TABLE 6.1 Summary of Basic Op Amp Circuits

Name	Circuit Schematic	Input-Output Relation
Inverting Amplifier		$v_{\text{out}} = -\dfrac{R_f}{R_1} v_{\text{in}}$
Noninverting Amplifier		$v_{\text{out}} = \left(1 + \dfrac{R_f}{R_1}\right) v_{\text{in}}$
Voltage Follower (also known as a Unity Gain Amplifier)		$v_{\text{out}} = v_{\text{in}}$
Summing Amplifier		$v_{\text{out}} = -\dfrac{R_f}{R}(v_1 + v_2 + v_3)$
Difference Amplifier		$v_{\text{out}} = v_2 - v_1$

ENGINEERING CIRCUIT ANALYSIS

ENGINEERING CIRCUIT ANALYSIS

SEVENTH EDITION

William H. Hayt, Jr. (late)
Purdue University

Jack E. Kemmerly (late)
California State University

Steven M. Durbin
University of Canterbury
Te Whare Wānanga o Waitaha

Higher Education

Boston   Burr Ridge, IL   Dubuque, IA   Madison, WI   New York   San Francisco   St. Louis
Bangkok   Bogotá   Caracas   Kuala Lumpur   Lisbon   London   Madrid   Mexico City
Milan   Montreal   New Delhi   Santiago   Seoul   Singapore   Sydney   Taipei   Toronto

Higher Education

ENGINEERING CIRCUIT ANALYSIS, SEVENTH EDITION

Published by McGraw-Hill, a business unit of The McGraw-Hill Companies, Inc., 1221 Avenue of the Americas, New York, NY 10020. Copyright © 2007 by The McGraw-Hill Companies, Inc. All rights reserved. No part of this publication may be reproduced or distributed in any form or by any means, or stored in a database or retrieval system, without the prior written consent of The McGraw-Hill Companies, Inc., including, but not limited to, in any network or other electronic storage or transmission, or broadcast for distance learning.

Some ancillaries, including electronic and print components, may not be available to customers outside the United States.

This book is printed on acid-free paper.

1 2 3 4 5 6 7 8 9 0 DOW/DOW 0 9 8 7 6

ISBN-13 978–0–07–286611–7
ISBN-10 0–07–286611–X

Publisher: *Suzanne Jeans*
Senior Sponsoring Editor: *Michael S. Hackett*
Senior Developmental Editor: *Melinda D. Bilecki*
Executive Marketing Manager: *Michael Weitz*
Senior Project Manager: *Kay J. Brimeyer*
Senior Production Supervisor: *Laura Fuller*
Lead Media Project Manager: *Audrey A. Reiter*
Associate Media Producer: *Christina Nelson*
Senior Coordinator of Freelance Design: *Michelle D. Whitaker*
Cover Designer: *Christopher Reese*
Interior Designer: *John Walker Design*
Senior Photo Research Coordinator: *John C. Leland*
Photo Research: *Jerry Marshall*
Supplement Producer: *Tracy L. Konrardy*
Compositor: *Interactive Composition Corporation*
Typeface: *10/12 Times Roman*
Printer: *R. R. Donnelley Willard, OH*

(USE) Cover Images: Aeroplane cockpit image: © *PhotoLink/Getty Images*
Wind turbines: *Russell Illig/Getty Images*
Fluke Graphic Multimeter: *Courtesy of Fluke Corporation*
Intel® Pentium® M Processor (Dothan) Die: *Courtesy of Intel Corporation*

Library of Congress Cataloging-in-Publication Data

Hayt, William Hart, 1920–
  Engineering circuit analysis / William H. Hayt, Jr., Jack E. Kemmerly, Steven M. Durbin. — 7th ed.
    p. cm.
  Includes index.
  ISBN 978–0–07–286611–7 — ISBN 0–07–286611–X (hard copy : alk. paper)
  1. Electric circuit analysis. 2. Electric network analysis. I. Kemmerly, Jack E. (Jack Ellsworth), 1924–.
II. Durbin, Steven M. III. Title.

TK454.H4   2007
621.319'2—dc22   2005033982
                                                        CIP

www.mhhe.com

To Sean and Kristi. The best part of every day.

OUR COMMITMENT TO ACCURACY

You have a right to expect an accurate textbook, and McGraw-Hill Engineering invests considerable time and effort to ensure that we deliver one. Listed below are the many steps we take in this process.

OUR ACCURACY VERIFICATION PROCESS

First Round

Step 1: Numerous **college engineering instructors** review the manuscript and report errors to the editorial team. The authors review their comments and make the necessary corrections in their manuscript.

Second Round

Step 2: An **expert in the field** works through every example and exercise in the final manuscript to verify the accuracy of the examples, exercises, and solutions. The authors review any resulting corrections and incorporate them into the final manuscript and solutions manual.

Step 3: The manuscript goes to a **copyeditor,** who reviews the pages for grammatical and stylistic considerations. At the same time, the expert in the field begins a second accuracy check. All corrections are submitted simultaneously to the **authors,** who review and integrate the editing, and then submit the manuscript pages for typesetting.

Third Round

Step 4: The **authors** review their page proofs for a dual purpose: 1) to make certain that any previous corrections were properly made, and 2) to look for any errors they might have missed.

Step 5: A **proofreader** is assigned to the project to examine the new page proofs, double check the authors' work, and add a fresh, critical eye to the book. Revisions are incorporated into a new batch of pages which the authors check again.

Fourth Round

Step 6: The **author team** submits the solutions manual to the **expert in the field,** who checks text pages against the solutions manual as a final review.

Step 7: The **project manager, editorial team,** and **author team** review the pages for a final accuracy check.

The resulting engineering textbook has gone through several layers of quality assurance and is verified to be as accurate and error-free as possible. Our authors and publishing staff are confident that through this process we deliver textbooks that are industry leaders in their correctness and technical integrity.

ABOUT THE AUTHORS

WILLIAM H. HAYT, Jr. received his B.S. and M.S. at Purdue University and his Ph.D. from the University of Illinois. After spending four years in industry, Professor Hayt joined the faculty of Purdue University, where he served as Professor and Head of the School of Electrical Engineering, and as Professor Emeritus after retiring in 1986. Besides *Engineering Circuit Analysis,* Professor Hayt authored three other texts, including *Engineering Electromagnetics,* now in its seventh edition with McGraw-Hill. Professor Hayt's professional society memberships included Eta Kapp Nu, Tau Beta Pi, Sigma Xi, Sigma Delta Chi, Fellow of IEEE, ASEE, and NAEB. While at Purdue, he received numerous teaching awards, including the university's Best Teacher Award. He is also listed in Purdue's Book of Great Teachers, a permanent wall display in the Purdue Memorial Union, dedicated on April 23, 1999. The book bears the names of the inaugural group of 225 faculty members, past and present, who have devoted their lives to excellence in teaching and scholarship. They were chosen by their students and their peers as Purdue's finest educators.

JACK E. KEMMERLY received his B.S. magna cum laude from The Catholic University of America, M.S. from University of Denver, and Ph.D. from Purdue University. Professor Kemmerly first taught at Purdue University and later worked as principal engineer at the Aeronutronic Division of Ford Motor Company. He then joined California State University, Fullerton, where he served as Professor, Chairman of the Faculty of Electrical Engineering, Chairman of the Engineering Division, and Professor Emeritus. Professor Kemmerly's professional society memberships included Eta Kappa Nu, Tau Beta Pi, Sigma Xi, ASEE, and IEEE (Senior Member). His pursuits outside of academe included being an officer in the Little League and a scoutmaster in the Boy Scouts.

STEVEN M. DURBIN received the B.S.E.E., M.S.E., and Ph.D. from Purdue University, West Lafayette, Indiana. After receiving the Ph.D., he joined the faculty of the Department of Electrical Engineering at Florida A&M University and The Florida State University. In August of 2000, he accepted a faculty position at the University of Canterbury, Christchurch, New Zealand, where he teaches circuits, electronics, and solid-state related courses and conducts research into novel electronic materials and device structures. He is a senior member of the IEEE and a member of Eta Kappa Nu, the Electron Devices Society, the American Physical Society, and the Royal Society of New Zealand.

PREFACE

Reading this book is intended to be an enjoyable experience, even though the text is by necessity scientifically rigorous and somewhat mathematical. We, the authors, are trying to share the idea that circuit analysis can be fun. Not only is it useful and downright essential to the study of engineering, it is a marvelous education in logical thinking, good even for those who may never analyze another circuit in their professional lifetime. Looking back after finishing the course, many students are truly amazed by all the excellent analytical tools that are derived from only **three simple scientific laws**—Ohm's law and Kirchhoff's voltage and current laws.

In many colleges and universities, the introductory course in electrical engineering will be preceded or accompanied by an introductory physics course in which the basic concepts of electricity and magnetism are introduced, most often from the field aspect. Such a background is not a prerequisite, however. Instead, several of the requisite basic concepts of electricity and magnetism are discussed (or reviewed) as needed. Only an introductory calculus course need be considered as a prerequisite—or possibly a corequisite—to the reading of the book. Circuit elements are introduced and defined here in terms of their circuit equations; only incidental comments are offered about the pertinent field relationships. In the past, we have tried introducing the basic circuit analysis course with three or four weeks of electromagnetic field theory, so as to be able to define circuit elements more precisely in terms of Maxwell's equations. The results, especially in terms of students' acceptance, were not good.

We intend that this text be one from which students may teach the science of circuit analysis to themselves. *It is written to the student, and not to the instructor,* because the student is probably going to spend more time than the instructor in reading it. If at all possible, each new term is clearly defined when it is first introduced. The basic material appears toward the beginning of each chapter and is explained carefully and in detail; numerical examples are used to introduce and suggest general results. Practice problems appear throughout each chapter; they are generally simple, and answers to the several parts are given in order. The more difficult problems appear at the ends of the chapters and follow the general order of presentation of the text material. These problems are occasionally used to introduce less important or more advanced topics through a guided step-by-step procedure, as well as to introduce topics that will appear in a subsequent chapter. The introduction and resulting repetition are both important to the learning process. In all, there are over 1200 end-of-chapter problems in addition to numerous practice problems and over 170 worked examples. Many of the exercises are new in this edition, and, with the assistance of several colleagues, each problem was solved by hand and checked by computer when appropriate.

If the book occasionally appears to be informal, or even lighthearted, it is because we feel that it is not necessary to be dry or pompous to be educational. Amused smiles on the faces of our students are seldom obstacles to their absorbing information. *If the writing of the text had its entertaining moments, then why not the reading too?* The presentation of the material in the text represents an evolutionary process through courses taught at Purdue University; the California State University, Fullerton; Fort Lewis College in Durango; the joint engineering program of Florida A&M University and The Florida State University; and the University of Canterbury (New Zealand). Those students saw it all first, and their frequent comments and suggestions are greatly appreciated.

It is a great privilege to serve as co-author to *Engineering Circuit Analysis,* first published in 1962. Now in its seventh edition, this book has seen both steady progression and significant change in the way circuit analysis is taught. I used it myself as an undergraduate engineering student at Purdue, where I was fortunate to take the course from Bill Hayt himself—without a doubt, one of the best professors I ever had.

There are several noteworthy features of *Engineering Circuit Analysis* that are responsible for its success. It is a very well structured, time-tested book—key concepts are presented in a logical format, but also interlinked seamlessly into a larger framework. Discussions are located well, interspersed with helpful examples and good practice problems. There is no timidity when it comes to presenting the theory underlying a specific topic, or pulling of punches in developing the mathematical underpinning. Everything, however, has been carefully designed to assist the student in learning how to perform circuit analysis for herself/himself; theory for the sake of theory has been left to other texts. Bill Hayt and Jack Kemmerly put a great deal of work into the creation of the First Edition, and their desire to impart some of their boundless enthusiasm to the reader comes through in every chapter.

NEW IN THE SEVENTH EDITION

When the decision to make the seventh edition four-color became official, everyone on the production team moved into high gear to make the most of this exciting opportunity. Countless (I'm sure somebody in accounting counted) drafts, revisions, models and templates crossed the ether(net) as we strove to make the color work to best advantage for the student. The final result of this team effort, I believe, would be hard to top. There are many other changes from the sixth edition, although great care was exercised to retain the key features, general flow, and overall content for the benefit of current instructors. We have therefore, once again, made use of several different icons:

 Provides a warning for common mistakes;

 Indicates a specific point worth noting;

Denotes a design problem to which there may be no unique answer; and

  Indicates a problem requiring computer-aided analysis.

With the mindset that engineering-oriented software packages can be of assistance in the learning process, but should not be used as a crutch, those end-of-chapter problems designated with are always phrased so that the appropriate software is used to *check* answers, not provide them.

Many instructors are hard-pressed to cover the required material for their specific circuits course, and may have to skip over some chapters in order to do so. This is particularly true for operational amplifiers, so this and subsequent chapters have been written so that the material can be omitted without loss of clarity or flow. The decision to place Chapter 6 immediately after completion of dc analysis was made so that op amp circuits can be used to reinforce those circuit analysis techniques studied in previous chapters. Transient effects and frequency response, with the exception of slew rate, are included at the end of the relevant chapters; this both avoids information overload and provides multiple opportunities for using op amps as practical examples for the circuit analysis concepts being studied.

The issue of complex frequency is also worth mentioning here. Bill Hayt was of the mindset that Laplace transforms should be presented as a special case of Fourier transforms—a straightforward exercise mathematically. Many programs, however, do not cover Fourier-based concepts until later courses in signals and systems, and so he and Jack Kemmerly introduced the student to the notion of complex frequency as an extension of phasors. This student-friendly approach has been retained and is a strong feature of the text, where competing treatments often start the Laplace analysis chapter by simply stating the integral transform.

CHANGES TO THE SEVENTH EDITION INCLUDE:

1. Numerous new and revised examples, particularly in the transient analysis portion of the text (Chapters 7, 8, and 9).

2. Substantial rewriting and expansion of the op amp material in Chapter 6. This material now includes discussions of using op amps to build current and voltage sources, and of slew rate, comparators, and instrumentation amplifiers. Several types of configurations are analyzed in detail, but a few variations are left for the students to tackle for themselves.

3. Addition of several hundred end-of-chapter problems.

4. Several new tables for easy reference.

5. Careful attention to each example to ensure concise explanations, adequate intermediate steps, and appropriate figures. As in the Sixth Edition, each example is phrased in a manner similar to that of a test question and designed to assist in problem-solving, as opposed to concept illustration.

6. In response to many student comments, a greater variety of end-of-chapter problems has been included, including straight-forward "confidence building" problems.

7. The "Goals & Objectives" section at the beginning of each chapter has been recast as "Key Concepts" to provide a quick reference for chapter content.

8. Several new Practical Application sections have been added, while existing ones were updated.

9. New photos, many in 4-color, to add a visual perspective to relevant topics.

10. New multimedia software to accompany the book, including a long-anticipated update to the COSMOS solutions manual system created for instructors.

The unexpected passing of Bill Hayt at the very beginning of Sixth Edition revision process was an enormous shock. I never had the opportunity to talk to him about the intended changes—I can only hope that the continued revisions have helped this book to speak to yet another generation of bright young engineering students. In the meantime, we (durbin@ieee.org and the editors at McGraw-Hill) welcome comments and feedback from both students and instructors. Positive and negative, they are all appreciated.

Of course, this project has been a team effort, and many people have participated and provided assistance. The ever-present support of the McGraw-Hill editorial and production staff, including Melinda Bilecki, Michelle Flomenhoft, Kalah Cavanaugh, Michael Hackett, Christina Nelson, Eric Weber, Phil Meek, and Kay Brimeyer is gratefully acknowledged. I would also like to thank my local McGraw-Hill representative, Nazier Hassan, who dropped by whenever on campus to grab a cup of coffee and ask how things were going. Working with these people has been incredible.

For the Seventh Edition, the following individuals deserve acknowledgment and a debt of gratitude for their time and energy in reviewing various versions of the manuscript:

Miroslav M. Begovic, *Georgia Institute of Technology*

Maqsood Chaudhry, *California State University, Fullerton*

Wade Enright, *Viva Technical Solutions, Ltd.*

Rick Fields, *TRW*

Victor Gerez, *Montana State University*

Dennis Goeckel, *Univeristy of Massachusetts, Amherst*

Paul M. Goggans, *University of Mississippi*

Riadh Habash, *University of Ottawa*

Jay H. Harris, *San Diego State University*

Archie Holmes, Jr. *University of Texas, Austin,*

Sheila Horan, *New Mexico State University*

Douglas E. Jussaume, *University of Tulsa*

James S. Kang, *California State Polytechnic University, Pomona*

Chandra Kavitha, *University of Massachusetts, Lowell*

Leon McCaughan, *University of Wisconsin*

John P. Palmer, *California State Polytechnic University, Pomona*

Craig S. Petrie, *Brigham Young University*

Mohammad Sarmadi, *The Pennsylvania State University*

A.C. Soudack, *University of British Columbia*

Earl Swartzlander, *University of Texas, Austin*

Val Tereski, *North Dakota State University*

Kamal Yacoub, *University of Miami.*

    The comments and suggestions from Drs. Jim Zheng, Reginald Perry, Rodney Roberts, and Tom Harrison of the Department of Electrical and Computer Engineering at Florida A&M University and The Florida State University are gratefully acknowledged, as is the incredible effort and enthusiasm of Bill Kennedy at the University of Canterbury, who once again proofread each chapter and provided many useful suggestions. Also, a special thanks to Ken Smart and Dermot Sallis for providing components for photographs, Duncan Shaw-Brown and Kristi Durbin for various photography services, Richard Blaikie for his help with the *h*-parameter Practical Application, Rick Millane for assistance with the image processing Practical Application, and Wade Enright for supplying numerous transformer photographs (nobody has more transformer pictures). Cadence and the Mathworks kindly provided assistance with computer-aided analysis software, which was much appreciated. Phillipa Haigh and Emily Hewat provided technical typing, photocopying, and proofing at various stages of the project, and certainly deserve written thanks for all their help. I would also like to thank my Department for granting sabbatical leave to start the revision process—meaning that my colleagues kindly stepped in and covered many of my regular duties.

    A number of people have influenced my teaching style over the years, including Profs. Bill Hayt, David Meyer, Alan Weitsman, and my thesis advisor, Jeffery Gray, but also the first electrical engineer I ever met—my father, Jesse Durbin, a graduate of the Indiana Institute of Technology. Support and encouragement from the other members of my family—including my mother, Roberta, brothers Dave, John, and James, as well as my parents-in-law Jack and Sandy—are also gratefully acknowledged. Finally and most importantly: thank you to my wife, Kristi, for your patience, your understanding, your support, and advice, and to our son, Sean, for making life so much fun.

Steven M. Durbin
Christchurch, New Zealand

GUIDED TOUR

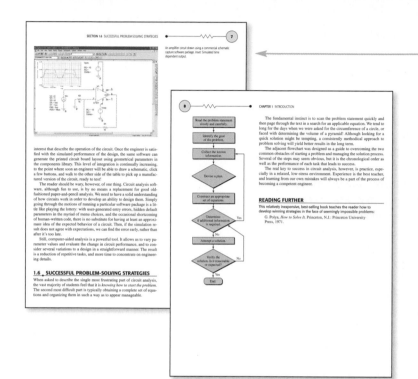

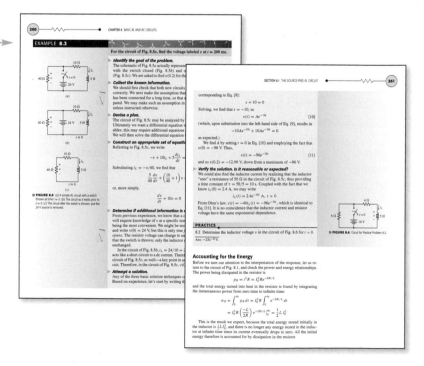

Focus on Problem Solving

Chapter one contains detailed information about the best way to approach an engineering circuit analysis problem and lays out the steps that one should work through to arrive at the correct answer.

A carefully chosen example in each subsequent chapter is labeled with these steps to continually reinforce effective problem solving skills.

Design Emphasis

The concept of design is introduced in Chapter 1; throughout the text, issues pertaining to design are interwoven with discussion of analysis procedures.

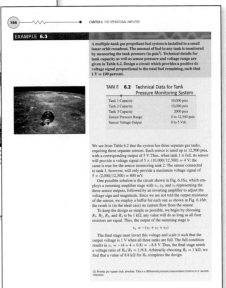

Special design problems are marked with a "D" icon.

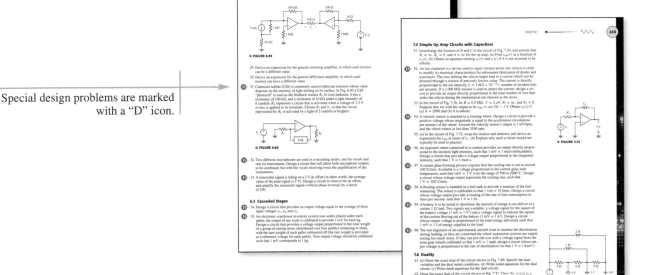

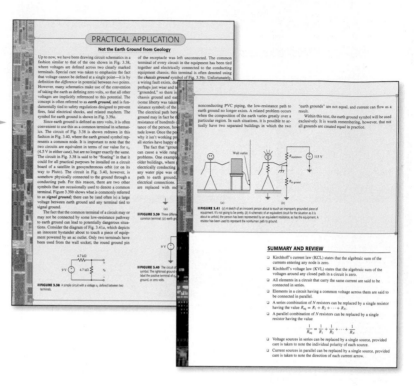

Real World Connections

Practical Application sections show how the material under discussion directly pertains to real-world situations.

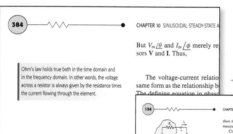

Extensive Margin Notes and Icons

Running margin notes provide tips, insights, and additional information about key aspects of the discussion.

A "take note" icon highlights specific points worth noting, and a "caution" icon identifies common sources of error for students.

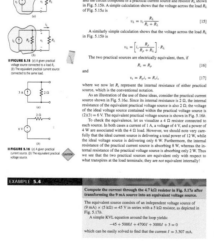

"Reading Further" Sections

At the end of each chapter, a list of additional readings on key chapter topics provides direction for students interested in further clarifying or deepening their knowledge of important circuit concepts.

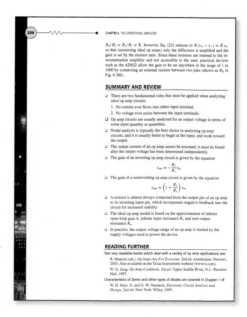

Summary and Review

An overview section at the end of each chapter offers students a chance to check their retention of the major ideas presented and serves as a reference for review and test preparation.

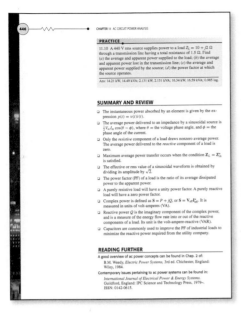

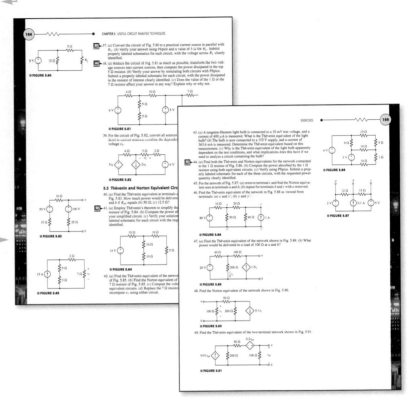

Computer Problems

PSpice and MATLAB examples have been introduced at appropriate places in many chapters. Computer-aided engineering is used as an aid to, not a substitute for, developing problem-solving skills. Computer-aided analysis is also introduced in selected homework problems to encourage students to compare hand calculations to simulation-generated results.

OUR COMMITMENT TO ACCURACY

You have a right to expect an accurate textbook, and McGraw-Hill Engineering invests considerable time and effort to ensure that we deliver one. Listed below are the many steps we take in this process.

OUR ACCURACY VERIFICATION PROCESS

First Round

Step 1: Numerous **college engineering instructors** review the manuscript and report errors to the editorial team. The authors review their comments and make the necessary corrections in their manuscript.

Second Round

Step 2: An **expert in the field** works through every example and exercise in the final manuscript to verify the accuracy of the examples, exercises, and solutions. The authors review any resulting corrections and incorporate them into the final manuscript and solutions manual.

Step 3: The manuscript goes to a **copyeditor,** who reviews the pages for grammatical and stylistic considerations. At the same time, the expert in the field begins a second accuracy check. All corrections are submitted simultaneously to the **authors,** who review and integrate the editing, and then submit the manuscript pages for typesetting.

Third Round

Step 4: The **authors** review their page proofs for a dual purpose: 1) to make certain that any previous corrections were properly made, and 2) to look for any errors they might have missed.

Step 5: A **proofreader** is assigned to the project to examine the new page proofs, double check the authors' work, and add a fresh, critical eye to the book. Revisions are incorporated into a new batch of pages which the authors check again.

Fourth Round

Step 6: The **author team** submits the solutions manual to the **expert in the field,** who checks text pages against the solutions manual as a final review.

Step 7: The **project manager, editorial team,** and **author team** review the pages for a final accuracy check.

The resulting engineering textbook has gone through several layers of quality assurance and is verified to be as accurate and error-free as possible. Our authors and publishing staff are confident that through this process we deliver textbooks that are industry leaders in their correctness and technical integrity.

Commitment to Accuracy

An Accuracy Statement describes the process the publisher and author have instituted to ensure accurate calculations.

McGraw-Hill's ARIS

McGraw-Hill's Assessment, Review, and Instruction System (ARIS) is a complete, online tutorial, electronic homework, and course management system, designed for greater ease of use than any other system available. ARIS includes a solutions manual, text image files, and transition guides for instructors as well as eProfessor video lectures, Network Analysis Tutorials, selected answers to text practice problems, additional problems and solutions, and FE Exam questions for students.

BRIEF CONTENTS

PREFACE   ix

1 ● INTRODUCTION   1

2 ● BASIC COMPONENTS AND ELECTRIC CIRCUITS   9

3 ● VOLTAGE AND CURRENT LAWS   35

4 ● BASIC NODAL AND MESH ANALYSIS   79

5 ● USEFUL CIRCUIT ANALYSIS TECHNIQUES   121

6 ● THE OPERATIONAL AMPLIFIER   173

7 ● CAPACITORS AND INDUCTORS   215

8 ● BASIC RL AND RC CIRCUITS   255

9 ● THE RLC CIRCUIT   319

10 ● SINUSOIDAL STEADY-STATE ANALYSIS   369

11 ● AC CIRCUIT POWER ANALYSIS   419

12 ● POLYPHASE CIRCUITS   457

13 ● MAGNETICALLY COUPLED CIRCUITS   491

14 ● COMPLEX FREQUENCY AND THE LAPLACE TRANSFORM   533

15 ● CIRCUIT ANALYSIS IN THE **s**-DOMAIN   571

16 ● FREQUENCY RESPONSE   627

17 ● TWO-PORT NETWORKS   691

18 ● FOURIER CIRCUIT ANALYSIS   735

Appendix 1 AN INTRODUCTION TO NETWORK TOPOLOGY   793

Appendix 2 SOLUTION OF SIMULTANEOUS EQUATIONS   805

Appendix 3 A PROOF OF THEVENIN'S THEOREM   813

Appendix 4 A PSPICE® TUTORIAL   815

Appendix 5 COMPLEX NUMBERS   821

Appendix 6 A BRIEF MATLAB® TUTORIAL   831

Appendix 7 ADDITIONAL LAPLACE TRANSFORM THEOREMS   837

INDEX   843

CONTENTS

CHAPTER 1

INTRODUCTION 1

1.1 Preamble 1
1.2 Overview of Text 2
1.3 Relationship of Circuit Analysis to Engineering 4
1.4 Analysis and Design 5
1.5 Computer-Aided Analysis 6
1.6 Successful Problem-Solving Strategies 7
READING FURTHER 8

CHAPTER 2

BASIC COMPONENTS AND ELECTRIC CIRCUITS 9

2.1 Units and Scales 9
2.2 Charge, Current, Voltage, and Power 11
2.3 Voltage and Current Sources 17
2.4 Ohm's Law 22
SUMMARY AND REVIEW 28
READING FURTHER 28
EXERCISES 29

CHAPTER 3

VOLTAGE AND CURRENT LAWS 35

3.1 Nodes, Paths, Loops, and Branches 35
3.2 Kirchhoff's Current Law 36
3.3 Kirchhoff's Voltage Law 38
3.4 The Single-Loop Circuit 42
3.5 The Single-Node-Pair Circuit 45
3.6 Series and Parallel Connected Sources 49
3.7 Resistors in Series and Parallel 51
3.8 Voltage and Current Division 57
SUMMARY AND REVIEW 62
READING FURTHER 63
EXERCISES 63

CHAPTER 4

BASIC NODAL AND MESH ANALYSIS 79

4.1 Nodal Analysis 80
4.2 The Supernode 89
4.3 Mesh Analysis 92

4.4 The Supermesh 98
4.5 Nodal vs. Mesh Analysis: A Comparison 101
4.6 Computer-Aided Circuit Analysis 103
SUMMARY AND REVIEW 108
READING FURTHER 108
EXERCISES 109

CHAPTER 5

USEFUL CIRCUIT ANALYSIS TECHNIQUES 121

5.1 Linearity and Superposition 121
5.2 Source Transformations 131
5.3 Thévenin and Norton Equivalent Circuits 139
5.4 Maximum Power Transfer 150
5.5 Delta-Wye Conversion 152
5.6 Selecting an Approach: A Summary of Various Techniques 155
SUMMARY AND REVIEW 156
READING FURTHER 156
EXERCISES 156

CHAPTER 6

THE OPERATIONAL AMPLIFIER 173

6.1 Background 173
6.2 The Ideal Op Amp: A Cordial Introduction 174
6.3 Cascaded Stages 182
6.4 Circuits for Voltage and Current Sources 186
6.5 Practical Considerations 190
6.6 Comparators and the Instrumentation Amplifier 201
SUMMARY AND REVIEW 204
READING FURTHER 204
EXERCISES 205

CHAPTER 7

CAPACITORS AND INDUCTORS 215

7.1 The Capacitor 215
7.2 The Inductor 224
7.3 Inductance and Capacitance Combinations 232
7.4 Consequences of Linearity 235
7.5 Simple Op Amp Circuits with Capacitors 238

7.6 Duality 240

7.7 Modeling Capacitors and Inductors with PSpice® 243

SUMMARY AND REVIEW 245

READING FURTHER 246

EXERCISES 246

CHAPTER 8
BASIC *RL* AND *RC* CIRCUITS 255

8.1 The Source-Free *RL* Circuit 255

8.2 Properties of the Exponential Response 262

8.3 The Source-Free *RC* Circuit 266

8.4 A More General Perspective 269

8.5 The Unit-Step Function 276

8.6 Driven *RL* Circuits 280

8.7 Natural and Forced Response 283

8.8 Driven *RC* Circuits 289

8.9 Predicting the Response of Sequentially Switched Circuits 294

SUMMARY AND REVIEW 300

READING FURTHER 302

EXERCISES 302

CHAPTER 9
THE *RLC* CIRCUIT 319

9.1 The Source-Free Parallel Circuit 319

9.2 The Overdamped Parallel *RLC* Circuit 324

9.3 Critical Damping 332

9.4 The Underdamped Parallel *RLC* Circuit 336

9.5 The Source-Free Series *RLC* Circuit 343

9.6 The Complete Response of the *RLC* Circuit 349

9.7 The Lossless *LC* Circuit 357

SUMMARY AND REVIEW 359

READING FURTHER 360

EXERCISES 360

CHAPTER 10
SINUSOIDAL STEADY-STATE ANALYSIS 369

10.1 Characteristics of Sinusoids 369

10.2 Forced Response to Sinusoidal Functions 372

10.3 The Complex Forcing Function 376

10.4 The Phasor 381

10.5 Phasor Relationships for *R*, *L*, and *C* 383

10.6 Impedance 387

10.7 Admittance 392

10.8 Nodal and Mesh Analysis 393

10.9 Superposition, Source Transformations, and Thévenin's Theorem 396

10.10 Phasor Diagrams 404

SUMMARY AND REVIEW 407

READING FURTHER 407

EXERCISES 408

CHAPTER 11
AC CIRCUIT POWER ANALYSIS 419

11.1 Instantaneous Power 420

11.2 Average Power 422

11.3 Effective Values of Current and Voltage 432

11.4 Apparent Power and Power Factor 437

11.5 Complex Power 440

11.6 Comparison of Power Terminology 445

SUMMARY AND REVIEW 446

READING FURTHER 446

EXERCISES 447

CHAPTER 12
POLYPHASE CIRCUITS 457

12.1 Polyphase Systems 458

12.2 Single-Phase Three-Wire Systems 460

12.3 Three-Phase Y-Y Connection 464

12.4 The Delta (Δ) Connection 470

12.5 Power Measurement in Three-Phase Systems 476

SUMMARY AND REVIEW 484

READING FURTHER 485

EXERCISES 485

CHAPTER 13
MAGNETICALLY COUPLED CIRCUITS 491

13.1 Mutual Inductance 491

13.2 Energy Considerations 499

13.3 The Linear Transformer 503

13.4 The Ideal Transformer 510

SUMMARY AND REVIEW 520

READING FURTHER 520

EXERCISES 521

CHAPTER 14
COMPLEX FREQUENCY AND THE LAPLACE TRANSFORM 533

14.1 Complex Frequency 533

14.2 The Damped Sinusoidal Forcing Function 537

14.3 Definition of the Laplace Transform 540

14.4 Laplace Transforms of Simple Time Functions 543

14.5 Inverse Transform Techniques 546

14.6 Basic Theorems for the Laplace Transform 553

14.7 The Initial-Value and Final-Value Theorems 561
SUMMARY AND REVIEW 564
READING FURTHER 564
EXERCISES 565

CHAPTER 15

CIRCUIT ANALYSIS IN THE **s**-DOMAIN 571

15.1 Z(s) and Y(s) 571
15.2 Nodal and Mesh Analysis in the **s**-Domain 578
15.3 Additional Circuit Analysis Techniques 585
15.4 Poles, Zeros, and Transfer Functions 588
15.5 Convolution 589
15.6 The Complex-Frequency Plane 598
15.7 Natural Response and the s-Plane 607
15.8 A Technique for Synthesizing the Voltage Ratio
$H(s) = V_{out}/V_{in}$ 612
SUMMARY AND REVIEW 616
READING FURTHER 616
EXERCISES 617

CHAPTER 16

FREQUENCY RESPONSE 627

16.1 Parallel Resonance 627
16.2 Bandwidth and High-Q Circuits 636
16.3 Series Resonance 641
16.4 Other Resonant Forms 645
16.5 Scaling 652
16.6 Bode Diagrams 656
16.7 Filters 672
SUMMARY AND REVIEW 680
READING FURTHER 681
EXERCISES 681

CHAPTER 17

TWO-PORT NETWORKS 691

17.1 One-Port Networks 691
17.2 Admittance Parameters 696
17.3 Some Equivalent Networks 703
17.4 Impedance Parameters 712
17.5 Hybrid Parameters 718
17.6 Transmission Parameters 720
SUMMARY AND REVIEW 724
READING FURTHER 725
EXERCISES 725

CHAPTER 18

FOURIER CIRCUIT ANALYSIS 735

18.1 Trigonometric Form of the Fourier Series 735
18.2 The Use of Symmetry 745
18.3 Complete Response to Periodic Forcing
Functions 750
18.4 Complex Form of the Fourier Series 752
18.5 Definition of the Fourier Transform 759
18.6 Some Properties of the Fourier Transform 763
18.7 Fourier Transform Pairs for Some Simple
Time Functions 766
18.8 The Fourier Transform of a General Periodic
Time Function 771
18.9 The System Function and Response
in the Frequency Domain 772
18.10 The Physical Significance of the System
Function 779
SUMMARY AND REVIEW 785
READING FURTHER 785
EXERCISES 785

APPENDIX 1 AN INTRODUCTION TO NETWORK
TOPOLOGY 793

APPENDIX 2 SOLUTION OF SIMULTANEOUS
EQUATIONS 805

APPENDIX 3 A PROOF OF THÉVENIN'S
THEOREM 813

APPENDIX 4 A PSPICE® TUTORIAL 815

APPENDIX 5 COMPLEX NUMBERS 821

APPENDIX 6 A BRIEF MATLAB® TUTORIAL 831

APPENDIX 7 ADDITIONAL LAPLACE TRANSFORM
THEOREMS 837

INDEX 843

1

Introduction

Facets to Circuit Analysis:
dc Analysis, Transient
Analysis, ac Analysis, and
Frequency Analysis

Analysis and Design

Computer-Aided Analysis

Problem-Solving Approaches

1.1 PREAMBLE

Today's engineering graduates are no longer employed solely for the technical design aspects of engineering problems. Their efforts now extend beyond the creation of better computers and communication systems to vigorous efforts to solve socioeconomic problems such as air and water pollution, urban planning, mass transportation, the discovery of new energy sources, and the conservation of existing natural resources, particularly oil and natural gas.

To contribute to the solution of these engineering problems an engineer must acquire many skills, one of which is a knowledge of electric circuit analysis. If we have already entered or intend to enter an electrical engineering program, then circuit analysis likely represents one of the introductory courses in our chosen field. If we are associated with another branch of engineering, then circuit analysis may represent a large fraction of our total study of electrical engineering—providing the basis for working with electronic instrumentation, electrically powered machines, and large-scale systems. Most important, however, is the possibility given to us to broaden our education and become more informed members of a team. Increasingly, such teams are multidisciplinary in composition, and effective communication within such a group can be achieved only if the language and definitions used are familiar to all.

In this chapter, just prior to launching into our agenda of technical discussions, we preview the topics which form the remainder of the text, pausing briefly to consider the relationship between analysis and design, and the evolving role computer tools play in modern circuit analysis.

Not all electrical engineers routinely make use of circuit analysis, but they often bring to bear analytical and problem-solving skills learned early on in their careers. A circuit analysis course is one of the first exposures to such concepts. (Solar Mirrors: © Corbis; Skyline: © Getty Images/PhotoLink; Oil Rig: © Getty Images; Dish: © Getty Images/J. Luke/PhotoLink)

1.2 OVERVIEW OF TEXT

The fundamental subject of this text is *linear circuit analysis*, which sometimes prompts a few readers to ask,

"Is there ever any *nonlinear* circuit analysis?"

Of course! We encounter nonlinear circuits every day: they capture and decode signals for our TVs and radios, perform calculations millions of times a second inside microprocessors, convert speech into electrical signals for transmission over phone lines, and execute many other functions outside our field of view. In designing, testing, and implementing such nonlinear circuits, detailed analysis is unavoidable.

"Then why study *linear* circuit analysis?"

Television sets include many nonlinear circuits. A great deal of them, however, can be understood and analyzed with the assistance of linear models. (© *Sony Electronics, Inc.*)

you might ask. An excellent question. The simple fact of the matter is that no physical system (including electrical circuits) is ever perfectly linear. Fortunately for us, however, a great many systems behave in a reasonably linear fashion over a limited range—allowing us to model them as linear systems if we keep the range limitations in mind.

For example, consider the common function

$$f(x) = e^x$$

A linear approximation to this function is

$$f(x) \approx 1 + x$$

Let's test this out. Table 1.1 shows both the exact value and the approximate value of $f(x)$ for a range of x. Interestingly, the linear approximation is exceptionally accurate up to about $x = 0.1$, when the relative error is still less than 1%. Although many engineers are rather quick on a calculator, it's hard to argue that any approach is faster than just adding 1.

TABLE 1.1 Comparison of a Linear Model for e^x to Exact Value

x	$f(x)$*	$1 + x$	Relative error**
0.0001	1.0001	1.0001	0.0000005%
0.001	1.0010	1.001	0.00005%
0.01	1.0101	1.01	0.005%
0.1	1.1052	1.1	0.5%
1.0	2.7183	2.0	26%

*Quoted to four significant figures.

**Relative error $\triangleq \left| 100 \times \dfrac{e^x - (1 + x)}{e^x} \right|$

Linear problems are inherently more easily solved than their nonlinear counterparts. For this reason, we often seek reasonably accurate linear approximations (or *models*) to physical situations. Furthermore, the linear models are more easily manipulated and understood—making design a more straightforward process.

The circuits we will encounter in subsequent chapters all represent linear approximations to physical electric circuits. Where appropriate, brief discussions of potential inaccuracies or limitations to these models are provided, but generally speaking we find them to be suitably accurate for most applications. When greater accuracy is required in practice, nonlinear models are employed, but with a considerable increase in solution complexity. A detailed discussion of what constitutes a *linear electric circuit* can be found in Chap. 2.

Linear circuit analysis can be separated into four broad categories: ***dc analysis***, ***transient analysis***, ***ac analysis***, and ***frequency response analysis***. We begin our journey with the topic of resistive circuits, which may include simple examples such as a flashlight or a toaster. This provides us with a perfect opportunity to learn a number of very powerful engineering circuit analysis techniques, such as *nodal analysis, mesh analysis, superposition, source transformation, Thévenin's theorem, Norton's theorem,* and several methods for simplifying networks of components connected in series or parallel. The single most redeeming feature of resistive circuits is that the

Modern trains are powered by electric motors. Their electrical systems are best analyzed using ac or phasor analysis techniques. (© Corbis)

Frequency-dependent circuits lie at the heart of many electronic devices, and they can be a great deal of fun to design. (© 1994–2005 Hewlett-Packard Company.)

time dependence of any quantity of interest does not affect our analysis procedure. In other words, if asked for an electrical quantity of a resistive circuit at several specific instants in time, we do not need to analyze the circuit more than once. As a result, we will spend most of our effort early on considering only dc circuits—those circuits whose electrical parameters do not vary with time.

Although dc circuits such as flashlights or automotive rear window defoggers are undeniably important in everyday life, things are often much more interesting when something happens suddenly (imagine a firecracker which takes 100 years to go from a quiet crackling to Bang!). In circuit analysis parlance, we refer to *transient analysis* as the suite of techniques used to study circuits which are suddenly energized or de-energized. To make such circuits interesting, we need to add elements that respond to the rate of change of electrical quantities, leading to circuit equations which include derivatives and integrals. Fortunately, we can obtain such equations using the simple techniques learned in the first part of our study.

Still, not all time-varying circuits are turned on and off suddenly. Air conditioners, fans, and fluorescent lights are only a few of the many examples we may see daily. In such situations, a calculus-based approach for every analysis can become tedious and time-consuming. Fortunately, there is a better alternative for situations where equipment has been allowed to run long enough for transient effects to die out, and this is commonly referred to as ac analysis, or sometimes *phasor analysis*.

The final leg of our journey deals with a subject known as *frequency response*. Working directly with the differential equations obtained in time-domain analysis helps us develop an intuitive understanding of the operation of circuits containing energy storage elements (e.g., capacitors and inductors). As we shall see, however, circuits with even a relatively small number of components can be somewhat onerous to analyze, and so much more straightforward methods have been developed. These methods, which include Laplace and Fourier analysis, allow us to transform differential equations into algebraic equations. Such methods also enable us to design circuits to respond in specific ways to particular frequencies. We make use of frequency-dependent circuits every day when we dial a telephone, select our favorite radio station, or connect to the Internet.

1.3 • RELATIONSHIP OF CIRCUIT ANALYSIS TO ENGINEERING

Whether we intend to pursue further circuit analysis at the completion of this course or not, it is worth noting that there are several layers to the concepts under study. Beyond the nuts and bolts of circuit analysis techniques lies the opportunity to develop a methodical approach to problem solving, the ability to determine the goal or goals of a particular problem, skill at collecting the information needed to effect a solution, and, perhaps equally importantly, opportunities for practice at verifying solution accuracy.

Students familiar with the study of other engineering topics such as fluid flow, automotive suspension systems, bridge design, supply chain management, or process control will recognize the general form of many of the equations we develop to describe the behavior of various circuits. We

A molecular beam epitaxy crystal growth facility. The equations governing its operation closely resemble those used to describe simple linear circuits.

simply need to learn how to "translate" the relevant variables (for example, replacing *voltage* with *force, charge* with *distance, resistance* with *friction coefficient,* etc.) to find that we already know how to work a new type of problem. Very often, if we have previous experience in solving a similar or related problem, our intuition can guide us through the solution of a totally new problem.

What we are about to learn regarding linear circuit analysis forms the basis for many subsequent electrical engineering courses. The study of electronics relies on the analysis of circuits with devices known as diodes and transistors, which are used to construct power supplies, amplifiers, and digital circuits. The skills which we will develop are typically applied in a rapid, methodical fashion by electronics engineers, who sometimes can analyze a complicated circuit without even reaching for a pencil! The time-domain and frequency-domain chapters of this text lead directly into discussions of signal processing, power transmission, control theory, and communications. We find that frequency-domain analysis in particular is an extremely powerful technique, easily applied to any physical system subjected to time-varying excitation.

An example of a robotic manipulator. The feedback control system can be modeled using linear circuit elements to determine situations in which the operation may become unstable. (*NASA Marshall Space Flight Center.*)

1.4 ANALYSIS AND DESIGN

Engineers take a fundamental understanding of scientific principles, combine this with practical knowledge often expressed in mathematical terms, and (frequently with considerable creativity) arrive at a solution to a given problem. *Analysis* is the process through which we determine the scope of a problem, obtain the information required to understand it, and compute the parameters of interest. *Design* is the process by which we synthesize something new as part of the solution to a problem. Generally speaking, there is an expectation that a problem requiring design will have no unique solution, whereas the analysis phase typically will. Thus, the last step in designing is always analyzing the result to see if it meets specifications.

Two proposed designs for a next-generation space shuttle. Although both contain similar elements, each is unique. (*NASA Dryden Flight Research Center.*)

This text is focused on developing our ability to analyze and solve problems because it is the starting point in every engineering situation. The philosophy of this book is that we need clear explanations, well-placed examples, and plenty of practice to develop such an ability. Therefore, elements of design are integrated into end-of-chapter problems and later chapters so as to be enjoyable rather than distracting.

1.5 COMPUTER-AIDED ANALYSIS

Solving the types of equations that result from circuit analysis can often become notably cumbersome for even moderately complex circuits. This of course introduces an increased probability that errors will be made, in addition to considerable time in performing the calculations. The desire to find a tool to help with this process actually predates electronic computers, with purely mechanical computers such as the Analytical Engine designed by Charles Babbage in the 1880s proposed as possible solutions. Perhaps the earliest successful electronic computer designed for solution of differential equations was the 1940s-era ENIAC, whose vacuum tubes filled a large room. With the advent of low-cost desktop computers, however, computer-aided circuit analysis has developed into an invaluable everyday tool which has become an integral part of not only analysis but design as well.

One of the most powerful aspects of computer-aided design is the relatively recent integration of multiple programs in a fashion transparent to the user. This allows the circuit to be drawn schematically on the screen, reduced automatically to the format required by an analysis program (such as SPICE, introduced in Chap. 4), and the resulting output smoothly transferred to a third program capable of plotting various electrical quantities of

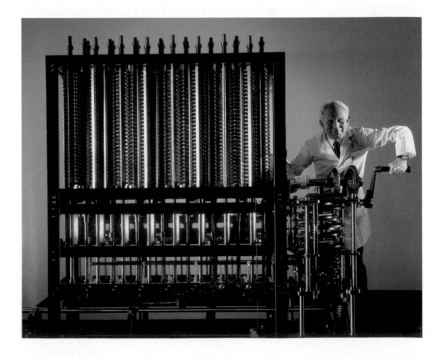

Charles Babbage's "Difference Engine Number 2," as completed by the Science Museum (London) in 1991. (© *Science Museum/Science & Society Picture Library.*)

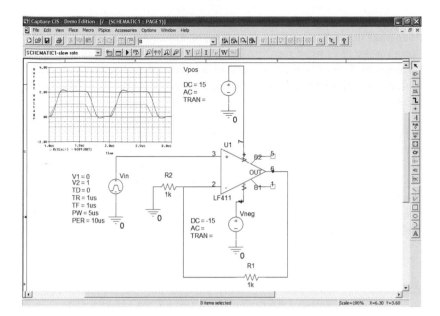

An amplifier circuit drawn using a commercial schematic capture software package. Inset: Simulated time dependent output.

interest that describe the operation of the circuit. Once the engineer is satisfied with the simulated performance of the design, the same software can generate the printed circuit board layout using geometrical parameters in the components library. This level of integration is continually increasing, to the point where soon an engineer will be able to draw a schematic, click a few buttons, and walk to the other side of the table to pick up a manufactured version of the circuit, ready to test!

The reader should be wary, however, of one thing. Circuit analysis software, although fun to use, is by no means a replacement for good old-fashioned paper-and-pencil analysis. We need to have a solid understanding of how circuits work in order to develop an ability to design them. Simply going through the motions of running a particular software package is a little like playing the lottery: with user-generated entry errors, hidden default parameters in the myriad of menu choices, and the occasional shortcoming of human-written code, there is no substitute for having at least an approximate idea of the expected behavior of a circuit. Then, if the simulation result does not agree with expectations, we can find the error early, rather than after it's too late.

Still, computer-aided analysis is a powerful tool. It allows us to vary parameter values and evaluate the change in circuit performance, and to consider several variations to a design in a straightforward manner. The result is a reduction of repetitive tasks, and more time to concentrate on engineering details.

1.6 SUCCESSFUL PROBLEM-SOLVING STRATEGIES

When asked to describe the single most frustrating part of circuit analysis, the vast majority of students feel that it is *knowing how to start the problem.* The second most difficult part is typically obtaining a complete set of equations and organizing them in such a way as to appear manageable.

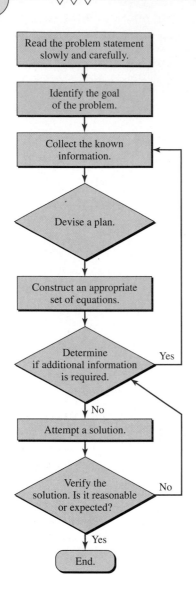

The fundamental instinct is to scan the problem statement quickly and then page through the text in a search for an applicable equation. We tend to long for the days when we were asked for the circumference of a circle, or faced with determining the volume of a pyramid! Although looking for a quick solution might be tempting, a consistently methodical approach to problem solving will yield better results in the long term.

The adjacent flowchart was designed as a guide to overcoming the two common obstacles of starting a problem and managing the solution process. Several of the steps may seem obvious, but it is the chronological order as well as the performance of each task that leads to success.

The real key to success in circuit analysis, however, is practice, especially in a relaxed, low-stress environment. Experience is the best teacher, and learning from our own mistakes will always be a part of the process of becoming a competent engineer.

READING FURTHER

This relatively inexpensive, best-selling book teaches the reader how to develop winning strategies in the face of seemingly impossible problems:

G. Polya, *How to Solve It*. Princeton, N.J.: Princeton University Press, 1971.

Basic Components and Electric Circuits

KEY CONCEPTS

Basic Electrical Quantities
and Associated Units:
Charge, Current, Voltage,
and Power

Current Direction and
Voltage Polarity

The Passive Sign Convention
for Calculating Power

Ideal Voltage and Current
Sources

Dependent Sources

Resistance and Ohm's Law

INTRODUCTION

The primary topic of this text is the analysis of electrical circuits and systems. In conducting a particular analysis, we often find ourselves seeking specific *currents, voltages,* or *powers,* so we begin with a brief description of these quantities. In terms of components that can be used to build electrical circuits, we have quite a few from which to choose. So as not to overwhelm, we initially focus on the *resistor,* a simple passive component, and a range of idealized active sources of voltage and current. As we move forward, new components will be added to the inventory to allow more complex (and useful) circuits to be considered.

One quick word of advice before we begin: Pay close attention to the role of "+" and "−" signs when labeling voltages, and the significance of the arrow in defining current; they often make the difference between wrong and right answers.

2.1 UNITS AND SCALES

In order to state the value of some measurable quantity, we must give both a *number* and a *unit,* such as "3 inches." Fortunately, we all use the same number system. This is not true for units, and a little time must be spent in becoming familiar with a suitable system. We must agree on a standard unit and be assured of its permanence and its general acceptability. The standard unit of length, for example, should not be defined in terms of the distance between two marks on a certain rubber band; this is not permanent, and furthermore everybody else is using another standard.

We have very little choice open to us with regard to a system of units. The one we will use was adopted by the National Bureau of Standards in 1964; it is used by all major professional engineering societies and is the language in which today's textbooks are written. This is the International System of Units (abbreviated **SI** in all languages), adopted by the General Conference on Weights and Measures in 1960. Modified several times since, the SI is built upon seven basic units: the *meter, kilogram, second, ampere, kelvin, mole,* and *candela* (see Table 2.1). This is a "metric system," some form of which is now in common use in most countries of the world, although it is not yet widely used in the United States. Units for other quantities such as volume, force, energy, etc., are derived from these seven base units.

TABLE   **2.1**   SI Base Units

Base Quantity	Name	Symbol
length	meter	m
mass	kilogram	kg
time	second	s
electric current	ampere	A
thermodynamic temperature	kelvin	K
amount of substance	mole	mol
luminous intensity	candela	cd

Units named after a person (e.g., the kelvin, after Lord Kelvin, a professor at the University of Glasgow) are written in lowercase, but abbreviated using an uppercase letter.

The "calorie" used with food, drink, and exercise is really a kilocalorie, 4.187 J.

The fundamental unit of work or energy is the **joule** (J). One joule (a kg m^2 s^{-2} in SI base units) is equivalent to 0.7376 foot pound-force (ft-lbf). Other energy units include the calorie (cal), equal to 4.187 J; the British thermal unit (Btu), which is 1055 J; and the kilowatthour (kWh), equal to 3.6×10^6 J. Power is defined as the *rate* at which work is done or energy is expended. The fundamental unit of power is the **watt** (W), defined as 1 J/s. One watt is equivalent to 0.7376 ft-lbf/s or, equivalently, 1/745.7 horsepower (hp).

The SI uses the decimal system to relate larger and smaller units to the basic unit, and employs prefixes to signify the various powers of 10. A list of prefixes and their symbols is given in Table 2.2; the ones most commonly encountered in engineering are highlighted.

TABLE   **2.2**   SI Prefixes

Factor	Name	Symbol	Factor	Name	Symbol
10^{-24}	yocto	y	10^{24}	yotta	Y
10^{-21}	zepto	z	10^{21}	zetta	Z
10^{-18}	atto	a	10^{18}	exa	E
10^{-15}	femto	f	10^{15}	peta	P
10^{-12}	pico	p	10^{12}	tera	T
10^{-9}	nano	n	10^{9}	giga	G
10^{-6}	micro	μ	10^{6}	mega	M
10^{-3}	milli	m	10^{3}	kilo	k
10^{-2}	centi	c	10^{2}	hecto	h
10^{-1}	deci	d	10^{1}	deka	da

These prefixes are worth memorizing, for they will appear often both in this text and in other technical work. Combinations of several prefixes, such as the millimicrosecond, are unacceptable. It is worth noting that in terms of distance, it is much more common to see "micron (μm)" as opposed to "micrometer," and often the angstrom (Å) is used for 10^{-10} meters. Also, in circuit analysis and engineering in general, it is fairly common to see numbers expressed in what are frequently termed "engineering units." In engineering notation, a quantity is represented by a number between 1 and 999 and an appropriate metric unit using a power divisible by 3. So, for example, it is preferable to express the quantity 0.048 W as 48 mW, instead of 4.8 cW, 4.8×10^{-2} W, or 48,000 μW.

PRACTICE

2.1 A krypton fluoride laser emits light at a wavelength of 248 nm. This is the same as: (*a*) 0.0248 mm; (*b*) 2.48 μm; (*c*) 0.248 μm; (*d*) 24,800 Å.

2.2 In a certain digital integrated circuit, a logic gate switches from the "on" state to the "off" state in 1 ns. This corresponds to: (*a*) 0.1 ps; (*b*) 10 ps; (*c*) 100 ps; (*d*) 1000 ps.

2.3 A typical incandescent reading lamp runs at 60 W. If it is left on constantly, how much energy (J) is consumed per day, and what is the weekly cost if energy is charged at a rate of 12.5 cents per kilowatthour?

Ans: 2.1 (c); 2.2 (d); 2.3 5.18 MJ, $1.26.

2.2 CHARGE, CURRENT, VOLTAGE, AND POWER

Charge

One of the most fundamental concepts in electric circuit analysis is that of charge conservation. We know from basic physics that there are two types of charge: positive (corresponding to a proton), and negative (corresponding to an electron). For the most part, this text is concerned with circuits in which only electron flow is relevant. There are many devices (such as batteries, diodes, and transistors) in which positive charge motion is important to understanding internal operation, but external to the device we typically concentrate on the electrons which flow through the connecting wires. Although we continuously transfer charges between different parts of a circuit, we do nothing to change the total amount of charge. In other words, we neither create nor destroy electrons (or protons) when running electric circuits.[1] Charge in motion represents a *current*.

In the SI system, the fundamental unit of charge is the **coulomb** (C). It is defined in terms of the **ampere** by counting the total charge that passes through an arbitrary cross section of a wire during an interval of one second; one coulomb is measured each second for a wire carrying a current of 1 ampere (Fig. 2.1). In this system of units, a single electron has a charge of -1.602×10^{-19} C and a single proton has a charge of $+1.602 \times 10^{-19}$ C.

As seen in Table 2.1, the base units of the SI are not derived from fundamental physical quantities. Instead, they represent historically agreed upon measurements, leading to definitions which occasionally seem backward. For example, it would make more sense physically to define the ampere based on electronic charge.

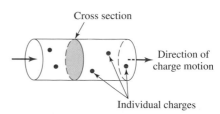

FIGURE 2.1 The definition of current illustrated using current flowing through a wire; 1 ampere corresponds to 1 coulomb of charge passing through the arbitrarily chosen cross section in 1 second.

---

(1) Although the occasional appearance of smoke may seem to suggest otherwise. . .

A quantity of charge that does not change with time is typically represented by Q. The instantaneous amount of charge (which may or may not be time-invariant) is commonly represented by $q(t)$, or simply q. This convention is used throughout the remainder of the text: capital letters are reserved for constant (time-invariant) quantities, whereas lowercase letters represent the more general case. Thus, a constant charge may be represented by *either* Q or q, but an amount of charge that changes over time *must* be represented by the lowercase letter q.

Current

The idea of "transfer of charge" or "charge in motion" is of vital importance to us in studying electric circuits because, in moving a charge from place to place, we may also transfer energy from one point to another. The familiar cross-country power-transmission line is a practical example of a device that transfers energy. Of equal importance is the possibility of varying the rate at which the charge is transferred in order to communicate or transfer information. This process is the basis of communication systems such as radio, television, and telemetry.

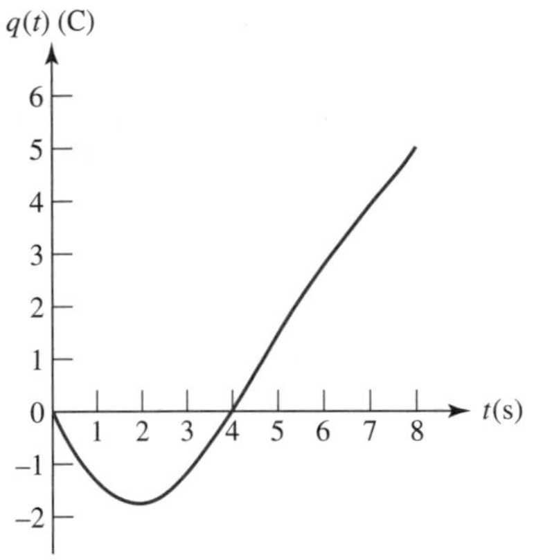

The current present in a discrete path, such as a metallic wire, has both a *numerical value* and a *direction* associated with it; it is a measure of the rate at which charge is moving past a given reference point in a specified direction.

Once we have specified a reference direction, we may then let $q(t)$ be the total charge that has passed the reference point since an arbitrary time $t = 0$, moving in the defined direction. A contribution to this total charge will be negative if negative charge is moving in the reference direction, or if positive charge is moving in the opposite direction. As an example, Fig. 2.2 shows a history of the total charge $q(t)$ that has passed a given reference point in a wire (such as the one shown in Fig. 2.1).

We define the current at a specific point and flowing in a specified direction as the instantaneous rate at which net positive charge is moving past that point in the specified direction. This, unfortunately, is the historical definition, which came into popular use before it was appreciated that current in wires is actually due to negative, not positive, charge motion. Current is symbolized by I or i, and so

$$i = \frac{dq}{dt} \qquad [1]$$

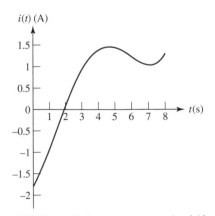

FIGURE 2.2 A graph of the instantaneous value of the total charge $q(t)$ that has passed a given reference point since $t = 0$.

The unit of current is the ampere (A), named after A. M. Ampère, a French physicist. It is commonly abbreviated as an "amp," although this is unofficial and somewhat informal. One ampere equals 1 coulomb per second.

Using Eq. [1], we compute the instantaneous current and obtain Fig. 2.3. The use of the lowercase letter i is again to be associated with an instantaneous value; an uppercase I would denote a constant (i.e., time-invariant) quantity.

The charge transferred between time t_0 and t may be expressed as a definite integral:

$$\int_{q(t_0)}^{q(t)} dq = \int_{t_0}^{t} i \, dt'$$

The total charge transferred over all time is thus given by

$$q(t) = \int_{t_0}^{t} i \, dt' + q(t_0) \qquad [2]$$

FIGURE 2.3 The instantaneous current $i = dq/dt$, where q is given in Fig. 2.2

Several different types of current are illustrated in Fig. 2.4. A current that is constant in time is termed a direct current, or simply dc, and is shown by Fig. 2.4a. We will find many practical examples of currents that vary sinusoidally with time (Fig. 2.4b); currents of this form are present in normal household circuits. Such a current is often referred to as alternating current, or ac. Exponential currents and damped sinusoidal currents (Fig. 2.4c and d) will also be encountered later.

We establish a graphical symbol for current by placing an arrow next to the conductor. Thus, in Fig. 2.5a the direction of the arrow and the value 3 A indicate either that a net positive charge of 3 C/s is moving to the right or that a net negative charge of -3 C/s is moving to the left each second. In Fig. 2.5b there are again two possibilities: either -3 A is flowing to the left or $+3$ A is flowing to the right. All four statements and both figures represent currents that are equivalent in their electrical effects, and we say that they are equal. A nonelectrical analogy that may be easier to visualize is to think in terms of a personal savings account: e.g., a deposit can be viewed as either a *negative* cash flow *out of* your account or a *positive* flow *into* your account.

It is convenient to think of current as the motion of positive charge, even though it is known that current flow in metallic conductors results from electron motion. In ionized gases, in electrolytic solutions, and in some semiconductor materials, positively charged elements in motion constitute part or all of the current. Thus, any definition of current can agree with the physical nature of conduction only part of the time. The definition and symbolism we have adopted are standard.

It is essential that we realize that the current arrow does not indicate the "actual" direction of current flow but is simply part of a convention that allows us to talk about "the current in the wire" in an unambiguous manner. The arrow is a fundamental part of the definition of a current! Thus, to talk about the value of a current $i_1(t)$ without specifying the arrow is to discuss an undefined entity. For example, Fig. 2.6a and b are meaningless representations of $i_1(t)$, whereas Fig. 2.6c is the proper definitive symbology.

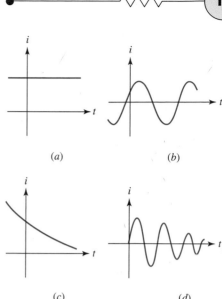

■ **FIGURE 2.4** Several types of current: (*a*) Direct current (dc). (*b*) Sinusoidal current (ac). (*c*) Exponential current. (*d*) Damped sinusoidal current.

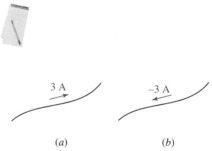

(*a*)   (*b*)

■ **FIGURE 2.5** Two methods of representation for the exact same current.

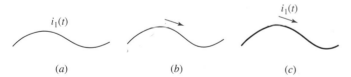

$i_1(t)$

(*a*)   (*b*)   (*c*)

■ **FIGURE 2.6** (*a, b*) Incomplete, improper, and incorrect definitions of a current. (*c*) the correct definition of $i_1(t)$.

PRACTICE
●

2.4 In the wire of Fig. 2.7, electrons are moving *left* to *right* to create a current of 1 mA. Determine I_1 and I_2.

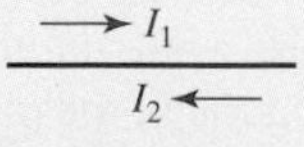

$\longrightarrow I_1$

$I_2 \longleftarrow$

■ **FIGURE 2.7**

Ans: $I_1 = -1$ mA; $I_2 = +1$ mA.

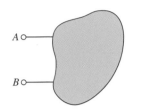

FIGURE 2.8 A general two-terminal circuit element.

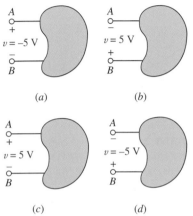

FIGURE 2.9 (*a, b*) Terminal *B* is 5 V positive with respect to terminal *A*; (*c, d*) terminal *A* is 5 V positive with respect to terminal *B*.

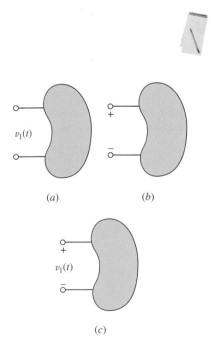

FIGURE 2.10 (*a, b*) These are inadequate definitions of a voltage. (*c*) A correct definition includes both a symbol for the variable and a plus-minus symbol pair.

Voltage

We must now begin to refer to a circuit element, something best defined in general terms to begin with. Such electrical devices as fuses, light bulbs, resistors, batteries, capacitors, generators, and spark coils can be represented by combinations of simple circuit elements. We begin by showing a very general circuit element as a shapeless object possessing two terminals at which connections to other elements may be made (Fig. 2.8).

There are two paths by which current may enter or leave the element. In subsequent discussions we will define particular circuit elements by describing the electrical characteristics that may be observed at their terminals.

In Fig. 2.8, let us suppose that a dc current is sent into terminal *A*, through the general element, and back out of terminal *B*. Let us also assume that pushing charge through the element requires an expenditure of energy. We then say that an electrical voltage (or a *potential difference*) exists between the two terminals, or that there is a voltage "across" the element. Thus, the voltage across a terminal pair is a measure of the work required to move charge through the element. The unit of voltage is the volt,[2] and 1 volt is the same as 1 J/C. Voltage is represented by V or *v*.

A voltage can exist between a pair of electrical terminals whether a current is flowing or not. An automobile battery, for example, has a voltage of 12 V across its terminals even if nothing whatsoever is connected to the terminals.

According to the principle of conservation of energy, the energy that is expended in forcing charge through the element must appear somewhere else. When we later meet specific circuit elements, we will note whether that energy is stored in some form that is readily available as electric energy or whether it changes irreversibly into heat, acoustic energy, or some other nonelectrical form.

We must now establish a convention by which we can distinguish between energy supplied *to* an element and energy that is supplied *by* the element itself. We do this by our choice of sign for the voltage of terminal *A* with respect to terminal *B*. If a positive current is entering terminal *A* of the element and an external source must expend energy to establish this current, then terminal *A* is positive with respect to terminal *B*. Alternatively, we may say that terminal *B* is negative with respect to terminal *A*.

The sense of the voltage is indicated by a plus-minus pair of algebraic signs. In Fig. 2.9*a*, for example, the placement of the + sign at terminal *A* indicates that terminal *A* is *v* volts positive with respect to terminal *B*. If we later find that *v* happens to have a numerical value of −5 V, then we may say either that *A* is −5 V positive with respect to *B* or that *B* is 5 V positive with respect to *A*. Other cases are shown in Fig. 2.9*b*, *c*, and *d*.

Just as we noted in our definition of current, it is essential to realize that the plus-minus pair of algebraic signs does not indicate the "actual" polarity of the voltage but is simply part of a convention that enables us to talk unambiguously about "the voltage across the terminal pair." Note: *The definition of any voltage must include a plus-minus sign pair*! Using a quantity $v_1(t)$ without specifying the location of the plus-minus sign pair is using an undefined term. Figure 2.10*a* and *b* do *not* serve as definitions of $v_1(t)$; Fig. 2.10*c* does.

(2) We are probably fortunate that the full name of the 18th century Italian physicist, *Alessandro Giuseppe Antonio Anastasio Volta,* is not used for our unit of potential difference!

PRACTICE

2.5 For the element in Fig. 2.11, $v_1 = 17$ V. Determine v_2.

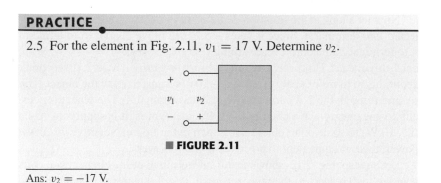

■ **FIGURE 2.11**

Ans: $v_2 = -17$ V.

Power

We have already defined power, and we will represent it by P or p. If one joule of energy is expended in transferring one coulomb of charge through the device in one second, then the rate of energy transfer is one watt. The absorbed power must be proportional both to the number of coulombs transferred per second (current) and to the energy needed to transfer one coulomb through the element (voltage). Thus,

$$p = vi \qquad [3]$$

Dimensionally, the right side of this equation is the product of joules per coulomb and coulombs per second, which produces the expected dimension of joules per second, or watts. The conventions for current, voltage, and power are shown in Fig. 2.12.

We now have an expression for the power being absorbed by a circuit element in terms of a voltage across it and current through it. Voltage was defined in terms of an energy expenditure, and power is the rate at which energy is expended. However, no statement can be made concerning energy transfer in any of the four cases shown in Fig. 2.9, for example, until the direction of the current is specified. Let us imagine that a current arrow is placed alongside each upper lead, directed to the right, and labeled "+2 A." First, consider the case shown in Fig. 2.9c. Terminal A is 5 V positive with respect to terminal B, which means that 5 J of energy is required to move each coulomb of positive charge into terminal A, through the object, and out terminal B. Since we are injecting +2 A (a current of 2 coulombs of positive charge per second) into terminal A, we are doing (5 J/C) × (2 C/s) = 10 J of work per second on the object. In other words, the object is absorbing 10 W of power from whatever is injecting the current.

We know from an earlier discussion that there is no difference between Fig. 2.9c and Fig. 2.9d, so we expect the object depicted in Fig. 2.9d to also be absorbing 10 W. We can check this easily enough: we are injecting +2 A into terminal A of the object, so +2 A flows out of terminal B. Another way of saying this is that we are injecting −2 A of current into terminal B. It takes −5 J/C to move charge from terminal B to terminal A, so the object is absorbing (−5 J/C) × (−2 C/s) = +10 W as expected. The only difficulty in describing this particular case is keeping the minus signs straight, but with a bit of care we see the correct answer can be obtained regardless of our choice of positive reference terminal (terminal A in Fig. 2.9c, and terminal B in Fig. 2.9d).

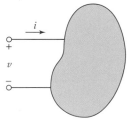

■ **FIGURE 2.12** The power absorbed by the element is given by the product $p = vi$. Alternatively, we can say that the element generates or supplies a power −vi.

Now let's look at the situation depicted in Fig. 2.9*a*, again with +2 A injected into terminal *A*. Since it takes −5 J/C to move charge from terminal *A* to terminal *B*, the object is absorbing (−5 J/C) × (2 C/s) = −10 W. What does this mean? How can anything absorb **negative** power? If we think about this in terms of energy transfer, −10 J is transferred to the object each second through the 2 A current flowing into terminal *A*. The object is actually losing energy—at a rate of 10 J/s. In other words, it is supplying 10 J/s (i.e. 10 W) to some other object not shown in the figure. Negative *absorbed* power, then, is equivalent to positive *supplied* power.

Let's recap. Fig. 2.12 shows that if one terminal of the element is *v* volts positive with respect to the other terminal, and if a current *i* is entering the element through that terminal, then a power *p* = *vi* is being *absorbed* by the element; it is also correct to say that a power *p* = *vi* is being *delivered* to the element. When the current arrow is directed into the element at the plus-marked terminal, we satisfy the ***passive sign convention.*** This convention should be studied carefully, understood, and memorized. In other words, it says that if the current arrow and the voltage polarity signs are placed such that the current enters that end of the element marked with the positive sign, then the power *absorbed* by the element can be expressed by the product of the specified current and voltage variables. If the numerical value of the product is negative, then we say that the element is absorbing negative power, or that it is actually generating power and delivering it to some external element. For example, in Fig. 2.12 with *v* = 5 V and *i* = −4 A, the element may be described as either absorbing −20 W or generating 20 W.

Conventions are only required when there is more than one way to do something, and confusion may result when two different groups try to communicate. For example, it is rather arbitrary to always place "North" at the top of a map; compass needles don't point "up," anyway. Still, if we were talking to people who had secretly chosen the opposite convention of placing "South" at the top of their maps, imagine the confusion that could result! In the same fashion, there is a general convention that always draws the current arrows pointing into the positive voltage terminal, irregardless of whether the element supplies or absorbs power. This convention is not incorrect but sometimes results in counterintuitive currents labeled on circuit schematics. The reason for this is that it simply seems more natural to refer to positive current flowing out of a voltage or current source that is supplying positive power to one or more circuit elements.

If the current arrow is directed into the "+" marked terminal of an element, then *p* = *vi* yields the *absorbed* power. A negative value indicates that power is actually being generated by the element; it might have been better to define a current flowing out of the "+" terminal.

If the current arrow is directed out of the "+" terminal of an element, then *p* = *vi* yields the *supplied* power. A negative value in this case indicates that power is being absorbed.

EXAMPLE 2.1

Compute the power absorbed by each part in Fig. 2.13.

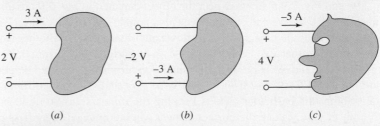

(a)   (b)   (c)

■ **FIGURE 2.13** (*a, b, c*) Three examples of two-terminal elements.

In Fig. 2.13*a*, we see that the reference current is defined consistent with the passive sign convention, which assumes that the element is absorbing power. With +3 A flowing into the positive reference terminal, we compute

$$P = (2 \text{ V})(3 \text{ A}) = 6 \text{ W}$$

of power absorbed by the element.

Fig. 2.13*b* shows a slightly different picture. Now, we have a current of −3 A flowing into the positive reference terminal. However, the voltage as defined is negative. This gives us an absorbed power

$$P = (-2 \text{ V})(-3 \text{ A}) = 6 \text{ W}$$

Thus, we see that the two cases are actually equivalent: A current of +3 A flowing into the top terminal is the same as a current of +3 A flowing out of the bottom terminal, or, equivalently, a current of −3 A flowing into the bottom terminal.

Referring to Fig. 2.13*c*, we again apply the passive sign convention rules and compute an absorbed power

$$P = (4 \text{ V})(-5 \text{ A}) = -20 \text{ W}$$

Since we computed a negative *absorbed* power, this tells us that the element in Fig. 2.13*c* is actually *supplying* +20 W (i.e., it's a source of energy).

PRACTICE

2.6 Find the power being absorbed by the circuit element in Fig. 2.14*a*.

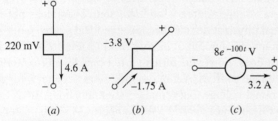

(a)   (b)   (c)

■ **FIGURE 2.14**

2.7 Find the power being generated by the circuit element in Fig. 2.14*b*.
2.8 Find the power being delivered to the circuit element in Fig. 2.14*c* at $t = 5$ ms.

Ans: 1.012 W; 6.65 W; −15.53 W.

2.3 • VOLTAGE AND CURRENT SOURCES

Using the concepts of current and voltage, it is now possible to be more specific in defining a *circuit element.*

In so doing, it is important to differentiate between the physical device itself and the mathematical model which we will use to analyze its behavior in a circuit. The model is only an approximation.

Let us agree that we will use the expression *circuit element* to refer to the mathematical model. The choice of a particular model for any real device must be made on the basis of experimental data or experience; we will usually assume that this choice has already been made. For simplicity, we initially consider circuits with idealized components represented by simple models.

All the simple circuit elements that we will consider can be classified according to the relationship of the current through the element to the voltage across the element. For example, if the voltage across the element is linearly proportional to the current through it, we will call the element a resistor. Other types of simple circuit elements have terminal voltages which are proportional to the *derivative* of the current with respect to time (an inductor), or to the *integral* of the current with respect to time (a capacitor). There are also elements in which the voltage is completely independent of the current, or the current is completely independent of the voltage; these are termed *independent sources*. Furthermore, we will need to define special kinds of sources for which either the source voltage or current depends upon a current or voltage elsewhere in the circuit; such sources are referred to as *dependent sources*. Dependent sources are used a great deal in electronics to model both dc and ac behavior of transistors, especially in amplifier circuits.

Independent Voltage Sources

The first element we will consider is the ***independent voltage source.*** The circuit symbol is shown in Fig. 2.15*a*; the subscript *s* merely identifies the voltage as a "source" voltage, and is common but not required. *An independent voltage source is characterized by a terminal voltage which is completely independent of the current through it.* Thus, if we are given an independent voltage source and are notified that the terminal voltage is 12 V, then we always assume this voltage, regardless of the current flowing.

The independent voltage source is an *ideal* source and does not represent exactly any real physical device, because the ideal source could theoretically deliver an infinite amount of energy from its terminals. This idealized voltage source does, however, furnish a reasonable approximation to several practical voltage sources. An automobile storage battery, for example, has a 12 V terminal voltage that remains essentially constant as long as the current through it does not exceed a few amperes. A small current may flow in either direction through the battery. If it is positive and flowing out of the positively marked terminal, then the battery is furnishing power to the headlights, for example; if the current is positive and flowing into the positive terminal, then the battery is charging by absorbing energy from the alternator.[3] An ordinary household electrical outlet also approximates an independent voltage source, providing a voltage $v_s = 115\sqrt{2}\cos 2\pi 60t$ V; this representation is valid for currents less than 20 A or so.

A point worth repeating here is that the presence of the plus sign at the upper end of the symbol for the independent voltage source in Fig. 2.15*a* does not necessarily mean that the upper terminal is numerically positive with respect to the lower terminal. Instead, it means that the upper terminal is v_s volts positive with respect to the lower. If at some instant v_s happens to be negative, then the upper terminal is actually negative with respect to the lower at that instant.

> By definition, a simple circuit element is the mathematical model of a two-terminal electrical device, and it can be completely characterized by its voltage-current relationship; it cannot be subdivided into other two-terminal devices.

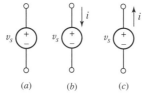

| (a) | (b) | (c) |

■ **FIGURE 2.15** Circuit symbol of the independent voltage source.

> If you've ever noticed the room lights dim when an air conditioner kicks on, it's because the sudden large current demand temporarily led to a voltage drop. After the motor starts moving, it takes less current to keep it in motion. At that point, the current demand is reduced, the voltage returns to its original value, and the wall outlet again provides a reasonable approximation of an ideal voltage source.

---

(3) Or the battery of a friend's car, if you accidentally left your headlights on. . . .

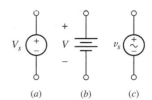

Consider a current arrow labeled "*i*" placed adjacent to the upper conductor of the source as in Fig. 2.15*b*. The current *i* is entering the terminal at which the positive sign is located, the passive sign convention is satisfied, and the source thus *absorbs* power $p = v_s i$. More often than not, a source is expected to deliver power to a network and not to absorb it. Consequently, we might choose to direct the arrow as in Fig. 2.15*c* so that $v_s i$ will represent the power *delivered* by the source. Technically, either arrow direction may be chosen; whenever possible, we will adopt the convention of Fig. 2.15*c* in this text for voltage and current sources, which are not usually considered passive devices.

An independent voltage source with a constant terminal voltage is often termed an independent dc voltage source and can be represented by either of the symbols shown in Fig. 2.16*a* and *b*. Note in Fig. 2.16*b* that when the physical plate structure of the battery is suggested, the longer plate is placed at the positive terminal; the plus and minus signs then represent redundant notation, but they are usually included anyway. For the sake of completeness, the symbol for an independent ac voltage source is shown in Fig. 2.16*c*.

FIGURE 2.16 (*a*) DC voltage source symbol; (*b*) battery symbol; (*c*) ac voltage source symbol.

Terms like dc voltage source and dc current source are commonly used. Literally, they mean "direct-current voltage source" and "direct-current current source," respectively. Although these terms may seem a little odd or even redundant, the terminology is so widely used there's no point in fighting it.

Independent Current Sources

Another ideal source which we will need is the **independent current source.** Here, the current through the element is completely independent of the voltage across it. The symbol for an independent current source is shown in Fig. 2.17. If i_s is constant, we call the source an independent dc current source. An ac current source is often drawn with a tilde through the arrow, similar to the ac voltage source shown in Fig. 2.16*c*.

Like the independent voltage source, the independent current source is at best a reasonable approximation for a physical element. In theory it can deliver infinite power from its terminals because it produces the same finite current for any voltage across it, no matter how large that voltage may be. It is, however, a good approximation for many practical sources, particularly in electronic circuits.

Although most students seem happy enough with an independent voltage source providing a fixed voltage but essentially any current, *it is a common mistake* to view an independent current source as having zero voltage across its terminals while providing a fixed current. In fact, we do not know a priori what the voltage across a current source will be—it depends entirely on the circuit to which it is connected.

FIGURE 2.17 Circuit symbol for the independent current source.

Dependent Sources

The two types of ideal sources that we have discussed up to now are called *independent* sources because the value of the source quantity is not affected in any way by activities in the remainder of the circuit. This is in contrast with yet another kind of ideal source, the *dependent,* or *controlled,* source, in which the source quantity is determined by a voltage or current existing at some other location in the system being analyzed. Sources such as these appear in the equivalent electrical models for many electronic devices, such as transistors, operational amplifiers, and integrated circuits. To distinguish between dependent and independent sources, we introduce the diamond symbols shown in Fig. 2.18. In Fig. 2.18*a* and *c*, *K* is a dimensionless scaling constant. In Fig.2.18*b*, *g* is a scaling factor with units of A/V; in Fig. 2.18*d*, *r* is a scaling factor with units of V/A. The controlling current i_x and the controlling voltage v_x must be defined in the circuit.

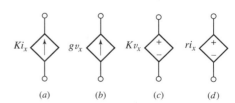

FIGURE 2.18 The four different types of dependent sources: (*a*) current-controlled current source; (*b*) voltage-controlled current source; (*c*) voltage-controlled voltage source; (*d*) current-controlled voltage source.

It does seem odd at first to have a current source whose value depends on a voltage, or a voltage source which is controlled by a current flowing through some other element. Even a voltage source depending on a remote voltage can appear strange. Such sources are invaluable for modeling complex systems, however, making the analysis algebraically straightforward. Examples include the drain current of a field effect transistor as a function of the gate voltage, or the output voltage of an analog integrated circuit as a function of differential input voltage. When encountered during circuit analysis, we write down the entire controlling expression for the dependent source just as we would if it was a numerical value attached to an independent source. This often results in the need for an additional equation to complete the analysis, unless the controlling voltage or current is already one of the specified unknowns in our system of equations.

EXAMPLE **2.2**

In the circuit of Fig. 2.19a, if v_2 is known to be 3 V, find v_L.

We have been provided with a partially labeled circuit diagram and the additional information that $v_2 = 3$ V. This is probably worth adding to our diagram, as shown in Fig. 2.19b.

Next we step back and look at the information collected. In examining the circuit diagram, we notice that the desired voltage v_L is the same as the voltage across the dependent source. Thus,

$$v_L = 5v_2$$

At this point, we would be done with the problem if only we knew v_2!

Returning to our diagram, we see that we actually do know v_2—it was specified as 3 V. We therefore write

$$v_2 = 3$$

We now have two (simple) equations in two unknowns, and solve to find $v_L = 15$ V.

An important lesson at this early stage of the game is that *the time it takes to completely label a circuit diagram is always a good investment.* As a final step, we should go back and check over our work to ensure that the result is correct.

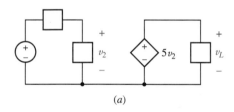

(a)

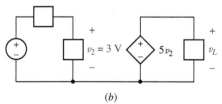

(b)

■ **FIGURE 2.19** (a) An example circuit containing a voltage-controlled voltage source. (b) The additional information provided is included on the diagram.

PRACTICE

2.9 Find the power *absorbed* by each element in the circuit in Fig. 2.20.

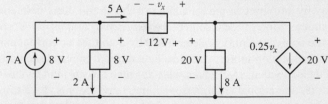

■ **FIGURE 2.20**

Ans: (left to right) −56 W; 16 W; −60 W; 160 W; −60 W.

Dependent and independent voltage and current sources are *active* elements; they are capable of delivering power to some external device. For the present we will think of a *passive* element as one which is capable only of receiving power. However, we will later see that several passive elements are able to store finite amounts of energy and then return that energy later to various external devices; since we still wish to call such elements passive, it will be necessary to improve upon our two definitions a little later.

Networks and Circuits

The interconnection of two or more simple circuit elements forms an electrical **network.** If the network contains at least one closed path, it is also an electric **circuit.** Note: Every circuit is a network, but not all networks are circuits (see Fig. 2.21)!

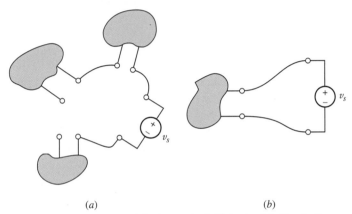

(a)   (b)

■ **FIGURE 2.21** (*a*) A network that is not a circuit. (*b*) A network that is a circuit.

A network that contains at least one active element, such as an independent voltage or current source, is an active network. A network that does not contain any active elements is a passive network.

We have now defined what we mean by the term **circuit element,** and we have presented the definitions of several specific circuit elements, the independent and dependent voltage and current sources. Throughout the remainder of the book we will define only five additional circuit elements: the resistor, inductor, capacitor, transformer, and the ideal operational amplifier ("op amp," for short). These are all ideal elements. They are important because we may combine them into networks and circuits that represent real devices as accurately as we require. Thus, the transistor shown in Fig. 2.22*a* and *b* may be modeled by the voltage terminals designated v_{gs} and the single dependent current source of Fig. 2.22*c*. Note that the dependent current source produces a current that depends on a voltage elsewhere in the circuit. The parameter g_m, commonly referred to as the transconductance, is calculated using transistor-specific details as well as the operating point determined by the circuit connected to the transistor. It is generally a small number, on the order of 10^{-2} to perhaps 10 A/V. This model works pretty well as long as the frequency of any sinusoidal source is neither very large nor very small; the model can be modified to account for frequency-dependent effects by including additional ideal circuit elements such as resistors and capacitors.

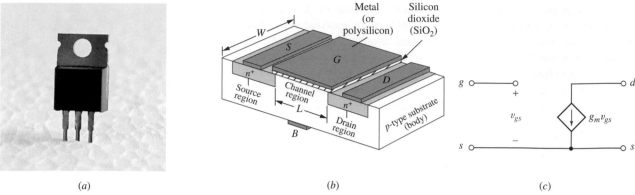

■ **FIGURE 2.22** The Metal Oxide Semiconductor Field Effect Transistor (MOSFET). (*a*) An IRF540 N-channel power MOSFET in a TO-220 package, rated at 100 V and 22 A; (*b*) cross-sectional view of a basic MOSFET (*R. Jaeger, Microelectronic Design, McGraw-Hill, 1997*); (*c*) equivalent circuit model for use in ac circuit analysis.

Similar (but much smaller) transistors typically constitute only one small part of an integrated circuit that may be less than 2 mm × 2 mm square and 200 μm thick and yet contains several thousand transistors plus various resistors and capacitors. Thus, we may have a physical device that is about the size of one letter on this page but requires a model composed of ten thousand ideal simple circuit elements. We use this concept of "circuit modeling" in a number of electrical engineering topics covered in other courses, including electronics, energy conversion, and antennas.

2.4 OHM'S LAW

So far, we have been introduced to both dependent and independent voltage and current sources and were cautioned that they were *idealized* active elements that could only be approximated in a real circuit. We are now ready to meet another idealized element, the linear resistor. The resistor is the simplest passive element, and we begin our discussion by considering the work of an obscure German physicist, Georg Simon Ohm, who published a pamphlet in 1827 that described the results of one of the first efforts to measure currents and voltages, and to describe and relate them mathematically. One result was a statement of the fundamental relationship we now call ***Ohm's law,*** even though it has since been shown that this result was discovered 46 years earlier in England by Henry Cavendish, a brilliant semirecluse. Ohm's pamphlet received much undeserved criticism and ridicule for several years after its first publication but was later accepted and served to remove the obscurity associated with his name.

Ohm's law states that the voltage across conducting materials is directly proportional to the current flowing through the material, or

$$v = R\,i \qquad\qquad [4]$$

where the constant of proportionality R is called the ***resistance.*** The unit of resistance is the *ohm,* which is 1 V/A and customarily abbreviated by a capital omega, Ω.

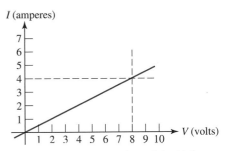

When this equation is plotted on *i*-versus-*v* axes, the graph is a straight line passing through the origin (Fig. 2.23). Equation [4] is a linear equation, and we will consider it as the definition of a *linear resistor*. Hence, if the ratio of the current and voltage associated with any simple current element is a constant, then the element is a linear resistor and has a resistance equal to the voltage-to-current ratio. Resistance is normally considered to be a positive quantity, although negative resistances may be simulated with special circuitry.

Again, it must be emphasized that the linear resistor is an idealized circuit element; it is only a mathematical model of a real, physical device. "Resistors" may be easily purchased or manufactured, but it is soon found that the voltage-current ratios of these physical devices are reasonably constant only within certain ranges of current, voltage, or power, and depend also on temperature and other environmental factors. We usually refer to a linear resistor as simply a resistor; any resistor that is nonlinear will always be described as such. Nonlinear resistors should not necessarily be considered undesirable elements. Although it is true that their presence complicates an analysis, the performance of the device may depend on or be greatly improved by the nonlinearity. For example, fuses for overcurrent protection and Zener diodes for voltage regulation are very nonlinear in nature, a fact that is exploited when using them in circuit design.

■ **FIGURE 2.23** Current-voltage relationship for an example 2 Ω linear resistor.

Power Absorption

Figure 2.24 shows several different resistor packages, as well as the most common circuit symbol used for a resistor. In accordance with the voltage, current, and power conventions already adopted, the product of *v* and *i*

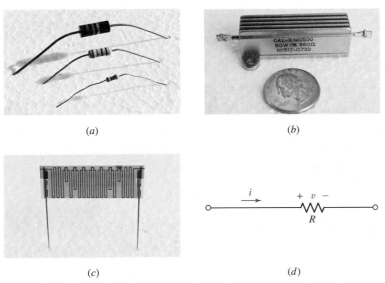

(*a*)

(*b*)

(*c*)

(*d*)

■ **FIGURE 2.24** (*a*) Several common resistor packages. (*b*) A 560 Ω power resistor rated at up to 500 W. (*c*) A 5% tolerance 10-Teraohm (10,000,000,000,000 Ω) resistor manufactured by Ohmcraft. (*d*) Circuit symbol for the resistor, applicable to all of the devices in (*a*) through (*c*).

gives the power absorbed by the resistor. That is, v and i are selected to satisfy the passive sign convention. The absorbed power appears physically as heat and/or light and is always positive; a (positive) resistor is a passive element that cannot deliver power or store energy. Alternative expressions for the absorbed power are

$$p = vi = i^2 R = v^2/R \qquad [5]$$

One of the authors (who prefers not to be identified further)[4] had the unfortunate experience of inadvertently connecting a 100 Ω, 2 W carbon resistor across a 110 V source. The ensuing flame, smoke, and fragmentation were rather disconcerting, demonstrating clearly that a practical resistor has definite limits to its ability to behave like the ideal linear model. In this case, the unfortunate resistor was called upon to absorb 121 W; since it was designed to handle only 2 W, its reaction was understandably violent.

EXAMPLE 2.3

The resistor shown in Fig. 2.24b is connected in a circuit that forces a current of 428 mA to flow through it. Calculate the voltage across its terminals and the power it is dissipating.

The voltage across a resistor is specified by Ohm's law, so

$$v = R\,i = (560)(0.428) = 239.7 \text{ V}$$

We may obtain the power dissipated by the resistor in several different ways. Since we have both the voltage across its terminals and the current flowing through it,

$$p = vi = (239.7)(0.428) = 102.6 \text{ W}$$

which is approximately 20% of its maximum rating of 500 W. We check our results using the two alternative equations:

$$p = v^2/R = (239.7)^2/560 = 102.6 \text{ W}$$
$$p = i^2 R = (0.428)^2\,560 = 102.6 \text{ W}$$

as expected.

PRACTICE

With reference to v and i defined in Fig. 2.25, compute the following quantities:

2.10 R if $i = -1.6$ mA and $v = -6.3$ V.

2.11 The absorbed power if $v = -6.3$ V and $R = 21\ \Omega$.

2.12 i if $v = -8$ V and R is absorbing 0.24 W.

Ans: 3.94 kΩ; 1.89 W; −30.0 mA.

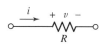

■ **FIGURE 2.25**

(4) Name gladly furnished upon written request to S.M.D.

Wire Gauge

Technically speaking, any material (except for a super-conductor) will provide resistance to current flow. As in all introductory circuits texts, however, we will tacitly assume that wires appearing in circuit diagrams have zero resistance. This implies that there is no potential difference between the ends of a wire, and hence no power absorbed or heat generated. Although this is usually not an unreasonable assumption, it does neglect practical considerations when choosing the appropriate wire diameter for a specific application.

Resistance is determined by (1) the inherent resistivity of a material and (2) the device geometry. **Resistivity,** represented by the symbol ρ, is a measure of the ease with which electrons can travel through a certain material. Since it is the ratio of the electric field (V/m) to the areal density of current flowing in the material (A/m^2), the general unit of ρ is an $\Omega \cdot$m, although metric prefixes are often employed. Every material has a different inherent resistivity, which depends on temperature. Some examples are shown in Table 2.3; as can be seen, there is a small variation between different types of copper (less than 1%) but a very large difference between different metals. In particular, although physically stronger than copper, steel wire is several times more resistive. In some technical discussions, it is more common to see the conductivity (symbolized by σ) of a

material quoted, which is simply the reciprocal of the resistivity.

The resistance of a particular object is obtained by multiplying the resistivity by the length ℓ of the resistor and dividing by the cross-sectional area (A) as in Eq. [6]; these parameters are illustrated in Fig. 2.26.

$$R = \rho \frac{\ell}{A} \qquad [6]$$

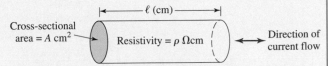

■ **FIGURE 2.26** Definition of geometrical parameters used to compute the resistance of a wire. The resistivity of the material is assumed to be spatially uniform.

We determine the resistivity when we select the material from which to fabricate a wire and measure the temperature of the application environment. Since a finite amount of power is absorbed by the wire due to its resistance, current flow leads to the production of heat. Thicker wires have lower resistance and also dissipate heat more easily but are heavier, take up a larger volume, and are more expensive. Thus, we are motivated by practical considerations to choose the smallest wire that

TABLE 2.3 Common Electrical Wire Materials and Resistivities*

ASTM Specification**	Temper and Shape	Resistivity at 20°C ($\mu\Omega \cdot$ cm)
B33	Copper, tinned soft, round	1.7654
B75	Copper, tube, soft, OF copper	1.7241
B188	Copper, hard bus tube, rectangular or square	1.7521
B189	Copper, lead-coated soft, round	1.7654
B230	Aluminum, hard, round	2.8625
B227	Copper-clad steel, hard, round, grade 40 HS	4.3971
B355	Copper, nickel-coated soft, round Class 10	1.9592
B415	Aluminum-clad steel, hard, round	8.4805

* C. B. Rawlins, "Conductor Materials," *Standard Handbook for Electrical Engineering,* 13th ed., D. G. Fink and H. W. Beaty, eds. New York: McGraw-Hill, 1993, pp. 4-4 to 4-8.

** American Society of Testing Materials

(Continued on next page)

can safely do the job, rather than simply choosing the largest diameter wire available in an effort to minimize resistive losses. The American Wire Gauge (AWG) is a standard system of specifying wire size. In selecting a wire gauge, smaller AWG corresponds to a larger wire diameter; an abbreviated table of common gauges is given in Table 2.4. Local fire and electrical safety codes typically dictate the required gauge for specific wiring applications, based on the maximum current expected as well as where the wires will be located.

TABLE 2.4 Some Common Wire Gauges and the Resistance of (Soft) Solid Copper Wire.*

Conductor Size (AWG)	Cross-Sectional Area (mm²)	Ohms per 1000 ft at 20°C
28	0.0804	65.3
24	0.205	25.7
22	0.324	16.2
18	0.823	6.39
14	2.08	2.52
12	3.31	1.59
6	13.3	0.3952
4	21.1	0.2485
2	33.6	0.1563

* C. B. Rawlins, et al., *Standard Handbook for Electrical Engineering*, 13th ed., D. G. Fink and H. W. Beaty, eds. New York: McGraw-Hill, 1993, p. 4-47.

EXAMPLE 2.4

A wire is run across a 2000 ft span to a high-power lamp that draws 100 A. If 4 AWG wire is used, how much power is dissipated (i.e., lost or wasted) within the wire?

The best place to begin this problem is to sketch a quick picture, as shown in Fig. 2.27. We see from Table 2.4 that 4 AWG wire is 0.2485 Ω per 1000 ft. The wire out to the lamp is 2000 ft long, and the wire back to the power source is also 2000 ft long, for a total of 4000 ft. Thus, the wire has a resistance of

$$R = (4000 \text{ ft})(0.2485 \text{ } \Omega/1000 \text{ ft}) = 0.994 \text{ } \Omega$$

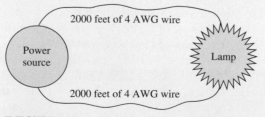

■ **FIGURE 2.27** A quick sketch of the lamp circuit for Example 2.4.

The dissipated power is given by $i^2 R$, where $i = 100$ A. Thus, 9940 W or 9.94 kW is dissipated by the wire. Even with less than 1 Ω total resistance, we find that a huge amount of power is wasted by the wire: this must also be supplied by the power source, but it never reaches the lamp!

PRACTICE

2.13 Faced with the significant power losses described in Example 2.4, your manager instructs you to have the 4 AWG wire replaced with 2 AWG. Calculate the power lost in the new wire, assuming the lamp is still drawing 100 A. Out of curiosity, how much more will the new wiring weigh (two times more, four times more, etc.)?

Ans: 6.25 kW, 1.59 times more.

Conductance

For a linear resistor the ratio of current to voltage is also a constant.

$$\frac{i}{v} = \frac{1}{R} = G \tag{7}$$

where G is called the *conductance*. The SI unit of conductance is the siemens (S), 1 A/V. An older, unofficial unit for conductance is the mho, abbreviated by an inverted capital omega, ℧. You will occasionally see it used on some circuit diagrams, as well as in catalogs and texts. The same circuit symbol (Fig. 2.24d) is used to represent both resistance and conductance. The absorbed power is again necessarily positive and may be expressed in terms of the conductance by

$$p = vi = v^2 G = \frac{i^2}{G} \tag{8}$$

Thus a 2 Ω resistor has a conductance of $\frac{1}{2}$ S, and if a current of 5 A is flowing through it, then a voltage of 10 V is present across the terminals and a power of 50 W is being absorbed.

All the expressions given so far in this section were written in terms of instantaneous current, voltage, and power, such as $v = iR$ and $p = vi$. We should recall that this is a shorthand notation for $v(t) = R i(t)$ and $p(t) = v(t) i(t)$. The current through and voltage across a resistor must both vary with time in the same manner. Thus, if $R = 10 \Omega$ and $v = 2 \sin 100t$ V, then $i = 0.2 \sin 100t$ A. Note that the power is given by $0.4 \sin^2 100t$ W, and a simple sketch will illustrate the different nature of its variation with time. Although the current and voltage are each negative during certain time intervals, the absorbed power is *never* negative!

Resistance may be used as the basis for defining two commonly used terms, *short circuit* and *open circuit*. We define a short circuit as a resistance of zero ohms; then, since $v = iR$, the voltage across a short circuit must be zero, although the current may have any value. In an analogous manner, we define an open circuit as an infinite resistance. It follows from Ohm's law that the current must be zero, regardless of the voltage across the open circuit.

Although real wires have a small resistance associated with them, we always assume them to have zero resistance unless otherwise specified. Thus, in all of our circuit schematics, wires are taken to be perfect short circuits.

SUMMARY AND REVIEW

- The system of units most commonly used in electrical engineering is the SI.

- The direction in which positive charges are moving is the direction of positive current flow; alternatively, positive current flow is in the direction opposite that of moving electrons.

- To define a current, both a value and a direction must be given. Currents are typically denoted by the uppercase letter "I" for constant (dc) values, and either $i(t)$ or simply i otherwise.

- To define a voltage across an element, it is necessary to label the terminals with "+" and "−" signs as well as to provide a value (either an algebraic symbol or a numerical value).

- Any element is said to supply positive power if positive current flows out of the positive voltage terminal. Any element absorbs positive power if positive current flows into the positive voltage terminal.

- There are six sources: the independent voltage source, the independent current source, the current-controlled dependent current source, the voltage-controlled dependent current source, the voltage-controlled dependent voltage source, and the current-controlled dependent voltage source.

- Ohm's law states that the voltage across a linear resistor is directly proportional to the current flowing through it; i.e., $v = R\,i$.

- The power dissipated by a resistor (which leads to the production of heat) is given by $p = vi = i^2 R = v^2/R$.

- Wires are typically assumed to have zero resistance in circuit analysis. When selecting a wire gauge for a specific application, however, local electrical and fire codes must be consulted.

Note that a current represented by i or $i(t)$ can be constant (dc) or time-varying, but currents represented by the symbol I must be non-time-varying.

READING FURTHER

A good book that discusses the properties and manufacture of resistors in considerable depth:

Felix Zandman, Paul-René Simon, and Joseph Szwarc, *Resistor Theory and Technology*. Raleigh, N.C.: SciTech Publishing, 2002.

A good all-purpose electrical engineering handbook:

Donald G. Fink and H. Wayne Beaty, *Standard Handbook for Electrical Engineers,* 13th ed., New York: McGraw-Hill, 1993.

In particular, pp. 1-1 to 1-51, 2-8 to 2-10, and 4-2 to 4-207 provide an in-depth treatment of topics related to those discussed in this chapter.

A detailed reference for the SI is available on the Web from the National Institute of Standards:

Barry N. Taylor, *Guide for the Use of the International System of Units (SI),* NIST Special Publication 811, 1995 Edition, www.nist.gov.

EXERCISES

2.1 Units and Scales

1. Convert the following to engineering notation:
   - (a) 1.2×10^{-5} s
   - (b) 750 mJ
   - (c) 1130 Ω
   - (d) 3,500,000,000 bits
   - (e) 0.0065 μm
   - (f) 13,560,000 Hz
   - (g) 0.039 nA
   - (h) 49,000 Ω
   - (i) 1.173×10^{-5} μA

2. Convert the following to engineering notation:
   - (a) 1,000,000 W
   - (b) 12.35 mm
   - (c) 47,000 W
   - (d) 0.00546 A
   - (e) 0.033 mJ
   - (f) 5.33×10^{-6} mW
   - (g) 0.000000001 s
   - (h) 5555 kW
   - (i) 32,000,000,000 pm

3. Convert the following to SI units. Be sure to use engineering notation, and retain four significant digits.
   - (a) 400 hp
   - (b) 12 ft
   - (c) 2.54 cm
   - (d) 67 Btu
   - (e) 285.4×10^{-15} s

4. A certain 15 V dry-cell battery, completely discharged, requires a current of 100 mA for 3 hr to completely recharge. What is the energy storage capacity of the battery, assuming the voltage does not depend on its charge status?

5. A zippy little electric car is equipped with a 175 hp motor.
   - (a) How many kW are required to run the motor if we assume 100 percent efficiency in converting electrical power to mechanical power?
   - (b) How much energy (in J) is expended if the motor is run continuously for 3 hours?
   - (c) If a single lead-acid battery has a 430-kilowatthour storage capacity, how many batteries are required for part (b)?

6. A KrF excimer laser generates 400 mJ laser pulses 20 ns in duration.
   - (a) What is the peak instantaneous power of the laser?
   - (b) If only 20 pulses can be generated per second, what is the average power output of the laser?

7. An amplified titanium:sapphire laser generates 1 mJ laser pulses 75 fs in duration.
   - (a) What is the peak instantaneous power of the laser?
   - (b) If only 100 pulses can be generated per second, what is the average power output of the laser?

8. The power supplied by a certain battery is a constant 6 W over the first 5 min, zero for the following 2 min, a value that increases linearly from zero to 10 W during the next 10 min, and a power that decreases linearly from 10 W to zero in the following 7 min. (a) What is the total energy in joules expended during this 24 min interval? (b) What is the average power in Btu/h during this time?

9. A new type of battery delivers 10 W of power for 8 hr without voltage or current fluctuation. After 8 hr, however, the power output drops linearly from 10 W to 0 in only 5 min. (a) What is the energy storage capacity of the battery? (b) How much energy is delivered during the last 5 min of the discharge cycle?

2.2 Charge, Current, Voltage, and Power

10. The total charge accumulated by a certain device is given as a function of time by $q = 18t^2 - 2t^4$ (in SI units). (a) What total charge is accumulated

at $t = 2$ s? (b) What is the maximum charge accumulated in the interval $0 \leq t \leq 3$ s, and when does it occur? (c) At what rate is charge being accumulated at $t = 0.8$ s? (d) Sketch curves of q versus t and i versus t in the interval $0 \leq t \leq 3$ s.

11. The current $i_1(t)$ shown in Fig. 2.6c is given as $-2 + 3e^{-5t}$ A for $t < 0$, and $-2 + 3e^{3t}$ A for $t > 0$. Find (a) $i_1(-0.2)$; (b) $i_1(0.2)$; (c) those instants at which $i_1 = 0$; (d) the total charge that has passed from left to right along the conductor in the interval $-0.8 < t < 0.1$ s.

12. The waveform shown in Fig. 2.28 has a period of 10 s. (a) What is the average value of the current over one period? (b) How much charge is transferred in the interval $1 < t < 12$ s? (c) If $q(0) = 0$, sketch $q(t)$, $0 < t < 16$ s.

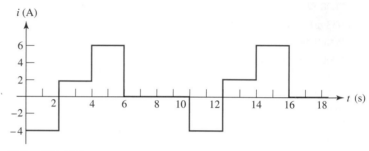

■ **FIGURE 2.28**

13. Consider a path with discrete points A, B, C, D, and E. It takes 2 pJ to move an electron from A to B or from B to C. It takes 3 pJ to move a proton from C to D. It takes no energy to move an electron from D to E.

   (a) What is the potential difference (in volts) between A and B? (Assume + reference at B.)

   (b) What is the potential difference (in volts) between D and E? (Assume + reference at E.)

   (c) What is the potential difference (in volts) between C and D? (Assume + reference at D.)

   (d) What is the potential difference (in volts) between D and B? (Assume + reference at D.)

14. An unmarked box is found in the back corner of a laboratory. It has two wires protruding from it, an orange wire and a purple wire. A voltmeter is connected to the two wires with the + reference on the purple wire. A voltage of -2.86 V is measured in this fashion. What would the voltage reading be if the voltmeter connections were reversed?

15. Determine the power being absorbed by each of the circuit elements shown in Fig. 2.29.

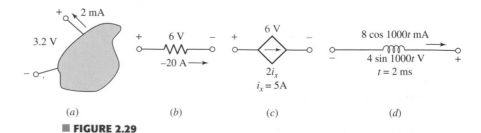

(a)   (b)   (c)   (d)

■ **FIGURE 2.29**

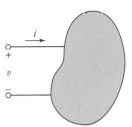

16. Let $i = 3te^{-100t}$ mA and $v = (0.006 - 0.6t)e^{-100t}$ V for the circuit element of Fig. 2.30. (*a*) What power is being absorbed by the circuit element at $t = 5$ ms? (*b*) How much energy is delivered to the element in the interval $0 < t < \infty$?

17. In Fig. 2.30, let $i = 3e^{-100t}$ A. Find the power being absorbed by the circuit element at $t = 8$ ms if v equals (*a*) $40i$; (*b*) $0.2 \, di/dt$; (*c*) $30 \int_0^t i \, dt + 20$ V.

18. The current-voltage characteristic of a silicon solar cell exposed to direct sunlight at noon in Florida during midsummer is given in Fig. 2.31. It is obtained by placing different-sized resistors across the two terminals of the device and measuring the resulting currents and voltages.

(*a*) What is the value of the short-circuit current?

(*b*) What is the value of the voltage at open circuit?

(*c*) Estimate the maximum power that can be obtained from the device.

■ FIGURE 2.30

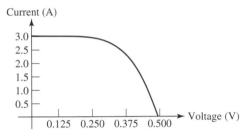

■ FIGURE 2.31

19. The current flowing into a certain circuit is monitored carefully over time. All voltages quoted assume the positive reference terminal is the top of the two circuit terminals. It is observed that for the first two hours, a current of 1 mA flows into the top terminal, while a voltage of $+5$ V is measured. Over the next 30 minutes, no current flows in or out. Then, for two hours, a current of 1 mA flows out of the top terminal, with a voltage of $+2$ V measured. After that, no current flows in or out again. Assuming the circuit had no energy initially stored, answer the following:

(*a*) How much power was delivered to the circuit during each of the three intervals?

(*b*) How much energy was supplied to the circuit during the first two hours of observation?

(*c*) How much energy remains in the circuit now?

2.3 Voltage and Current Sources

20. Determine which of the five sources in Fig. 2.32 are being charged (absorbing positive power), and show that the algebraic sum of the five absorbed power values is zero.

21. Refer to the circuit of Fig. 2.32. Multiply each current and voltage by 4, and determine which of the five sources are acting as sources of energy (i.e., supplying positive power to other elements).

22. In the simple circuit shown in Fig. 2.33, the same current flows through each element. If $V_x = 1$ V and $V_R = 9$ V, compute:

(*a*) the power absorbed by element *A*;

(*b*) the power supplied by each of the two sources.

(*c*) Does the total power supplied equal the total power absorbed? Is your finding reasonable? Why (or why not)?

23. For the circuit in Fig. 2.34, if $v_2 = 1000i_2$ and $i_2 = 5$ mA, determine v_S.

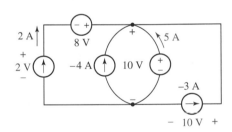

■ FIGURE 2.32

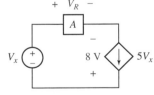

■ FIGURE 2.33

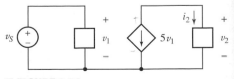

■ FIGURE 2.34

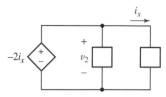

■ **FIGURE 2.35**

24. For the circuit of Fig. 2.35, if $i_x = -1$ mA, calculate the voltage v_2.

25. A simple circuit is formed using a 12 V lead-acid battery and an automobile headlight. If the battery delivers a total energy of 460.8 watt-hours over an 8-hour discharge period,

   (a) how much power is delivered to the headlight?

   (b) what is the current flowing through the bulb? (Assume the battery voltage remains constant while discharging.)

26. A fuse must be selected for a certain application. You may choose from fuses rated to "blow" when the current exceeds 1.5 A, 3 A, 4.5 A, or 5 A. If the supply voltage is 110 V and the maximum allowable power dissipation is 500 W, which fuse should be chosen, and why?

2.4 Ohm's Law

27. A 10% tolerance 1 kΩ resistor may in fact have a value anywhere in the range of 900 to 1100 Ω. If 5.0 V is applied across it, (a) what is the range of currents that might be measured? (b) What is the range of power that might be measured?

28. A current of 2 mA is forced to flow through a 5% tolerance 470 Ω resistor. What power rating should the resistor have, and why? (Note that "5% tolerance" means that the resistor could in reality have a value anywhere in the range of 446.5 Ω and 493.5 Ω.)

29. Let $R = 1200$ Ω for the resistor shown in Fig. 2.24d. Find the power being absorbed by R at $t = 0.1$ s if (a) $i = 20e^{-12t}$ mA; (b) $v = 40 \cos 20t$ V; (c) $vi = 8t^{1.5}$ VA.

30. A certain voltage is $+10$ V for 20 ms and -10 V for the succeeding 20 ms and continues oscillating back and forth between these two values at 20 ms intervals. The voltage is present across a 50 Ω resistor. Over any 40 ms interval find (a) the maximum value of the voltage; (b) the average value of the voltage; (c) the average value of the resistor current; (d) the maximum value of the absorbed power; (e) the average value of the absorbed power.

31. In the circuit in Fig. 2.36, the same current must flow through all three components as a result of conservation laws. Using the fact that the total power supplied equals the total power absorbed, show that the voltage across resistor R_2 is given by:

$$V_{R_2} = V_S \frac{R_2}{R_1 + R_2}$$

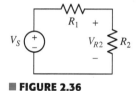

■ **FIGURE 2.36**

32. The following experimental measurements were made on a two-terminal device by setting the voltage using a variable power supply and measuring the resulting current flow into one of the terminals.

Voltage (V)	Current (mA)
−1.5	−3.19
−0.3	−0.638
0.0	1.01×10^{-8}
1.2	2.55
2.5	5.32

   (a) Plot the current vs. voltage characteristic.

   (b) Compute the effective conductance and resistance of the device.

   (c) On a different graph, plot the current vs. voltage characteristic if the device resistance is increased by a factor of 3.

33. For each of the circuits in Fig. 2.37, find the current I and compute the power absorbed by the resistor.

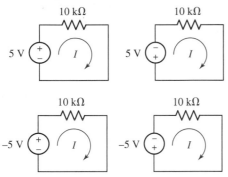

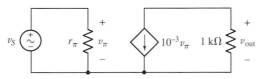

■ **FIGURE 2.37**

34. It is not uncommon to see a variety of subscripts on voltages, currents, and resistors in circuit diagrams. In the circuit in Fig. 2.38, the voltage v_π appears across the resistor named r_π. Compute v_{out} if $v_s = 0.01 \cos 1000t$ V.

■ **FIGURE 2.38**

35. The circuit of Fig. 2.38 is constructed so that $v_S = 2 \sin 5t$ V, and $r_\pi = 80\ \Omega$. Calculate v_{out} at $t = 0$ and $t = 314$ ms.

36. A length of 18 AWG solid copper wire is run along the side of a road to connect a sensor to a central computer system. If the wire is known to have a resistance of 53 Ω, what is the total length of the wire? (Assume the temperature is $\sim 20°$C.)

37. You're stranded on a desert island, and the air temperature is $108°$F. After realizing that your transmitter is not working, you trace the problem to a broken 470 Ω resistor. Fortunately, you notice that a large spool of 28 AWG solid copper wire also washed ashore. How many feet of wire will you require to use as a replacement for the 470 Ω resistor? Note that because the island is in the tropics, it is a little balmier than the $20°$C used to quote the wire resistance in Table 2.4. You may use the following relationship[5] to correct the values in Table 2.4:

$$\frac{R_2}{R_1} = \frac{234.5 + T_2}{234.5 + T_1}$$

where T_1 = reference temperature ($20°$C in this case)
              R_1 = resistance at the reference temperature
              T_2 = new temperature (in degrees Celsius)
              R_2 = resistance at the new temperature.

38. The resistance of a conductor having a length l and a uniform cross-sectional area A is given by $R = l/\sigma A$, where σ (sigma) is the electrical conductivity. If $\sigma = 5.8 \times 10^7$ S/m for copper: (*a*) what is the resistance of a #18 copper wire (diameter = 1.024 mm) that is 50 ft long? (*b*) If a circuit board has a copper-foil conducting ribbon 33 μm thick and 0.5 mm wide that can carry 3 A safely at $50°$C, find the resistance of a 15 cm length of this ribbon and the power delivered to it by the 3 A current.

(5) D. G. Fink and H. W. Beaty, *Standard Handbook for Electrical Engineers,* 13th ed. New York: McGraw-Hill, 1993, p. 2–9.

39. Table 2.3 lists several types of copper wire standards, with a resistivity of approximately 1.7 $\mu\Omega \cdot$ cm. Use the information in Table 2.4 for 28 AWG wire to extract the resistivity of the corresponding soft copper wire. Is your value consistent with Table 2.3?

40. (a) List three examples of "nonlinear" resistors. (b) Imagine a battery connected to a resistor. Energy is transferred from the battery to the resistor until the battery is completely discharged. Keeping in mind the physical principle of energy conservation, where exactly did the energy initially stored in the battery go?

41. If B33 copper is used to make round wire having a diameter of 1 mm, how much power would be dissipated in 100 m of wire carrying a current of 1.5 A?

D 42. Based on the information in Table 2.4, design a mechanical device that acts as a continuously variable resistor. (Hint: a coil might be of help.)

43. The diode, a very common two-terminal nonlinear device, can be modeled using the following current-voltage relationship:

$$I = 10^{-9}(e^{39V} - 1)$$

(a) Sketch the current-voltage characteristic for $V = -0.7$ to 0.7 V.

(b) What is the effective resistance of the diode at $V = 0.55$ V?

(c) At what current does the diode have a resistance of 1 Ω?

D 44. A resistance of 10 Ω is required to repair a voltage regulator circuit for a portable application. The only available materials are 10,000 ft spools of each wire gauge listed in Table 2.4. Design a suitable resistor.

D 45. The resistivity of "n-type" crystalline silicon is given by $\rho = 1/qN_D\mu_n$, where q, the charge per electron, is 1.602×10^{-19} C, $N_D =$ the number of phosphorus impurity atoms per cm^3, and $\mu_n =$ the electron mobility (in units of cm^2 V^{-1}s^{-1}). The mobility and impurity concentration are related by Fig. 2.39. Assuming a 6 in–diameter silicon wafer 250 μm thick, design a 100 Ω resistor by specifying a phosphorus concentration in the range of $10^{15} \leq N_D \leq 10^{18}$ atoms/cm^3, and a suitable device geometry.

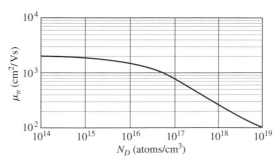

■ **FIGURE 2.39**

Voltage and Current Laws

INTRODUCTION

In Chap. 2 we were introduced to the resistor as well as to several types of sources. After defining a few new circuit terms, we will be ready to begin analyzing simple circuits constructed from these devices. The techniques we will learn are based on two relatively simple laws: Kirchhoff's current law (KCL) and Kirchhoff's voltage law (KVL). KCL is based on the principle of conservation of charge, and KVL is based on the principle of conservation of energy—both fundamental physical laws. Once familiar with basic analysis, we make further use of KCL and KVL to reduce series and parallel combinations of resistors, voltage sources, or current sources, and we develop the important concepts of voltage and current division. In subsequent chapters, we learn additional techniques that allow us to efficiently analyze even more complex networks.

3.1 NODES, PATHS, LOOPS, AND BRANCHES

We now focus our attention on the current-voltage relationships in simple networks of two or more circuit elements. The elements will be connected together by wires (sometimes referred to as "leads"), which have zero resistance. Since the network then appears as a number of simple elements and a set of connecting leads, it is called a ***lumped-parameter network.*** A more difficult analysis problem arises when we are faced with a ***distributed-parameter network,*** which contains an essentially infinite number of vanishingly small elements. We will concentrate on lumped-parameter networks in this text.

In circuits assembled in the real world, the wires will always have finite resistance. However, this resistance is typically so small compared to other resistances in the circuit that we can neglect it without introducing significant error. In our idealized circuits, we will therefore refer to "zero resistance" wires from now on.

◇ CAUTION

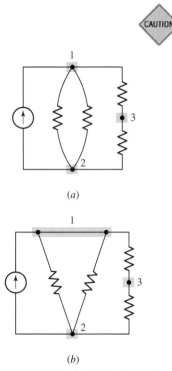

(a)

(b)

■ **FIGURE 3.1** (a) A circuit containing three nodes and five branches. (b) Node 1 is redrawn to look like two nodes; it is still one node.

A point at which two or more elements have a common connection is called a **node.** For example, Fig. 3.1a shows a circuit containing three nodes. Sometimes networks are drawn so as to trap an unwary student into believing that there are more nodes present than is actually the case. This occurs when a node, such as node 1 in Fig. 3.1a, is shown as two separate junctions connected by a (zero-resistance) conductor, as in Fig. 3.1b. However, all that has been done is to spread the common point out into a common zero-resistance line. Thus, we must necessarily consider all of the perfectly conducting leads or portions of leads attached to the node as part of the node. Note also that every element has a node at each of its ends.

Suppose that we start at one node in a network and move through a simple element to the node at the other end. We then continue from that node through a different element to the next node, and continue this movement until we have gone through as many elements as we wish. If no node was encountered more than once, then the set of nodes and elements that we have passed through is defined as a **path.** If the node at which we started is the same as the node on which we ended, then the path is, by definition, a closed path or a **loop.**

For example, in Fig. 3.1a, if we move from node 2 through the current source to node 1, and then through the upper right resistor to node 3, we have established a path; since we have not continued on to node 2 again, we have not made a loop. If we proceeded from node 2 through the current source to node 1, down through the left resistor to node 2, and then up through the central resistor to node 1 again, we do not have a path, since a node (actually two nodes) was encountered more than once; we also do not have a loop, because a loop must be a path.

Another term whose use will prove convenient is **branch.** We define a branch as a single path in a network, composed of one simple element and the node at each end of that element. Thus, a path is a particular collection of branches. The circuit shown in Fig. 3.1a and b contains five branches.

3.2 • KIRCHHOFF'S CURRENT LAW

We are now ready to consider the first of the two laws named for Gustav Robert Kirchhoff (two h's and two f's), a German university professor who was born about the time Ohm was doing his experimental work. This axiomatic law is called Kirchhoff's current law (abbreviated KCL), and it simply states that:

> The algebraic sum of the currents entering any node is zero.

This law represents a mathematical statement of the fact that charge cannot accumulate at a node. *A node is not a circuit element,* and it certainly cannot store, destroy, or generate charge. Hence, the currents must sum to zero. A hydraulic analogy is sometimes useful here: for example, consider three water pipes joined in the shape of a Y. We define three "currents" as flowing *into* each of the three pipes. If we insist that water is always flowing, then obviously we cannot have three positive water currents, or the pipes would burst. This is a result of our defining currents independent of

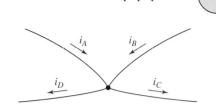

the direction that water is actually flowing. Therefore, the value of either one or two of the currents as defined must be negative.

Consider the node shown in Fig. 3.2. The algebraic sum of the four currents entering the node must be zero:

$$i_A + i_B + (-i_C) + (-i_D) = 0$$

It is evident that the law could be equally well applied to the algebraic sum of the currents *leaving* the node:

$$(-i_A) + (-i_B) + i_C + i_D = 0$$

We might also wish to equate the sum of the currents having reference arrows directed into the node to the sum of those directed out of the node:

$$i_A + i_B = i_C + i_D$$

which simply states that the sum of the currents going in must equal the sum of the currents going out.

A compact expression for Kirchhoff's current law is

$$\sum_{n=1}^{N} i_n = 0 \qquad [1]$$

which is just a shorthand statement for

$$i_1 + i_2 + i_3 + \cdots + i_N = 0 \qquad [2]$$

When Eq. [1] or Eq. [2] is used, it is understood that the N current arrows are either all directed toward the node in question, or are all directed away from it.

■ FIGURE 3.2 Example node to illustrate the application of Kirchhoff's current law.

EXAMPLE 3.1

For the circuit in Fig. 3.3*a*, compute the current through resistor R_3 if it is known that the voltage source supplies a current of 3 A.

▶ **Identify the goal of the problem.**
The current through resistor R_3, labeled as i on the circuit diagram.

▶ **Collect the known information.**
This current flows from the top node of R_3, which is connected to three other branches. The current flowing into the node from each branch will add to form the current i.

▶ **Devise a plan.**
If we label the current through R_1 (Fig. 3.3*b*), we may write a KCL equation at the top node of resistors R_2 and R_3.

▶ **Construct an appropriate set of equations.**
Summing the currents flowing into the node:

$$i_{R_1} - 2 - i + 5 = 0$$

The currents flowing into this node are shown in the expanded diagram of Fig. 3.3*c* for clarity.

(Continued on next page)

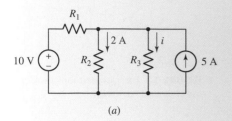

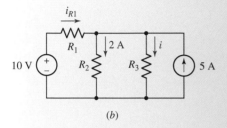

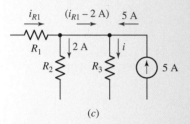

■ **FIGURE 3.3** (a) Simple circuit for which the current through resistor R_3 is desired. (b) The current through resistor R_1 is labeled so that a KCL equation can be written. (c) The currents into the top node of R_3 are redrawn for clarity.

▶ **Determine if additional information is required.**

We see that we have one equation but two unknowns, which means we need to obtain an additional equation. At this point, the fact that we know the 10 V source is supplying 3 A comes in handy: KCL shows us that this is also the current i_{R_1}.

▶ **Attempt a solution.**

Substituting, we find that $i = 3 - 2 + 5 = 6$ A.

▶ **Verify the solution. Is it reasonable or expected?**

It is always worth the effort to recheck our work. Also, we can attempt to evaluate whether at least the magnitude of the solution is reasonable. In this case, we have two sources—one supplies 5 A, and the other supplies 3 A. There are no other sources, independent or dependent. Thus, we would not expect to find any current in the circuit in excess of 8 A.

PRACTICE

3.1 Count the number of branches and nodes in the circuit in Fig. 3.4. If $i_x = 3$ A and the 18 V source delivers 8 A of current, what is the value of R_A? (Hint: You need Ohm's law as well as KCL.)

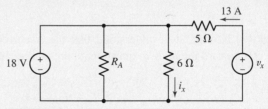

■ **FIGURE 3.4**

Ans: 5 branches, 3 nodes, 1Ω.

3.3 KIRCHHOFF'S VOLTAGE LAW

Current is related to the charge flowing *through* a circuit element, whereas voltage is a measure of potential energy difference *across* the element. There is a single unique value for any voltage in circuit theory. Thus, the energy required to move a unit charge from point A to point B in a circuit must have a value independent of the path chosen to get from A to B (there is often more than one such path). We may assert this fact through Kirchhoff's voltage law (abbreviated **KVL**):

The algebraic sum of the voltages around any closed path is zero.

In Fig. 3.5, if we carry a charge of 1 C from A to B through element 1, the reference polarity signs for v_1 show that we do v_1 joules of work.[1] Now

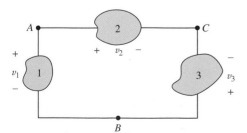

■ **FIGURE 3.5** The potential difference between points A and B is independent of the path selected.

(1) Note that we chose a 1 C charge for the sake of numerical convenience: therefore, we did (1 C)(v_1 J/C) = v_1 joules of work.

if, instead, we choose to proceed from A to B via node C, then we expend $v_2 - v_3$ joules of energy. The work done, however, is independent of the path in a circuit, and these values must be equal. Any route must lead to the same value for the voltage. In other words,

$$v_1 = v_2 - v_3 \qquad [3]$$

It follows that if we trace out a closed path, the algebraic sum of the voltages across the individual elements around it must be zero. Thus, we may write

$$v_1 + v_2 + v_3 + \cdots + v_N = 0$$

or, more compactly,

$$\sum_{n=1}^{N} v_n = 0 \qquad [4]$$

We may apply KVL to a circuit in several different ways. One method that leads to fewer equation-writing errors than others consists of moving mentally around the closed path in a clockwise direction and writing down directly the voltage of each element whose (+) terminal is entered, and writing down the negative of every voltage first met at the (−) sign. Applying this to the single loop of Fig. 3.5, we have

$$-v_1 + v_2 - v_3 = 0$$

which agrees with our previous result, Eq. [3].

EXAMPLE 3.2

In the circuit of Fig. 3.6, find v_x and i_x.

We know the voltage across two of the three elements in the circuit. Thus, KVL can be applied immediately to obtain v_x.

Beginning with the bottom node of the 5 V source, we apply KVL clockwise around the loop:

$$-5 - 7 + v_x = 0$$

so $v_x = 12$ V.

KCL applies to this circuit, but only tells us that the same current (i_x) flows through all three elements. We now know the voltage across the 100 Ω resistor, however.

Invoking Ohm's law,

$$i_x = \frac{v_x}{100} = \frac{12}{100} \text{ A} = 120 \text{ mA}$$

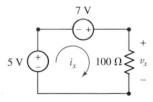

■ **FIGURE 3.6** A simple circuit with two voltage sources and a single resistor.

PRACTICE

3.2 Determine i_x and v_x in the circuit of Fig. 3.7.

Ans: $v_x = -4$ V; $i_x = -400$ mA.

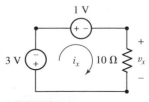

■ **FIGURE 3.7**

EXAMPLE 3.3

In the circuit of Fig. 3.8 there are eight circuit elements; voltages with plus-minus pairs are shown across each element. Find v_{R2} (the voltage across R_2) and the voltage labeled v_x.

The best approach for finding v_{R2} in this situation is to look for a loop to which we can apply KVL. There are several options, but after looking at the circuit carefully we see that the leftmost loop offers a straightforward route, as two of the voltages are clearly specified. Thus, we find v_{R2} by writing a KVL equation around the loop on the left, starting at point c:

$$4 - 36 + v_{R2} = 0$$

which leads to $v_{R2} = 32$ V.

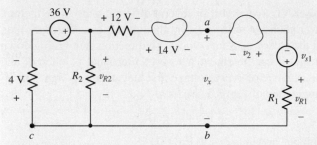

■ FIGURE 3.8 A circuit with eight elements for which we desire v_{R2} and v_x.

To find v_x, we might think of this as the (algebraic) sum of the voltages across the three elements on the right. However, since we do not have values for these quantities, such an approach would not lead to a numerical answer. Instead, we apply KVL beginning at point c, moving up and across the top to a, through v_x to b, and through the conducting lead to the starting point:

$$+4 - 36 + 12 + 14 + v_x = 0$$

so that

$$v_x = 6 \text{ V}$$

An alternative approach: Knowing v_{R2}, we might have taken the shortcut through R_2:

$$-32 + 12 + 14 + v_x = 0$$

yielding $v_x = 6$ V once again.

Points b and c, as well as the wire between them, are all part of the same node.

As we have just seen, the key to correctly analyzing a circuit is to first methodically label all voltages and currents on the diagram. This way, carefully written KCL or KVL equations will yield correct relationships, and Ohm's law can be applied as necessary if more unknowns than equations are obtained initially. We illustrate these principles with a more detailed example.

EXAMPLE **3.4**

Determine v_x in the circuit of Fig. 3.9a.

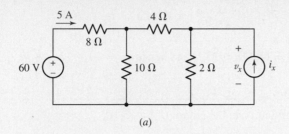

(a)

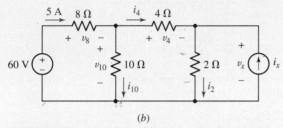

(b)

■ **FIGURE 3.9** (a) A circuit for which v_x is to be determined using KVL. (b) Circuit with voltages and currents labeled.

We begin by labeling voltages and currents on the rest of the elements in the circuit (Fig. 3.9b). Note that v_x appears across the 2 Ω resistor and the source i_x as well.

If we can obtain the current through the 2 Ω resistor, Ohm's law will yield v_x. Writing the appropriate KCL equation, we see that

$$i_2 = i_4 + i_x$$

Unfortunately, we do not have values for any of these three quantities. Our solution has (temporarily) stalled.

Since we were given the current flowing from the 60 V source, perhaps we should consider starting from that side of the circuit. Instead of finding v_x using i_2, it might be possible to find v_x directly using KVL. Working from this perspective, we can write the following KVL equations:

$$-60 + v_8 + v_{10} = 0$$

and

$$-v_{10} + v_4 + v_x = 0 \qquad [5]$$

This is progress: we now have two equations in four unknowns, a small improvement over one equation in which *all* terms were unknown. In fact, we know that $v_8 = 40$ V through Ohm's law, as we were told that 5 A flows through the 8 Ω resistor. Thus, $v_{10} = 0 + 60 - 40 = 20$ V, so that Eq. [5] reduces to

$$v_x = 20 - v_4$$

If we can determine v_4, the problem is solved.

(Continued on next page)

The best route to finding a numerical value for the voltage v_4 in this case is to employ Ohm's law, which requires a value for i_4. From KCL, we see that

$$i_4 = 5 - i_{10} = 5 - \frac{v_{10}}{10} = 5 - \frac{20}{10} = 3$$

so that $v_4 = (4)(3) = 12$ V and hence $v_x = 20 - 12 = 8$ V.

PRACTICE

3.3 Determine v_x in the circuit of Fig. 3.10.

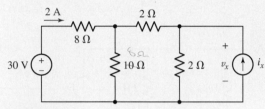

■ **FIGURE 3.10**

Ans: $v_x = 12.8$ V.

3.4 THE SINGLE-LOOP CIRCUIT

We have seen that repeated use of KCL and KVL in conjunction with Ohm's law can be applied to nontrivial circuits containing several loops and a number of different elements. Before proceeding further, this is a good time to focus on the concept of series (and, in the next section, parallel) circuits, as they form the basis of any network we will encounter in the future.

All of the elements in a circuit that carry the same current are said to be connected in **series.** As an example, consider the circuit of Fig. 3.9. The 60 V source is in series with the 8 Ω resistor; they carry the same 5 A current. However, the 8 Ω resistor is not in series with the 4 Ω resistor; they carry different currents. Note that elements may carry equal currents and not be in series; two 100 W light bulbs in neighboring houses may very well carry equal currents, but they certainly do not carry the same current and are *not* connected in series.

Figure 3.11*a* shows a simple circuit consisting of two batteries and two resistors. Each terminal, connecting lead, and solder glob is assumed to have zero resistance; together they constitute an individual node of the circuit diagram in Fig. 3.11*b*. Both batteries are modeled by ideal voltage sources; any internal resistances they may have are assumed to be small enough to neglect. The two resistors are assumed to be replaceable by ideal (linear) resistors.

We seek the current *through* each element, the voltage *across* each element, and the power *absorbed* by each element. Our first step in the analysis is the assumption of reference directions for the unknown currents. Arbitrarily, let us select a clockwise current i which flows out of the upper terminal of the voltage source on the left. This choice is indicated by an arrow labeled i at that point in the circuit, as shown in Fig. 3.11*c*. A trivial

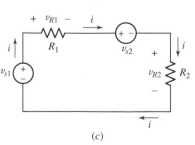

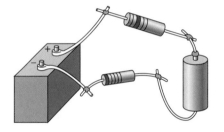

■ **FIGURE 3.11** (*a*) A single-loop circuit with four elements. (*b*) The circuit model with source voltages and resistance values given. (*c*) Current and voltage reference signs have been added to the circuit.

application of Kirchhoff's current law assures us that this same current must also flow through every other element in the circuit; we emphasize this fact this one time by placing several other current symbols about the circuit.

Our second step in the analysis is a choice of the voltage reference for each of the two resistors. The passive sign convention requires that the resistor current and voltage variables be defined so that the current enters the terminal at which the positive voltage reference is located. Since we already (arbitrarily) selected the current direction, v_{R1} and v_{R2} are defined as in Fig. 3.11c.

The third step is the application of Kirchhoff's voltage law to the only closed path. Let us decide to move around the circuit in the clockwise direction, beginning at the lower left corner, and to write down directly every voltage first met at its positive reference, and to write down the negative of every voltage encountered at the negative terminal. Thus,

$$-v_{s1} + v_{R1} + v_{s2} + v_{R2} = 0 \qquad [6]$$

We then apply Ohm's law to the resistive elements:

$$v_{R1} = R_1 i \quad \text{and} \quad v_{R2} = R_2 i$$

Substituting into Eq. [6] yields

$$-v_{s1} + R_1 i + v_{s2} + R_2 i = 0$$

Since i is the only unknown, we find that

$$i = \frac{v_{s1} - v_{s2}}{R_1 + R_2}$$

The voltage or power associated with any element may now be obtained by applying $v = Ri$, $p = vi$, or $p = i^2 R$.

PRACTICE

3.4 In the circuit of Fig. 3.11b, $v_{s1} = 120$ V, $v_{s2} = 30$ V, $R_1 = 30\ \Omega$, and $R_2 = 15\ \Omega$. Compute the power absorbed by each element.

Ans: $p_{120V} = -240$ W; $p_{30V} = +60$ W; $p_{30\Omega} = 120$ W; $p_{15\Omega} = 60$ W.

EXAMPLE **3.5**

Compute the power absorbed in each element for the circuit shown in Fig. 3.12a.

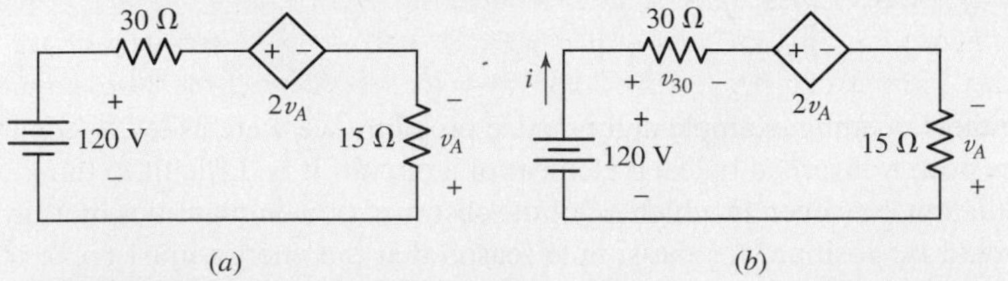

(a)   (b)

■ **FIGURE 3.12** (a) A single-loop circuit containing a dependent source. (b) The current i and voltage v_{30} are assigned.

(Continued on next page)

We first assign a reference direction for the current i and a reference polarity for the voltage v_{30} as shown in Fig. 3.12b. There is no need to assign a voltage to the 15 Ω resistor, since the controlling voltage v_A for the dependent source is already available. (It is worth noting, however, that the reference signs for v_A are reversed from those we would have assigned based on the passive sign convention.)

This circuit contains a dependent voltage source, the value of which remains unknown until we determine v_A. However, its algebraic value $2v_A$ can be used in the same fashion as if a numerical value were available. Thus, applying KVL around the loop:

$$-120 + v_{30} + 2v_A - v_A = 0 \qquad [7]$$

Using Ohm's law to introduce the known resistor values:

$$v_{30} = 30i \quad \text{and} \quad v_A = -15i$$

Note that the negative sign is required since i flows into the negative terminal of v_A.

Substituting into Eq. [7] yields

$$-120 + 30i - 30i + 15i = 0$$

and so we find that

$$i = 8 \text{ A}$$

Computing the power *absorbed* by each element:

$$
\begin{aligned}
p_{120\text{V}} &= (120)(-8) = -960 \text{ W} \\
p_{30\Omega} &= (8)^2(30) &= 1920 \text{ W} \\
p_{\text{dep}} &= (2v_A)(8) &= 2[(-15)(8)](8) \\
&&= -1920 \text{ W} \\
p_{15\Omega} &= (8)^2(15) &= 960 \text{ W}
\end{aligned}
$$

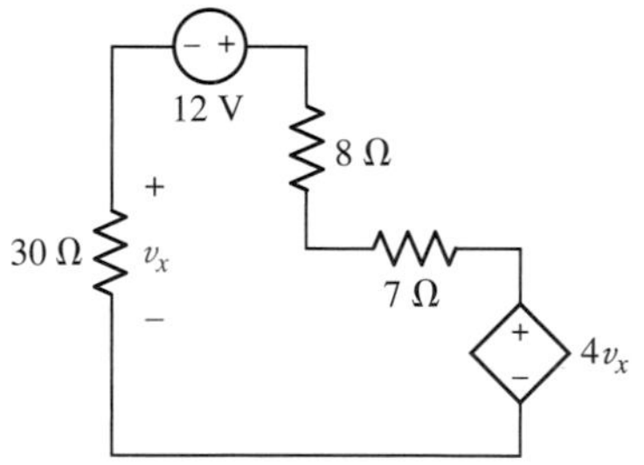

PRACTICE

3.5 In the circuit of Fig. 3.13, find the power absorbed by each of the five elements in the circuit.

Ans: (CW from left) 0.768 W, 1.92 W, 0.2048 W, 0.1792 W, −3.072 W

■ **FIGURE 3.13** A simple loop circuit.

In the preceding example and practice problem, we were asked to compute the power absorbed by each element of a circuit. It is difficult to think of a situation, however, in which *all* of the absorbed power quantities of a circuit would be positive, for the simple reason that the energy must come from somewhere. Thus, from simple conservation of energy, we expect that ***the sum of the absorbed power for each element of a circuit should be zero.*** In other words, at least one of the quantities should be negative (neglecting the trivial case where the circuit is not operating). Stated another way, the sum of the supplied power for each element should be zero. More pragmatically, *the sum of the absorbed power equals the sum of the supplied power,* which seems reasonable enough at face value.

Let's test this with the circuit of Fig. 3.12 from Example 3.5, which consists of two sources (one dependent and one independent) and two resistors. Adding the power absorbed by each element, we find

$$\sum_{\text{all elements}} p_{\text{absorbed}} = -960 + 1920 - 1920 + 960 = 0$$

In reality (our indication is the sign associated with the absorbed power) the 120 V source *supplies* $+960$ W, and the dependent source supplies $+1920$ W. Thus, the sources supply a total of $960 + 1920 = 2880$ W. The resistors are expected to absorb positive power, which in this case sums to a total of $1920 + 960 = 2880$ W. Thus, if we take into account each element of the circuit,

$$\sum p_{\text{absorbed}} = \sum p_{\text{supplied}}$$

as we expect.

Turning our attention to Practice Problem 3.5, the solution to which the reader might want to verify, we see that the absorbed powers sum to $0.768 + 1.92 + 0.2048 + 0.1792 - 3.072 = 0$. Interestingly enough, the 12 V independent voltage source is absorbing $+1.92$ W, which means it is *dissipating* power, not supplying it. Instead, the dependent voltage source appears to be supplying all the power in this particular circuit. Is such a thing possible? We usually expect a source to supply positive power, but since we are employing idealized sources in our circuits, it is in fact possible to have a net power flow into any source. If the circuit is changed in some way, the same source might then be found to supply positive power. The result is not known until a circuit analysis has been completed.

3.5 • THE SINGLE-NODE-PAIR CIRCUIT

The companion of the single-loop circuit discussed in Sec. 3.4 is the single-node-pair circuit, in which any number of simple elements are connected between the same pair of nodes. An example of such a circuit is shown in Fig. 3.14a. The two current sources and the resistance values are known. First, assume a voltage across any element, assigning an arbitrary reference polarity. KVL then forces us to recognize that the voltage across each branch is the same as that across any other branch. *Elements in a circuit having a common voltage across them are said to be connected in **parallel.***

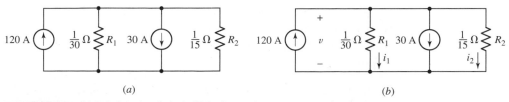

■ **FIGURE 3.14** (*a*) A single-node-pair circuit. (*b*) A voltage and two currents are assigned.

EXAMPLE 3.6

Find the voltage, current, and power associated with each element in the circuit of Fig. 3.14a.

We first define a voltage v and arbitrarily select its polarity as shown in Fig. 3.14b. Two currents, flowing in the resistors, are selected in conformance with the passive sign convention, as shown in Fig. 3.14b.

Determining either current i_1 or i_2 will enable us to obtain a value for v. Thus, our next step is to apply KCL to either of the two nodes in the circuit. Equating the algebraic sum of the currents leaving the upper node to zero:

$$-120 + i_1 + 30 + i_2 = 0$$

Writing both currents in terms of the voltage v using Ohm's law,

$$i_1 = 30v \quad \text{and} \quad i_2 = 15v$$

we obtain

$$-120 + 30v + 30 + 15v = 0$$

Solving this equation for v results in

$$v = 2 \text{ V}$$

and invoking Ohm's law then gives

$$i_1 = 60 \text{ A} \quad \text{and} \quad i_2 = 30 \text{ A}$$

The absorbed power in each element can now be computed. In the two resistors,

$$p_{R1} = 30(2)^2 = 120 \text{ W} \quad \text{and} \quad p_{R2} = 15(2)^2 = 60 \text{ W}$$

and for the two sources,

$$p_{120A} = 120(-2) = -240 \text{ W} \quad \text{and} \quad p_{30A} = 30(2) = 60 \text{ W}$$

Since the 120 A source absorbs negative 240 W, it is actually *supplying* power to the other elements in the circuit. In a similar fashion, we find that the 30 A source is actually *absorbing* power rather than *supplying* it.

PRACTICE

3.6 Determine v in the circuit of Fig. 3.15.

■ **FIGURE 3.15**

Ans: 50 V.

EXAMPLE **3.7**

Determine the value of v and the power supplied by the independent current source in Fig. 3.16.

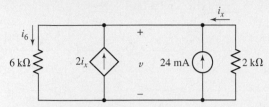

■ **FIGURE 3.16** A voltage v and a current i_6 are assigned in a single-node-pair circuit containing a dependent source.

By KCL, the sum of the currents leaving the upper node must be zero, so that

$$i_6 - 2i_x - 0.024 - i_x = 0$$

Again, note that the value of the dependent source $(2i_x)$ is treated the same as any other current would be, even though its exact value is not known until the circuit has been analyzed.

We next apply Ohm's law to each resistor:

$$i_6 = \frac{v}{6000} \quad \text{and} \quad i_x = \frac{-v}{2000}$$

Therefore,

$$\frac{v}{6000} - 2\left(\frac{-v}{2000}\right) - 0.024 - \left(\frac{-v}{2000}\right) = 0$$

and so $v = (600)(0.024) = 14.4$ V.

Any other information we may want to find for this circuit is now easily obtained, usually in a single step. For example, the power supplied by the independent source is $p_{24} = 14.4(0.024) = 0.3456$ W (345.6 mW).

PRACTICE

3.7 For the single-node-pair circuit of Fig. 3.17, find i_A, i_B, and i_C.

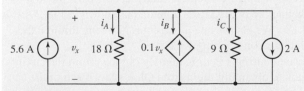

■ **FIGURE 3.17**

Ans: 3 A; −5.4 A; 6 A.

EXAMPLE 3.8

For the circuit of Fig. 3.18*a*, find i_1, i_2, i_3, and i_4.

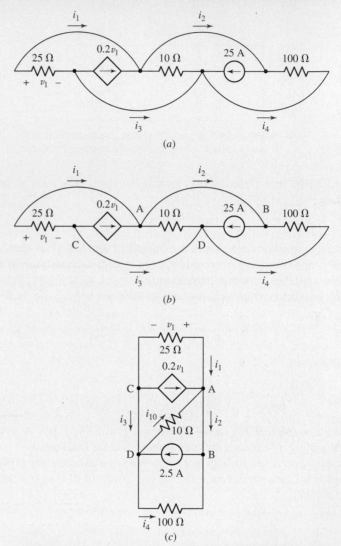

■ **FIGURE 3.18** (*a*) A single-node-pair circuit. (*b*) Circuit with points labeled to assist in redrawing. (*c*) Redrawn circuit.

As drawn, this circuit is a little difficult to analyze, so we decide first to redraw it after labeling the points *A*, *B*, *C*, and *D* as shown in Fig. 3.18*b* and finally Fig. 3.18*c*. We also define a current i_{10} flowing through the 10 Ω resistor in anticipation of using KCL.

None of the desired currents is immediately obvious from the circuit diagram, so we look to obtaining them from Ohm's law. Each of the three resistors has the same voltage (v_1) across it, and we simply sum the currents flowing into the rightmost node:

$$-\frac{v_1}{100} - 2.5 - \frac{v_1}{10} + 0.2\,v_1 - \frac{v_1}{25} = 0$$

Solving, we find $v_1 = 250/5 = 50$ V.

Looking to the bottom of the circuit, we see that

$$i_4 = \frac{-v_1}{100} = -\frac{50}{100} = -0.5 \text{ A}$$

In a similar fashion, we determine that $i_1 = -2$ A and $i_{10} = -5$ A. The remaining two currents i_2 and i_3 are found by using KCL to independently sum the known currents into the right-hand and left-hand nodes. Thus,

$$i_2 = i_1 + 0.2v_1 + i_{10} = -2 + 10 - 5 = 3 \text{ A}$$

and

$$i_3 = i_{10} - 2.5 + i_4 = -5 - 2.5 - 0.5 = -8 \text{ A}$$

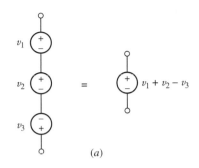

3.6 SERIES AND PARALLEL CONNECTED SOURCES

It turns out that some of the equation writing that we have been doing for series and parallel circuits can be avoided by combining sources. Note, however, that all the current, voltage, and power relationships in the remainder of the circuit will be unchanged. For example, several voltage sources in series may be replaced by an equivalent voltage source having a voltage equal to the algebraic sum of the individual sources (Fig. 3.19a). Parallel current sources may also be combined by algebraically adding the individual currents, and the order of the parallel elements may be rearranged as desired (Fig. 3.19b).

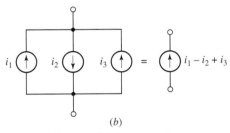

■ **FIGURE 3.19** (a) Series-connected voltage sources can be replaced by a single source. (b) Parallel current sources can be replaced by a single source.

EXAMPLE **3.9**

Find the current through the 470 Ω resistor in Fig. 3.20a by first combining the four sources into a single voltage source.

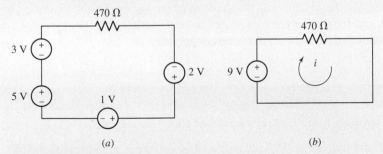

(a)   (b)

■ **FIGURE 3.20** (a) A simple loop circuit containing four voltage sources in series. (b) Equivalent circuit.

We have four voltage sources connected in series. Choosing to replace them with a single voltage source having its "+" reference terminal at the top, we begin at the "+" reference terminal of the 3 V source and write:

$$+3 + 5 - 1 + 2 = 9 \text{ V}$$

(Continued on next page)

The equivalent circuit is shown in Fig. 3.20b. We now find i using Ohm's law:

$$i = \frac{9}{470} = 19.15 \text{ mA}$$

Typically, there is very little to be gained from including a dependent source in either a voltage or current source combination, but it is not incorrect to do so.

PRACTICE

3.8 Determine v in the circuit of Fig. 3.21 by first combining the three current sources.

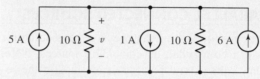

■ **FIGURE 3.21**

Ans: 50 V.

To conclude the discussion of parallel and series source combinations, we should consider the parallel combination of two voltage sources and the series combination of two current sources. For instance, what is the equivalent of a 5 V source in parallel with a 10 V source? By the definition of a voltage source, the voltage across the source cannot change; by Kirchhoff's voltage law, then, 5 equals 10 and we have hypothesized a physical impossibility. Thus, *ideal* voltage sources in parallel are permissible only when each has the same terminal voltage at every instant. In a similar way, two current sources may not be placed in series unless each has the same current, including sign, for every instant of time.

EXAMPLE **3.10**

Determine which of the circuits of Fig. 3.22 are valid.

The circuit of Fig. 3.22a consists of two voltage sources in parallel. The value of each source is different, so this circuit violates KVL. For example, if a resistor is placed in parallel with the 5 V source, it is also in parallel with the 10 V source. The actual voltage across it is therefore ambiguous, and clearly the circuit cannot be constructed as indicated. If we attempt to build such a circuit in real life, we will find it impossible to locate "ideal" voltage sources—all real-world sources have an internal resistance. The presence of such resistance allows a voltage difference

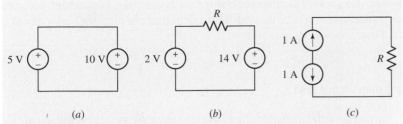

■ FIGURE 3.22 (*a*) to (*c*) Examples of circuits with multiple sources, some of which violate Kirchhoff's laws.

between the two *real* sources. Along these lines, the circuit of Fig. 3.22*b* is perfectly valid.

The circuit of Fig. 3.22*c* violates KCL: it is unclear what current actually flows through the resistor *R*.

PRACTICE

3.9 Determine whether the circuit of Fig. 3.23 violates either of Kirchhoff's laws.

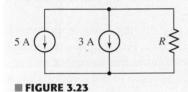

■ FIGURE 3.23

Ans: No. If the resistor were removed, however, the resulting circuit would.

3.7 RESISTORS IN SERIES AND PARALLEL

It is often possible to replace relatively complicated resistor combinations with a single equivalent resistor. This is useful when we are not specifically interested in the current, voltage, or power associated with any of the individual resistors in the combinations. *All the current, voltage, and power relationships in the remainder of the circuit will be unchanged.*

Consider the series combination of *N* resistors shown in Fig. 3.24*a*. We want to simplify the circuit with replacing the *N* resistors with a single resistor R_{eq} so that the remainder of the circuit, in this case only the voltage

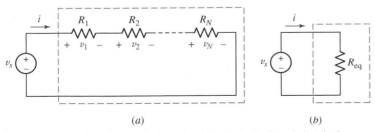

■ FIGURE 3.24 (*a*) Series combination of *N* resistors. (*b*) Electrically equivalent circuit.

source, does not realize that any change has been made. The current, voltage, and power of the source must be the same before and after the replacement.

First, apply KVL:

$$v_s = v_1 + v_2 + \cdots + v_N$$

and then Ohm's law:

$$v_s = R_1 i + R_2 i + \cdots + R_N i = (R_1 + R_2 + \cdots + R_N)i$$

Now compare this result with the simple equation applying to the equivalent circuit shown in Fig. 3.24b:

$$v_s = R_{eq} i$$

Thus, the value of the equivalent resistance for N series resistors is

$$\boxed{R_{eq} = R_1 + R_2 + \cdots + R_N} \tag{8}$$

We are therefore able to replace a two-terminal network consisting of N series resistors with a single two-terminal element R_{eq} that has the same v-i relationship.

It should be emphasized again that we might be interested in the current, voltage, or power of one of the original elements. For example, the voltage of a dependent voltage source may depend upon the voltage across R_3. Once R_3 is combined with several series resistors to form an equivalent resistance, then it is gone and the voltage across it cannot be determined until R_3 is identified by removing it from the combination. In that case, it would have been better to look ahead and not make R_3 a part of the combination initially.

Another tip: Inspection of the KVL equation for a series circuit shows that the order in which elements are placed makes no difference.

EXAMPLE 3.11

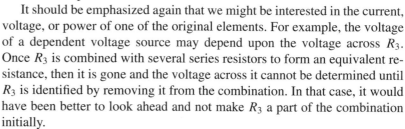

Use resistance and source combinations to determine the current i in Fig. 3.25a and the power delivered by the 80 V source.

We first interchange the element positions in the circuit, being careful to preserve the proper sense of the sources, as shown in Fig. 3.25b. The next step is to then combine the three voltage sources into an equivalent 90 V source, and the four resistors into an equivalent 30 Ω resistance, as in Fig. 3.25c. Thus, instead of writing

$$-80 + 10i - 30 + 7i + 5i + 20 + 8i = 0$$

we have simply

$$-90 + 30i = 0$$

and so we find that

$$i = 3 \text{ A}$$

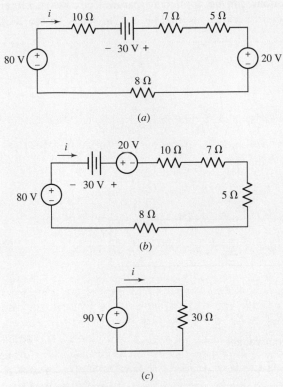

■ **FIGURE 3.25** (*a*) A series circuit with several sources and resistors.
(*b*) The elements are rearranged for the sake of clarity. (*c*) A simpler
equivalent.

In order to calculate the power delivered to the circuit by the 80 V
source appearing in the given circuit, it is necessary to return to
Fig. 3.25*a* with the knowledge that the current is 3 A. The desired
power is then 80 V × 3 A = 240 W.

It is interesting to note that no element of the original circuit remains
in the equivalent circuit.

PRACTICE

3.10 Determine *i* in the circuit of Fig. 3.26.

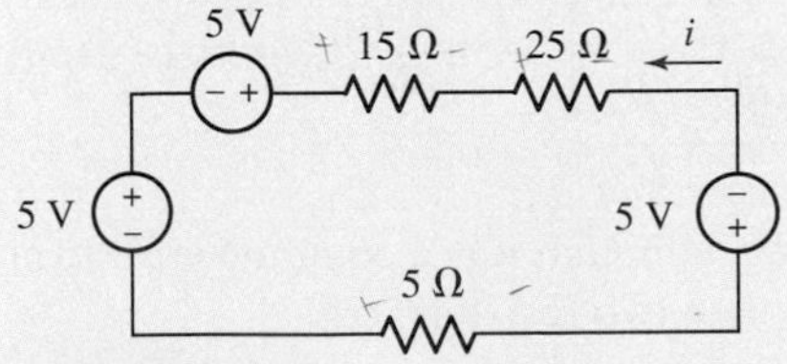

■ **FIGURE 3.26**

Ans: −333 mA.

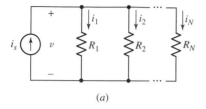

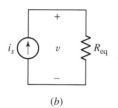

■ **FIGURE 3.27** (*a*) A circuit with *N* resistors in parallel. (*b*) Equivalent circuit.

Similar simplifications can be applied to parallel circuits. A circuit containing N resistors in parallel, as in Fig. 3.27*a*, leads to the KCL equation

$$i_s = i_1 + i_2 + \cdots + i_N$$

or

$$i_s = \frac{v}{R_1} + \frac{v}{R_2} + \cdots + \frac{v}{R_N}$$

$$= \frac{v}{R_{eq}}$$

Thus,

$$\boxed{\frac{1}{R_{eq}} = \frac{1}{R_1} + \frac{1}{R_2} + \cdots + \frac{1}{R_N}} \qquad [9]$$

which can be written as

$$R_{eq}^{-1} = R_1^{-1} + R_2^{-1} + \cdots + R_N^{-1}$$

or, in terms of conductances, as

$$G_{eq} = G_1 + G_2 + \cdots + G_N$$

The simplified (equivalent) circuit is shown in Fig. 3.27*b*.

A parallel combination is routinely indicated by the following shorthand notation:

$$R_{eq} = R_1 \| R_2 \| R_3$$

The special case of only two parallel resistors is encountered fairly often, and is given by

$$R_{eq} = R_1 \| R_2$$

$$= \frac{1}{\dfrac{1}{R_1} + \dfrac{1}{R_2}}$$

Or, more simply,

$$\boxed{R_{eq} = \frac{R_1 R_2}{R_1 + R_2}} \qquad [10]$$

The last form is worth memorizing, although it is a common error to attempt to generalize Eq. [10] to more than two resistors, e.g.,

$$R_{eq} \neq \frac{R_1 R_2 R_3}{R_1 + R_2 + R_3}$$

A quick look at the units of this equation will immediately show that the expression cannot possibly be correct.

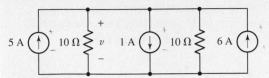
PRACTICE

3.11 Determine v in the circuit of Fig. 3.28 by first combining the three current sources, and then the two 10 Ω resistors.

5 A $\uparrow$ 10 Ω $+$ v $-$ 1 A $\downarrow$ 10 Ω   6 A $\uparrow$

■ **FIGURE 3.28**

Ans: 50 V.

EXAMPLE 3.12

Calculate the power and voltage of the dependent source in Fig. 3.29a.

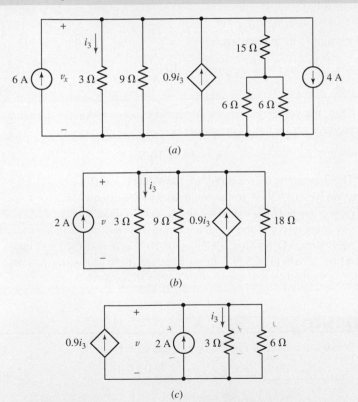

(a)

(b)

(c)

■ **FIGURE 3.29** (a) A multinode circuit. (b) The two independent current sources are combined into a 2 A source, and the 15 Ω resistor in series with the two parallel 6 Ω resistors are replaced with a single 18 Ω resistor. (c) A simplified equivalent circuit.

We will seek to simplify the circuit before analyzing it, but take care not to include the dependent source since its voltage and power characteristics are of interest.

(Continued on next page)

Despite not being drawn adjacent to one another, the two independent current sources are in fact in parallel, so we replace them with a 2 A source.

The two 6 Ω resistors are in parallel and can be replaced with a single 3 Ω resistor in series with the 15 Ω resistor. Thus, the two 6 Ω resistors and the 15 Ω resistor are replaced by an 18 Ω resistor (Fig. 3.29*b*).

No matter how tempting, *we should not combine the remaining three resistors;* the controlling variable i_3 depends on the 3 Ω resistor and so that resistor must remain untouched. The only further simplification, then, is 9 $\Omega \| 18 \Omega = 6 \Omega$, as shown in Fig. 3.29*c*.

Applying KCL at the top node of Fig. 3.29*c*, we have

$$-0.9i_3 - 2 + i_3 + \frac{v}{6} = 0$$

Employing Ohm's law,

$$v = 3i_3$$

which allows us to compute

$$i_3 = \frac{10}{3} \text{ A}$$

Thus, the voltage across the dependent source (which is the same as the voltage across the 3 Ω resistor) is

$$v = 3i_3 = 10 \text{ V}$$

The dependent source therefore furnishes $v \times 0.9i_3 = 10(0.9)(10/3) = 30$ W to the remainder of the circuit.

Now if we are later asked for the power dissipated in the 15 Ω resistor, we must return to the original circuit. This resistor is in series with an equivalent 3 Ω resistor; a voltage of 10 V is across the 18 Ω total; therefore, a current of 5/9 A flows through the 15 Ω resistor and the power absorbed by this element is $(5/9)^2(15)$ or 4.63 W.

PRACTICE

3.12 For the circuit of Fig. 3.30, find the voltage v.

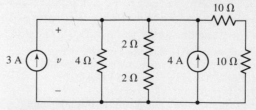

■ **FIGURE 3.30**

Ans: 12.73 V.

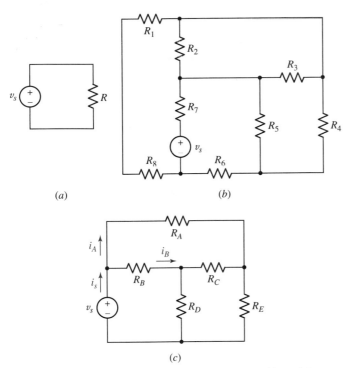

■ **FIGURE 3.31** These two circuit elements are both in series and in parallel.
(b) R_2 and R_3 are in parallel, and R_1 and R_8 are in series. (c) There are no circuit
elements either in series or in parallel with one another.

Three final comments on series and parallel combinations might be
helpful. The first is illustrated by referring to Fig. 3.31a and asking, "*Are v_s
and R in series or in parallel?*" The answer is "*Both.*" The two elements
carry the same current and are therefore in series; they also enjoy the same
voltage and consequently are in parallel.

The second comment is a word of caution. Circuits can be drawn by in-
experienced students or insidious instructors in such a way as to make series
or parallel combinations difficult to spot. In Fig. 3.31b, for example, the
only two resistors in parallel are R_2 and R_3, while the only two in series are
R_1 and R_8.

The final comment is simply that a simple circuit element need not be in
series or parallel with any other simple circuit element in a circuit. For example, R_4 and R_5 in Fig. 3.31b are not in series or parallel with any other simple
circuit element, and there are no simple circuit elements in Fig. 3.31c that are
in series or parallel with any other simple circuit element. In other words, we
cannot simplify that circuit further using any of the techniques discussed in
this chapter.

3.8 VOLTAGE AND CURRENT DIVISION

By combining resistances and sources, we have found one method of short-
ening the work of analyzing a circuit. Another useful shortcut is the application of the ideas of voltage and current division. Voltage division is used
to express the voltage across one of several series resistors in terms of the

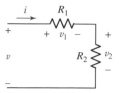

■ **FIGURE 3.32** An illustration of voltage division.

voltage across the combination. In Fig. 3.32, the voltage across R_2 is found via KVL and Ohm's law:

$$v = v_1 + v_2 = iR_1 + iR_2 = i(R_1 + R_2)$$

so

$$i = \frac{v}{R_1 + R_2}$$

Thus,

$$v_2 = iR_2 = \left(\frac{v}{R_1 + R_2}\right) R_2$$

or

$$v_2 = \frac{R_2}{R_1 + R_2} v$$

and the voltage across R_1 is, similarly,

$$v_1 = \frac{R_1}{R_1 + R_2} v$$

If the network of Fig. 3.32 is generalized by removing R_2 and replacing it with the series combination of $R_2, R_3, \ldots, R_N$, then we have the general result for voltage division across a string of N series resistors,

$$v_k = \frac{R_k}{R_1 + R_2 + \cdots + R_N} v \qquad [11]$$

which allows us to compute the voltage v_k that appears across an arbitrary resistor R_k of the series.

EXAMPLE 3.13

Determine v_x in the circuit of Fig. 3.33a.

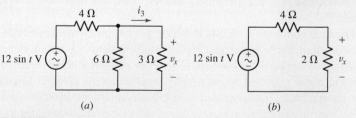

■ **FIGURE 3.33** A numerical example illustrating resistance combination and voltage division. (a) Original circuit. (b) Simplified circuit.

We first combine the 6 Ω and 3 Ω resistors, replacing them with $(6)(3)/(6 + 3) = 2$ Ω.

Since v_x appears across the parallel combination, our simplification has not lost this quantity. However, further simplification of the circuit by replacing the series combination of the 4 Ω resistor with our new 2 Ω resistor would.

Thus, we proceed by simply applying voltage division to the circuit in Fig. 3.33*b*:

$$v_x = (12 \sin t) \frac{2}{4+2} = 4 \sin t \qquad \text{volts}$$

PRACTICE

3.13 Use voltage division to determine v_x in the circuit of Fig. 3.34.

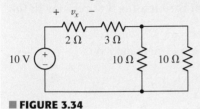

■ **FIGURE 3.34**

Ans: 2 V.

The dual[2] of voltage division is current division. We are now given a total current supplied to several parallel resistors, as shown in the circuit of Fig. 3.35.

The current flowing through R_2 is

$$i_2 = \frac{v}{R_2} = \frac{i(R_1 \| R_2)}{R_2} = \frac{i}{R_2} \frac{R_1 R_2}{R_1 + R_2}$$

or

$$\boxed{i_2 = i \frac{R_1}{R_1 + R_2}} \qquad [12]$$

and, similarly,

$$\boxed{i_1 = i \frac{R_2}{R_1 + R_2}} \qquad [13]$$

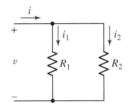

■ **FIGURE 3.35** An illustration of current division.

Nature has not smiled on us here, for these last two equations have a factor which differs subtly from the factor used with voltage division, and some effort is going to be needed to avoid errors. Many students look on the expression for voltage division as "obvious" and that for current division as being "different." It helps to realize that the larger of two parallel resistors always carries the smaller current.

For a parallel combination of *N* resistors, the current through resistor R_k is

$$\boxed{i_k = i \frac{\dfrac{1}{R_k}}{\dfrac{1}{R_1} + \dfrac{1}{R_2} + \cdots + \dfrac{1}{R_N}}} \qquad [14]$$

(2) The principle of duality is encountered often in engineering. We will consider the topic briefly in Chap. 7 when we compare inductors and capacitors.

Written in terms of conductances,

$$i_k = i \frac{G_k}{G_1 + G_2 + \cdots + G_N}$$

which strongly resembles Eq. [11] for voltage division.

EXAMPLE 3.14

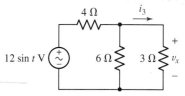

FIGURE 3.36 A circuit used as an example of current division. The wavy line in the voltage source symbol indicates a sinusoidal variation with time.

Write an expression for the current through the 3 Ω resistor in the circuit of Fig. 3.36.

The total current flowing into the 3 Ω–6 Ω combination is

$$i(t) = \frac{12 \sin t}{4 + 3\|6} = \frac{12 \sin t}{4 + 2} = 2 \sin t \quad \text{A}$$

and thus the desired current is given by current division:

$$i_3(t) = (2 \sin t)\left(\frac{6}{6+3}\right) = \frac{4}{3} \sin t \quad \text{A}$$

Unfortunately, current division is sometimes applied when it is not applicable. As one example, let us consider again the circuit shown in Fig. 3.31c, a circuit that we have already agreed contains no circuit elements that are in series or in parallel. Without parallel resistors, there is no way that current division can be applied. Even so, there are too many students who take a quick look at resistors R_A and R_B and try to apply current division, writing an incorrect equation such as

$$i_A \neq i_S \frac{R_B}{R_A + R_B}$$

Remember, *parallel resistors must be branches between the same pair of nodes.*

PRACTICE

3.14 In the circuit of Fig. 3.37, use resistance combination methods and current division to find i_1, i_2, and v_3.

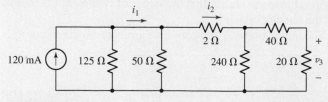

FIGURE 3.37

Ans: 100 mA; 50 mA; 0.8 V.

Not the Earth Ground from Geology

Up to now, we have been drawing circuit schematics in a fashion similar to that of the one shown in Fig. 3.38, where voltages are defined across two clearly marked terminals. Special care was taken to emphasize the fact that voltage cannot be defined at a single point—it is by definition the *difference* in potential between *two* points. However, many schematics make use of the convention of taking the earth as defining zero volts, so that all other voltages are implicitly referenced to this potential. The concept is often referred to as ***earth ground,*** and is fundamentally tied to safety regulations designed to prevent fires, fatal electrical shocks, and related mayhem. The symbol for earth ground is shown in Fig. 3.39*a*.

Since earth ground is defined as zero volts, it is often convenient to use this as a common terminal in schematics. The circuit of Fig. 3.38 is shown redrawn in this fashion in Fig. 3.40, where the earth ground symbol represents a common node. It is important to note that the two circuits are equivalent in terms of our value for v_a (4.5 V in either case), but are no longer exactly the same. The circuit in Fig. 3.38 is said to be "floating" in that it could for all practical purposes be installed on a circuit board of a satellite in geosynchronous orbit (or on its way to Pluto). The circuit in Fig. 3.40, however, is somehow physically connected to the ground through a conducting path. For this reason, there are two other symbols that are occasionally used to denote a common terminal. Figure 3.39*b* shows what is commonly referred to as ***signal ground;*** there can be (and often is) a large voltage between earth ground and any terminal tied to signal ground.

The fact that the common terminal of a circuit may or may not be connected by some low-resistance pathway to earth ground can lead to potentially dangerous situations. Consider the diagram of Fig. 3.41*a*, which depicts an innocent bystander about to touch a piece of equipment powered by an ac outlet. Only two terminals have been used from the wall socket; the round ground pin

of the receptacle was left unconnected. The common terminal of every circuit in the equipment has been tied together and electrically connected to the conducting equipment chassis; this terminal is often denoted using the ***chassis ground*** symbol of Fig. 3.39*c*. Unfortunately, a wiring fault exists, due to either poor manufacturing or perhaps just wear and tear. At any rate, the chassis is not "grounded," so there is a very large resistance between chassis ground and earth ground. A pseudo-schematic (some liberty was taken with the person's equivalent resistance symbol) of the situation is shown in Fig. 3.41*b*. The electrical path between the conducting chassis and ground may in fact be the table, which could represent a resistance of hundreds of megaohms or more. The resistance of the person, however, is many orders of magnitude lower. Once the person taps on the equipment to see why it isn't working properly . . . well, let's just say not all stories have happy endings.

The fact that "ground" is not always "earth ground" can cause a wide range of safety and electrical noise problems. One example is occasionally encountered in older buildings, where plumbing originally consisted of electrically conducting copper pipes. In such buildings, any water pipe was often treated as a low-resistance path to earth ground, and therefore used in many electrical connections. However, when corroded pipes are replaced with more modern and cost-effective

(a)   *(b)*   *(c)*

■ **FIGURE 3.39** Three different symbols used to represent a ground or common terminal: (*a*) earth ground; (*b*) signal ground; (*c*) chassis ground.

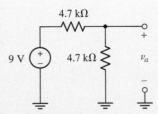

■ **FIGURE 3.40** The circuit of Fig. 3.38, redrawn using the earth ground symbol. The rightmost ground symbol is redundant; it is only necessary to label the positive terminal of v_a; the negative reference is then implicitly ground, or zero volts.

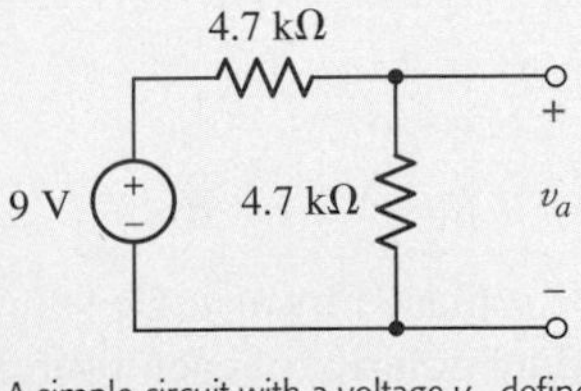

■ **FIGURE 3.38** A simple circuit with a voltage v_a defined between two terminals.

(Continued on next page)

nonconducting PVC piping, the low-resistance path to earth ground no longer exists. A related problem occurs when the composition of the earth varies greatly over a particular region. In such situations, it is possible to actually have two separated buildings in which the two "earth grounds" are not equal, and current can flow as a result.

Within this text, the earth ground symbol will be used exclusively. It is worth remembering, however, that not all grounds are created equal in practice.

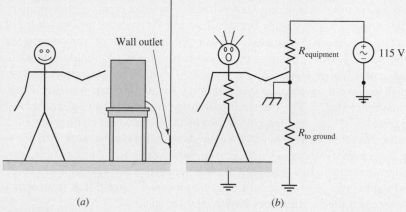

■ **FIGURE 3.41** (*a*) A sketch of an innocent person about to touch an improperly grounded piece of equipment. It's not going to be pretty. (*b*) A schematic of an equivalent circuit for the situation as it is about to unfold; the person has been represented by an equivalent resistance, as has the equipment. A resistor has been used to represent the nonhuman path to ground.

SUMMARY AND REVIEW

- ❏ Kirchhoff's current law (KCL) states that the algebraic sum of the currents entering any node is zero.
- ❏ Kirchhoff's voltage law (KVL) states that the algebraic sum of the voltages around any closed path in a circuit is zero.
- ❏ All elements in a circuit that carry the same current are said to be connected in series.
- ❏ Elements in a circuit having a common voltage across them are said to be connected in parallel.
- ❏ A series combination of N resistors can be replaced by a single resistor having the value $R_{eq} = R_1 + R_2 + \cdots + R_N$.
- ❏ A parallel combination of N resistors can be replaced by a single resistor having the value

$$\frac{1}{R_{eq}} = \frac{1}{R_1} + \frac{1}{R_2} + \cdots + \frac{1}{R_N}$$

- ❏ Voltage sources in series can be replaced by a single source, provided care is taken to note the individual polarity of each source.
- ❏ Current sources in parallel can be replaced by a single source, provided care is taken to note the direction of each current arrow.

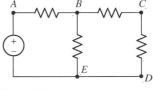

❑ Voltage division allows us to calculate what fraction of the total voltage across a series string of resistors is dropped across any one resistor (or group of resistors).

❑ Current division allows us to calculate what fraction of the total current into a parallel string of resistors flows through any one of the resistors.

READING FURTHER

A discussion of the principles of conservation of energy and conservation of charge, as well as Kirchhoff's laws, can be found in

R. Feynman, R. B. Leighton, and M. L. Sands, *The Feynman Lectures on Physics*. Reading, Mass.: Addison-Wesley, 1989, pp. 4-1, 4-7, and 25-9.

A very detailed discussion of grounding practices consistent with the 1996 National Electrical Code® can be found in

J. F. McPartland and B. J. McPartland, *McGraw-Hill's National Electrical Code® Handbook*, 22nd ed. New York: McGraw-Hill, 1996, pp. 337–485.

EXERCISES

3.1 Nodes, Paths, Loops, and Branches

1. Redraw the circuit of Fig. 3.42, consolidating nodes into the minimum number possible.

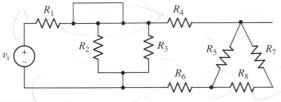

■ FIGURE 3.42

2. In the circuit of Fig. 3.42, count the number of (*a*) nodes; (*b*) branches.

3. In Fig. 3.43,
   (*a*) How many nodes are there?
   (*b*) How many branches are there?
   (*c*) If we move from *A* to *B* to *E* to *D* to *C* to *B*, have we formed a path? A loop?

4. In Fig. 3.44,
   (*a*) How many nodes are there?
   (*b*) How many branches are there?
   (*c*) If we move from *B* to *F* to *E* to *C*, have we formed a path? A loop?

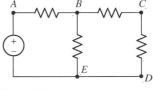

■ FIGURE 3.43

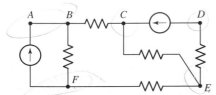

■ FIGURE 3.44

5. Referring to the circuit depicted in Fig. 3.43,

   (a) If a second wire is connected between points E and D of the circuit, how many nodes does the new circuit have?

   (b) If a resistor is added to the circuit so that one terminal is connected to point C and the other terminal is left floating, how many nodes does the new circuit have?

   (c) Which of the following represent loops?

       (i) Moving from A to B to C to D to E to A.

       (ii) Moving from B to E to A.

       (iii) Moving from B to C to D to E to B.

       (iv) Moving from A to B to C.

       (v) Moving from A to B to C to B to A.

3.2 Kirchhoff's Current Law

6. (a) Determine the current labeled i_z in the circuit shown in Fig. 3.45. (b) If the resistor carrying 3 A has a value of 1 Ω, what is the value of the resistor carrying -5 A?

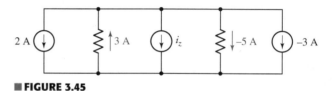

■ **FIGURE 3.45**

7. Find i_x in each of the circuits in Fig. 3.46.

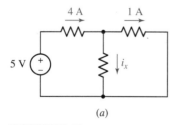

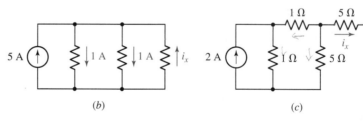

■ **FIGURE 3.46**

8. Referring to Fig. 3.47,

   (a) Find i_x if $i_y = 2$ A and $i_z = 0$ A. (b) Find i_y if $i_x = 2$ A and $i_z = 2\,i_y$.

   (c) Find i_z if $i_x = i_y = i_z$.

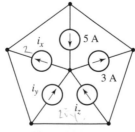

■ **FIGURE 3.47**

9. Find i_x and i_y in the circuit of Fig. 3.48.

10. A 100 W light bulb, a 60 W light bulb, and a 40 W light bulb are connected in parallel to each other and to a standard North American household 115 V supply. Compute the current flowing through each light bulb and the total current delivered by the voltage supply.

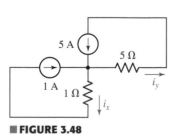

■ **FIGURE 3.48**

11. The digital multimeter (DMM) is a device commonly used to measure voltages. It is equipped with two leads (usually red for the positive reference and black for the negative reference) and an LCD display. Let's suppose a DMM is connected to the circuit of Fig. 3.46b with the positive lead at the top node and the negative lead on the bottom node. Using KCL, explain why we would ideally want a DMM used in this way to have an infinite resistance as opposed to zero resistance.

12. A local restaurant has a neon sign constructed from 12 separate bulbs; when a bulb fails, it appears as an infinite resistance and cannot conduct current. In wiring the sign, the manufacturer offers two options (Fig. 3.49). From what you've learned about KCL, which one should the restaurant owner select? Explain.

■ **FIGURE 3.49**

13. In the circuit of Fig. 3.50,
    (a) Calculate v_y if $i_z = -3$ A.
    (b) What voltage would need to replace the 5 V source to obtain $v_y = -6$ V if $i_z = 0.5$ A?

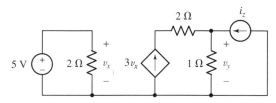

■ **FIGURE 3.50**

14. Referring to the circuit in Fig. 3.51a,
    (a) If $i_x = 5$ A, find v_1 and i_y. (b) If $v_1 = 3$ V, find i_x and i_y.
    (c) What value of i_s will lead to $v_1 \neq v_2$?

15. Find R and G in the circuit of Fig. 3.51b if the 5 A source is supplying 100 W and the 40 V source is supplying 500 W.

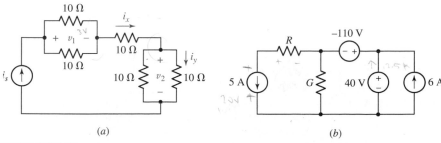

(a)   (b)

■ **FIGURE 3.51**

3.3 Kirchhoff's Voltage Law

16. In the circuits of Fig. 3.52a and b, determine the current labeled i.

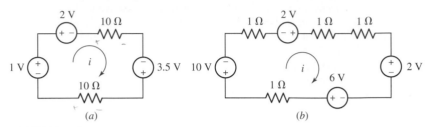

■ **FIGURE 3.52**

17. Calculate the value of i in each circuit of Fig. 3.53.

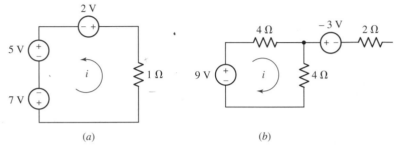

■ **FIGURE 3.53**

18. Consider the simple circuit shown in Fig. 3.54. Using KVL, derive the expressions

$$v_1 = v_s \frac{R_1}{R_1 + R_2} \quad \text{and} \quad v_2 = v_s \frac{R_2}{R_1 + R_2}$$

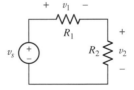

■ **FIGURE 3.54**

19. The circuit shown in Fig. 3.55 includes a device known as an op amp. This device has two unusual properties in the circuit shown: (1) $V_d = 0$ V, and (2) no current can flow into either input terminal (marked "−" and "+" inside the symbol), but it *can* flow through the output terminal (marked "OUT"). This seemingly impossible situation—in direct conflict with KCL—is a result of power leads to the device that are not included in the symbol. Based on this information, calculate V_{out}. (Hint: two KVL equations are required, both involving the 5 V source.)

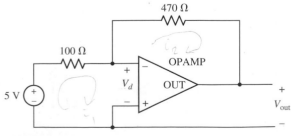

■ **FIGURE 3.55**

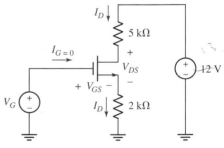

20. Use Ohm's and Kirchhoff's laws on the circuit of Fig. 3.56 to find (a) v_x; (b) i_{in}; (c) I_s; (d) the power provided by the dependent source.

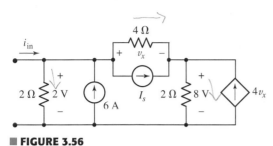

■ **FIGURE 3.56**

21. (a) Use Kirchhoff's and Ohm's laws in a step-by-step procedure to evaluate all the currents and voltages in the circuit of Fig. 3.57. (b) Calculate the power absorbed by each of the five circuit elements and show that the sum is zero.

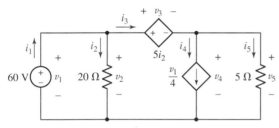

■ **FIGURE 3.57**

22. With reference to the circuit shown in Fig. 3.58, find the power absorbed by each of the seven circuit elements.

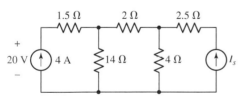

■ **FIGURE 3.58**

23. A certain circuit contains six elements and four nodes, numbered 1, 2, 3, and 4. Each circuit element is connected between a different pair of nodes. The voltage v_{12} (+ reference at first-named node) is 12 V, and $v_{34} = -8$ V. Find v_{13}, v_{23}, and v_{24} if v_{14} equals (a) 0; (b) 6 V; (c) −6 V.

24. Refer to the transistor circuit shown in Fig. 3.59. Keep in mind that although we do not know the current-voltage relationship for the device, it still obeys both KCL and KVL. (a) If $I_D = 1.5$ mA, compute V_{DS}. (b) If $I_D = 2$ mA and $V_G = 3$ V, compute V_{GS}.

3.4 The Single-Loop Circuit

25. Find the power being absorbed by element X in Fig. 3.60 if it is a (a) 100 Ω resistor; (b) 40 V independent voltage source, + reference on top; (c) dependent voltage source labeled $25i_x$, + reference on top; (d) dependent voltage source labeled $0.8v_1$, + reference on top; (e) 2 A independent current source, arrow directed upward.

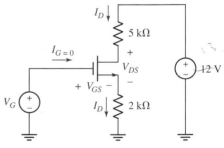

■ **FIGURE 3.59**

■ **FIGURE 3.60**

26. Find i_1 in the circuit of Fig. 3.61 if the dependent voltage source is labeled: (a) $2v_2$; (b) $1.5v_3$; (c) $-15i_1$.

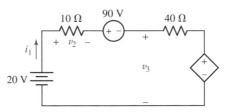

■ **FIGURE 3.61**

27. Refer to the circuit of Fig. 3.61 and label the dependent source $1.8v_3$. Find v_3 if (a) the 90 V source generates 180 W; (b) the 90 V source absorbs 180 W; (c) the dependent source generates 100 W; (d) the dependent source absorbs 100 W of power.

28. For the battery charger modeled by the circuit of Fig. 3.62, find the value of the adjustable resistor R so that: (a) a charging current of 4 A flows; (b) a power of 25 W is delivered to the battery (0.035 Ω and 10.5 V); (c) a voltage of 11 V is present at the terminals of the battery (0.035 Ω and 10.5 V).

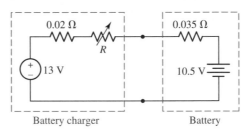

Battery charger   Battery

■ **FIGURE 3.62**

29. The circuit of Fig. 3.62 is modified by installing a dependent voltage source in series with the battery. Break the top wire, place the + reference at the right and let the control be $0.05i$, where i is the clockwise loop current. Find this current and the terminal voltage of the battery, including the dependent source, if $R = 0.5$ Ω.

30. Find the power absorbed by each of the six circuit elements in Fig. 3.63, and show that they sum to zero.

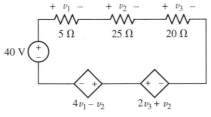

■ **FIGURE 3.63**

31. For the circuit of Fig. 3.64,

   (a) Determine the resistance R that will result in the 25 kΩ resistor absorbing 2 mW.

   (b) Determine the resistance R that results in the 12 V source delivering 3.6 mW to the circuit.

   (c) Replace the resistor R with a voltage source such that no power is absorbed by either resistor; draw the circuit, indicating the voltage polarity of the new source.

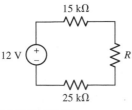

■ **FIGURE 3.64**

32. Referring to Table 2.4, if the bottom wire segment in the circuit of Fig. 3.65 is 22 AWG solid copper and 3000 ft long, compute the current i.

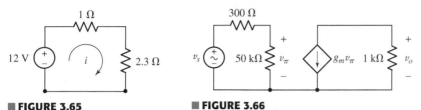

■ **FIGURE 3.65**   ■ **FIGURE 3.66**

33. In Fig. 3.66, if $g_m = 25 \times 10^{-3}$ siemens and $v_s = 10 \cos 5t$ mV, find $v_o(t)$.

34. Kirchhoff's laws apply whether or not Ohm's law applies to a particular element. The *I-V* characteristic of a diode, for example, is given by

$$I_D = I_S \left(e^{V_D/V_T} - 1 \right)$$

where $V_T = 27$ mV at room temperature and I_S can vary from 10^{-12} to 10^{-3} A. In the circuit of Fig. 3.67, use KVL/KCL to obtain V_D if $I_S = 3 \ \mu$A. (*Note: This problem results in a transcendental equation, requiring an iterative approach to obtaining a numerical solution. Most scientific calculators will perform such a function.*)

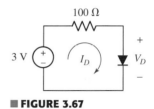

■ **FIGURE 3.67**

3.5 The Single-Node-Pair Circuit

35. Find the power absorbed by each circuit element of Fig. 3.68 if the control for the dependent source is (*a*) $0.8i_x$; (*b*) $0.8i_y$. In each case, demonstrate that the absorbed power quantities sum to zero.

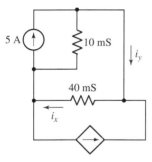

■ **FIGURE 3.68**   ■ **FIGURE 3.69**

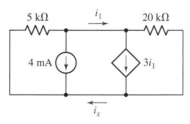

36. Find i_x in the circuit of Fig. 3.69.

37. Find the power absorbed by each element in the single-node-pair circuit of Fig. 3.70, and show that the sum is equal to zero.

38. Find the power absorbed by element X in the circuit of Fig. 3.71 if it is a (*a*) 4 kΩ resistor; (*b*) 20 mA independent current source, reference arrow downward; (*c*) dependent current source, reference arrow downward, labeled $2i_x$; (*d*) 60 V independent voltage source, + reference at top.

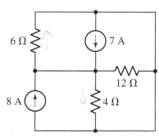

■ **FIGURE 3.70**

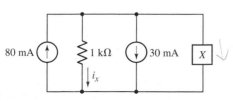

■ **FIGURE 3.71**

39. (a) Let element X in Fig. 3.72 be an independent current source, arrow directed upward, labeled i_s. What is i_s if none of the four circuit elements absorbs any power? (b) Let element X be an independent voltage source, + reference on top, labeled v_s. What is v_s if the voltage source absorbs no power?

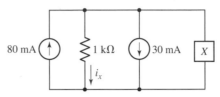

80 mA   1 kΩ   30 mA   X

i_x

■ **FIGURE 3.72**

40. (a) Apply the techniques of single-node-pair analysis to the upper right node in Fig. 3.73 and find i_x. (b) Now work with the upper left node and find v_8. (c) How much power is the 5 A source generating?

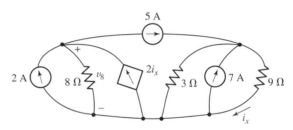

5 A

2 A   8 Ω v_8 $2i_x$ 3 Ω   7 A   9 Ω

i_x

■ **FIGURE 3.73**

41. Find the power absorbed by the 5 Ω resistor in Fig. 3.74.

42. Compute the power supplied by each element shown in Fig. 3.75, and show that their sum is equal to zero.

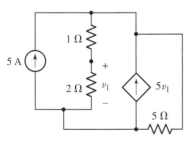

5 A   1 Ω   2 Ω v_1 $5v_1$ 5 Ω

■ **FIGURE 3.74**

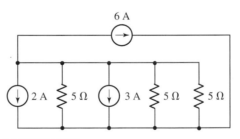

6 A

2 A   5 Ω   3 A   5 Ω   5 Ω

■ **FIGURE 3.75**

43. Referring to Table 2.4, how many miles of 28 AWG solid copper wire is required for the labelled wire segment of Fig. 3.76 to obtain $i_1 = 5$ A?

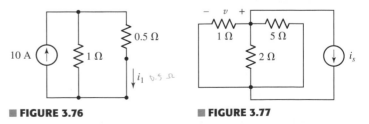

10 A   1 Ω   0.5 Ω i_1 0.5 Ω

$-$ v $+$ 1 Ω   5 Ω   2 Ω i_s

■ **FIGURE 3.76**   ■ **FIGURE 3.77**

44. In the circuit of Fig. 3.77, if $v = 6$ V, find i_s.

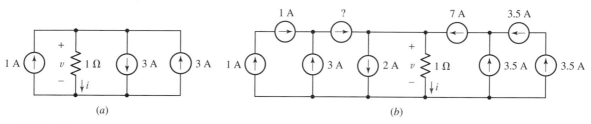

3.6 Series and Parallel Connected Sources

45. Using combinations of sources, compute i for both circuits in Fig. 3.78.

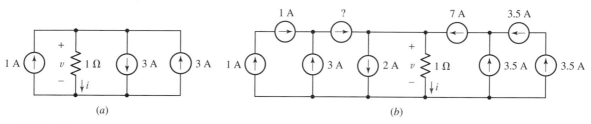

(a)   (b)

■ **FIGURE 3.78**

46. Compute v for each of the circuits in Fig. 3.78 by first combining sources.
47. Compute the current labeled i in each of the circuits in Fig. 3.79.

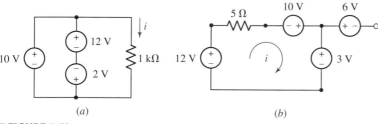

(a)   (b)

■ **FIGURE 3.79**

48. Compute the power absorbed by each element of the circuit shown in Fig. 3.80, and verify that their sum is zero.

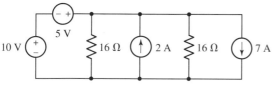

■ **FIGURE 3.80**

49. For the circuit in Fig. 3.81, compute i if:
    (a) $v_1 = v_2 = 10$ V and $v_3 = v_4 = 6$ V.
    (b) $v_1 = v_3 = 3$ V and $v_2 = v_4 = 2.5$ V.
    (c) $v_1 = -3$ V, $v_2 = 1.5$ V, $v_3 = -0.5$ V, and $v_4 = 0$ V.
50. In the circuit of Fig. 3.82, choose v_1 to obtain a current i_x of 2 A.

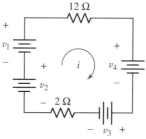

■ **FIGURE 3.81**

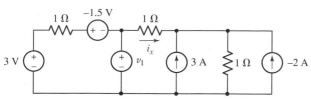

■ **FIGURE 3.82**

51. Find the voltage v in the circuit of Fig. 3.83.

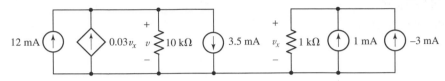

FIGURE 3.83

52. The circuit shown in Fig. 3.84 contains several examples of independent current and voltage sources connected in series and in parallel. (*a*) Find the power absorbed by each source. (*b*) To what value should the 4 V source be changed to reduce the power supplied by the -5 A source to zero?

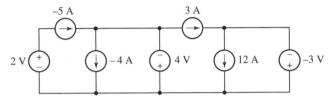

FIGURE 3.84

3.7 Resistors in Series and Parallel

53. Compute the equivalent resistance as indicated in Fig. 3.85 if each resistor is 1 kΩ.

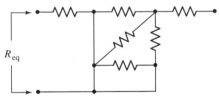

FIGURE 3.85

54. For the circuit in Fig. 3.86,

(*a*) Compute the equivalent resistance.

(*b*) Derive an expression for the equivalent resistance if the circuit is extended using N branches, each branch having one more resistor than the branch to its left.

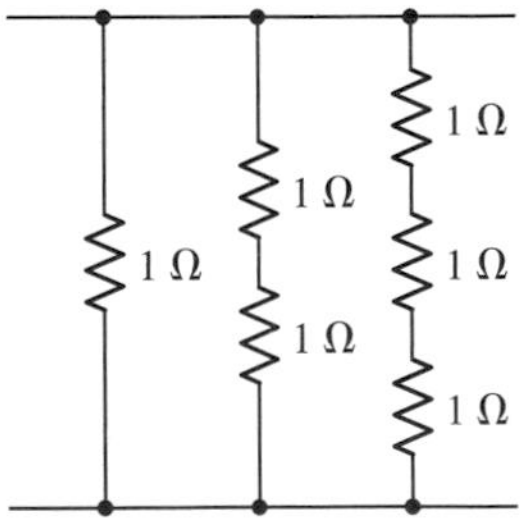

55. Given three 10 kΩ resistors, three 47 kΩ resistors, and three 1 kΩ resistors, find a combination (not all resistors need to be used) that yields:

(*a*) 5 kΩ

(*b*) 57,333 Ω

(*c*) 29.5 kΩ

56. Simplify the networks in Fig. 3.87 using resistor and source combinations.

FIGURE 3.86

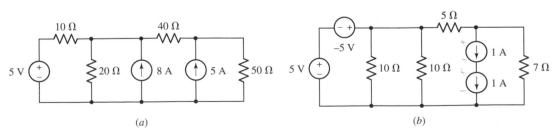

(*a*)   (*b*)

FIGURE 3.87

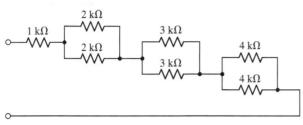

57. Compute the equivalent resistance of the circuit in Fig. 3.88.

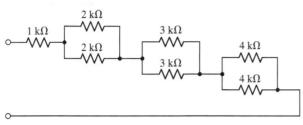

■ **FIGURE 3.88**

58. Find R_{eq} for each of the resistive networks shown in Fig. 3.89.

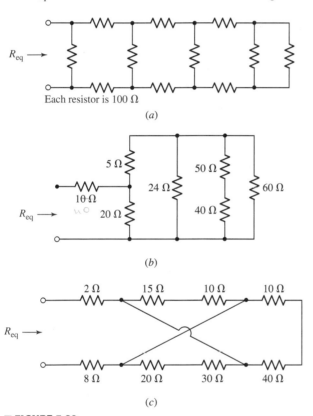

Each resistor is 100 Ω

(a)

(b)

(c)

■ **FIGURE 3.89**

59. In the network shown in Fig. 3.90: (a) let $R = 80\ \Omega$ and find R_{eq}; (b) find R if $R_{eq} = 80\ \Omega$; (c) find R if $R = R_{eq}$.

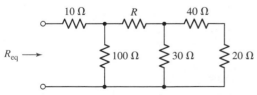

■ **FIGURE 3.90**

60. Show how to combine four 100 Ω resistors to obtain an equivalent resistance of (a) 25 Ω; (b) 60 Ω; (c) 40 Ω.

61. Find the power absorbed by each of the resistors in the circuit of Fig. 3.91.

62. Use source- and resistor-combination techniques as a help in finding v_x and i_x in the circuit of Fig. 3.92.

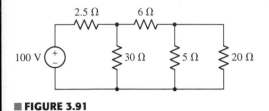

■ **FIGURE 3.91**

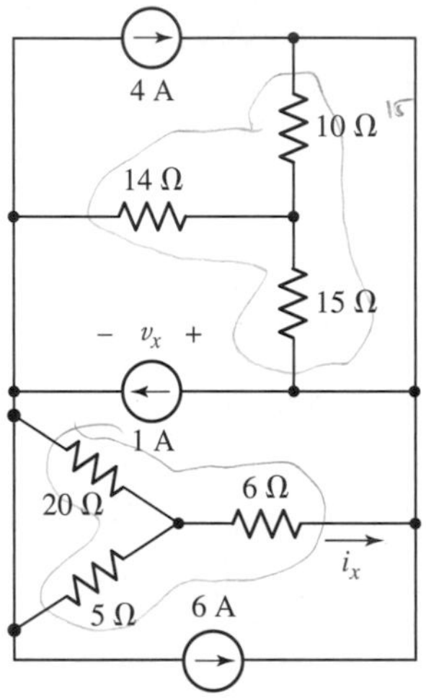

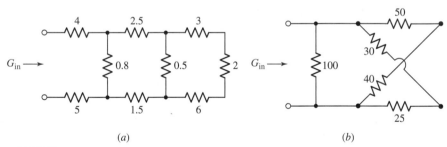

■ **FIGURE 3.92**

63. Determine G_{in} for each network shown in Fig. 3.93. Values are all given in millisiemens.

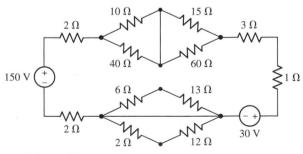

(a)   (b)

■ **FIGURE 3.93**

3.8 Voltage and Current Division

64. Use both resistance and source combinations, as well as current division, in the circuit of Fig. 3.94 to find the power absorbed by the 1 Ω, 10 Ω, and 13 Ω resistors.

■ **FIGURE 3.94**

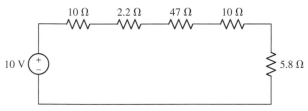

65. The **Wheatstone bridge** (Fig. 3.95) is one of the most well-known electrical circuits and is used in resistance measurement. The resistor with an arrow through its symbol (R_3) is a variable resistor, sometimes referred to as a potentiometer; its value can be changed by simply rotating a knob. The ammeter, symbolized by a circle with a diagonal arrow in the center, measures the current through the center wire. We assume this ammeter to be ideal, so that it has zero internal resistance.

Operation is simple. The values of R_1, R_2, and R_3 are known, and the value of R is desired. Resistor R_3 is adjusted until $i_m = 0$; in other words, until no current flows through the ammeter. At this point the bridge is said to be "balanced."

Using KCL and KVL, show that $R = \dfrac{R_2}{R_1} R_3$. (Hints: The value of V_s is irrelevant; with $i_m = 0$, $i_1 = i_3$ and $i_2 = i_R$; and there is no voltage dropped across the ammeter.)

■ **FIGURE 3.95**

66. The circuit of Fig. 3.96 consists of several resistors connected in a series string. Use voltage division to calculate how much voltage is dropped across the smallest resistor and across the largest resistor, respectively.

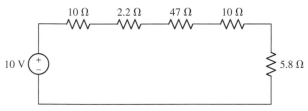

■ **FIGURE 3.96**

67. Employ voltage division to calculate the voltage across the 47 kΩ resistor of Fig. 3.97.

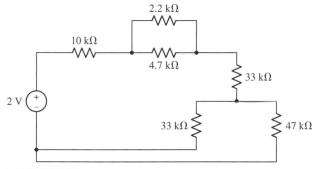

■ **FIGURE 3.97**

68. Referring to the circuit depicted in Fig. 3.98, use current division to calculate the current flowing downward through (*a*) the 33 Ω resistor; (*b*) the rightmost 134 Ω resistor.

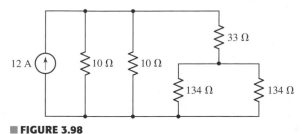

■ **FIGURE 3.98**

69. It appears that despite the large number of components in the circuit of Fig. 3.99, only the voltage across the 15 Ω resistor is of interest. Use current division to assist in calculating the correct value.

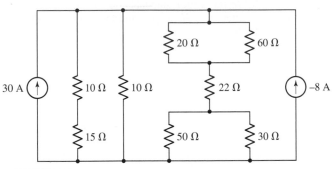

■ **FIGURE 3.99**

D 70. Choosing from the following resistor values (they may be used more than once), set v_s, R_1, and R_2 in Fig. 3.100 to obtain $v_x = 5.5$ V. [1 kΩ, 3.3 kΩ, 4.7 kΩ, 10 kΩ]

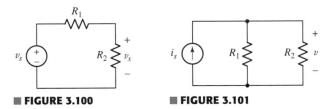

■ **FIGURE 3.100**   ■ **FIGURE 3.101**

D 71. Choosing from the following resistor values (they may be used more than once), set i_s, R_1, and R_2 in Fig. 3.101 to obtain $v = 5.5$ V. [1 kΩ, 3.3 kΩ, 4.7 kΩ, 10 kΩ]

72. Determine the power dissipated by (absorbed by) the 15 kΩ resistor in Fig. 3.102.

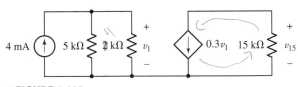

■ **FIGURE 3.102**

73. For the circuit in Fig. 3.103, determine i_x, and compute the power dissipated by (absorbed by) the 15 kΩ resistor.

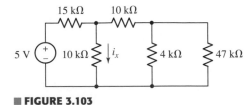

■ **FIGURE 3.103**

74. For the circuit in Fig. 3.104, find i_x, i_y, and the power dissipated by (absorbed by) the 3 Ω resistor.

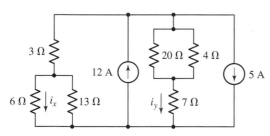

■ FIGURE 3.104

75. What is the power dissipated by (absorbed by) the 47 kΩ resistor in Fig. 3.105?

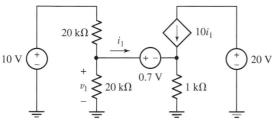

■ FIGURE 3.105

76. Explain why voltage division cannot be used to determine v_1 in Fig. 3.106.

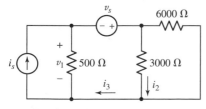

■ FIGURE 3.106

77. Use current and voltage division on the circuit of Fig. 3.107 to find an expression for (a) v_2; (b) v_1; (c) i_4.

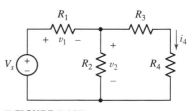

■ FIGURE 3.107

78. With reference to the circuit shown in Fig. 3.108: (a) let $v_s = 40$ V, $i_s = 0$, and find v_1; (b) let $v_s = 0$, $i_s = 3$ mA, and find i_2 and i_3.

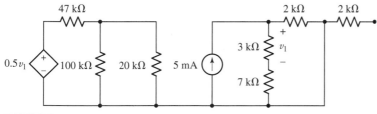

■ FIGURE 3.108

79. In Fig. 3.109: (a) let $v_x = 10$ V and find I_s; (b) let $I_s = 50$ A and find v_x; (c) calculate the ratio v_x/I_s.

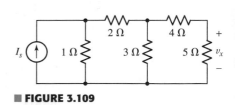

■ FIGURE 3.109

80. Determine how much power is absorbed by R_x in the circuit of Fig. 3.110.

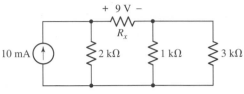

■ **FIGURE 3.110**

81. Use current and voltage division to help obtain an expression for v_5 in Fig. 3.111.

82. With reference to the circuit of Fig. 3.112, find (a) I_x if $I_1 = 12$ mA; (b) I_1 if $I_x = 12$ mA; (c) I_x if $I_2 = 15$ mA; (d) I_x if $I_s = 60$ mA.

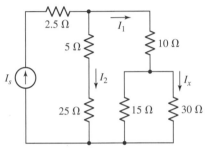

■ **FIGURE 3.111**

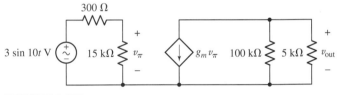

■ **FIGURE 3.112**

83. The circuit in Fig. 3.113 is a commonly used equivalent circuit used to model the ac behavior of a MOSFET amplifier circuit. If $g_m = 4$ m℧, compute v_{out}.

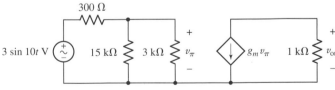

■ **FIGURE 3.113**

84. The circuit in Fig. 3.114 is a commonly used equivalent circuit used to model the ac behavior of a bipolar junction transistor amplifier circuit. If $g_m = 38$ m℧, compute v_{out}.

■ **FIGURE 3.114**

Basic Nodal and Mesh Analysis

KEY CONCEPTS

Nodal Analysis

The Supernode Technique

Mesh Analysis

The Supermesh Technique

Choosing Between Nodal and Mesh Analysis

Computer-Aided Analysis, Including PSpice and MATLAB

INTRODUCTION

Armed with the trio of Ohm's and Kirchhoff's laws, analyzing a simple linear circuit to obtain useful information such as the current, voltage, or power associated with a particular element is perhaps starting to seem a straightforward enough venture. Still, for the moment at least, every circuit seems unique, requiring (to some degree) a measure of creativity in approaching the analysis. In this chapter, we learn two basic circuit analysis techniques—*nodal analysis* and *mesh analysis*—both of which allow us to investigate many different circuits with a consistent, methodical approach. The result is a streamlined analysis, a more uniform level of complexity in our equations, fewer errors and, perhaps most importantly, a reduced occurrence of "*I don't know how to even start!*"

Most of the circuits we have seen up to now have been rather simple and (to be honest) of questionable practical use. Such circuits are valuable, however, in helping us to learn to apply fundamental techniques. Although the more complex circuits appearing in this chapter may represent a variety of electrical systems including control circuits, communication networks, motors, or integrated circuits, as well as electric circuit models of nonelectrical systems, we believe it best not to dwell on such specifics at this early stage. Rather, it is important to initially focus on the *methodology of problem solving* that we will continue to develop throughout the book.

4.1 NODAL ANALYSIS

We begin our study of general methods for methodical circuit analysis by considering a powerful method based on KCL, namely *nodal analysis.* In Chap. 3 we considered the analysis of a simple circuit containing only two nodes. We found that the major step of the analysis was obtaining a single equation in terms of a single unknown quantity—the voltage between the pair of nodes.

We will now let the number of nodes increase and correspondingly provide one additional unknown quantity and one additional equation for each added node. Thus, a three-node circuit should have two unknown voltages and two equations; a 10-node circuit will have nine unknown voltages and nine equations; an N-node circuit will need $(N - 1)$ voltages and $(N - 1)$ equations. Each equation is a simple KCL equation.

To illustrate the basic mechanics of the technique, consider the three-node circuit shown in Fig. 4.1a, redrawn in Fig. 4.1b to emphasize the fact that there are only three nodes, numbered accordingly. Our goal will be to determine the voltage across each element, and the next step in the analysis is critical. We designate one node as a *reference node;* it will be the negative terminal of our $N - 1 = 2$ nodal voltages, as shown in Fig. 4.1c.

A little simplification in the resultant equations is obtained if the node connected to the greatest number of branches is identified as the reference node. If there is a ground node, it is usually most convenient to select it as the reference node, although many people seem to prefer selecting the bottom node of a circuit as the reference, especially if no explicit ground is noted.

The voltage of node 1 *relative to the reference node* is defined as v_1, and v_2 is defined as the voltage of node 2 with respect to the reference node.

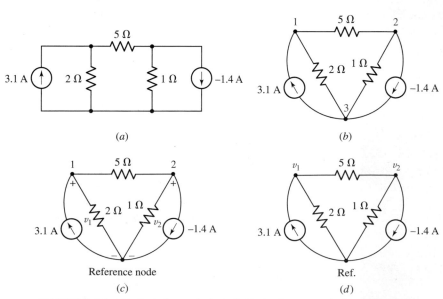

■ **FIGURE 4.1** (*a*) A simple three-node circuit. (*b*) Circuit redrawn to emphasize nodes. (*c*) Reference node selected and voltages assigned. (*d*) Shorthand voltage references. If desired, an appropriate ground symbol may be substituted for "Ref."

These two voltages are sufficient, as the voltage between any other pair of nodes may be found in terms of them. For example, the voltage of node 1 with respect to node 2 is $v_1 - v_2$. The voltages v_1 and v_2 and their reference signs are shown in Fig. 4.1c. It is common practice once a reference node has been labeled to omit the reference signs for the sake of clarity; the node labeled with the voltage is taken to be the positive terminal (Fig. 4.1d). This is understood to be a type of shorthand voltage notation.

We now apply KCL to nodes 1 and 2. We do this by equating the total current leaving the node through the several resistors to the total source current entering the node. Thus,

$$\frac{v_1}{2} + \frac{v_1 - v_2}{5} = 3.1 \tag{1}$$

or

$$0.7v_1 - 0.2v_2 = 3.1 \tag{2}$$

At node 2 we obtain

$$\frac{v_2}{1} + \frac{v_2 - v_1}{5} = -(-1.4) \tag{3}$$

or

$$-0.2v_1 + 1.2v_2 = 1.4 \tag{4}$$

Equations [2] and [4] are the desired two equations in two unknowns, and they may be solved easily. The results are $v_1 = 5$ V and $v_2 = 2$ V.

From this, it is straightforward to determine the voltage across the 5 Ω resistor: $v_{5\Omega} = v_1 - v_2 = 3$ V. The currents and absorbed powers may also be computed in one step.

We should note at this point that there is more than one way to write the KCL equations for nodal analysis. For example, the reader may prefer to sum all the currents entering a given node and set this quantity to zero. Thus, for node 1 we might have written

$$3.1 - \frac{v_1}{2} - \frac{v_1 - v_2}{5} = 0$$

or

$$3.1 + \frac{-v_1}{2} + \frac{v_2 - v_1}{5} = 0$$

either of which is equivalent to Eq. [1]. *Is one way better than any other?* Every instructor and every student develops a personal preference, and at the end of the day the most important thing is to be consistent. The authors prefer constructing KCL equations for nodal analysis in such a way as to end up with all current source terms on one side and all resistor terms on the other. Specifically,

> Σ currents entering the node from current sources
> = Σ currents leaving the node through resistors

There are several advantages to such an approach. First, there is never any confusion regarding whether a term should be "$v_1 - v_2$" or "$v_2 - v_1$;"

The reference node in a schematic is implicitly defined as zero volts. However, it is important to remember that any terminal can be designated as the reference terminal. Thus, the reference node is at zero volts with respect to the other defined nodal voltages, and not necessarily with respect to *earth* ground.

the first voltage in every resistor current expression corresponds to the node for which a KCL equation is being written, as seen in Eqs. [1] and [3]. Second, it allows a quick check that a term has not been accidentally omitted. Simply count the current sources connected to a node and then the resistors; grouping them in the stated fashion makes the comparison a little easier.

EXAMPLE 4.1

Determine the current flowing left to right through the 15 Ω resistor of Fig. 4.2a.

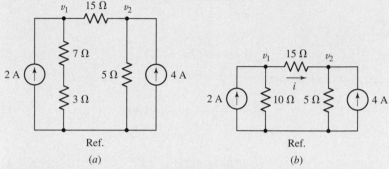

■ **FIGURE 4.2** (*a*) A four-node circuit containing two independent current sources. (*b*) The two resistors in series are replaced with a single 10 Ω resistor, reducing the circuit to three nodes.

Nodal analysis will directly yield numerical values for the nodal voltages v_1 and v_2, and the desired current is given by $i = (v_1 - v_2)/15$.

Before launching into nodal analysis, however, we first note that no details regarding either the 7 Ω resistor or the 3 Ω resistor are of interest. Thus, we may replace their series combination with a 10 Ω resistor as in Fig. 4.2b. The result is a reduction in the number of equations to solve.

Writing an appropriate KCL equation for node 1,

$$2 = \frac{v_1}{10} + \frac{v_1 - v_2}{15} \qquad [5]$$

and for node 2,

$$4 = \frac{v_2}{5} + \frac{v_2 - v_1}{15} \qquad [6]$$

Rearranging, we obtain

$$5v_1 - 2v_2 = 60$$

and

$$-v_1 + 4v_2 = 60$$

Solving, we find that $v_1 = 20$ V and $v_2 = 20$ so that $v_1 - v_2 = 0$. In other words, *zero current* is flowing through the 15 Ω resistor in this circuit!

PRACTICE

4.1 For the circuit of Fig. 4.3, determine the nodal voltages v_1 and v_2.

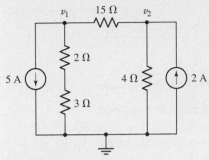

■ **FIGURE 4.3**

Ans: $v_1 = -145/8$ V, $v_2 = 5/2$ V.

Now let us increase the number of nodes so that we may use this technique to work a slightly more difficult problem.

EXAMPLE 4.2

Find the node voltages in the circuit of Fig. 4.4a.

▶ ***Identify the goal of the problem.***
There are four nodes in this circuit. Selecting the bottom node as our reference, we label the other three nodes as shown in Fig. 4.4b. The circuit has also been redrawn slightly for convenience.

▶ ***Collect the known information.***
We have three unknown voltages, v_1, v_2, and v_3. All current sources and resistors have designated values, which are marked on the schematic.

▶ ***Devise a plan.***
This problem is well suited to the recently introduced technique of nodal analysis, as three independent KCL equations may be written in terms of the current sources and the current through each resistor.

▶ ***Construct an appropriate set of equations.***
We begin by writing a KCL equation for node 1:

$$-8 - 3 = \frac{v_1 - v_2}{3} + \frac{v_1 - v_3}{4}$$

or

$$0.5833v_1 - 0.3333v_2 - 0.25v_3 = -11 \qquad [7]$$

At node 2:

$$-(-3) = \frac{v_2 - v_1}{3} + \frac{v_2}{1} + \frac{v_2 - v_3}{7}$$

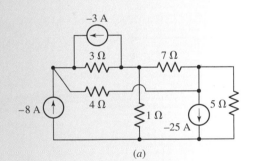

(a)

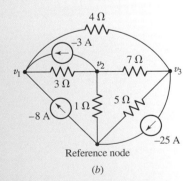

Reference node

(b)

■ **FIGURE 4.4** (a) A four-node circuit. (b) Redrawn circuit with reference node chosen and voltages labeled.

(Continued on next page)

or

$$-0.3333v_1 + 1.4762v_2 - 0.1429v_3 = 3 \qquad [8]$$

And, at node 3:

$$-(-25) = \frac{v_3}{5} + \frac{v_3 - v_2}{7} + \frac{v_3 - v_1}{4}$$

or, more simply,

$$-0.25v_1 - 0.1429v_2 + 0.5929v_3 = 25 \qquad [9]$$

▶ **Determine if additional information is required.**

We have three equations in three unknowns. Provided that they are independent, this is sufficient to determine the three voltages.

▶ **Attempt a solution.**

Equations [7] through [9] may be solved by successive elimination of variables, by matrix methods, or by *Cramer's rule* and *determinants*. Using the latter method, described in App. 2, we have

$$v_1 = \frac{\begin{vmatrix} -11 & -0.3333 & -0.2500 \\ 3 & 1.4762 & -0.1429 \\ 25 & -0.1429 & 0.5929 \end{vmatrix}}{\begin{vmatrix} 0.5833 & -0.3333 & -0.2500 \\ -0.3333 & 1.4762 & -0.1429 \\ -0.2500 & -0.1429 & 0.5929 \end{vmatrix}} = \frac{1.714}{0.3167} = 5.412 \text{ V}$$

Similarly,

$$v_2 = \frac{\begin{vmatrix} 0.5833 & -11 & -0.2500 \\ -0.3333 & 3 & -0.1429 \\ -0.2500 & 25 & 0.5929 \end{vmatrix}}{0.3167} = \frac{2.450}{0.3167} = 7.736 \text{ V}$$

and

$$v_3 = \frac{\begin{vmatrix} 0.5833 & -0.3333 & -11 \\ -0.3333 & 1.4762 & 3 \\ -0.2500 & -0.1429 & 25 \end{vmatrix}}{0.3167} = \frac{14.67}{0.3167} = 46.32 \text{ V}$$

▶ **Verify the solution. Is it reasonable or expected?**

One way to check part of our solution is to solve the three equations using another technique. Beyond that, is it possible to determine whether these voltages are "reasonable" values? We have a maximum possible current of $3 + 8 + 25 = 36$ amperes anywhere in the circuit. The largest resistor is $7\ \Omega$, so we do not expect any voltage magnitude greater than $7 \times 36 = 252$ V.

There are, of course, numerous methods available for the solution of linear systems of equations, and we describe several in App. 2 in detail. Prior to the advent of the scientific calculator, Cramer's rule as seen in Example 4.2 was very common in circuit analysis, although occasionally tedious to implement. It is, however, straightforward to use on a simple four-function calculator and so an awareness of the technique can be valuable. MATLAB, on the other hand, although not likely to be available during an examination, is a powerful software package that can greatly simplify the solution process; a brief tutorial on getting started is provided in App. 6.

For the situation encountered in Example 4.2, there are several options available through MATLAB. First, we can represent Eqs. [7]–[9] in *matrix form:*

$$
\begin{bmatrix}
0.5833 & -0.3333 & -0.25 \\
-0.3333 & 1.4762 & -0.1429 \\
-0.25 & -0.1429 & 0.5929
\end{bmatrix}
\begin{bmatrix}
v_1 \\ v_2 \\ v_3
\end{bmatrix}
=
\begin{bmatrix}
-11 \\ 3 \\ 25
\end{bmatrix}
$$

so that

$$
\begin{bmatrix}
v_1 \\ v_2 \\ v_3
\end{bmatrix}
=
\begin{bmatrix}
0.5833 & -0.3333 & -0.25 \\
-0.3333 & 1.4762 & -0.1429 \\
-0.25 & -0.1429 & 0.5929
\end{bmatrix}^{-1}
\begin{bmatrix}
-11 \\ 3 \\ 25
\end{bmatrix}
$$

In MATLAB, we write

```
>> a = [0.5833 -0.3333 -0.25; -0.3333 1.4762 -0.1429; -0.25 -0.1429 0.5929];
>> c = [-11; 3; 25];
>> b = a^-1 * c

b =

   5.4124
   7.7375
  46.3127

>>
```

where spaces separate elements along rows, and a semicolon separates rows. The matrix named **b**, which can also be referred to as a *vector* as it has only one column, is our solution. Thus, $v_1 = 5.412$ V, $v_2 = 7.738$ V, and $v_3 = 46.31$ V (some rounding error has been incurred).

We could also use the KCL equations as we wrote them initially if we employ the symbolic processor of MATLAB.

```
>> eqn1 = '-8 -3 = (v1 - v2)/ 3 + (v1 - v3)/ 4';
>> eqn2 = '-(-3) = (v2 - v1)/ 3 + v2/ 1 + (v2 - v3)/ 7';
>> eqn3 = '-(-25) = v3/ 5 + (v3 - v2)/ 7 + (v3 - v1)/ 4';
>> answer = solve(eqn1, eqn2, eqn3, 'v1', 'v2', 'v3');
>> answer.v1

ans =

720/133

>> answer.v2

ans =

147/19

>> answer.v3

ans =

880/19

>>
```

which results in exact answers, with no rounding errors. The *solve()* routine is invoked with the list of symbolic equations we named eqn1, eqn2, and eqn3, but the variables v1, v2 and v3 must also be specified. If *solve()* is called with fewer variables than equations, an algebraic solution is returned. The form of the solution is worth a quick comment; it is returned in what is referred to in programming parlance as a *structure;* in this case, we called our structure "answer." Each component of the structure is accessed separately by name as shown.

PRACTICE

4.2 For the circuit of Fig. 4.5, compute the voltage across each current source.

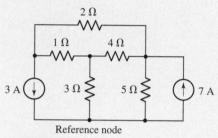

■ **FIGURE 4.5**

Ans: $v_{3A} = 5.235$ V; $v_{7A} = 11.47$ V.

The previous examples have demonstrated the basic approach to nodal analysis, but it is worth considering what happens if dependent sources are present as well.

EXAMPLE 4.3

Determine the power supplied by the dependent source of Fig. 4.6a.

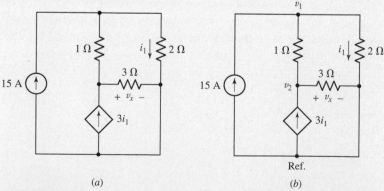

(a)   (b)

■ **FIGURE 4.6** (*a*) A four-node circuit containing a dependent current source. (*b*) Circuit labeled for nodal analysis.

We choose the bottom node as our reference, since it has the largest number of branch connections, and proceed to label the nodal voltages v_1 and v_2 as shown in Fig. 4.6b. The quantity labeled v_x is actually equal to v_2.

At node 1, we write

$$15 = \frac{v_1 - v_2}{1} + \frac{v_1}{2} \qquad [10]$$

and at node 2

$$3i_1 = \frac{v_2 - v_1}{1} + \frac{v_2}{3} \qquad [11]$$

Unfortunately, we have only two equations but three unknowns; *this is a direct result of the presence of the dependent current source, since it is not controlled by a nodal voltage*. Thus, we must develop an additional equation that relates i_1 to one or more nodal voltages.

In this case, we find that

$$i_1 = \frac{v_1}{2} \qquad [12]$$

which upon substitution into Eq. [11] yields (with a little rearranging)

$$3v_1 - 2v_2 = 30 \qquad [13]$$

and Eq. [10] simplifies to

$$-15v_1 + 8v_2 = 0 \qquad [14]$$

Solving, we find that $v_1 = -40$ V, $v_2 = -75$ V, and $i_1 = 0.5v_1 = -20$ A. Thus, the power supplied by the dependent source is equal to $(3i_1)(v_2) = (-60)(-75) = 4.5$ kW.

We see that the presence of a dependent source will create the need for an additional equation in our analysis if the controlling quantity is not a nodal voltage. Now let's look at the same circuit, but with the controlling variable of the dependent current source changed to a different quantity—the voltage across the 3 Ω resistor, which is in fact a nodal voltage. We will find that only *two* equations are required to complete the analysis.

EXAMPLE **4.4**

Determine the power supplied by the dependent source of Fig. 4.7a.

We select the bottom node as our reference and label the nodal voltages as shown in Fig. 4.7b. We have labeled the nodal voltage v_x explicitly for clarity, but this redundancy is of course not necessary. Note that our choice of reference node is important in this case; it led to the quantity v_x being a nodal voltage.

Our KCL equation for node 1 is

$$15 = \frac{v_1 - v_x}{1} + \frac{v_1}{2} \qquad [15]$$

(Continued on next page)

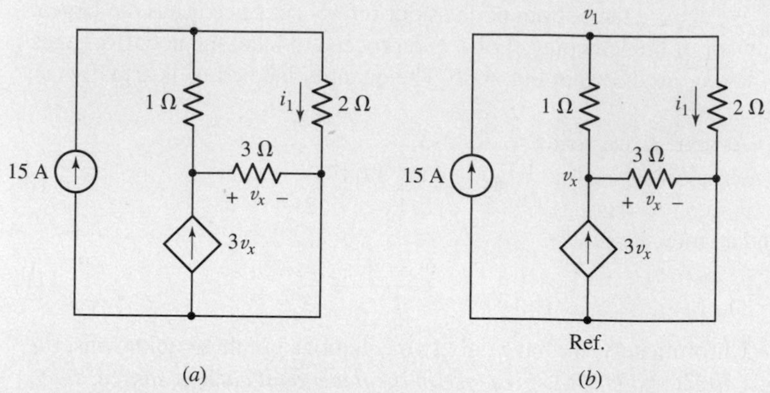

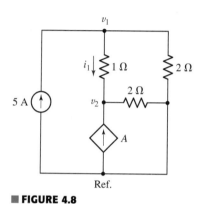

■ **FIGURE 4.7** (*a*) A four-node circuit containing a dependent current source. (*b*) Circuit labeled for nodal analysis.

and for node *x* is

$$3v_x = \frac{v_x - v_1}{1} + \frac{v_2}{3} \qquad [16]$$

Grouping terms and solving, we find that $v_1 = \frac{50}{7}$ V and $v_x = -\frac{30}{7}$ V. Thus, the dependent source in this circuit generates $(3v_x)(v_x) = 55.1$ W.

PRACTICE

4.3 For the circuit of Fig. 4.8, determine the nodal voltage v_1 if A is (*a*) $2i_1$; (*b*) $2v_1$.

Ans: (*a*) $\frac{70}{9}$ V; (*b*) −10 V.

■ **FIGURE 4.8**

Summary of Basic Nodal Analysis Procedure

1. **Count the number of nodes** (N).

2. **Designate a reference node.** The number of terms in your nodal equations can be minimized by selecting the node with the greatest number of branches connected to it.

3. **Label the nodal voltages** (there are $N - 1$ of them).

4. **Write a KCL equation for each of the nonreference nodes.** Sum the currents flowing *into* a node from sources on one side of the equation. On the other side, sum the currents flowing *out of* the node through resistors. Pay close attention to "−" signs.

5. **Express any additional unknowns such as currents or voltages other than nodal voltages in terms of appropriate nodal voltages.** This situation can occur if voltage sources or dependent sources appear in our circuit.

6. **Organize the equations.** Group terms according to nodal voltages.

7. **Solve the system of equations for the nodal voltages** (there will be $N - 1$ of them).

These seven basic steps will work on any circuit we ever encounter, although the presence of voltage sources will require extra care. Such situations are discussed in Sec. 4.2.

4.2 THE SUPERNODE

We next consider how voltage sources affect the strategy of nodal analysis.

As a typical example, consider the circuit shown in Fig. 4.9a. The original four-node circuit of Fig. 4.4 has been changed by replacing the 7 Ω resistor between nodes 2 and 3 with a 22 V voltage source. We still assign the same node-to-reference voltages v_1, v_2, and v_3. Previously, the next step was the application of KCL at each of the three nonreference nodes. If we try to do that once again, we see that we will run into some difficulty at both nodes 2 and 3, for we do not know what the current is in the branch with the voltage source. There is no way by which we can express the current as a function of the voltage, for the definition of a voltage source is exactly that the voltage is independent of the current.

There are two ways out of this dilemma. The more difficult approach is to assign an unknown current to the branch which contains the voltage source, proceed to apply KCL three times, and then apply KVL ($v_3 - v_2 = 22$) once between nodes 2 and 3; the result is four equations in four unknowns for this example.

The easier method is to treat node 2, node 3, and the voltage source together as a sort of *supernode* and apply KCL to both nodes at the same time; the supernode is indicated by the region enclosed by the broken line in Fig. 4.9a. This is certainly possible because, if the total current leaving node 2 is zero and the total current leaving node 3 is zero, then the total current leaving the combination of the two nodes is zero. This concept is depicted graphically in the expanded view of Fig. 4.9b.

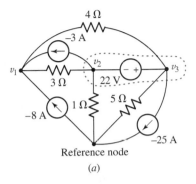

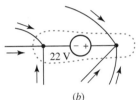

FIGURE 4.9 (a) The circuit of Example 4.2 with a 22 V source in place of the 7 Ω resistor. (b) Expanded view of the region defined as a supernode; KCL requires that all currents flowing into the region must sum to zero, or we would pile up or run out of electrons.

EXAMPLE 4.5

Determine the value of the unknown node voltage v_1 in the circuit of Fig. 4.9a.

The KCL equation at node 1 is unchanged from Example 4.2:

$$-8 - 3 = \frac{v_1 - v_2}{3} + \frac{v_1 - v_3}{4}$$

or

$$0.5833v_1 - 0.3333v_2 - 0.2500v_3 = -11 \qquad [17]$$

Next we consider the 2, 3 supernode. Two current sources are connected, and four resistors. Thus,

$$3 + 25 = \frac{v_2 - v_1}{3} + \frac{v_3 - v_1}{4} + \frac{v_3}{5} + \frac{v_2}{1}$$

or

$$-0.5833v_1 + 1.3333v_2 + 0.45v_3 = 28 \qquad [18]$$

(Continued on next page)

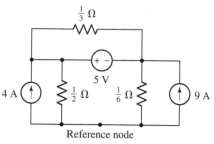

FIGURE 4.10

Since we have three unknowns, we need one additional equation, and it must utilize the fact that there is a 22 V voltage source between nodes 2 and 3:

$$v_2 - v_3 = -22 \qquad [19]$$

Solving Eqs. [17] to [19], the solution for v_1 is 1.071 V.

PRACTICE

4.4 For the circuit of Fig. 4.10, compute the voltage across each current source.

Ans: 5.375 V, 375 mV.

The presence of a voltage source thus reduces by one the number of non-reference nodes at which we must apply KCL, regardless of whether the voltage source extends between two nonreference nodes or is connected between a node and the reference. We should be careful in analyzing circuits such as that of Practice Prob. 4.4. Since both ends of the resistor are part of the supernode, there must technically be *two* corresponding current terms in the KCL equation, but they cancel each other out. We can summarize the supernode method as follows:

Summary of Supernode Analysis Procedure

1. **Count the number of nodes** (*N*).

2. **Designate a reference node.** The number of terms in your nodal equations can be minimized by selecting the node with the greatest number of branches connected to it.

3. **Label the nodal voltages** (there are $N - 1$ of them).

4. **If the circuit contains voltage sources, form a supernode about each one.** This is done by enclosing the source, its two terminals, and any other elements connected between the two terminals within a broken-line enclosure.

5. **Write a KCL equation for each nonreference node and for each supernode *that does not contain the reference node.*** Sum the currents flowing *into* a node/supernode from current sources on one side of the equation. On the other side, sum the currents flowing *out* of the node/supernode through resistors. Pay close attention to "−" signs.

6. **Relate the voltage across each voltage source to nodal voltages.** This is accomplished by simple application of KVL; one such equation is needed for each supernode defined.

7. **Express any additional unknowns (i.e., currents or voltages other than nodal voltages) in terms of appropriate nodal voltages.** This situation can occur if dependent sources appear in our circuit.

8. **Organize the equations.** Group terms according to nodal voltages.

9. **Solve the system of equations for the nodal voltages** (there will be $N - 1$ of them).

We see that we have added two additional steps from our general nodal analysis procedure. In reality, however, application of the supernode technique to a circuit containing voltage sources not connected to the reference node will result in a reduction in the number of KCL equations required. With this in mind, let's consider the circuit of Fig. 4.11, which contains all four types of sources and has five nodes.

EXAMPLE **4.6**

Determine the node-to-reference voltages in the circuit of Fig. 4.11.

After establishing a supernode about each *voltage* source, we see that we need to write KCL equations only at node 2 and at the supernode containing the dependent voltage source. By inspection, it is clear that $v_1 = -12$ V.

At node 2,

$$\frac{v_2 - v_1}{0.5} + \frac{v_2 - v_3}{2} = 14 \qquad [20]$$

while at the 3-4 supernode,

$$0.5v_x = \frac{v_3 - v_2}{2} + \frac{v_4}{1} + \frac{v_4 - v_1}{2.5} \qquad [21]$$

We next relate the source voltages to the node voltages:

$$v_3 - v_4 = 0.2v_y \qquad [22]$$

and

$$0.2v_y = 0.2(v_4 - v_1) \qquad [23]$$

Finally, we express the dependent current source in terms of the assigned variables:

$$0.5v_x = 0.5(v_2 - v_1) \qquad [24]$$

Five nodes requires *four* KCL equations in general nodal analysis, but we have reduced this requirement to *only two,* as we formed two separate supernodes. Each supernode required a KVL equation (Eq. [22] and $v_1 = -12$, the latter written by inspection). Neither dependent source was controlled by a nodal voltage, so two additional equations were needed as a result.

With this done, we can now eliminate v_x and v_y to obtain a set of four equations in the four node voltages:

$$
\begin{aligned}
-2v_1 + 2.5v_2 - 0.5v_3 \qquad\qquad &= \ 14 \\
0.1v_1 - \ v_2 + 0.5v_3 + 1.4v_4 &= \ 0 \\
v_1 \qquad\qquad\qquad\qquad &= -12 \\
0.2v_1 \qquad + \ v_3 - 1.2v_4 &= \ 0
\end{aligned}
$$

Solving, $v_1 = -12$ V, $v_2 = -4$ V, $v_3 = 0$ V, and $v_4 = -2$ V.

PRACTICE

4.5 Determine the nodal voltages in the circuit of Fig. 4.12.

Ans: $v_1 = 3$ V, $v_2 = 5.09$ V, $v_3 = 1.28$ V, $v_4 = 1.68$ V.

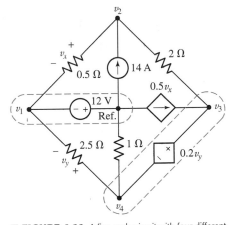

■ **FIGURE 4.11** A five-node circuit with four different types of sources.

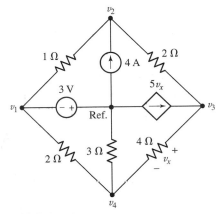

■ **FIGURE 4.12**

4.3 • MESH ANALYSIS

The technique of nodal analysis described in the preceding section is completely general and can always be applied to any electrical network. An alternative method that is sometimes easier to apply to certain circuits is known as *mesh analysis.* Even though this technique is not applicable to every network, it can be applied to many of the networks we will need to analyze. Mesh analysis is applicable only to those networks which are planar, a term we hasten to define.

If it is possible to draw the diagram of a circuit on a plane surface in such a way that no branch passes over or under any other branch, then that circuit is said to be a *planar circuit.* Thus, Fig. 4.13*a* shows a planar network, Fig. 4.13*b* shows a nonplanar network, and Fig. 4.13*c* also shows a planar network, although it is drawn in such a way as to make it appear nonplanar at first glance.

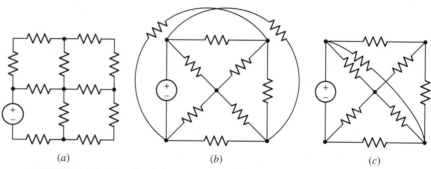

(a)　　　　(b)　　　　(c)

■ **FIGURE 4.13** Examples of planar and nonplanar networks; crossed wires without a solid dot are not in physical contact with each other.

We should mention that mesh-type analysis can be applied to nonplanar circuits, but since it is not possible to define a complete set of unique meshes for such circuits, assignment of unique mesh currents is not possible.

In Sec. 3.1, the terms *path, closed path,* and *loop* were defined. Before we define a mesh, let us consider the sets of branches drawn with heavy lines in Fig. 4.14. The first set of branches is not a path, since four branches are connected to the center node, and it is of course also not a loop. The second set of branches does not constitute a path, since it is traversed only by passing through the central node twice. The remaining four paths are all loops. The circuit contains 11 branches.

The mesh is a property of a planar circuit and is undefined for a nonplanar circuit. We define a *mesh* as a loop that does not contain any other loops within it. Thus, the loops indicated in Fig. 4.14*c* and *d* are not meshes, whereas those of parts *e* and *f* are meshes. Once a circuit has been drawn neatly in planar form, it often has the appearance of a multipaned window; the boundary of each pane in the window may be considered to be a mesh.

If a network is planar, mesh analysis can be used to accomplish the analysis. This technique involves the concept of a *mesh current,* which we introduce by considering the analysis of the two-mesh circuit of Fig. 4.15*a*.

As we did in the single-loop circuit, we will begin by defining a current through one of the branches. Let us call the current flowing to the right through the 6 Ω resistor i_1. We will apply KVL around each of the two meshes, and the two resulting equations are sufficient to determine two unknown currents. We next define a second current i_2 flowing to the right in

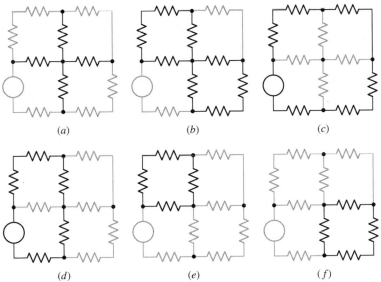

(a)   (b)   (c)

(d)   (e)   (f)

■ **FIGURE 4.14** (a) The set of branches identified by the heavy lines is neither a path nor a loop. (b) The set of branches here is not a path, since it can be traversed only by passing through the central node twice. (c) This path is a loop but not a mesh, since it encloses other loops. (d) This path is also a loop but not a mesh. (e, f) Each of these paths is both a loop and a mesh.

the 4 Ω resistor. We might also choose to call the current flowing downward through the central branch i_3, but it is evident from KCL that i_3 may be expressed in terms of the two previously assumed currents as $(i_1 - i_2)$. The assumed currents are shown in Fig. 4.15b.

Following the method of solution for the single-loop circuit, we now apply KVL to the left-hand mesh,

$$-42 + 6i_1 + 3(i_1 - i_2) = 0$$

or

$$9i_1 - 3i_2 = 42 \qquad [25]$$

Applying KVL to the right-hand mesh,

$$-3(i_1 - i_2) + 4i_2 - 10 = 0$$

or

$$-3i_1 + 7i_2 = 10 \qquad [26]$$

Equations [25] and [26] are independent equations; one cannot be derived from the other. There are two equations and two unknowns, and the solution is easily obtained:

$$i_1 = 6 \text{ A} \qquad i_2 = 4 \text{ A} \qquad \text{and} \qquad (i_1 - i_2) = 2 \text{ A}$$

If our circuit contains M meshes, then we expect to have M mesh currents and therefore will be required to write M independent equations.

Now let us consider this same problem in a slightly different manner by using mesh currents. We define a **mesh current** as a current that flows only around the perimeter of a mesh. One of the greatest advantages in the use of mesh currents is the fact that Kirchhoff's current law is automatically satisfied. If a mesh current flows *into* a given node, it flows *out* of it also.

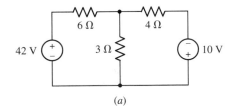

(a)

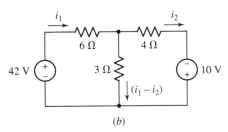

(b)

■ **FIGURE 4.15** (a, b) A simple circuit for which currents are required.

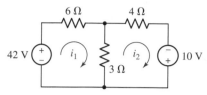

■ **FIGURE 4.16** The same circuit considered in Fig. 4.15*b*, but viewed a slightly different way.

A mesh current may often be identified as a branch current, as i_1 and i_2 have been identified in this example. This is not always true, however, for consideration of a square nine-mesh network soon shows that the central mesh current cannot be identified as the current in any branch.

If we label the left-hand mesh of our problem as mesh 1, then we may establish a mesh current i_1 flowing in a clockwise direction about this mesh. A mesh current is indicated by a curved arrow that almost closes on itself and is drawn inside the appropriate mesh, as shown in Fig. 4.16. The mesh current i_2 is established in the remaining mesh, again in a clockwise direction. Although the directions are arbitrary, we will always choose clockwise mesh currents because a certain error-minimizing symmetry then results in the equations.

We no longer have a current or current arrow shown directly on each branch in the circuit. The current through any branch must be determined by considering the mesh currents flowing in every mesh in which that branch appears. This is not difficult, because no branch can appear in more than two meshes. For example, the 3 Ω resistor appears in both meshes, and the current flowing downward through it is $i_1 - i_2$. The 6 Ω resistor appears only in mesh 1, and the current flowing to the right in that branch is equal to the mesh current i_1.

For the left-hand mesh,

$$-42 + 6i_1 + 3(i_1 - i_2) = 0$$

while for the right-hand mesh,

$$3(i_2 - i_1) + 4i_2 - 10 = 0$$

and these two equations are equivalent to Eqs. [25] and [26].

EXAMPLE 4.7

Determine the power supplied by the 2 V source of Fig. 4.17*a*.

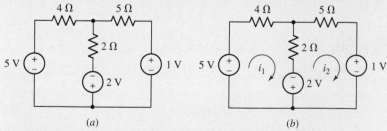

(a)   *(b)*

■ **FIGURE 4.17** (*a*) A two-mesh circuit containing three sources. (*b*) Circuit labeled for mesh analysis.

We first define two clockwise mesh currents as shown in Fig. 4.17*b*.

Beginning at the bottom left node of mesh 1, we write the following KVL equation as we proceed clockwise through the branches:

$$-5 + 4i_1 + 2(i_1 - i_2) - 2 = 0$$

Doing the same for mesh 2, we write

$$+2 + 2(i_2 - i_1) + 5i_2 + 1 = 0$$

Rearranging and grouping terms,

$$6i_1 - 2i_2 = 7$$

and

$$-2i_1 + 7i_2 = -3$$

Solving, $i_1 = \dfrac{43}{38} = 1.132$ A and $i_2 = -\dfrac{2}{19} = -0.1053$ A.

The current flowing out of the positive reference terminal of the 2 V source is $i_1 - i_2$. Thus, the 2 V source supplies $(2)(1.237) = 2.474$ W.

PRACTICE

4.6 Determine i_1 and i_2 in the circuit in Fig. 4.18.

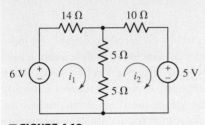

■ **FIGURE 4.18**

Ans: $i_1 = +184.2$ mA; -157.9 mA.

Let us next consider the five-node, seven-branch, three-mesh circuit shown in Fig. 4.19. This is a slightly more complicated problem because of the additional mesh.

EXAMPLE 4.8

Use mesh analysis to determine the three mesh currents in the circuit of Fig. 4.19.

The three required mesh currents are assigned as indicated in Fig. 4.19, and we methodically apply KVL about each mesh:

$$-7 + 1(i_1 - i_2) + 6 + 2(i_1 - i_3) = 0$$
$$1(i_2 - i_1) + 2i_2 + 3(i_2 - i_3) = 0$$
$$2(i_3 - i_1) - 6 + 3(i_3 - i_2) + 1i_3 = 0$$

Simplifying,

$$3i_1 - i_2 - 2i_3 = 1$$
$$-i_1 + 6i_2 - 3i_3 = 0$$
$$-2i_1 - 3i_2 + 6i_3 = 6$$

and solving, we obtain $i_1 = 3$ A, $i_2 = 2$ A, and $i_3 = 3$ A.

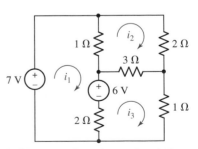

■ **FIGURE 4.19** A five-node, seven-branch, three-mesh circuit.

PRACTICE

4.7 Determine i_1 and i_2 in the circuit of Fig 4.20.

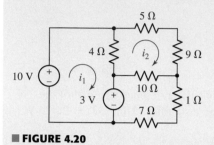

■ **FIGURE 4.20**

Ans: 2.220 A, 470.0 mA.

The previous examples dealt with circuits powered exclusively by independent voltage sources. If a current source is included in the circuit, it may either simplify or complicate the analysis, as discussed in Sec. 4.4. As seen in our study of the nodal analysis technique, dependent sources generally require an additional equation besides the M mesh equations, unless the controlling variable is a mesh current (or sum of mesh currents). We explore this in the following example.

EXAMPLE 4.9

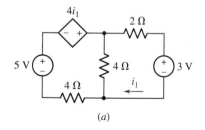

(a)

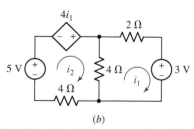

(b)

■ **FIGURE 4.21** (a) A two-mesh circuit containing a dependent source. (b) Circuit labeled for mesh analysis.

Determine the current i_1 in the circuit of Fig. 4.21a.

The current i_1 is actually a mesh current, so rather than redefine it we label the rightmost mesh current i_1 and define a clockwise mesh current i_2 for the left mesh, as shown in Fig. 4.21b.

For the left mesh, KVL yields

$$-5 - 4i_1 + 4(i_2 - i_1) + 4i_2 = 0 \qquad [27]$$

and for the right mesh we find

$$4(i_1 - i_2) + 2i_1 + 3 = 0 \qquad [28]$$

Grouping terms, these equations may be written more compactly as

$$-8i_1 + 8i_2 = 5$$

and

$$6i_1 - 4i_2 = -3$$

Solving, $i_2 = 375$ mA, so $i_1 = -250$ mA.

Since the dependent source of Fig. 4.21 is controlled by a mesh current (i_1), only two equations—Eqs. [27] and [28]—were required to analyze the two-mesh circuit. In the following example, we explore the situation that arises if the controlling variable is *not* a mesh current.

EXAMPLE **4.10**

Determine the current i_1 in the circuit of Fig. 4.22a.

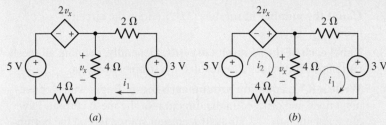

(a)   (b)

■ **FIGURE 4.22** (a) A circuit with a dependent source controlled by a voltage. (b) Circuit labeled for mesh analysis.

In order to draw comparisons to Example 4.9 we use the same mesh current definitions, as shown in Fig. 4.22b.

For the left mesh, KVL now yields

$$-5 - 2v_x + 4(i_2 - i_1) + 4i_2 = 0 \qquad [29]$$

and for the right mesh we find the same as before, namely

$$4(i_1 - i_2) + 2i_1 + 3 = 0 \qquad [30]$$

Since the dependent source is controlled by the unknown voltage v_x, we are faced with *two* equations in *three* unknowns. The way out of our dilemma is to simply construct an equation for v_x in terms of mesh currents, such as

$$v_x = 4(i_2 - i_1) \qquad [31]$$

We simplify this system of equations by substituting Eq. [31] into Eq. [29], resulting in

$$4i_1 = 5$$

Solving, we find that $i_1 = 1.25$ A. In this particular instance, Eq. [30] is not needed unless a value for i_2 is desired.

PRACTICE
●———————————————————————————

4.8 Determine i_1 in the circuit of Fig. 4.23 if the controlling quantity A is equal to: (a) $2i_2$; (b) $2v_x$.

Ans: (a) 1.35 A; (b) 546 mA.

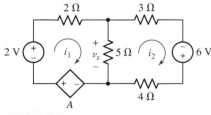

■ **FIGURE 4.23**

The mesh analysis procedure can be summarized by the seven basic steps that follow. It will work on any *planar* circuit we ever encounter, although the presence of current sources will require extra care. Such situations are discussed in Sec. 4.4.

Summary of Basic Mesh Analysis Procedure

1. **Determine if the circuit is a planar circuit.** If not, perform nodal analysis instead.

2. **Count the number of meshes** (*M*). Redraw the circuit if necessary.

3. **Label each of the *M* mesh currents.** Generally, defining all mesh currents to flow clockwise results in a simpler analysis.

4. **Write a KVL equation around each mesh.** Begin with a convenient node and proceed in the direction of the mesh current. Pay close attention to "−" signs. If a current source lies on the periphery of a mesh, no KVL equation is needed and the mesh current is determined by inspection.

5. **Express any additional unknowns such as voltages or currents other than mesh currents in terms of appropriate mesh currents.** This situation can occur if current sources or dependent sources appear in our circuit.

6. **Organize the equations.** Group terms according to mesh currents.

7. **Solve the system of equations for the mesh currents** (there will be *M* of them).

4.4 • THE SUPERMESH

How must we modify this straightforward procedure when a current source is present in the network? Taking our lead from nodal analysis, we should feel that there are two possible methods. First, we could assign an unknown voltage across the current source, apply KVL around each mesh as before, and then relate the source current to the assigned mesh currents. This is generally the more difficult approach.

A better technique is one that is quite similar to the supernode approach in nodal analysis. There we formed a supernode, completely enclosing the voltage source inside the supernode and reducing the number of non-reference nodes by 1 for each voltage source. Now we create a kind of **"*supermesh*"** from two meshes that have a current source as a common element; the current source is in the interior of the supermesh. We thus reduce the number of meshes by 1 for each current source present. If the current source lies on the *perimeter* of the circuit, then the single mesh in which it is found is ignored. Kirchhoff's voltage law is thus applied only to those meshes or supermeshes in the reinterpreted network.

EXAMPLE 4.11

Use the technique of mesh analysis to evaluate the three mesh currents in Fig. 4.24*a*.

Here we note that a 7 A independent current source is in the common boundary of two meshes. Mesh currents i_1, i_2, and i_3 have already been assigned, and the current source leads us to create a supermesh whose interior is that of meshes 1 and 3 as shown in Fig. 4.24*b*. Applying KVL about this loop,

$$-7 + 1(i_1 - i_2) + 3(i_3 - i_2) + 1i_3 = 0$$

or

$$i_1 - 4i_2 + 4i_3 = 7 \qquad [32]$$

and around mesh 2,

$$1(i_2 - i_1) + 2i_2 + 3(i_2 - i_3) = 0$$

or

$$-i_1 + 6i_2 - 3i_3 = 0 \qquad [33]$$

Finally, the independent-source current is related to the assumed mesh currents,

$$i_1 - i_3 = 7 \qquad [34]$$

Solving Eqs. [32] through [34], we find $i_1 = 9$ A, $i_2 = 2.5$ A, and $i_3 = 2$ A.

PRACTICE

4.9 Determine the current i_1 in the circuit of Fig. 4.25.

Ans: −1.93 A.

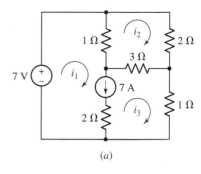

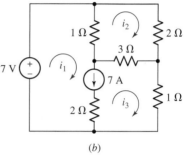

■ **FIGURE 4.24** (*a*) A three-mesh circuit with an independent current source. (*b*) A supermesh is defined by the colored line.

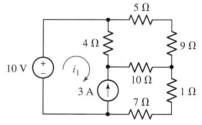

■ **FIGURE 4.25**

The presence of one or more dependent sources merely requires each of these source quantities and the variable on which it depends to be expressed in terms of the assigned mesh currents. In Fig. 4.26, for example, we note that both a dependent and an independent current source are included in the network. Let's see how their presence affects the analysis of the circuit and actually simplifies it.

EXAMPLE **4.12**

Use mesh analysis to evaluate the three unknown currents in the circuit of Fig. 4.26.

The current sources appear in meshes 1 and 3. Since the 15 A source is located on the perimeter of the circuit, we may eliminate mesh 1 from consideration—it is clear that $i_1 = 15$ A.

We find that because we now know one of the two mesh currents relevant to the dependent current source, there is no need to write a supermesh equation about meshes 1 and 3. Instead, we simply relate i_1 and i_3 to the current from the dependent source using KCL:

$$\frac{v_x}{9} = i_3 - i_1 = \frac{3(i_3 - i_2)}{9}$$

(Continued on next page)

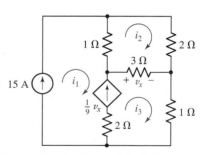

■ **FIGURE 4.26** A three-mesh circuit with one dependent and one independent current source.

which can be written more compactly as

$$-i_1 + \frac{1}{3}i_2 + \frac{2}{3}i_3 = 0 \quad \text{or} \quad \frac{1}{3}i_2 + \frac{2}{3}i_3 = 15 \qquad [35]$$

With one equation in two unknowns, all that remains is to write a KVL equation about mesh 2

$$1(i_2 - i_1) + 2i_2 + 3(i_2 - i_3) = 0$$

or

$$6i_2 - 3i_3 = 15 \qquad [36]$$

Solving Eqs. [35] and [36], we find that $i_2 = 11$ A and $i_3 = 17$ A; we already determined that $i_1 = 15$ A by inspection.

PRACTICE

4.10 Use mesh analysis to find v_3 in the circuit of Fig. 4.27.

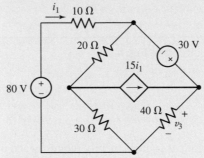

■ **FIGURE 4.27**

Ans: 104.2 V

We can now summarize the general approach to writing mesh equations, whether or not dependent sources, voltage sources, and/or current sources are present, provided that the circuit can be drawn as a planar circuit:

Summary of Supermesh Analysis Procedure

1. **Determine if the circuit is a planar circuit.** If not, perform nodal analysis instead.

2. **Count the number of meshes** (M). Redraw the circuit if necessary.

3. **Label each of the M mesh currents.** Generally, defining all mesh currents to flow clockwise results in a simpler analysis.

4. **If the circuit contains current sources shared by two meshes, form a supermesh to enclose both meshes.** A highlighted enclosure helps when writing KVL equations.

5. **Write a KVL equation around each mesh/supermesh.** Begin with a convenient node and proceed in the direction of the mesh current. Pay close attention to "−" signs. If a current source lies

on the periphery of a mesh, no KVL equation is needed and the mesh current is determined by inspection.

6. **Relate the current flowing from each current source to mesh currents.** This is accomplished by simple application of KCL; one such equation is needed for each supermesh defined.

7. **Express any additional unknowns such as voltages or currents other than mesh currents in terms of appropriate mesh currents.** This situation can occur if dependent sources appear in our circuit.

8. **Organize the equations.** Group terms according to nodal voltages.

9. **Solve the system of equations for the nodal voltages** (there will be M of them).

4.5 • NODAL VS. MESH ANALYSIS: A COMPARISON

Now that we have examined two distinctly different approaches to circuit analysis, it seems logical to ask if there is ever any advantage to using one over the other. If the circuit is nonplanar, then there is no choice: only nodal analysis may be applied.

Provided that we are indeed considering the analysis of a *planar* circuit, however, there are situations where one technique has a small advantage over the other. If we plan to use nodal analysis, then a circuit with N nodes will lead to at most $N - 1$ KCL equations. Each supernode defined will further reduce this number by one. If the same circuit has M distinct meshes, then we will obtain at most M KVL equations; each supermesh will reduce this number by one. Based on these facts, we should select the approach that will result in the smaller number of simultaneous equations.

If one or more dependent sources are included in the circuit, then each controlling quantity may influence our choice of nodal or mesh analysis. For example, a dependent voltage source controlled by a nodal voltage does not require an additional equation when we perform nodal analysis. Likewise, a dependent current source controlled by a mesh current does not require an additional equation when we perform mesh analysis. *What about the situation where a dependent voltage source is controlled by a current? Or the converse, where a dependent current source is controlled by a voltage?* Provided that the controlling quantity can be easily related to mesh currents, we might expect mesh analysis to be the more straightforward option. Likewise, if the controlling quantity can be easily related to nodal voltages, nodal analysis may be preferable. One final point in this regard is to keep in mind the *location* of the source; current sources which lie on the periphery of a mesh, whether dependent or independent, are easily treated in mesh analysis; voltage sources connected to the reference terminal are easily treated in nodal analysis.

When either method results in essentially the same number of equations, it may be worthwhile to also consider what quantities are being sought. Nodal analysis results in direct calculation of nodal voltages, whereas mesh analysis provides currents. If we are asked to find currents through a set of resistors, for example, after performing nodal analysis, we must still invoke Ohm's law at each resistor to determine the current.

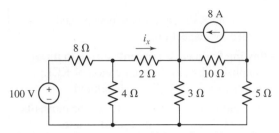

■ **FIGURE 4.28** A planar circuit with five nodes and four meshes.

As an example, consider the circuit in Fig. 4.28. We wish to determine the current i_x.

We choose the bottom node as the reference node, and note that there are four nonreference nodes. Although this means that we can write four distinct equations, there is no need to label the node between the 100 V source and the 8 Ω resistor, since that node voltage is clearly 100 V. Thus, we label the remaining node voltages v_1, v_2, and v_3 as in Fig. 4.29.

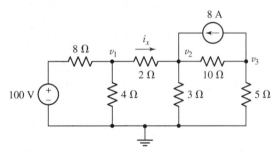

■ **FIGURE 4.29** The circuit of Fig. 4.28 with node voltages labeled. Note that an earth ground symbol was chosen to designate the reference terminal.

We write the following three equations:

$$\frac{v_1 - 100}{8} + \frac{v_1}{4} + \frac{v_1 - v_2}{2} = 0 \quad \text{or} \quad 0.875v_1 - 0.5v_2 \qquad\qquad = 12.5 \quad [37]$$

$$\frac{v_2 - v_1}{2} + \frac{v_2}{3} + \frac{v_2 - v_3}{10} - 8 = 0 \quad \text{or} \quad -0.5v_1 - 0.9333v_2 - 0.1v_3 = 8 \quad [38]$$

$$\frac{v_3 - v_2}{10} + \frac{v_3}{5} + 8 = 0 \qquad\qquad \text{or} \qquad -0.1v_2 \quad + 0.3v_3 = -8 \quad [39]$$

Solving, we find that $v_1 = 25.89$ V and $v_2 = 20.31$ V. We determine the current i_x by application of Ohm's law:

$$i_x = \frac{v_1 - v_2}{2} = 2.79 \text{ A} \quad [40]$$

Next, we consider the same circuit using mesh analysis. We see in Fig. 4.30 that we have four distinct meshes, although it is obvious that $i_4 = -8$ A; we therefore need to write three distinct equations.

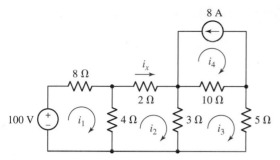

■ FIGURE 4.30 The circuit of Fig. 4.28 with mesh currents labeled.

Writing a KVL equation for meshes 1, 2, and 3:

$$-100 + 8i_1 + 4(i_1 - i_2) = 0 \qquad \text{or} \qquad 12i_1 - 4i_2 \qquad\quad = 100 \qquad [41]$$

$$4(i_2 - i_1) + 2i_2 + 3(i_2 - i_3) = 0 \qquad \text{or} \qquad -4i_1 + 9i_2 - 3i_3 = 0 \qquad [42]$$

$$3(i_3 - i_2) + 10(i_3 + 8) + 5i_3 = 0 \qquad \text{or} \qquad\qquad -3i_2 + 18i_3 = -80 \qquad [43]$$

Solving, we find that i_2 ($= i_x$) = 2.79 A. For this particular problem, mesh analysis proved to be simpler. Since either method is valid, however, working the same problem both ways can also serve as a means to check our answers.

4.6 • COMPUTER-AIDED CIRCUIT ANALYSIS

We have seen that it does not take many components at all to create a circuit of respectable complexity. As we continue to examine even more complex circuits, it will become obvious rather quickly that it is easy to make errors during the analysis, and verifying solutions by hand can be time-consuming. A powerful computer software package known as PSpice is commonly employed for rapid analysis of circuits, and the schematic capture tools are typically integrated with either a printed circuit board or integrated circuit layout tool. Originally developed in the early 1970s at the University of California at Berkeley, SPICE (*Simulation Program with Integrated Circuit Emphasis*) is now an industry standard. MicroSim Corporation introduced PSpice in 1984, which built intuitive graphical interfaces around the core SPICE program. Depending on the type of circuit application being considered, there are now several companies offering variations of the basic SPICE package.

Although computer-aided analysis is a relatively quick means of determining voltages and currents in a circuit, we should be careful not to allow simulation packages to completely replace traditional "paper and pencil" analysis. There are several reasons for this. First, in order to design we must be able to analyze. Overreliance on software tools can inhibit the development of necessary analytical skills, similar to introducing calculators too early in grade school. Second, it is virtually impossible to use a complicated software package over a long period of time without making some type of data-entry error. If we have no basic intuition as to what type of answer to expect from a simulation, then there is no way to determine whether or not it is valid. Thus, the generic name really is a fairly accurate description: computer-*aided* analysis. Human brains are not obsolete. Not yet, anyway.

As an example, consider the circuit of Fig. 4.15*b*, which includes two dc voltage sources and three resistors. We wish to simulate this circuit using

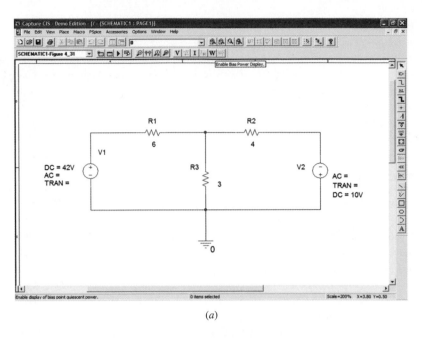

(a)

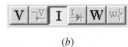

(b)

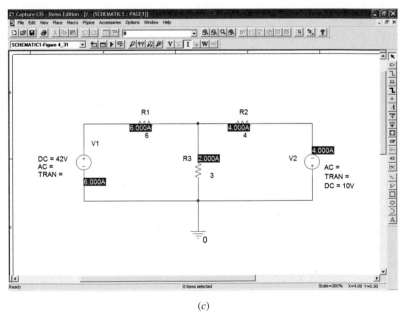

(c)

■ **FIGURE 4.31** (a) Circuit of Fig. 4.15a drawn using Orcad schematic capture software. (b) Current, voltage, and power display buttons. (c) Circuit after simulation run, with current display enabled.

PSpice so that we may determine the currents i_1 and i_2. Figure 4.31a shows the circuit as drawn using a schematic capture program.[1]

In order to determine the mesh currents, we need only run a bias point simulation. Under **PSpice,** select **New Simulation Profile.** Type in First

(1) Refer to Appendix 4 for a brief tutorial on PSpice and schematic capture.

Example (or your personal preference) and click on **Create.** Under the **Analysis Type:** pull-down menu, select **Bias Point,** then click on **OK.** Returning to the original schematic window, under **PSpice** select **Run** (or use either of the two shortcuts: pressing the F11 key or clicking on the blue "Play" symbol). To see the currents calculated by PSpice, make sure the current button is selected (Fig. 4.31*b*). The results of our simulation are shown in Fig. 4.31*c*. We see that the two currents i_1 and i_2 are 6 A and 4 A, respectively, as we found previously.

As a further example, consider the circuit shown in Fig. 4.32*a*. It contains a dc voltage source, a dc current source, and a voltage-controlled current source. We are interested in the three nodal voltages, which from either nodal or mesh analysis are found to be 82.91 V, 69.9 V, and 59.9 V, respectively as we move from left to right across the top of the circuit. Figure 4.32*b* shows this circuit, drawn using a schematic capture tool, after the simulation was performed. The three nodal voltages are indicated directly on the schematic. Note that in drawing a dependent source using the schematic capture tool, we must *explicitly* link two terminals of the source to the controlling voltage or current.

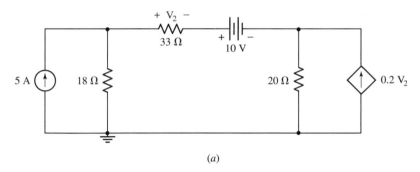

(*a*)

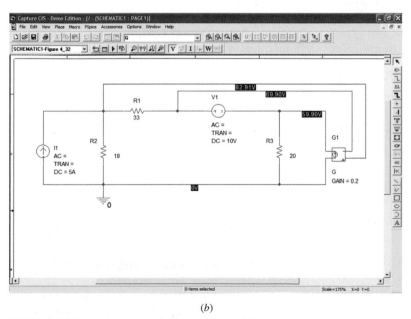

(*b*)

■ **FIGURE 4.32** (*a*) Circuit with dependent current source. (*b*) Circuit drawn using schematic capture tool, with simulation results presented directly on the schematic.

PRACTICAL APPLICATION

Node-Based PSpice Schematic Creation

The most common method of describing a circuit in conjunction with computer-aided circuit analysis is with some type of graphical schematic drawing package, an example output of which was shown in Fig. 4.32. SPICE, however, was written before the advent of such software, and as such requires circuits to be described in a specific text-based format. The format has its roots in the syntax used for punch cards, which gives it a somewhat distinct appearance. The basis for circuit description is the definition of elements, each terminal of which is assigned a node number. So, although we have just studied two different generalized circuit analysis methods—the nodal and mesh techniques—it is interesting that SPICE and PSpice were written using a clearly defined nodal analysis approach.

Even though modern circuit analysis is largely done using graphics-oriented interactive software, when errors are generated (usually due to a mistake in drawing the schematic or in selecting a combination of analysis options), the ability to read the text-based "input deck" generated by the schematic capture tool can be invaluable in tracking down the specific problem. The easiest way to develop such an ability is to learn how to run PSpice directly from a user-written input deck.

Consider, for example, the sample input deck below (lines beginning with an asterisk are comments, and are skipped by SPICE).

```
* Example input deck for a simple voltage divider.

.OP                        (Asks SPICE to determine the dc operating point of the circuit.)

R1 1 2 1k                  (R1 is defined between nodes 1 and 2; it has a value of 1 kΩ.)
R2 2 0 1k                  (R2 is defined between nodes 2 and 0; it has a value of 1 kΩ.)
V1 1 0 DC 5                (V1 is defined between nodes 1 and 0; it has a value of 5 V dc.)

* End of input deck.
```

We can create the input deck by using the Notepad program from Windows or our favorite text editor. Saving the file under the name example.cir, we next invoke, PSpice A/D (see App. 4). Under **File,** we choose **Open,** locate the directory in which we saved our file example.cir, and for **Files of Type:** select **Circuit Files (*.cir).** After selecting our file and clicking **Open,** we see the PSpice A/D window with our circuit file loaded (Fig. 4.33a). A netlist such as this, containing instruc-

tions for the simulation to be performed, can be created by schematic capture software or created manually as in this example.

We run the simulation by either clicking the blue "play" symbol at the top right, or selecting **Run** under **Simulation.** In the lower left corner of the main window, a smaller summary window informs us that the simulation ran successfully (Fig. 4.33b). To view the results, we select **Output File** from under the **View** menu and see:

```
**** 02/18/04 09:53:57 ************* PSpice Lite (Jan 2003) *****************

 * Example input deck for a simple voltage divider.

 ****    CIRCUIT DESCRIPTION

 ***********************************************************************

 .OP

 R1 1 2 1k
 R2 2 0 1k
 V1 1 0 DC 5

 * End of input deck.
```

The input deck is repeated in the output for reference and to assist in error checking.

```
**** 02/18/04 09:53:57 ************** PSpice Lite (Jan 2003) *****************

* Example input deck for a simple voltage divider.

****      SMALL SIGNAL BIAS SOLUTION       TEMPERATURE =   27.000 DEG C

***********************************************************************

 NODE    VOLTAGE       NODE   VOLTAGE     NODE   VOLTAGE      NODE   VOLTAGE

(    1)    5.0000   (    2)    2.5000
```

In the output summary, we are given the voltage between each node and node 0. Our 5 V source is connected between nodes 1 and 0; resistor R2, connected between nodes 2 and 0, has 2.5 V across it as expected.

```
        VOLTAGE SOURCE CURRENTS
        NAME           CURRENT
```

Also note a quirk of SPICE: The current provided by our source is quoted using the passive sign convention (i.e., −2.5 mA).

```
        V1             -2.500E-03


        TOTAL POWER DISSIPATION   1.25E-02  WATTS
```

As we see, using the text-based approach to describing circuits is somewhat less user-friendly than schematic capture tools. In particular, it is very easy to introduce simple (but significant) errors into a simulation by mistakenly numbering the nodes incorrectly, as there is no direct visualization of the input deck beyond what is written on paper. However, interpreting the output is very straightforward, and practicing reading a few such files is well worth the effort.

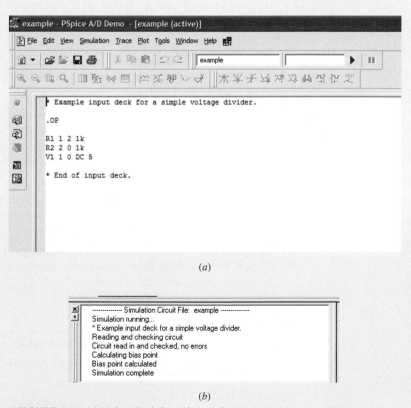

(a)

(b)

■ **FIGURE 4.33** (a) PSpice A/D window with circuit file loaded. (b) Simulation activity summary.

At this point, the real power of computer-aided analysis begins to be apparent: Once you have the circuit drawn in the schematic capture program, it is easy to experiment by simply changing component values and observing the effect on currents and voltages. To gain a little experience at this point, try simulating any of the circuits shown in previous examples and practice problems.

SUMMARY AND REVIEW

- ❏ Prior to beginning an analysis, make a neat, simple circuit diagram. Indicate all element and source values.
- ❏ If nodal analysis is the chosen approach,
  - ❏ Choose one of the nodes as a reference node. Then label the node voltages $v_1, v_2, \ldots, v_{N-1}$, remembering that each is understood to be measured with respect to the reference node.
  - ❏ If the circuit contains only current sources, apply KCL at each nonreference node.
  - ❏ If the circuit contains voltage sources, form a supernode about each one, and then proceed to apply KCL at all nonreference nodes and supernodes.
- ❏ If you are considering mesh analysis, first make certain that the network is a planar network.
  - ❏ Assign a clockwise mesh current in each mesh: $i_1, i_2, \ldots, i_M$.
  - ❏ If the circuit contains only voltage sources, apply KVL around each mesh.
  - ❏ If the circuit contains current sources, create a supermesh for each one that is common to two meshes, and then apply KVL around each mesh and supermesh.
- ❏ Dependent sources will add an additional equation to nodal analysis if the controlling variable is a current, but not if the controlling variable is a nodal voltage. Conversely, a dependent source will add an additional equation to mesh analysis if the controlling variable is a voltage, but not if the controlling variable is a mesh current.
- ❏ In deciding whether to use nodal or mesh analysis for a planar circuit, a circuit with fewer nodes/supernodes than meshes/supermeshes will result in fewer equations using nodal analysis.
- ❏ Computer-aided analysis is useful for checking results and analyzing circuits with large numbers of elements. However, common sense must be used to check simulation results.

READING FURTHER

A detailed treatment of nodal and mesh analysis can be found in:

R. A. DeCarlo and P. M. Lin, *Linear Circuit Analysis,* 2nd ed. New York: Oxford University Press, 2001.

A solid guide to SPICE is

P. Tuinenga, *SPICE: A Guide to Circuit Simulation and Analysis Using PSPICE,* 3rd ed. Upper Saddle River, N. J.: Prentice-Hall, 1995.

EXERCISES

4.1 Nodal Analysis

1. (a) Find v_2 if $0.1v_1 - 0.3v_2 - 0.4v_3 = 0$, $-0.5v_1 + 0.1v_2 = 4$, and $-0.2v_1 - 0.3v_2 + 0.4v_3 = 6$. (b) Evalute the determinant:

$$\begin{vmatrix} 2 & 3 & 4 & 1 \\ 3 & 4 & 1 & 2 \\ 4 & 1 & 2 & 3 \\ 1 & -2 & 3 & 0 \end{vmatrix}$$

2. (a) Find v_A, v_B, and v_C if $v_A + v_B + v_C = 27$, $2v_B + 16 = v_A - 3v_C$, and $4v_C + 2v_A + 6 = 0$. (b) Evaluate the determinant:

$$\begin{vmatrix} 0 & 1 & 2 & 3 \\ 1 & 2 & 3 & 4 \\ 2 & 3 & 4 & 1 \\ 3 & 4 & 1 & 2 \end{vmatrix}$$

3. (a) Solve the following system of equations:

$$4 = v_1/100 + (v_1 - v_2)/20 + (v_1 - v_x)/50$$
$$10 - 4 - (-2) = (v_x - v_1)/50 + (v_x - v_2)/40$$
$$-2 = v_2/25 + (v_2 - v_x)/40 + (v_2 - v_1)/20$$

(b) Verify your solution using MATLAB.

4. Determine the value of the voltage labeled v_1 in Fig. 4.34.

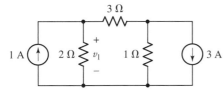

FIGURE 4.34

5. Determine the value of the voltage labeled v_1 in Fig. 4.35.

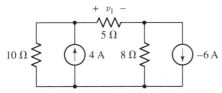

FIGURE 4.35

6. For the circuit of Fig. 4.36, determine the value of the voltage labeled v_1 and the current labeled i_1.

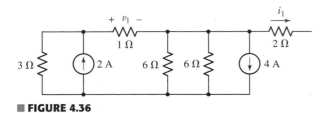

FIGURE 4.36

7. Use nodal analysis to find v_P in the circuit shown in Fig. 4.37.

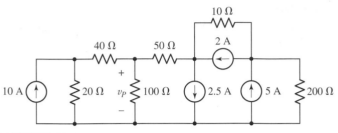

■ **FIGURE 4.37**

8. Use nodal analysis to find v_x in the circuit of Fig. 4.38.

9. For the circuit of Fig. 4.39 (*a*) use nodal analysis to determine v_1 and v_2.
   (*b*) Compute the power absorbed by the 6 Ω resistor.

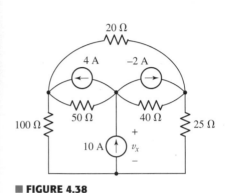

■ **FIGURE 4.38**

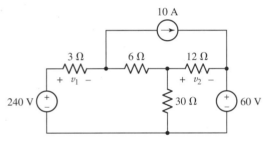

■ **FIGURE 4.39**

10. Employ nodal analysis techniques to find v_1 and i_2 in the circuit of Fig. 4.40.

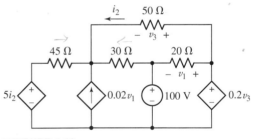

■ **FIGURE 4.40**

11. Referring to the circuit depicted in Fig. 4.41, use nodal analysis to determine the value of V_2 that will result in $v_1 = 0$.

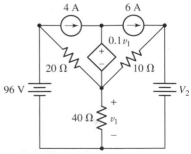

■ **FIGURE 4.41**

12. For the circuit of Fig. 4.42, use nodal analysis to determine the current i_5.

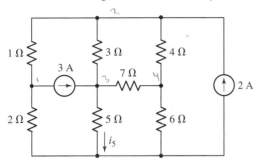

■ **FIGURE 4.42**

13. Employ nodal analysis to obtain a value for v_x as indicated in Fig. 4.43.

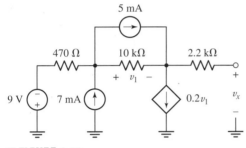

■ **FIGURE 4.43**

14. Determine the voltage labeled v in the circuit of Fig. 4.44 using nodal analysis techniques.

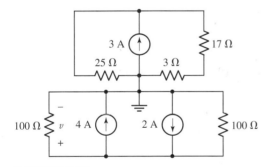

■ **FIGURE 4.44**

15. Determine the nodal voltages indicated in the circuit of Fig. 4.45.

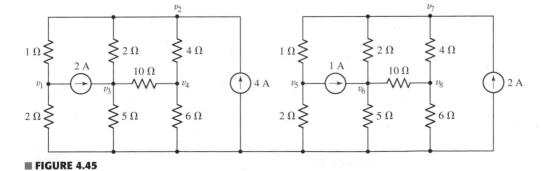

■ **FIGURE 4.45**

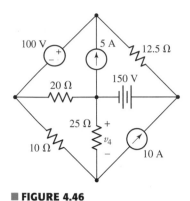

■ **FIGURE 4.46**

4.2 The Supernode

16. Use nodal analysis to find v_4 in the circuit shown in Fig. 4.46.

17. With the help of nodal analysis on the circuit of Fig. 4.47, find (*a*) v_A; (*b*) the power dissipated in the 2.5 Ω resistor.

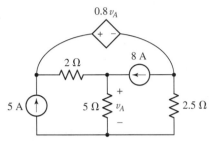

■ **FIGURE 4.47**

18. Use nodal analysis to determine v_1 and the power being supplied by the dependent current source in the circuit shown in Fig. 4.48.

19. In Fig. 4.49, use nodal analysis to find the value of k that will cause v_y to be zero.

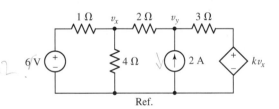

■ **FIGURE 4.49**

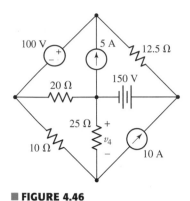

■ **FIGURE 4.48**

20. Consider the circuit of Fig. 4.50. Determine the current labeled i_1.

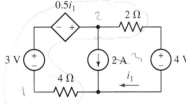

■ **FIGURE 4.50**

21. Make use of the supernode concept to assist in the determination of the voltage labeled v_{20} in Fig. 4.51. Crossed wires not marked by a solid dot are not in physical contact.

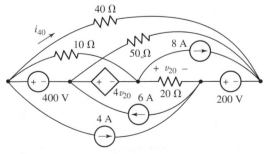

■ **FIGURE 4.51**

22. For the circuit of Fig. 4.52, determine all four nodal voltages.

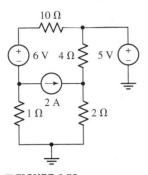

■ **FIGURE 4.52**

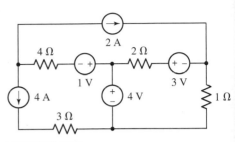

■ **FIGURE 4.53**

23. Determine the power supplied by the 2 A source in the circuit of Fig. 4.53.

24. Determine the power supplied by the 2 A source in the circuit of Fig. 4.54.

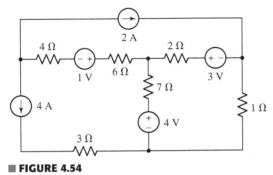

■ **FIGURE 4.54**

25. Determine the nodal voltages characterizing the circuit of Fig. 4.55.

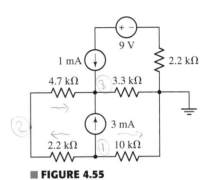

■ **FIGURE 4.55**

4.3 Mesh Analysis

26. Determine the mesh currents i_1 and i_2 as shown in the circuit of Fig. 4.56.

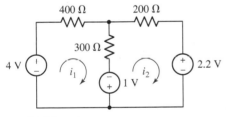

■ **FIGURE 4.56**

27. Regarding the circuit of Fig. 4.57, employ mesh analysis to determine (*a*) the current i_y; (*b*) the power supplied by the 10 V source.

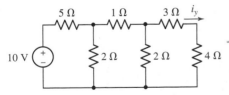

■ **FIGURE 4.57**

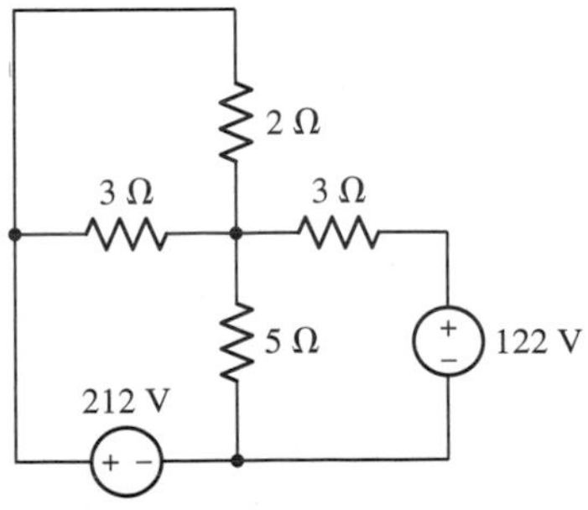

28. Employ mesh analysis to determine the current flowing in the circuit of Fig. 4.58 through (*a*) the 2 Ω resistor; (*b*) the 5 Ω resistor.

■ **FIGURE 4.58**

29. For the circuit of Fig. 4.59, use mesh analysis to determine (*a*) the current labeled i_x; (*b*) the power absorbed by the 25 Ω resistor.

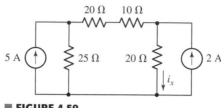

■ **FIGURE 4.59**

30. Use mesh analysis to determine the current labeled *i* in the circuit of Fig. 4.60.
31. Use mesh analysis to find i_x in the circuit shown in Fig. 4.61.

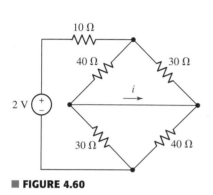

■ **FIGURE 4.60**

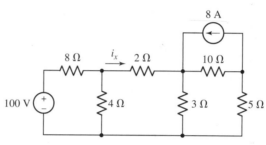

■ **FIGURE 4.61**

32. Calculate the power being dissipated in the 2 Ω resistor for the circuit of Fig. 4.62.

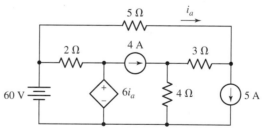

■ **FIGURE 4.62**

33. Use mesh analysis on the circuit shown in Fig. 4.48 to find the power being supplied by the dependent voltage source.
34. Use mesh analysis to find i_x in the circuit shown in Fig. 4.63.

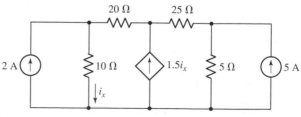

■ **FIGURE 4.63**

35. Determine the clockwise mesh currents for the circuit of Fig. 4.64.

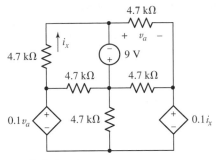

■ **FIGURE 4.64**

36. Determine each mesh current in the circuit of Fig. 4.65.

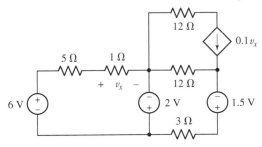

■ **FIGURE 4.65**

37. (a) Referring to the circuit of Fig. 4.66, determine the value of R if it is known that the mesh current $i_1 = 1.5$ mA. (b) Is the value of R necessarily unique? Explain.

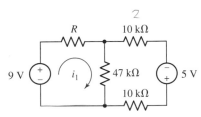

■ **FIGURE 4.66**

38. For the circuit of Fig. 4.67, employ the mesh analysis technique to find the power absorbed by each resistor.

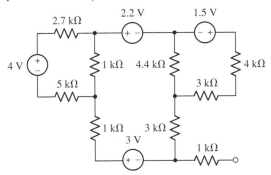

■ **FIGURE 4.67**

39. The circuit shown in Fig. 4.68 is the equivalent circuit of a common-base bipolar junction transistor amplifier. The input source has been shorted, and a

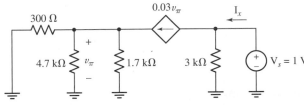

■ **FIGURE 4.68**

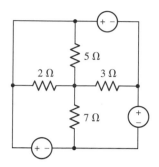

FIGURE 4.69

1 V source has been substituted for the output device. (*a*) Use mesh analysis to find I_x. (*b*) Verify your solution to part (*a*) using nodal analysis. (*c*) What is the physical significance of the quantity V_s/I_x?

40. Choose nonzero values for the three voltage sources of Fig. 4.69 so that no current flows through any resistor in the circuit.

4.4 The Supermesh

41. Use mesh analysis to help find the power generated by each of the five sources in Fig. 4.70.

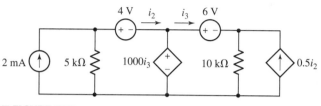

FIGURE 4.70

42. Find i_A in the circuit of Fig. 4.71.

43. Use the supermesh concept to determine the power supplied by the 2.2 V source of Fig. 4.72.

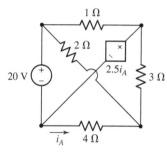

FIGURE 4.71

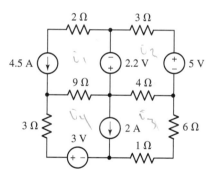

FIGURE 4.72

44. Determine the voltage across the 2 mA source in Fig. 4.73, assuming its bottom node is ground.

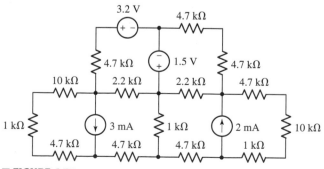

FIGURE 4.73

45. Employ mesh analysis to obtain the voltage across the 2.5 Ω resistor of Fig. 4.74.

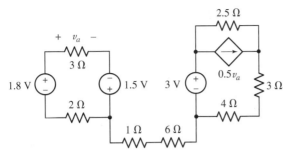

■ **FIGURE 4.74**

46. Calculate the mesh currents for the circuit of Fig. 4.75.

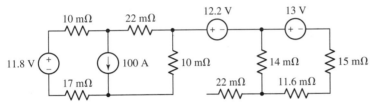

■ **FIGURE 4.75**

47. For the circuit of Fig. 4.76, determine the value of resistor **X** if $i_2 = 2.273$ A.

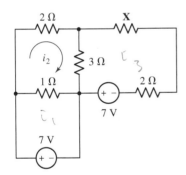

■ **FIGURE 4.76**

48. Consider the circuit of Fig. 4.77. Compute the three mesh currents indicated.

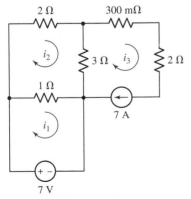

■ **FIGURE 4.77**

4.5 Nodal vs. Mesh Analysis: A Comparison

49. Determine the voltage labeled v_x in each of the circuits of Fig. 4.78.

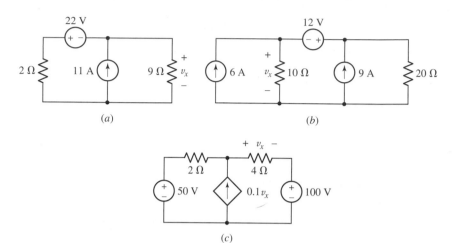

(a)   (b)

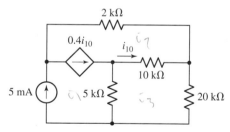

(c)

■ **FIGURE 4.78**

50. Find v_3 in the circuit of Fig. 4.79 if element A is (a) a short circuit; (b) a 9 V independent voltage source, with positive reference on the left; (c) a dependent current source, arrow head on the left, labeled $5i_1$.

51. Determine the currents i_1 and i_2 in the circuit of Fig. 4.79 if element A is a 12 Ω resistor. Explain the logic behind your choice of either nodal or mesh analysis.

52. Obtain a value for the current labeled i_{10} in the circuit of Fig. 4.80.

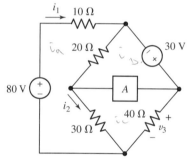

■ **FIGURE 4.79**

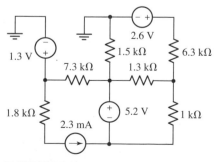

■ **FIGURE 4.80**

53. Determine the two currents labeled in the circuit of Fig. 4.81.

54. For the circuit of Fig. 4.82, determine the voltage of the center node.

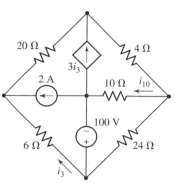

■ **FIGURE 4.81**　　　　　　■ **FIGURE 4.82**

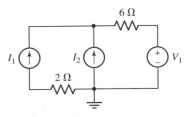

55. Determine the current through each branch of the circuit in Fig. 4.83.

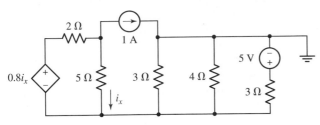

■ **FIGURE 4.83**

56. Determine the voltage across the 2 mA current source of Fig. 4.84.

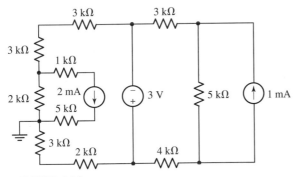

■ **FIGURE 4.84**

57. For the circuit of Fig. 4.85, let A be a 5 V voltage source with positive reference at the top, let B represent a 3 A current source with the arrow pointing toward ground, let C be a 3 Ω resistor, let D be a 2 A current source with arrow pointing toward ground, let F be a 1 V voltage source with negative reference on the right, and let E be a 4 Ω resistor. Compute i_1.

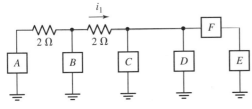

■ **FIGURE 4.85**

58. Choose any nonzero values for I_1, I_2, and V_1 so that 6 W is dissipated by the 6 Ω resistor in the circuit of Fig. 4.86.

59. Referring to the circuit of Fig. 4.84, replace the 2 mA current source with a 2 V voltage source, "+" reference terminal at the bottom, and the 3 V source with a 7 mA current source, arrow pointing down. Determine the mesh currents for the new circuit.

60. In the circuit of Fig. 4.85, A is a dependent current source with arrow pointing down and labeled $5i_1$. Let B and E be 2 Ω resistors, let C be a 2 A current source with arrow pointing toward ground, let F be a 2 V voltage source with negative reference on the right, and let D be a 3 A current source with the arrow at the top. Determine the nodal voltages and all mesh currents.

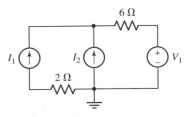

■ **FIGURE 4.86**

4.6 Computer-Aided Circuit Analysis

61. Use PSpice to verify the solution of Exer. 52. Submit a printout of a properly labeled schematic. Include hand calculations.

62. Use PSpice to verify the solution of Exer. 54. Submit a printout of a properly labeled schematic. Include hand calculations.

63. Use PSpice to verify the solution of Exer. 56. Submit a printout of a properly labeled schematic. Include hand calculations.

64. Use PSpice to verify the solution of Exer. 58. Submit a printout of a properly labeled schematic. Include hand calculations.

65. Use PSpice to verify the solution of Exer. 60. Submit a printout of a properly labeled schematic. Include hand calculations.

66. Construct a circuit consisting of a 5 V source in series with a 100 Ω resistor, connected to a network containing at least one 3 A source, three different resistors, and a voltage-controlled current source that depends on the voltage across the 100 Ω resistor. (*a*) Determine all node voltages and all branch currents. (*b*) Use PSpice to verify your results.

67. Construct a circuit using a 10 V battery, a 3 A source, and as many 1 Ω resistors as necessary to obtain a potential of 5 V across the 3 A source. Verify your hand calculations using PSpice.

68. Write an appropriate input deck for SPICE to find v_5 in the circuit of Fig. 4.87. Submit a printout of the output file, with the solution highlighted.

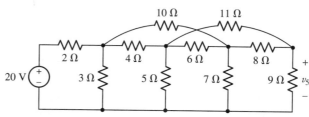

■ **FIGURE 4.87**

69. Design a circuit using only 9 V batteries and resistors that will provide nodal voltages of 4 V, 3 V, and 2 V. Write an appropriate input deck for SPICE to simulate your solution, and submit a printout of the output file with the desired voltages highlighted. Draw a labeled schematic on the printout for reference, with node numbers identified.

70. A very long string of multicolored outdoor lights is installed on a house. After applying power, the homeowner notices that two bulbs are burned out. (*a*) Are the lights connected in series or parallel? (*b*) Write a SPICE input deck to simulate the lights, assuming 20 AWG wire, 115 V ac power supply, and an individual bulb rating of 1 W. There are 400 lights in the string; simulate an electrically equivalent circuit with fewer than 25 components for simplicity. Submit a printout of the output file, with the power supplied by the wall socket highlighted. (*c*) After replacing the burned-out bulbs, the homeowner notices that the lights closest to the outlet are approximately 10 percent brighter than the lights at the far end of the string. Provide a possible explanation, keeping in mind that nothing in the string is zero ohms.

Useful Circuit Analysis Techniques

INTRODUCTION

The techniques of nodal and mesh analysis described in Chap. 4 are reliable and extremely powerful methods. However, both require that we develop a complete set of equations to describe a particular circuit as a general rule, even if only one current, voltage, or power quantity is of interest. In this chapter, we investigate several different techniques for isolating specific parts of a circuit in order to simplify the analysis. After examining the usage of these techniques, we focus on how one might go about selecting one method over another.

5.1 • LINEARITY AND SUPERPOSITION

All of the circuits which we plan to analyze can be classified as *linear circuits,* so this is a good time to be more specific in defining exactly what we mean by that. Having done this, we can then consider the most important consequence of linearity, the principle of **superposition.** This principle is very basic and will appear repeatedly in our study of linear circuit analysis. As a matter of fact, the nonapplicability of superposition to nonlinear circuits is the very reason they are so difficult to analyze!

> The principle of superposition states that the *response* (a desired current or voltage) in a linear circuit having more than one independent source can be obtained by adding the responses caused by the separate independent sources *acting alone.*

Linear Elements and Linear Circuits

Let us first define a **linear element** as a passive element that has a linear voltage-current relationship. By a "linear voltage-current

KEY CONCEPTS

Superposition as a Means of Determining the *Individual Contributions* of Different Sources to Any Current or Voltage

Source Transformation as a Means of Simplifying Circuits

Thévenin's Theorem

Norton's Theorem

Thévenin and Norton Equivalent Networks

Maximum Power Transfer

$\Delta \leftrightarrow Y$ Transformations for Resistive Networks

Selecting a Particular Combination of Analysis Techniques

Performing dc Sweep Simulations Using PSpice

relationship" we simply mean that multiplication of the current through the element by a constant K results in the multiplication of the voltage across the element by the same constant K. At this time, only one passive element has been defined (the resistor) and its voltage-current relationship

$$v(t) = Ri(t)$$

is clearly linear. As a matter of fact, if $v(t)$ is plotted as a function of $i(t)$, the result is a straight line.

We must also define a ***linear dependent source*** as a dependent current or voltage source whose output current or voltage is proportional only to the first power of a specified current *or* voltage variable in the circuit (or to the *sum* of such quantities). For example, the dependent voltage source $v_s = 0.6i_1 - 14v_2$ is linear, but $v_s = 0.6i_1^2$ and $v_s = 0.6i_1 v_2$ are not.

We may now define a ***linear circuit*** as a circuit composed entirely of independent sources, linear dependent sources, and linear elements. From this definition, it is possible to show[1] that "the response is proportional to the source," or that multiplication of all independent source voltages and currents by a constant K increases all the current and voltage responses by the same factor K (including the dependent source voltage or current outputs).

The Superposition Principle

The most important consequence of linearity is superposition. Let us develop the superposition principle by considering first the circuit of Fig. 5.1, which contains two independent sources, the current generators that force the currents i_a and i_b into the circuit. Sources are often called *forcing functions* for this reason, and the nodal voltages that they produce can be termed *response functions*, or simply *responses*. Both the forcing functions and the responses may be functions of time. The two nodal equations for this circuit are

$$0.7v_1 - 0.2v_2 = i_a \qquad [1]$$

$$-0.2v_1 + 1.2v_2 = i_b \qquad [2]$$

Now let us perform experiment x. We change the two forcing functions to i_{ax} and i_{bx}; the two unknown voltages will now be different, so we will call them v_{1x} and v_{2x}. Thus,

$$0.7v_{1x} - 0.2v_{2x} = i_{ax} \qquad [3]$$

$$-0.2v_{1x} + 1.2v_{2x} = i_{bx} \qquad [4]$$

We next perform experiment y by changing the source currents to i_{ay} and i_{by} and measure the responses v_{1y} and v_{2y}:

$$0.7v_{1y} - 0.2v_{2y} = i_{ay} \qquad [5]$$

$$-0.2v_{1y} + 1.2v_{2y} = i_{by} \qquad [6]$$

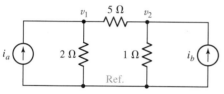

■ **FIGURE 5.1** A circuit with two independent current sources.

---

(1) The proof involves first showing that the use of nodal analysis on the linear circuit can produce only linear equations of the form

$$a_1 v_1 + a_2 v_2 + \cdots + a_N v_N = b$$

where the a_i are constants (combinations of resistance or conductance values, constants appearing in dependent source expressions, 0, or ± 1), the v_i are the unknown node voltages (responses), and b is an independent source value or a sum of independent source values. Given a set of such equations, if we multiply all the b's by K, then it is evident that the solution of this new set of equations will be the node voltages $Kv_1, Kv_2, \ldots, Kv_N$.

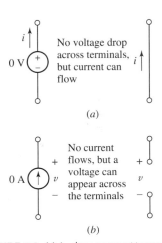

These three sets of equations describe the same circuit with three different sets of source currents. Let us *add* or "*superpose*" the last two sets of equations. Adding Eqs. [3] and [5],

$$(0.7v_{1x} + 0.7v_{1y}) - (0.2v_{2x} + 0.2v_{2y}) = i_{ax} + i_{ay} \qquad [7]$$

$$0.7v_1 \quad - \quad 0.2v_2 \quad = \quad i_a \qquad [1]$$

and adding Eqs. [4] and [6],

$$-(0.2v_{1x} + 0.2v_{1y}) + (1.2v_{2x} + 1.2v_{2y}) = i_{bx} + i_{by} \qquad [8]$$

$$-0.2v_1 \quad + \quad 1.2v_2 \quad = \quad i_b \qquad [2]$$

where Eq. [1] has been written immediately below Eq. [7] and Eq. [2] below Eq. [8] for easy comparison.

The linearity of all these equations allows us to compare Eq. [7] with Eq. [1] and Eq. [8] with Eq. [2] and draw an interesting conclusion. If we select i_{ax} and i_{ay} such that their sum is i_a and select i_{bx} and i_{by} such that their sum is i_b, then the desired responses v_1 and v_2 may be found by adding v_{1x} to v_{1y} and v_{2x} to v_{2y}, respectively. In other words, we can perform experiment x and note the responses, perform experiment y and note the responses, and finally add the two sets of responses. This leads to the fundamental concept involved in the superposition principle: to look at each independent source (and the response it generates) one at a time with the other independent sources "turned off" or "zeroed out."

If we reduce a voltage source to zero volts, we have effectively created a short circuit (Fig. 5.2*a*). If we reduce a current source to zero amps, we have effectively created an open circuit (Fig. 5.2*b*). Thus, the **superposition theorem** can be stated as:

> In any linear resistive network, the voltage across or the current through any resistor or source may be calculated by adding algebraically all the individual voltages or currents caused by the separate independent sources acting alone, with all other independent voltage sources replaced by short circuits and all other independent current sources replaced by open circuits.

Thus, if there are N independent sources we must perform N experiments, each having only one of the independent sources active and the others inactive/turned off/zeroed out. Note that *dependent* sources are in general active in every experiment.

The circuit we have just used as an example, however, should indicate that a much stronger theorem might be written; a *group* of independent sources may be made active and inactive collectively, if we wish. For example, suppose there are three independent sources. The theorem states that we may find a given response by considering each of the three sources acting alone and adding the three results. Alternatively, we may find the response due to the first and second sources operating with the third inactive, and then add to this the response caused by the third source acting alone. This amounts to treating several sources collectively as a sort of "supersource."

There is also no reason that an independent source must assume only its given value or a zero value in the several experiments; it is necessary only for the sum of the several values to be equal to the original value. An inactive source almost always leads to the simplest circuit, however.

FIGURE 5.2 (*a*) A voltage source set to zero acts like a short circuit. (*b*) A current source set to zero acts like an open circuit.

Let us illustrate the application of the superposition principle by considering an example in which both types of independent source are present.

EXAMPLE 5.1

For the circuit of Fig. 5.3a, use superposition to write an expression for the unknown branch current i_x.

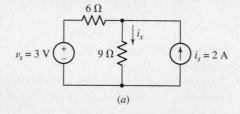

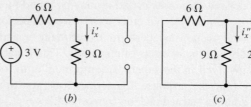

■ FIGURE 5.3 (a) An example circuit with two independent sources for which the branch current i_x is desired; (b) same circuit with current source open-circuited; (c) original circuit with voltage source short-circuited.

We first set the current source equal to zero and redraw the circuit as shown in Fig. 5.3b. The portion of i_x due to the voltage source has been designated i'_x to avoid confusion and is easily found to be 0.2 A.

We next set the voltage source in Fig. 5.3a to zero and again redraw the circuit, as shown in Fig. 5.3c. Routine application of current division allows us to determine that i''_x (the portion of i_x due to the 2 A current source) is 0.8 A.

We may now compute the complete current i_x as the sum of the two individual components:

$$i_x = i_x|_{3\,V} + i_x|_{2\,A} = i'_x + i''_x$$

or

$$i_x = \frac{3}{6+9} + 2\left(\frac{6}{6+9}\right) = 0.2 + 0.8 = 1.0\ A$$

Another way of looking at Example 5.1 is that the 3 V source and the 2 A source are each performing work on the circuit, resulting in a total current i_x flowing through the 9 Ω resistor. *However, the contribution of the 3 V source to i_x does not depend on the contribution of the 2 A source, and vice versa.* For example, if we double the output of the 2 A source to 4 A, it will now contribute 1.6 A to the total current i_x flowing through the 9 Ω resistor. However, the 3 V source would still contribute only 0.2 A to i_x, for a new total current of $0.2 + 1.6 = 1.8$ A.

V = iR.

SECTION 5.1 LINEARITY AND SUPERPOSITION

PRACTICE

5.1 For the circuit of Fig. 5.4, use superposition to compute the current i_x.

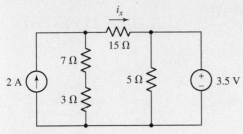

■ FIGURE 5.4

Ans: 660 mA.

As we will see, superposition does not generally reduce our workload when considering a particular circuit, since it leads to the analysis of several new circuits to obtain the desired response. However, it is particularly useful in identifying the significance of various parts of a more complex circuit. It also forms the basis of phasor analysis, which is introduced in Chap. 10.

EXAMPLE 5.2

Referring to the circuit of Fig. 5.5*a*, determine the maximum *positive* current to which the source I_X can be set before any resistor exceeds its power rating and overheats.

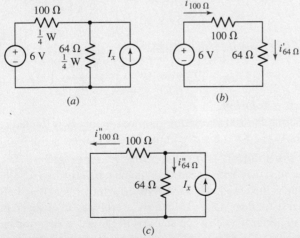

■ FIGURE 5.5 (*a*) A circuit with two resistors each rated at $\frac{1}{4}$ W. (*b*) Circuit with only the 6 V source active. (*c*) Circuit with the source I_x active.

▶ **Identify the goal of the problem.**
Each resistor is rated to a maximum of 250 mW. If the circuit allows this value to be exceeded (by forcing too much current through either resistor), excessive heating will occur—possibly leading to

(Continued on next page)

an accident. The 6 V source cannot be changed, so we are looking for an equation involving I_x and the maximum current through each resistor.

▶ **Collect the known information.**

Based on its 250 mW power rating, the maximum current the 100 Ω resistor can tolerate is

$$\sqrt{\frac{P_{max}}{R}} = \sqrt{\frac{0.250}{100}} = 50 \text{ mA}$$

and, similarly, the current through the 64 Ω resistor must be less than 62.5 mA.

▶ **Devise a plan.**

Either nodal or mesh analysis may be applied to the solution of this problem, but superposition may give us a slight edge, since we are primarily interested in the effect of the current source.

▶ **Construct an appropriate set of equations.**

Using superposition, we redraw the circuit as in Fig. 5.5b and find that the 6 V source contributes a current

$$i'_{100\Omega} = \frac{6}{100 + 64} = 36.59 \text{ mA}$$

to the 100 Ω resistor and, since the 64 Ω resistor is in series, $i'_{64\Omega} = 36.59$ mA as well.

Recognizing the current divider in Fig. 5.5c, we note that $i''_{64\Omega}$ will *add* to $i'_{64\Omega}$, but $i''_{100\Omega}$ is *opposite* in direction to $i'_{100\Omega}$. I_X can therefore safely contribute $62.5 - 36.59 = 25.91$ mA to the 64 Ω resistor current, and $50 - (-36.59) = 86.59$ mA to the 100 Ω resistor current.

The 100 Ω resistor therefore places the following constraint on I_X:

$$I_X < (86.59 \times 10^{-3}) \left(\frac{100 + 64}{64} \right)$$

and the 64 Ω resistor requires that

$$I_X < (25.91 \times 10^{-3}) \left(\frac{100 + 64}{100} \right)$$

▶ **Attempt a solution.**

Considering the 100 Ω resistor first, we see that I_X is limited to $I_X < 221.9$ mA. The 64 Ω resistor limits I_X such that $I_X < 42.49$ mA.

▶ **Verify the solution. Is it reasonable or expected?**

In order to satisfy both constraints, I_X must be less than 42.49 mA. If the value is increased, the 64 Ω resistor will overheat long before the 100 Ω resistor does. One particularly useful way to evaluate our solution is to perform a dc sweep analysis in PSpice as described after the next example. An interesting question, however, is whether we would have expected the 64 Ω resistor to overheat first.

Originally we found that the 100 Ω resistor has a smaller maximum current, so it might be reasonable to expect it to limit I_X. However, because I_X *opposes* the current sent by the 6 V source through the 100 Ω resistor but *adds* to the 6 V source's contribution to the current through the 64 Ω resistor, it turns out to work the other way—it's the 64 Ω resistor that sets the limit on I_X.

EXAMPLE **5.3**

In the circuit of Fig. 5.6a, use the superposition principle to determine the value of i_x.

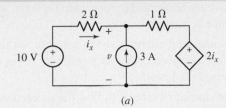

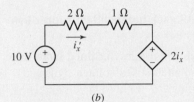 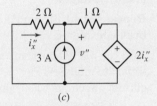

(b)   (c)

■ **FIGURE 5.6** (a) An example circuit with two independent sources and one dependent source for which the branch current i_x is desired. (b) Circuit with the 3 A source open-circuited. (c) Original circuit with the 10 V source short-circuited.

We first open-circuit the 3 A source (Fig. 5.6b). The single mesh equation is

$$-10 + 2i'_x + 1i'_x + 2i'_x = 0$$

so that

$$i'_x = 2 \text{ A}$$

Next, we short-circuit the 10 V source (Fig. 5.6c) and write the single-node equation

$$\frac{v''}{2} + \frac{v'' - 2i''_x}{1} = 3$$

and relate the dependent-source-controlling quantity to v'':

$$v'' = 2(-i''_x)$$

We find

$$i''_x = -0.6 \text{ A}$$

and, thus,

$$i_x = i'_x + i''_x = 2 + (-0.6) = 1.4 \text{ A}$$

Note that in redrawing each subcircuit, we are always careful to use some type of notation to indicate that we are not working with the original variables. This prevents the possibility of rather disastrous errors when we add the individual results.

PRACTICE

5.2 For the circuit of Fig. 5.7, use superposition to obtain the voltage across each current source.

Ans: $v_1|_{2A} = 9.180$ V, $v_2|_{2A} = -1.148$ V, $v_1|_{3V} = 1.967$ V, $v_2|_{3V} = -0.246$ V; $v_1 = 11.147$ V, $v_2 = -1.394$ V.

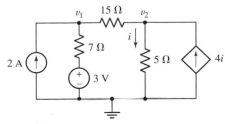

■ **FIGURE 5.7**

> **Summary of Basic Superposition Procedure**
>
> 1. **Select one of the independent sources. Set all other independent sources to zero.** This means voltage sources are replaced with short circuits and current sources are replaced with open circuits. Leave dependent sources alone.
>
> 2. **Relabel voltages and currents using suitable notation** (e.g., v', i_2''). Be sure to relabel controlling variables of dependent sources to avoid confusion.
>
> 3. **Analyze the simplified circuit to find the desired currents and/or voltages.**
>
> 4. **Repeat steps 1 through 3 until each independent source has been considered.**
>
> 5. **Add the partial currents and/or voltages obtained from the separate analyses.** Pay careful attention to voltage signs and current directions when summing.
>
> 6. **Do not add power quantities.** If power quantities are required, calculate only after partial voltages and/or currents have been summed.

Note that step 1 may be altered in several ways. First, independent sources can be considered in groups as opposed to individually if it simplifies the analysis, as long as no independent source is included in more than one subcircuit. Second, it is technically not necessary to set sources to zero, although this is almost always the best route. For example, a 3 V source may appear in two subcircuits as a 1.5 V source, since $1.5 + 1.5 = 3$ V just as $0 + 3 = 3$ V. Because it is unlikely to simplify our analysis, however, there is little point to such an exercise.

COMPUTER-AIDED ANALYSIS

Although PSpice is extremely useful in verifying that we have analyzed a complete circuit correctly, it can also assist us in determining the contribution of each source to a particular response. To do this, we employ what is known as a *dc parameter sweep*.

Consider the circuit presented in Example 5.2, when we were asked to determine the maximum positive current that could be obtained from the current source without exceeding the power rating of either resistor in the circuit. The circuit is shown redrawn using the Orcad Capture CIS schematic tool in Fig. 5.8. Note that no value has been assigned to the current source.

After the schematic has been entered and saved, the next step is to specify the dc sweep parameters. This option allows us to specify a range of values for a voltage or current source (in the present case, the current source I_x), rather than a specific value. Selecting **New Simulation Profile** under **PSpice,** we provide a name for our profile and are then provided with the dialog box shown in Fig. 5.9.

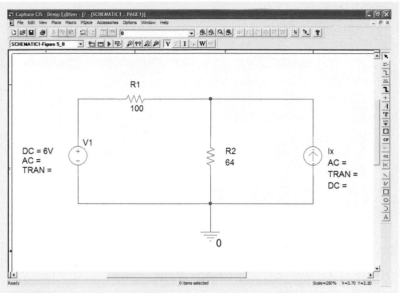

■ **FIGURE 5.8** The circuit from Example 5.2.

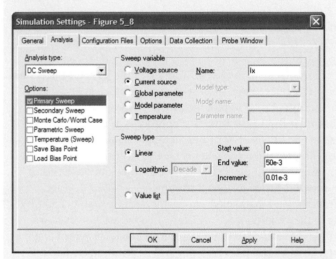

■ **FIGURE 5.9** DC Sweep dialog box shown with I_x selected as the sweep variable.

Under **Analysis Type,** we pull down the **DC Sweep** option, specify the "sweep variable" as **Current Source,** and then type in I_x in the **Name** box. There are several options under Sweep Type: **Linear, Logarithmic,** and **Value List.** The last option allows us to specify each value to assign to I_x. In order to generate a smooth plot, however, we choose to perform a **Linear** sweep, with a **Start Value** of 0 mA, an **End Value** of 50 mA, and a value of 0.01 mA for the **Increment.**

After we perform the simulation, the graphical output package Probe is automatically launched. When the window appears, the horizontal axis (corresponding to our variable, I_x) is displayed, but the vertical axis variable must be chosen. Selecting **Add Trace** from the **Trace**

(Continued on next page)

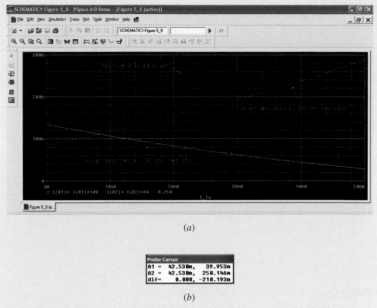

(a)

```
Probe Cursor
A1 =   42.530m,    39.953m
A2 =   42.530m,   250.146m
dif=    0.000,  -210.193m
```

(b)

■ **FIGURE 5.10** (a) Probe output with text labels identifying the power absorbed by the two resistors individually. A horizontal line indicating 250 mW has also been included, as well as text labels to improve clarity. (b) Cursor dialog box.

menu, we click on **I(R1),** then type an asterisk in the **Trace Expression** box, click on **I(R1)** once again, insert yet another asterisk, and finally type in 100. This asks Probe to plot the power absorbed by the 100 Ω resistor. In a similar fashion, we repeat the process to add the power absorbed by the 64 Ω resistor, resulting in a plot similar to that shown in Fig. 5.10a. A horizontal reference line at 250 mW was also added to the plot by typing 0.250 in the **Trace Expression** box after selecting **Add Trace** from the **Trace** menu a third time.

We see from the plot that the 64 Ω resistor *does* exceed its 250 mW power rating in the vicinity of $I_x = 43$ mA. In contrast, however, we see that regardless of the value of the current source I_x (provided that it is between 0 and 50 mA), the 100 Ω resistor will never dissipate 250 mW; in fact, the absorbed power *decreases* with increasing current from the current source. If we desire a more precise answer we can make use of the cursor tool, which is invoked by selecting **Trace, Cursor, Display** from the menu bar. Figure 5.10b shows the result of dragging both cursors to 42.53 mA; the 64 Ω resistor has just barely exceeded its rating at this current level. Increased precision can be obtained by decreasing the increment value used in the dc sweep.

This technique is very useful in analyzing electronic circuits, where we might need, for example, to determine what input voltage is required to a complicated amplifier circuit in order to obtain a zero output voltage. We also notice that there are several other types of parameter sweeps that we can perform, including a dc voltage sweep. The ability to vary temperature is useful only when dealing with component models that have a temperature parameter built in, such as diodes and transistors.

Unfortunately, it usually turns out that little if any time is saved in analyzing a circuit containing one or more dependent sources by use of the superposition principle, for there must always be at least two sources in operation: one independent source and all the dependent sources.

We must constantly be aware of the limitations of superposition. It is applicable only to linear responses, and thus the most common nonlinear response—power—is not subject to superposition. For example, consider two 1 V batteries in series with a 1 Ω resistor. The power delivered to the resistor is obviously 4 W, but if we mistakenly try to apply superposition we might say that each battery alone furnished 1 W and thus the total power is 2 W. This is incorrect, but a surprisingly easy mistake to make.

CAUTION

5.2 • SOURCE TRANSFORMATIONS

Practical Voltage Sources

Up to now we have been working exclusively with *ideal* voltage and current sources; it is now time to take a step closer to reality by considering *practical* sources. These sources will enable us to make more realistic representations of physical devices. Once we have defined practical sources, we will see that *practical* current and voltage sources may be interchanged without affecting the remainder of the circuit. Such sources will be called *equivalent* sources. Our methods will be applicable to both independent and dependent sources, although we will find that they are not as useful with dependent sources.

The ideal voltage source was defined as a device whose terminal voltage is independent of the current through it. A 1 V dc source produces a current of 1 A through a 1 Ω resistor, and a current of 1,000,000 A through a 1 $\mu\Omega$ resistor; it can provide an unlimited amount of power. No such device exists practically, of course, and we agreed previously that a *real* physical voltage source could be represented by an *ideal* voltage source only as long as relatively small currents, or powers, were drawn from it. For example, a car battery may be approximated by an ideal 12 V dc voltage source if its current is limited to a few amperes (Fig. 5.11*a*). However, anyone who has ever tried to start an automobile with the headlights on must have observed that the lights dimmed perceptibly when the battery was asked to deliver the heavy starter current, 100 A or more, in addition to the headlight current. Under these conditions, an ideal voltage source is not really an adequate representation of the battery.

To better approximate the behavior of a real device, the ideal voltage source must be modified to account for the lowering of its terminal voltage when large currents are drawn from it. Let us suppose that we observe experimentally that our car battery has a terminal voltage of 12 V when no current is flowing through it, and a reduced voltage of 11 V when 100 A is flowing. How could we model this behavior? Well, a more accurate model might be an ideal voltage source of 12 V in series with a resistor across which 1 V appears when 100 A flows through it. A quick calculation shows that the resistor must be 1 V/100 A = 0.01 Ω, and the ideal voltage source and this series resistor constitute a ***practical voltage source*** (Fig. 5.11*b*). Thus, we are using the series combination of two ideal circuit elements, an independent voltage source and a resistor, to model a real device.

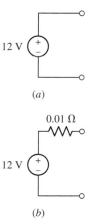

(a)

(b)

■ **FIGURE 5.11** (*a*) An ideal 12 V dc voltage source used to model a car battery. (*b*) A more accurate model that accounts for the observed reduction in terminal voltage at large currents.

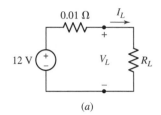

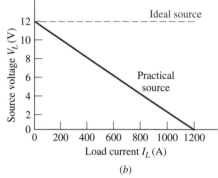

FIGURE 5.12 (*a*) A practical source, which approximates the behavior of a certain 12 V automobile battery, is shown connected to a load resistor R_L. (*b*) The relationship between I_L and V_L is linear.

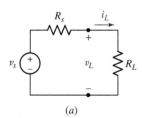

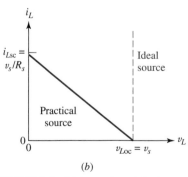

FIGURE 5.13 (*a*) A general practical voltage source connected to a load resistor R_L. (*b*) The terminal voltage of a practical voltage source decreases as i_L increases and $R_L = v_L/i_L$ decreases. The terminal voltage of an ideal voltage source (also plotted) remains the same for any current delivered to a load.

We do not expect to find such an arrangement of ideal elements inside our car battery, of course. Any real device is characterized by a certain current-voltage relationship at its terminals, and our problem is to develop some combination of ideal elements that can furnish a similar current-voltage characteristic, at least over some useful range of current, voltage, or power.

In Fig. 5.12*a*, we show our two-piece practical model of the car battery now connected to some load resistor R_L. The terminal voltage of the practical source is the same as the voltage across R_L and is marked[2] V_L. Figure 5.12*b* shows a plot of load voltage V_L as a function of the load current I_L for this practical source. The KVL equation for the circuit of Fig. 5.12*a* may be written in terms of I_L and V_L:

$$12 = 0.01I_L + V_L$$

and thus

$$V_L = -0.01I_L + 12$$

This is a linear equation in I_L and V_L, and the plot in Fig. 5.12*b* is a straight line. Each point on the line corresponds to a different value of R_L. For example, the midpoint of the straight line is obtained when the load resistance is equal to the internal resistance of the practical source, or $R_L = 0.01 \ \Omega$. Here, the load voltage is exactly one-half the ideal source voltage.

When $R_L = \infty$ and no current whatsoever is being drawn by the load, the practical source is open-circuited and the terminal voltage, or open-circuit voltage, is $V_{Loc} = 12$ V. If, on the other hand, $R_L = 0$, thereby short-circuiting the load terminals, then a load current or short-circuit current, $I_{Lsc} = 1200$ A, would flow. (*In practice, such an experiment would probably result in the destruction of the short circuit, the battery, and any measuring instruments incorporated in the circuit!*)

Since the plot of V_L versus I_L is a straight line for this practical voltage source, we should note that the values of V_{Loc} and I_{Lsc} uniquely determine the entire $V_L - I_L$ curve.

The horizontal broken line of Fig. 5.12*b* represents the $V_L - I_L$ plot for an *ideal* voltage source; the terminal voltage remains constant for any value of load current. For the practical voltage source, the terminal voltage has a value near that of the ideal source only when the load current is relatively small.

Let us now consider a *general* practical voltage source, as shown in Fig. 5.13*a*. The voltage of the ideal source is v_s, and a resistance R_s, called an *internal resistance* or *output resistance,* is placed in series with it. Again, we must note that the resistor is not really present as a separate component but merely serves to account for a terminal voltage that decreases as the load current increases. Its presence enables us to model the behavior of a physical voltage source more closely.

The linear relationship between v_L and i_L is

$$v_L = v_s - R_s i_L \qquad [9]$$

---

(2) From this point on we will endeavor to adhere to the standard convention of referring to strictly dc quantities using capital letters, whereas lowercase letters denote a quantity that we know to possess some time-varying component. However, in describing general theorems which apply to either dc or ac, we will continue to use lowercase to emphasize the general nature of the concept.

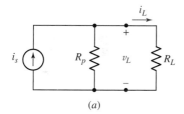

and this is plotted in Fig. 5.13b. The open-circuit voltage ($R_L = \infty$, so $i_L = 0$) is

$$v_{Loc} = v_s \qquad [10]$$

and the short-circuit current ($R_L = 0$, so $v_L = 0$) is

$$i_{Lsc} = \frac{v_s}{R_s} \qquad [11]$$

Once again, these values are the intercepts for the straight line in Fig. 5.13b, and they serve to define it completely.

Practical Current Sources

An ideal current source is also nonexistent in the real world; there is no physical device that will deliver a constant current regardless of the load resistance to which it is connected or the voltage across its terminals. Certain transistor circuits will deliver a constant current to a wide range of load resistances, but the load resistance can always be made sufficiently large that the current through it becomes very small. Infinite power is simply never available (unfortunately).

A practical current source is defined as an ideal current source in parallel with an internal resistance R_p. Such a source is shown in Fig. 5.14a, and the current i_L and voltage v_L associated with a load resistance R_L are indicated. Application of KCL yields

$$i_L = i_s - \frac{v_L}{R_p} \qquad [12]$$

which is again a linear relationship. The open-circuit voltage and the short-circuit current are

$$v_{Loc} = R_p i_s \qquad [13]$$

and

$$i_{Lsc} = i_s \qquad [14]$$

The variation of load current with changing load voltage may be investigated by changing the value of R_L as shown in Fig. 5.14b. The straight line is traversed from the short-circuit, or "northwest," end to the open-circuit termination at the "southeast" end by increasing R_L from zero to infinite ohms. The midpoint occurs for $R_L = R_p$. The load current i_L and the ideal source current are approximately equal only for small values of load voltage, which are obtained with values of R_L that are small compared to R_p.

Equivalent Practical Sources

Having defined both practical sources, we are now ready to discuss their equivalence. We will define two sources as being *equivalent* if they produce identical values of v_L and i_L when they are connected to identical values of R_L, no matter what the value of R_L may be. Since $R_L = \infty$ and $R_L = 0$ are two such values, equivalent sources provide the same open-circuit voltage and short-circuit current. In other words, if we are given two equivalent sources, one a practical voltage source and the other a practical current source, each enclosed in a black box with only a single pair of terminals,

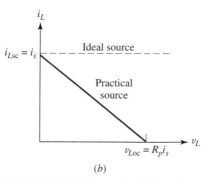

FIGURE 5.14 (*a*) A general practical current source connected to a load resistor R_L. (*b*) The load current provided by the practical current source is shown as a function of the load voltage.

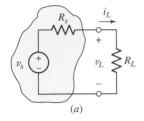

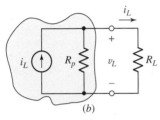

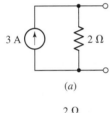

■ **FIGURE 5.15** (a) A given practical voltage source connected to a load R_L. (b) The equivalent practical current source connected to the same load.

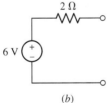

■ **FIGURE 5.16** (a) A given practical current source. (b) The equivalent practical voltage source.

CAUTION

then there is no way in which we can tell which source is in which box by measuring current or voltage in a resistive load.

Consider the practical voltage source and resistor R_L shown in Fig. 5.15a, and the circuit composed of a practical current source and resistor R_L shown in Fig. 5.15b. A simple calculation shows that the voltage across the load R_L of Fig. 5.15a is

$$v_L = v_s \frac{R_L}{R_s + R_L} \qquad [15]$$

A similarly simple calculation shows that the voltage across the load R_L in Fig. 5.15b is

$$v_L = \left[i_s \frac{R_p}{R_p + R_L} \right] \cdot R_L$$

The two practical sources are electrically equivalent, then, if

$$R_s = R_p \qquad [16]$$

and

$$v_s = R_p i_s = R_s i_s \qquad [17]$$

where we now let R_s represent the internal resistance of either practical source, which is the conventional notation.

As an illustration of the use of these ideas, consider the practical current source shown in Fig. 5.16a. Since its internal resistance is 2 Ω, the internal resistance of the equivalent practical voltage source is also 2 Ω; the voltage of the ideal voltage source contained within the practical voltage source is $(2)(3) = 6$ V. The equivalent practical voltage source is shown in Fig. 5.16b.

To check the equivalence, let us visualize a 4 Ω resistor connected to each source. In both cases a current of 1 A, a voltage of 4 V, and a power of 4 W are associated with the 4 Ω load. However, we should note very carefully that the ideal current source is delivering a total power of 12 W, while the ideal voltage source is delivering only 6 W. Furthermore, the internal resistance of the practical current source is absorbing 8 W, whereas the internal resistance of the practical voltage source is absorbing only 2 W. Thus we see that the two practical sources are equivalent only with respect to what transpires at the load terminals; they are *not* equivalent internally!

EXAMPLE 5.4

Compute the current through the 4.7 kΩ resistor in Fig. 5.17a after transforming the 9 mA source into an equivalent voltage source.

The equivalent source consists of an independent voltage source of $(9 \text{ mA}) \times (5 \text{ k}\Omega) = 45$ V in series with a 5 kΩ resistor, as depicted in Fig. 5.17b.

A simple KVL equation around the loop yields:

$$-45 + 5000I + 4700I + 3000I + 3 = 0$$

which can be easily solved to find that the current $I = 3.307$ mA.

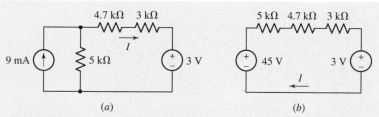

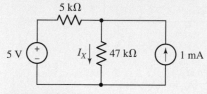

■ **FIGURE 5.17** (*a*) A circuit with both a voltage source and a current source. (*b*) The circuit after the 9 mA source is transformed into an equivalent voltage source.

PRACTICE

5.3 For the circuit of Fig. 5.18, compute the current I_X through the 47 kΩ resistor after performing a source transformation on the voltage source.

■ **FIGURE 5.18**

Ans: 192 μA.

EXAMPLE **5.5**

Calculate the current through the 2 Ω resistor in Fig. 5.19*a* on the next page by making use of source transformations to first simplify the circuit.

We begin by transforming each current source into a voltage source (Fig. 5.19*b*), the strategy being to convert the circuit into a simple loop.

We must be careful to retain the 2 Ω resistor for two reasons: first, the dependent source controlling variable appears across it, and second, we desire the current flowing through it. However, we can combine the 17 Ω and 9 Ω resistors, since they appear in series. We also see that the 3 Ω and 4 Ω resistors may be combined into a single 7 Ω resistor, which can then be used to transform the 15 V source into a 15/7 A source as in Fig. 5.19*c*.

As a final simplification, we note that the two 7 Ω resistors can be combined into a single 3.5 Ω resistor, which may be used to transform the 15/7 A current source into a 7.5 V voltage source. The result is a simple loop circuit, shown in Fig. 5.19*d*.

The current I can now be found using KVL:

$$-7.5 + 3.5I - 51V_x + 28I + 9 = 0$$

where

$$V_x = 2I$$

Thus,

$$I = 21.28 \text{ mA}$$

(*Continued on next page*)

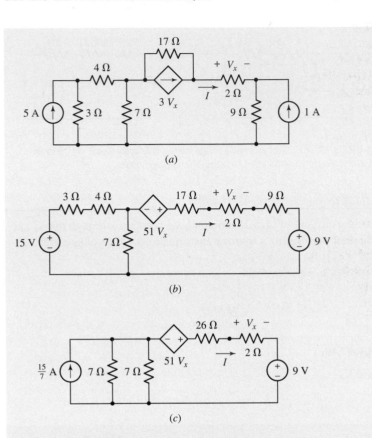

■ **FIGURE 5.19** (*a*) A circuit with two independent current sources and one dependent source. (*b*) The circuit after each source is transformed into a voltage source. (*c*) The circuit after further combinations. (*d*) The final circuit.

PRACTICE

5.4 For the circuit of Fig. 5.20, compute the voltage *V* across the 1 MΩ resistor using repeated source transformations.

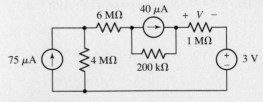

■ **FIGURE 5.20**

Ans: 27.23 V.

Several Key Points

We conclude our discussion of practical sources and source transformations with a few specialized observations. First, when we transform a voltage source, we must be sure that the source is in fact *in series* with the resistor under consideration. For example, in the circuit shown in Fig. 5.21, it is perfectly valid to perform a source transformation on the voltage source using the 10 Ω resistor, as they are in series. However, it would be incorrect to attempt a source transformation using the 60 V source and the 30 Ω resistor—a very common type of error.

In a similar fashion, when we transform a current source and resistor combination, we must be sure that they are in fact *in parallel*. Consider the current source shown in Fig. 5.22a. We may perform a source transformation including the 3 Ω resistor, as they are in parallel, but after the transformation there may be some ambiguity as to where to place the resistor. In such circumstances, it is helpful to first redraw the components to be transformed as in Fig. 5.22b. Then, the transformation to a voltage source in series with a resistor may be drawn correctly as shown in Fig. 5.22c; the resistor may in fact be drawn above or below the voltage source.

It is also worthwhile to consider the unusual case of a current source in series with a resistor and its dual, the case of a voltage source in parallel

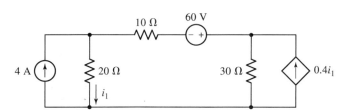

■ FIGURE 5.21 An example circuit to illustrate how to determine if a source transformation can be performed.

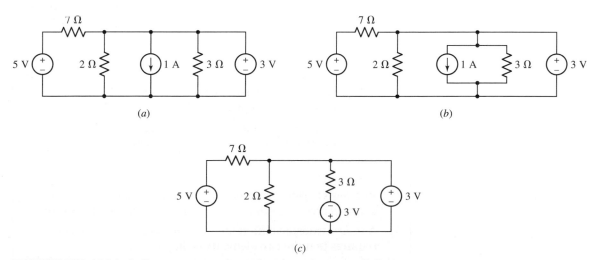

(a)

(b)

(c)

■ FIGURE 5.22 (a) A circuit with a current source to be transformed to a voltage source. (b) Circuit redrawn so as to avoid errors. (c) Transformed source/resistor combination.

with a resistor. Let's start with the simple circuit of Fig. 5.23a, where we are interested only in the voltage across the resistor marked R_2. We note that regardless of the value of resistor R_1, $V_{R_2} = I_x R_2$. Although we might be tempted to perform an inappropriate source transformation on such a circuit, in fact *we may simply omit resistor R_1* (provided that it is of no interest to us itself). A similar situation arises with a voltage source in parallel with a resistor, as depicted in Fig. 5.23b. Again, if we are only interested in some quantity regarding resistor R_2, we may find ourselves tempted to perform some strange (and incorrect) source transformation on the voltage source and resistor R_1. In reality, we may omit resistor R_1 from our circuit as far as resistor R_2 is concerned—its presence does not alter either the voltage across, the current through, or the power dissipated by resistor R_2.

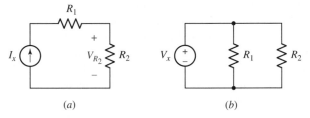

(a) (b)

■ **FIGURE 5.23** (a) Circuit with a resistor R_1 in series with a current source. (b) A voltage source in parallel with two resistors.

Summary of Source Transformation

1. **A common goal in source transformation is to end up with either all current sources or all voltage sources in the circuit.** This is especially true if it makes nodal or mesh analysis easier.

2. **Repeated source transformations can be used to simplify a circuit by allowing resistors and sources to eventually be combined.**

3. **The resistor value does not change during a source transformation, but it is not the same resistor.** This means that currents or voltages associated with the original resistor are irretrievably lost when we perform a source transformation.

4. **If the voltage or current associated with a particular resistor is used as a controlling variable for a dependent source, it should not be included in any source transformation.** The original resistor must be retained in the final circuit, untouched.

5. **If the voltage or current associated with a particular element is of interest, that element should not be included in any source transformation.** The original element must be retained in the final circuit, untouched.

6. **In a source transformation, the head of the current source arrow corresponds to the "+" terminal of the voltage source.**

7. **A source transformation on a current source and resistor requires that the two elements be in parallel.**

8. **A source transformation on a voltage source and resistor requires that the two elements be in series.**

5.3 • THÉVENIN AND NORTON EQUIVALENT CIRCUITS

Now that we have been introduced to source transformations and the super-position principle, it is possible to develop two more techniques that will greatly simplify the analysis of many linear circuits. The first of these theorems is named after M. L. Thévenin, a French engineer working in telegraphy who published the theorem in 1883; the second may be considered a corollary of the first and is credited to E. L. Norton, a scientist with the Bell Telephone Laboratories.

Let us suppose that we need to make only a partial analysis of a circuit. For example, perhaps we need to determine the current, voltage, and power delivered to a single "load" resistor by the remainder of the circuit, which may consist of a sizable number of sources and resistors (Fig. 5.24a). Or, perhaps we wish to find the response for different values of the load resistance. Thévenin's theorem tells us that it is possible to replace everything except the load resistor with an independent voltage source in series with a resistor (Fig. 5.24b); the response measured *at the load resistor* will be unchanged. Using Norton's theorem, we obtain an equivalent composed of an independent current source in parallel with a resistor (Fig. 5.24c).

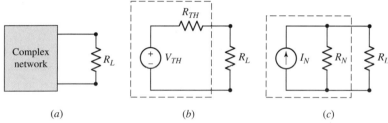

(a)   (b)   (c)

■ **FIGURE 5.24** (a) A complex network including a load resistor R_L. (b) A Thévenin equivalent network connected to the load resistor R_L. (c) A Norton equivalent network connected to the load resistor R_L.

It should thus be apparent that one of the main uses of Thévenin's and Norton's theorems is the replacement of a large part of a circuit, often a complicated and uninteresting part, with a very simple equivalent. The new, simpler circuit enables us to make rapid calculations of the voltage, current, and power which the original circuit is able to deliver to a load. It also helps us to choose the best value of this load resistance. In a transistor power amplifier, for example, the Thévenin or Norton equivalent enables us to determine the maximum power that can be taken from the amplifier and delivered to the speakers.

EXAMPLE **5.6**

Consider the circuit shown in Fig. 5.25a on the next page. Determine the Thévenin equivalent of network A, and compute the power delivered to the load resistor R_L.

The dashed regions separate the circuit into networks A and B; our main interest is in network B, which consists only of the load resistor R_L. Network A may be simplified by making repeated source transformations.

(Continued on next page)

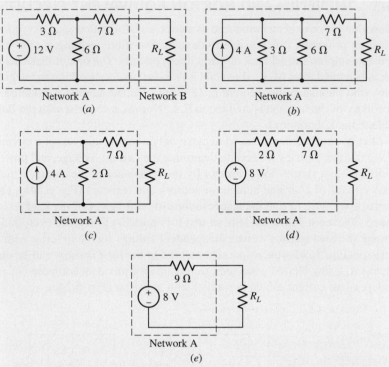

■ **FIGURE 5.25** (*a*) A circuit separated into two networks. (*b*)–(*d*) Intermediate steps to simplifying network *A*. (*e*) The Thévenin equivalent circuit.

We first treat the 12 V source and the 3 Ω resistor as a practical voltage source and replace it with a practical current source consisting of a 4 A source in parallel with 3 Ω (Fig. 5.25*b*). The parallel resistances are then combined into 2 Ω (Fig. 5.25*c*), and the practical current source that results is transformed back into a practical voltage source (Fig. 5.25*d*). The final result is shown in Fig. 5.25*e*.

From the viewpoint of the load resistor R_L, this network A (the Thévenin equivalent) is equivalent to the original network A; from our viewpoint, the circuit is much simpler, and we can now easily compute the power delivered to the load:

$$P_L = \left(\frac{8}{9 + R_L}\right)^2 R_L$$

Furthermore, we can see from the equivalent circuit that the maximum voltage that can be obtained across R_L is 8 V and corresponds to $R_L = \infty$. A quick transformation of network A to a practical current source (the Norton equivalent) indicates that the maximum current that may be delivered to the load is 8/9 A, which occurs when $R_L = 0$. Neither of these facts is readily apparent from the original circuit.

PRACTICE

5.5 Using repeated source transformations, determine the Norton equivalent of the highlighted network in the circuit of Fig. 5.26.

Ans: 1 A, 5 Ω.

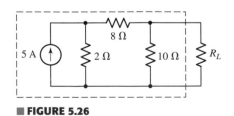

■ **FIGURE 5.26**

Thévenin's Theorem

Using the technique of source transformation to find a Thévenin or Norton equivalent network worked well enough in Example 5.6, but it can rapidly become impractical in situations where dependent sources are present or the circuit is composed of a large number of elements. An alternative is to employ Thévenin's theorem (or Norton's theorem) instead. We will state the theorem as a somewhat formal procedure and then proceed to consider various ways to make the approach more practical depending on the situation we face.

A Statement of Thévenin's Theorem[3]

1. **Given any linear circuit, rearrange it in the form of two networks, *A* and *B*, connected by two wires.** *A* is the network to be simplified; *B* will be left untouched.

2. **Disconnect network *B*.** Define a voltage v_{oc} as the voltage now appearing across the terminals of network *A*.

3. **Turn off or "zero out" every independent source in network *A* to form an inactive network.** Leave dependent sources unchanged.

4. **Connect an independent voltage source with value v_{oc} in series with the inactive network.** Do not complete the circuit; leave the two terminals disconnected.

5. **Connect network *B* to the terminals of the new network *A*.** All currents and voltages in *B* will remain unchanged.

Note that if either network contains a dependent source, *its control variable must be in the same network.*

Let us see if we can apply Thévenin's theorem successfully to the circuit we considered in Fig. 5.25. We have already found the Thévenin equivalent of the circuit to the left of R_L in Example 5.6, but we want to see if there is an easier way to obtain the same result.

EXAMPLE **5.7**

Use Thévenin's theorem to determine the Thévenin equivalent for that part of the circuit in Fig. 5.25*a* to the left of R_L.

We begin by disconnecting R_L, and note that no current flows through the 7 Ω resistor in the resulting partial circuit shown in Fig. 5.27*a* on the next page. Thus, V_{oc} appears across the 6 Ω resistor (with no current through the 7 Ω resistor there is no voltage drop across it), and voltage division enables us to determine that

$$V_{oc} = 12 \left(\frac{6}{3 + 6} \right) = 8 \text{ V}$$

(Continued on next page)

---

(3) A proof of Thévenin's theorem in the form in which we have stated it is rather lengthy, and therefore it has been placed in App. 3, where the curious may peruse it.

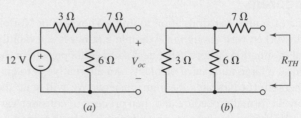

■ **FIGURE 5.27**   (a) The circuit of Fig. 5.25a with network B (the resistor R_L) disconnected and the voltage across the connecting terminals labeled as V_{oc}. (b) The independent source in Fig. 5.25a has been killed, and we look into the terminals where network B was connected to determine the effective resistance of network A.

Killing network A (i.e., replacing the 12 V source with a short circuit), we see looking back into the dead network a 7 Ω resistor connected in series with the parallel combination of 6 Ω and 3 Ω (Fig. 5.27b).

Thus, the dead network can be represented here by a 9 Ω resistor, referred to as the ***Thévenin equivalent resistance*** of network A. The Thévenin equivalent then is V_{oc} in series with a 9 Ω resistor, which agrees with our previous result.

PRACTICE

5.6 Use Thévenin's theorem to find the current through the 2 Ω resistor in the circuit of Fig. 5.28. (Hint: Designate the 2 Ω resistor as network B.)

Ans: $V_{TH} = 2.571$ V, $R_{TH} = 7.857$ Ω, $I_{2\Omega} = 260.8$ mA.

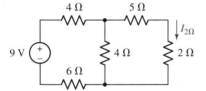

■ **FIGURE 5.28**

A Few Key Points

The equivalent circuit we have learned how to obtain is completely independent of network B, because we have been instructed first to remove network B and then measure the open-circuit voltage produced by network A, an operation that certainly does not depend on network B in any way. The B network is mentioned in the statement of the theorem only to indicate that an equivalent for A may be obtained no matter what arrangement of elements is connected to the A network; the B network represents this general network.

There are several points about the theorem which deserve emphasis.

- The only restriction that we must impose on A or B is that all *dependent* sources in A have their control variables in A, and similarly for B.

- No restrictions were imposed on the complexity of A or B; either one may contain any combination of independent voltage or current sources, linear dependent voltage or current sources, resistors, or any other circuit elements which are linear.

- The dead network A can be represented by a single equivalent resistance R_{TH}, which we will call the Thévenin equivalent resistance.

This holds true whether or not dependent sources exist in the dead *A* network, an idea we will explore shortly.

- A Thévenin equivalent consists of two components: a voltage source in series with a resistance. Either may be zero, although this is not usually the case.

Norton's Theorem

Norton's theorem bears a close resemblance to Thévenin's theorem and may be stated as follows:

A Statement of Norton's Theorem

1. **Given any linear circuit, rearrange it in the form of two networks, *A* and *B*, connected by two wires.** *A* is the network to be simplified; *B* will be left untouched. As before, if either network contains a dependent source, *its controlling variable must be in the same network.*

2. **Disconnect network *B*, and short the terminals of *A*.** Define a current i_{sc} as the current now flowing through the shorted terminals of network *A*.

3. **Turn off or "zero out" every independent source in network *A* to form an inactive network.** Leave dependent sources unchanged.

4. **Connect an independent current source with value i_{sc} in parallel with the inactive network.** Do not complete the circuit; leave the two terminals disconnected.

5. **Connect network *B* to the terminals of the new network *A*.** All currents and voltages in *B* will remain unchanged.

The Norton equivalent of a linear network is the Norton current source i_{sc} in parallel with the Thévenin resistance R_{TH}. Thus, we see that in fact it is possible to obtain the Norton equivalent of a network by performing a source transformation on the Thévenin equivalent. This results in a direct relationship between v_{oc}, i_{sc}, and R_{TH}:

$$v_{oc} = R_{TH}i_{sc} \qquad [18]$$

In circuits containing dependent sources, we will often find it more convenient to determine either the Thévenin or Norton equivalent by finding both the open-circuit voltage and the short-circuit current and then determining the value of R_{TH} as their quotient. It is therefore advisable to become adept at finding both open-circuit voltages and short-circuit currents, even in the simple problems that follow. If the Thévenin and Norton equivalents are determined independently, Eq. [18] can serve as a useful check.

Let us consider three different examples of the determination of a Thévenin or Norton equivalent circuit.

EXAMPLE 5.8

Find the Thévenin and Norton equivalent circuits for the network faced by the 1 kΩ resistor in Fig. 5.29a.

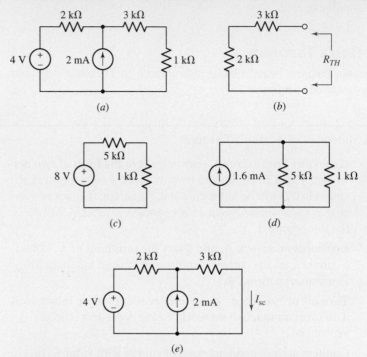

■ **FIGURE 5.29** (a) A given circuit in which the 1 kΩ resistor is identified as network B. (b) Network A with all independent sources killed. (c) The Thévenin equivalent is shown for network A. (d) The Norton equivalent is shown for network A. (e) Circuit for determining I_{sc}.

From the way the problem statement is worded, we know that network B is the 1 kΩ resistor, and network A is the remainder of the circuit. The circuit contains no dependent sources, and the easiest way to find the Thévenin equivalent is to determine R_{TH} for the dead network directly, followed by a calculation of either V_{oc} or I_{sc}.

We first determine the open-circuit voltage; in this case it is easily found by superposition. With only the 4 V source operating, the open-circuit voltage is 4 V; when only the 2 mA source is on, the open-circuit voltage is 2 mA × 2 kΩ = 4 V (no current flows through the 3 kΩ resistor with the 1 kΩ resistor disconnected). With both independent sources on, we see that $V_{oc} = 4 + 4 = 8$ V.

We next kill both independent sources to determine the form of the dead A network. With the 4 V source short-circuited and the 2 mA source open-circuited as in Fig. 5.29b, the result is the series combination of a 2 kΩ and a 3 kΩ resistor, or the equivalent, a 5 kΩ resistor.

This determines the Thévenin equivalent, shown in Fig. 5.29c, and from it the Norton equivalent of Fig. 5.29d can be drawn quickly. As a check, let us determine I_{sc} for the given circuit (Fig. 5.29e). We use superposition and a little current division:

$$I_{sc} = I_{sc}|_{4\,V} + I_{sc}|_{2\,mA} = \frac{4}{2+3} + (2)\frac{2}{2+3}$$
$$= 0.8 + 0.8 = 1.6 \text{ mA}$$

which completes the check.[4]

PRACTICE

5.7 Determine the Thévenin and Norton equivalents of the circuit of Fig. 5.30.

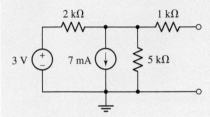

■ **FIGURE 5.30**

Ans: −7.857 V, −3.235 mA, 2.429 kΩ.

When Dependent Sources Are Present

Technically speaking, there does not always have to be a "network *B*" for us to invoke either Thévenin's theorem or Norton's theorem; we could instead be asked to find the equivalent of a network with two terminals not yet connected to another network. If there *is* a network *B* that we do not want to involve in the simplification procedure, however, we must use a little caution if it contains dependent sources. In such situations, the controlling variable and the associated element(s) must be included in network *B* and excluded from network *A*. Otherwise, there will be no way to analyze the final circuit because the controlling quantity will be lost.

If network *A* contains a dependent source, then again we must ensure that the controlling variable and its associated element(s) cannot be in network *B*. Up to now, we have only considered circuits with resistors and independent sources. Although technically speaking it is correct to leave a dependent source in the "dead" or "inactive" network when creating a Thévenin or Norton equivalent, in practice this does not result in any kind of simplification. What we really want is an independent voltage source in series with a single resistor, or an independent current source in parallel with a single resistor—in other words, a two-component equivalent. In the following examples, we consider various means of reducing networks with dependent sources and resistors into a single resistance.

(4) Note: If we use resistance in kΩ throughout our equation and voltage is expressed in volts, then the current will always automatically be in mA.

EXAMPLE **5.9**

Determine the Thévenin equivalent of the circuit in Fig. 5.31a.

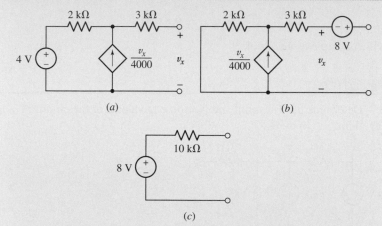

■ **FIGURE 5.31** (a) A given network whose Thévenin equivalent is desired. (b) A possible, but rather useless, form of the Thévenin equivalent. (c) The best form of the Thévenin equivalent for this linear resistive network.

To find V_{oc} we note that $v_x = V_{\mathrm{oc}}$ and that the dependent source current must pass through the 2 kΩ resistor, since no current can flow through the 3 kΩ resistor. Using KVL around the outer loop:

$$-4 + 2 \times 10^3 \left(-\frac{v_x}{4000}\right) + 3 \times 10^3(0) + v_x = 0$$

and

$$v_x = 8 \text{ V} = V_{\mathrm{oc}}$$

By Thévenin's theorem, then, the equivalent circuit could be formed with the dead A network in series with an 8 V source, as shown in Fig. 5.31b. This is correct, but not very simple and not very helpful; in the case of linear resistive networks, we should certainly show a much simpler equivalent for the inactive A network, namely, R_{TH}.

The presence of the dependent source prevents us from determining R_{TH} directly for the inactive network through resistance combination; we therefore seek I_{sc}. Upon short-circuiting the output terminals in Fig. 5.31a, it is apparent that $V_x = 0$ and the dependent current source is dead. Hence, $I_{\mathrm{sc}} = 4/(5 \times 10^3) = 0.8$ mA. Thus,

$$R_{TH} = \frac{V_{\mathrm{oc}}}{I_{\mathrm{sc}}} = \frac{8}{(0.8 \times 10^{-3})} = 10 \text{ k}\Omega$$

and the acceptable Thévenin equivalent of Fig. 5.31c is obtained.

PRACTICE

5.8 Find the Thévenin equivalent for the network of Fig. 5.32. (Hint: A quick source transformation on the dependent source might help.)

Ans: -502.5 mV, -100.5 Ω.

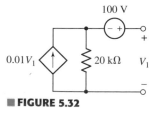

■ **FIGURE 5.32**

As our final example, let us consider a network having a dependent source but no independent source.

EXAMPLE **5.10**

Find the Thévenin equivalent of the circuit shown in Fig. 5.33a.

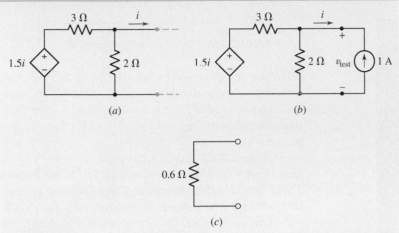

■ FIGURE 5.33 (*a*) A network with no independent sources. (*b*) A hypothetical measurement to obtain R_{TH}. (*c*) The Thévenin equivalent to the original circuit.

Since the rightmost terminals are already open-circuited, $i = 0$. Consequently, the dependent source is dead, so $v_{oc} = 0$.

We next seek the value of R_{TH} represented by this two-terminal network. However, we cannot find v_{oc} and i_{sc} and take their quotient, for there is no independent source in the network and both v_{oc} and i_{sc} are zero. Let us, therefore, be a little tricky.

We apply a 1 A source externally, measure the voltage v_{test} that results, and then set $R_{TH} = v_{test}/1$. Referring to Fig. 5.33*b*, we see that $i = -1$ A. Applying nodal analysis,

$$\frac{v_{test} - 1.5(-1)}{3} + \frac{v_{test}}{2} = 1$$

so that

$$v_{test} = 0.6 \text{ V}$$

and thus

$$R_{TH} = 0.6 \ \Omega$$

The Thévenin equivalent is shown in Fig. 5.33*c*.

A Quick Recap of Procedures

We have now looked at three examples in which we determined a Thévenin or Norton equivalent circuit. The first example (Fig. 5.29) contained only independent sources and resistors, and several different methods could have been applied to it. One would involve calculating R_{TH} for the dead network and then V_{oc} for the live network. We could also have found R_{TH} and I_{sc}, or V_{oc} and I_{sc}.

The Digital Multimeter

One of the most common pieces of electrical test equipment is the DMM, or digital multimeter (Fig. 5.34), which is designed to measure voltage, current, and resistance values.

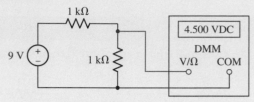

■ **FIGURE 5.35** A DMM connected to measure voltage.

reference terminal—often referred to as the *common terminal*—is typically designated by "COM." The typical convention is to use a red-colored lead for the positive reference terminal and a black lead for the common terminal.

From our discussion of Thévenin and Norton equivalents, it may now be apparent that the DMM has its own Thévenin equivalent resistance. This Thévenin equivalent resistance will appear in parallel with our circuit, and its value can affect the measurement (Fig. 5.36). The DMM does not supply power to the circuit to measure voltage, so its Thévenin equivalent consists of only a resistance, which we will name R_{DMM}.

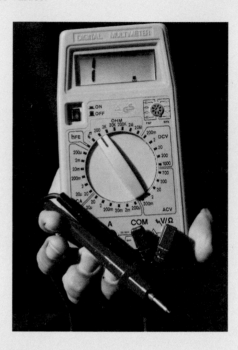

■ **FIGURE 5.34** A handheld digital multimeter.

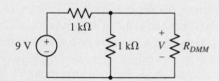

■ **FIGURE 5.36** DMM in Fig. 5.35 shown as its Thévenin equivalent resistance, R_{DMM}.

In a voltage measurement, two leads from the DMM are connected across the appropriate circuit element, as depicted in Fig. 5.35. The positive reference terminal of the meter is typically marked "V/Ω," and the negative

The input resistance of a good DMM is typically 10 MΩ or more. The measured voltage V thus appears

In the second example (Fig. 5.31), both independent and dependent sources were present, and the method we used required us to find V_{oc} and I_{sc}. We could not easily find R_{TH} for the dead network because the dependent source could not be made inactive.

The last example did not contain any independent sources, and therefore the Thévenin and Norton equivalents do not contain an independent source. We found R_{TH} by applying 1 A and finding $v_{\text{test}} = 1 \times R_{TH}$. We could also apply 1 V and determine $i = 1/R_{TH}$. These two related techniques can be applied to any circuit with dependent sources, *as long as all independent sources are set to zero first*.

Two other methods have a certain appeal because they can be used for any of the three types of networks considered. In the first, simply replace network B with a voltage source v_s, define the current leaving its positive terminal as i, then analyze network A to obtain i, and put the equation in the form $v_s = ai + b$. Then, $a = R_{TH}$ and $b = v_{\text{oc}}$.

across 1 kΩ‖10 MΩ = 999.9 Ω. Using voltage division, we find that V = 4.4998 volts, slightly less than the expected value of 4.5 volts. Thus, the finite input resistance of the voltmeter introduces a small error in the measured value.

To measure currents, the DMM must be placed in series with a circuit element, generally requiring that we cut a wire (Fig. 5.37). One DMM lead is connected to the common terminal of the meter, and the other lead is placed in a connector usually marked "A" to signify current measurement. Again, the DMM does not supply power to the circuit in this type of measurement.

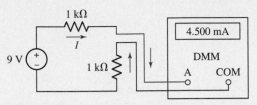

■ **FIGURE 5.37** A DMM connected to measure current.

We see from this figure that the Thévenin equivalent resistance (R_{DMM}) of the DMM is in series with our circuit, so its value can affect the measurement. Writing a simple KVL equation around the loop,

$$-9 + 1000I + R_{DMM}I + 1000I = 0$$

Note that since we have reconfigured the meter to perform a current measurement, the Thévenin equivalent resistance is not the same as when the meter is configured to measure voltages. In fact, we would ideally like R_{DMM} to be 0 Ω for current measurements, and ∞ for voltage measurements. If R_{DMM} is now 0.1 Ω, we see

that the measured current I is 4.4998 mA, which is only slightly different from the expected value of 4.5 mA. Depending on the number of digits that can be displayed by the meter, we may not even notice the effect of nonzero DMM resistance on our measurement.

The same meter can be used to determine resistance, provided no independent sources are active during the measurement. Internally, a known current is passed through the resistor being measured, and the voltmeter circuitry is used to measure the resulting voltage. Replacing the DMM with its Norton equivalent (which now includes an active independent current source to generate the predetermined current), we see that R_{DMM} appears in parallel with our unknown resistor R (Fig. 5.38).

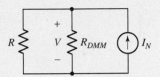

■ **FIGURE 5.38** DMM in resistance measurement configuration replaced by its Norton equivalent, showing R_{DMM} in parallel with the unknown resistor R to be measured.

As a result, the DMM actually measures $R‖R_{DMM}$. If R_{DMM} = 10 MΩ and R = 10 Ω, $R_{measured}$ = 9.99999 Ω, which is more than accurate enough for most purposes. However, if R = 10 MΩ, $R_{measured}$ = 5 MΩ. The input resistance of a DMM therefore places a practical upper limit on the values of resistance that can be measured, and special techniques must be used to measure larger resistances. We should note that if a digital multimeter is *programmed* with knowledge of R_{DMM}, it is possible to compensate and allow measurement of larger resistances.

We could also apply a current source i_s, let its voltage be v, and then determine $i_s = cv - d$, where $c = 1/R_{TH}$ and $d = i_{sc}$ (the minus sign arises from assuming both current source arrows are directed into the same node). Both of these last two procedures are universally applicable, but some other method can usually be found that is easier and more rapid.

Although we are devoting our attention almost entirely to the analysis of linear circuits, it is good to know that Thévenin's and Norton's theorems are both valid if network B is nonlinear; only network A must be linear.

PRACTICE

5.9 Find the Thévenin equivalent for the network of Fig. 5.39. (Hint: Try a 1 V test source.)

Ans: I_{test} = 50 mA so R_{TH} = 20 Ω.

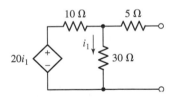

■ **FIGURE 5.39** See Practice Problem 5.9.

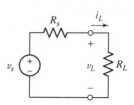

■ FIGURE 5.40 A practical voltage source connected to a load resistor R_L.

5.4 • MAXIMUM POWER TRANSFER

A very useful power theorem may be developed with reference to a practical voltage or current source. For the practical voltage source (Fig. 5.40), the power delivered to the load R_L is

$$p_L = i_L^2 R_L = \frac{v_s^2 R_L}{(R_s + R_L)^2} \qquad [19]$$

To find the value of R_L that absorbs a maximum power from the given practical source, we differentiate with respect to R_L:

$$\frac{d \, p_L}{d \, R_L} = \frac{(R_s + R_L)^2 v_s^2 - v_s^2 R_L (2)(R_s + R_L)}{(R_s + R_L)^4}$$

and equate the derivative to zero, obtaining

$$2R_L(R_s + R_L) = (R_s + R_L)^2$$

or

$$R_s = R_L$$

Since the values $R_L = 0$ and $R_L = \infty$ both give a minimum ($p_L = 0$), and since we have already developed the equivalence between practical voltage and current sources, we have therefore proved the following **maximum power transfer theorem:**

> An independent voltage source in series with a resistance R_s, or an independent current source in parallel with a resistance R_s, delivers a maximum power to that load resistance R_L for which $R_L = R_s$.

It may have occurred to the reader that an alternative way to view the maximum power theorem is possible in terms of the Thévenin equivalent resistance of a network:

> A network delivers the maximum power to a load resistance R_L when R_L is equal to the Thévenin equivalent resistance of the network.

Thus, the maximum power transfer theorem tells us that a 2 Ω resistor draws the greatest power (4.5 W) from either practical source of Fig. 5.16, whereas a resistance of 0.01 Ω receives the maximum power (3.6 kW) in Fig. 5.11.

There is a distinct difference between *drawing* maximum power from a *source* and *delivering* maximum power to a *load*. If the load is sized such that its Thévenin resistance is equal to the Thévenin resistance of the network to which it is connected, it will receive maximum power from that network. *Any change to the load resistance will reduce the power delivered to the load.* However, consider just the Thévenin equivalent of the network itself. We draw the maximum possible power from the voltage source by drawing the maximum possible current—which is achieved by shorting the network terminals! However, in this extreme example *we deliver zero power* to the "load"—a short circuit in this case—as $p = i^2 R$, and we just set $R = 0$ by shorting the network terminals.

A minor amount of algebra applied to Eq. [19] coupled with the maximum power transfer requirement that $R_L = R_s = R_{TH}$ will provide

$$p_{\max}|_{\text{delivered to load}} = \frac{v_s^2}{4R_s} = \frac{v_{TH}^2}{4R_{TH}}$$

where v_{TH} and R_{TH} recognize that the practical voltage source of Fig. 5.40 can also be viewed as a Thévenin equivalent of some specific source.

It is also not uncommon for the maximum power theorem to be misinterpreted. It is designed to help us select an optimum load in order to maximize power absorption. If the load resistance is already specified, however, the maximum power theorem is of no assistance. If for some reason we can affect the size of the Thévenin equivalent resistance of the network connected to our load, setting it equal to the load does not guarantee maximum power transfer to our predetermined load. A quick consideration of the power lost in the Thévenin resistance will clarify this point.

EXAMPLE 5.11

The circuit shown in Fig. 5.41 is a model for the common-emitter bipolar junction transistor amplifier. Choose a load resistance so that maximum power is transferred to it from the amplifier, and calculate the actual power absorbed.

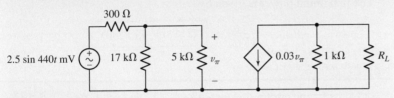

■ **FIGURE 5.41** A small-signal model of the common-emitter amplifier, with the load resistance unspecified.

Since it is the load resistance we are asked to determine, the maximum power theorem applies. The first step is to find the Thévenin equivalent of the rest of the circuit.

We first determine the Thévenin equivalent resistance, which requires that we remove R_L and short-circuit the independent source as in Fig. 5.42a.

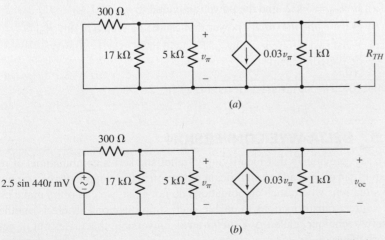

■ **FIGURE 5.42** (a) Circuit with R_L removed and independent source short-circuited. (b) Circuit for determining v_{TH}.

(Continued on next page)

Since $v_\pi = 0$, the dependent current source is an open circuit, so $R_{TH} = 1$ kΩ. This can be verified by connecting an independent 1 A current source across the 1 kΩ resistor; v_π will still be zero, so the dependent source remains inactive and hence contributes nothing to R_{TH}.

In order to obtain maximum power delivered into the load, R_L should be set to $\boxed{R_{TH} = 1 \text{ k}\Omega.}$

To find v_{TH} we consider the circuit shown in Fig. 5.42b, which is Fig. 5.41 with R_L removed. We may write

$$v_{oc} = -0.03v_\pi(1000) = -30v_\pi$$

where the voltage v_π may be found from simple voltage division:

$$v_\pi = (2.5 \times 10^{-3}\sin 440t)\left(\frac{3864}{300 + 3864}\right)$$

so that our Thévenin equivalent is a voltage $-69.6\sin 440t$ mV in series with 1 kΩ.

The maximum power is given by

$$p_{max} = \frac{v_{TH}^2}{4R_{TH}} = \boxed{1.211\sin^2 440t \ \mu\text{W}}$$

PRACTICE

5.10 Consider the circuit of Fig. 5.43.

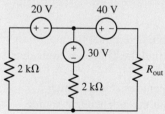

■ **FIGURE 5.43**

(a) If $R_{out} = 3$ kΩ, find the power delivered to it.

(b) What is the maximum power that can be delivered to any R_{out}?

(c) What two different values of R_{out} will have exactly 20 mW delivered to them?

Ans: 230 mW; 306 mW; 59.2 kΩ and 16.88 Ω.

5.5 DELTA-WYE CONVERSION

We saw previously that identifying parallel and series combinations of resistors can often lead to a significant reduction in the complexity of a circuit. In situations where such combinations do not exist, we can often make use of source transformations to enable such simplifications. There is another useful technique, called Δ-**Y (delta-wye)** conversion, that arises out of network theory.

Consider the circuits in Fig. 5.44. There are no series or parallel combinations that can be made to further simplify any of the circuits (note that 5.44a and 5.44b are identical, as are 5.44c and 5.44d), and without any

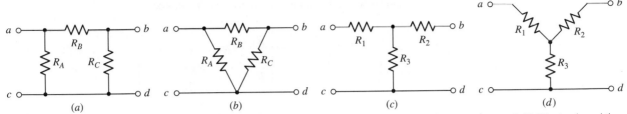

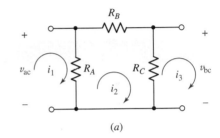

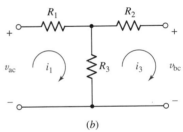

FIGURE 5.44 (*a*) Π network consisting of three resistors and three unique connections. (*b*) Same network drawn as a Δ network. (*c*) A T network consisting of three resistors. (*d*) Same network drawn as a Y network.

sources present, no source transformations can be performed. However, it is possible to convert between these two types of networks.

We first define two voltages v_{ab} and v_{cd}, and three currents i_1, i_2, and i_3 as depicted in Fig. 5.45. If the two networks are equivalent, then the terminal voltages and currents must be equal (there is no current i_2 in the T-connected network). A set of relationships between R_A, R_B, R_C and R_1, R_2, and R_3 can now be defined simply by performing mesh analysis. For example, for the network of Fig. 5.45*a* we may write

$$R_A i_1 - R_A i_2 \qquad\qquad = v_{ac} \qquad [20]$$

$$-R_A i_1 + (R_A + R_B + R_C)i_2 - R_C i_3 = 0 \qquad [21]$$

$$-R_C i_2 \qquad\qquad + R_C i_3 = -v_{bc} \qquad [22]$$

and for the network of Fig. 5.45*b* we have

$$(R_1 + R_3)i_1 - R_3 i_3 \qquad = v_{ac} \qquad [23]$$

$$-R_3 i_1 + (R_2 + R_3)i_3 \qquad = -v_{bc} \qquad [24]$$

We next remove i_2 from Eqs. [20] and [22] using Eq. [21], resulting in

$$\left(R_A - \frac{R_A^2}{R_A + R_B + R_C}\right)i_1 - \frac{R_A R_C}{R_A + R_B + R_C}i_3 = v_{ac} \qquad [25]$$

and

$$-\frac{R_A R_C}{R_A + R_B + R_C}i_1 + \left(R_C - \frac{R_C^2}{R_A + R_B + R_C}\right)i_3 = -v_{bc} \qquad [26]$$

Comparing terms between Eq. [25] and Eq. [23], we see that

$$R_3 = \frac{R_A R_C}{R_A + R_B + R_C}$$

In a similar fashion, we may find expressions for R_1 and R_2 in terms of R_A, R_B, and R_C, as well as expressions for R_A, R_B, and R_C in terms of R_1, R_2, and R_3; we leave the remainder of the derivations as an exercise for the reader. Thus, to convert from a Y network to a Δ network, the new resistor values are calculated using

$$R_A = \frac{R_1 R_2 + R_2 R_3 + R_3 R_1}{R_2}$$

$$R_B = \frac{R_1 R_2 + R_2 R_3 + R_3 R_1}{R_3}$$

$$R_C = \frac{R_1 R_2 + R_2 R_3 + R_3 R_1}{R_1}$$

FIGURE 5.45 (*a*) Labeled Π network; (*b*) labeled T network.

and to convert from a Δ network to a Y network,

$$
\begin{array}{l}
R_1 = \dfrac{R_A R_B}{R_A + R_B + R_C} \\[2ex]
R_2 = \dfrac{R_B R_C}{R_A + R_B + R_C} \\[2ex]
R_3 = \dfrac{R_C R_A}{R_A + R_B + R_C}
\end{array}
$$

Application of these equations is straightforward, although identifying the actual networks sometimes requires a little concentration.

EXAMPLE 5.12

Use the technique of Δ-Y conversion to find the Thévenin equivalent resistance of the circuit in Fig. 5.46a.

We see that the network in Fig. 5.46a is composed of two Δ-connected networks that share the 3 Ω resistor. We must be careful at this point not to be too eager, attempting to convert both Δ-connected networks to two Y-connected networks. The reason for this may be more obvious after we convert the top network consisting of the 1, 4, and 3 Ω resistors into a Y-connected network (Fig. 5.46b).

Note that in converting the upper network to a Y-connected network, we have removed the 3 Ω resistor. As a result, there is no way to convert the original Δ-connected network consisting of the 2, 5, and 3 Ω resistors into a Y-connected network.

We proceed by combining the $\frac{3}{8}$ Ω and 2 Ω resistors and the $\frac{3}{2}$ Ω and 5 Ω resistors (Fig. 5.46c). We now have a $\frac{19}{8}$ Ω resistor in parallel with a $\frac{13}{2}$ Ω resistor, and this parallel combination is in series with the $\frac{1}{2}$ Ω resistor. Thus, we can replace the original network of Fig. 5.46a with a single $\frac{159}{71}$ Ω resistor (Fig. 5.46d).

PRACTICE

5.11 Use the technique of Y-Δ conversion to find the Thévenin equivalent resistance of the circuit of Fig. 5.47.

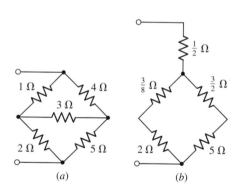

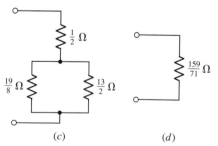

(a) (b) (c) (d)

■ FIGURE 5.46 (a) A given resistive network whose input resistance is desired. (b) The upper Δ network is replaced by an equivalent Y network. (c, d) Series and parallel combinations result in a single resistance value.

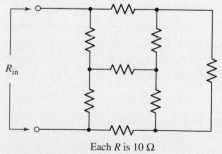

Each R is 10 Ω

■ FIGURE 5.47

Ans: 11.43 Ω.

5.6 • SELECTING AN APPROACH: A SUMMARY OF VARIOUS TECHNIQUES

In Chap. 3, we were introduced to Kirchhoff's current law (KCL) and Kirchhoff's voltage law (KVL). These two laws apply to any circuit we will ever encounter, provided that we take care to consider the entire system that the circuits represent. The reason for this is that KCL and KVL enforce charge and energy conservation, respectively, which are very fundamental principles. Based on KCL, we developed the very powerful method of nodal analysis. A similar technique based on KVL (unfortunately only applicable to planar circuits) is known as mesh analysis and is also a useful circuit analysis approach.

For the most part, this text is concerned with developing analytical skills that apply to *linear* circuits. If we know a circuit is constructed of only linear components (in other words, all voltages and currents are related by linear functions), then we can often simplify circuits prior to employing either mesh or nodal analysis. Perhaps the most important result that comes from the knowledge that we are dealing with a completely linear system is that the principle of superposition applies. Given a number of independent sources acting on our circuit, we can add the contribution of each source independently of the other sources. This technique is extremely pervasive throughout the field of engineering, and we will encounter it often. In many real situations, we will find that although several "sources" are acting simultaneously on our "system," typically one of them dominates the system response. Superposition allows us to quickly identify that source, provided that we have a reasonably accurate linear model of the system.

However, from a circuit analysis standpoint, unless we are asked to find which independent source contributes the most to a particular response, we find that rolling up our sleeves and launching straight into either nodal or mesh analysis is often a more straightforward tactic. The reason for this is that applying superposition to a circuit with 12 independent sources will require us to redraw the original circuit 12 times, and often we will have to apply nodal or mesh analysis to each partial circuit, anyway.

The technique of source transformations, however, is often a very useful tool in circuit analysis. Performing source transformations can allow us to consolidate resistors or sources that are not in series or parallel in the original circuit. Source transformations may also allow us to convert all or at least most of the sources in the original circuit to the same type (either all voltage sources or all current sources), so nodal or mesh analysis is more straightforward.

Thévenin's theorem is extremely important for a number of reasons. In working with electronic circuits, we are always aware of the Thévenin equivalent resistance of different parts of our circuit, especially the input and output resistances of amplifier stages. The reason for this is that matching of resistances is frequently the best route to optimizing the performance of a given circuit. We have seen a small preview of this in our discussion of maximum power transfer, where the load resistance should be chosen to match the Thévenin equivalent resistance of the network to which the load is connected. In terms of day-to-day circuit analysis, however, we find that converting part of a circuit to its Thévenin or Norton equivalent is almost as much work as analyzing the complete circuit. Therefore, as in the case of superposition, Thévenin's and Norton's theorems are typically applied only when we require specialized information about part of our circuit.

SUMMARY AND REVIEW

- ❏ The principle of superposition states that the *response* in a linear circuit can be obtained by adding the individual responses caused by the separate *independent* sources *acting alone*.

- ❏ Superposition is most often used when it is necessary to determine the individual contribution of each source to a particular response.

- ❏ A practical model for a real voltage source is a resistor in series with an independent voltage source. A practical model for a real current source is a resistor in parallel with an independent current source.

- ❏ Source transformations allow us to convert a practical voltage source into a practical current source, and vice versa.

- ❏ Repeated source transformations can greatly simplify analysis of a circuit by providing the means to combine resistors and sources.

- ❏ The Thévenin equivalent of a network is a resistor in series with an independent voltage source. The Norton equivalent is the same resistor in parallel with an independent current source.

- ❏ There are several ways to obtain the Thévenin equivalent resistance, depending on whether or not dependent sources are present in the network.

- ❏ Maximum power transfer occurs when the load resistor matches the Thévenin equivalent resistance of the network to which it is connected.

- ❏ When faced with a Δ-connected resistor network, it is straightforward to convert it to a Y-connected network. This can be useful in simplifying the network prior to analysis. Conversely, a Y-connected resistor network can be converted to a Δ-connected network to assist in simplification of the network.

READING FURTHER

A book about battery technology, including characteristics of built-in resistance:

D. Linden, *Handbook of Batteries,* 2nd ed. New York: McGraw-Hill, 1995.

An excellent discussion of pathological cases and various circuit analysis theorems can be found in:

R. A. DeCarlo and P. M. Lin, *Linear Circuit Analysis,* 2nd ed. New York: Oxford University Press, 2001.

EXERCISES

5.1 Linearity and Superposition

1. The concept of linearity is very important, as linear systems are much more easily analyzed than nonlinear systems. Unfortunately, most practical systems that we encounter are nonlinear in nature. It is possible, however, to create a linear model for a nonlinear system that is valid over a small range of the controlling variable. As an example of this, consider the simple exponential

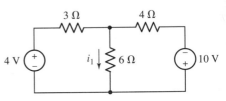

function e^x. The Taylor series representation of this function is

$$e^x \approx 1 + x + \frac{x^2}{2} + \frac{x^3}{6} + \cdots$$

Construct a linear model of this function by truncating it after the linear term (x^1). Evaluate your new function at $x = 0.001, 0.005, 0.01, 0.05, 0.10, 0.5$, 1.0, and 5.0. For which values of x does the linear model give a "reasonable" approximation to e^x?

2. In the circuit of Fig. 5.48, (a) determine the contribution of the 4 V source to the current labeled i_1; (b) determine the contribution of the 10 V source to i_1; and (c) determine i_1.

■ **FIGURE 5.48**

3. Referring to the two-source circuit depicted in Fig. 5.49, determine the contribution of the 1 A source to v_1, and calculate the total current flowing through the 7 Ω resistor.

4. Employ the principle of superposition to determine the current labeled i_y in the circuit of Fig. 5.50 by considering each source individually.

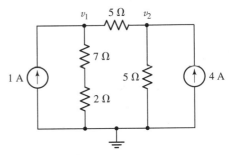

■ **FIGURE 5.49**

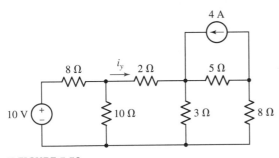

■ **FIGURE 5.50**

5. For the circuit shown in Fig. 5.48, change only the value of the sources to obtain a factor of 10 increase in the current i_1; *both* source values must be changed and *neither* may be set to zero.

6. With sources i_A and v_B on in the circuit of Fig. 5.51 and $v_C = 0$, $i_x = 20$ A; with i_A and v_C on and $v_B = 0$, $i_x = -5$ A; and finally, with all three sources on, $i_x = 12$ A. Find i_x if the only source operating is (a) i_A; (b) v_B; (c) v_C. (d) Find i_x if i_A and v_C are doubled in amplitude and v_B is reversed.

7. Use superposition to find the value of v_x in the circuit of Fig. 5.52.

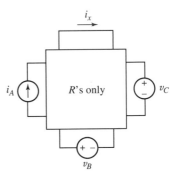

■ **FIGURE 5.51**

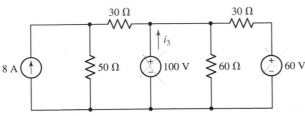

■ **FIGURE 5.52**

8. Apply superposition to the circuit of Fig. 5.53 to find i_3.

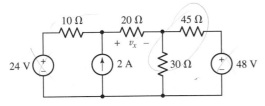

■ **FIGURE 5.53**

9. (*a*) Use the superposition theorem to find i_2 in the circuit shown in Fig. 5.54. (*b*) Calculate the power absorbed by each of the five circuit elements.

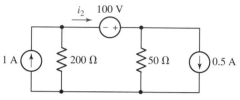

■ **FIGURE 5.54**

10. Use superposition on the circuit shown in Fig. 5.55 to find the voltage v. Note that there is a dependent source present.

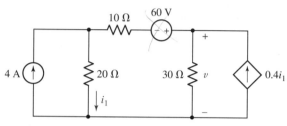

■ **FIGURE 5.55**

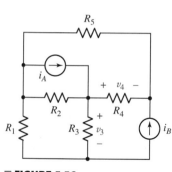

■ **FIGURE 5.56**

11. In the circuit shown in Fig. 5.56: (*a*) if $i_A = 10$ A and $i_B = 0$, then $v_3 = 80$ V; find v_3 if $i_A = 25$ A and $i_B = 0$. (*b*) If $i_A = 10$ A and $i_B = 25$ A, then $v_4 = 100$ V, while $v_4 = -50$ V if $i_A = 25$ A and $i_B = 10$ A; find v_4 if $i_A = 20$ A and $i_B = -10$ A.

12. Use superposition to determine the voltage across the current source in Fig. 5.57.

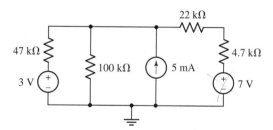

■ **FIGURE 5.57**

13. Use superposition to find the power dissipated by the 500 kΩ resistor in Fig. 5.58.

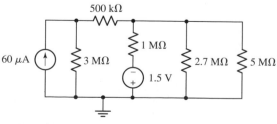

■ **FIGURE 5.58**

14. Employ superposition to determine the voltage across the 17 kΩ resistor in Fig. 5.59. If the maximum power rating of the resistor is 250 mW, what is the maximum positive voltage to which the 5 V source can be increased before the resistor overheats?

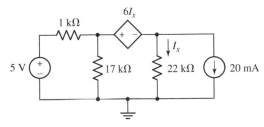

■ **FIGURE 5.59**

15. Which source in Fig. 5.60 contributes the most to the power dissipated in the 2 Ω resistor? The least? What *is* the power dissipated in the 2 Ω resistor?

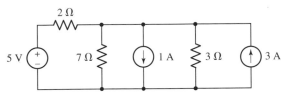

■ **FIGURE 5.60**

16. Use superposition to find i_B in the circuit of Fig. 5.61, which is a commonly used model circuit for a bipolar junction transistor amplifier.

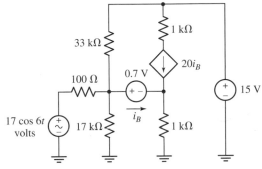

■ **FIGURE 5.61**

 17. For the circuit shown in Fig. 5.62,
   (*a*) Use superposition to compute V_x.
   (*b*) Verify the contribution of each source to V_x using a dc sweep PSpice analysis. Submit a labeled schematic, relevant probe output, and a brief summary of the results.

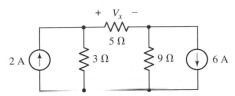

■ **FIGURE 5.62**

18. For the circuit shown in Fig. 5.63,
   (a) Use superposition to compute V_x.
   (b) Verify the contribution of each source to V_x using a dc sweep PSpice analysis. Submit a labeled schematic, relevant probe output, and a brief summary of the results.

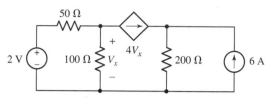

■ **FIGURE 5.63**

19. Consider the three circuits shown in Fig. 5.64. Analyze each circuit, and demonstrate that $V_x = V_x' + V_x''$ (i.e., *superposition is most useful when sources are set to zero, but the principle is in fact much more general than that*).

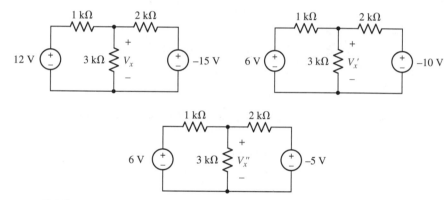

■ **FIGURE 5.64**

5.2 Source Transformations

20. With the assistance of the method of source transformations, (a) convert the circuit of Fig. 5.65a to a single independent voltage source in series with an appropriately sized resistor; and (b) convert the circuit of Fig. 5.65b to a single independent current source in parallel with an appropriately sized resistor. For both (a) and (b), leave the right-hand terminals in your final circuit.

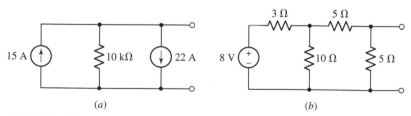

(a)   (b)

■ **FIGURE 5.65**

21. (a) Use the method of source transformations to reduce the circuit of Fig. 5.66 to a practical voltage source in series with the 10 Ω resistor.
   (b) Calculate v. (c) Explain why the 10 Ω resistor should not be included in a source transformation.

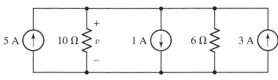

■ FIGURE 5.66

22. Use source transformations and resistance combinations to simplify both of the networks of Fig. 5.67 until only two elements remain to the left of terminals *a* and *b*.

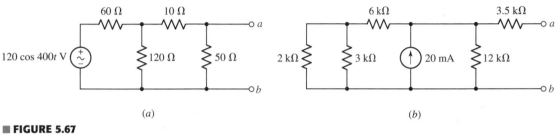

(a)   (b)

■ FIGURE 5.67

23. Using source transformation to first simplify the circuit, determine the power dissipated by the 5.8 kΩ resistor in Fig. 5.68.

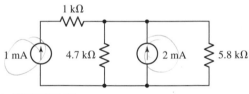

■ FIGURE 5.68

24. Using source transformation, determine the power dissipated by the 5.8 kΩ resistor in Fig. 5.69.

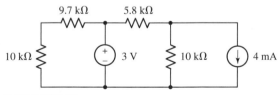

■ FIGURE 5.69

25. Determine the power dissipated by the 1 MΩ resistor using source transformation to first simplify the circuit shown in Fig. 5.70.

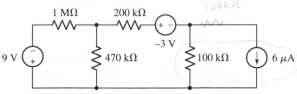

■ FIGURE 5.70

26. Determine I_1 using source transformation to first simplify the circuit of Fig. 5.71.

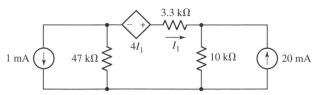

■ **FIGURE 5.71**

27. (a) Find V_1 in the circuit of Fig. 5.72 using source transformation to obtain a simplified equivalent circuit first.

(b) Verify your analysis by performing a PSpice analysis of the circuit in Fig. 5.72. Submit a schematic with V_1 clearly labeled.

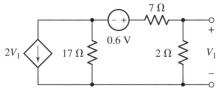

■ **FIGURE 5.72**

28. (a) Use repeated source transformation to determine the current I_x as indicated in Fig. 5.73.

(b) Verify your analysis by performing a PSpice analysis of the circuit in Fig. 5.73. Submit a schematic with I_X clearly labeled.

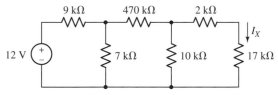

■ **FIGURE 5.73**

29. Use repeated source transformation to determine the current I_X in the circuit of Fig. 5.74.

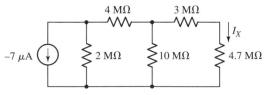

■ **FIGURE 5.74**

30. Convert the circuit in Fig. 5.75 to a single current source in parallel with a single resistor.

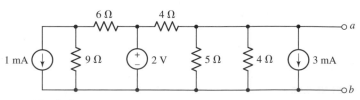

■ **FIGURE 5.75**

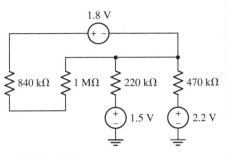

31. Use source transformation to convert the circuit in Fig. 5.76 to a single current source in parallel with a single resistor.

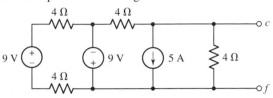

■ **FIGURE 5.76**

32. Determine the power dissipated by the 1 MΩ resistor in the circuit of Fig. 5.77.

■ **FIGURE 5.77**

33. The measurements in Table 5.1 are made on a 1.5 V alkaline battery. Use the information to construct a simple two-component practical voltage source model for the battery that is relatively accurate for currents in the range of 1 to 20 mA. Note that besides the obvious experimental error, the "internal resistance" of the battery is significantly different over the current range measured in the experiment.

TABLE **5.1** Measured current-voltage characteristics of a 1.5 V alkaline battery connected to a variable load resistance

Current Output (mA)	Terminal Voltage (V)
0.0000589	1.584
0.3183	1.582
1.4398	1.567
7.010	1.563
12.58	1.558

34. Use the data in Table 5.1 to construct a simple two-component practical current source model for the battery that is relatively accurate for currents in the range of 1 to 20 mA. Note that besides the obvious experimental error, the "internal resistance" of the battery is significantly different over the current range measured in the experiment.

35. Reduce the circuit in Fig. 5.78 to a single voltage source in series with a single resistor. Leave the right-hand terminals in your final circuit.

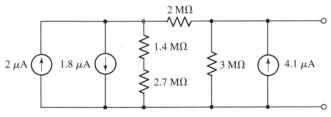

■ **FIGURE 5.78**

36. Find the power absorbed by the 5 Ω resistor in Fig. 5.79.

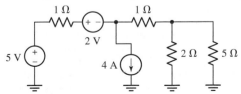

■ **FIGURE 5.79**

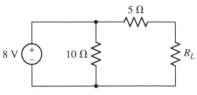

■ FIGURE 5.80

37. (*a*) Convert the circuit of Fig. 5.80 to a practical current source in parallel with R_L. (*b*) Verify your answer using PSpice and a value of 5 Ω for R_L. Submit properly labeled schematics for each circuit, with the voltage across R_L clearly identified.

38. (*a*) Reduce the circuit of Fig. 5.81 as much as possible, transform the two voltage sources into current sources, then compute the power dissipated in the top 5 Ω resistor. (*b*) Verify your answer by simulating both circuits with PSpice. Submit a properly labeled schematic for each circuit, with the power dissipated in the resistor of interest clearly identified. (*c*) Does the value of the 1 Ω or the 7 Ω resistor affect your answer in any way? Explain why or why not.

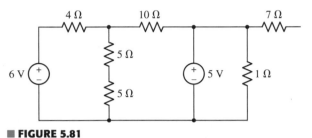

■ FIGURE 5.81

39. For the circuit of Fig. 5.82, convert all sources (both dependent and independent) to current sources, combine the dependent sources, and calculate the voltage v_3.

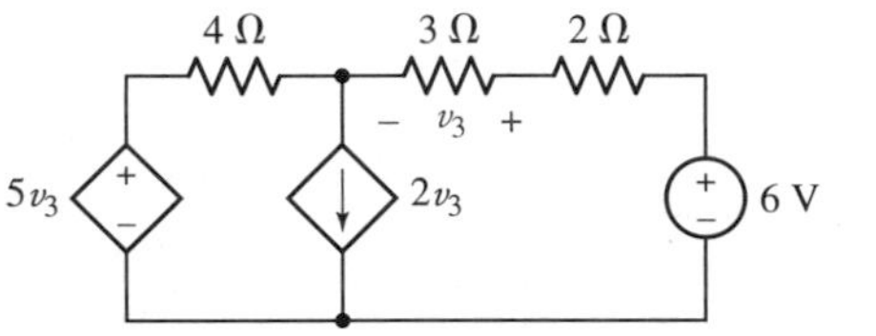

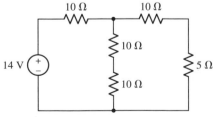

■ FIGURE 5.82

5.3 Thévenin and Norton Equivalent Circuits

40. (*a*) Find the Thévenin equivalent at terminals *a* and *b* for the network shown in Fig. 5.83. How much power would be delivered to a resistor connected to *a* and *b* if R_{ab} equals (*b*) 50 Ω; (*c*) 12.5 Ω?

41. (*a*) Employ Thévenin's theorem to simplify the network connected to the 5 Ω resistor of Fig. 5.84. (*b*) Compute the power absorbed by the 5 Ω resistor using your simplified circuit. (*c*) Verify your solution with PSpice. Submit a properly labeled schematic for each circuit with the requested power quantity clearly identified.

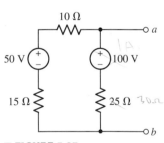

■ FIGURE 5.83

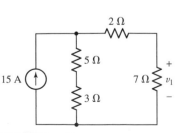

■ FIGURE 5.85

■ FIGURE 5.84

42. (*a*) Find the Thévenin equivalent of the network connected to the 7 Ω resistor of Fig. 5.85. (*b*) Find the Norton equivalent of the network connected to the 7 Ω resistor of Fig. 5.85. (*c*) Compute the voltage v_1 using both of your equivalent circuits. (*d*) Replace the 7 Ω resistor with a 1 Ω resistor, and recompute v_1 using either circuit.

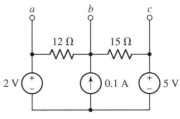

43. (*a*) A tungsten-filament light bulb is connected to a 10 mV test voltage, and a current of 400 μA is measured. What is the Thévenin equivalent of the light bulb? (*b*) The bulb is now connected to a 110 V supply, and a current of 363.6 mA is measured. Determine the Thévenin equivalent based on this measurement. (*c*) Why is the Thévenin equivalent of the light bulb apparently dependent on the test conditions, and what implications does this have if we need to analyze a circuit containing the bulb?

44. (*a*) Find both the Thévenin and Norton equivalents for the network connected to the 1 Ω resistor of Fig. 5.86. (*b*) Compute the power absorbed by the 1 Ω resistor using both equivalent circuits. (*c*) Verify using PSpice. Submit a properly labeled schematic for each of the three circuits, with the requested power quantity clearly identified.

45. For the network of Fig. 5.87: (*a*) remove terminal *c* and find the Norton equivalent seen at terminals *a* and *b*; (*b*) repeat for terminals *b* and *c* with *a* removed.

46. Find the Thévenin equivalent of the network in Fig. 5.88 as viewed from terminals: (*a*) *x* and *x'*; (*b*) *y* and *y'*.

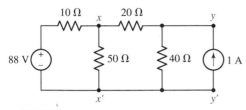

■ FIGURE 5.88

■ FIGURE 5.86

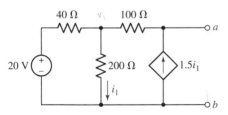

■ FIGURE 5.87

47. (*a*) Find the Thévenin equivalent of the network shown in Fig. 5.89. (*b*) What power would be delivered to a load of 100 Ω at *a* and *b*?

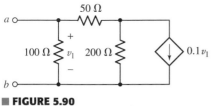

■ FIGURE 5.89

48. Find the Norton equivalent of the network shown in Fig. 5.90.

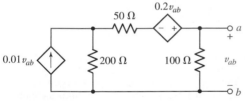

■ FIGURE 5.90

49. Find the Thévenin equivalent of the two-terminal network shown in Fig. 5.91.

■ FIGURE 5.91

50. Find the Thévenin equivalent of the circuit in Fig. 5.92.

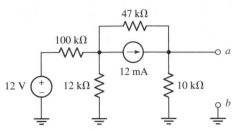

■ **FIGURE 5.92**

51. For the network in Fig. 5.93, determine: (*a*) the Thévenin equivalent; (*b*) the Norton equivalent.

52. For the circuit in Fig. 5.94*a*, determine the Norton equivalent of the network connected to R_L. For the circuit in Fig. 5.94*b*, determine the Thévenin equivalent of the network connected to R_L.

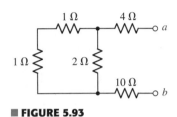

■ **FIGURE 5.93**

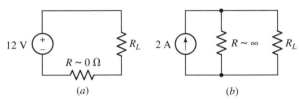

■ **FIGURE 5.94**

53. Determine the Thévenin and Norton equivalents of the network shown in Fig. 5.95.

54. Determine the Thévenin and Norton equivalents of the network shown in Fig. 5.96.

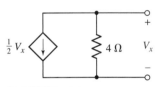

■ **FIGURE 5.95**

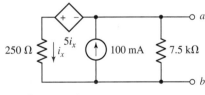

■ **FIGURE 5.96**

55. Determine the Thévenin and Norton equivalents of the network shown in Fig. 5.97.

56. Find the Thévenin equivalent resistance seen by the 2 kΩ resistor in the circuit of Fig. 5.98. Ignore the dashed line in the figure.

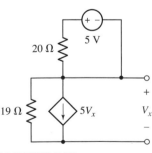

■ **FIGURE 5.97**

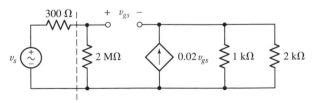

■ **FIGURE 5.98**

57. Referring to the circuit of Fig. 5.98, determine the Thévenin equivalent resistance of the circuit to the right of the dashed line. This circuit is a common-source transistor amplifier, and you are calculating its input resistance.

58. Referring to the circuit of Fig. 5.99, determine the Thévenin equivalent resistance of the circuit to the right of the dashed line. This circuit is a common-collector transistor amplifier, and you are calculating its input resistance.

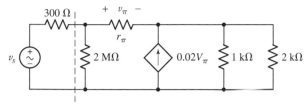

■ **FIGURE 5.99**

59. The circuit shown in Fig. 5.100 is a reasonably accurate model of an operational amplifier. In cases where R_i and A are very large and $R_o \sim 0$, a resistive load (such as a speaker) connected between ground and the terminal labeled v_{out} will see a voltage $-R_f/R_1$ times larger than the input signal v_{in}. Find the Thévenin equivalent of the circuit, taking care to label v_{out}.

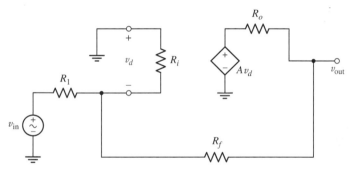

■ **FIGURE 5.100**

5.4 Maximum Power Transfer

60. Assuming that we can determine the Thévenin equivalent resistance of our wall socket, why don't toaster, microwave oven, and TV manufacturers match each appliance's Thévenin equivalent resistance to this value? Wouldn't it permit maximum power transfer from the utility company to our household appliances?

61. If any value whatsoever may be selected for R_L in the circuit of Fig. 5.101, what is the maximum power that could be dissipated in R_L?

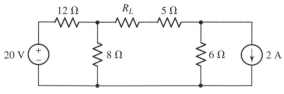

■ **FIGURE 5.101**

62. (a) Find the Thévenin equivalent at terminals a and b for the network shown in Fig. 5.102. How much power would be delivered to a resistor connected to a and b if R_{ab} equals (b) 10 Ω; (c) 75 Ω?

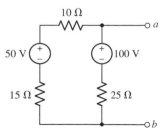

■ **FIGURE 5.102**

63. (*a*) Determine the Thévenin equivalent of the network shown in Fig. 5.103, and (*b*) find the maximum power that can be drawn from it.

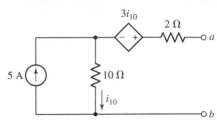

■ **FIGURE 5.103**

64. With reference to the circuit of Fig. 5.104: (*a*) determine that value of R_L to which a maximum power can be delivered, and (*b*) calculate the voltage across R_L then (+ reference at top).

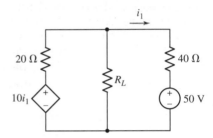

■ **FIGURE 5.104**

65. A certain practical dc voltage source can provide a current of 2.5 A when it is (momentarily) short-circuited, and can provide a power of 80 W to a 20 Ω load. Find (*a*) the open-circuit voltage and (*b*) the maximum power it could deliver to a well-chosen R_L. (*c*) What is the value of that R_L?

66. A practical current source provides 10 W to a 250 Ω load and 20 W to an 80 Ω load. A resistance R_L, with voltage v_L and current i_L, is connected to it. Find the values of R_L, v_L, and i_L if (*a*) $v_L i_L$ is a maximum; (*b*) v_L is a maximum; (*c*) i_L is a maximum.

67. A certain battery can accurately be modeled as a 9 V independent source in series with a 1.2 Ω resistor over the current range of interest. No current flows if an infinite resistance load is connected to the battery. We also know that maximum power will be transferred to a resistor of 1.2 Ω, and less power transferred to either a 1.1 Ω or 1.3 Ω resistor. However, if we simply short the terminals of the battery together (not recommended!), we will obtain *much more* current than for a 1.2 Ω resistive load. Doesn't this conflict with what we derived previously for maximum power transfer (*after all, isn't power proportional to i^2*)? Explain.

68. The circuit in Fig. 5.105 is part of an audio amplifier. If we want to transfer maximum power to the 8 Ω speaker, what value of R_E is needed? Verify your solution with PSpice.

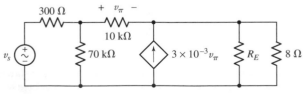

■ **FIGURE 5.105**

69. The circuit shown in Fig. 5.106 depicts a circuit separated into two stages. Select R_1 so that maximum power is transferred from stage 1 to stage 2.

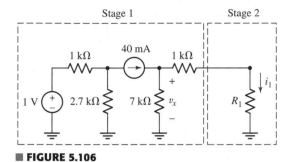

FIGURE 5.106

5.5 Delta-Wye Conversion

70. Convert the network in Fig. 5.107 to a Y-connected network.

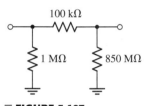

FIGURE 5.107

71. Convert the network in Fig. 5.108 to a △-connected network.
72. Find R_{in} for the network shown in Fig. 5.109.

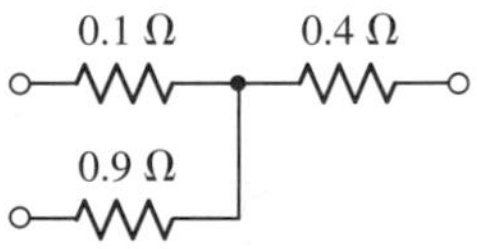

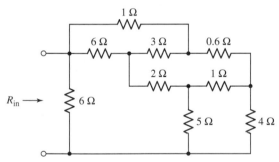

FIGURE 5.109

FIGURE 5.108

73. Use Y-△ and △-Y transformations to find the input resistance of the network shown in Fig. 5.110.

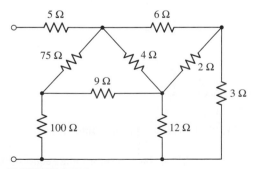

FIGURE 5.110

74. Find R_{in} for the circuit of Fig. 5.111.

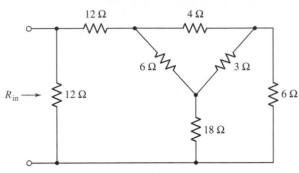

■ **FIGURE 5.111**

75. Find the Thévenin equivalent of the circuit in Fig. 5.112.

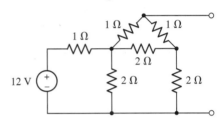

■ **FIGURE 5.112**

76. Find the Norton equivalent of the circuit in Fig. 5.113.

77. If all resistors in Fig. 5.114 are 10 Ω, determine the Thévenin equivalent for the circuit.

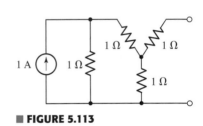

■ **FIGURE 5.113**

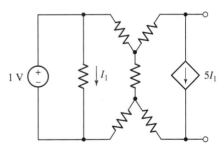

■ **FIGURE 5.114**

78. (a) Replace the network in Fig. 5.115 with an equivalent three-resistor Y network.

(b) Perform a PSpice analysis to verify that your answer is in fact equivalent. (Hint: Try adding a load resistor.)

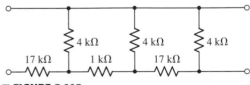

■ **FIGURE 5.115**

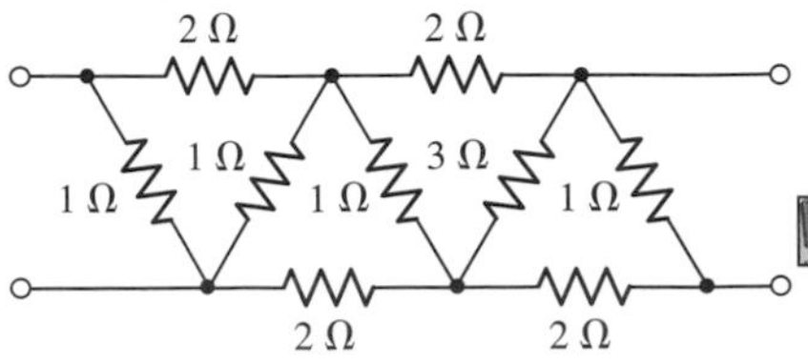

79. (a) Replace the network in Fig. 5.116 with an equivalent three-resistor Δ network.

(b) Perform a PSpice analysis to verify that your answer is in fact equivalent. (Hint: Try adding a load resistor.)

■ **FIGURE 5.116**

EXERCISES

171

5.6 Selecting an Approach: A Summary of Various Techniques

80. The circuit shown in Fig. 5.117 is a reasonably accurate model for a bipolar junction transistor operating in what is known as the *forward active region*. Determine the collector current I_C. Verify your answer with PSpice.

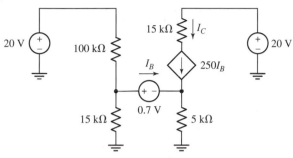

■ **FIGURE 5.117**

81. The load resistor in Fig. 5.118 can safely dissipate up to 1 W before overheating and bursting into flame. The lamp can be treated as a 10.6 Ω resistor if less than 1 A flows through it and a 15 Ω resistor if more than 1 A flows through it. What is the maximum permissible value of I_S? Verify your answer with PSpice.

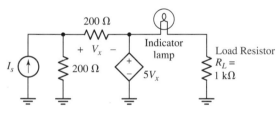

■ **FIGURE 5.118**

82. The human ear can detect sound waves in the frequency range of about 20 Hz to 20 kHz. If each 8 Ω resistor in Fig. 5.119 is a loudspeaker, which of the signal generators (modeled as practical voltage sources) produces the most sound? (Take "loudness" as proportional to power delivered to a speaker.)

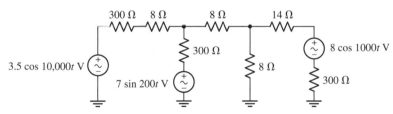

■ **FIGURE 5.119**

83. A DMM is connected to a resistor circuit as shown in Fig. 5.120. If the input resistance of the DMM is 1 MΩ, what value will be displayed if the DMM is measuring resistance?

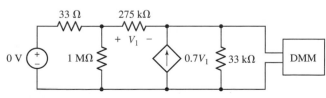

■ **FIGURE 5.120**

84. A metallic substance is extracted from a meteorite found in rural Indiana. The substance is found to have a resistivity of 50 Ω · cm and is fabricated into a simple cylinder. The cylinder is connected into the circuit of Fig. 5.121 and is found to have a temperature dependence of $T = 200P^{0.25}$ °C, where P is the power delivered to the cylinder in watts. Interestingly enough, the resistivity of the substance does not appear to depend on temperature. If $R = 10$ Ω and is absorbing maximum power in the circuit shown, what is the temperature of the cylinder?

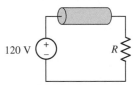

■ **FIGURE 5.121**

Ⓓ 85. As part of a security system, a very thin 100 Ω wire is attached to a window using nonconducting epoxy. Given only a box of twelve rechargeable 1.5 V AAA batteries, one thousand 1 Ω resistors, and a 2900 Hz piezo buzzer that draws 15 mA at 6 V, design a circuit with no moving parts that will set off the buzzer if the window is broken (and hence the thin wire as well). Note that the buzzer requires a dc voltage of at least 6 V (maximum 28 V) to operate.

Ⓓ 86. Three 45 W light bulbs originally wired in a Y network configuration with a 120 V ac source connected across each port are rewired as a Δ network. The neutral, or center, connection is not used. If the intensity of each light is proportional to the power it draws, design a new 120 V ac power circuit so that the three lights have the same intensity in the Δ configuration as they did when connected in a Y configuration. Verify your design using PSpice by comparing the power drawn by each light in your circuit (modeled as an appropriately chosen resistor value) with the power each would draw in the original Y-connected circuit.

Ⓓ 87. A certain red LED has a maximum current rating of 35 mA, and if this value is exceeded, overheating and catastrophic failure will result. The resistance of the LED is a nonlinear function of its current, but the manufacturer warrants a minimum resistance of 47 Ω and a maximum resistance of 117 Ω. Only 9 V batteries are available to power the LED. Design a suitable circuit to deliver the maximum power possible to the LED without damaging it. Use only combinations of the standard resistor values given in the inside front cover.

The Operational Amplifier

KEY CONCEPTS

Ideal Characteristics of
Op Amps

Inverting and Noninverting
Amplifiers

Summing and Difference
Amplifier Circuits

Cascaded Op Amp Stages

Using Op Amps to Build
Voltage and Current Sources

Nonideal Characteristics of
Op Amps

Voltage Gain and Feedback

Basic Comparator and
Instrumentation Amplifier
Circuits

INTRODUCTION

We have now been introduced to enough basic laws and circuit analysis techniques that we should be able to apply them successfully to some interesting practical circuits. In this chapter, we focus on a very useful everyday electrical device called the *operational amplifier,* or ***op amp*** for short.

6.1 BACKGROUND

The origins of the operational amplifier date back to the 1940s, when basic circuits were constructed using vacuum tubes to perform mathematical operations such as addition, subtraction, multiplication, division, differentiation, and integration. This enabled the construction of analog (as opposed to digital) computers tasked with the solution of complex differential equations. The first commercially available op amp *device* is generally considered to be the K2-W, manufactured by Philbrick Researches, Inc. of Boston from about 1952 through the early 1970s (Fig. 6.1a). These early vacuum tube devices weighed 3 oz (85 g), measured $1^{33}/_{64}"\times 2^{9}/_{64}"\times 4^{7}/_{64}"$ (3.8 cm $\times$ 5.4 cm $\times$ 10.4 cm), and sold for about US$22. In contrast, modern integrated circuit (IC) op amps such as the Fairchild KA741 weigh less than 500 mg, measure 5.7 mm $\times$ 4.9 mm $\times$ 1.8 mm, and sell for approximately US$0.22.

Compared to op amps based on vacuum tubes, modern IC op amps are constructed using perhaps 25 or more transistors all on the same silicon "chip," as well as resistors and capacitors needed to obtain the desired performance characteristics. As a result, they run at much lower dc supply voltages ($\pm$18 V, for example, as opposed to $\pm$300 V for the K2-W), are more reliable, and considerably smaller (Fig. 6.1b,c). In some cases, the IC may contain several op amps. In

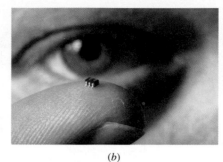

(a)   (b)   (c)

■ **FIGURE 6.1** (*a*) A Philbrick K2-W op amp, based on a matched pair of 12AX7A vacuum tubes. (*b*) LMV321 op amp, used in a variety of phone and game applications. (*c*) LMC6035 operational amplifier, which packs 114 transistors into a package so small that it fits on the head of a pin. (*Photos courtesy of National Semiconductor Corporation.*)

Offset null V^-

Input {

Output

Offset null V^+

(a)

(b)

■ **FIGURE 6.2** (*a*) Electrical symbol for the op amp. (*b*) Minimum required connections to be shown on a circuit schematic.

addition to the output pin and the two inputs, other pins enable power to be supplied to run the transistors, and for external adjustments to be made to balance and compensate the op amp. The symbol commonly used for an op amp is shown in Fig. 6.2*a*. At this point, we are not concerned with the internal circuitry of the op amp or the IC, but only with the voltage and current relationships that exist between the input and output terminals. Thus, for the time being we will use a simpler electrical symbol, shown in Fig. 6.2*b*. Two input terminals are shown on the left, and a single output terminal appears at the right. The terminal marked by a "+" is referred to as the **noninverting input,** and the "−" marked terminal is called the **inverting input.**

6.2 • THE IDEAL OP AMP: A CORDIAL INTRODUCTION

When designing an op amp, the integrated circuit engineer works very hard to ensure that the device has nearly ideal characteristics. In practice, we find that most op amps perform so well that we can often make the assumption that we are dealing with an "ideal" op amp. The characteristics of an **ideal op amp** form the basis for two fundamental rules that at first may seem somewhat unusual:

Ideal Op Amp Rules

1. No current ever flows into either input terminal.
2. There is no voltage difference between the two input terminals.

In a real op amp, a very small leakage current will flow into the input (sometimes as low as 40 femtoamps). It is also possible to maintain a very small voltage across the two input terminals. However, compared to other voltages and currents in most circuits, such values are so small that including them in the analysis does not typically affect our calculations.

When analyzing op amp circuits, we should keep one other point in mind. As opposed to the circuits that we have studied so far, an op amp circuit always has an *output* that depends on some type of *input*. Therefore, we will analyze op amp circuits with the goal of obtaining an expression for the output in terms of the input quantities. *We will find that it is usually a good idea to begin the analysis of an op amp circuit at the input, and proceed from there.*

The circuit shown in Fig. 6.3 is known as an ***inverting amplifier.*** We choose to analyze this circuit using KVL, beginning with the input voltage source. The current labeled i flows only through the two resistors R_1 and R_f; ideal op amp rule 1 states that no current flows into the inverting input terminal. Thus, we can write

$$-v_{in} + R_1 i + R_f i + v_{out} = 0$$

which can be rearranged to obtain an equation that relates the output to the input:

$$v_{out} = v_{in} - (R_1 + R_f)i \qquad [1]$$

However, we are now in the situation of having one equation in two unknowns, since we were only given $v_{in} = 5 \sin 3t$ mV, $R_1 = 4.7$ kΩ, and $R_f = 47$ kΩ. To compute the output voltage, we therefore require an additional equation that expresses i only in terms of v_{out}, v_{in}, R_1, and/or R_f.

This is a good time to mention that we have not yet made use of ideal op amp rule 2. Since the noninverting input is grounded, it is at zero volts. By ideal op amp rule 2, the inverting input is therefore also at zero volts. *This does not mean that the two inputs are physically shorted together, and we should be careful not to make such an assumption.* Rather, the two input voltages simply track each other: If we try to change the voltage at one pin, the other pin will be driven by internal circuitry to the same value. Thus, we can write one more KVL equation:

$$-v_{in} + R_1 i + 0 = 0$$

or

$$i = \frac{v_{in}}{R_1} \qquad [2]$$

Combining Eq. [2] with Eq. [1], we obtain an expression for v_{out} in terms of v_{in}:

$$v_{out} = -\frac{R_f}{R_1} v_{in} \qquad [3]$$

Substituting $v_{in} = 5 \sin 3t$ mV, $R_1 = 4.7$ kΩ, and $R_f = 47$ kΩ,

$$v_{out} = -50 \sin 3t \qquad \text{mV}$$

Since we were given $R_f > R_1$, this circuit amplifies the input voltage signal v_{in}. If we choose $R_f < R_1$, the signal will be attenuated instead. We also note that the output voltage has the opposite sign of the input voltage,[1] hence the name "inverting amplifier." The output is sketched in Fig. 6.4, along with the input waveform for comparison.

At this point, it is worth mentioning that the ideal op amp seems to be violating KCL. Specifically, in the above circuit no current flows into or out of either input terminal, but somehow current is able to flow into the output pin! This would imply that the op amp is somehow able to either create electrons out of nowhere, or store them forever (depending on the direction of current flow). Obviously, this is not possible. The conflict arises because we have been treating the op amp the same way we treated passive elements

(1) Or, "*the output is 180° out of phase with the input,*" which sounds more impressive.

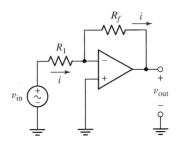

■ **FIGURE 6.3** An op amp used to construct an inverting amplifier circuit, where $v_{in} = 5 \sin 3t$ mV, $R_1 = 4.7$ kΩ, and $R_f = 47$ kΩ.

The fact that the inverting input terminal finds itself at zero volts in this type of circuit configuration leads to what is often referred to as a "virtual ground." This does not mean that the pin is actually grounded, which is sometimes a source of confusion for students. The op amp makes whatever internal adjustments are necessary to prevent a voltage difference between the input terminals. The input terminals are not shorted together.

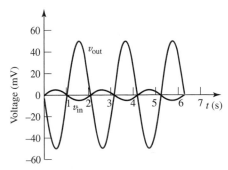

■ **FIGURE 6.4** Input and output waveforms of the inverting amplifier circuit.

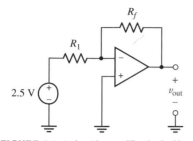

■ FIGURE 6.5 An inverting amplifier circuit with a 2.5 V input.

such as the resistor. In reality, however, the op amp cannot function unless it is connected to external power sources. It is through those power sources that we can direct current flow through the output terminal.

Although we have shown that the inverting amplifier circuit of Fig. 6.3 can amplify an ac signal (a sine wave in this case having a frequency of 3 rad/s and an amplitude of 5 mV), it works just as well with dc inputs. We consider this type of situation in Fig. 6.5, where values for R_1 and R_f are to be selected to obtain an output voltage of -10 V.

This is the same circuit as shown in Fig. 6.3, but with a 2.5 V dc input. Since no other change has been made, the expression we presented as Eq. [3] is valid for this circuit as well. To obtain the desired output, we seek a ratio of R_f to R_1 of 10/2.5 or 4. Since it is only the ratio that is important here, we simply need to pick a convenient value for one resistor, and the other resistor value is then fixed at the same time. For example, we could choose $R_1 = 100\ \Omega$ (so $R_f = 400\ \Omega$), or even $R_f = 8\ \text{M}\Omega$ (so $R_1 = 2\ \text{M}\Omega$). In practice, other constraints (such as bias current) may limit our choices.

This circuit configuration therefore acts as a convenient type of voltage amplifier (or **attenuator,** if the ratio of R_f to R_1 is less than 1), but does have the sometimes inconvenient property of inverting the sign of the input. There is an alternative, however, which is analyzed just as easily—the noninverting amplifier shown in Fig. 6.6. We examine such a circuit in the following example.

EXAMPLE 6.1

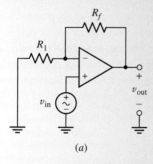

(a)

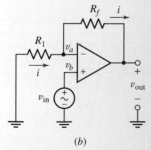

(b)

■ FIGURE 6.6 (a) An op amp used to construct a noninverting amplifier circuit. (b) Circuit with the current through R_1 and R_f defined, as well as both input voltages labeled.

Sketch the output waveform of the noninverting amplifier circuit in Fig. 6.6a. Use $v_{in} = 5\sin 3t$ mV, $R_1 = 4.7$ kΩ, and $R_f = 47$ kΩ.

▶ **Identify the goal of the problem.**
We require an expression for v_{out} that only depends on the known quantities v_{in}, R_1, and R_f.

▶ **Collect the known information.**
Since values have been specified for the resistors and the input waveform, we begin by labeling the current i and the two input voltages as shown in Fig. 6.6b. We will assume that the op amp is an ideal op amp.

▶ **Devise a plan.**
Although mesh analysis is a favorite technique of students, it turns out to be more practical in most op amp circuits to apply nodal analysis, since there is no direct way to determine the current flowing out of the op amp output.

▶ **Construct an appropriate set of equations.**
Note that we are using ideal op amp rule 1 implicitly by defining the same current through both resistors: no current flows into the inverting input terminal. Employing nodal analysis to obtain our expression for v_{out} in terms of v_{in}, we thus find that

At node a:

$$0 = \frac{v_a}{R_1} + \frac{v_a - v_{out}}{R_f} \qquad [4]$$

At node b:

$$v_b = v_{in} \qquad [5]$$

▶ Determine if additional information is required.

Our goal is to obtain a single expression that relates the input and output voltages, although neither Eq. [4] nor Eq. [5] appears to do so. However, we have not yet employed ideal op amp rule 2, and we will find that in almost every op amp circuit *both* rules need to be invoked in order to obtain such an expression.

Thus, we recognize that $v_a = v_b = v_{in}$, and Eq. [4] becomes

$$0 = \frac{v_{in}}{R_1} + \frac{v_{in} - v_{out}}{R_f}$$

▶ Attempt a solution.

Rearranging, we obtain an expression for the output voltage in terms of the input voltage v_{in}:

$$v_{out} = \left(1 + \frac{R_f}{R_1}\right) v_{in} = 11\, v_{in} = 55 \sin 3t \quad mV$$

▶ Verify the solution. Is it reasonable or expected?

The output waveform is sketched in Fig. 6.7, along with the input waveform for comparison. In contrast to the output waveform of the inverting amplifier circuit, we note that the input and output are in phase for the noninverting amplifier. This should not be entirely unexpected: It is implicit in the name "noninverting amplifier."

PRACTICE

6.1 Derive an expression for v_{out} in terms of v_{in} for the circuit shown in Fig. 6.8.

Ans: $v_{out} = v_{in}$. The circuit is known as a "*voltage follower*," since the output voltage tracks or "*follows*" the input voltage.

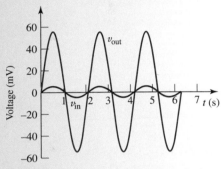

■ **FIGURE 6.7** Input and output waveforms for the noninverting amplifier circuit.

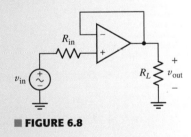

■ **FIGURE 6.8**

Just like the inverting amplifier, the noninverting amplifier works with dc as well as ac inputs, but has a voltage gain of $v_{out}/v_{in} = 1 + (R_f/R_1)$. Thus, if we set $R_f = 9\ \Omega$ and $R_1 = 1\ \Omega$, we obtain an output v_{out} 10 times larger than the input voltage v_{in}. In contrast to the inverting amplifier, the output and input of the noninverting amplifier always have the same sign, and the output voltage cannot be less than the input; the minimum gain is 1. Which amplifier we choose depends on the application we are considering. In the special case of the voltage follower circuit shown in Fig. 6.8, which represents a noninverting amplifier with R_1 set to ∞ and R_f set to zero, the output is identical to the input both in sign *and* magnitude. This may seem rather pointless as a general type of circuit, but we should keep in mind that *the voltage follower draws no current from the input* (in the ideal case)—it therefore can act as a **buffer** between the voltage v_{in} and some resistive load R_L connected to the output of the op amp.

We mentioned earlier that the name "operational amplifier" originates from using such devices to perform arithmetical operations on analog (i.e., nondigitized, real-time, real-world) signals. As we see in the following two circuits, this includes both addition and subtraction of input voltage signals.

EXAMPLE **6.2**

Obtain an expression for v_{out} in terms of v_1, v_2, and v_3 for the op amp circuit in Fig. 6.9, also known as a *"summing amplifier."*

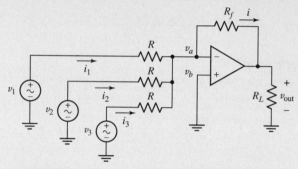

■ **FIGURE 6.9** Basic summing amplifier circuit with three inputs.

We first note that this circuit is similar to the inverting amplifier circuit of Fig. 6.3. Again, the goal is to obtain an expression for v_{out} (which in this case appears across a load resistor R_L) in terms of the inputs (v_1, v_2, and v_3).

Since no current can flow into the inverting input terminal, we know that

$$i = i_1 + i_2 + i_3$$

Therefore, we can write the following equation at the node labeled v_a:

$$0 = \frac{v_a - v_{out}}{R_f} + \frac{v_a - v_1}{R} + \frac{v_a - v_2}{R} + \frac{v_a - v_3}{R}$$

This equation contains both v_{out} and the input voltages, but unfortunately it also contains the nodal voltage v_a. To remove this unknown quantity from our expression, we need to write an additional equation that relates v_a to v_{out}, the input voltages, R_f, and/or R. At this point, we remember that we have not yet used ideal op amp rule 2, and that we will almost certainly require the use of both rules when analyzing an op amp circuit. Thus, since $v_a = v_b = 0$, we can write the following:

$$0 = \frac{v_{out}}{R_f} + \frac{v_1}{R} + \frac{v_2}{R} + \frac{v_3}{R}$$

Rearranging, we obtain the following expression for v_{out}:

$$v_{\text{out}} = -\frac{R_f}{R}(v_1 + v_2 + v_3) \qquad [6]$$

In the special case where $v_2 = v_3 = 0$, we see that our result agrees with Eq. [3], which was derived for essentially the same circuit.

There are several interesting features about the result we have just derived. First, if we select R_f so that it is equal to R, then the output is the (negative of the) sum of the three input signals v_1, v_2, and v_3. Further, we can select the ratio of R_f to R to multiply this sum by a fixed constant. So, for example, if the three voltages represented signals from three separate scales calibrated so that -1 V $= 1$ lb, we could set $R_f = R/2.205$ to obtain a voltage signal that represented the combined weight in kilograms (to within about 1 percent accuracy due to our conversion factor).

Also, we notice that R_L did not appear in our final expression. As long as its value is not too low, the operation of the circuit will not be affected; at present, we have not considered a detailed enough model of an op amp to predict such an occurrence. This resistor represents the Thévenin equivalent of whatever we use to monitor the amplifier output. If our output device is a simple voltmeter, then R_L represents the Thévenin equivalent resistance seen looking into the voltmeter terminals (typically 10 MΩ or more). Or, our output device might be a speaker (typically 8 Ω), in which case we hear the sum of the three separate sources of sound; v_1, v_2, and v_3 might represent microphones in that case.

One word of caution: It is frequently tempting to assume that the current labeled i in Fig. 6.9 flows not only through R_f but through R_L also. Not true! It is very possible that current is flowing through the output terminal of the op amp as well, so that *the currents through the two resistors are not the same*. It is for this reason that we almost universally avoid writing KCL equations at the output pin of an op amp, which leads to the preference of nodal over mesh analysis when working with most op amp circuits.

CAUTION

PRACTICE

6.2 Derive an expression for v_{out} in terms of v_1 and v_2 for the circuit shown in Fig. 6.10, also known as a **difference amplifier**.

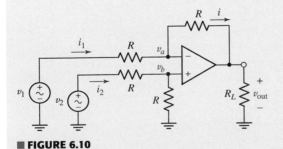

■ **FIGURE 6.10**

Ans: $v_{\text{out}} = v_2 - v_1$. Hint: Use voltage division to obtain v_b.

TABLE 6.1 Summary of Basic Op Amp Circuits

Name	Circuit Schematic	Input-Output Relation
Inverting Amplifier		$v_{\text{out}} = -\dfrac{R_f}{R_1} v_{\text{in}}$
Noninverting Amplifier		$v_{\text{out}} = \left(1 + \dfrac{R_f}{R_1}\right) v_{\text{in}}$
Voltage Follower (also known as a Unity Gain Amplifier)		$v_{\text{out}} = v_{\text{in}}$
Summing Amplifier		$v_{\text{out}} = -\dfrac{R_f}{R}(v_1 + v_2 + v_3)$
Difference Amplifier		$v_{\text{out}} = v_2 - v_1$

A Fiber Optic Intercom

A point-to-point intercom system can be constructed using a number of different approaches, depending on the intended application environment. Low-power radio frequency (RF) systems work very well and are generally cost-effective, but are subject to interference from other RF sources and are also prone to eavesdropping. Use of a simple wire to connect the two intercom systems instead can eliminate a great deal of the RF interference as well as increase privacy. However, wires are subject to corrosion and short circuits when the plastic insulation wears, and their weight can be a concern in aircraft and related applications (Fig. 6.11).

FIGURE 6.11 The application environment often dictates design constraints.
™ & © Boeing. Used under license.

An alternative design would be to convert the electrical signal from the microphone to an optical signal, which could then be transmitted through a thin (~50 μm diameter) optical fiber. The optical signal is then converted back to an electrical signal, which is amplified and delivered to a speaker. A schematic diagram of such a system is shown in Fig. 6.12; two such systems would be needed for two-way communication.

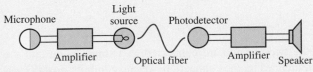

FIGURE 6.12 Schematic diagram of one-half of a simple fiber optic intercom.

We can consider the design of the transmission and reception circuits separately, since the two circuits are in fact electrically independent. Figure 6.13 shows a simple

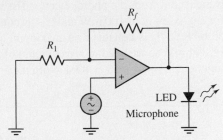

FIGURE 6.13 Circuit used to convert the electrical microphone signal into an optical signal for transmission through a fiber.

signal generation circuit consisting of a microphone, a light-emitting diode (LED), and an op amp used in a noninverting amplifier circuit to drive the LED; not shown are the power connections required for the op amp itself. The light output of the LED is roughly proportional to its current, although less so for very small and very large values of current.

We know the gain of the amplifier is given by

$$\frac{v_{\text{out}}}{v_{\text{in}}} = 1 + \frac{R_f}{R_1}$$

which is independent of the resistance of the LED. In order to select values for R_f and R_1, we need to know the input voltage from the microphone and the necessary output voltage to power the LED. A quick measurement indicates that the typical voltage output of the microphone peaks at 40 mV when someone is using a normal speaking voice. The LED manufacturer recommends operating at approximately 1.6 V, so we design for a gain of 1.6/0.04 = 40. Arbitrarily choosing $R_1 = 1$ kΩ leads to a required value of 39 kΩ for R_f.

The circuit of Fig. 6.14 is the receiver part of our one-way intercom system. It converts the optical signal from the fiber into an electrical signal, amplifying it so that an audible sound emanates from the speaker.

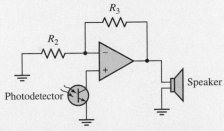

FIGURE 6.14 Receiver circuit used to convert the optical signal into an audio signal.

(Continued on next page)

After coupling the LED output of the transmitting circuit to the optical fiber, a signal of approximately 10 mV is measured from the photodetector. The speaker is rated for a maximum of 100 mW and has an equivalent resistance of 8 Ω. This equates to a maximum speaker voltage of 894 mV, so we need to select values of R_2 and R_3 to obtain a gain of $894/10 = 89.4$. With the arbitrary selection of $R_2 = 10$ kΩ, we find that a value of 884 kΩ completes our design.

This circuit will work in practice, although the non-linear characteristics of the LED lead to a noticeable distortion of the audio signal. We leave improved designs for more advanced texts.

6.3 CASCADED STAGES

Although the op amp is an extremely versatile device, there are numerous applications in which a single op amp will not suffice. In such instances, it is often possible to meet application requirements by cascading several individual op amps together in the same circuit. An example of this is shown in Fig. 6.15, which consists of the summing amplifier circuit of Fig. 6.9 with only two input sources, and the output fed into a simple inverting amplifier. The result is a two-stage op amp circuit.

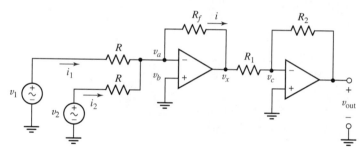

■ **FIGURE 6.15** A two-stage op amp circuit consisting of a summing amplifier cascaded with an inverting amplifier circuit.

We have already analyzed each of these op amp circuits separately. Based on our previous experience, if the two op amp circuits were disconnected, we would expect

$$v_x = -\frac{R_f}{R}(v_1 + v_2) \qquad [7]$$

and

$$v_{out} = -\frac{R_2}{R_1}v_x \qquad [8]$$

In fact, since the two circuits are connected at a single point and the voltage v_x is not influenced by the connection, we can combine Eqs. [7] and [8] to obtain

$$v_{out} = \frac{R_2}{R_1}\frac{R_f}{R}(v_1 + v_2) \qquad [9]$$

which describes the input-output characteristics of the circuit shown in Fig. 6.15. We may not always be able to reduce such a circuit to familiar stages, however, so it is worth seeing how the two-stage circuit of Fig. 6.15 can be analyzed as a whole.

When analyzing cascaded circuits, it is sometimes helpful to begin with the last stage and work backward toward the input stage. Referring to ideal op amp rule 1, the same current flows through R_1 and R_2. Writing the appropriate nodal equation at the node labeled v_c yields

$$0 = \frac{v_c - v_x}{R_1} + \frac{v_c - v_{out}}{R_2} \qquad [10]$$

Applying ideal op amp rule 2, we can set $v_c = 0$ in Eq. [10], resulting in

$$0 = \frac{v_x}{R_1} + \frac{v_{out}}{R_2} \qquad [11]$$

Since our goal is an expression for v_{out} in terms of v_1 and v_2, we proceed to the first op amp in order to obtain an expression for v_x in terms of the two input quantities.

Applying ideal op amp rule 1 at the inverting input of the first op amp,

$$0 = \frac{v_a - v_x}{R_f} + \frac{v_a - v_1}{R} + \frac{v_a - v_2}{R} \qquad [12]$$

Ideal op amp rule 2 allows us to replace v_a in Eq. [12] with zero, since $v_a = v_b = 0$. Thus, Eq. [12] becomes

$$0 = \frac{v_x}{R_f} + \frac{v_1}{R} + \frac{v_2}{R} \qquad [13]$$

We now have an equation for v_{out} in terms of v_x (Eq. [11]), and an equation for v_x in terms of v_1 and v_2 (Eq. [13]). These equations are identical to Eqs. [7] and [8] respectively, which means that cascading the two separate circuits as in Fig. 6.15 did not affect the input-output relationship of either stage. Combining Eqs. [11] and [13], we find that the input-output relationship for the cascaded op amp circuit is

$$v_{out} = \frac{R_2}{R_1} \frac{R_f}{R} (v_1 + v_2) \qquad [14]$$

which is identical to Eq. [9].

Thus, the cascaded circuit acts as a summing amplifier, but without a phase reversal between the input and output. By choosing the resistor values carefully, we can either amplify or attenuate the sum of the two input voltages. If we select $R_2 = R_1$ and $R_f = R$, we can also obtain an amplifier circuit where $v_{out} = v_1 + v_2$, if desired.

EXAMPLE **6.3**

A multiple-tank gas propellant fuel system is installed in a small lunar orbit runabout. The amount of fuel in any tank is monitored by measuring the tank pressure (in psia[2]). Technical details for tank capacity as well as sensor pressure and voltage range are given in Table 6.2. Design a circuit which provides a positive dc voltage signal proportional to the total fuel remaining, such that 1 V = 100 percent.

© Corbis

TABLE 6.2 Technical Data for Tank Pressure Monitoring System

Tank 1 Capacity	10,000 psia
Tank 2 Capacity	10,000 psia
Tank 3 Capacity	2000 psia
Sensor Pressure Range	0 to 12,500 psia
Sensor Voltage Output	0 to 5 Vdc

We see from Table 6.2 that the system has three separate gas tanks, requiring three separate sensors. Each sensor is rated up to 12,500 psia, with a corresponding output of 5 V. Thus, when tank 1 is full, its sensor will provide a voltage signal of $5 \times (10,000/12,500) = 4$ V; the same is true for the sensor monitoring tank 2. The sensor connected to tank 3, however, will only provide a maximum voltage signal of $5 \times (2,000/12,500) = 800$ mV.

One possible solution is the circuit shown in Fig. 6.16a, which employs a summing amplifier stage with v_1, v_2, and v_3 representing the three sensor outputs, followed by an inverting amplifier to adjust the voltage sign and magnitude. Since we are not told the output resistance of the sensor, we employ a buffer for each one as shown in Fig. 6.16b; the result is (in the ideal case) no current flow from the sensor.

To keep the design as simple as possible, we begin by choosing R_1, R_2, R_3, and R_4 to be 1 kΩ; any value will do as long as all four resistors are equal. Thus, the output of the summing stage is

$$v_x = -(v_1 + v_2 + v_3)$$

The final stage must invert this voltage and scale it such that the output voltage is 1 V when all three tanks are full. The full condition results in $v_x = -(4 + 4 + 0.8) = -8.8$ V. Thus, the final stage needs a voltage ratio of $R_6/R_5 = 1/8.8$. Arbitrarily choosing $R_6 = 1$ kΩ, we find that a value of 8.8 kΩ for R_5 completes the design.

---

(2) Pounds per square inch, absolute. This is a differential pressure measurement relative to a vacuum reference.

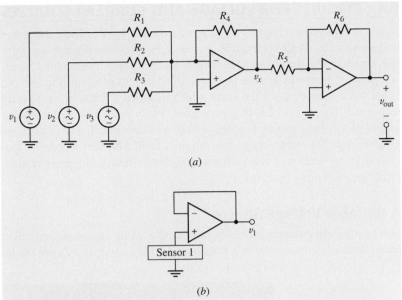

(a)

(b)

■ **FIGURE 6.16** (a) A proposed circuit to provide a total fuel remaining readout. (b) Buffer design to avoid errors associated with the internal resistance of the sensor and limitations on its ability to provide current. One such buffer is used for each sensor, providing the inputs v_1, v_2, and v_3 to the summing amplifier stage.

PRACTICE

6.3 An historic bridge is showing signs of deterioration. Until renovations can be performed, it is decided that only cars weighing less than 1600 kg will be allowed across. To monitor this, a four-pad weighing system is designed. There are four independent voltage signals, one from each wheel pad, with 1 mV = 1 kg. Design a circuit to provide a positive voltage signal to be displayed on a DMM (digital multimeter) that represents the total weight of a vehicle, such that 1 mV = 1 kg. You may assume there is no need to buffer the wheel pad voltage signals.

Ans: See Fig. 6.17.

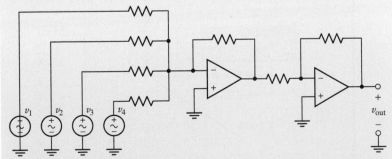

■ **FIGURE 6.17** One possible solution to Practice Problem 6.3; all resistors are 10 kΩ (although any value will do as long as they are all equal). Input voltages v_1, v_2, v_3, and v_4 represent the voltage signals from the four wheel pad sensors, and v_{out} is the output signal to be connected to the positive input terminal of the DMM. All five voltages are referenced to ground, and the common terminal of the DMM should be connected to ground as well.

6.4 ● CIRCUITS FOR VOLTAGE AND CURRENT SOURCES

In this and previous chapters we have often made use of ideal current and voltage sources, which we assume provide the same value of current or voltage, respectively, regardless of how they are connected in a circuit. Our assumption of independence has its limits, of course, as mentioned in Sec. 5.2 when we discussed practical sources which included a "built-in" or inherent resistance. The effect of such a resistance was a reduction of the voltage output of a voltage source as more current was demanded, or a diminished current output as more voltage was required from a current source. As discussed in this section, it is possible to construct circuits with more reliable characteristics using op amps.

A Reliable Voltage Source

One of the most common means of providing a stable and consistent reference voltage is to make use of a nonlinear device known as a **Zener diode.**

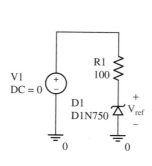

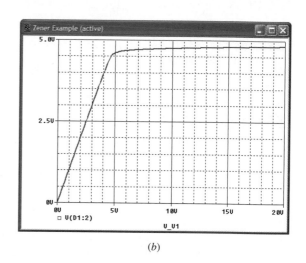

(a)　　　　　　　　　　　　　　　　(b)

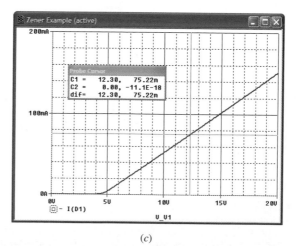

(c)

■ **FIGURE 6.18** (a) PSpice schematic of a simple voltage reference circuit based on the 1N750 Zener diode. (b) Simulation of the circuit showing the diode voltage V_{ref} as a function of the driving voltage V1. (c) Simulation of the diode current, showing that its maximum rating is exceeded when V1 exceeds 12.3 V. (Note that performing this calculation assuming an *ideal* Zener diode yields 12.2 V.)

Its symbol is a triangle with a Z-like line across the top of the triangle, as shown for a 1N750 in the circuit of Fig. 6.18a.

Although a two-terminal element, a Zener diode behaves very differently than a simple (linear) resistor. In particular, whereas a resistor is a symmetric device, a diode is not. Instead, its two terminals are labeled the **anode** (the flat part of the triangle) and the **cathode** (the point of the triangle), and very different behavior is obtained depending on which way the diode is inserted in a circuit. A *Zener* diode is a special type of diode designed to be used with a positive voltage at the cathode with respect to the anode; when connected this way, the diode is said to be *reverse biased*. For low voltages, the diode acts like a resistor with a small linear increase in current flow as the voltage is increased. Once a certain voltage (V_{BR}) is reached, however—known as the *reverse breakdown voltage* or **Zener voltage** of the diode—the voltage does not significantly increase further, but essentially any current can flow up to the maximum rating of the diode (75 mA for a 1N750, whose Zener voltage is 4.7 V).

Let's consider the simulation result presented in Fig. 6.18b, which shows the voltage V_{ref} across the diode as the voltage source V1 is swept from 0 to 20 V. Provided V1 remains above 5 V, *the voltage across our diode is essentially constant*. Thus, we could replace V1 with a 9 V battery, and not be too concerned with changes in our voltage reference as the battery voltage begins to drop as it discharges. The purpose of R1 in this circuit is simply to provide the necessary voltage drop between the battery and the diode; its value should be chosen to ensure that the diode is operating at its Zener voltage but below its maximum rated current. For example, Fig. 6.18c shows that the 75 mA rating is exceeded in our circuit if the source voltage V1 is much greater than 12 V. Thus, the value of resistor R1 should be sized corresponding to the source voltage available, as we explore in Example 6.4.

EXAMPLE 6.4

Design a circuit based on the 1N750 Zener diode that runs on a single 9 V battery and provides a reference voltage of 4.7 V.

The 1N750 has a maximum current rating of 75 mA, and a Zener voltage of 4.7 V. The voltage of a 9 V battery can vary slightly depending on its state of charge, but we neglect this for the present design.

A simple circuit such as the one shown in Fig. 6.19a is adequate for our purposes; the only issue is determining a suitable value for the resistor R_{ref}.

If 4.7 V is dropped across the diode, then $9 - 4.7 = 4.3$ V must be dropped across R_{ref}. Thus,

$$R_{ref} = \frac{9 - V_{ref}}{I_{ref}} = \frac{4.3}{I_{ref}}$$

We determine R_{ref} by specifying a current value. We know that I_{ref} should not be allowed to exceed 75 mA for this diode, and that larger currents will discharge the battery more quickly. However, as seen in Fig. 6.19b, we cannot simply select I_{ref} arbitrarily; very low currents

(Continued on next page)

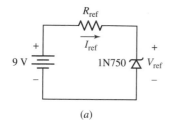

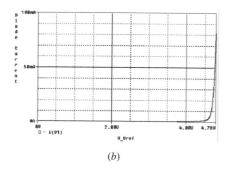

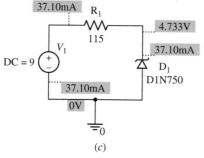

FIGURE 6.19 (a) A voltage reference circuit based on the 1N750 Zener diode. (b) Diode I-V relationship. (c) PSpice simulation of the final design.

do not allow the diode to operate in the Zener breakdown region. In the absence of a detailed equation for the diode's current-voltage relationship (which is clearly nonlinear), we design for 50 percent of the maximum rated current as a rule of thumb. Thus,

$$R_{ref} = \frac{4.3}{0.0375} = 115 \ \Omega$$

Detailed "tweaking" can be obtained by performing a PSpice simulation of the final circuit, although we see from Fig. 6.19c that our first pass is reasonably close (within 1 percent) to our target value.

The basic Zener diode voltage reference circuit of Fig. 6.18a works very well in many situations, but we are limited somewhat in the value of the voltage depending on which Zener diodes are available. Also, we often find that the circuit shown is not well suited to applications requiring more than a few milliamperes of current. In such instances, we may use the Zener reference circuit in conjunction with a simple amplifier stage, as shown in Fig. 6.20. The result is a stable voltage that can be controlled by adjusting the value of either R_1 or R_f, without having to switch to a different Zener diode.

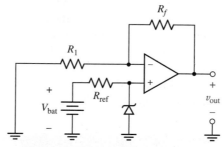

FIGURE 6.20 An op amp–based voltage source based on a Zener voltage reference.

PRACTICE

6.4 Design a circuit to provide a reference voltage of 6 V using a 1N750 Zener diode and a noninverting amplifier.

Ans: Using the circuit topology shown in Fig. 6.20, choose $V_{bat} = 9$ V, $R_{ref} = 115 \ \Omega$, $R_1 = 1$ kΩ, and $R_f = 268 \ \Omega$.

A Reliable Current Source

Consider the circuit shown in Fig. 6.21a, where V_{ref} is provided by a regulated voltage source such as the one shown in Fig. 6.19a. The reader may recognize this circuit as a simple inverting amplifier configuration, assuming we tap the output pin of the op amp. We can also use this circuit as a current source, however, where R_L represents a resistive load.

The input voltage V_{ref} appears across reference resistor R_{ref}, since the noninverting input of the op amp is connected to ground. With no current

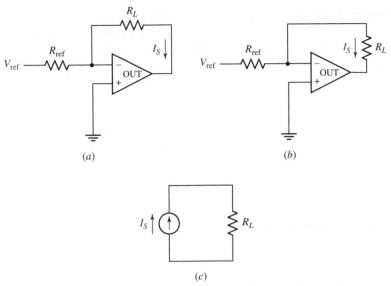

FIGURE 6.21 (a) An op amp–based current source, controlled by the reference voltage V_{ref}. (b) Circuit redrawn to highlight load. (c) Circuit model. Resistor R_L represents the Norton equivalent of an unknown passive load circuit.
Jung, Walter G.: IC OP-AMP COOKBOOK, 3rd Edition, © 1986. Reprinted by permission of Pearson Education, Inc., Upper Saddle River, NJ.

flowing into the inverting input, the current flowing through the load resistor R_L is simply

$$I_s = \frac{V_{ref}}{R_{ref}}$$

In other words, the current supplied to R_L does not depend on its resistance—the primary attribute of an ideal current source. It is also worth noting that we are not tapping the output voltage of the op amp here as a quantity of interest. Instead, we may view the load resistor R_L as the Norton (or Thévenin) equivalent of some unknown passive load circuit, which receives power from the op amp circuit. Redrawing the circuit slightly as in Fig. 6.21b, we see that it has a great deal in common with the more familiar circuit of Fig. 6.21c. In other words, we may use this op amp circuit as an independent

EXAMPLE **6.5**

Design a current source that will deliver 1 mA to an arbitrary resistive load.

Basing our design on the circuits of Fig. 6.20 and Fig. 6.21a, we know that the current through our load R_L will be given by

$$I_s = \frac{V_{ref}}{R_{ref}}$$

where values for V_{ref} and R_{ref} must be selected, and a circuit to provide V_{ref} must also be designed. If we use a 1N750 Zener diode in series with a 9 V

(Continued on next page)

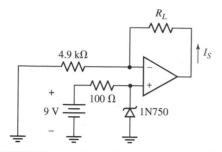

■ FIGURE 6.22 One possible design for the desired current source. Note the change in current direction from Fig. 6.21b.

battery and a 100 Ω resistor, we know from Fig. 6.18b that a voltage of 4.9 V will exist across the diode. Thus, $V_{ref} = 4.9$ V, dictating a value of $4.9/10^{-3} = 4.9$ kΩ for R_{ref}. The complete circuit is shown in Fig. 6.22.

Note that if we had assumed a diode voltage of 4.7 V instead, the error in our designed current would only be a few percent, well within the typical 5 to 10 percent tolerance in resistor values we might expect.

The only issue remaining is whether 1 mA can in fact be provided to any value of R_L. For the case of $R_L = 0$, the output of the op amp will be 4.9 V, which is not unreasonable. As the load resistor is increased, however, the op amp output voltage increases. Eventually we must reach some type of limit, as discussed in Sec. 6.5.

PRACTICE

6.5 Design a current source capable of providing 500 μA to a resistive load.

Ans: See Fig. 6.23 for one possible solution.

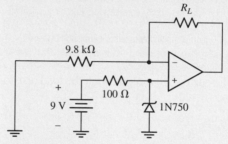

■ FIGURE 6.23 One possible solution to Practice Problem 6.5.

current source with essentially ideal characteristics, up to the maximum rated output current of the op amp selected.

6.5 PRACTICAL CONSIDERATIONS

A More Detailed Op Amp Model

Reduced to its essentials, the op amp can be thought of as a voltage-controlled dependent voltage source. The dependent voltage source provides the output of the op amp, and the voltage on which it depends is applied to the input terminals. A schematic diagram of a reasonable model for a practical op amp is shown in Fig. 6.24; it includes a dependent voltage source with voltage gain A, an output resistance R_o, and an input resistance R_i. Table 6.3 gives typical values for these parameters for several types of commercially available op amps.

The parameter A is referred to as the ***open-loop voltage gain*** of the op amp, and is typically in the range of 10^5 to 10^6. We notice that all of the op amps listed in Table 6.3 have extremely large open-loop voltage gain, especially compared to the voltage gain of 11 that characterized the noninverting

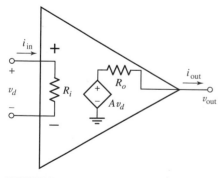

■ FIGURE 6.24 A more detailed model for the op amp.

TABLE 6.3 Typical Parameter Values for Several Types of Op Amps

Part Number	μA741	LM324	LF411	AD549K	OPA690
Description	General purpose	Low-power quad	Low-offset, low-drift JFET input	Ultralow input bias current	Wideband video frequency op amp
Open loop gain A	2×10^5 V/V	10^5 V/V	2×10^5 V/V	10^6 V/V	2800 V/V
Input resistance	2 MΩ	*	1 TΩ	10 TΩ	190 kΩ
Output resistance	75 Ω	*	~1 Ω	~15 Ω	*
Input bias current	80 nA	45 nA	50 pA	75 fA	3 μA
Input offset voltage	1.0 mV	2.0 mV	0.8 mV	0.150 mV	±1.0 mV
CMRR	90 dB	85 dB	100 dB	100 dB	65 dB
Slew Rate	0.5 V/μs	*	15 V/μs	3 V/μs	1800 V/μs
PSpice Model	✓	✓	✓		

* Not provided by manufacturer.
✓ Indicates that a PSpice model is included in Orcad Capture CIS Version 10.0.

amplifier circuit of Example 6.1. It is important to remember the distinction between the open-loop voltage gain of the op amp itself, and the **closed-loop voltage gain** that characterizes a particular op amp circuit. The "loop" in this case refers to an *external* path between the output pin and the inverting input pin; it can be a wire, a resistor, or another type of element, depending on the application.

The μA741 is a very common op amp, originally produced by Fairchild Corporation in 1968. It is characterized by an open-loop voltage gain of 200,000, an input resistance of 2 MΩ, and an output resistance of 75 Ω. In order to evaluate how well the ideal op amp model approximates the behavior of this particular device, let's revisit the inverting amplifier circuit of Fig. 6.3.

EXAMPLE 6.6

Using the appropriate values for the μA741 op amp in the model of Fig. 6.24, reanalyze the inverting amplifier circuit of Fig. 6.3.

We begin by replacing the ideal op amp symbol of Fig. 6.3 with the detailed model, resulting in the circuit shown in Fig. 6.25.
Note that we can no longer invoke the ideal op amp rules, since we are not using the ideal op amp model. Thus, we write two nodal equations:

$$0 = \frac{-v_d - v_{in}}{R_1} + \frac{-v_d - v_{out}}{R_f} + \frac{-v_d}{R_i}$$

$$0 = \frac{v_{out} + v_d}{R_f} + \frac{v_{out} - Av_d}{R_o}$$

Performing some straightforward but rather lengthy algebra, we eliminate v_d and combine these two equations to obtain the following

(Continued on next page)

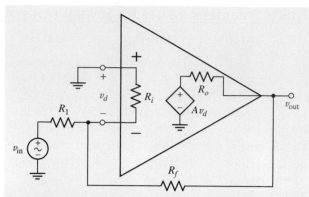

■ **FIGURE 6.25** Inverting amplifier circuit drawn using detailed op amp model.

expression for v_{out} in terms of v_{in}:

$$v_{\text{out}} = \left[\frac{(R_o + R_f)}{R_o - AR_f} \left(\frac{1}{R_1} + \frac{1}{R_f} + \frac{1}{R_i} \right) - \frac{1}{R_f} \right]^{-1} \frac{v_{\text{in}}}{R_1} \qquad [15]$$

Substituting $v_{\text{in}} = 5 \sin 3t\,\text{mV}$, $R_1 = 4.7\,\text{k}\Omega$, $R_f = 47\,\text{k}\Omega$, $R_o = 75\,\Omega$, $R_i = 2\,\text{M}\Omega$, and $A = 2 \times 10^5$, we obtain

$$v_{\text{out}} = -9.999448 v_{\text{in}} = -49.99724 \sin 3t \qquad \text{mV}$$

Upon comparing this to the expression found assuming an ideal op amp ($v_{\text{out}} = -10 v_{\text{in}} = -50 \sin 3t$ mV), we see that the ideal op amp is indeed a reasonably accurate model. Further, assuming an ideal op amp leads to a significant reduction in the algebra required to perform the circuit analysis. Note that if we allow $A \to \infty$, $R_o \to 0$, and $R_i \to \infty$, Eq. [15] reduces to

$$v_{\text{out}} = -\frac{R_f}{R_1} v_{\text{in}}$$

which is what we derived earlier for the inverting amplifier when assuming the op amp was ideal.

PRACTICE

6.6 Assuming a finite open-loop gain (A), a finite input resistance (R_i), and zero output resistance (R_o), derive an expression for v_{out} in terms of v_{in} for the op amp circuit of Fig. 6.3.

Ans: $v_{\text{out}}/v_{\text{in}} = -AR_f R_i / [(1 + A) R_1 R_i + R_1 R_f + R_f R_i]$.

Derivation of the Ideal Op Amp Rules

We have seen that the ideal op amp is a pretty accurate model for the behavior of practical devices. However, using our more detailed model which includes a finite open-loop gain, finite input resistance, and nonzero output resistance, it is actually straightforward to derive the two ideal op amp rules.

Referring to Fig. 6.24, we see that the open circuit output voltage of a practical op amp can be expressed as

$$v_{\text{out}} = A v_d \qquad [16]$$

Rearranging this equation, we find that v_d, sometimes referred to as the *differential input voltage,* can be written as

$$v_d = \frac{v_{out}}{A} \qquad [17]$$

As we might expect, there are practical limits to the output voltage v_{out} that can be obtained from a real op amp. As described in the next section, we must connect our op amp to external dc voltage supplies in order to power the internal circuitry. These external voltage supplies represent the maximum value of v_{out}, and are typically in the range of 5 to 24 V. If we divide 24 V by the open-loop gain of the μA741 (2×10^5), we obtain $v_d = 120 \, \mu V$. Although this is mode the same as zero volts, such a small value compared to the output voltage of 24 V is *practically* zero. An ideal op amp would have infinite open-loop gain, resulting in $v_d = 0$ regardless of v_{out}; this leads to ideal op amp rule 2.

Ideal op amp rule 1 states that, "*No current ever flows into either input terminal.*" Referring to Fig. 6.23, the input current of an op amp is simply

$$i_{in} = \frac{v_d}{R_i}$$

We have just determined that v_d is typically a very small voltage. As we can see from Table 6.3, the input resistance of a typical op amp is very large, ranging from the mega-ohms to the tera-ohms! Using the value of $v_d = 120 \, \mu V$ from above and $R_i = 2 \, M\Omega$, we compute an input current of 60 pA. This is an extremely small current, and we would require a specialized ammeter (known as a picoammeter) to measure it. We see from Table 6.3 that the typical input current (more accurately termed the **input bias current**) of a μA741 is 80 nA, three orders of magnitude larger than our estimate. This is a shortcoming of the op amp model we are using, which is not designed to provide accurate values for input bias current. Compared to the other currents flowing in a typical op amp circuit, however, either value is essentially zero. More modern op amps (such as the AD549) have even lower input bias currents. Thus, we conclude that ideal op amp rule 1 is a fairly reasonable assumption.

From our discussion, it is clear that an ideal op amp has infinite open-loop voltage gain, and infinite input resistance. However, we have not yet considered the output resistance of the op amp and its possible effects on our circuit. Referring to Fig. 6.24, we see that

$$v_{out} = Av_d - R_o i_{out}$$

where i_{out} flows from the output pin of the op amp. Thus, a nonzero value of R_o acts to reduce the output voltage, an effect which becomes more pronounced as the output current increases. For this reason, an *ideal* op amp has an output resistance of zero ohms. The μA741 has a maximum output resistance of 75 Ω, and more modern devices such as the AD549 have even lower output resistance.

Common-Mode Rejection

The op amp is occasionally referred to as a *difference amplifier,* since the output is proportional to the voltage difference between the two input

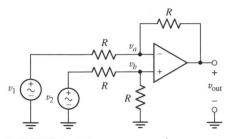

■ **FIGURE 6.26** An op amp connected as a difference amplifier.

terminals. This means that if we apply identical voltages to both input terminals, we expect the output voltage to be zero. This ability of the op amp is one of its most attractive qualities, and is known as ***common-mode rejection.*** The circuit shown in Fig. 6.26 is connected to provide an output voltage

$$v_{out} = v_2 - v_1$$

If $v_1 = 2 + 3 \sin 3t$ volts and $v_2 = 2$ volts, we would expect the output to be $-3 \sin 3t$ volts; the 2 V component common to v_1 and v_2 would not be amplified, nor does it appear in the output.

For practical op amps, we do in fact find a small contribution to the output in response to common mode signals. In order to compare one op amp type to another, it is often helpful to express the ability of an op amp to reject common mode signals through a parameter known as the common mode rejection ratio, or **CMRR.** Defining $v_{o_{CM}}$ as the output obtained when both inputs are equal ($v_1 = v_2 = v_{CM}$), we can determine A_{CM}, the common mode gain of the op amp

$$A_{CM} = \left| \frac{v_{o_{CM}}}{v_{CM}} \right|$$

We then define CMRR in terms of the ratio of differential-mode gain A to the common mode gain A_{CM}, or

$$\text{CMRR} \equiv \left| \frac{A}{A_{CM}} \right| \qquad [18]$$

although this is often expressed in decibels (dB), a logarithmic scale:

$$\text{CMRR}_{(dB)} \equiv 20 \log_{10} \left| \frac{A}{A_{CM}} \right| \text{ dB} \qquad [19]$$

Typical values for several different op amps are provided in Table 6.3; a value of 100 dB corresponds to an absolute ratio of 10^5 for A to A_{CM}.

Negative Feedback

We have seen that the open-loop gain of an op amp is very large, ideally infinite. In practical situations, however, its exact value can vary from the value specified by the manufacturer as typical. Temperature, for example, can have a number of significant effects on the performance of an op amp, so that the operating behavior in $-20°C$ weather may be significantly different from the behavior observed on a warm sunny day. Also, there are typically small variations between devices fabricated at different times. If we design a circuit in which the output voltage is the open-loop gain times the voltage at one of the input terminals, the output voltage could therefore be difficult to predict with a reasonable degree of precision, and might be expected to change depending on the ambient temperature.

A solution to such potential problems is to employ the technique of ***negative feedback,*** which is the process of subtracting a small portion of the output from the input. If some event changes the characteristics of the amplifier such that the output tries to increase, the input is decreasing at the same time. Too much negative feedback will prevent any useful amplification, but a small amount provides stability. An example of negative feedback is the unpleasant sensation we feel as our hand draws near a flame. The closer we move toward the flame, the larger the negative signal sent from our hand. Overdoing the proportion of negative feedback, however, might cause us to abhor heat, and eventually freeze to death. ***Positive feedback*** is the process where some fraction of the output signal is added back to the input. A common example is when a microphone is directed toward a speaker—a very soft sound is rapidly amplified over and over until the system "screams." Positive feedback generally leads to an unstable system.

All of the circuits considered in this chapter incorporate negative feedback through the presence of a resistor between the output pin and the inverting input. The resulting loop between the output and the input reduces the dependency of the output voltage on the actual value of the open-loop gain (as seen in Example 6.6). This obviates the need to measure the precise open-loop gain of each op amp we use, as small variations in A will not significantly impact the operation of the circuit. Negative feedback also provides increased stability in situations where A is sensitive to the op amp's surroundings. For example, if A suddenly increases in response to a change in the ambient temperature, a larger feedback voltage is added to the inverting input. This acts to reduce the differential input voltage v_d, and therefore the change in output voltage Av_d is smaller. We should note that the closed-loop circuit gain is always less than the open-loop device gain; this is the price we pay for stability and reduced sensitivity to parameter variations.

Saturation

So far, we have treated the op amp as a purely linear device, assuming that its characteristics are independent of the way in which it is connected in a circuit. In reality, it is necessary to supply power to an op amp in order to run the internal circuitry, as shown in Fig. 6.27. A positive supply, typically in the range of 5 to 24 V dc, is connected to the terminal marked V^+, and a negative supply of equal magnitude is connected to the terminal marked V^-. There are also a number of applications where a single voltage supply is acceptable, as well as situations where the two voltage magnitudes may be unequal. The op amp manufacturer will usually specify a maximum power supply voltage, beyond which damage to the internal transistors will occur.

The power supply voltages are a critical choice when designing an op amp circuit, because they represent the maximum possible output voltage of the op amp.[3] For example, consider the op amp circuit shown in Fig. 6.26, now connected as a noninverting amplifier having a gain of 10. As shown in the PSpice simulation in Fig. 6.28, we do in fact observe linear behavior from the op amp, but only in the range of ±1.71 V for the input voltage. Outside of this range, the output voltage is no longer proportional to the input, reaching a peak magnitude of 17.6 V. This important nonlinear effect is known as ***saturation,*** which refers to the fact that further increases in the input voltage do not result in a change in the output voltage. This phenomenon refers to the fact that the output of a real op amp cannot exceed its

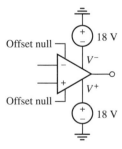

■ **FIGURE 6.27** Op amp with positive and negative voltage supplies connected. Two 18 V supplies are used as an example; note the polarity of each source.

(3) In practice, we find the maximum output voltage is slightly less than the supply voltage by as much as a volt or so.

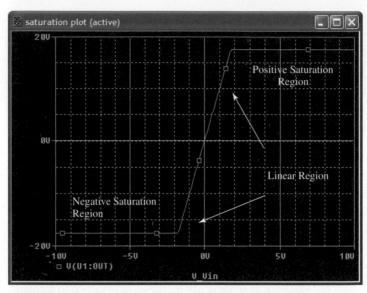

■ **FIGURE 6.28** Simulated input-output characteristics of a μA741 connected as a noninverting amplifier with a gain of 10, and powered by ±18 V supplies.

supply voltages. For example, if we choose to run the op amp with a +9 V supply and a −5 V supply, then our output voltage will be limited to the range of −5 to +9 V. The output of the op amp is a linear response bounded by the positive and negative saturation regions, and as a general rule, we try to design our op amp circuits so that we do not accidentally enter the saturation region. This requires us to select the operating voltage carefully based on the closed-loop gain and maximum expected input voltage.

Input Offset Voltage

As we are discovering, there are a number of practical considerations to keep in mind when working with op amps. One particular nonideality worth mentioning is the tendency for real op amps to have a nonzero output even when the two input terminals are shorted together. The value of the output under such conditions is known as the offset voltage, and the input voltage required to reduce the output to zero is referred to as the ***input offset voltage.*** Referring to Table 6.3 we see that typical values for the input offset voltage are on the order of a few millivolts or less.

Most op amps are provided with two pins marked either "offset null" or "balance." These terminals can be used to adjust the output voltage by connecting them to a variable resistor. A variable resistor is a three-terminal device commonly used for such applications as volume controls on radios. The device comes with a knob that can be rotated to select the actual value of resistance, and has three terminals. If the variable resistor is connected using only the two extreme terminals, its resistance is fixed regardless of the position of the knob. Using the middle terminal and one of the end terminals creates a resistor whose value depends on the knob position. Figure 6.29 shows a typical circuit used to adjust the output voltage of an op amp; the manufacturer's data sheet may suggest alternative circuitry for a particular device.

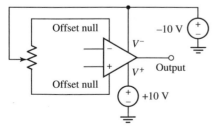

■ **FIGURE 6.29** Suggested external circuitry for obtaining a zero output voltage. The ±10 V supplies are shown as an example; the actual supply voltages used in the final circuit would be chosen in practice.

Slew Rate

Up to now, we have tacitly assumed that the op amp will respond equally well to signals of any frequency, although perhaps we would not be surprised to find that in practice there is some type of limitation in this regard. Since we know that op amp circuits work well at dc, which is essentially zero frequency, it is the performance as the signal frequency is *increased* that we must consider. One measure of the frequency performance of an op amp is its ***slew rate,*** which is the rate at which the output voltage can respond to changes in the input; it is most often expressed in V/μs. The typical slew rate specification for several commercially available devices is provided in Table 6.3, showing values on the order of a few volts per microsecond. One notable exception is the OPA690, which is designed as a high-speed op amp for video applications requiring operation at several hundred MHz. As can be seen, a respectable slew rate of 1800 V/μs is not unrealistic for this device, although its other parameters, particularly input bias current and CMRR, suffer somewhat as a result.

The PSpice simulations shown in Fig. 6.30 illustrate the degradation in performance of an op amp due to slew rate limitations. The circuit simulated is an LF411 configured as a noninverting amplifier with a gain of 2 and powered by ±15 V supplies. The input waveform is shown in green, and has

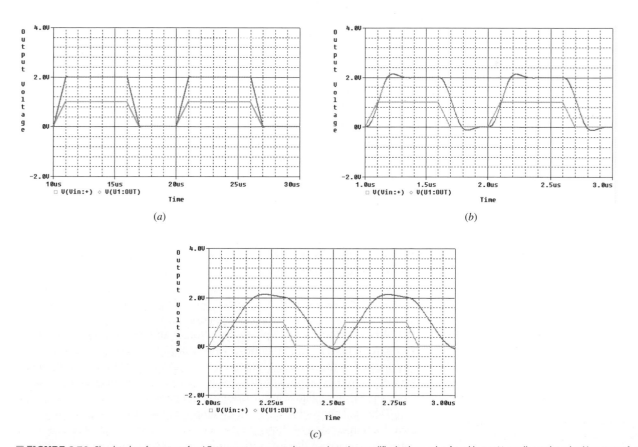

■ **FIGURE 6.30** Simulated performance of an LF411 op amp connected as a noninverting amplifier having a gain of 2, with ±15 V supplies and a pulsed input waveform. (*a*) Rise and fall times = 1 μs, pulse width = 5 μs; (*b*) Rise and fall times = 100 ns, pulse width = 500 ns; (*c*) Rise and fall times = 50 ns, pulse width = 250 ns.

a peak voltage of 1 V; the output voltage is shown in red. The simulation of Fig. 6.30a corresponds to a rise and fall time of 1 μs, which although a short time span for humans, is easily coped with by the LF411. As the rise and fall times are decreased by a factor of 10 to 100 ns (Fig. 6.30b), we begin to see that the LF411 is having a small difficulty in tracking the input. In the case of a 50 ns rise and fall time (Fig. 6.30c), we see that not only is there a significant delay between the output and the input, but the waveform is noticeably distorted as well—not a good feature of an amplifier. This observed behavior is consistent with the typical slew rate of 15 V/μs specified in Table 6.3, which indicates that the output might be expected to require roughly 130 ns to change from 0 to 2 V (or 2 V to 0 V).

Packaging

Modern op amps are available in a number of different types of packages. Some styles are better suited to high temperatures, and there are a variety of different ways to mount ICs on printed-circuit boards. Figure 6.31 shows several different styles of the LM741, manufactured by National Semiconductor. The label "NC" next to a pin means "no connection." The package styles shown in the figure are standard configurations, and are used for a large number of different integrated circuits; occasionally there are more pins available on a package than required.

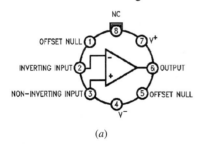

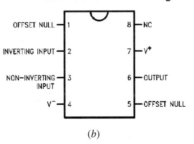

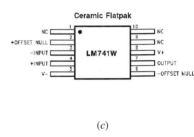

(a) (b) (c)

■ **FIGURE 6.31** Several different package styles for the LM741 op amp. (*a*) metal can; (*b*) dual-in-line package; (*c*) ceramic flatpak. (© *2000 National Semiconductor Corporation/www.national.com*).

COMPUTER-AIDED ANALYSIS

As we have just seen, PSpice can be enormously helpful in predicting the output of an op amp circuit, especially in the case of time-varying inputs. We will find, however, that our ideal op amp model agrees fairly well with PSpice simulations as a general rule.

When performing a PSpice simulation of an op amp circuit, we must be careful to remember that positive and negative dc supplies must be connected to the device. Although the model shows the offset null pins used to zero the output voltage, PSpice does not build in any offset, so these pins are typically left floating (unconnected).

Table 6.3 shows the different op amp part numbers available in the Evaluation version of PSpice; other models are available in the commercial version of the software and from some manufacturers.

EXAMPLE **6.7**

Simulate the circuit of Fig. 6.3 using PSpice. Determine the point(s) at which saturation begins if ±15 V dc supplies are used to power the device. Compare the gain calculated by PSpice to what was predicted using the ideal op amp model.

We begin by drawing the inverting amplifier circuit of Fig. 6.3 using the schematic capture tool as shown in Fig. 6.32. Note that two separate 15 V dc supplies are required to power the op amp.

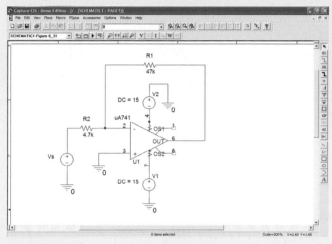

■ **FIGURE 6.32** The inverting amplifier of Fig. 6.3 drawn using a μA741 op amp.

Our previous analysis using an ideal op amp model predicted a gain of -10. With an input of $5 \sin 3t$ mV, this led to an output voltage of $-50 \sin 3t$ mV. However, an implicit assumption in the analysis was that *any* voltage input would be amplified by a factor of -10. Based on practical considerations, we expect this to be true for *small* input voltages, but the output will eventually saturate to a value comparable to the corresponding power supply voltage.

We perform a dc sweep from -2 to $+2$ volts, as shown in Fig. 6.33; this is a slightly larger range than the supply voltage divided by the gain, so we expect our results to include the positive and negative saturation regions.

As we can see using the cursor tool on the simulation results shown in Fig. 6.34a (expanded in Fig. 6.34b for clarity), the input-output characteristic of the amplifier is indeed linear over a wide input range, corresponding approximately to $-1.45 < V_s < +1.45$ V. This range is slightly less than the range defined by dividing the positive and negative supply voltages by the gain. Outside this range, the output of the op amp saturates, with only a slight dependence on the input voltage. In the two saturation regions, then, the circuit does not perform as a linear amplifier.

Increasing the number of cursor digits (**Tools**, **Options**, **Number of Cursor Digits**) to 10, we find that at an input voltage of $V_s = 1.0$ V, the

(Continued on next page)

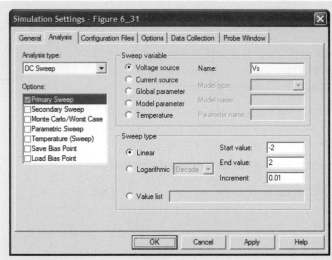

■ **FIGURE 6.33** DC sweep setup window.

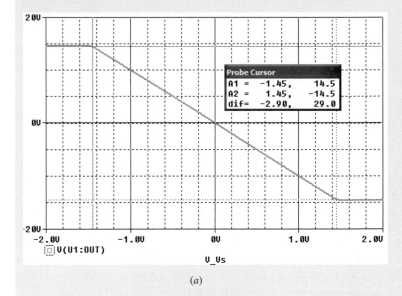

(*a*)

```
Probe Cursor
A1 =   -1.45,      14.5
A2 =    1.45,     -14.5
dif=   -2.91,      29.0
```

(*b*)

■ **FIGURE 6.34** (*a*) Output voltage of the inverting amplifer circuit, with the onset of saturation identified with the cursor tool. (*b*) Close-up of the cursor window.

output voltage is −9.99548340, slightly less than the value of −10 predicted from the ideal op amp model, and slightly different from the value of −9.999448 obtained in Example 6.6 using an analytical model. Still, the results predicted by the PSpice μA741 model are within a few hundredths of a percent of either analytical model, demonstrating that the ideal op amp model is indeed a remarkably accurate approximation for modern operational amplifier integrated circuits.

PRACTICE

6.7 Simulate the remaining op amp circuits described in this chapter, and compare the results to those predicted using the ideal op amp model.

6.6 COMPARATORS AND THE INSTRUMENTATION AMPLIFIER

The Comparator

Every op amp circuit we have discussed up to now has featured an electrical connection between the output pin and the inverting input pin. This is known as *closed-loop* operation, and is used to provide negative feedback as discussed previously. Closed loop is the preferred method of using an op amp as an amplifier, as it serves to isolate the circuit performance from variations in the open-loop gain that arise from changes in temperature or manufacturing differences. There are a number of applications, however, where it is advantageous to use an op amp in an *open-loop* configuration. Devices intended for such applications are frequently referred to as **comparators,** as they are designed slightly differently from regular op amps in order to improve their speed in open-loop operation.

Figure 6.35a shows a simple comparator circuit where a 2.5 V reference voltage is connected to the noninverting input, and the voltage being compared (v_{in}) is connected to the inverting input. Since the op amp has a very large open-loop gain A (10^5 or greater, typically, as seen in Table 6.3), it does not take a large voltage difference between the input terminals to drive it into saturation. In fact, a differential input voltage as small as the supply voltage divided by A is required—approximately $\pm120\ \mu$V in the case of the circuit in Fig. 6.35a and $A = 10^5$. The distinctive output of the comparator circuit is shown in Fig. 6.35b, where the response swings

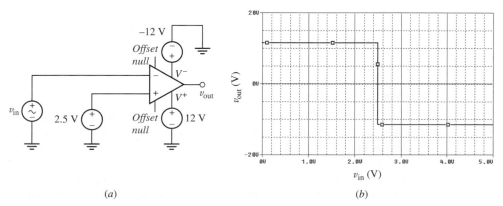

(a)   (b)

■ **FIGURE 6.35** (a) An example comparator circuit with a 2.5 V reference voltage. (b) Graph of input-output characteristic.

between positive and negative saturation, with essentially no linear "amplification" region. Thus, a positive 12 V output from the comparator indicates that the input voltage is *less than* the reference voltage, and a negative 12 V output indicates an input voltage *greater than* the reference. Opposite behavior is obtained if we connect the reference voltage to the inverting input instead.

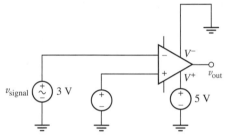

EXAMPLE **6.8**

Design a circuit that provides a "logic 1" 5 V output if a certain voltage signal drops below 3 V, and zero volts otherwise.

Since we want the output of our comparator to swing between 0 and 5 V, we will use an op amp with a single-ended +5 V supply, connected as shown in Fig. 6.36. We connect a +3 V reference voltage to the noninverting input, which may be provided by two 1.5 V batteries in series, or a suitable Zener diode reference circuit. The input voltage signal (designated v_{signal}) is then connected to the inverting input. In reality, the saturation voltage range of a comparator circuit will be slightly less than that of the supply voltages, so that some adjustment may be required in conjunction with simulation or testing.

■ **FIGURE 6.36** One possible design for the required circuit.

PRACTICE

6.8 Design a circuit that provides a 12 V output if a certain voltage (v_{signal}) exceeds 0 V, and a −2 V output otherwise.

Ans: One possible solution is shown in Fig. 6.37.

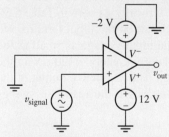

■ **FIGURE 6.37** One possible solution to Practice Problem 6.8.

The Instrumentation Amplifier

The basic comparator circuit acts on the voltage difference between the two input terminals to the device, although it does not technically amplify signals as the output is not proportional to the input. The difference amplifier of Fig. 6.10 also acts on the voltage difference between the inverting and noninverting inputs, and as long as care is taken to avoid saturation, *does* provide an output directly proportional to this difference. When dealing with a very small input voltage, however, a better alternative is a device known as an ***instrumentation amplifier,*** which is actually three op amp devices in a single package.

An example of the common instrumentation amplifier configuration is shown in Fig. 6.38a, and its symbol is shown in Fig. 6.38b. Each input is fed directly into a voltage follower stage, and the output of both voltage followers is fed into a difference amplifier stage. It is particularly well-suited to applications where the input voltage signal is very small (for example, on the order of millivolts), such as that produced by thermocouples or strain gauges, and where a significant common mode noise signal of several volts may be present.

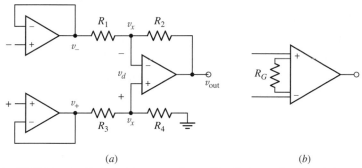

(a)   (b)

■ **FIGURE 6.38** (a) The basic instrumentation amplifier. (b) Commonly used symbol.

If components of the instrumentation amplifier are fabricated all on the same silicon "chip," then it is possible to obtain well-matched device characteristics and to achieve precise ratios for the two sets of resistors. In order to maximize the CMRR of the instrumentation amplifier, we expect $R_4/R_3 = R_2/R_1$, so that equal amplification of common-mode components of the input signals is obtained. To explore this further, we identify the voltage at the output of the top voltage follower as "v_-," and the voltage at the output of the bottom voltage follower as "v_+." Assuming all three op amps are ideal and naming the voltage at either input of the difference stage v_x, we may write the following nodal equations:

$$\frac{v_x - v_-}{R_1} + \frac{v_x - v_{out}}{R_2} = 0 \qquad [20]$$

and

$$\frac{v_x - v_+}{R_3} + \frac{v_x}{R_4} = 0 \qquad [21]$$

Solving Eq. [21] for v_x we find that

$$v_x = \frac{v_+}{1 + R_3/R_4} \qquad [22]$$

and upon substituting into Eq. [20] obtain an expression for v_{out} in terms of the input:

$$v_{out} = \frac{R_4}{R_3}\left(\frac{1 + R_2/R_1}{1 + R_4/R_3}\right)v_+ - \frac{R_2}{R_1}v_- \qquad [23]$$

From Eq. [23] it is clear that the general case allows amplification of common-mode components to the two inputs. In the specific case where

$R_4/R_3 = R_2/R_1 = K$, however, Eq. [23] reduces to $K(v_+ - v_-) = Kv_d$, so that (asssuming ideal op amps) only the difference is amplified and the gain is set by the resistor ratio. Since these resistors are internal to the instrumentation amplifier and not accessible to the user, practical devices such as the AD622 allow the gain to be set anywhere in the range of 1 to 1000 by connecting an external resistor between two pins (shown as R_G in Fig. 6.38b).

SUMMARY AND REVIEW

❑ There are two fundamental rules that must be applied when analyzing *ideal* op amp circuits:

   1. No current ever flows into either input terminal.

   2. No voltage ever exists between the input terminals.

❑ Op amp circuits are usually analyzed for an output voltage in terms of some input quantity or quantities.

❑ Nodal analysis is typically the best choice in analyzing op amp circuits, and it is usually better to begin at the input, and work toward the output.

❑ The output current of an op amp cannot be assumed; it must be found after the output voltage has been determined independently.

❑ The gain of an inverting op amp circuit is given by the equation

$$v_{\text{out}} = -\frac{R_f}{R_1}v_{\text{in}}$$

❑ The gain of a noninverting op amp circuit is given by the equation

$$v_{\text{out}} = \left(1 + \frac{R_f}{R_1}\right)v_{\text{in}}$$

❑ A resistor is almost always connected from the output pin of an op amp to its inverting input pin, which incorporates negative feedback into the circuit for increased stability.

❑ The ideal op amp model is based on the approximation of infinite open-loop gain A, infinite input resistance R_i, and zero output resistance R_o.

❑ In practice, the output voltage range of an op amp is limited by the supply voltages used to power the device.

READING FURTHER

Two very readable books which deal with a variety of op amp applications are:

R. Mancini (ed.), *Op Amps Are For Everyone,* 2nd ed. Amsterdam: Newnes, 2003. Also available on the Texas Instruments website (www.ti.com).

W. G. Jung, *Op Amp Cookbook,* 3rd ed. Upper Saddle River, N.J.: Prentice Hall, 1997.

Characteristics of Zener and other types of diodes are covered in Chapter 1 of

W. H. Hayt, Jr. and G. W. Neudeck, *Electronic Circuit Analysis and Design,* 2nd ed. New York: Wiley, 1995.

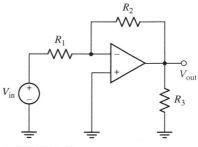

One of the first reports of the implementation of an "operational amplifier" can be found in

> J. R. Ragazzini, R. M. Randall, and F. A. Russell, "Analysis of problems in dynamics by electronic circuits," *Proceedings of the IRE* **35**(5), 1947, pp. 444–452.

And an early applications guide for the op amp can be found on the Analog Devices, Inc. website (www.analog.com):

> George A. Philbrick Researches, Inc., *Applications Manual for Computing Amplifiers for Modelling, Measuring, Manipulating & Much Else.* Norwood, Mass.: Analog Devices, 1998.

EXERCISES

6.2 The Ideal Op Amp

1. For the op amp circuit of Fig. 6.39, calculate V_{out} if (a) $V_{in} = 3$ V, $R_1 = 10$ Ω and $R_2 = 100$ Ω; (b) $V_{in} = 2.5$ V, $R_1 = 1$ MΩ and $R_2 = 1$ MΩ; (c) $V_{in} = -1$ V, $R_1 = 3.3$ kΩ and $R_2 = 4.7$ kΩ.

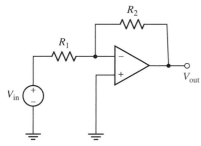

■ **FIGURE 6.39**

2. For the op amp circuit of Fig. 6.40, calculate V_{out} if (a) $V_{in} = 1.5$ V, $R_3 = 10$ Ω, $R_1 = 10$ Ω, and $R_2 = 47$ Ω; (b) $V_{in} = -9$ V, $R_3 = 1$ kΩ, $R_1 = 1$ MΩ, and $R_2 = 1$ MΩ; (c) $V_{in} = 100$ mV, $R_3 = 330$ Ω, $R_1 = 1$ kΩ, and $R_2 = 6.8$ kΩ.

3. Sketch the output voltage v_{out} of the op amp circuit shown in Fig. 6.41 if (a) $v_{in} = 2 \sin 5t$ V; (b) $v_{in} = 1 + 0.5 \sin 5t$ V.

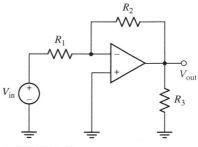

■ **FIGURE 6.40**

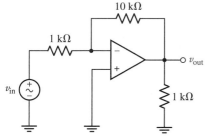

■ **FIGURE 6.41**

4. Sketch the output voltage v_{out} of the op amp circuit shown in Fig. 6.42 if (a) $v_{in} = 10 \cos 4t$ V; (b) $v_{in} = 15 + 4 \cos 4t$ V.

Ⓓ 5. Design a circuit that delivers -9 V to a 47 kΩ load, if only ± 5 V supplies are available. (For the purposes of this problem, there is no need to include the power supplies that actually power the op amp, which are not restricted to ± 5 V.)

Ⓓ 6. Design a circuit to deliver $+20$ V to a 1 kΩ load, if only ± 5 V supplies are available. (For the purposes of this problem, there is no need to include the power supplies that actually power the op amp, which are not restricted to ± 5 V.)

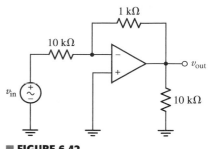

■ **FIGURE 6.42**

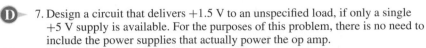

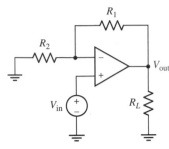

FIGURE 6.43

(D) 7. Design a circuit that delivers $+1.5$ V to an unspecified load, if only a single $+5$ V supply is available. For the purposes of this problem, there is no need to include the power supplies that actually power the op amp.

(D) 8. Design a circuit to deliver $+3$ V to an unspecified load, if only a single $+9$ V supply is available. For the purposes of this problem, there is no need to include the power supplies that actually power the op amp.

9. For the op amp circuit of Fig. 6.43, calculate V_{out} if (a) $V_{in} = 300$ mV, $R_2 = 10$ Ω, and $R_1 = 47$ Ω; (b) $V_{in} = 1.5$ V, $R_1 = 1$ MΩ, and $R_2 = 1$ MΩ; (c) $V_{in} = -1$ V, $R_1 = 4.7$ kΩ, and $R_2 = 3.3$ kΩ.

10. For the op amp circuit of Fig. 6.44, calculate V_{out} if (a) $V_{in} = 200$ mV, $R_L = 10$ kΩ, $R_1 = 10$ Ω, and $R_2 = 47$ Ω; (b) $V_{in} = -9$ V, $R_L = 1$ kΩ, $R_1 = 1$ MΩ, and $R_2 = 1$ MΩ; (c) $V_{in} = 100$ mV, $R_L = 330$ Ω, $R_1 = 1$ kΩ, and $R_2 = 6.8$ kΩ.

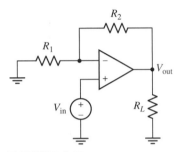

FIGURE 6.44

11. In the op amp circuit shown in Fig. 6.45, $R_1 = R_f = 1$ kΩ. Sketch the output voltage v_{out} if (a) $v_{in} = 4 \sin 10t$ V; (b) $v_{in} = 1 + 0.25 \sin 10t$ V.

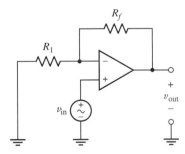

FIGURE 6.45

12. In the op amp circuit shown in Fig. 6.45, $R_1 = 2$ kΩ and $R_f = 1$ kΩ. Sketch the output voltage v_{out} if (a) $v_{in} = 2 \cos 2t$ V; (b) $v_{in} = 4 + \cos 2t$ V.

13. Referring to Fig. 6.46, calculate the voltage v_{out}.

14. In the circuit of Fig. 6.47, what value of R is required so that 150 mW is delivered to the 10 kΩ resistor?

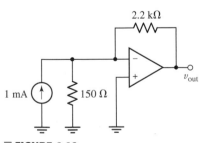

FIGURE 6.46

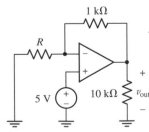

FIGURE 6.47

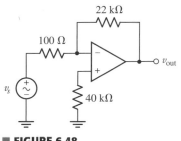

D 15. A certain microphone is capable of delivering 0.5 V when someone claps their hands at a distance of 20 ft. A particular electronic switch has a Thévenin equivalent resistance of 670 Ω, and requires 100 mA to energize. Design a circuit that will connect the microphone to the electronic switch in such a way that the switch is activated by someone clapping their hands.

16. For the circuit of Fig. 6.48, derive an expression for v_{out} in terms of v_s.

17. For the circuit of Fig. 6.49, calculate the voltage V_1.

■ FIGURE 6.48

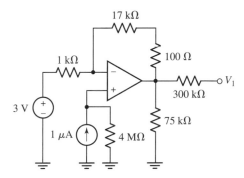

■ FIGURE 6.49

18. For the circuit of Fig. 6.50, calculate the voltage V_2.

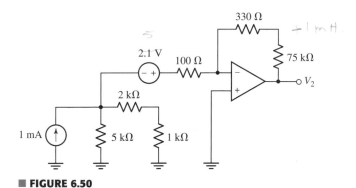

■ FIGURE 6.50

19. Find an expression for v_{out} in the circuit of Fig. 6.51, and evaluate it at $t = 3$ seconds.

20. What value of V_{in} will lead to an output voltage of 18 V in the circuit of Fig. 6.52?

■ FIGURE 6.51

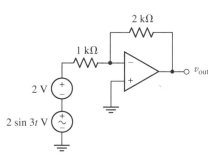

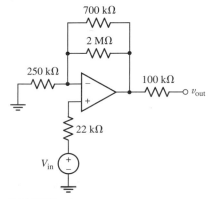

■ FIGURE 6.52

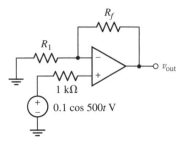

■ **FIGURE 6.53**

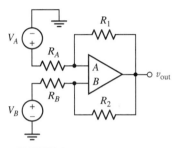

■ **FIGURE 6.55**

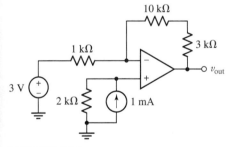

■ **FIGURE 6.56**

(D) 21. Choose R_1 and R_f in Fig. 6.53 to obtain $v_{out} = 23.7 \cos 500t$ volts.

22. Derive an expression for v_{out} for the circuit of Fig. 6.54 without using source transformations.

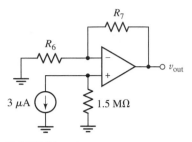

■ **FIGURE 6.54**

23. Referring to the circuit of Fig. 6.55,
    (a) If $V_A = 0$, $V_B = 1$ V, $R_A = R_B = 10$ kΩ, $R_1 = 70$ kΩ, $R_2 = \infty$, and $v_{out} = 8$ V, which terminal (A or B) is the noninverting input? Explain.
    (b) $V_A = 10$ V and $V_B = 0$ V. If B is the inverting input, select R_A, R_B, R_1, and R_2 to obtain an output voltage of 20 V.
    (c) $V_A = V_B = 1$ V, $R_1 = 0$, and $R_2 = \infty$. If V_{out} is found to be 1 V, which terminal (A or B) is the inverting input? Explain.

24. For the op amp circuit of Fig. 6.56, calculate v_{out}.

25. If $v_s = 5 \sin 3t$ mV in the circuit of Fig. 6.57, calculate v_{out} at $t = 0.25$ s.

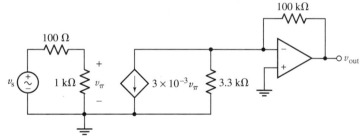

■ **FIGURE 6.57**

26. Use appropriate circuit analysis techniques to calculate v_{out} for the circuit of Fig. 6.58.

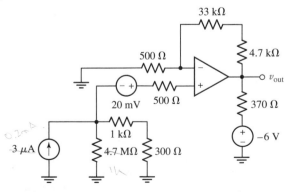

■ **FIGURE 6.58**

27. In Fig. 6.58, replace the 3μA source with a 27μA source and compute v_{out}.

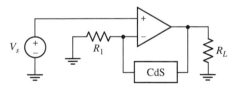

28. Compute v_x for the multiple op amp circuit of Fig. 6.59.

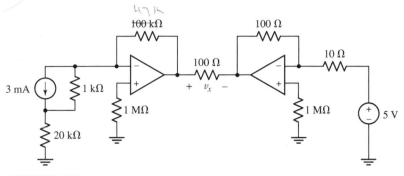

■ **FIGURE 6.59**

29. Derive an expression for the general summing amplifier, in which each resistor can be a different value.

30. Derive an expression for the general difference amplifier, in which each resistor can have a different value.

(D) 31. Cadmium sulfide (CdS) is commonly used to fabricate resistors whose value depends on the intensity of light shining on its surface. In Fig. 6.60 a CdS "photocell" is used as the feedback resistor R_f. In total darkness, it has a resistance of 100 kΩ, and a resistance of 10 kΩ under a light intensity of 6 candela. R_L represents a circuit that is activated when a voltage of 1.5 V or less is applied to its terminals. Choose R_1 and V_s so that the circuit represented by R_L is activated by a light of 2 candela or brighter.

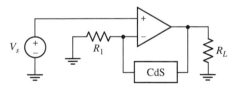

■ **FIGURE 6.60**

(D) 32. Two different microphones are used in a recording studio, one for vocals and one for instruments. Design a circuit that will allow both microphone outputs to be combined, but with the vocals receiving twice the amplification of the instruments.

(D) 33. A sinusoidal signal is riding on a 2 V dc offset (in other words, the average value of the total signal is 2 V). Design a circuit to remove the dc offset, and amplify the sinusoidal signal (without phase reversal) by a factor of 100.

6.3 Cascaded Stages

(D) 34. Design a circuit that provides an output voltage equal to the average of three input voltages v_1, v_2, and v_3.

(D) 35. An electronic warehouse inventory system uses scales placed under each pallet; the output of any scale is calibrated to provide 1 mV for each kg. Design a circuit that provides a voltage output proportional to the total weight of a group of similar items (distributed over four pallets) remaining in stock, with the tare weight of each pallet subtracted off (the tare weight is provided as a reference voltage for each pallet). Your output voltage should be calibrated such that 1 mV corresponds to 1 kg.

D 36. A certain manufacturer's radar gun for measuring vehicular speed provides a voltage output proportional to the speed of the object being targeted, such that 10 mV = 1 mph. If the speedometer of the police vehicle is tapped to provide a signal proportional to its speed as well, also such that 10 mV = 1 mph, design a multistage circuit that (*a*) provides a voltage signal equal to the difference in velocity of the speeding car and the police vehicle (such that +10 mV = police vehicle slower by 1 mph) and (*b*) provides a voltage signal for each of the three quantities in kph, such that 10 mV = 1 kph.

37. Compute v_{out} for the circuit of Fig. 6.61.

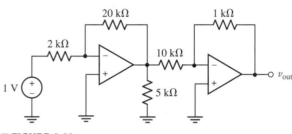

■ **FIGURE 6.61**

38. Derive an expression for V_{out} in terms of V_1 and V_2 for the circuit of Fig. 6.62.

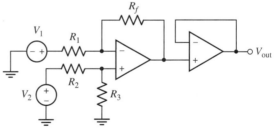

■ **FIGURE 6.62**

39. For the cascaded op amp circuit shown in Fig. 6.63, compute the output voltage of each stage.

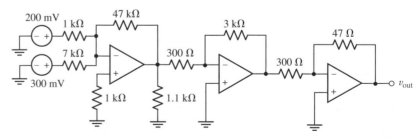

■ **FIGURE 6.63**

40. Referring to the op amp circuit of Fig. 6.64, what value of R is required to obtain $V_{out} = 10$ V?

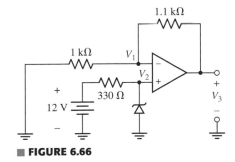

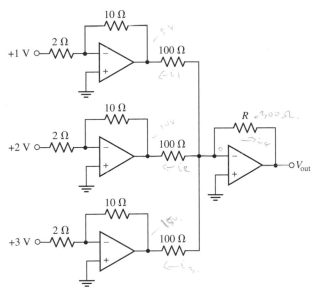

FIGURE 6.64

41. Compute v_{out} for the two-stage op amp circuit of Fig. 6.65.

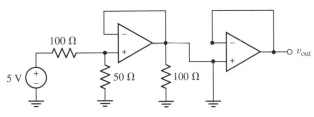

FIGURE 6.65

6.4 Circuits for Voltage and Current Sources

42. (a) The circuit in Fig. 6.66 uses a 1N750 diode, which is characterized by a Zener voltage of 4.7 V. Determine the voltages labeled V_1, V_2, and V_3. (b) Verify your analysis with an appropriate PSpice simulation. Submit a properly labeled schematic and comment on the possible sources of any differences between the two analyses.

43. Design a circuit to supply a +5.1 V reference voltage as the input to a voltage follower if only 9 V batteries are available. Use a 1N4733 diode, which has a Zener voltage of 5.1 V at a current of 76 mA.

44. Design a circuit to supply a −2.5 V reference voltage as the input to a voltage follower if only 9 V batteries are available. Use a 1N4740 diode, which has a Zener voltage of 10 V at a current of 25 mA.

45. Design a circuit to supply a +12 V reference voltage as the input to a voltage follower if only 9 V batteries are available. Use a 1N4747 diode, which has a Zener voltage of 20 V at a current of 12.5 mA.

46. (a) Design a circuit to supply a −5 V reference voltage as the input to a voltage follower if only 9 V batteries are available. Use a 1N4728 diode, which has a Zener voltage of 3.3 V at a current of 76 mA. (b) Modify your design to provide a +2.2 V reference voltage instead.

47. Design a current source circuit that can provide 25 mA to an unspecified load. Use a 1N4740 Zener diode, which has a breakdown voltage of 10 V at a current of 25 mA.

FIGURE 6.66

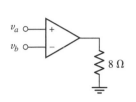

■ FIGURE 6.67

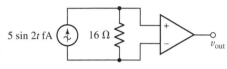

■ FIGURE 6.68

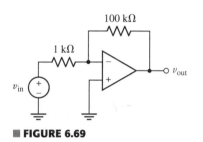

■ FIGURE 6.69

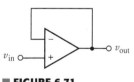

■ FIGURE 6.71

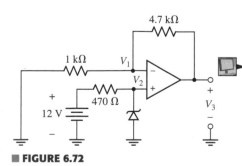

■ FIGURE 6.72

(D) 48. Design a current source circuit that can provide 12.5 mA to an unspecified load. Use a 1N4747 Zener diode, which has a breakdown voltage of 20 V at a current of 12.5 mA.

(D) 49. Design a current source circuit that can provide 75 mA to an unspecified load. Use a 1N4747 Zener diode, which has a breakdown voltage of 20 V at a current of 12.5 mA. If the amplifier is powered using ± 15 V supplies, what range of loads is possible with your design?

50. Using the detailed model for a μA741, determine the power delivered to the 8 Ω resistor of Fig. 6.67 if

    (a) $v_a = v_b = 1$ nV; (b) $v_a = 0$, $v_b = 1$ nV; (c) $v_a = 2$ pV, $v_b = 1$ fV; (d) $v_a = 50$ μV, $v_b = -4$ μV.

51. An inverting op amp circuit is constructed with an AD549. If $R_1 = 270$ kΩ and $R_f = 1$ MΩ, what input bias current is expected for:

    (a) $V_S = 1$ mV; (b) $V_S = -7.5$ mV; (c) $V_S = 1$ V?

52. Compute v_{out} for the circuit of Fig. 6.68 if

    (a) $A = 10^5$, $R_i = 100$ MΩ, and $R_o = 0$

    (b) $A = 10^6$, $R_i = 1$ TΩ, and $R_o = 0$

53. For the circuit of Fig. 6.69(a) derive an expression for $v_{\text{out}}/v_{\text{in}}$ if $R_i = \infty$, $R_o = 0$, and A is finite. (b) What value of open-loop gain A is required for the closed-loop gain to be within 1 percent of its ideal value?

54. For the circuit of Fig. 6.70, compute the power dissipated by the 8 Ω resistor if $\delta = $ (a) 0 V; (b) 1 nV; (c) 2.5 μV.

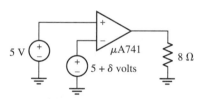

■ FIGURE 6.70

55. Using the parameters for an AD549, calculate v_{out} for the circuit of Fig. 6.71 if $v_{\text{in}} = -16$ mV.

6.5 Practical Considerations

56. Derive an expression for the output voltage of a voltage follower in terms of the input voltage v_{in} for the case of finite open-loop gain and input resistance and nonzero output resistance. Verify that your expression reduces to $v_{\text{out}} = v_{\text{in}}$ for the case of an ideal op amp.

57. (a) Construct a detailed op amp model that includes a common-mode gain A_{CM} contribution to the output voltage. (b) Use your model with $A = 10^5$, $R_i = \infty$, $R_o = 0$, and $A_{\text{CM}} = 10$ to analyze the circuit of Fig. 6.25 with $v_1 = 5 + 2\sin t$ V and $v_2 = 5$. (c) Compare your answer to what would be obtained if $A_{\text{CM}} = 0$.

58. Define *slew rate,* and explain its significance to the output waveform of an op amp circuit.

59. The circuit in Fig. 6.72 uses a 1N750 diode, which is characterized by a Zener voltage of 4.7 V. (a) Determine the voltages labeled V_1, V_2, and V_3. (b) Verify your answer by performing a PSpice simulation, employing a μA741 op amp and ± 18 V power supplies. Submit a properly labeled schematic with your results. If your simulation does not agree exactly with your hand calculations, attempt to determine the origin of the discrepancy. (c) What is the minimum value to which the 12 V source can be reduced before the Zener circuit stops performing its function?

60. Perform a PSpice simulation of an inverting op amp circuit using a μA741, $\pm$15 V supplies, $R_1 = 10$ kΩ, and $R_f = 1$ MΩ. Plot the input-output characteristics, and label the linear and positive/negative saturation regions. Is the gain predicted by simulation in agreement with that predicted by the ideal op amp model?

61. If we know that our applications will only require inverting op amp configurations, and are not concerned with being able to trim the output voltage, what is the minimum number of pins required for our op amp package? List each by name.

62. Use PSpice to simulate the circuit in Fig. 6.73 using a(n) (a) μA741; (b) LM324; (c) LF411. Determine the differential input voltage V_S required to saturate each type of op amp using $\pm$15 V supplies. (d) Compare your results to what you would expect based on the information in Table 6.3.

63. Perform a PSpice simulation of a noninverting op amp circuit using a μA741, $\pm$15 V supplies, $R_1 = 4.7$ kΩ, and $R_f = 1$ MΩ. Plot the input-output characteristics, and label the linear and positive/negative saturation regions. Is the gain in agreement with that predicted by the ideal op amp model?

64. Use PSpice and the circuit shown in Fig. 6.74 to determine the output resistance of the μA741 and the LF411. Vary the supply voltage, and determine if it affects the simulation results. Do the simulations agree with the values of Table 6.3?

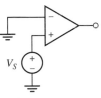

V_S

■ **FIGURE 6.73**

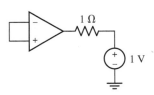

1 Ω

1 V

■ **FIGURE 6.74**

65. Simulate the circuit of Fig. 6.75 using an LF411. Determine the input bias current and the differential input voltage. Compare these results to values predicted using the detailed model and the values listed in Table 6.3.

66. A sensor provides a signal voltage between -30 mV and $+75$ mV. (a) If an inverting amplifier is used with a voltage gain of $|v_{out}/v_{in}| = 1000$ and $\pm$15 V dc supplies, what is the expected output voltage range? (b) If a noninverting amplifier is used with $\pm$15 V dc supplies, what is the maximum resistance ratio R_f/R_1 that can be used without the op amp saturating?

67. (a) Simulate the circuit shown in Fig. 6.76 using a μA741 over the range -10 V $\leq V_{in} \leq +10$ V. Determine the precise voltages at which saturation begins using the cursor tool. Compare your results to what would be predicted using Table 6.3.

(b) A real μA741 op amp is capable of providing up to 35 mA of current under continuous short-circuit conditions. Determine the maximum possible short-circuit current allowed by the model used in PSpice.

68. In an attempt to improve the security of transmission, a chaotic time-varying signal is added to an audio signal prior to its being broadcast. The same chaotic time-varying signal is also broadcast on a separate frequency. Assuming that any receiving antenna can be modeled as a time-varying voltage source in series with a 300 Ω resistance, design a circuit to separate the two signals, discard the chaotic signal, amplify the audio signal by a factor of 10, and deliver the result to an 8 Ω speaker.

69. Design an op amp circuit that will provide an output voltage equal to the average of three input voltages. You may assume that the input voltages will be confined to the range -10 V $\leq V_{in} \leq +10$ V. Verify your design with PSpice and a suitable set of input voltages.

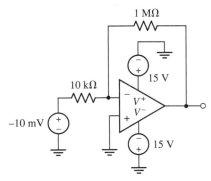

1 MΩ

15 V

10 kΩ

-10 mV

15 V

■ **FIGURE 6.75**

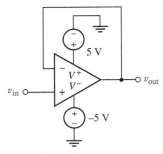

5 V

v_{in}

v_{out}

-5 V

■ **FIGURE 6.76**

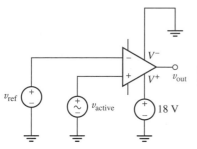

■ FIGURE 6.77

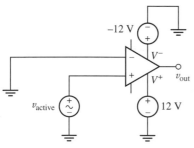

■ FIGURE 6.78

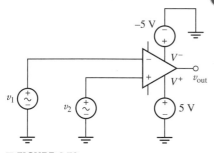

■ FIGURE 6.79

6.6 Comparators and the Instrumentation Amplifier

70. For the circuit depicted in Fig. 6.77, sketch the expected output voltage v_{out} as a function of v_{active} for $-5\text{ V} \le v_{active} \le +5\text{ V}$, if v_{ref} is equal to (a) -3 V; (b) $+3$ V.

71. For the circuit depicted in Fig. 6.78, sketch the expected output voltage v_{out} as a function of v_{active}, if $-2\text{ V} \le v_{active} \le +2\text{ V}$. Verify your solution using a μA741 (although it is slow compared to op amps designed specifically for use as comparators, its PSpice model works well, and as this is a dc application speed is not an issue). Submit a properly labeled schematic with your results.

72. For the circuit depicted in Fig. 6.79, (a) sketch the expected output voltage v_{out} as a function of v_1 for $-5\text{ V} \le v_1 \le +5\text{ V}$, if $v_2 = +2\text{ V}$; (b) sketch the expected output voltage v_{out} as a function of v_2 for $-5\text{ V} \le v_2 \le +5\text{ V}$, if $v_1 = +2\text{ V}$.

73. In digital logic applications, a $+5$ V signal represents a logic "1" state, and a 0 V signal represents a logic "0" state. In order to process real-world information using a digital computer, some type of interface is required, which typically includes an analog-to-digital (A/D) converter—a device that converts analog signals into digital signals. Design a circuit that acts as a simple 1-bit A/D, with any signal less than 1.5 V resulting in a logic "0" and any signal greater than 1.5 V resulting in a logic "1".

74. For the instrumentation amplifier shown in Fig. 6.38a, assume that the three internal op amps are ideal, and determine the CMRR of the circuit if (a) $R_1 = R_3$ and $R_2 = R_4$; (b) all four resistors are different values.

75. A common application for instrumentation amplifiers is measuring voltages in resistive strain gauge circuits. These strain sensors work by exploiting the changes in resistance that result from geometric distortions, as in Eq. [6] of Chap. 2. They are often part of a bridge circuit, as shown in Fig. 6.80a, where the strain gauge is identified as R_G. (a) Show that

$$V_{out} = V_{in}\left[\frac{R_2}{R_1+R_2} - \frac{R_3}{R_3+R_{Gauge}}\right].$$ (b) Verify that $V_{out} = 0$ when the three

fixed-value resistors R_1, R_2, and R_3 are all chosen to be equal to the unstrained gauge resistance R_{Gauge}. (c) For the intended application, the gauge selected has an unstrained resistance of 5 kΩ, and a maximum resistance increase of 50 mΩ is expected. Only ±12 V supplies are available. Using the instrumentation amplifier of Fig. 6.80b, design a circuit that will provide a voltage signal of $+1$ V when the strain gauge is at its maximum loading.

AD622 Specifications

Amplifier gain G can be varied from 2 to 1000 by connecting a resistor between pins 1 and 8 with a value calculated by $R = \dfrac{50.5}{G-1}$ kΩ.

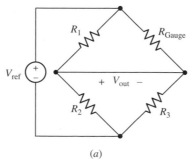

(a)

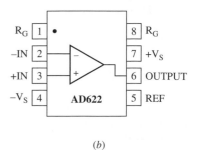

(b)

■ FIGURE 6.80

© Analog Devices.

Capacitors and Inductors

KEY CONCEPTS

The Voltage-Current Relationship of an Ideal Capacitor

The Current-Voltage Relationship of an Ideal Inductor

Calculating Energy Stored in Capacitors and Inductors

Analysis of Capacitor and Inductor Response to Time-Varying Waveforms

Series and Parallel Capacitor Combinations

Series and Parallel Inductor Combinations

Op Amp Circuits Using Capacitors

PSpice Modeling of Energy Storage Elements

INTRODUCTION

In this chapter we introduce two new passive circuit elements, the *capacitor* and the *inductor,* each of which has the ability to both store and deliver *finite* amounts of energy. They differ from ideal sources in this respect, since they cannot sustain a finite average power flow over an infinite time interval. Although classed as linear elements, the current-voltage relationships for these new elements are time-dependent, leading to many interesting circuits. As we are about to see, the range of capacitance and inductance values we might encounter can be huge, so that at times they may dominate circuit behavior, and at other times be essentially insignificant. Such issues continue to be relevant in modern circuit applications, particularly as computer and communication systems move to increasingly higher operating frequencies and component densities.

7.1 THE CAPACITOR

Ideal Capacitor Model

We previously termed the independent and dependent voltage and current sources active elements and the linear resistor a passive element, although our definitions of active and passive are still slightly fuzzy and need to be brought into sharper focus. We now define an *active element* as an element that is capable of furnishing an average power greater than zero to some external device, where the average is taken over an infinite time interval. Ideal sources are active elements, and the operational amplifier is also an active device. A *passive element,* however, is defined as an element that cannot supply an average power that is greater than zero over an infinite time interval. The resistor falls into this category; the energy it receives is usually transformed into heat, and it never supplies energy.

We now introduce a new passive circuit element, the **capacitor.** We define capacitance C by the voltage-current relationship

$$i = C\frac{dv}{dt}$$

[1]

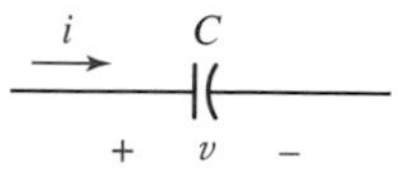

■ FIGURE 7.1 Electrical symbol and current-voltage conventions for a capacitor.

where v and i satisfy the conventions for a passive element, as shown in Fig. 7.1. We should bear in mind that v and i are functions of time; if needed, we can emphasize this fact by writing $v(t)$ and $i(t)$, instead. From Eq. [1], we may determine the unit of capacitance as an ampere-second per volt, or coulomb per volt. We will now define the *farad*[1] (F) as one coulomb per volt, and use this as our unit of capacitance.

The ideal capacitor defined by Eq. [1] is only a mathematical model of a real device. A capacitor consists of two conducting surfaces on which charge may be stored, separated by a thin insulating layer that has a very large resistance. If we assume that this resistance is sufficiently large that it may be considered infinite, then equal and opposite charges placed on the capacitor "plates" can never recombine, at least by any path within the element. The construction of the physical device is suggested by the circuit symbol shown in Fig. 7.1.

Let us visualize some external device connected to this capacitor and causing a positive current to flow into one plate of the capacitor and out of the other plate. Equal currents are entering and leaving the two terminals, and this is no more than we expect for any circuit element. Now let us examine the interior of the capacitor. The positive current entering one plate represents positive charge moving toward that plate through its terminal lead; this charge cannot pass through the interior of the capacitor, and it therefore accumulates on the plate. As a matter of fact, the current and the increasing charge are related by the familiar equation

$$i = \frac{dq}{dt}$$

Now let us consider this plate as an overgrown node and apply Kirchhoff's current law. It apparently does not hold; current is approaching the plate from the external circuit, but it is not flowing out of the plate into the "internal circuit." This dilemma bothered a famous Scottish scientist, James Clerk Maxwell, more than a century ago. The unified electromagnetic theory that he subsequently developed hypothesizes a "displacement current" that is present wherever an electric field or a voltage is varying with time. The displacement current flowing internally between the capacitor plates is exactly equal to the conduction current flowing in the capacitor leads; Kirchhoff's current law is therefore satisfied if we include both conduction and displacement currents. However, circuit analysis is not concerned with this internal displacement current, and since it is fortunately equal to the conduction current, we may consider Maxwell's hypothesis as relating the conduction current to the changing voltage across the capacitor.

A capacitor constructed of two parallel conducting plates of area A, separated by a distance d, has a capacitance $C = \varepsilon A/d$, where ε is the permittivity, a constant of the insulating material between the plates; this assumes

(1) Named in honor of Michael Faraday.

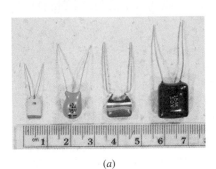
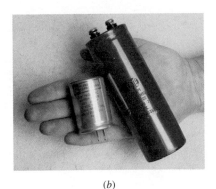
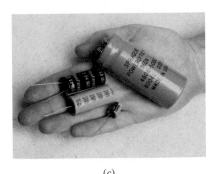

(a)   (b)   (c)

■ **FIGURE 7.2** Several examples of commercially available capacitors. (a) Left to right: 270 pF ceramic, 20 μF tantalum, 15 nF polyester, 150 nF polyester. (b) Left: 2000 μF 40 VDC rated electrolytic, 25,000 μF 35 VDC rated electrolytic. (c) Clockwise from smallest: 100 μF 63 VDC rated electrolytic, 2200 μF 50 VDC rated electrolytic, 55 F 2.5 VDC rated electrolytic, and 4800 μF 50 VDC rated electrolytic. Note that generally speaking larger capacitance values require larger packages, with one notable exception above. What was the tradeoff in that case?

the linear dimensions of the conducting plates are all very much greater than d. For air or vacuum, $\varepsilon = \varepsilon_0 = 8.854$ pF/m. Most capacitors employ a thin dielectric layer with a larger permittivity than air in order to minimize the device size. Examples of various types of commercially available capacitors are shown in Fig. 7.2, although we should remember that any two conducting surfaces not in direct contact with each other may be characterized by a nonzero (although probably small) capacitance. We should also note that a capacitance of several hundred *microfarads* (μF) is considered "large."

Several important characteristics of our new mathematical model can be discovered from the defining equation, Eq. [1]. A constant voltage across a capacitor results in zero current passing through it; a capacitor is thus an "*open circuit to dc.*" This fact is pictorially represented by the capacitor symbol. It is also apparent that a sudden jump in the voltage requires an infinite current. Since this is physically impossible, we will therefore prohibit the voltage across a capacitor to change in zero time.

EXAMPLE 7.1

Determine the current i flowing through the capacitor of Fig. 7.1 for the two voltage waveforms of Fig. 7.3 if $C = 2$ F.

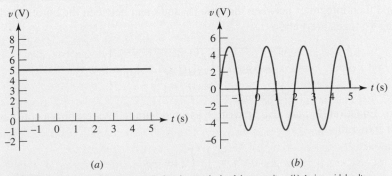

(a)   (b)

■ **FIGURE 7.3** (a) A dc voltage applied to the terminals of the capacitor. (b) A sinusoidal voltage waveform applied to the capacitor terminals.

(Continued on next page)

The current i is related to the voltage v across the capacitor by Eq. [1]:

$$i = C\frac{dv}{dt}$$

For the voltage waveform depicted in Fig. 7.3a, $dv/dt = 0$, so $i = 0$; the result is plotted in Fig. 7.4a. For the case of the sinusoidal waveform of Fig. 7.3b, we expect a cosine current waveform to flow in response, having the same frequency and twice the magnitude (since $C = 2$ F). The result is plotted in Fig. 7.4b.

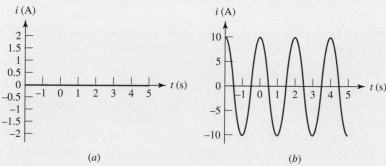

(a)   (b)

■ **FIGURE 7.4** (a) $i = 0$ as the voltage applied is dc. (b) The current has a cosine form in response to a sine-wave voltage.

PRACTICE

7.1 Determine the current flowing through a 5 mF capacitor in response to a voltage $v = :$ (a) -20 V; (b) $2e^{-5t}$ V.

Ans: (a) 0 A; (b) $-50e^{-5t}$ mA.

Integral Voltage-Current Relationships

The capacitor voltage may be expressed in terms of the current by integrating Eq. [1]. We first obtain

$$dv = \frac{1}{C}i(t)\,dt$$

and then integrate[2] between the times t_0 and t and between the corresponding voltages $v(t_0)$ and $v(t)$:

$$v(t) = \frac{1}{C}\int_{t_0}^{t} i(t')\,dt' + v(t_0) \qquad [2]$$

Equation [2] may also be written as an indefinite integral plus a constant of integration:

$$v(t) = \frac{1}{C}\int i\,dt + k$$

(2) Note that we are employing the mathematically correct procedure of defining a *dummy variable t'* in situations where the integration variable *t* is also a limit.

Finally, in many real problems we will find that $v(t_0)$, the voltage initially across the capacitor, is not able to be discerned. In such instances it will be mathematically convenient to set $t_0 = -\infty$ and $v(-\infty) = 0$, so that

$$v(t) = \frac{1}{C} \int_{-\infty}^{t} i\, dt'$$

Since the integral of the current over any time interval is the charge accumulated in that period on the capacitor plate into which the current is flowing, we may also define capacitance as

$$q(t) = Cv(t)$$

where $q(t)$ and $v(t)$ represent instantaneous values of the charge on either plate and the voltage between the plates, respectively.

EXAMPLE 7.2

Find the capacitor voltage that is associated with the current shown graphically in Fig. 7.5a.

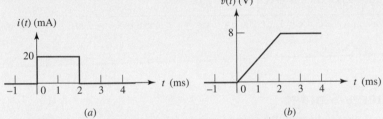

FIGURE 7.5 (a) The current waveform applied to a 5 μF capacitor. (b) The resultant voltage waveform obtained by graphical integration.

Interpreting Eq. [2] graphically, we know that the difference between the values of the voltage at t and t_0 is proportional to the area under the current curve between these same two values of time. The proportionality constant is $1/C$. The area can be obtained from Fig. 7.5a by inspection for desired values of t_0 and t. Let us select our starting point t_0 prior to zero time. For simplicity, the first interval of t is selected between $-\infty$ and 0, and since our waveform implies that no current has ever been applied to this capacitor since the beginning of time,

$$v(t_0) = v(-\infty) = 0$$

Referring to Eq. [2], the integral of the current between $t_0 = -\infty$ and 0 is simply zero, since $i = 0$ in that interval. Thus,

$$v(t) = 0 + v(-\infty) \qquad -\infty \le t \le 0$$

or

$$v(t) = 0 \qquad t \le 0$$

(Continued on next page)

If we now consider the time interval represented by the rectangular pulse, we obtain

$$v(t) = \frac{1}{5 \times 10^{-6}} \int_0^t 20 \times 10^{-3} \, dt' + v(0)$$

Since $v(0) = 0$,

$$v(t) = 4000t \qquad 0 \le t \le 2 \text{ ms}$$

For the semi-infinite interval following the pulse, the integral of $i(t)$ is once again zero, so that

$$v(t) = 8 \qquad t \ge 2 \text{ ms}$$

The results are expressed much more simply in a sketch than by these analytical expressions, as shown in Fig. 7.5b.

PRACTICE

7.2 Determine the current through a 100 pF capacitor if its voltage as a function of time is given by Fig. 7.6.

Ans: $0 \text{ A}, -\infty \le t \le 1 \text{ ms}; 200 \text{ nA}, 1 \text{ ms} \le t \le 2 \text{ ms}; 0 \text{ A}, t \ge 2 \text{ ms}$.

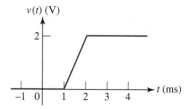

$v(t)$ (V)

■ **FIGURE 7.6**

Energy Storage

The power delivered to a capacitor is

$$p = vi = Cv\frac{dv}{dt}$$

and the change in energy stored in its electric field is therefore

$$\int_{t_0}^t p \, dt' = C \int_{t_0}^t v\frac{dv}{dt'} \, dt' = C \int_{v(t_0)}^{v(t)} v' \, dv' = \frac{1}{2}C\left\{[v(t)]^2 - [v(t_0)]^2\right\}$$

and thus

$$w_C(t) - w_C(t_0) = \tfrac{1}{2}C\left\{[v(t)]^2 - [v(t_0)]^2\right\} \qquad [3]$$

where the stored energy is $w_C(t_0)$ in joules (J) and the voltage at t_0 is $v(t_0)$. If we select a zero-energy reference at t_0, implying that the capacitor voltage is also zero at that instant, then

$$\boxed{w_C(t) = \tfrac{1}{2}C\,v^2} \qquad [4]$$

Let us consider a simple numerical example. As sketched in Fig. 7.7, a sinusoidal voltage source is in parallel with a 1 MΩ resistor and a 20 μF capacitor. The parallel resistor may be assumed to represent the finite resistance of the dielectric between the plates of the physical capacitor (an *ideal* capacitor has infinite resistance).

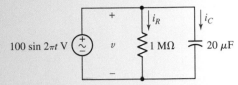

EXAMPLE **7.3**

Find the maximum energy stored in the capacitor of Fig. 7.7 and the energy dissipated in the resistor over the interval $0 < t < 0.5$ s.

▶ **Identify the goal of the problem.**

The energy stored in the capacitor varies with time; we are asked for the *maximum* value over a specific time interval. We are also asked to find the *total* amount of energy dissipated by the resistor over this interval. These are actually two completely different questions.

▶ **Collect the known information.**

The only source of energy in the circuit is the independent voltage source, which has a value of $100 \sin 2\pi t$ V. We are only interested in the time interval of $0 < t < 0.5$ s. The circuit is properly labeled.

▶ **Devise a plan.**

We will determine the energy in the capacitor by evaluating the voltage. To find the energy dissipated in the resistor during the same time interval, we need to integrate the dissipated *power*, $p_R = i_R^2 \cdot R$.

▶ **Construct an appropriate set of equations.**

The energy stored in the capacitor is simply

$$w_C(t) = \tfrac{1}{2}Cv^2 = 0.1 \sin^2 2\pi t \qquad \text{J}$$

We obtain an expression for the power dissipated by the resistor in terms of the current i_R:

$$i_R = \frac{v}{R} = 10^{-4} \sin 2\pi t \qquad \text{A}$$

and so

$$p_R = i_R^2 R = (10^{-4})(10^6) \sin^2 2\pi t$$

so that the energy dissipated in the resistor between 0 and 0.5 s is

$$w_R = \int_0^{0.5} p_R \, dt = \int_0^{0.5} 10^{-2} \sin^2 2\pi t \, dt \qquad \text{J}$$

▶ **Determine if additional information is required.**

We have an expression for the energy stored in the capacitor; a sketch is shown in Fig. 7.8. The expression derived for the energy dissipated by the resistor does not involve any unknown quantities, and so may also be readily evaluated.

▶ **Attempt a solution.**

From our sketch of the expression for the energy stored in the capacitor, we see that it increases from zero at $t = 0$ to a maximum of 100 mJ at $t = \frac{1}{4}$ s, and then decreases to zero in another $\frac{1}{4}$ s. Thus, $w_{C_{\max}} = 100$ mJ. Evaluating our integral expression for the energy dissipated in the resistor, we find that $w_R = 2.5$ mJ.

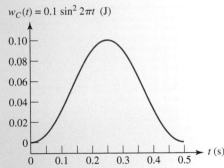

■ FIGURE 7.7 A sinusoidal voltage source is applied to a parallel *RC* network. The 1 MΩ resistor might represent the finite resistance of the "real" capacitor's dielectric layer.

$w_C(t) = 0.1 \sin^2 2\pi t$ (J)

■ FIGURE 7.8 A sketch of the energy stored in the capacitor as a function of time.

(Continued on next page)

▶ **Verify the solution. Is it reasonable or expected?**
We do not expect to calculate a *negative* stored energy, which is borne out in our sketch. Further, since the maximum value of $\sin 2\pi t$ is 1, the maximum energy expected would be $(1/2)(20 \times 10^{-6})(100)^2 = 100$ mJ.

The resistor dissipated 2.5 mJ in the period of 0 to 500 ms, although the capacitor stored a maximum of 100 mJ at one point during that interval. What happened to the "other" 97.5 mJ? To answer this, we compute the capacitor current:

$$i_C = 20 \times 10^{-6}\frac{dv}{dt} = 0.004\pi \cos 2\pi t$$

and the current i_s defined as flowing *into* the voltage source

$$i_s = -i_C - i_R$$

both of which are plotted in Fig. 7.9. We observe that the current flowing through the resistor is a small fraction of the source current; not entirely surprising as 1 MΩ is a relatively large resistance value. As current flows from the source, a small amount is diverted to the resistor, with the rest flowing into the capacitor as it charges. After $t = 250$ ms, the source current is seen to change sign; current is now flowing from the capacitor back into the source. Most of the energy stored in the capacitor is being returned to the ideal voltage source, except for the small fraction dissipated in the resistor.

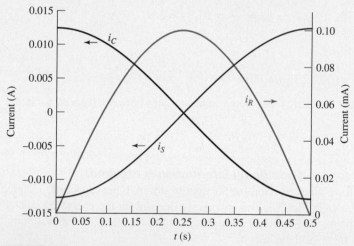

■ **FIGURE 7.9** Plot of the resistor, capacitor, and source currents during the interval of 0 to 500 ms.

PRACTICE

7.3 Calculate the energy stored in a 1000 μF capacitor at $t = 50\,\mu$s if the voltage across it is 1.5 $\cos 10^5 t$ volts.

Ans: 90.52 μJ.

Ultracapacitors

Digital cellular and satellite system telephones have three basic operating modes: *standby, receiving,* and *transmitting.* Signal reception and standby do not typically demand a great deal of current from the batteries, but transmission does (Fig. 7.10). However, the time spent in transmission is typically a small fraction of the total time the device is drawing power, as indicated in the figure.

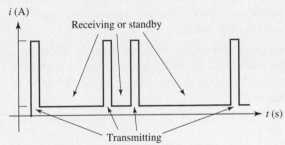

■ **FIGURE 7.10** Typical duty cycle of a cellular telephone.

As we explored in Chap. 5, batteries maintain a constant voltage only for small currents. Thus, when the current demand increases, the battery voltage decreases (Fig. 7.11). This can lead to problems, since most circuits have a minimum voltage, or *cutoff voltage,* below which they can no longer function properly.

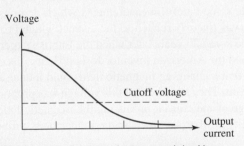

■ **FIGURE 7.11** Example battery voltage-current relationship.

If the peak current draw of the circuit is such that the battery voltage can drop below the cutoff voltage, a much larger battery is really needed. However, this is usually undesirable in portable applications, where small, light-weight batteries are typically desired. One alternative to using batteries exclusively is to employ a hybrid device

composed of a standard battery and a specially designed capacitor (sometimes referred to as an *electrochemical capacitor* or an *ultracapacitor*). An example of a commercially available device is shown in Fig. 7.12.

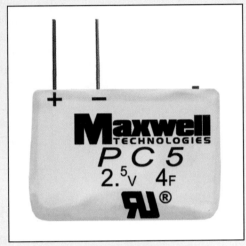

■ **FIGURE 7.12** A commercially available ultracapacitor. Photo courtesy of Maxwell Technologies Inc.

The principle behind the hybrid device is that while the battery is keeping up with the current demanded by the circuit (for example, while the phone is in *receive* mode), the capacitor stores energy from the battery ($\frac{1}{2}CV^2$). If the current demand suddenly increases (for example, when the phone is transmitting), the battery voltage will try to decrease. At this point, current will flow out of the charged capacitor in response to the resulting dv/dt. Provided that the Thévenin equivalent resistance of the circuit is much smaller than the internal resistance of the battery, the current will flow through the phone circuit rather than into the battery. The charge will leave the capacitor very quickly, so that this current "boost" is short-lived. However, if the *transmit* operation is of a short duration, the capacitor can effectively assist the battery and prevent the circuit from cutting off. In Chap. 8, we will learn how to predict how long such a capacitor can assist a battery provided that we know the Thévenin resistance of both the battery and the circuit.

Important Characteristics of an Ideal Capacitor

1. There is no current through a capacitor if the voltage across it is not changing with time. A capacitor is therefore an *open circuit to dc*.

2. A finite amount of energy can be stored in a capacitor even if the current through the capacitor is zero, such as when the voltage across it is constant.

3. It is impossible to change the voltage across a capacitor by a finite amount in zero time, for this requires an infinite current through the capacitor. A capacitor resists an abrupt change in the voltage across it in a manner analogous to the way a spring resists an abrupt change in its displacement.

4. A capacitor never dissipates energy, but only stores it. Although this is true for the *mathematical model,* it is not true for a *physical* capacitor due to finite resistances associated with the dielectric as well as the packaging.

7.2 • THE INDUCTOR

Ideal Inductor Model

Although we shall define an **inductor** strictly from a circuit point of view, that is, by a voltage-current equation, a few comments about the development of magnetic field theory may provide a better understanding of the definition. In the early 1800s the Danish scientist Oersted showed that a current-carrying conductor produced a magnetic field (compass needles were affected in the presence of a wire when current was flowing). Shortly thereafter, Ampère made some careful measurements which demonstrated that this magnetic field was *linearly* related to the current which produced it. The next step occurred some 20 years later when the English experimentalist Michael Faraday and the American inventor Joseph Henry discovered almost simultaneously[3] that a changing magnetic field could induce a voltage in a neighboring circuit. They showed that this voltage was proportional to the time rate of change of the current producing the magnetic field. The constant of proportionality is what we now call the inductance, symbolized by L, and therefore

$$v = L \frac{di}{dt} \qquad [5]$$

where we must realize that v and i are both functions of time. When we wish to emphasize this, we may do so by using the symbols $v(t)$ and $i(t)$.

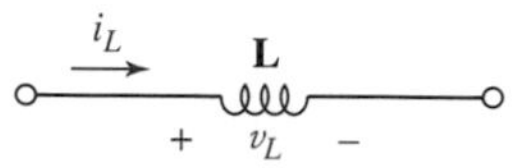

The circuit symbol for the inductor is shown in Fig. 7.13, and it should be noted that the passive sign convention is used, just as it was with the resistor and the capacitor. The unit in which inductance is measured is the **henry** (H), and the defining equation shows that the henry is just a shorter expression for a volt-second per ampere.

■ **FIGURE 7.13** Electrical symbol and current-voltage conventions for an inductor.

(3) Faraday won.

The inductor whose inductance is defined by Eq. [5] is a mathematical model; it is an *ideal* element which we may use to approximate the behavior of a *real* device. A physical inductor may be constructed by winding a length of wire into a coil. This serves effectively to increase the current that is causing the magnetic field and also to increase the "number" of neighboring circuits into which Faraday's voltage may be induced. The result of this twofold effect is that the inductance of a coil is approximately proportional to the square of the number of complete turns made by the conductor out of which it is formed. For example, an inductor or "coil" that has the form of a long helix of very small pitch is found to have an inductance of $\mu N^2 A/s$, where A is the cross-sectional area, s is the axial length of the helix, N is the number of complete turns of wire, and μ (mu) is a constant of the material inside the helix, called the permeability. For free space (and very closely for air), $\mu = \mu_0 = 4\pi \times 10^{-7}$ H/m $= 4\pi$ nH/cm. Several examples of commercially available inductors are shown in Fig. 7.14.

Let us now scrutinize Eq. [5] to determine some of the electrical characteristics of the mathematical model. This equation shows that the voltage across an inductor is proportional to the time rate of change of the current through it. In particular, it shows that there is no voltage across an inductor carrying a constant current, regardless of the magnitude of this current. Accordingly, we may view an inductor as a "*short circuit to dc.*"

Another fact that can be obtained from Eq. [5] is that a sudden or discontinuous change in the current must be associated with an infinite voltage across the inductor. In other words, if we wish to produce an abrupt change in an inductor current, we must apply an infinite voltage. Although an infinite-voltage forcing function might be acceptable theoretically, it can never be a part of the phenomena displayed by a real physical device. As we

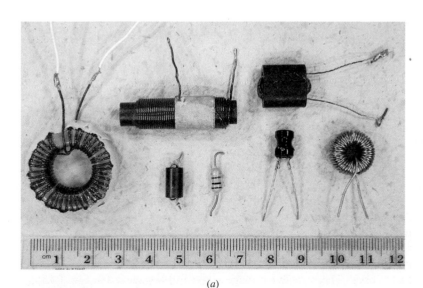

(a)

(b)

■ **FIGURE 7.14** (a) Several different types of commercially available inductors, sometimes also referred to as "chokes." Clockwise, starting from far left: 287 μH ferrite core toroidal inductor, 266 μH ferrite core cylindrical inductor, 215 μH ferrite core inductor designed for VHF frequencies, 85 μH iron powder core toroidal inductor, 10 μH bobbin-style inductor, 100 μH axial lead inductor, and 7 μH lossy-core inductor used for RF suppression. (b) An 11 H inductor, measuring 10 cm (tall) × 8 cm (wide) × 8 cm (deep).

shall see shortly, an abrupt change in the inductor current also requires an abrupt change in the energy stored in the inductor, and this sudden change in energy requires infinite power at that instant; infinite power is again not a part of the real physical world. In order to avoid infinite voltage and infinite power, an inductor current must not be allowed to jump *instantaneously* from one value to another.

If an attempt is made to open-circuit a physical inductor through which a finite current is flowing, an arc may appear across the switch. This is useful in the ignition system of some automobiles, where the current through the spark coil is interrupted by the distributor and the arc appears across the spark plug. Although this does not occur instantaneously, it happens in a very short timespan, leading to the creation of a large voltage. The presence of a large voltage across a short distance equates to a very large electric field; the stored energy is dissipated in ionizing the air in the path of the arc.

Equation [5] may also be interpreted (and solved, if necessary) by graphical methods, as seen in Example 7.4.

EXAMPLE **7.4**

Given the waveform of the current in a 3 H inductor as shown in Fig. 7.15a, determine the inductor voltage and sketch it.

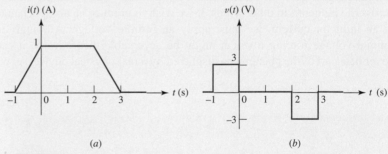

■ **FIGURE 7.15** (*a*) The current waveform in a 3 H inductor. (*b*) The corresponding voltage waveform, $v = 3 \, di/dt$.

Provided that the voltage v and the current i are defined to satisfy the passive sign convention, we may obtain v from Fig. 7.15a using Eq. [5]:

$$v = 3\frac{di}{dt}$$

Since the current is zero for $t < -1$ s, the voltage is zero in this interval. The current then begins to increase at the linear rate of 1 A/s, and thus a constant voltage of $L \, di/dt = 3$ V is produced. During the following 2 s interval, the current is constant and the voltage is therefore zero. The final decrease of the current results in $di/dt = -1$ A/s, yielding $v = -3$ V. For $t > 3$ s, $i(t)$ is a constant (zero), so that $v(t) = 0$ for that interval. The complete voltage waveform is sketched in Fig. 7.15b.

Let us now investigate the effect of a more rapid rise and decay of the current between the zero and 1 A values.

EXAMPLE **7.5**

Find the inductor voltage that results from applying the current waveform shown in Fig. 7.16a to the inductor of Example 7.4.

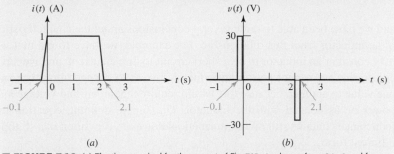

(a)   (b)

■ **FIGURE 7.16** (a) The time required for the current of Fig. 7.15a to change from 0 to 1 and from 1 to 0 is decreased by a factor of 10. (b) The resultant voltage waveform. The pulse widths are exaggerated for clarity.

Note that the intervals for the rise and fall have decreased to 0.1 s. Thus, the magnitude of each derivative will be 10 times larger; this condition is shown in the current and voltage sketches of Fig. 7.16a and b. In the voltage waveforms of Fig. 7.15b and 7.16b, it is interesting to note that the area under each voltage pulse is 3 V-s.

Just for curiosity's sake, let's continue in the same vein for a moment. A further decrease in the rise and fall times of the current waveform will produce a proportionally larger voltage magnitude, but only within the interval in which the current is increasing or decreasing. An abrupt change in the current will cause the infinite voltage "spikes" (each having an area of 3 V-s) that are suggested by the waveforms of Fig. 7.17a and b; or, from the equally valid but opposite point of view, these infinite voltage spikes are required to produce the abrupt changes in the current.

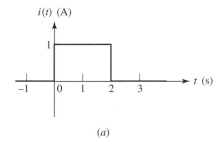

(a)

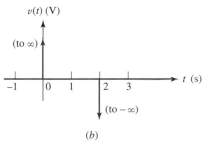

(b)

■ **FIGURE 7.17** (a) The time required for the current of Fig. 7.16a to change from 0 to 1 and from 1 to 0 is decreased to zero; the rise and fall are abrupt. (b) The resultant voltage across the 3 H inductor consists of a positive and a negative infinite spike.

PRACTICE
●
7.4 The current through a 200 mH inductor is shown in Fig. 7.18. Assume the passive sign convention, and find v_L at t equal to: (a) 0; (b) 2 ms; (c) 6 ms.

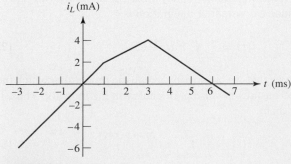

■ **FIGURE 7.18**

Ans: 0.4 V; 0.2 V; −0.267 V.

Integral Voltage-Current Relationships

We have defined inductance by a simple differential equation,

$$v = L\frac{di}{dt}$$

and we have been able to draw several conclusions about the characteristics of an inductor from this relationship. For example, we have found that we may consider an inductor to be a short circuit to direct current, and we have agreed that we cannot permit an inductor current to change abruptly from one value to another, because this would require that an infinite voltage and power be associated with the inductor. The simple defining equation for inductance contains still more information, however. Rewritten in a slightly different form,

$$di = \frac{1}{L}v\,dt$$

it invites integration. Let us first consider the limits to be placed on the two integrals. We desire the current i at time t, and this pair of quantities therefore provides the upper limits on the integrals appearing on the left and right side of the equation, respectively; the lower limits may also be kept general by merely assuming that the current is $i(t_0)$ at time t_0. Thus,

$$\int_{i(t_0)}^{i(t)} di' = \frac{1}{L}\int_{t_0}^{t} v(t')\,dt'$$

which leads to the equation

$$i(t) - i(t_0) = \frac{1}{L}\int_{t_0}^{t} v\,dt'$$

or

$$\boxed{i(t) = \frac{1}{L}\int_{t_0}^{t} v\,dt' + i(t_0)} \qquad [6]$$

Equation [5] expresses the inductor voltage in terms of the current, whereas Eq. [6] gives the current in terms of the voltage. Other forms are also possible for the latter equation. We may write the integral as an indefinite integral and include a constant of integration k:

$$i(t) = \frac{1}{L}\int v\,dt + k \qquad [7]$$

We also may assume that we are solving a realistic problem in which the selection of t_0 as $-\infty$ ensures no current or energy in the inductor. Thus, if $i(t_0) = i(-\infty) = 0$, then

$$i(t) = \frac{1}{L}\int_{-\infty}^{t} v\,dt' \qquad [8]$$

Let us investigate the use of these several integrals by working a simple example where the voltage across an inductor is specified.

EXAMPLE **7.6**

The voltage across a 2 H inductor is known to be 6 cos 5t V. Determine the resulting inductor current if $i(t = -\pi/2) = 1$ A.

From Eq. [6],

$$i(t) = \frac{1}{2} \int_{t_0}^{t} 6 \cos 5t' \, dt' + i(t_0)$$

or

$$i(t) = \frac{1}{2} \left(\frac{6}{5}\right) \sin 5t - \frac{1}{2} \left(\frac{6}{5}\right) \sin 5t_0 + i(t_0)$$

$$= 0.6 \sin 5t - 0.6 \sin 5t_0 + i(t_0)$$

The first term indicates that the inductor current varies sinusoidally; the second and third terms together merely represent a constant which becomes known when the current is numerically specified at some instant of time. Using the fact that the current is 1 A at $t = -\pi/2$ s, we identify t_0 as $-\pi/2$ with $i(t_0) = 1$, and find that

$$i(t) = 0.6 \sin 5t - 0.6 \sin(-2.5\pi) + 1$$

or

$$i(t) = 0.6 \sin 5t + 1.6$$

We may obtain the same result from Eq. [7]. We have

$$i(t) = 0.6 \sin 5t + k$$

and we establish the numerical value of k by forcing the current to be 1 A at $t = -\pi/2$:

$$1 = 0.6 \sin(-2.5\pi) + k$$

or

$$k = 1 + 0.6 = 1.6$$

and so, as before,

$$i(t) = 0.6 \sin 5t + 1.6$$

Equation [8] is going to cause trouble with this particular voltage. We based the equation on the assumption that the current was zero when $t = -\infty$. To be sure, this must be true in the real, physical world, but we are working in the land of the mathematical model; our elements and forcing functions are all idealized. The difficulty arises after we integrate, obtaining

$$i(t) = 0.6 \sin 5t' \Big|_{-\infty}^{t}$$

and attempt to evaluate the integral at the lower limit:

$$i(t) = 0.6 \sin 5t - 0.6 \sin(-\infty)$$

The sine of $\pm\infty$ is indeterminate, and therefore we cannot evaluate our expression. Equation [8] is only useful if we are evaluating functions which approach zero as $t \rightarrow -\infty$.

We should not make any snap judgments, however, as to which single form of Eqs. [6], [7], and [8] we are going to use forever after; each has its advantages, depending on the problem and the application. Equation [6] represents a long, general method, but it shows clearly that the constant of integration is a current. Equation [7] is a somewhat more concise expression of Eq. [6], but the nature of the integration constant is suppressed. Finally, Eq. [8] is an excellent expression, since no constant is necessary; however, it applies only when the current is zero at $t = -\infty$ and when the analytical expression for the current is not indeterminate there.

Energy Storage

Let us now turn our attention to power and energy. The absorbed power is given by the current-voltage product

$$p = vi = Li\frac{di}{dt}$$

The energy w_L accepted by the inductor is stored in the magnetic field around the coil. The change in this energy is expressed by the integral of the power over the desired time interval:

$$\int_{t_0}^{t} p\,dt' = L\int_{t_0}^{t} i\frac{di}{dt'}\,dt' = L\int_{i(t_0)}^{i(t)} i'\,di'$$

$$= \frac{1}{2}L\left\{[i(t)]^2 - [i(t_0)]^2\right\}$$

Thus,

$$w_L(t) - w_L(t_0) = \tfrac{1}{2}L\left\{[i(t)]^2 - [i(t_0)]^2\right\} \qquad [9]$$

where we have again assumed that the current is $i(t_0)$ at time t_0. In using the energy expression, it is customary to assume that a value of t_0 is selected at which the current is zero; it is also customary to assume that the energy is zero at this time. We then have simply

$$\boxed{w_L(t) = \tfrac{1}{2}Li^2} \qquad [10]$$

where we now understand that our reference for zero energy is any time at which the inductor current is zero. At any subsequent time at which the current is zero, we also find no energy stored in the coil. Whenever the current is not zero, and regardless of its direction or sign, energy is stored in the inductor. It follows, therefore, that energy may be delivered to the inductor for a part of the time and recovered from the inductor later. All the stored energy may be recovered from an ideal inductor; there are no storage charges or agent's commissions in the mathematical model. A physical coil,

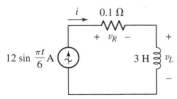

FIGURE 7.19 A sinusoidal current is applied as a forcing function to a series *RL* circuit. The 0.1 Ω represents the inherent resistance of the wire from which the inductor is fabricated.

however, must be constructed out of real wire and thus will always have a resistance associated with it. Energy can no longer be stored and recovered without loss.

These ideas may be illustrated by a simple example. In Fig. 7.19, a 3 H inductor is shown in series with a 0.1 Ω resistor and a sinusoidal current source, $i_s = 12 \sin \frac{\pi t}{6}$ A. The resistor should be interpreted as the resistance of the wire which must be associated with the physical coil.

EXAMPLE 7.7

Find the maximum energy stored in the inductor of Fig. 7.19, and calculate how much energy is dissipated in the resistor in the time during which the energy is being stored in and then recovered from the inductor.

The energy stored in the inductor is

$$w_L = \frac{1}{2}Li^2 = 216 \sin^2 \frac{\pi t}{6} \quad \text{J}$$

and this energy increases from zero at $t = 0$ to 216 J at $t = 3$ s. Thus, the maximum energy stored in the inductor is 216 J.

After reaching its peak value at $t = 3$ s, the energy leaves the inductor completely. Let us see what price we have paid in this coil for the privilege of storing and removing 216 J in these 6 seconds. The power dissipated in the resistor is easily found as

$$p_R = i^2 R = 14.4 \sin^2 \frac{\pi t}{6} \quad \text{W}$$

and the energy converted into heat in the resistor within this 6 s interval is therefore

$$w_R = \int_0^6 p_R \, dt = \int_0^6 14.4 \sin^2 \frac{\pi}{6} t \, dt$$

or

$$w_R = \int_0^6 14.4 \left(\frac{1}{2}\right) \left(1 - \cos \frac{\pi}{3} t\right) dt = 43.2 \text{ J}$$

Thus, we have expended 43.2 J in the process of storing and then recovering 216 J in a 6 s interval. This represents 20 percent of the maximum stored energy, but it is a reasonable value for many coils having this large an inductance. For coils having an inductance of about 100 μH, we might expect a figure closer to 2 or 3 percent.

PRACTICE

7.6 Let $L = 25$ mH for the inductor of Fig. 7.20. (a) Find v at $t = 12$ ms if $i = 10te^{-100t}$ A. (b) Find i at $t = 0.1$ s if $v = 6e^{-12t}$ V and $i(0) = 10$ A. If $i = 8(1 - e^{-40t})$ mA, find: (c) the power being delivered to the inductor at $t = 50$ ms, and (d) the energy stored in the inductor at $t = 40$ ms.

Ans: -15.06 mV; 24.0 A; 7.49 μW; 0.510 μJ.

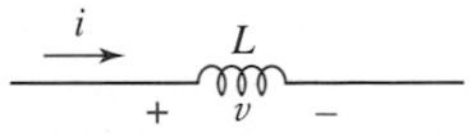

$i \quad\quad L$

$+ \quad v \quad -$

■ FIGURE 7.20

Let us now recapitulate by listing four key characteristics of an inductor which result from its defining equation $v = L\, di/dt$:

Important Characteristics of an Ideal Inductor

1. There is no voltage across an inductor if the current through it is not changing with time. An inductor is therefore a *short circuit to dc*.

2. A finite amount of energy can be stored in an inductor even if the voltage across the inductor is zero, such as when the current through it is constant.

3. It is impossible to change the current through an inductor by a finite amount in zero time, for this requires an infinite voltage across the inductor. An inductor resists an abrupt change in the current through it in a manner analogous to the way a mass resists an abrupt change in its velocity.

4. The inductor never dissipates energy, but only stores it. Although this is true for the *mathematical* model, it is not true for a *physical* inductor due to series resistances.

It is interesting to anticipate our discussion of **duality** in Sec. 7.6 by rereading the previous four statements with certain words replaced by their "duals." If *capacitor* and *inductor, capacitance* and *inductance, voltage* and *current, across* and *through, open circuit* and *short circuit, spring* and *mass,* and *displacement* and *velocity* are interchanged (in either direction), the four statements previously given for capacitors are obtained.

7.3 • INDUCTANCE AND CAPACITANCE COMBINATIONS

Now that we have added the inductor and capacitor to our list of passive circuit elements, we need to decide whether or not the methods we have developed for resistive circuit analysis are still valid. It will also be convenient to learn how to replace series and parallel combinations of either of these elements with simpler equivalents, just as we did with resistors in Chap. 3.

We look first at Kirchhoff's two laws, both of which are axiomatic. However, when we hypothesized these two laws, we did so with no restrictions as to the types of elements constituting the network. Both, therefore, remain valid.

Inductors in Series

Now we may extend the procedures we have derived for reducing various combinations of resistors into one equivalent resistor to the analogous cases of inductors and capacitors. We shall first consider an ideal voltage source applied to the series combination of N inductors, as shown in Fig. 7.21a. We desire a single equivalent inductor, with inductance L_{eq}, which may replace the series combination so that the source current $i(t)$ is unchanged.

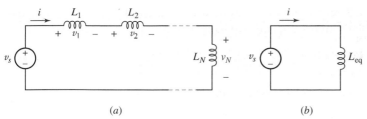

FIGURE 7.21 (a) A circuit containing N inductors in series. (b) The desired equivalent circuit, in which $L_{eq} = L_1 + L_2 + \cdots + L_N$.

The equivalent circuit is sketched in Fig. 7.21b. Applying KVL to the original circuit,

$$v_s = v_1 + v_2 + \cdots + v_N$$

$$= L_1 \frac{di}{dt} + L_2 \frac{di}{dt} + \cdots + L_N \frac{di}{dt}$$

$$= (L_1 + L_2 + \cdots + L_N) \frac{di}{dt}$$

or, written more concisely,

$$v_s = \sum_{n=1}^{N} v_n = \sum_{n=1}^{N} L_n \frac{di}{dt} = \frac{di}{dt} \sum_{n=1}^{N} L_n$$

But for the equivalent circuit we have

$$v_s = L_{eq} \frac{di}{dt}$$

and thus the equivalent inductance is

$$L_{eq} = (L_1 + L_2 + \cdots + L_N)$$

or

$$L_{eq} = \sum_{n=1}^{N} L_n \qquad [11]$$

The inductor which is equivalent to several inductors connected in series is one whose inductance is the sum of the inductances in the original circuit. *This is exactly the same result we obtained for resistors in series.*

Inductors in Parallel

The combination of a number of parallel inductors is accomplished by writing the single nodal equation for the original circuit, shown in Fig. 7.22a,

$$i_s = \sum_{n=1}^{N} i_n = \sum_{n=1}^{N} \left[\frac{1}{L_n} \int_{t_0}^{t} v \, dt' + i_n(t_0) \right]$$

$$= \left(\sum_{n=1}^{N} \frac{1}{L_n} \right) \int_{t_0}^{t} v \, dt' + \sum_{n=1}^{N} i_n(t_0)$$

and comparing it with the result for the equivalent circuit of Fig. 7.22b,

$$i_s = \frac{1}{L_{eq}} \int_{t_0}^{t} v \, dt' + i_s(t_0)$$

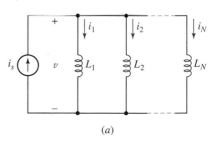

(a)

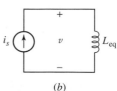

(b)

FIGURE 7.22 (a) The parallel combination of N inductors. (b) The equivalent circuit, where $L_{eq} = [1/L_1 + 1/L_2 + \cdots + 1/L_N]^{-1}$.

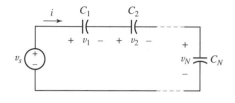

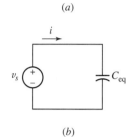

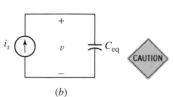

■ **FIGURE 7.23** (*a*) A circuit containing *N* capacitors in series. (*b*) The desired equivalent circuit, where $C_{eq} = [1/C_1 + 1/C_2 + \cdots + 1/C_N]^{-1}$.

Since Kirchhoff's current law demands that $i_s(t_0)$ be equal to the sum of the branch currents at t_0, the two integral terms must also be equal; hence,

$$L_{eq} = \frac{1}{1/L_1 + 1/L_2 + \cdots + 1/L_N} \quad [12]$$

For the special case of two inductors in parallel,

$$L_{eq} = \frac{L_1 L_2}{L_1 + L_2} \quad [13]$$

and we note that inductors in parallel combine exactly as do resistors in parallel.

Capacitors in Series

In order to find a capacitor that is equivalent to N capacitors in series, we use the circuit of Fig. 7.23a and its equivalent in Fig. 7.23b to write

$$v_s = \sum_{n=1}^{N} v_n = \sum_{n=1}^{N} \left[\frac{1}{C_n} \int_{t_0}^{t} i \, dt' + v_n(t_0) \right]$$

$$= \left(\sum_{n=1}^{N} \frac{1}{C_n} \right) \int_{t_0}^{t} i \, dt' + \sum_{n=1}^{N} v_n(t_0)$$

and

$$v_s = \frac{1}{C_{eq}} \int_{t_0}^{t} i \, dt' + v_s(t_0)$$

However, Kirchhoff's voltage law establishes the equality of $v_s(t_0)$ and the sum of the capacitor voltages at t_0; thus

$$C_{eq} = \frac{1}{1/C_1 + 1/C_2 + \cdots + 1/C_N} \quad [14]$$

and capacitors in series combine as do conductances in series, or resistors in parallel. The special case of two capacitors in series, of course, yields

$$C_{eq} = \frac{C_1 C_2}{C_1 + C_2} \quad [15]$$

Capacitors in Parallel

Finally, the circuits of Fig. 7.24 enable us to establish the value of the capacitor which is equivalent to N parallel capacitors as

$$C_{eq} = C_1 + C_2 + \cdots + C_N \quad [16]$$

and it is no great source of amazement to note that capacitors in parallel combine in the same manner in which we combine resistors in series, that is, by simply adding all the individual capacitances.

These formulas are well worth memorizing. The formulas applying to series and parallel combinations of inductors are identical to those for resistors, so they typically seem "obvious." Care should be exercised, however, in the case of the corresponding expressions for series and parallel combinations of capacitors, as they are opposite those of resistors and inductors, frequently leading to errors when calculations are made too hastily.

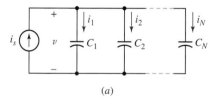

■ **FIGURE 7.24** (*a*) The parallel combination of *N* capacitors. (*b*) The equivalent circuit, where $C_{eq} = C_1 + C_2 + \cdots + C_N$.

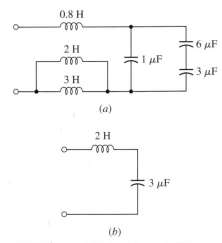

EXAMPLE 7.8

Simplify the network of Fig. 7.25a using series/parallel combinations.

The 6 and 3 μF series capacitors are first combined into a 2 μF equivalent, and this capacitor is then combined with the 1 μF element with which it is in parallel to yield an equivalent capacitance of 3 μF. In addition, the 3 and 2 H inductors are replaced by an equivalent 1.2 H inductor, which is then added to the 0.8 H element to give a total equivalent inductance of 2 H. The much simpler (and probably less expensive) equivalent network is shown in Fig. 7.25b.

■ **FIGURE 7.25** (a) A given *LC* network. (b) A simpler equivalent circuit.

PRACTICE

7.7 Find C_{eq} for the network of Fig. 7.26.

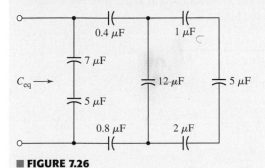

■ **FIGURE 7.26**

Ans: 3.18 μF.

The network shown in Fig. 7.27 contains three inductors and three capacitors, but no series or parallel combinations of either the inductors or the capacitors can be achieved. Simplification of this network cannot be accomplished using the techniques presented here.

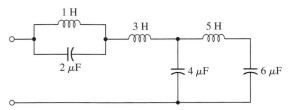

■ **FIGURE 7.27** An *LC* network in which no series or parallel combinations of either the inductors or the capacitors can be made.

7.4 CONSEQUENCES OF LINEARITY

Next let us turn to nodal and mesh analysis. Since we already know that we may safely apply Kirchhoff's laws, we should have little difficulty in writing a set of equations that are both sufficient and independent. They will be constant-coefficient linear integrodifferential equations, however, which are hard enough to pronounce, let alone solve. Consequently, we shall write them now to gain familiarity with the use of Kirchhoff's laws in *RLC* circuits and discuss the solution of the simpler cases in subsequent chapters.

EXAMPLE 7.9

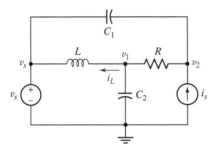

■ **FIGURE 7.28** A four-node *RLC* circuit with node voltages assigned.

Write appropriate nodal equations for the circuit of Fig. 7.28.

Node voltages are chosen as indicated, and we sum currents leaving the central node:

$$\frac{1}{L}\int_{t_0}^{t}(v_1 - v_s)\,dt' + i_L(t_0) + \frac{v_1 - v_2}{R} + C_2\frac{dv_1}{dt} = 0$$

where $i_L(t_0)$ is the value of the inductor current at the time the integration begins. At the right-hand node,

$$C_1\frac{d(v_2 - v_s)}{dt} + \frac{v_2 - v_1}{R} - i_s = 0$$

Rewriting these two equations, we have

$$\frac{v_1}{R} + C_2\frac{dv_1}{dt} + \frac{1}{L}\int_{t_0}^{t}v_1\,dt' - \frac{v_2}{R} = \frac{1}{L}\int_{t_0}^{t}v_s\,dt' - i_L(t_0)$$

$$-\frac{v_1}{R} + \frac{v_2}{R} + C_1\frac{dv_2}{dt} = C_1\frac{dv_s}{dt} + i_s$$

These are the promised integrodifferential equations, and we may note several interesting points about them. First, the source voltage v_s happens to enter the equations as an integral and as a derivative, but not simply as v_s. Since both sources are specified for all time, we should be able to evaluate the derivative or integral. Second, the initial value of the inductor current, $i_L(t_0)$, acts as a (constant) source current at the center node.

PRACTICE

7.8 If $v_C(t) = 4\cos 10^5 t$ V in the circuit in Fig. 7.29, find $v_s(t)$.

Ans: $-2.4\cos 10^5 t$ V.

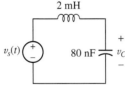

■ **FIGURE 7.29**

We will not attempt the solution of integrodifferential equations at this time. It is worthwhile pointing out, however, that when the voltage forcing functions are sinusoidal functions of time, it will be possible to define a voltage-current ratio (called ***impedance***) or a current-voltage ratio (called ***admittance***) for each of the three passive elements. The factors operating on the two node voltages in the preceding equations will then become simple multiplying factors, and the equations will be linear algebraic equations once again. These we may solve by determinants or a simple elimination of variables as before.

We may also show that the benefits of linearity apply to *RLC* circuits as well. In accordance with our previous definition of a linear circuit, these circuits are also linear because the voltage-current relationships for the inductor and capacitor are linear relationships. For the inductor, we have

$$v = L\frac{di}{dt}$$

and multiplication of the current by some constant K leads to a voltage that is also greater by a factor K. In the integral formulation,

$$i(t) = \frac{1}{L}\int_{t_0}^{t} v\,dt' + i(t_0)$$

it can be seen that, if each term is to increase by a factor of K, then the initial value of the current must also increase by this same factor.

A corresponding investigation of the capacitor shows that it, too, is linear. Thus, a circuit composed of independent sources, linear dependent sources, and linear resistors, inductors, and capacitors is a linear circuit.

In this linear circuit the response is again proportional to the forcing function. The proof of this statement is accomplished by first writing a general system of integrodifferential equations. Let us place all the terms having the form of Ri, $L\,di/dt$, and $1/C\int i\,dt$ on the left side of each equation, and keep the independent source voltages on the right side. As a simple example, one of the equations might have the form

$$Ri + L\frac{di}{dt} + \frac{1}{C}\int_{t_0}^{t} i\,dt' + v_C(t_0) = v_s$$

If every independent source is now increased by a factor K, then the right side of each equation is greater by the factor K. Now each term on the left side is either a linear term involving some loop current or an initial capacitor voltage. In order to cause all there sponses (loop currents) to increase by a factor K, it is apparent that we must also increase the initial capacitor voltages by a factor K. That is, we must treat the initial capacitor voltage as an independent source voltage and increase it also by a factor K. In a similar manner, initial inductor currents appear as independent source currents in nodal analysis.

The principle of proportionality between source and response is thus extensible to the general *RLC* circuit, and it follows that the principle of superposition is also applicable. It should be emphasized that initial inductor currents and capacitor voltages must be treated as independent sources in applying the superposition principle; each initial value must take its turn in being rendered inactive. In Chap. 5 we learned that the principle of superposition is a natural consequence of the linear nature of resistive circuits. The resistive circuits are linear because the voltage-current relationship for the resistor is linear and Kirchhoff's laws are linear.

Before we can apply the superposition principle to *RLC* circuits, however, it is first necessary to develop methods of solving the equations describing these circuits when only one independent source is present. At this time we should feel convinced that a linear circuit will possess a response whose amplitude is proportional to the amplitude of the source. We should be prepared to apply superposition later, considering an inductor current or capacitor voltage specified at $t = t_0$ as a source that must be killed when its turn comes.

Thévenin's and Norton's theorems are based on the linearity of the initial circuit, the applicability of Kirchhoff's laws, and the superposition principle. The general *RLC* circuit conforms perfectly to these requirements, and it follows, therefore, that all linear circuits that contain any combinations of independent voltage and current sources, linear dependent voltage and current sources, and linear resistors, inductors, and capacitors may be analyzed with the use of these two theorems, if we wish. It is not necessary to repeat the theorems here, for they were previously stated in a manner that is equally applicable to the general *RLC* circuit.

7.5 • SIMPLE OP AMP CIRCUITS WITH CAPACITORS

In Chap. 6 we were introduced to several different types of amplifier circuits based on the ideal op amp. In almost every case, we found that the output was related to the input voltage by some combination of resistance ratios. If we replace one or more of these resistors with a capacitor, it is possible to obtain some interesting circuits in which the output is proportional to either the derivative or integral of the input voltage. Such circuits find widespread use in practice. For example, a velocity sensor can be connected to an op amp circuit that provides a signal proportional to the acceleration, or an output signal can be obtained that represents the total charge incident on a metal electrode during a specific period of time by simply integrating the measured current.

To create an integrator using an ideal op amp, we ground the noninverting input, install an ideal capacitor as a feedback element from the output back to the inverting input, and connect a signal source v_s to the inverting input through an ideal resistor as shown in Fig. 7.30.

Performing nodal analysis at the inverting input,

$$0 = \frac{v_a - v_s}{R_1} + i$$

We can relate the current i to the voltage across the capacitor,

$$i = C_f \frac{dv_{C_f}}{dt}$$

Resulting in

$$0 = \frac{v_a - v_s}{R_1} + C_f \frac{dv_{C_f}}{dt}$$

Invoking ideal op amp rule 2, we know that $v_a = v_b = 0$, so

$$0 = \frac{-v_s}{R_1} + C_f \frac{dv_{C_f}}{dt}$$

Integrating and solving for v_{out}, we obtain

$$v_{C_f} = v_a - v_{\text{out}} = 0 - v_{\text{out}} = \frac{1}{R_1 C_f} \int_0^t v_s \, dt' + v_{C_f}(0)$$

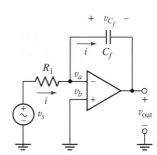

■ **FIGURE 7.30** An ideal op amp connected as an integrator.

or

$$v_{\text{out}} = -\frac{1}{R_1 C_f} \int_0^t v_s \, dt' - v_{C_f}(0) \qquad [17]$$

We therefore have combined a resistor, a capacitor, and an op amp to form an integrator. Note that the first term of the output is $1/RC$ times the negative of the integral of the input from $t' = 0$ to t, and the second term is the negative of the initial value of v_{C_f}. The value of $(RC)^{-1}$ can be made equal to unity if we wish by choosing $R = 1 \text{ M}\Omega$ and $C = 1 \text{ }\mu\text{F}$, for example; other selections may be made that will increase or decrease the output voltage.

Before we leave the integrator circuit, we might anticipate a question from an inquisitive reader, "*Could we use an inductor in place of the capacitor and obtain a differentiator?*" Indeed we could, but circuit designers usually avoid the use of inductors whenever possible because of their size, weight, cost, and associated resistance and capacitance. Instead, it is possible to interchange the positions of the resistor and capacitor in Fig. 7.30 and obtain a differentiator.

EXAMPLE 7.10

Derive an expression for the output voltage of the op amp circuit shown in Fig. 7.31.

We begin by writing a nodal equation at the inverting input pin, with $v_{C_1} \triangleq v_a - v_s$:

$$0 = C_1 \frac{dv_{C_1}}{dt} + \frac{v_a - v_{\text{out}}}{R_f}$$

Invoking ideal op amp rule 2, $v_a = v_b = 0$. Thus,

$$C_1 \frac{dv_{C_1}}{dt} = \frac{v_{\text{out}}}{R_f}$$

Solving for v_{out},

$$v_{\text{out}} = R_f C_1 \frac{dv_{C_1}}{dt}$$

Since $v_{C_1} = v_a - v_s = -v_s$,

$$v_{\text{out}} = -R_f C_1 \frac{dv_s}{dt}$$

So, simply by swapping the resistor and capacitor in the circuit of Fig. 7.30, we obtain a differentiator instead of an integrator.

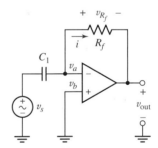

■ **FIGURE 7.31** An ideal op amp connected as a differentiator.

PRACTICE

7.9 Derive an expression for v_{out} in terms of v_{in} for the circuit shown in Fig. 7.32.

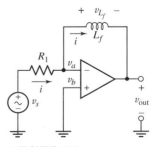

■ **FIGURE 7.32**

Ans: $v_{\text{out}} = -L_f/R_1 \, dv_s/dt$.

7.6 DUALITY

The concept of **duality** applies to many fundamental engineering concepts. In this section, we shall define duality in terms of the circuit equations. Two circuits are "duals" if the mesh equations that characterize one of them have the *same mathematical form* as the nodal equations that characterize the other. They are said to be exact duals if each mesh equation of one circuit is numerically identical with the corresponding nodal equation of the other; the current and voltage variables themselves cannot be identical, of course. Duality itself merely refers to any of the properties exhibited by dual circuits.

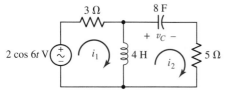

FIGURE 7.33 A given circuit to which the definition of duality may be applied to determine the dual circuit. Note that $v_c(0) = 10$ V.

Let us interpret the definition and use it to construct an exact dual circuit by writing the two mesh equations for the circuit shown in Fig. 7.33. Two mesh currents i_1 and i_2 are assigned, and the mesh equations are

$$3i_1 + 4\frac{di_1}{dt} - 4\frac{di_2}{dt} = 2\cos 6t \tag{18}$$

$$-4\frac{di_1}{dt} + 4\frac{di_2}{dt} + \frac{1}{8}\int_0^t i_2\,dt' + 5i_2 = -10 \tag{19}$$

We may now construct the two equations that describe the exact dual of our circuit. We wish these to be nodal equations, and thus begin by replacing the mesh currents i_1 and i_2 in Eqs. [18] and [19] by two node-to-reference voltages v_1 and v_2. We obtain

$$3v_1 + 4\frac{dv_1}{dt} - 4\frac{dv_2}{dt} = 2\cos 6t \tag{20}$$

$$-4\frac{dv_1}{dt} + 4\frac{dv_2}{dt} + \frac{1}{8}\int_0^t v_2\,dt' + 5v_2 = -10 \tag{21}$$

and we now seek the circuit represented by these two nodal equations.

Let us first draw a line to represent the reference node, and then we may establish two nodes at which the positive references for v_1 and v_2 are located. Equation [20] indicates that a current source of $2\cos 6t$ A is connected between node 1 and the reference node, oriented to provide a current entering node 1. This equation also shows that a 3 S conductance appears between node 1 and the reference node. Turning to Eq. [21], we first consider the nonmutual terms, i.e., those terms which do not appear in Eq. [20], and they instruct us to connect an 8 H inductor and a 5 S conductance (in parallel) between node 2 and the reference. The two similar terms in Eqs. [20] and [21] represent a 4 F capacitor present mutually at nodes 1 and 2; the circuit is completed by connecting this capacitor between the two nodes. The constant term on the right side of Eq. [21] is the value of the inductor current at $t = 0$; in other words, $i_L(0) = 10$ A. The dual circuit is shown in Fig. 7.34; since the two sets of equations are numerically identical, the circuits are exact duals.

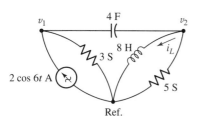

FIGURE 7.34 The exact dual of the circuit of Fig. 7.33.

Dual circuits may be obtained more readily than by this method, for the equations need not be written. In order to construct the dual of a given circuit, we think of the circuit in terms of its mesh equations. With each mesh we must associate a nonreference node, and, in addition, we must supply the reference node. On a diagram of the given circuit we therefore place a node in the center of each mesh and supply the reference node as a line near the diagram or a loop enclosing the diagram. Each element that appears jointly in two meshes is a mutual element and gives rise to identical terms, except for sign, in the two corresponding mesh equations. It must be replaced by an element that supplies the dual term in the two corresponding nodal equations. This dual element must therefore be connected directly between the two nonreference nodes that are within the meshes in which the given mutual element appears.

The nature of the dual element itself is easily determined; the mathematical form of the equations will be the same only if inductance is replaced by capacitance, capacitance by inductance, conductance by resistance, and resistance by conductance. Thus, the 4 H inductor which is common to meshes 1 and 2 in the circuit of Fig. 7.33 appears as a 4 F capacitor connected directly between nodes 1 and 2 in the dual circuit.

Elements that appear only in one mesh must have duals that appear between the corresponding node and the reference node. Referring again to Fig. 7.33, the voltage source 2 cos 6t V appears only in mesh 1; its *dual* is a current source 2 cos 6t A, which is connected only to node 1 and the reference node. Since the voltage source is clockwise-sensed, the current source must be into-the-nonreference-node-sensed. Finally, provision must be made for the dual of the initial voltage present across the 8 F capacitor in the given circuit. The equations have shown us that the dual of this initial voltage across the capacitor is an initial current through the inductor in the dual circuit; the numerical values are the same, and the correct sign of the initial current may be determined most readily by considering both the initial voltage in the given circuit and the initial current in the dual circuit as sources. Thus, if v_C in the given circuit is treated as a source, it would appear as $-v_C$ on the right side of the mesh equation; in the dual circuit, treating the current i_L as a source would yield a term $-i_L$ on the right side of the nodal equation. Since each has the same sign when treated as a source, then, if $v_C(0) = 10$ V, $i_L(0)$ must be 10 A.

The circuit of Fig. 7.33 is repeated in Fig. 7.35, and its exact dual is constructed on the circuit diagram itself by merely drawing the dual of each given element between the two nodes that are inside the two meshes that are common to the given element. A reference node that surrounds the given circuit may be helpful. After the dual circuit is redrawn in more standard form, it appears as shown in Fig. 7.34.

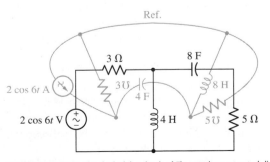

■ **FIGURE 7.35** The dual of the circuit of Fig. 7.33 is constructed directly from the circuit diagram.

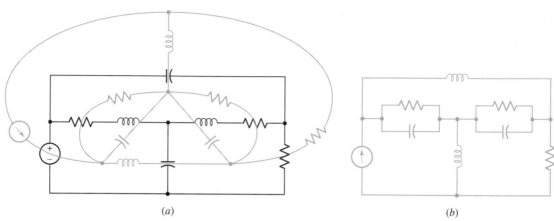

(a)   (b)

■ **FIGURE 7.36** (a) The dual (in gray) of a given circuit (in black) is constructed on the given circuit.
(b) The dual circuit is drawn in more conventional form for comparison to the original.

An additional example of the construction of a dual circuit is shown in Fig. 7.36a and b. Since no particular element values are specified, these two circuits are duals, but not necessarily exact duals. The original circuit may be recovered from the dual by placing a node in the center of each of the five meshes of Fig. 7.36b and proceeding as before.

The concept of duality may also be carried over into the language by which we describe circuit analysis or operation. For example, if we are given a voltage source in series with a capacitor, we might wish to make the important statement, "*The voltage source causes a current to flow through the capacitor*"; the dual statement is, "*The current source causes a voltage to exist across the inductor.*" The dual of a less carefully worded statement, such as "*The current goes round and round the series circuit,*" may require a little inventiveness.[4]

Practice in using dual language can be obtained by reading Thévenin's theorem in this sense; Norton's theorem should result.

We have spoken of dual elements, dual language, and dual circuits. What about a dual network? Consider a resistor R and an inductor L in series. The dual of this two-terminal network exists and is most readily obtained by connecting some ideal source to the given network. The dual circuit is then obtained as the dual source in parallel with a conductance G with the same magnitude as R, and a capacitance C having the same magnitude as L. We consider the dual network as the two-terminal network that is connected to the dual source; it is thus a pair of terminals between which G and C are connected in parallel.

Before leaving the definition of duality, it should be pointed out that duality is defined on the basis of mesh and nodal equations. Since nonplanar circuits cannot be described by a system of mesh equations, a circuit that cannot be drawn in planar form does not possess a dual.

We shall use duality principally to reduce the work that we must do to analyze the simple standard circuits. After we have analyzed the series RL circuit the parallel RC circuit requires less attention, not because it is less important, but because the analysis of the dual network is already known. Since the analysis of some complicated circuit is not apt to be well known, duality will usually not provide us with any quick solution.

(4) Someone suggested, "The voltage is across all over the parallel circuit."

PRACTICE

7.10 Write the single nodal equation for the circuit of Fig. 7.37a, and show, by direct substitution, that $v = -80e^{-10^6 t}$ mV is a solution. Knowing this, find (a) v_1; (b) v_2; and (c) i for the circuit of Fig. 7.37b.

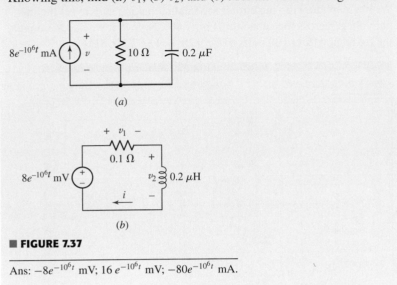

(a)

(b)

■ **FIGURE 7.37**

Ans: $-8e^{-10^6 t}$ mV; $16\,e^{-10^6 t}$ mV; $-80e^{-10^6 t}$ mA.

7.7 MODELING CAPACITORS AND INDUCTORS WITH PSPICE

When using PSpice to analyze circuits containing inductors and capacitors, it is frequently necessary to be able to specify the initial condition of each element [i.e., $v_C(0)$ and $i_L(0)$]. This is achieved by double-clicking on the element symbol, resulting in the dialog box shown in Fig. 7.38a. At the far right (not shown), we find the value of the capacitance, which defaults to 1 nF. We can also specify the initial condition (**IC**), set to 2 V in Fig. 7.38a. Clicking on the right mouse button and selecting Display results in the dialog box shown in Fig. 7.38b, which allows the initial condition to be displayed on the schematic. The procedure for setting the initial condition of an inductor is essentially the same. We should also note that when a capacitor is first placed in the schematic, it appears horizontally; the positive reference terminal for the initial voltage is the *left* terminal.

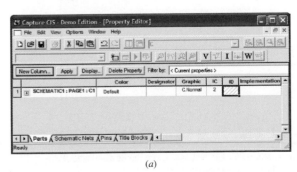

(a)

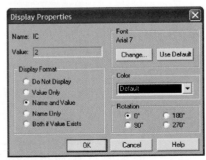

(b)

■ **FIGURE 7.38** (a) Capacitor property editor window. (b) Display Properties dialog box.

EXAMPLE 7.11

Simulate the output voltage waveform of the circuit in Fig. 7.39 if $v_s = 1.5 \sin 100t$ V, $R_1 = 10$ kΩ, $C_f = 4.7$ μF, and $v_C(0) = 2$ V.

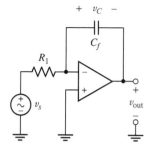

■ FIGURE 7.39 An integrating op amp circuit.

We begin by drawing the circuit schematic, making sure to set the initial voltage across the capacitor (Fig. 7.40). Note that we had to convert the frequency from 100 rad/s to $100/2\pi = 15.92$ Hz.

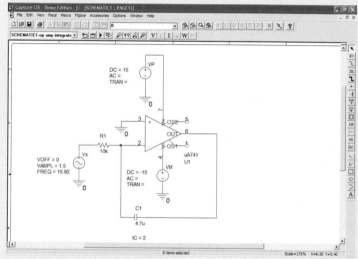

■ FIGURE 7.40 The schematic representation of the circuit shown in Fig. 7.39, with the initial capacitor voltage set to 2 V.

In order to obtain time-varying voltages and currents, we need to perform what is referred to as a *transient analysis*. Under the **PSpice** menu, we create a **New Simulation Profile** named **op amp integrator,** which leads to the dialog box re-created in Fig. 7.41. **Run to time**

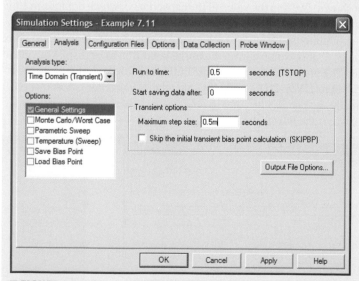

■ FIGURE 7.41 Dialog box for setting up a transient analysis. We choose a final time of 0.5 s to obtain several periods of the output waveform ($1/15.92 \approx 0.06$ s).

represents the time at which the simulation is terminated; PSpice will select its own discrete times at which to calculate the various voltages and currents. Occasionally we obtain an error message stating that the transient solution could not converge, or the output waveform does not appear as smooth as we would like. In such situations, it is useful to set a value for **Maximum step size,** which has been set to 0.5 ms in this example.

From our earlier analysis and Eq. [17], we expect the output to be proportional to the negative integral of the input waveform, i.e., $v_{out} = 0.319 \cos 100t - 2.319$ V, as shown in Fig. 7.42. The initial condition of 2 V across the capacitor has combined with a constant term from the integration to result in *a nonzero average value* for the output, unlike the input which has an average value of zero.

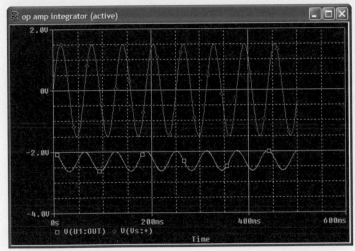

■ **FIGURE 7.42** Probe output for the simulated integrator circuit along with the input waveform for comparison.

SUMMARY AND REVIEW

❏ The current through a capacitor is given by $i = C \, dv/dt$.

❏ The voltage across a capacitor is related to its current by

$$v(t) = \frac{1}{C} \int_{t_0}^{t} i(t') \, dt' + v(t_0)$$

❏ A capacitor is an *open circuit* to dc currents.

❏ The voltage across an inductor is given by $v = L \, di/dt$.

❏ The current through an inductor is related to its voltage by

$$i(t) = \frac{1}{L} \int_{t_0}^{t} v \, dt' + i(t_0)$$

❏ An inductor is a *short circuit* to dc currents.

❑ The energy presently stored in a capacitor is given by $\frac{1}{2} C v^2$, whereas the energy presently stored in an inductor is given by $\frac{1}{2} L i^2$; both are referenced to a time at which no energy was stored.

❑ Series and parallel combinations of inductors can be combined using the same equations as for resistors.

❑ Series and parallel combinations of capacitors work the *opposite* way that they do for resistors.

❑ A capacitor as the feedback element in an inverting op amp leads to an output voltage proportional to the *integral* of the input voltage. Swapping the input resistor and the feedback capacitor leads to an output voltage proportional to the *derivative* of the input voltage.

❑ Since capacitors and inductors are linear elements, KVL, KCL, superposition, Thévenin's and Norton's theorems, and nodal and mesh analysis apply to their circuits as well.

❑ The concept of duality provides another perspective on the relationship between circuits with inductors and circuits with capacitors.

❑ PSpice allows us to set the initial voltage across a capacitor, and the initial current through an inductor. A transient analysis provides details of the time-dependent response of circuits containing these types of elements.

READING FURTHER

A detailed guide to characteristics and selection of various capacitor and inductor types can be found in:

H. B. Drexler, *Passive Electronic Component Handbook,* 2nd ed., C. A. Harper, ed. New York: McGraw-Hill, 2003, pp. 69–203.

C. J. Kaiser, *The Inductor Handbook,* 2nd ed. Olathe, Kans.: C.J. Publishing, 1996.

Two books that describe capacitor-based op amp circuits are:

R. Mancini, (ed.). *Op Amps Are For Everyone,* 2nd ed. Amsterdam: Newnes, 2003.

W. G. Jung, *Op Amp Cookbook,* 3rd ed. Upper Saddle River, N.J.: Prentice Hall, 1997.

EXERCISES

7.1 The Capacitor

1. Calculate the current flowing through a 10 μF capacitor if the voltage across its terminals is: (*a*) 5 V; (*b*) $115\sqrt{2} \cos 120\pi t$ V; (*c*) $4e^{-t}$ mV.

2. Sketch the current flowing through a 4.7 μF capacitor in response to the voltage waveform shown in Fig. 7.43. Assume the current and voltage are defined consistently with the passive sign convention.

3. Calculate the current flowing through a 1 mF capacitor in response to a voltage v across its terminals if v equals: (*a*) $30te^{-t}$ V; (*b*) $4e^{-5t} \sin 100t$ V.

4. What is the maximum amount of energy that can be stored in each of the electrolytic capacitors of Fig. 7.2*b* and *c*? Explain your answer.

5. A capacitor is fabricated from two thin aluminum disks 1 cm in diameter which are separated by a distance of 100 μm (0.1 mm). (*a*) Compute the

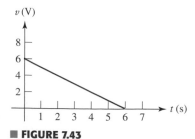

FIGURE 7.43

capacitance, assuming only air between the metal plates. (*b*) Determine the voltage that must be applied to store a measly 1 mJ of energy in the capacitor. (*c*) If the capacitor is needed to store 2.5 μJ of energy in an application that can supply up to 100 V, what value of relative permittivity $\varepsilon/\varepsilon_0$ would be required for the region between the plates?

6. A silicon *pn* junction diode is characterized by a junction capacitance defined as

$$C_j = \frac{K_s \varepsilon_0 A}{W}$$

where $K_s = 11.8$ for silicon, ε_0 is the vacuum permittivity, $A =$ the cross-sectional area of the junction, and W is known as the depletion width of the junction. W depends on not only how the diode is fabricated, but also on the voltage applied to its two terminals. It can be computed using

$$W = \sqrt{\frac{2K_s\varepsilon_0}{qN}\,(V_{bi} - V_A)}$$

Thus, diodes are frequently used in electronic circuits, since they can be thought of as voltage-controlled capacitors. Assuming parameter values of $N = 10^{18}$ cm^{-3}, $V_{bi} = 0.57$ V, and using $q = 1.6 \times 10^{-19}$ C, calculate the capacitance of a diode with cross-sectional area $A = 1\ \mu$m $\times 1\ \mu$m at applied voltages of $V_A = -1, -5,$ and -10 volts.

D 7. Design a capacitor whose capacitance can be varied manually between 100 pF and 1 nF by rotating a knob. Include appropriately labeled diagrams to explain your design.

8. A voltage $v(t) = \begin{cases} 3\text{ V} & t < 0 \\ 3e^{-t/5}\text{ V}, & t \geq 0 \end{cases}$ is applied to a 300 μF capacitor. (*a*) Compute the energy stored in the capacitor at $t = 2$ ms. (*b*) At what time has the energy stored in the capacitor dropped to 37 percent of its maximum value? (*Round to the nearest second.*) (*c*) Determine the current flowing through the capacitor at $t = 1.2$ s. (*d*) Calculate the power delivered by the capacitor to the external circuit at $t = 2$ s.

9. The current through a 47 μF capacitor is shown in Fig. 7.44. Calculate the voltage across the device at (*a*) $t = 2$ ms; (*b*) $t = 4$ ms; (*c*) $t = 5$ ms.

10. The current through a capacitor is given by $i(t) = 7 \sin \pi t$ mA. If the energy stored at $t = 200$ ms is 3 μJ, what is the value of the capacitance?

11. (*a*) If the capacitor shown in Fig. 7.1 has a capacitance of 0.2 μF, let $v_C = 5 + 3\cos^2 200t$ V, and find $i_C(t)$. (*b*) What is the maximum energy stored in the capacitor? (*c*) If $i_C = 0$ for $t < 0$ and $i_C = 8e^{-100t}$ mA for $t > 0$, find $v_C(t)$ for $t > 0$. (*d*) If $i_C = 8e^{-100t}$ mA for $t > 0$ and $v_C(0) = 100$ V, find $v_C(t)$ for $t > 0$.

12. The current waveform shown for $t > 0$ in Fig. 7.45 is applied to a 2 mF capacitor. Given that $v_C(0) = 250$ V, and assuming the passive sign convention, during what time interval is the value of v_C between 2000 and 2100 V?

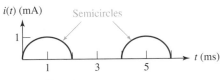

$i(t)$ (mA) Semicircles

■ **FIGURE 7.44**

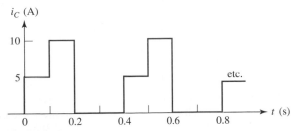

■ **FIGURE 7.45**

13. A resistance R is connected in parallel with a 1 μF capacitor. For any $t \le 0$, the energy stored in the capacitor is $20e^{-1000t}$ mJ. (a) Find R. (b) By integration, show that the energy dissipated in R over the interval $0 \le t < \infty$ is 0.02 J.

14. For the circuits in Fig. 7.46, (a) compute the voltage across each capacitor. (b) Verify your answers with PSpice. Submit a properly labeled schematic with your simulation results.

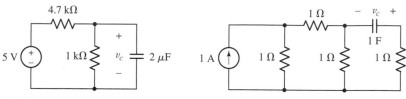

■ **FIGURE 7.46**

7.2 The Inductor

15. Calculate the voltage across a 10 nH inductor if the current into the "+" reference terminal is: (a) 5 mA; (b) $115\sqrt{2} \cos 120\pi t$ A; (c) $4e^{-6t}$ mA.

16. Sketch the voltage appearing across a 1 pH inductor in response to the current waveform shown in Fig. 7.47. Assume the current and voltage are defined consistently with the passive sign convention.

17. Calculate the voltage that develops across a 5 μH inductor in response to a current i flowing into its "+" reference terminal if i equals: (a) $30te^{-t}$ nA; (b) $4e^{-5t} \sin 100t$ mA.

18. What is the maximum amount of energy that can be stored in a 5 mH inductor if the wire is rated for a maximum current of 1.5 A? Explain your answer.

19. With reference to Fig. 7.48: (a) sketch v_L as a function of time, $0 < t < 60$ ms; (b) find the value of time at which the inductor is absorbing a maximum power; (c) find the value of time at which it is supplying a maximum power; and (d) find the energy stored in the inductor at $t = 40$ ms.

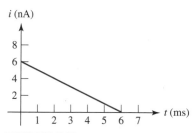

■ **FIGURE 7.47**

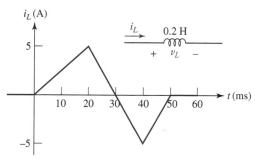

■ **FIGURE 7.48**

20. In Fig. 7.13, let $L = 50$ mH, with $i_L = 0$ for $t < 0$ and $80te^{-100t}$ mA for $t > 0$. Find the maximum values of $|i_L|$ and $|v|$, and the time at which each maximum occurs.

21. (a) If $i_s = 0.4t^2$ A for $t > 0$ in the circuit of Fig. 7.49a, find and sketch $v_{in}(t)$ for $t > 0$. (b) If $v_s = 40t$ V for $t > 0$ and $i_L(0) = 5$ A, find and sketch $i_{in}(t)$ for $t > 0$ in the circuit of Fig. 7.49b.

22. The voltage 20 cos 1000t V is applied to a 25 mH inductor. If the inductor current is zero at $t = 0$, find and sketch for $(0 \le t \le 2\pi$ ms): (a) the power being absorbed by the inductor; (b) the energy stored in the inductor.

23. The voltage v_L across a 0.2 H inductor is 100 V for $0 < t \le 10$ ms; decreases linearly to zero in the interval $10 < t < 20$ ms; is 0 for $20 \le t < 30$ ms; is 100 V for $30 < t < 40$ ms; and is zero thereafter. Assume the passive sign

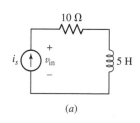

(a)

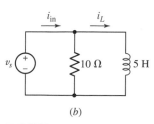

(b)

■ **FIGURE 7.49**

convention for v_L and i_L. (*a*) Calculate i_L at $t = 8$ ms if $i_L(0) = -2$ A. (*b*) Determine the stored energy at $t = 22$ ms if $i_L(0) = 0$.

24. The circuit depicted in Fig. 7.50 has been connected for an interminably long time. Calculate the current i_x.

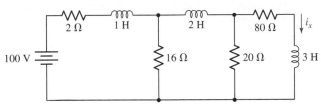

■ **FIGURE 7.50**

25. The voltage across a 5 H inductor is $v_L = 10(e^{-t} - e^{-2t})$ V. If $i_L(0) = 80$ mA and v_L and i_L satisfy the passive sign convention, find (*a*) $v_L(1$ s$)$; (*b*) $i_L(1$ s$)$; and (*c*) $i_L(\infty)$.

26. A long time after all connections have been made in the circuit shown in Fig. 7.51, find v_x if (*a*) a capacitor is present between x and y and (*b*) an inductor is present between x and y.

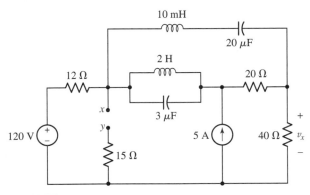

■ **FIGURE 7.51**

27. With reference to the circuit shown in Fig. 7.52, find (*a*) w_L; (*b*) w_C; (*c*) the voltage across each circuit element; (*d*) the current in each circuit element.

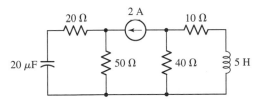

■ **FIGURE 7.52**

28. Let $v_s = 400t^2$ V for $t > 0$ and $i_L(0) = 0.5$ A in the circuit of Fig. 7.53. At $t = 0.4$ s, find the values of energy: (*a*) stored in the capacitor; (*b*) stored in the inductor; and (*c*) dissipated by the resistor since $t = 0$.

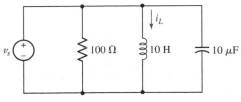

■ **FIGURE 7.53**

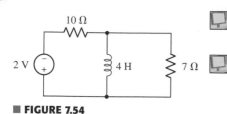

■ FIGURE 7.54

29. For the circuit of Fig. 7.54, (a) compute the power dissipated by the 7 Ω and 10 Ω resistors, respectively. (b) Verify your answers with PSpice. Submit a properly labeled schematic with your simulation results.

30. (a) Determine the Thévenin equivalent of the network connected to the inductor of Fig. 7.55. (b) Compute the current through the inductor. (c) Verify your answer with PSpice. Submit a properly labeled schematic with your simulation results.

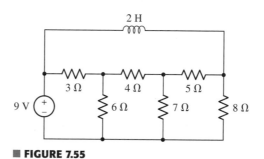

■ FIGURE 7.55

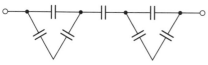

■ FIGURE 7.56

7.3 Inductance and Capacitance Combinations

31. Determine the equivalent capacitance of the network in Fig. 7.56 if all capacitors are 10 μF.

32. Determine the equivalent inductance of the network in Fig. 7.57 if all inductors are 77 pH.

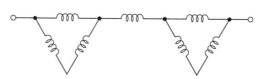

■ FIGURE 7.57

33. For the circuit of Fig. 7.58, (a) reduce the circuit to the fewest possible components using series/parallel combinations; (b) determine v_x if all resistors are 10 kΩ, all capacitors are 50 μF, and all inductors are 1 mH.

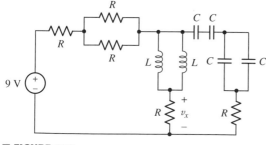

■ FIGURE 7.58

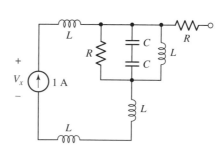

■ FIGURE 7.59

34. For the circuit of Fig. 7.59, (a) redraw the circuit using the fewest possible components by employing series/parallel combination rules; (b) determine V_x if all resistors are 1 Ω, all capacitors are 50 μF, and all inductors are 10 nH.

35. Reduce the network of Fig. 7.60 to a single equivalent capacitance as seen looking into terminals a and b.

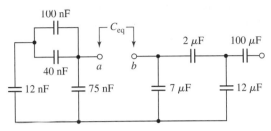

■ **FIGURE 7.60**

36. Reduce the network of Fig. 7.61 to a single equivalent inductance as seen looking into terminals a and b.

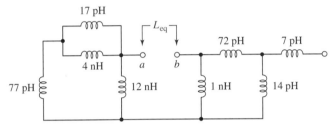

■ **FIGURE 7.61**

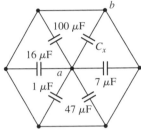

■ **FIGURE 7.62**

37. The network of Fig. 7.62 stores 534.8 μJ of energy when a voltage of 2.5 V is connected to terminals a and b. What is the value of C_x?

38. The network of Fig. 7.63 consists of three stages in series, with each stage containing a corresponding number of inductors in parallel. (a) Find the equivalent inductance if all inductors are 1.5 H. (b) Derive an expression for a general network of this type having N stages.

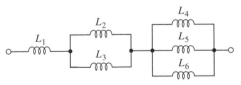

■ **FIGURE 7.63**

39. For the network of Fig. 7.63, $L_1 = 1$ H, $L_2 = L_3 = 2$ H, $L_4 = L_5 = L_6 = 3$ H. (a) Find the equivalent inductance. (b)Derive an expression for a general network of this type having N stages, assuming stage N is composed of N inductors, each having inductance N henrys.

40. Extend the concept of Δ-Y transformations to simplify the network of Fig. 7.64 if each element is a 2 pF capacitor.

41. Extend the concept of Δ-Y transformations to simplify the network of Fig. 7.64 if each element is a 1 nH inductor.

42. Given a box full of 1 μH inductors, show how (using as few components as possible) one may obtain an equivalent inductance of (a) 2.25 μH; (b) 750 nH; (c) 450 nH.

43. Refer to the network shown in Fig. 7.65 and find (a) R_{eq} if each element is a 10 Ω resistor; (b) L_{eq} if each element is a 10 H inductor; and (c) C_{eq} if each element is a 10 F capacitor.

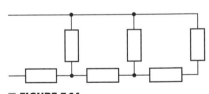

■ **FIGURE 7.64**

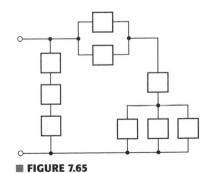

■ **FIGURE 7.65**

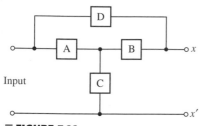

■ **FIGURE 7.66**

44. In Fig. 7.66, let elements A, B, C, and D be (a) 1 H, 2 H, 3 H, and 4 H inductors, respectively, and find the input inductance with x-x' first open-circuited and then short-circuited; (b) 1 F, 2 F, 3 F, and 4 F capacitors, respectively, and find the input capacitance with x-x' first open-circuited and then short-circuited.

45. Given a boxful of 1 nF capacitors, and using as few capacitors as possible, show how it is possible to obtain an equivalent capacitance of (a) 2.25 nF; (b) 0.75 nF; (c) 0.45 nF.

7.4 Consequences of Linearity

46. In the circuit shown in Fig. 7.67, let $i_s = 60e^{-200t}$ mA with $i_1(0) = 20$ mA. (a) Find $v(t)$ for all t. (b) Find $i_1(t)$ for $t \geq 0$. (c) Find $i_2(t)$ for $t \geq 0$.

47. Let $v_s = 100e^{-80t}$ V and $v_1(0) = 20$ V in the circuit of Fig. 7.68. (a) Find $i(t)$ for all t. (b) Find $v_1(t)$ for $t \geq 0$. (c) Find $v_2(t)$ for $t \geq 0$.

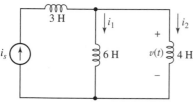

■ **FIGURE 7.67**

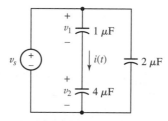

■ **FIGURE 7.68**

48. (a) Write nodal equations for the circuit of Fig. 7.69. (b) Write mesh equations for the same circuit.

49. If it is assumed that all the sources in the circuit of Fig. 7.70 have been connected and operating for a very long time, use the superposition principle to find $v_C(t)$ and $v_L(t)$.

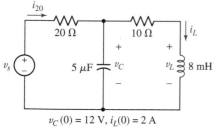

$v_C(0) = 12$ V, $i_L(0) = 2$ A

■ **FIGURE 7.69**

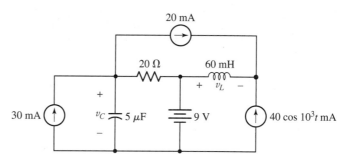

■ **FIGURE 7.70**

50. For the circuit of Fig. 7.71, assume no energy is stored at $t = 0$, and write a complete set of nodal equations.

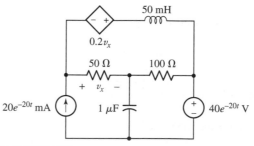

■ **FIGURE 7.71**

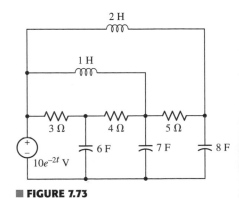

7.5 Simple Op Amp Circuits with Capacitors

51. Interchange the location of R and C in the circuit of Fig. 7.30, and assume that $R_i = \infty$, $R_o = 0$, and $A = \infty$ for the op amp. (a) Find $v_{out}(t)$ as a function of $v_s(t)$. (b) Obtain an equation relating $v_o(t)$ and $v_s(t)$ if A is not assumed to be infinite.

52. An ion implanter is a device used to inject ionized atoms into silicon in order to modify its electrical characteristics for subsequent fabrication of diodes and transistors. The ions striking the silicon target lead to a current which can be directed through a resistor of precisely known value. The current is directly proportional to the ion intensity ($i = 1.602 \times 10^{-19} \times$ number of incident ions per second). If a 1.000 MΩ resistor is used to detect the current, design a circuit to provide an output directly proportional to the total number of ions that strike the silicon during the implantation run (known as the *dose*).

53. In the circuit of Fig. 7.30, let $R = 0.5$ MΩ, $C = 2\,\mu$F, $R_i = \infty$, and $R_o = 0$. Suppose that we wish the output to be $v_{out} = \cos 10t - 1$ V. Obtain $v_s(t)$ if (a) $A = 2000$ and (b) A is infinite.

54. A velocity sensor is attached to a rotating wheel. Design a circuit to provide a positive voltage whose magnitude is equal to the acceleration (revolutions per minute) of the wheel. Assume the velocity sensor's output is 1 mV/rpm, and the wheel rotates at less than 3500 rpm.

55. (a) In the circuit of Fig. 7.72, swap the resistor and inductor, and derive an expression for v_{out} in terms of v_s. (b) Explain why such a circuit would not typically be used in practice.

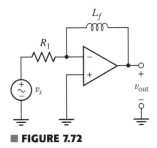

■ FIGURE 7.72

56. An exposure meter connected to a camera provides an output directly proportional to the incident light intensity, such that 1 mV = 1 mcd (millicandela). Design a circuit that provides a voltage output proportional to the integrated intensity, such that 1 V = 1 mcd-s.

57. A certain glass-forming process requires that the cooling rate is not to exceed 100°C/min. Available is a voltage proportional to the current glass melt temperature, such that 1mV = 1°C over the range of 500 to 2000°C. Design a circuit whose voltage output represents the cooling rate, such that 1 V = 100°C/min.

58. A floating sensor is installed in a fuel tank to provide a measure of the fuel remaining. The sensor is calibrated so that 1 volt = 10 liters. Design a circuit whose voltage output provides a reading of the rate of fuel consumption in liters per second, such that 1 V = 1 l/s.

59. A battery is to be tested to determine the amount of energy it can deliver to a certain 1 Ω load. Two signals are available: a voltage signal for the square of the battery voltage (1 mV = 1 V²) and a voltage signal to indicate the square of the current flowing out of the battery (1 mV = 1 A²). Design a circuit whose output voltage is proportional to the total energy delivered, such that 1 mV = 1 J of energy supplied to the load.

60. The test engineers of an experimental aircraft want to monitor the deceleration during landing, as they are concerned the wheel suspension systems are experiencing too much stress. If they can provide you with a voltage signal from the nose gear wheels calibrated so that 1 mV = 1 mph, design a circuit whose output voltage is proportional to the rate of deceleration (so that 1 V = 1 km/s²).

7.6 Duality

61. (a) Draw the exact dual of the circuit shown in Fig. 7.69. Specify the dual variables and the dual initial conditions. (b) Write nodal equations for the dual circuit. (c) Write mesh equations for the dual circuit.

62. Draw the exact dual of the circuit shown in Fig. 7.51. Draw the circuit in a neat, clean form with square corners, a recognizable reference node, and no crossovers.

63. Draw the exact dual of the circuit shown in Fig. 7.73. Keep it neat!

■ FIGURE 7.73

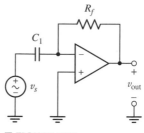

■ FIGURE 7.74

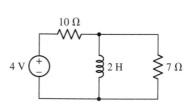

■ FIGURE 7.75

64. (*a*) Draw the exact dual of the circuit given for Exer. 47, including variables. (*b*) Write the dual of the problem statement for Exer. 47. (*c*) Solve your new Exer. 47.

65. Determine the dual of the circuit in Fig. 7.74, and obtain an expression for i_{out} in terms of i_s. (Hint: Use the detailed model for an op amp.)

7.7 Modeling Capacitors and Inductors with PSpice

66. Calculate the energy stored in the inductor of Fig. 7.75. Verify your solution using PSpice; submit a properly labeled schematic with your simulation results.

67. Calculate the energy stored in the inductor of Fig. 7.76. Verify your solution using PSpice; submit a properly labeled schematic with your simulation results.

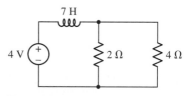

■ FIGURE 7.76

68. Calculate the energy stored in the capacitor of Fig. 7.77. Verify your solution using PSpice; submit a properly labeled schematic with your simulation results.

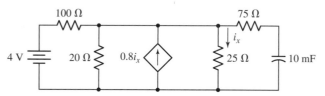

■ FIGURE 7.77

69. Calculate the energy stored in the capacitor of Fig. 7.78. Verify your solution using PSpice; submit a properly labeled schematic with your simulation results.

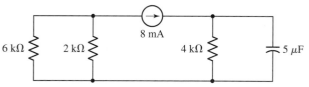

■ FIGURE 7.78

70. For the op amp differentiator of Fig. 7.31, set $C_1 = 5$ nF and $R_f = 100$ MΩ. (*a*) Predict the output if $v_s(t) = 3 \sin 10t$ V. (*b*) Verify your solution with a PSpice simulation. Submit a properly labeled schematic with your simulation results.

71. Use PSpice to verify that the energy stored in a 33 μF capacitor is 221 μJ at $t = 10^{-2}$ s when connected to a voltage source $v(t) = 5 \cos 75t$ V. (Hint: Use the component VSIN.)

72. Use PSpice to verify that the energy stored in a 100 pH inductor is 669 pJ at $t = 0.01$ s when connected to a current source $i(t) = 5 \cos 75t$ A. (Hint: Use the component ISIN.)

73. For the circuit of Fig. 7.72, select R_1 and L_f so that the output is twice the derivative of the input voltage if $v_s = A \cos 2\pi 10^3 t$ V. Verify your design using PSpice.

74. Work Exer. 71, but with $v(t) = 5 \cos 75t - 7$ V instead.

75. Work Exer. 72, but with $i(t) = 5 \cos 75t - 7$ A instead.

Basic *RL* and *RC* Circuits

KEY CONCEPTS

RL and *RC* Time Constants

Natural and Forced Response

Calculating the Time-Dependent Response to DC Excitation

How to Determine Initial Conditions and Their Effect on the Circuit Response

Analyzing Circuits with Step Function Input and with Switches

Construction of Pulse Waveforms Using Unit Step Functions

The Response of Sequentially Switched Circuits

INTRODUCTION

In Chap. 7 we wrote equations governing the response of several circuits containing both inductance and capacitance, but we did not solve any of them. At this time we are ready to proceed with the solution of the simpler circuits, restricting our attention to those which contain only resistors and inductors or only resistors and capacitors.

Although the circuits that we are about to consider have a very elementary appearance, they are also of practical importance. Networks of this form find use in electronic amplifiers, automatic control systems, operational amplifiers, communications equipment, and many other applications. A familiarity with these simple circuits will enable us to predict the accuracy with which the output of an amplifier can follow an input that is changing rapidly with time, or to predict how quickly the speed of a motor will change in response to a change in its field current. Our understanding of simple *RL* and *RC* circuits will also enable us to suggest modifications to the amplifier or motor in order to obtain a more desirable response.

8.1 THE SOURCE-FREE *RL* CIRCUIT

The analysis of circuits containing inductors and/or capacitors is dependent upon the formulation and solution of the integrodifferential equations that characterize the circuits. We will call the special type of equation we obtain a *homogeneous linear differential equation*, which is simply a differential equation in which every term is of the first degree in the dependent variable or one of its derivatives. A solution is obtained when we have found an expression for the dependent variable that satisfies both the differential equation and also the prescribed energy distribution in the inductors or capacitors at a prescribed instant of time, usually $t = 0$.

The solution of the differential equation represents a response of the circuit, and it is known by many names. Since this response depends upon the general "nature" of the circuit (the types of elements, their sizes, the interconnection of the elements), it is often called a ***natural response.*** However, any real circuit we construct cannot store energy forever; the resistances intrinsically associated with inductors and capacitors will eventually convert all stored energy into heat. The response must eventually die out, and for this reason it is frequently referred to as the ***transient response.*** Finally, we should also be familiar with the mathematicians' contribution to the nomenclature; they call the solution of a homogeneous linear differential equation a *complementary function.*

When we consider independent sources acting on a circuit, part of the response will resemble the nature of the particular source (or *forcing function*) used; this part of the response, called the *particular solution,* the *steady-state response,* or the ***forced response,*** will be "complemented" by the complementary response produced in the source-free circuit. The complete response of the circuit will then be given by the sum of the complementary function and the particular solution. In other words, the complete response is the sum of the natural response and the forced response. The source-free response may be called the *natural* response, the *transient* response, the *free* response, or the *complementary function,* but because of its more descriptive nature, we will most often call it the *natural* response.

We will consider several different methods of solving these differential equations. The mathematical manipulation, however, is not circuit analysis. Our greatest interest lies in the solutions themselves, their meaning, and their interpretation, and we will try to become sufficiently familiar with the form of the response that we are able to write down answers for new circuits by just plain thinking. Although complicated analytical methods are needed when simpler methods fail, a well-developed intuition is an invaluable resource in such situations.

We begin our study of transient analysis by considering the simple series *RL* circuit shown in Fig. 8.1. Let us designate the time-varying current as $i(t)$; we will represent the value of $i(t)$ at $t = 0$ as I_0; in other words, $i(0) = I_0$. We therefore have

$$Ri + v_L = Ri + L\frac{di}{dt} = 0$$

or

$$\frac{di}{dt} + \frac{R}{L}i = 0 \qquad [1]$$

Our goal is an expression for $i(t)$ which satisfies this equation *and* also has the value I_0 at $t = 0$. The solution may be obtained by several different methods.

A Direct Approach

One very direct method of solving a differential equation consists of writing the equation in such a way that the variables are separated, and then

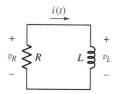

FIGURE 8.1 A series *RL* circuit for which $i(t)$ is to be determined, subject to the initial condition that $i(0) = I_0$.

It may seem pretty strange to discuss a time-varying current flowing in a circuit with no sources! Keep in mind that we only know the current at the time specified as $t = 0$; we don't know the current prior to that time. In the same vein, we don't know what the circuit looked like prior to $t = 0$, either. In order for a current to be flowing, a source had to have been present at some point, but we are not privy to this information. Fortunately, it is not required in order to analyze the circuit we are given.

integrating each side of the equation. The variables in Eq. [1] are i and t, and it is apparent that the equation may be multiplied by dt, divided by i, and arranged with the variables separated:

$$\frac{di}{i} = -\frac{R}{L}\,dt \qquad [2]$$

Since the current is I_0 at $t = 0$ and $i(t)$ at time t, we may equate the two definite integrals which are obtained by integrating each side between the corresponding limits:

$$\int_{I_0}^{i(t)} \frac{di'}{i'} = \int_{0}^{t} -\frac{R}{L}\,dt'$$

Performing the indicated integration,

$$\ln i' \Big|_{I_0}^{i} = -\frac{R}{L} t' \Big|_{0}^{t}$$

which results in

$$\ln i - \ln I_0 = -\frac{R}{L}(t - 0)$$

After a little manipulation, we find that the current $i(t)$ is given by

$$i(t) = I_0 e^{-Rt/L} \qquad [3]$$

We check our solution by first showing that substitution of Eq. [3] in Eq. [1] yields the identity $0 = 0$, and then showing that substitution of $t = 0$ in Eq. [3] produces $i(0) = I_0$. Both steps are necessary; the solution must satisfy the differential equation which characterizes the circuit, and it must also satisfy the initial condition.

CAUTION

EXAMPLE 8.1

If the inductor of Fig. 8.2 has a current $i_L = 2$ A at $t = 0$, find an expression for $i_L(t)$ valid for $t > 0$, and its value at $t = 200\ \mu s$.

This is the identical type of circuit just considered, so we expect an inductor current of the form

$$i_L(t) = I_0 e^{-R/L\,t}$$

where $R = 200\ \Omega$, $L = 50$ mH and I_0 is the initial current flowing through the inductor at $t = 0$. Thus,

$$i_L(t) = 2e^{-4000t}$$

Substituting $t = 200 \times 10^{-6}$ s, we find that $i_L(t) = 898.7$ mA, less than half the initial value.

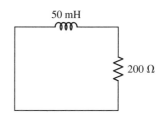

■ **FIGURE 8.2** A simple *RL* circuit in which energy is stored in the inductor at $t = 0$.

PRACTICE

8.1 Determine the current i_R through the resistor of Fig. 8.3 at $t = 1$ ns if $i_R(0) = 6$ A.

Ans: 812 mA.

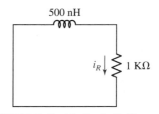

■ **FIGURE 8.3** Circuit for Practice Problem 8.1.

An Alternative Approach

The solution may also be obtained by a slight variation of the method we have described. After separating the variables, we may obtain the indefinite integral of each side of Eq. [2] if we also include a constant of integration. Thus,

$$\int \frac{di}{i} = -\int \frac{R}{L} dt + K$$

and integration gives us

$$\ln i = -\frac{R}{L}t + K \qquad [4]$$

The constant K cannot be evaluated by substitution of Eq. [4] in the original differential equation [1]; the identity $0 = 0$ will result, because Eq. [4] is a solution of Eq. [1] for *any* value of K (try it out on your own). The constant of integration must be selected to satisfy the initial condition $i(0) = I_0$. Thus, at $t = 0$, Eq. [4] becomes

$$\ln I_0 = K$$

and we use this value for K in Eq. [4] to obtain the desired response

$$\ln i = -\frac{R}{L}t + \ln I_0$$

or

$$i(t) = I_0 e^{-Rt/L}$$

as before.

A More General Solution Approach

Either of these methods can be used when the variables are separable, but this is not always the situation. In the remaining cases we will rely on a very powerful method, the success of which will depend upon our intuition or experience. We simply guess or assume a form for the solution and then test our assumptions, first by substitution in the differential equation, and then by applying the given initial conditions. Since we cannot be expected to guess the exact numerical expression for the solution, we will assume a solution containing several unknown constants and select the values for these constants in order to satisfy the differential equation and the initial conditions. Many of the differential equations encountered in circuit analysis have a solution which may be represented by the exponential function or by the sum of several exponential functions. Let us assume a solution of Eq. [1] in exponential form,

$$i(t) = A e^{s_1 t} \qquad [5]$$

where A and s_1 are constants to be determined. After substituting this assumed solution in Eq. [1], we have

$$A s_1 e^{s_1 t} + A \frac{R}{L} e^{s_1 t} = 0$$

or

$$\left(s_1 + \frac{R}{L}\right) A e^{s_1 t} = 0 \qquad [6]$$

In order to satisfy this equation for all values of time, it is necessary that either $A = 0$, or $s_1 = -\infty$, or $s_1 = -R/L$. But if $A = 0$ or $s_1 = -\infty$, then every response is zero; neither can be a solution to our problem. Therefore, we must choose

$$s_1 = -\frac{R}{L} \qquad [7]$$

and our assumed solution takes on the form

$$i(t) = A e^{-Rt/L}$$

The remaining constant must be evaluated by applying the initial condition $i(0) = I_0$. Thus, $A = I_0$, and the final form of the assumed solution is (again)

$$i(t) = I_0 e^{-Rt/L}$$

A summary of the basic approach is outlined in Fig. 8.4.

In fact, there is a more direct route that we can take. In obtaining Eq. [7], we solved

$$s_1 + \frac{R}{L} = 0 \qquad [8]$$

which is known as the ***characteristic equation.*** We can obtain the characteristic equation directly from the differential equation, without the need for substitution of our trial solution. Consider the general first-order differential equation

$$a\frac{df}{dt} + bf = 0$$

where a and b are constants. We substitute s^1 for df/dt and s^0 for f, resulting in

$$a\frac{df}{dt} + bf = (as + b)f = 0$$

From this we may directly obtain the characteristic equation

$$as + b = 0$$

which has the single root $s = -b/a$. The solution to our differential equation is then

$$f = A e^{-bt/a}$$

This basic procedure is easily extended to second-order differential equations, as we will explore in Chap. 9.

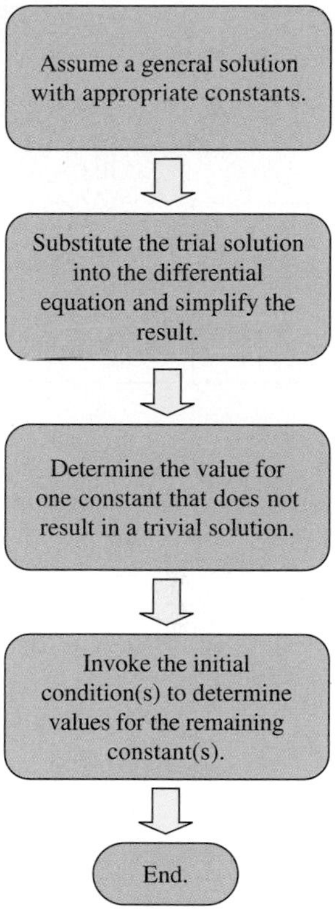

Assume a general solution with appropriate constants.

Substitute the trial solution into the differential equation and simplify the result.

Determine the value for one constant that does not result in a trivial solution.

Invoke the initial condition(s) to determine values for the remaining constant(s).

End.

■ **FIGURE 8.4** Flowchart for the general approach to solution of first-order differential equations where, based on experience, we can guess the form of the solution.

EXAMPLE 8.2

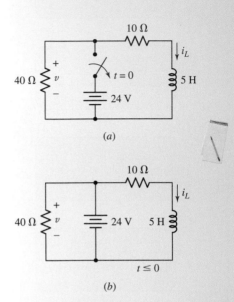

For the circuit of Fig. 8.5a, find the voltage labeled v at $t = 200$ ms.

▶ **Identify the goal of the problem.**

The schematic of Fig. 8.5a actually represents *two different* circuits: one with the switch closed (Fig. 8.5b) and one with the switch open (Fig. 8.5c). We are asked to find $v(0.2)$ for the circuit shown in Fig. 8.5c.

▶ **Collect the known information.**

We should first check that both new circuits are drawn and labeled correctly. We next make the assumption that the circuit in Fig. 8.5b has been connected for a long time, so that any transients have dissipated. We may make such an assumption in these circumstances unless instructed otherwise.

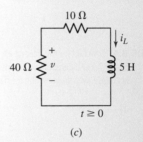

(b)

▶ **Devise a plan.**

The circuit of Fig. 8.5c may be analyzed by writing a KVL equation. Ultimately we want a differential equation with only v and t as variables; this may require additional equations and some substitution. We will then solve the differential equation for $v(t)$.

▶ **Construct an appropriate set of equations.**

Referring to Fig. 8.5c, we write

$$-v + 10i_L + 5\frac{di_L}{dt} = 0$$

Substituting $i_L = -v/40$, we find that

$$\frac{5}{40}\frac{dv}{dt} + \left(\frac{10}{40} + 1\right)v = 0$$

or, more simply,

$$\frac{dv}{dt} + 10v = 0 \qquad [9]$$

■ **FIGURE 8.5** (*a*) A simple *RL* circuit with a switch thrown at time $t = 0$. (*b*) The circuit as it exists prior to $t = 0$. (*c*) The circuit after the switch is thrown, and the 24 V source is removed.

▶ **Determine if additional information is required.**

From previous experience, we know that a complete expression for v will require knowledge of v at a specific instant of time, with $t = 0$ being the most convenient. We might be tempted to look at Fig. 8.5b and write $v(0) = 24$ V, but this is only true *just before the switch opens*. The resistor voltage can change to any value in the instant that the switch is thrown; only the inductor current must remain unchanged.

In the circuit of Fig. 8.5b, $i_L = 24/10 = 2.4$ A since the inductor acts like a short circuit to a dc current. Therefore, $i_L(0) = 2.4$ A in the circuit of Fig. 8.5c, as well—a key point in analyzing this type of circuit. Therefore, in the circuit of Fig. 8.5c, $v(0) = (40)(-2.4) = -96$ V.

▶ **Attempt a solution.**

Any of the three basic solution techniques can be brought to bear. Based on experience, let's start by writing the characteristic equation corresponding to Eq. [9]:

$$s + 10 = 0$$

Solving, we find that $s = -10$, so

$$v(t) = Ae^{-10t} \qquad [10]$$

(which, upon substitution into the left-hand side of Eq. [9], results in

$$-10Ae^{-10t} + 10Ae^{-10t} = 0$$

as expected.)

We find A by setting $t = 0$ in Eq. [10] and employing the fact that $v(0) = -96$ V. Thus,

$$v(t) = -96e^{-10t} \qquad [11]$$

and so $v(0.2) = -12.99$ V, down from a maximum of -96 V.

▶ **Verify the solution. Is it reasonable or expected?**
We could also find the inductor current by realizing that the inductor "sees" a resistance of 50 Ω in the circuit of Fig. 8.5c, thus providing a time constant of $\tau = 50/5 = 10$ s. Coupled with the fact that we know $i_L(0) = 2.4$ A, we may write

$$i_L(t) = 2.4e^{-10t} \text{ A}, \ t > 0$$

From Ohm's law, $v(t) = -40i_L(t) = -96e^{-10t}$, which is identical to Eq. [11]. It is no coincidence that the inductor current and resistor voltage have the same exponential dependence.

PRACTICE

8.2 Determine the inductor voltage v in the circuit of Fig. 8.6 for $t > 0$.

Ans: $-25e^{-2t}$ V.

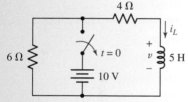

■ **FIGURE 8.6** Circuit for Practice Problem 8.2.

Accounting for the Energy

Before we turn our attention to the interpretation of the response, let us return to the circuit of Fig. 8.1, and check the power and energy relationships. The power being dissipated in the resistor is

$$p_R = i^2 R = I_0^2 R e^{-2Rt/L}$$

and the total energy turned into heat in the resistor is found by integrating the instantaneous power from zero time to infinite time:

$$w_R = \int_0^{\infty} p_R \, dt = I_0^2 R \int_0^{\infty} e^{-2Rt/L} \, dt$$

$$= I_0^2 R \left(\frac{-L}{2R} \right) e^{-2Rt/L} \Big|_0^{\infty} = \frac{1}{2} L I_0^2$$

This is the result we expect, because the total energy stored initially in the inductor is $\frac{1}{2}L I_0^2$, and there is no longer any energy stored in the inductor at infinite time since its current eventually drops to zero. All the initial energy therefore is accounted for by dissipation in the resistor.

8.2 • PROPERTIES OF THE EXPONENTIAL RESPONSE

Let us now consider the nature of the response in the series *RL* circuit. We have found that the inductor current is represented by

$$i(t) = I_0 e^{-Rt/L}$$

At $t = 0$, the current has value I_0, but as time increases, the current decreases and approaches zero. The shape of this decaying exponential is seen by the plot of $i(t)/I_0$ versus t shown in Fig. 8.7. Since the function we are plotting is $e^{-Rt/L}$, the curve will not change if R/L remains unchanged. Thus, the same curve must be obtained for every series *RL* circuit having the same R/L or L/R ratio. Let us see how this ratio affects the shape of the curve.

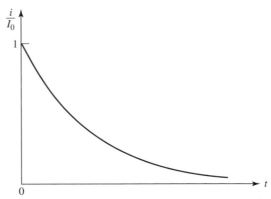

■ **FIGURE 8.7** A plot of $e^{-Rt/L}$ versus t.

If we double the ratio of L to R, the exponent will be unchanged only if t is also doubled. In other words, the original response will occur at a later time, and the new curve is obtained by moving each point on the original curve twice as far to the right. With this larger L/R ratio, the current takes longer to decay to any given fraction of its original value. We might have a tendency to say that the "width" of the curve is doubled, or that the width is proportional to L/R. However, we find it difficult to define our term *width,* because each curve extends from $t = 0$ to ∞! Instead, let us consider the time that would be required for the current to drop to zero *if it continued to drop at its initial rate.*

The initial rate of decay is found by evaluating the derivative at zero time:

$$\frac{d}{dt}\frac{i}{I_0}\bigg|_{t=0} = -\frac{R}{L}e^{-Rt/L}\bigg|_{t=0} = -\frac{R}{L}$$

We designate the value of time it takes for i/I_0 to drop from unity to zero, assuming a constant rate of decay, by the Greek letter τ (tau). Thus,

$$\left(\frac{R}{L}\right)\tau = 1$$

or

$$\boxed{\tau = \frac{L}{R}}$$ [12]

The ratio L/R has the units of seconds, since the exponent $-Rt/L$ must be dimensionless. This value of time τ is called the **time constant** and is shown pictorially in Fig. 8.8. The time constant of a series RL circuit may be found graphically from the response curve; it is necessary only to draw the tangent to the curve at $t = 0$ and determine the intercept of this tangent line with the time axis. This is often a convenient way of approximating the time constant from the display on an oscilloscope.

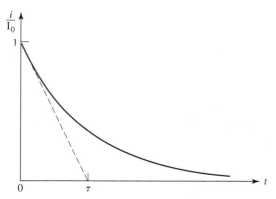

■ **FIGURE 8.8** The time constant τ is L/R for a series RL circuit. It is the time required for the response curve to drop to zero if it decays at a constant rate equal to its initial rate of decay.

An equally important interpretation of the time constant τ is obtained by determining the value of $i(t)/I_0$ at $t = \tau$. We have

$$\frac{i(\tau)}{I_0} = e^{-1} = 0.3679 \qquad \text{or} \qquad i(\tau) = 0.3679 I_0$$

Thus, in one time constant the response has dropped to 36.8 percent of its initial value; the value of τ may also be determined graphically from this fact, as indicated by Fig. 8.9. It is convenient to measure the decay of the current at intervals of one time constant, and recourse to a hand calculator or a table of negative exponentials shows that $i(t)/I_0$ is 0.3679 at $t = \tau$, 0.1353 at $t = 2\tau$, 0.04979 at $t = 3\tau$, 0.01832 at $t = 4\tau$, and 0.006738 at $t = 5\tau$. At some point three to five time constants after zero time, most of

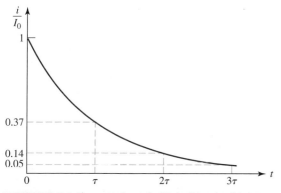

■ **FIGURE 8.9** The current in a series RL circuit is reduced to 37 percent of its initial value at $t = \tau$, 14 percent at $t = 2\tau$, and 5 percent at $t = 3\tau$.

us would agree that the current is a negligible fraction of its former self. Thus, if we are asked, "*How long does it take for the current to decay to zero?*" our answer might be, "*About five time constants.*" At that point, the current is less than 1 percent of its original value!

PRACTICE

8.3 In a source-free series *RL* circuit, find the numerical value of the ratio: (*a*) $i(2\tau)/i(\tau)$, (*b*) $i(0.5\tau)/i(0)$, and (*c*) t/τ if $i(t)/i(0) = 0.2$; (*d*) t/τ if $i(0) - i(t) = i(0)\ln 2$.

Ans: 0.368; 0.607; 1.609; 1.181.

COMPUTER-AIDED ANALYSIS

The transient analysis capability of PSpice is very useful when considering the response of source-free circuits. In this example, we make use of a special feature that allows us to vary a component parameter, similar to the way we varied the dc voltage in other simulations. We do this by adding the component **PARAM** to our schematic; it may be placed anywhere, as we will not wire it into the circuit. Our complete *RL* circuit is shown in Fig. 8.10, which includes an initial inductor current of 1 mA.

In order to relate our resistor value to the proposed parameter sweep, we must perform three tasks. First, we provide a name for our parameter, which we choose to call Resistance for the sake of simplicity. This is accomplished by double-clicking on the **PARAMETERS:** label in the schematic, which opens the Property Editor for this pseudocomponent. Clicking on **New Column** results in the dialog box shown

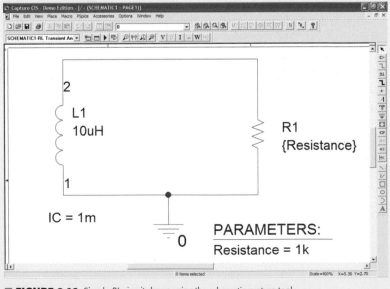

■ **FIGURE 8.10** Simple *RL* circuit drawn using the schematic capture tool.

in Fig. 8.11*a*, in which we enter *Resistance* under **Name** and a place-holder value of 1k under **Value.** Our second task consists of linking the value of R1 to our parameter sweep, which we accomplish by double-clicking on the default value of R1 on the schematic, resulting in the dialog box of Fig. 8.11*b*. Under **Value,** we simply enter *{Resistance}*. (Note the curly brackets are required.)

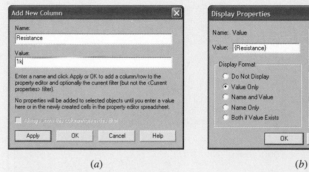

(*a*)   (*b*)

■ **FIGURE 8.11** (*a*) Add New Column dialog box in the Property Editor for PARAM. (*b*) Resistor value dialog box.

Our third task consists of setting up the simulation, which includes setting transient analysis parameters as well as the values we desire for R1. Under **PSpice** we select **New Simulation Profile** (Fig. 8.12*a*), in which we select **Time Domain (Transient)** for **Analysis type,** 300 ns for **Run to time,** and tick the **Parametric Sweep box** under **Options.** This last action results in the dialog box shown in Fig. 8.12*b*, in which we select **Global parameter** for **Sweep variable** and enter *Resistance* for **Parameter name.** The final setup step required is to select **Logarithmic** under **Sweep type,** a **Start value** of 10, an **End value** of 1000, and 1 **Points/Decade;** alternatively we could list the desired resistor values using **Value list.**

After running the simulation, the notification box shown in Fig. 8.13 appears, listing the available data sets for plotting (**Resistance** = 10,

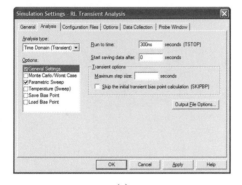

(*a*)

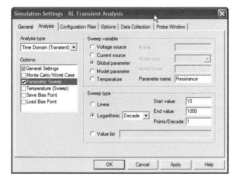

(*b*)

■ **FIGURE 8.12** (*a*) Simulation dialog box. (*b*) Parameter sweep dialog box.

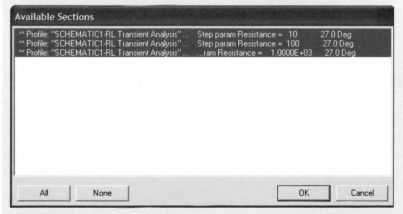

■ **FIGURE 8.13** Available data sections dialog box.

(*Continued on next page*)

100, and 1000 in this case). A particular data set is selected by high-lighting it; we select all three for this example, resulting in the Probe output of Fig. 8.14.

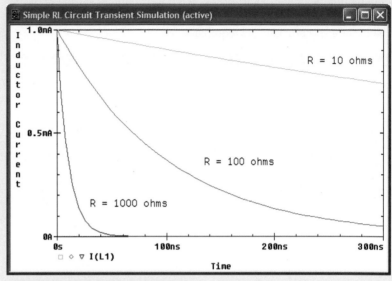

■ **FIGURE 8.14** Probe output for the three resistances.

Why does a larger value of the time constant L/R produce a response curve that decays more slowly? Let us consider the effect of each element.

In terms of the time constant τ, the response of the series RL circuit may be written simply as

$$i(t) = I_0 e^{-t/\tau}$$

An increase in L allows a greater energy storage for the same initial current, and this larger energy requires a longer time to be dissipated in the resistor. We may also increase L/R by reducing R. In this case, the power flowing into the resistor is less for the same initial current; again, a greater time is required to dissipate the stored energy. This effect is seen clearly in our simulation result of Fig. 8.14.

8.3 • THE SOURCE-FREE *RC* CIRCUIT

Circuits based on resistor-capacitor combinations are more common than their resistor-inductor analogs. The principal reasons for this are the smaller losses present in a physical capacitor, the lower cost, the better agreement between the simple mathematical model and the actual device behavior, as well as smaller size and lighter weight, both of which are particularly important for integrated circuit applications.

Let us see how closely the analysis of the parallel (or is it series?) *RC* circuit shown in Fig. 8.15 corresponds to that of the *RL* circuit. We will assume an initial stored energy in the capacitor by selecting

$$v(0) = V_0$$

■ **FIGURE 8.15** A parallel *RC* circuit for which $v(t)$ is to be determined, subject to the initial condition that $v(0) = V_0$.

The total current leaving the node at the top of the circuit diagram must be zero, so we may write

$$C\frac{dv}{dt} + \frac{v}{R} = 0$$

Division by C gives us

$$\frac{dv}{dt} + \frac{v}{RC} = 0 \qquad [13]$$

Equation [13] has a familiar form; comparison with Eq. [1],

$$\frac{di}{dt} + \frac{R}{L}i = 0 \qquad [1]$$

shows that the replacement of i by v and L/R by RC produces the identical equation we considered previously. It should, for the RC circuit we are now analyzing is the dual of the RL circuit we considered first. This *duality* forces $v(t)$ for the RC circuit and $i(t)$ for the RL circuit to have identical expressions if the resistance of one circuit is equal to the reciprocal of the resistance of the other circuit, and if L is numerically equal to C. Thus, the response of the RL circuit,

$$i(t) = i(0)e^{-Rt/L} = I_0 e^{-Rt/L}$$

enables us to immediately write

$$v(t) = v(0)e^{-t/RC} = V_0 e^{-t/RC} \qquad [14]$$

for the RC circuit.

Suppose instead that we had selected the current i as our variable in the RC circuit, rather than the voltage v. Applying Kirchhoff's voltage law,

$$\frac{1}{C}\int_{t_0}^{t} i\,dt' - v_0(t_0) + Ri = 0$$

we obtain an integral equation as opposed to a differential equation. However, taking the time derivative of both sides of this equation,

$$\frac{i}{C} + R\frac{di}{dt} = 0 \qquad [15]$$

and replacing i with v/R, we obtain Eq. [13] again:

$$\frac{v}{RC} + \frac{dv}{dt} = 0$$

Equation [15] could have been used as our starting point, but the application of duality principles would not have been as natural.

Let us discuss the physical nature of the voltage response of the RC circuit as expressed by Eq. [14]. At $t = 0$ we obtain the correct initial condition, and as t becomes infinite the voltage approaches zero. This latter result agrees with our thinking that if there were any voltage remaining across the capacitor, then energy would continue to flow into the resistor and be dissipated as heat. *Thus, a final voltage of zero is necessary.* The time constant of the RC circuit may be found by using the duality relationships on the expression for the time constant of the RL circuit, or it may be found by simply noting the time at which the response has dropped to 37 percent of its initial value:

$$\frac{\tau}{RC} = 1$$

so that

$$\boxed{\tau = RC} \qquad [16]$$

Our familiarity with the negative exponential and the significance of the time constant τ enables us to sketch the response curve readily (Fig. 8.16). Larger values of R or C provide larger time constants and slower dissipation of the stored energy. A larger resistance will dissipate a smaller power with a given voltage across it, thus requiring a greater time to convert the stored energy into heat; a larger capacitance stores a larger energy with a given voltage across it, again requiring a greater time to lose this initial energy.

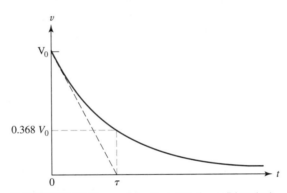

■ **FIGURE 8.16** The capacitor voltage $v(t)$ in the parallel *RC* circuit is plotted as a function of time. The initial value of $v(t)$ is V_0.

EXAMPLE **8.3**

For the circuit of Fig. 8.17a, find the voltage labeled v at $t = 200\ \mu s$.

To find the requested voltage, we will need to draw and analyze two separate circuits: one corresponding to before the switch is thrown (Fig. 8.17b), and one corresponding to after the switch is thrown (Fig. 8.17c).

The sole purpose of analyzing the circuit of Fig. 8.17b is to obtain an initial capacitor voltage; we assume any transients in that circuit died out long ago, leaving a purely dc circuit. With no current through either the capacitor or the 4 Ω resistor, then,

$$v(0) = 9\ \text{V} \qquad [17]$$

We next turn our attention to the circuit of Fig. 8.17c, recognizing that

$$\tau = RC = (2 + 4)(10 \times 10^{-6}) = 60 \times 10^{-6}\ \text{s}$$

Thus, from Eq. [14],

$$v(t) = v(0)e^{-t/RC} = v(0)e^{-t/60 \times 10^{-6}} \qquad [18]$$

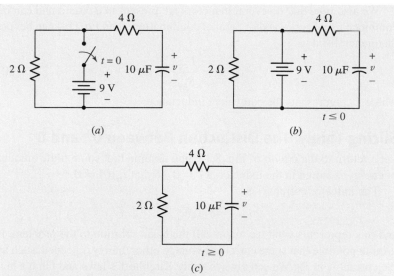

FIGURE 8.17 (*a*) A simple *RC* circuit with a switch thrown at time $t = 0$. (*b*) The circuit as it exists prior to $t = 0$. (*c*) The circuit after the switch is thrown, and the 9 V source is removed.

The capacitor voltage must be the same in both circuits at $t = 0$; no such restriction is placed on any other voltage or current. Substituting Eq. [17] into Eq. [18],

$$v(t) = 9e^{-t/60 \times 10^{-6}} \text{ V}$$

so that $v(200 \times 10^{-6}) = 321.1$ mV (less than 4 percent of its maximum value).

PRACTICE

8.4 Find $v(0)$ and $v(2 \text{ ms})$ for the circuit of Fig. 8.18.

Ans: 50 V, 14.33 V.

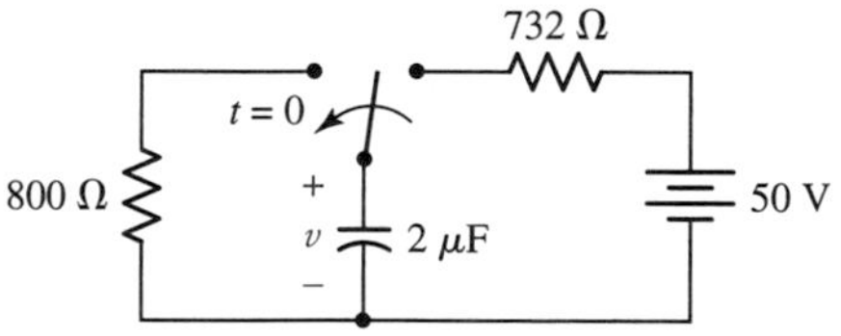

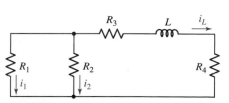

FIGURE 8.18

8.4 A MORE GENERAL PERSPECTIVE

As we have seen in Examples 8.2 and 8.3, it is not difficult to extend the results obtained for the series *RL* circuit to a circuit containing any number of resistors and one inductor. Likewise, we can generalize our results for the *RC* circuit to a circuit with any number of resistors and one capacitor. It is even possible to consider circuits containing dependent sources.

General *RL* Circuits

As an example, consider the circuit shown in Fig. 8.19. The equivalent resistance the inductor faces is

$$R_{eq} = R_3 + R_4 + \frac{R_1 R_2}{R_1 + R_2}$$

and the time constant is therefore

$$\tau = \frac{L}{R_{eq}} \qquad [19]$$

FIGURE 8.19 A source-free circuit containing one inductor and several resistors is analyzed by determining the time constant $\tau = L/R_{eq}$.

We could also state this as

$$\tau = \frac{L}{R_{TH}},$$

where R_{TH} is the Thévenin equivalent resistance "seen" by the inductor L.

We also note that if several inductors are present in a circuit and can be combined using series and/or parallel combination, then Eq. [19] can be further generalized to

$$\tau = \frac{L_{eq}}{R_{eq}} \qquad [20]$$

where L_{eq} represents the equivalent inductance.

Slicing Thinly: The Distinction Between 0^+ and 0^-

Let's return to the circuit of Fig. 8.19, and assume that some finite amount of energy is stored in the inductor at $t = 0$, so that $i_L(0) \neq 0$.

The inductor current i_L is

$$i_L = i_L(0)e^{-t/\tau}$$

and this represents what we might call the basic solution to the problem. It is quite possible that some current or voltage other than i_L is needed, such as the current i_2 in R_2. We can always apply Kirchhoff's laws and Ohm's law to the resistive portion of the circuit without any difficulty, but current division provides the quickest answer in this circuit:

$$i_2 = -\frac{R_1}{R_1 + R_2}[i_L(0)e^{-t/\tau}]$$

It may also happen that we know the initial value of some current *other* than the inductor current. Since *the current in a resistor may change instantaneously,* we will indicate the instant after any change that might have occurred at $t = 0$ by the use of the symbol 0^+; in more mathematical language, $i_1(0^+)$ is the limit from the right of $i_1(t)$ as t approaches zero.[1] Thus, if we are given the initial value of i_1 as $i_1(0^+)$, then the initial value of i_2 is

$$i_2(0^+) = i_1(0^+)\frac{R_1}{R_2}$$

Note that $i_L(0^+)$ is *always* equal to $i_L(0^-)$. This is not necessarily true for the inductor voltage or any resistor voltage or current, since they may change in zero time.

From these values, we obtain the necessary initial value of $i_L(0)$:

$$i_L(0^+) = -[i_1(0^+) + i_2(0^+)] = -\frac{R_1 + R_2}{R_2}i_1(0^+)$$

and the expression for i_2 becomes

$$i_2 = i_1(0^+)\frac{R_1}{R_2}e^{-t/\tau}$$

Let us see if we can obtain this last expression more directly. Since the inductor current decays exponentially as $e^{-t/\tau}$, *every current throughout the circuit must follow the same functional behavior.* This is made clear by considering the inductor current as a source current that is being applied to a resistive network. Every current and voltage in the resistive network must have the same time dependence. Using these ideas, we therefore express i_2 as

$$i_2 = Ae^{-t/\tau}$$

where

$$\tau = \frac{L}{R_{eq}}$$

---

(1) Note that this is a notational convenience *only.* When faced with $t = 0^+$ or its companion $t = 0^-$ in an equation, we simply use the value zero. This notation allows us to clearly differentiate between the time before and after an event, such as a switch opening or closing, or a power supply being turned on or off.

and A must be determined from a knowledge of the initial value of i_2. Since $i_1(0^+)$ is known, the voltage across R_1 and R_2 is known, and

$$R_2 i_2(0^+) = R_1 i_1(0^+)$$

leads to

$$i_2(0^+) = i_1(0^+)\frac{R_1}{R_2}$$

Therefore,

$$i_2(t) = i_1(0^+)\frac{R_1}{R_2}e^{-t/\tau}$$

A similar sequence of steps will provide a rapid solution to a large number of problems. We first recognize the time dependence of the response as an exponential decay, determine the appropriate time constant by combining resistances, write the solution with an unknown amplitude, and then determine the amplitude from a given initial condition.

This same technique can be applied to any circuit with one inductor and any number of resistors, as well as to those special circuits containing two or more inductors and also two or more resistors that may be simplified by resistance or inductance combination to one inductor and one resistor.

EXAMPLE 8.4

Determine both i_1 and i_L in the circuit shown in Fig. 8.20a for $t > 0$.

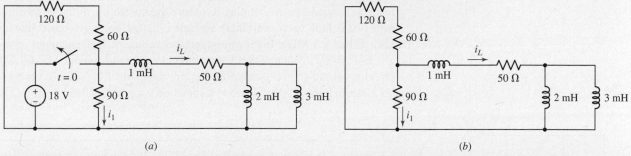

(a)   (b)

■ **FIGURE 8.20** (a) A circuit with multiple resistors and inductors. (b) After $t = 0$, the circuit simplifies to an equivalent resistance of 110 Ω in series with $L_{eq} = 2.2$ mH.

After $t = 0$, when the voltage source is disconnected as shown in Fig. 8.20b, we easily calculate an equivalent inductance,

$$L_{eq} = \frac{2 \times 3}{2 + 3} + 1 = 2.2 \text{ mH}$$

an equivalent resistance, in series with the equivalent inductance,

$$R_{eq} = \frac{90(60 + 120)}{90 + 180} + 50 = 110 \ \Omega$$

(*Continued on next page*)

and the time constant,

$$\tau = \frac{L_{\text{eq}}}{R_{\text{eq}}} = \frac{2.2 \times 10^{-3}}{110} = 20 \ \mu\text{s}$$

Thus, the form of the natural response is $Ke^{-50,000t}$, where K is an unknown constant. Considering the circuit just prior to the switch opening ($t = 0^-$), $i_L = 18/50$ A. Since $i_L(0^+) = i_L(0^-)$, we know that $i_L = 18/50$ A or 360 mA at $t = 0^+$ and so

$$i_L = \begin{cases} 360 \ \text{mA}, & t < 0 \\ 360e^{-50,000t} \ \text{mA}, & t \geq 0 \end{cases}$$

There is no restriction on i_1 changing instantaneously at $t = 0$, so its value at $t = 0^-$ (18/90 A or 200 mA) is not relevant to finding i_1 for $t > 0$. Instead, we must find $i_1(0^+)$ through our knowledge of $i_L(0^+)$. Using current division,

$$i_1(0^+) = -i_L(0^+)\frac{120 + 60}{120 + 60 + 90} = -240 \ \text{mA}$$

Hence,

$$i_1 = \begin{cases} 200 \ \text{mA}, & t < 0 \\ -240e^{-50,000t} \ \text{mA}, & t \geq 0 \end{cases}$$

We can verify our analysis using PSpice and the switch model **Sw_tOpen,** although it should be remembered that this part is actually just two resistance values: one corresponding to before the switch opens at the specified time (the default value is 10 mΩ), and one for after the switch opens (the default value is 1 MΩ). If the equivalent resistance of the remainder of the circuit is comparable to either value, the values should be edited by double-clicking on the switch symbol in the circuit schematic. Note that there is also a switch model that closes at a specified time: **Sw_tClose.**

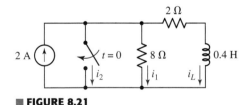

■ **FIGURE 8.21**

PRACTICE
●

8.5 At $t = 0.15$ s in the circuit of Fig. 8.21, find the value of (*a*) i_L; (*b*) i_1; (*c*) i_2.

Ans: 0.756 A; 0; 1.244 A.

We have now considered the task of finding the natural response of any circuit which can be represented by an equivalent inductor in series with an equivalent resistor. *A circuit containing several resistors and several inductors does not always possess a form which allows either the resistors or the inductors to be combined into single equivalent elements.* In such instances, there is no single negative exponential term or single time constant associated with the circuit. Rather, there will, in general, be several negative exponential terms, the number of terms being equal to the number of inductors that remain after all possible inductor combinations have been made. We consider this situation further in Chap. 9.

General RC Circuits

Many of the RC circuits for which we would like to find the natural response contain more than a single resistor and capacitor. Just as we did for the RL circuits, we first consider those cases in which the given circuit may be reduced to an equivalent circuit consisting of only one resistor and one capacitor.

Let us suppose first that we are faced with a circuit containing a single capacitor, but any number of resistors. It is possible to replace the two-terminal resistive network which is across the capacitor terminals with an equivalent resistor, and we may then write down the expression for the capacitor voltage immediately. In such instances, the circuit has an effective time constant given by

$$\tau = R_{eq}C$$

where R_{eq} is the equivalent resistance of the network. An alternative perspective is that R_{eq} is in fact the Thévenin equivalent resistance "seen" by the capacitor.

If the circuit has more than one capacitor, but they may be replaced somehow using series and/or parallel combinations with an equivalent capacitance C_{eq}, then the circuit has an effective time constant given by

$$\tau = RC_{eq}$$

with the general case expressed as

$$\tau = R_{eq}C_{eq}$$

It is worth noting, however, that parallel capacitors replaced by an equivalent capacitance would have to have identical initial conditions.

EXAMPLE **8.5**

Find $v(0^+)$ and $i_1(0^+)$ for the circuit shown in Fig. 8.22a if $v(0^-) = V_0$.

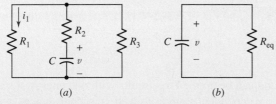

(a)   (b)

■ **FIGURE 8.22** (a) A given circuit containing one capacitor and several resistors. (b) The resistors have been replaced by a single equivalent resistor; the time constant is simply $\tau = R_{eq}C$.

We first simplify the circuit of Fig. 8.22a to that of Fig. 8.22b, enabling us to write

$$v = V_0 e^{-t/R_{eq}C}$$

(Continued on next page)

where

$$v(0^+) = v(0^-) = V_0 \qquad \text{and} \qquad R_{\text{eq}} = R_2 + \frac{R_1 R_3}{R_1 + R_3}$$

Every current and voltage in the resistive portion of the network must have the form $Ae^{-t/R_{\text{eq}}C}$, where A is the initial value of that current or voltage. Thus, the current in R_1, for example, may be expressed as

$$i_1 = i_1(0^+)e^{-t/\tau}$$

where

$$\tau = \left(R_2 + \frac{R_1 R_3}{R_1 + R_3} \right) C$$

and $i_1(0^+)$ remains to be determined from the initial condition. Any current flowing in the circuit at $t = 0^+$ must come from the capacitor. Therefore, since v cannot change instantaneously, $v(0^+) = v(0^-) = V_0$ and

$$i_1(0^+) = \frac{V_0}{R_2 + R_1 R_3/(R_1 + R_3)} \frac{R_3}{R_1 + R_3}$$

PRACTICE

8.6 Find values of v_C and v_o in the circuit of Fig. 8.23 at t equal to: (*a*) 0^-; (*b*) 0^+; (*c*) 1.3 ms.

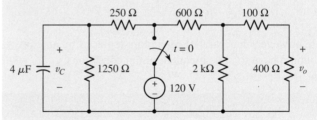

■ **FIGURE 8.23**

Ans: 100 V, 38.4 V; 100 V, 25.6 V; 59.5 V, 15.22 V.

Our method can be applied to circuits with one energy storage element and one or more dependent sources as well. In such instances, we may write an appropriate KCL or KVL equation along with any necessary supporting equations, distill this down into a single differential equation, and extract the characteristic equation to find the time constant. Alternatively, we may begin by finding the Thévenin equivalent resistance of the network connected to the capacitor or inductor, and use this in calculating the appropriate *RL* or *RC* time constant—unless the dependent source is controlled by a voltage or current associated with the energy storage element, in which case the Thévenin approach cannot be used. We explore this in the following example.

EXAMPLE **8.6**

For the circuit of Fig. 8.24*a*, find the voltage labeled v_C for $t > 0$ if $v_C(0^-) = 2$ V.

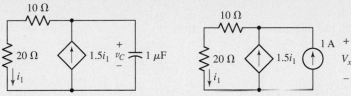

■ **FIGURE 8.24** (*a*) A simple *RC* circuit containing a dependent source not controlled by a capacitor voltage or current. (*b*) Circuit for finding the Thévenin equivalent of the network connected to the capacitor.

The dependent source is not controlled by a capacitor voltage or current, so we can start by finding the Thévenin equivalent of the network to the left of the capacitor. Connecting a 1 A reference source as in Fig. 8.24*b*,

$$V_x = (1 + 1.5i_1)(30)$$

where

$$i_1 = \left(\frac{1}{20}\right)\frac{20}{10 + 20}V_x = \frac{V_x}{30}$$

Performing a little algebra, we find that $V_x = -60$ V, so the network has a Thévenin equivalent resistance of $-60\ \Omega$ (unusual, but not impossible when dealing with a dependent source). Our circuit therefore has a *negative* time constant

$$\tau = -60(1 \times 10^{-6}) = -60\ \mu\text{s}$$

The capacitor voltage is therefore

$$v_C(t) = Ae^{t/60 \times 10^{-6}}\ \text{V}$$

where $A = v_C(0^+) = v_C(0^-) = 2$ V. Thus,

$$\boxed{v_C(t) = 2e^{t/60 \times 10^{-6}}\ \text{V}} \qquad [21]$$

which, interestingly enough is unstable: it grows exponentially with time. This cannot continue indefinitely; one or more elements in the circuit will eventually fail.

Alternatively, we could write a simple KCL equation for the top node of Fig. 8.24*a*

$$v_C = 30\left(1.5i_1 - 10^{-6}\frac{dv_C}{dt}\right) \qquad [22]$$

where

$$i_1 = \frac{v_C}{30} \qquad [23]$$

(Continued on next page)

Substituting Eq. [23] into Eq. [22] and performing some algebra, we obtain

$$\frac{dv_C}{dt} - \frac{1}{60 \times 10^{-6}} v_C = 0$$

which has the characteristic equation

$$s - \frac{1}{60 \times 10^{-6}} = 0$$

Thus,

$$s = \frac{1}{60 \times 10^{-6}}$$

and so

$$v_C(t) = Ae^{t/60 \times 10^{-6}} \text{ V}$$

as we found before. Substitution of $A = v_C(0^+) = 2$ results in Eq. [21], our expression for the capacitor voltage for $t > 0$.

■ **FIGURE 8.25** Circuit for Practice Problem 8.7.

PRACTICE

8.7 (*a*) Regarding the circuit of Fig. 8.25, determine the voltage $v_C(t)$ for $t > 0$ if $v_C(0^-) = 11$ V. (*b*) Is the circuit "stable?"

Ans: (*a*) $v_C(t) = 11e^{-2 \times 10^3 t/3}$ V, $t > 0$. (*b*) Yes; it decays (exponentially) rather than grows with time.

Some circuits containing a number of both resistors and capacitors may be replaced by an equivalent circuit containing only one resistor and one capacitor; it is necessary that the original circuit be one which can be broken into two parts, one containing all resistors and the other containing all capacitors, such that the two parts are connected by only two ideal conductors. This is not generally the case, however, so that multiple time constants will more than likely be required to describe a circuit with several resistors and capacitors.

As a parting comment, we should be wary of certain situations involving only ideal elements which are suddenly connected together. For example, we may imagine connecting two ideal capacitors in series having unequal voltages prior to $t = 0$. This poses a problem using our mathematical model of an ideal capacitor; however, real capacitors have resistances associated with them through which energy can be dissipated.

8.5 THE UNIT-STEP FUNCTION

We have been studying the response of *RL* and *RC* circuits when no sources or forcing functions were present. We termed this response the *natural response,* because its form depends only on the nature of the circuit. The reason that any response at all is obtained arises from the presence of initial energy storage within the inductive or capacitive elements in the circuit. In some cases we were confronted with circuits containing sources and switches; we were informed that certain switching operations were performed at $t = 0$ in order to remove all the sources from the circuit, while leaving known amounts of energy stored here and there. In other words, we have been solving problems in which energy sources are suddenly *removed* from the circuit; now we must consider that type of response which results when energy sources are suddenly *applied* to a circuit.

We will focus on the response which occurs when the energy sources suddenly applied are dc sources. Since every electrical device is intended to be energized at least once, and since most devices are turned on and off many times in the course of their lifetimes, our study applies to many practical cases. Even though we are now restricting ourselves to dc sources, there are still innumerable cases in which these simpler examples correspond to the operation of physical devices. For example, the first circuit we will analyze could represent the buildup of the current when a dc motor is started. The generation and use of the rectangular voltage pulses needed to represent a number or a command in a microprocessor provide many examples in the field of electronic or transistor circuitry. Similar circuits are found in the synchronization and sweep circuits of television receivers, in communication systems using pulse modulation, and in radar systems, to name but a few examples.

We have been speaking of the "sudden application" of an energy source, and by this phrase we imply its application in zero time.[2] The operation of a switch in series with a battery is thus equivalent to a forcing function which is zero up to the instant that the switch is closed and is equal to the battery voltage thereafter. The forcing function has a break, or discontinuity, at the instant the switch is closed. Certain special forcing functions which are discontinuous or have discontinuous derivatives are called *singularity functions*, the two most important of these singularity functions being the *unit-step function* and the *unit-impulse function*.

We define the unit-step forcing function as a function of time which is zero for all values of its argument less than zero and which is unity for all positive values of its argument. If we let $(t - t_0)$ be the argument and represent the unit-step function by u, then $u(t - t_0)$ must be zero for all values of t less than t_0, and it must be unity for all values of t greater than t_0. At $t = t_0$, $u(t - t_0)$ changes *abruptly* from 0 to 1. Its value at $t = t_0$ is not defined, but its value is known for all instants of time that are arbitrarily close to $t = t_0$. We often indicate this by writing $u(t_0^-) = 0$ and $u(t_0^+) = 1$. The concise mathematical definition of the unit-step forcing function is

$$u(t - t_0) = \begin{cases} 0 & t < t_0 \\ 1 & t > t_0 \end{cases}$$

and the function is shown graphically in Fig. 8.26. Note that a vertical line of unit length is shown at $t = t_0$. Although this "riser" is not strictly a part of the definition of the unit step, it is usually shown in each drawing.

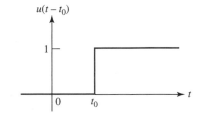

■ **FIGURE 8.26** The unit-step forcing function, $u(t - t_0)$.

---

(2) Of course, this is not physically possible. However, if the time scale over which such an event occurs is very short compared to all other relevant time scales that describe the operation of a circuit, this is approximately true, and mathematically convenient.

We also note that the unit step need not be a time function. For example, $u(x - x_0)$ could be used to denote a unit-step function where x might be a distance in meters, for example, or a frequency.

Very often in circuit analysis a discontinuity or a switching action takes place at an instant that is defined as $t = 0$. In that case $t_0 = 0$, and we then represent the corresponding unit-step forcing function by $u(t - 0)$, or more simply $u(t)$. This is shown in Fig. 8.27. Thus

$$u(t) = \begin{cases} 0 & t < 0 \\ 1 & t > 0 \end{cases}$$

The unit-step forcing function is in itself dimensionless. If we wish it to represent a voltage, it is necessary to multiply $u(t - t_0)$ by some constant voltage, such as 5 V. Thus, $v(t) = 5u(t - 0.2)$ V is an ideal voltage source which is zero before $t = 0.2$ s and a constant 5 V after $t = 0.2$ s. This forcing function is shown connected to a general network in Fig. 8.28a.

Physical Sources and the Unit-Step Function

We should now logically ask what physical source is the equivalent of this discontinuous forcing function. By equivalent, we mean simply that the voltage-current characteristics of the two networks are identical. For the step-voltage source of Fig. 8.28a, the voltage-current characteristic is quite simple: The voltage is zero prior to $t = 0.2$ s, it is 5 V after $t = 0.2$ s, and the current may be any (finite) value in either time interval. Our first thoughts might produce the attempt at an equivalent shown in Fig. 8.28b, a 5 V dc source in series with a switch which closes at $t = 0.2$ s. This network is not equivalent for $t < 0.2$ s, however, because the voltage across the battery and switch is completely unspecified in this time interval. The "equivalent" source is an open circuit, and the voltage across it *may be anything*. After $t = 0.2$ s, the networks are equivalent, and if this is the only time interval in which we are interested, and if the initial currents which flow from the two networks are identical at $t = 0.2$ s, then Fig. 8.28b becomes a useful equivalent of Fig. 8.28a.

In order to obtain an exact equivalent for the voltage-step forcing function, we may provide a single-pole double-throw switch. Before $t = 0.2$ s, the switch serves to ensure zero voltage across the input terminals of the general network. After $t = 0.2$ s, the switch is thrown to provide a constant input voltage of 5 V. At $t = 0.2$ s, the voltage is indeterminate (as is the step forcing function), and the battery is momentarily short-circuited (it is

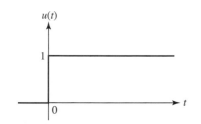

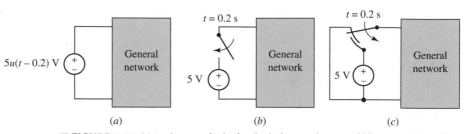

FIGURE 8.28 (a) A voltage-step forcing function is shown as the source driving a general network. (b) A simple circuit which, although not the exact equivalent of part (a), may be used as its equivalent in many cases. (c) An exact equivalent of part (a).

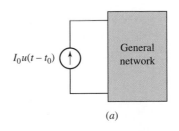

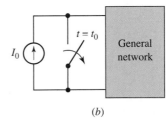

fortunate that we are dealing with mathematical models!). This exact equivalent of Fig. 8.28*a* is shown in Fig. 8.28*c*.

Figure 8.29*a* shows a current-step forcing function driving a general network. If we attempt to replace this circuit by a dc source in parallel with a switch (which opens at $t = t_0$), we must realize that the circuits are equivalent after $t = t_0$ but that the responses after $t = t_0$ are alike only if the initial conditions are the same. The circuit in Fig. 8.29*b* implies no voltage exists across the current source terminals for $t < t_0$. This is not the case for the circuit of Fig. 8.29*a*. However, we may often use the circuits of Fig. 8.29*a* and *b* interchangeably. The exact equivalent of Fig. 8.29*a* is the dual of the circuit of Fig. 8.28*c*; the exact equivalent of Fig. 8.29*b* cannot be constructed with current- and voltage-step forcing functions alone.[3]

FIGURE 8.29 (*a*) A current-step forcing function is applied to a general network. (*b*) A simple circuit which, although not the exact equivalent of part (*a*), may be used as its equivalent in many cases.

The Rectangular Pulse Function

Some very useful forcing functions may be obtained by manipulating the unit-step forcing function. Let us define a rectangular voltage pulse by the following conditions:

$$v(t - t_0) = \begin{cases} 0 & t < t_0 \\ V_0 & t_0 < t < t_1 \\ 0 & t > t_1 \end{cases}$$

The pulse is drawn in Fig. 8.30. Can this pulse be represented in terms of the unit-step forcing function? Let us consider the difference of the two unit steps, $u(t - t_0) - u(t - t_1)$. The two step functions are shown in Fig. 8.31*a*, and their difference is a rectangular pulse. The source $V_0 u(t - t_0) - V_0 u(t - t_1)$ which provides us with the desired voltage is indicated in Fig. 8.31*b*.

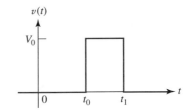

FIGURE 8.30 A useful forcing function, the rectangular voltage pulse.

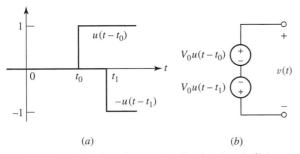

FIGURE 8.31 (*a*) The unit steps $u(t - t_0)$ and $-u(t - t_1)$. (*b*) A source which yields the rectangular voltage pulse of Fig. 8.30.

If we have a sinusoidal voltage source $V_m \sin \omega t$ which is suddenly connected to a network at $t = t_0$, then an appropriate voltage forcing function would be $v(t) = V_m u(t - t_0) \sin \omega t$. If we wish to represent one burst of energy from the transmitter for a radio-controlled car operating at 47 MHz (295 Mrad/s), we may turn the sinusoidal source off 70 ns later by a second unit-step forcing function.[4] The voltage pulse is thus

$$v(t) = V_m [u(t - t_0) - u(t - t_0 - 7 \times 10^{-8})] \sin(295 \times 10^6 t)$$

This forcing function is sketched in Fig. 8.32.

(3) The equivalent can be drawn if the current through the switch prior to $t = t_0$ is known.
(4) Apparently, we're pretty good at the controls of this car. A reaction time of 70 ns?

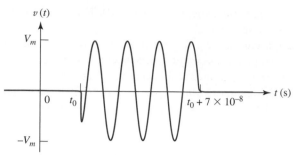

■ **FIGURE 8.32** A 47 MHz radio-frequency pulse, described by
$v(t) = V_m[u(t - t_0) - u(t - t_0 - 7 \times 10^{-8})] \sin(295 \times 10^6 t)$.

PRACTICE

8.8 Evaluate each of the following at $t = 0.8$: (*a*) $3u(t) - 2u(-t) + 0.8u(1 - t)$; (*b*) $[4u(t)]u(-t)$; (*c*) $2u(t) \sin \pi t$.

Ans: 3.8; 0; 1.176.

8.6 · DRIVEN *RL* CIRCUITS

We are now ready to subject a simple network to the sudden application of a dc source. The circuit consists of a battery whose voltage is V_0 in series with a switch, a resistor R, and an inductor L. The switch is closed at $t = 0$, as indicated on the circuit diagram of Fig. 8.33*a*. It is evident that the current $i(t)$ is zero before $t = 0$, and we are therefore able to replace the battery and switch by a voltage-step forcing function $V_0 u(t)$, which also produces no response prior to $t = 0$. After $t = 0$, the two circuits are clearly identical. Hence, we seek the current $i(t)$ either in the given circuit of Fig. 8.33*a* or in the equivalent circuit of Fig. 8.33*b*.

We will find $i(t)$ at this time by writing the appropriate circuit equation and then solving it by separation of the variables and integration. After we obtain the answer and investigate the two parts of which it is composed, we will see that there is physical significance to each of these two terms. With a more intuitive understanding of how each term originates, we will be able to produce more rapid and more meaningful solutions to every problem involving the sudden application of any source. Let us now proceed with the more formal method of solution.

Applying Kirchhoff's voltage law to the circuit of Fig. 8.33*b*, we have

$$Ri + L\frac{di}{dt} = V_0 u(t)$$

Since the unit-step forcing function is discontinuous at $t = 0$, we will first consider the solution for $t < 0$ and then for $t > 0$. The application of zero voltage since $t = -\infty$ forces a zero response, so that

$$i(t) = 0 \qquad t < 0$$

For positive time, however, $u(t)$ is unity and we must solve the equation

$$Ri + L\frac{di}{dt} = V_0 \qquad t > 0$$

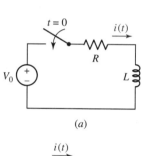

(*a*)

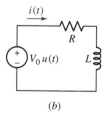

(*b*)

■ **FIGURE 8.33** (*a*) The given circuit. (*b*) An equivalent circuit, possessing the same response *i(t)* for all time.

The variables may be separated in several simple algebraic steps, yielding

$$\frac{L\,di}{V_0 - Ri} = dt$$

and each side may be integrated directly:

$$-\frac{L}{R}\ln(V_0 - Ri) = t + k$$

In order to evaluate k, an initial condition must be invoked. Prior to $t = 0$, $i(t)$ is zero, and thus $i(0^-) = 0$. Since the current in an inductor cannot change by a finite amount in zero time without being associated with an infinite voltage, we thus have $i(0^+) = 0$. Setting $i = 0$ at $t = 0$, we obtain

$$-\frac{L}{R}\ln V_0 = k$$

and, hence,

$$-\frac{L}{R}[\ln(V_0 - Ri) - \ln V_0] = t$$

Rearranging,

$$\frac{V_0 - Ri}{V_0} = e^{-Rt/L}$$

or

$$i = \frac{V_0}{R} - \frac{V_0}{R}e^{-Rt/L} \qquad t > 0 \qquad\qquad [24]$$

Thus, an expression for the response valid for all t would be

$$i = \left(\frac{V_0}{R} - \frac{V_0}{R}e^{-Rt/L}\right)u(t) \qquad\qquad [25]$$

A More Direct Procedure

This is the desired solution, but it has not been obtained in the simplest manner. In order to establish a more direct procedure, let us try to interpret the two terms appearing in Eq. [25]. The exponential term has the functional form of the natural response of the *RL* circuit; it is a negative exponential, it approaches zero as time increases, and it is characterized by the time constant L/R. The functional form of this part of the response is thus identical with that which is obtained in the source-free circuit. However, the amplitude of this exponential term depends on the source voltage V_0. We might generalize, then, that the response will be the sum of two terms, where one term has a functional form identical to that of the source-free response, but has an amplitude that depends on the forcing function. But what of the other term?

Equation [25] also contains a constant term, V_0/R. Why is it present? The answer is simple: The natural response approaches zero as the energy is gradually dissipated, but the total response must not approach zero. Eventually the circuit behaves as a resistor and an inductor in series with a battery. Since the inductor looks like a short circuit to dc, the only current now flowing is V_0/R. This current is a part of the response that is directly attributable to the forcing function, and we call it the *forced response*. It is the response that is present a long time after the switch is closed.

The complete response is composed of two parts, the *natural response* and the *forced response*. The natural response is a characteristic of the circuit and not of the sources. Its form may be found by considering the source-free circuit, and it has an amplitude that depends on both the initial amplitude of the source and the initial energy storage. The forced response has the characteristics of the forcing function; it is found by pretending that all switches were thrown a long time ago. Since we are presently concerned only with switches and dc sources, the forced response is merely the solution of a simple dc circuit problem.

EXAMPLE 8.7

For the circuit of Fig. 8.34, find $i(t)$ for $t = \infty, 3^-, 3^+$, and 100 μs after the source changes value.

Long after any transients have died out ($t \to \infty$), the circuit is a simple dc circuit driven by a 12 V voltage source. The inductor appears as a short circuit, so

$$i(\infty) = \frac{12}{1000} = 12 \text{ mA}$$

What is meant by $i(3^-)$? This is simply a notational convenience to indicate the instant before the voltage source changes value. For $t < 3$, $u(t - 3) = 0$. Thus, $i(3^-) = 0$ as well.

At $t = 3^+$, the forcing function $12u(t - 3) = 12$ V. However, since the inductor current cannot change in zero time, $i(3^+) = i(3^-) = 0$.

The most straightforward approach to analyzing the circuit for $t > 3$ s is to rewrite Eq. [25] as

$$i(t') = \left(\frac{V_0}{R} - \frac{V_0}{R}e^{-Rt'/L}\right)u(t')$$

and note that this equation applies to our circuit as well if we shift the time axis such that

$$t' = t - 3$$

Therefore, with $V_0/R = 12$ mA and $R/L = 20,000$ s^{-1},

$$i(t - 3) = \left(12 - 12e^{-20,000(t-3)}\right)u(t - 3) \text{ mA} \qquad [26]$$

which can be written more simply as

$$i(t) = \left(12 - 12e^{-20,000(t-3)}\right)u(t - 3) \text{ mA} \qquad [27]$$

since the unit step function forces a zero value for $t < 3$, as required. Substituting $t = 3.0001$ s into Eq. [26] or [27], we find that $i = 10.38$ mA at a time 100 μs after the source changes value.

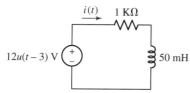

■ **FIGURE 8.34** A simple *RL* circuit driven by a voltage-step forcing function.

Developing an Intuitive Understanding

The reason for the two responses, forced and natural, may be seen from physical arguments. We know that our circuit will eventually assume the forced response. However, at the instant the switches are thrown, the initial inductor currents (or, in *RC* circuits, the voltages across the capacitors) will have values that depend only on the energy stored in these elements. These currents or voltages cannot be expected to be the same as the currents and voltages demanded by the forced response. Hence, there must be a transient period during which the currents and voltages change from their given initial values to their required final values. The portion of the response that provides the transition from initial to final values is the natural response (often called the *transient* response, as we found earlier). If we describe the response of the simple source-free *RL* circuit in these terms, then we should say that the forced response is zero and that the natural response serves to connect the initial response dictated by the stored energy with the zero value of the forced response.

This description is appropriate only for those circuits in which the natural response eventually dies out. This always occurs in physical circuits where some resistance is associated with every element, but there are a number of "pathologic" circuits in which the natural response is nonvanishing as time becomes infinite. Those circuits in which trapped currents circulate around inductive loops, or voltages are trapped in series strings of capacitors, are examples.

8.7 NATURAL AND FORCED RESPONSE

There is also an excellent mathematical reason for considering the complete response to be composed of two parts—the forced response and the natural response. The reason is based on the fact that the solution of any linear differential equation may be expressed as the sum of two parts: the *complementary solution* (natural response) and the *particular solution* (forced response). Without delving into the general theory of differential equations, let us consider a general equation of the type met in the previous section:

$$\frac{di}{dt} + Pi = Q$$

or

$$di + Pi\,dt = Q\,dt \qquad [28]$$

We may identify Q as a forcing function and express it as $Q(t)$ to emphasize its general time dependence. Let us simplify the discussion by assuming that P is a positive constant. Later, we will also assume that Q is constant, thus restricting ourselves to dc forcing functions.

In any standard text on elementary differential equations, it is shown that if both sides of Eq. [28] are multiplied by a suitable "integrating factor," then each side becomes an exact differential that can be integrated directly to obtain the solution. We are not separating the variables, but merely arranging them in such a way that integration is possible. For this equation, the integrating factor is $e^{\int P\,dt}$ or simply e^{Pt}, since P is a constant. We multiply each side of the equation by this integrating factor and obtain

$$e^{Pt}\,di + iPe^{Pt}\,dt = Qe^{Pt}\,dt \qquad [29]$$

The form of the left side may be simplified by recognizing it as the exact differential of ie^{Pt}:

$$d(ie^{Pt}) = e^{Pt}\,di + iPe^{Pt}\,dt$$

so that Eq. [29] becomes

$$d(ie^{Pt}) = Qe^{Pt}\,dt$$

Integrating each side,

$$ie^{Pt} = \int Qe^{Pt}\,dt + A$$

where A is a constant of integration. Multiplication by e^{-Pt} produces the solution for $i(t)$,

$$i = e^{-Pt}\int Qe^{Pt}\,dt + Ae^{-Pt} \qquad [30]$$

If our forcing function $Q(t)$ is known, then we can obtain the functional form of $i(t)$ by evaluating the integral. We will not evaluate such an integral for each problem, however; instead, we are interested in using Eq. [30] to draw several very general conclusions.

The Natural Response

We note first that, for a source-free circuit, Q must be zero, and the solution is the natural response

$$i_n = Ae^{-Pt} \qquad [31]$$

We will find that the constant P is never negative for a circuit with only resistors, inductors, and capacitors; its value depends only on the passive circuit elements[5] and their interconnection in the circuit. *The natural response therefore approaches zero as time increases without limit.* This must be the case for the simple *RL* circuit, because the initial energy is gradually dissipated in the resistor, leaving the circuit in the form of heat. There are also idealized circuits in which P is zero; in these circuits the natural response does not die out.

We therefore find that one of the two terms making up the complete response has the form of the natural response; it has an amplitude which will

---

(5) If the circuit contains a dependent source or a negative resistance, P may be negative.

depend on (but *not* always be equal to) the initial value of the complete response and thus on the initial value of the forcing function also.

The Forced Response

We next observe that the first term of Eq. [30] depends on the functional form of $Q(t)$, the forcing function. Whenever we have a circuit in which the natural response dies out as t becomes infinite, this first term must completely describe the form of the response after the natural response has disappeared. This term is typically called the *forced response*; it is also called the *steady-state response,* the *particular solution,* or the *particular integral.*

For the present, we have elected to consider only those problems involving the sudden application of dc sources, and $Q(t)$ will therefore be a constant for all values of time. If we wish, we can now evaluate the integral in Eq. [30], obtaining the forced response

$$i_f = \frac{Q}{P} \tag{32}$$

and the complete response

$$i(t) = \frac{Q}{P} + Ae^{-Pt} \tag{33}$$

For the *RL* series circuit, Q/P is the constant current V_0/R and $1/P$ is the time constant τ. We should see that the forced response might have been obtained without evaluating the integral, because it must be the complete response at infinite time; it is merely the source voltage divided by the series resistance. The forced response is thus obtained by inspection of the final circuit.

Determination of the Complete Response

Let us use the simple *RL* series circuit to illustrate how to determine the complete response by the addition of the natural and forced responses. The circuit shown in Fig. 8.35 was analyzed earlier, but by a longer method. The desired response is the current $i(t)$, and we first express this current as the sum of the natural and the forced current,

$$i = i_n + i_f$$

The functional form of the natural response must be the same as that obtained without any sources. We therefore replace the step-voltage source by a short circuit and recognize the old *RL* series loop. Thus,

$$i_n = Ae^{-Rt/L}$$

where the amplitude A is yet to be determined; since the initial condition applies to the *complete* response, we cannot simply assume $A = i(0)$.

We next consider the forced response. In this particular problem the forced response must be constant, because the source is a constant V_0 for all positive values of time. After the natural response has died out, therefore, there can be no voltage across the inductor; hence, a voltage V_0 appears across R, and the forced response is simply

$$i_f = \frac{V_0}{R}$$

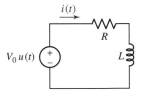

■ **FIGURE 8.35** A series *RL* circuit that is used to illustrate the method by which the complete response is obtained as the sum of the natural and forced responses.

Note that the forced response is determined completely; there is no unknown amplitude. We next combine the two responses to obtain

$$i = Ae^{-Rt/L} + \frac{V_0}{R}$$

and apply the initial condition to evaluate A. The current is zero prior to $t = 0$, and it cannot change value instantaneously since it is the current flowing through an inductor. Thus, the current is zero immediately after $t = 0$, and

$$0 = A + \frac{V_0}{R}$$

and so

$$i = \frac{V_0}{R}(1 - e^{-Rt/L}) \qquad [34]$$

Note carefully that A is not the initial value of i, since $A = -V_0/R$, while $i(0) = 0$. In considering source-free circuits, we found that A was the initial value of the response. When forcing functions are present, however, we must first find the initial value of the response and then substitute this in the equation for the complete response to find A.

This response is plotted in Fig. 8.36, and we can see the manner in which the current builds up from its initial value of zero to its final value of V_0/R. The transition is effectively accomplished in a time 3τ. If our circuit represents the field coil of a large dc motor, we might have $L = 10\,\text{H}$ and $R = 20\,\Omega$, obtaining $\tau = 0.5\,\text{s}$. The field current is thus established in about 1.5 s. In one time constant, the current has attained 63.2 percent of its final value.

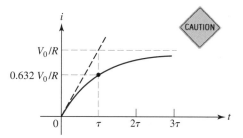

■ **FIGURE 8.36** The current flowing through the inductor of Fig. 8.35 is shown graphically. A line extending the initial slope meets the constant forced response at $t = \tau$.

EXAMPLE **8.8**

Determine $i(t)$ for all values of time in the circuit of Fig. 8.37.

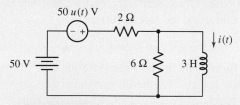

■ **FIGURE 8.37** The circuit of Example 8.8.

The circuit contains a dc voltage source as well as a step-voltage source. We might choose to replace everything to the left of the inductor by the Thévenin equivalent, but instead let us merely recognize the form of that equivalent as a resistor in series with some voltage source. The circuit contains only one energy-storage element, the inductor. We first note that

$$\tau = \frac{L}{R_{\text{eq}}} = \frac{3}{1.5} = 2\,\text{s}$$

and recall that

$$i = i_f + i_n$$

The natural response is therefore a negative exponential as before:

$$i_n = Ke^{-t/2} \text{ A} \qquad t > 0$$

Since the forcing function is a dc source, the forced response will be a constant current. The inductor acts like a short circuit to dc, so that

$$i_f = \tfrac{100}{2} = 50 \text{ A}$$

Thus,

$$i = 50 + Ke^{-0.5t} \qquad \text{A} \qquad t > 0$$

In order to evaluate K, we must establish the initial value of the inductor current. Prior to $t = 0$, this current is 25 A, and it cannot change instantaneously. Thus,

$$25 = 50 + K$$

or

$$K = -25$$

Hence,

$$i = 50 - 25e^{-0.5t} \qquad \text{A} \qquad t > 0$$

We complete the solution by also stating

$$i = 25 \text{ A} \qquad t < 0$$

or by writing a single expression valid for all t,

$$i = 25 + 25(1 - e^{-0.5t})u(t) \qquad \text{A}$$

The complete response is sketched in Fig. 8.38. Note how the natural response serves to connect the response for $t < 0$ with the constant forced response.

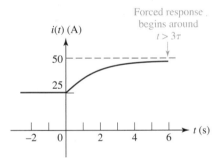

■ **FIGURE 8.38** The response $i(t)$ of the circuit shown in Fig. 8.37 is sketched for values of times less and greater than zero.

PRACTICE

8.10 A voltage source, $v_s = 20u(t)$ V, is in series with a 200 Ω resistor and a 4 H inductor. Find the magnitude of the inductor current at t equal to (a) 0^-; (b) 0^+; (c) 8 ms; (d) 15 ms.

Ans: 0; 0; 33.0 mA; 52.8 mA.

As a final example of this method by which the complete response of any circuit subjected to a transient may be written down *almost by inspection,* let us again consider the simple *RL* series circuit, but subjected to a voltage pulse.

EXAMPLE **8.9**

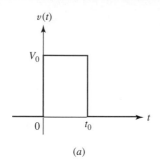

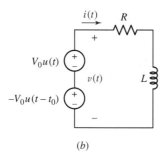

■ FIGURE 8.39 (*a*) A rectangular voltage pulse which is to be used as the forcing function in a simple series *RL* circuit. (*b*) The series *RL* circuit, showing the representation of the forcing function by the series combination of two independent voltage-step sources. The current *i*(*t*) is desired.

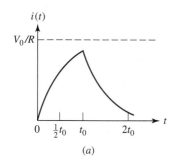

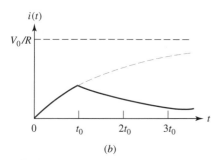

■ FIGURE 8.40 Two possible response curves are shown for the circuit of Fig. 8.39*b*. (*a*) τ is selected as $t_0/2$. (*b*) τ is selected as $2t_0$.

Find the current response in a simple series *RL* circuit when the forcing function is a rectangular voltage pulse of amplitude V_0 and duration t_0.

We represent the forcing function as the sum of two step-voltage sources $V_0 u(t)$ and $-V_0 u(t - t_0)$, as indicated in Fig. 8.39*a* and *b*, and we plan to obtain the response by using the superposition principle. Let $i_1(t)$ designate that part of $i(t)$ which is due to the upper source $V_0 u(t)$ acting alone, and let $i_2(t)$ represent that part due to $-V_0 u(t - t_0)$ acting alone. Then,

$$i(t) = i_1(t) + i_2(t)$$

Our object is now to write each of the partial responses i_1 and i_2 as the sum of a natural and a forced response. The response $i_1(t)$ is familiar; this problem was solved in Eq. [34]:

$$i_1(t) = \frac{V_0}{R}(1 - e^{-Rt/L}), \qquad t > 0$$

Note that this solution is only valid for $t > 0$ as indicated; $i_1 = 0$ for $t < 0$.

We now turn our attention to the other source and its response $i_2(t)$. Only the polarity of the source and the time of its application are different. There is no need therefore to determine the form of the natural response and the forced response; the solution for $i_1(t)$ enables us to write

$$i_2(t) = -\frac{V_0}{R}[1 - e^{-R(t-t_0)/L}], \qquad t > t_0$$

where the applicable range of t, $t > t_0$, must again be indicated; and $i_2 = 0$ for $t < t_0$.

We now add the two solutions, but do so carefully, since each is valid over a different interval of time. Thus,

$$i(t) = 0, \qquad t < 0 \tag{35}$$

$$i(t) = \frac{V_0}{R}(1 - e^{-Rt/L}), \qquad 0 < t < t_0 \tag{36}$$

and

$$i(t) = \frac{V_0}{R}(1 - e^{-Rt/L}) - \frac{V_0}{R}(1 - e^{-R(t-t_0)/L}), \qquad t > t_0$$

or more compactly,

$$i(t) = \frac{V_0}{R}e^{-Rt/L}(e^{Rt_0/L} - 1), \qquad t > t_0 \tag{37}$$

Although Eqs. [35] through [37] completely describe the response of the circuit in Fig. 8.39*b* to the pulse waveform of Fig. 8.39*a*, the current waveform itself is sensitive to both the circuit time constant τ and the voltage pulse duration t_0. Two possible curves are shown in Fig. 8.40.

The left curve is drawn for the case where the time constant is only one-half as large as the length of the applied pulse; the rising portion of the exponential has therefore almost reached V_0/R before the decaying exponential begins. The opposite situation is shown to the right; there, the time constant is twice t_0 and the response never has a chance to reach the larger amplitudes.

The procedure we have been using to find the response of an *RL* circuit after dc sources have been switched on or off (or in or out of the circuit) at some instant of time is summarized in the following. We assume that the circuit is reducible to a single equivalent resistance R_{eq} in series with a single equivalent inductance L_{eq} when all independent sources are set equal to zero. The response we seek is represented by $f(t)$.

---

1. With all independent sources zeroed out, simplify the circuit to determine R_{eq}, L_{eq}, and the time constant $\tau = L_{eq}/R_{eq}$.

2. Viewing L_{eq} as a short circuit, use dc analysis methods to find $i_L(0^-)$, the inductor current just prior to the discontinuity.

3. Again viewing L_{eq} as a short circuit, use dc analysis methods to find the forced response. This is the value approached by $f(t)$ as $t \to \infty$; we represent it by $f(\infty)$.

4. Write the total response as the sum of the forced and natural responses: $f(t) = f(\infty) + Ae^{-t/\tau}$.

5. Find $f(0^+)$ by using the condition that $i_L(0^+) = i_L(0^-)$. If desired, L_{eq} may be replaced by a current source $i_L(0^+)$ [an open circuit if $i_L(0^+) = 0$] for this calculation. With the exception of inductor currents (and capacitor voltages), other currents and voltages in the circuit may change abruptly.

6. $f(0^+) = f(\infty) + A$ and $f(t) = f(\infty) + [f(0^+) - f(\infty)]\, e^{-t/\tau}$, or total response $=$ final value $+$ (initial value $-$ final value) $e^{-t/\tau}$.

PRACTICE

8.11 The circuit shown in Fig. 8.41 has been in the form shown for a very long time. The switch opens at $t = 0$. Find i_R at t equal to
(a) 0^-; (b) 0^+; (c) ∞; (d) 1.5 ms.

Ans: 0; 10 mA; 4 mA; 5.34 mA.

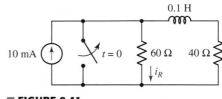

■ **FIGURE 8.41**

8.8 DRIVEN *RC* CIRCUITS

The complete response of any *RC* circuit may also be obtained as the sum of the natural and the forced response. Since the procedure is virtually identical to what we have already discussed in detail for *RL* circuits, the best approach at this stage is to illustrate it by working a relevant example completely, where the goal is not just a capacitor-related quantity but the current associated with a resistor as well.

EXAMPLE 8.10

Find the capacitor voltage $v_C(t)$ and the current $i(t)$ in the 200 Ω resistor of Fig. 8.42 for all time.

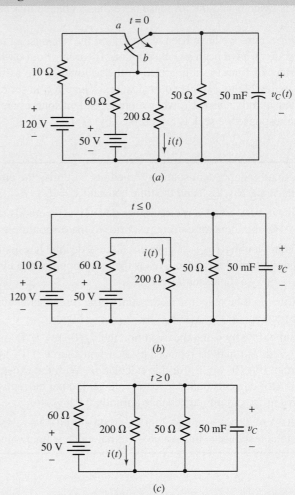

■ **FIGURE 8.42** (*a*) An *RC* circuit in which the complete responses v_C and *i* are obtained by adding a forced response and a natural response. (*b*) Circuit for $t \le 0$. (*c*) Circuit for $t \ge 0$.

We begin by considering the state of the circuit at $t < 0$, corresponding to the switch at position *a* as represented in Fig. 8.42*b*. As per usual, we assume no transients are present, so that only a forced response due to the 120 V source is relevant to finding $v_C(0^-)$. Simple voltage division then gives us the initial voltage,

$$v_C(0) = \frac{50}{50 + 10}(120) = 100 \text{ V}$$

Since the capacitor voltage cannot change instantaneously, this voltage is equally valid at $t = 0^-$ and $t = 0^+$.

The switch is now thrown to *b*, and the complete response is

$$v_C = v_{Cf} + v_{Cn}$$

The corresponding circuit has been redrawn in Fig. 8.42*c* for convenience. The form of the natural response is obtained by replacing the 50 V source by a short circuit and evaluating the equivalent resistance to find the time constant (in other words, we are finding the Thévenin equivalent resistance "seen" by the capacitor):

$$R_{eq} = \frac{1}{\frac{1}{50} + \frac{1}{200} + \frac{1}{60}} = 24 \ \Omega$$

Thus,

$$v_{Cn} = Ae^{-t/R_{eq}C} = Ae^{-t/1.2}$$

In order to evaluate the forced response with the switch at *b*, we wait until all the voltages and currents have stopped changing, thus treating the capacitor as an open circuit, and use voltage division once more:

$$v_{C_f} = 50 \left(\frac{200 \parallel 50}{60 + 200 \parallel 50} \right)$$

$$= 50 \left(\frac{(50)(200)/250}{60 + (50)(200)/250} \right) = 20 \text{ V}$$

Thus,

$$v_C = 20 + Ae^{-t/1.2} \qquad \text{V}$$

and from the initial condition already obtained,

$$100 = 20 + A$$

or

$$v_C = 20 + 80e^{-t/1.2} \qquad \text{V}, \qquad t \geq 0$$

and

$$v_C = 100 \text{ V}, \qquad t < 0$$

This response is sketched in Fig. 8.43*a*; again the natural response is seen to form a transition from the initial to the final response.

Next we attack $i(t)$. This response need not remain constant during the instant of switching. With the contact at *a*, it is evident that $i = 50/260 = 192.3$ milliamperes. When the switch moves to position *b*, the forced response for this current becomes

$$i_f = \frac{50}{60 + (50)(200)/(50 + 200)} \left(\frac{50}{50 + 200} \right) = 0.1 \text{ amperes}$$

The form of the natural response is the same as that which we already determined for the capacitor voltage:

$$i_n = Ae^{-t/1.2}$$

Combining the forced and natural responses, we obtain

$$i = 0.1 + Ae^{-t/1.2} \qquad \text{amperes}$$

(Continued on next page)

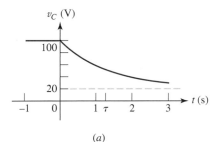

(a)

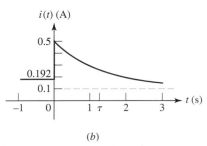

(b)

■ **FIGURE 8.43** The responses (*a*) v_C and (*b*) *i* are plotted as functions of time for the circuit of Fig. 8.42.

To evaluate A, we need to know $i(0^+)$. This is found by fixing our attention on the energy-storage element (the capacitor). The fact that v_C must remain 100 V during the switching interval is the governing condition which establishes the other currents and voltages at $t = 0^+$. Since $v_C(0^+) = 100$ V, and since the capacitor is in parallel with the 200 Ω resistor, we find $i(0^+) = 0.5$ ampere, $A = 0.4$ ampere, and thus

$$i(t) = 0.1923 \text{ ampere} \qquad t < 0$$

$$i(t) = 0.1 + 0.4e^{-t/1.2} \qquad \text{ampere} \qquad t > 0$$

or

$$i(t) = 0.1923 + (-0.0923 + 0.4e^{-t/1.2})u(t) \qquad \text{amperes}$$

where the last expression is correct for all t.

The complete response for all t may also be written concisely by using $u(-t)$, which is unity for $t < 0$ and 0 for $t > 0$. Thus,

$$i(t) = 0.1923u(-t) + (0.1 + 0.4e^{-t/1.2})u(t) \qquad \text{amperes}$$

This response is sketched in Fig. 8.43b. Note that only four numbers are needed to write the functional form of the response for this single-energy-storage-element circuit, or to prepare the sketch: the constant value prior to switching (0.1923 ampere), the instantaneous value just after switching (0.5 ampere), the constant forced response (0.1 ampere), and the time constant (1.2 s). The appropriate negative exponential function is then easily written or drawn.

PRACTICE

8.12 For the circuit of Fig. 8.44, find $v_C(t)$ at t equal to (a) 0^-; (b) 0^+; (c) ∞; (d) 0.08 s.

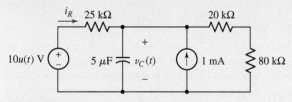

■ **FIGURE 8.44**

Ans: 20 V; 20 V; 28 V; 24.4 V.

We conclude by listing the duals of the statements given at the end of Sec. 8.7.

The procedure we have been using to find the response of an *RC* circuit after dc sources have been switched on or off, or in or out of the circuit, at some instant of time, say $t = 0$, is summarized in the following. We assume that the circuit is reducible to a single equivalent resistance R_{eq} in parallel with a single equivalent capacitance C_{eq} when all independent sources are set equal to zero. The response we seek is represented by $f(t)$.

1. With all independent sources zeroed out, simplify the circuit to determine R_{eq}, C_{eq}, and the time constant $\tau = R_{eq}C_{eq}$.

2. Viewing C_{eq} as an open circuit, use dc analysis methods to find $v_C(0^-)$, the capacitor voltage just prior to the discontinuity.

3. Again viewing C_{eq} as an open circuit, use dc analysis methods to find the forced response. This is the value approached by $f(t)$ as $t \to \infty$; we represent it by $f(\infty)$.

4. Write the total response as the sum of the forced and natural responses: $f(t) = f(\infty) + Ae^{-t/\tau}$.

5. Find $f(0^+)$ by using the condition that $v_C(0^+) = v_C(0^-)$. If desired, C_{eq} may be replaced by a voltage source $v_C(0^+)$ [a short circuit if $v_C(0^+) = 0$] for this calculation. With the exception of capacitor voltages (and inductor currents), other voltages and currents in the circuit may change abruptly.

6. $f(0^+) = f(\infty) + A$ and $f(t) = f(\infty) + [f(0^+) - f(\infty)]e^{-t/\tau}$, or total response = final value + (initial value − final value) $e^{-t/\tau}$.

As we have just seen, the same basic steps that apply to the analysis of *RL* circuits can be applied to *RC* circuits as well. Up to now, we have confined ourselves to the analysis of circuits with dc forcing functions only, despite the fact that Eq. [30] holds for more general functions such as $Q(t) = 9\cos(5t - 7°)$ or $Q(t) = 2e^{-5t}$. Before concluding this section, we explore one such non-dc scenario.

EXAMPLE 8.11

Determine an expression for $v(t)$ in the circuit of Fig. 8.45 valid for $t > 0$.

Based on experience, we expect a complete response of the form
$$v(t) = v_f + v_n$$
where v_f will likely resemble our forcing function and v_n will have the form $Ae^{-t/\tau}$.

What *is* the circuit time constant τ? We replace our source with an open circuit and find the Thévenin equivalent resistance in parallel with the capacitor:
$$R_{eq} = 4.7 + 10 = 14.7\ \Omega$$

Thus, our time constant is $\tau = R_{eq}C = 323.4\ \mu s$, or equivalently $1/\tau = 3.092 \times 10^3\ s^{-1}$.

There are several ways to proceed, although perhaps the most straightforward is to perform a source transformation, resulting in a voltage source $23.5e^{-2000t}u(t)$ V in series with $14.7\ \Omega$ and $22\ \mu F$. (*Note this does not change the time constant.*)

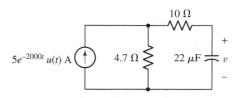

■ **FIGURE 8.45** A simple *RC* circuit driven by an exponentially decaying forcing function.

(Continued on next page)

Writing a simple KVL equation for $t > 0$, we find that

$$23.5e^{-2000t} = (14.7)(22 \times 10^{-6})\frac{dv}{dt} + v$$

A little rearranging results in

$$\frac{dv}{dt} + 3.092 \times 10^3 v = 72.67 \times 10^3\, e^{-2000t}$$

which, upon comparison with Eqs. [28] and [30] allows us to write the complete response as

$$v(t) = e^{-Pt} \int Qe^{Pt}\,dt + Ae^{-Pt}$$

where in our case $P = 1/\tau = 3.092 \times 10^3$ and $Q(t) = 72.67 \times 10^3 e^{-2000t}$. We therefore find that

$$v(t) = e^{-3092t} \int 72.67 \times 10^3 e^{-2000t} e^{3092t}\,dt + Ae^{-3092t}\ \text{V}$$

Performing the indicated integration,

$$v(t) = 66.55e^{-2000t} + Ae^{-3092t}\ \text{V} \tag{38}$$

Our only source is controlled by a step function with zero value for $t < 0$, so we know that $v(0^-) = 0$. Since v is a capacitor voltage, $v(0^+) = v(0^-)$, and we therefore find our initial condition $v(0) = 0$ easily enough. Substituting this into Eq. [38], we find $A = -66.55$ V and so

$$v(t) = 66.55(e^{-2000t} - e^{-3092t})\ \text{V}, \qquad t > 0$$

PRACTICE

8.13 Determine the capacitor voltage v in the circuit of Fig. 8.46 for $t > 0$.

Ans: $23.5\cos 3t + 22.8 \times 10^{-3}\sin 3t - 23.5e^{-3092t}$ V.

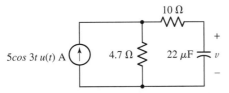

■ **FIGURE 8.46** A simple *RC* circuit driven by a sinusoidal forcing function.

8.9 • PREDICTING THE RESPONSE OF SEQUENTIALLY SWITCHED CIRCUITS

In Ex. 8.9 we briefly considered the response of an *RL* circuit to a pulse waveform, in which a source was effectively switched into and subsequently switched out of the circuit. This type of situation is very common in practice, as few circuits are designed to be energized only once (passenger vehicle airbag triggering circuits, for example). In predicting the response of simple *RL* and *RC* circuits subjected to pulses and series of pulses—sometimes referred to as *sequentially switched circuits*—the key is the relative size of the circuit time constant to the various times that define the pulse sequence. The underlying principle behind the analysis will be whether the energy storage element has time to fully charge before the pulse ends, and whether it has time to fully discharge before the next pulse begins.

Consider the circuit shown in Fig. 8.47*a*, which is connected to a pulsed voltage source described by seven separate parameters defined in Fig. 8.47*b*. The waveform is bounded by two values, **V1** and **V2**. The time t_r required to change from **V1** to **V2** is called the *rise time (TR)*, and the time t_f required to change from **V2** to **V1** is called the *fall time (TF)*.

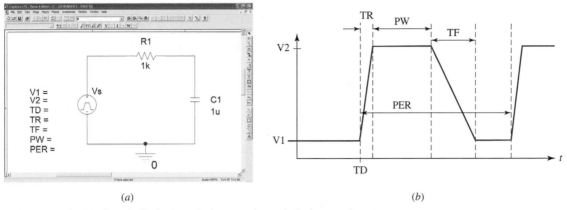

■ **FIGURE 8.47** (*a*) Schematic of a simple *RC* circuit connected to a pulsed voltage waveform.
(*b*) Diagram of the SPICE **VPULSE** parameter definitions.

The duration W_p of the pulse is referred to as the **pulse width (PW),** and the **period** T of the waveform **(PER)** is the time it takes for the pulse to repeat. Note also that SPICE allows a time delay **(TD)** before the pulse train begins, which can be useful in allowing initial transient responses to decay for some circuit configurations.

For the purposes of this discussion, we set a zero time delay, $V1 = 0$, and $V2 = 9$ V. The circuit time constant is $\tau = RC = 1$ ms, so we set the rise and fall times to be 1 ns. Although SPICE will not allow a voltage to change in zero time since it solves the differential equations using discrete time intervals, compared to our circuit time constant 1 ns is a reasonable approximation to "instantaneous."

We will consider four basic cases, summarized in Table 8.1. In the first two cases, the pulse width W_p is much longer than the circuit time constant τ, so we expect the transients resulting from the beginning of the pulse to die out before the pulse is over. In the latter two cases, the opposite is true: the pulse width is so short that the capacitor does not have time to fully charge before the pulse ends. A similar issue arises when we consider the response of the circuit when the time between pulses $(T - W_p)$ is either short (Case II) or long (Case III) compared to the circuit time constant.

TABLE 8.1 Four Separate Cases of Pulse Width and Period Relative to the Circuit Time Constant of 1 ms

Case	Pulse Width W_p	Period T
I	10 ms ($\tau \ll W_p$)	20 ms ($\tau \ll T - W_p$)
II	10 ms ($\tau \ll W_p$)	10.1 ms ($\tau \gg T - W_p$)
III	0.1 ms ($\tau \gg W_p$)	10.1 ms ($\tau \ll T - W_p$)
IV	0.1 ms ($\tau \gg W_p$)	0.2 ms ($\tau \gg T - W_p$)

We qualitatively sketch the circuit response for each of the four cases in Fig. 8.48, arbitrarily selecting the capacitor voltage as the quantity of interest as any voltage or current is expected to have the same time dependence.

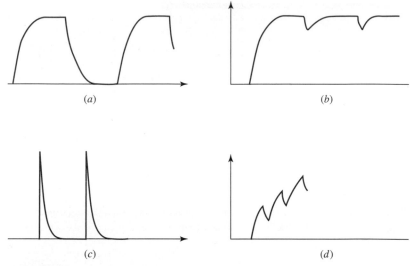

■ **FIGURE 8.48**   Capacitor voltage for the *RC* circuit, with pulse width and period as in (*a*) Case I;
(*b*) Case II; (*c*) Case III; and (*d*) Case IV.

In Case I, the capacitor has time to both fully charge and fully discharge (Fig. 8.48*a*), whereas in Case II (Fig. 8.48*b*), when the time between pulses is reduced, it no longer has time to fully discharge. In contrast, the capacitor does not have time to fully charge in either Case III (Fig. 8.48*c*) or Case IV (Fig. 8.49*d*).

Case I: Time Enough to Fully Charge and Fully Discharge

We can obtain exact values for the response in each case, of course, by performing a series of analyses. We consider Case I first. Since the capacitor has time to fully charge, the forced response will correspond to the 9 V dc driving voltage. The complete response to the first pulse is therefore

$$v_C(t) = 9 + Ae^{-1000t} \text{ V}$$

With $v_C(0) = 0$, $A = -9$ V and so

$$v_C(t) = 9(1 - e^{-1000t}) \text{ V} \qquad [39]$$

in the interval of $0 < t < 10$ ms. At $t = 10$ ms, the source drops suddenly to 0 V, and the capacitor begins to discharge through the resistor. In this time interval we are faced with a simple "source-free" *RC* circuit, and we can write the response as

$$v_C(t) = Be^{-1000(t-0.01)}, \qquad 10 < t < 20 \text{ ms} \qquad [40]$$

where $B = 8.99959$ V is found by substituting $t = 10$ ms in Eq. [39]; we will be pragmatic here and round this to 9 V, noting that the value calculated is consistent with our assumption that the initial transient dissipates before the pulse ends.

At $t = 20$ ms, the voltage source jumps immediately back to 9 V. The capacitor voltage just prior to this event is given by substituting $t = 20$ ms in Eq. [40], leading to $v_C(20\,\text{ms}) = 408.6\ \mu\text{V}$, essentially zero compared to the peak value of 9 V.

If we keep to our convention of rounding to four significant digits, the capacitor voltage at the beginning of the second pulse is zero, which is the same as our starting point. Thus, Eqs. [39] and [40] form the basis of the response for all subsequent pulses, and we may write

$$v_C(t) = \begin{cases} 9(1 - e^{-1000t}) \text{ V,} & 0 \le t \le 10 \text{ ms} \\ 9e^{-1000(t-0.01)} \text{ V,} & 10 < t \le 20 \text{ ms} \\ 9(1 - e^{-1000(t-0.02)}) \text{ V,} & 20 < t \le 30 \text{ ms} \\ 9e^{-1000(t-0.03)} \text{ V,} & 30 < t \le 40 \text{ ms} \end{cases}$$

and so on.

Case II: Time Enough to Fully Charge But Not Fully Discharge

Next we consider what happens if the capacitor is not allowed to completely discharge (Case II). Eq. [39] still describes the situation in the interval of $0 < t < 10$ ms, and Eq. [40] describes the capacitor voltage in the interval between pulses, which has been reduced to $10 < t < 10.1$ ms.

Just prior to the onset of the second pulse at $t = 10.1$ ms, v_C is now 8.144 V; the capacitor has only had 0.1 ms to discharge, and therefore still retains 82 percent of its maximum energy when the next pulse begins. Thus, in the next interval,

$$v_C(t) = 9 + Ce^{-1000(t-10.1\times 10^{-3})} \text{ V,} \quad 10.1 < t < 20.1 \text{ ms}$$

where $v_C(10.1 \text{ ms}) = 9 + C = 8.144$ V, so $C = -0.856$ V and

$$v_C(t) = 9 - 0.856e^{-1000(t-10.1\times 10^{-3})} \text{ V,} \quad 10.1 < t < 20.1 \text{ ms}$$

which reaches the peak value of 9 V much more quickly than for the previous pulse.

Case III: No Time to Fully Charge But Time to Fully Discharge

What if it isn't clear that the transient will dissipate before the end of the voltage pulse? In fact, this situation arises in Case III. Just as we wrote for Case I,

$$v_C(t) = 9 + Ae^{-1000t} \text{ V} \qquad [41]$$

still applies to this situation, but now only in the interval $0 < t < 0.1$ ms. Our initial condition has not changed, so $A = -9$ V as before. Now, however, just before this first pulse ends at $t = 0.1$ ms, we find that $v_C = 0.8565$ V. This is a far cry from the maximum of 9 V possible if we allow the capacitor time to fully charge, and is a direct result of the pulse lasting only one-tenth of the circuit time constant.

The capacitor now begins to discharge, so that

$$v_C(t) = Be^{-1000(t-1\times 10^{-4})} \text{ V,} \qquad 0.1 < t < 10.1 \text{ ms} \qquad [42]$$

We have already determined that $v_C(0.1^- \text{ ms}) = 0.8565$ V, so $v_C(0.1^+ \text{ ms}) = 0.8565$ V and substitution into Eq. [42] yields $B = 0.8565$ V. Just prior to the onset of the second pulse at $t = 10.1$ ms, the capacitor voltage has decayed to essentially 0 V; this is the initial condition at the start of the second pulse and so Eq. [41] can be rewritten as

$$v_C(t) = 9 - 9e^{-1000(t-10.1\times 10^{-3})} \text{ V,} \qquad 10.1 < t < 10.2 \text{ ms} \qquad [43]$$

to describe the corresponding response.

Case IV: No Time to Fully Charge or Even Fully Discharge

In the last case, we consider the situation where the pulse width and period are so short that the capacitor can neither fully charge nor fully discharge in any one period. Based on experience, we can write

$$v_C(t) = 9 - 9e^{-1000t} \text{ V}, \qquad\qquad 0 < t < 0.1 \text{ ms} \qquad [44]$$

$$v_C(t) = 0.8565e^{-1000(t-1\times10^{-4})} \text{ V}, \qquad 0.1 < t < 0.2 \text{ ms} \qquad [45]$$

$$v_C(t) = 9 + Ce^{-1000(t-2\times10^{-4})} \text{ V}, \qquad 0.2 < t < 0.3 \text{ ms} \qquad [46]$$

$$v_C(t) = De^{-1000(t-3\times10^{-4})} \text{ V}, \qquad 0.3 < t < 0.4 \text{ ms} \qquad [47]$$

Just prior to the onset of the second pulse at $t = 0.2$ ms, the capacitor voltage has decayed to $v_C = 0.7750$ V; with insufficient time to fully discharge, it retains a large fraction of the little energy it had time to store initially. For the interval of $0.2 < t < 0.3$ ms, substitution of $v_C(0.2^+) = v_C(0.2^-) = 0.7750$ V into Eq. [46] yields $C = -8.225$ V. Continuing, we evaluate Eq. [46] at $t = 0.3$ ms and calculate $v_C = 1.558$ V just prior to the end of the second pulse. Thus, $D = 1.558$ V and our

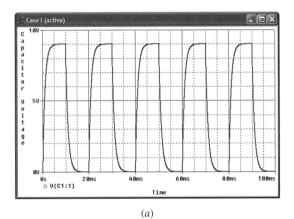

(a)

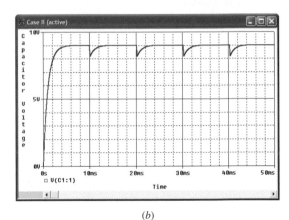

(b)

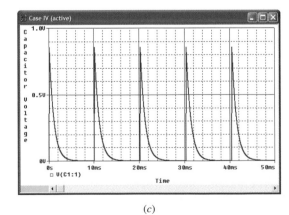

(c)

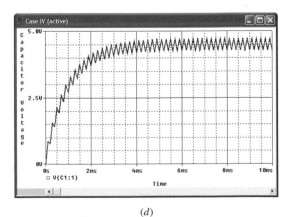

(d)

■ **FIGURE 8.49** PSpice simulation results corresponding to (*a*) Case I; (*b*) Case II; (*c*) Case III; (*d*) Case IV.

capacitor is slowly charging to ever increase voltage levels over several pulses. At this stage it might be useful if we plot the detailed responses, so we show the PSpice simulation results of Cases I through IV in Fig. 8.49. Note in particular that in Fig. 8.49d, the small charge/discharge transient response similar in shape to that shown in Figs. 8.49a–c is superimposed on a charging-type response of the form $(1 - e^{-t/\tau})$. Thus, it takes about 3 to 5 circuit time constants for the capacitor to charge to its maximum value in situations where a single period does not allow it to fully charge or discharge!

What we have not yet done is predict the behavior of the response for $t \gg 5\tau$, although we would be interested in doing so, especially if it was not necessary to consider a very long sequence of pulses one at a time. We note that the response of Fig. 8.49d has an *average* value of 4.50 V from about 4 ms onwards. This is exactly half the value we would expect if the voltage source pulse width allowed the capacitor to fully charge. In fact, this long-term average value can be computed by multiplying the dc capacitor voltage by the ratio of the pulse width to the period.

PRACTICE

8.14 Sketch $i_L(t)$ in the range of $0 < t < 6$ s for (a) $v_S(t) = 3u(t) - 3u(t-2) + 3u(t-4) - 3u(t-6) + \cdots$; (b) $v_S(t) = 3u(t) - 3u(t-2) + 3u(t-2.1) - 3u(t-4.1) + \cdots$.

Ans: (b) See Fig. 8.50a; (c) See Fig. 8.50b.

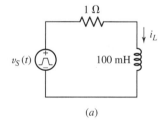

(a)

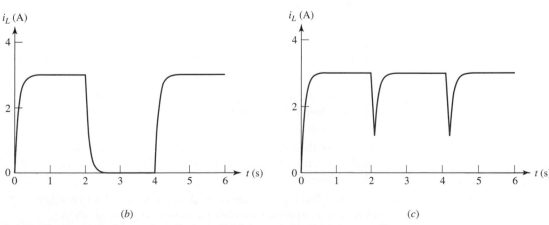

(b)   (c)

■ **FIGURE 8.50** (a) Circuit for Practice Problem 8.14; (b) Solution to part (a); (c) Solution to part (b).

Frequency Limits in Digital Integrated Circuits

Modern digital integrated circuits such as programmable array logic (PALs) and microprocessors (Fig. 8.51) are composed of interconnected transistor circuits known as *gates*.

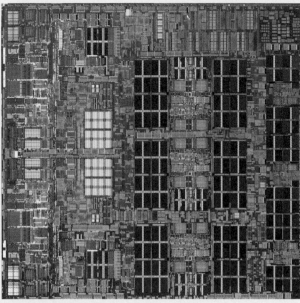

■ FIGURE 8.51 IBM Power Chip.

Digital signals are represented symbolically by combinations of ones and zeroes, and can be either data or instructions (such as "add" or "subtract"). Electrically, we represent a logic "1" by a "high" voltage, and a logic "0"

by a "low" voltage. In practice, there is a range of voltages that correspond to each; for example, in the 7400 series of TTL logic integrated circuits, any voltage between 2 and 5 V will be interpreted as a logic "1," and any voltage between 0 and 0.8 V will be interpreted as a logic "0." Voltages between 0.8 and 2 V do not correspond to either logic state, as shown in Fig. 8.52.

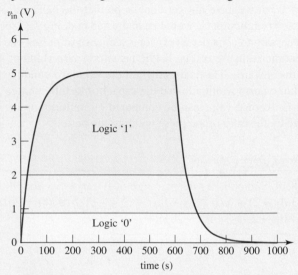

■ FIGURE 8.52 Charge/discharge characteristic of a pathway capacitance identifying the TTL voltage ranges for logic "1" and logic "0," respectively.

A key parameter in digital circuits is the speed at which we can effectively use them. In this sense, "speed" refers to how quickly we can switch a gate from one

SUMMARY AND REVIEW

❑ The response of a circuit having sources suddenly switched in or out of a circuit containing capacitors and inductors will always be composed of two parts: a *natural* response and a *forced* response.

❑ The form of the natural response (also referred to as the *transient response*) depends only on the component values and the way they are wired together.

❑ The form of the forced response mirrors the form of the forcing function. Therefore, a dc forcing function always leads to a constant forced response.

❑ A circuit reduced to a single equivalent inductance L and a single equivalent resistance R will have a natural response given by $i(t) = I_0 e^{-t/\tau}$, where $\tau = L/R$ is the circuit time constant.

logic state to another (either logic "0" to logic "1" or vice versa), and the time delay required to convey the output of one gate to the input of the next gate. Although transistors contain "built-in" capacitances that affect their switching speed, it is the *interconnect pathways* that presently limit the speed of the fastest digital integrated circuits. We can model the interconnect pathway between two logic gates using a simple *RC* circuit (although as feature sizes continue to decrease in modern designs, more detailed models are required to accurately predict circuit performance). For example, consider a 2000 μm-long pathway 2 μm wide. We can model this pathway in a typical silicon-based integrated circuit as having a capacitance of 0.5 pF and a resistance of 100 Ω, shown schematically in Fig. 8.53.

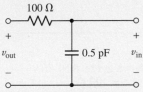

■ **FIGURE 8.53** Circuit model for an integrated circuit pathway.

Let's assume the voltage v_{out} represents the output voltage of a gate that is changing from a logic "0" state to a logic "1" state. The voltage v_{in} appears across the input of a second gate, and we are interested in how long it takes v_{in} to reach the same value as v_{out}.

Assuming the 0.5 pF capacitance that characterizes the interconnect pathway is initially discharged [i.e., $v_{in}(0) = 0$], calculating the *RC* time constant for our pathway as $\tau = RC = 50$ ps, and defining $t = 0$ as when v_{out} changes, we obtain the expression

$$v_{in}(t) = Ae^{-t/\tau} + v_{out}(0)$$

Setting $v_{in}(0) = 0$, we find that $A = -v_{out}(0)$ so that

$$v_{in}(t) = v_{out}(0)[1 - e^{-t/\tau}]$$

Upon examining this equation, we see that v_{in} will reach the value $v_{out}(0)$ after $\sim 5\tau$ or 250 ps. If the voltage v_{out} changes again before this transient time period is over, then the capacitance does not have sufficient time to fully charge. In such situations, v_{in} will be less than $v_{out}(0)$. Assuming that $v_{out}(0)$ equals the minimum logic "1" voltage, for example, this means that v_{in} will not correspond to a logic "1." If v_{out} now suddenly changes to 0 V (logic "0"), the capacitance will begin to discharge so that v_{in} decreases further. Thus, by switching our logic states too quickly, we are unable to transfer the information from one gate to another.

The fastest speed at which we can change logic states is therefore $(5\tau)^{-1}$. This can be expressed in terms of the maximum operating frequency:

$$f_{max} = \frac{1}{2(5\tau)} = 2 \text{ GHz}$$

where the factor of 2 represents a charge/discharge period. If we desire to operate our integrated circuit at a higher frequency so that calculations can be performed faster, we need to reduce the interconnect capacitance and/or the interconnect resistance.

❑ A circuit reduced to a single equivalent capacitance C and a single equivalent resistance R will have a natural response given by $v(t) = V_0 e^{-t/\tau}$, where $\tau = RC$ is the circuit time constant.

❑ The unit-step function is a useful way to model the closing or opening of a switch, provided we are careful to keep an eye on the initial conditions.

❑ The *complete* response of an *RL* or *RC* circuit excited by a dc source will have the form $f(0^+) = f(\infty) + A$ and $f(t) = f(\infty) + [f(0^+) - f(\infty)]e^{-t/\tau}$, or total response = final value + (initial value − final value)$e^{-t/\tau}$.

❑ The complete response for an *RL* or *RC* circuit may also be determined by writing a single differential equation for the quantity of interest and solving.

❏ When dealing with sequentially switched circuits, or circuits connected to pulsed waveforms, the relevant issue is whether the energy storage element has sufficient time to fully charge and to fully discharge, as measured relative to the circuit time constant.

READING FURTHER

A guide to solution techniques for differential equations can be found in:

W. E. Boyce and R. C. DiPrima, *Elementary Differential Equations and Boundary Value Problems,* 7th ed. New York: Wiley, 2002.

A detailed description of transients in electric circuits is given in:

E. Weber, *Linear Transient Analysis Volume I.* New York: Wiley, 1954. (Out of print, but in many university libraries.)

EXERCISES

8.1 The Source-Free *RL* Circuit

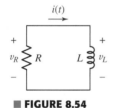

■ **FIGURE 8.54**

1. Consider the simple *RL* circuit shown in Fig. 8.54. If $R = 4.7$ kΩ, $L = 1$ μH, and $i(0) = 2$ mA, calculate (*a*) *i* at $t = 100$ ps; (*b*) *i* at $t = 212.8$ ps; (*c*) v_R at 75 ps; (*d*) v_L at 75 ps.

2. The circuit shown in Fig. 8.54 consists of a resistance $R = 1$ Ω and an inductance $L = 2$ H. At $t = 0$, the inductance is storing 100 mJ of energy. Calculate (*a*) *i* at $t = 1$ s; (*b*) *i* at $t = 5$ s; (*c*) *i* at $t = 10$ s; (*d*) the energy remaining in the inductance at $t = 2$ s.

3. For the simple *RL* circuit shown in Fig. 8.54, *R* is known to be 100 Ω. If $i(0) = 2$ mA and $i(50\,\mu s) = 735.8\,\mu$A, determine the value of the inductance *L*.

4. For the simple *RL* circuit shown in Fig. 8.54, *L* is known to be 3 mH. If $i(0) = 1.5$ A and $i(2$ s$) = 551.8$ mA, determine the value of the resistor *R*.

5. The 3 mH inductance in the circuit of Fig. 8.54 is storing 1 J of energy at $t = 0$, and 100 mJ at $t = 1$ ms. Calculate *R*.

6. The switch in the circuit of Fig. 8.55 has been closed since dinosaurs last walked the earth. If the switch is opened at $t = 0$, find (*a*) i_L the instant after the switch changes; (*b*) *v* the instant after the switch changes.

7. The switch in the circuit of Fig. 8.56 is a single-pole, double-throw switch that is drawn to indicate that it closes one circuit before opening the other; this type of switch is often referred to as a "make before break" switch. Assuming the switch has been in the position drawn in the figure for a long time, determine the value of *v* and i_L (*a*) the instant just *prior* to the switch changing; (*b*) the instant just *after* the switch changes.

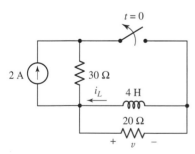

■ **FIGURE 8.55**

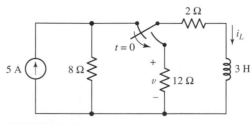

■ **FIGURE 8.56**

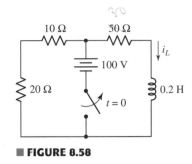

8. After being in the configuration shown for hours, the switch in the circuit of Fig. 8.57 is closed at $t = 0$. At $t = 5\ \mu s$, calculate: (a) i_L; (b) i_{SW}.

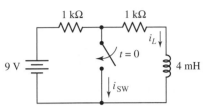

■ **FIGURE 8.57**

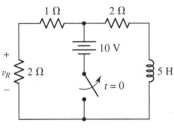

■ **FIGURE 8.58**

9. After having been closed for a long time, the switch in the circuit of Fig. 8.58 is opened at $t = 0$. (a) Find $i_L(t)$ for $t > 0$. (b) Evaluate i_L. (c) Find t_1 if $i_L(t_1) = 0.5i_L(0)$.

10. For the circuit depicted in Fig. 8.59, (a) write the differential equation that describes the resistor voltage v_R for $t > 0$. (b) Solve the characteristic equation. (c) Compute v_R just *before* the switch opens, just *after* the switch opens, and at $t = 1$ s.

8.2 Properties of the Exponential Response

11. Figure 8.7 shows a plot of i/I_0 as a function of t. (a) Determine the values of t/τ at which i/I_0 is 0.1, 0.01, and 0.001. (b) If a tangent to the curve is drawn at the point where $t/\tau = 1$, where will it intersect the t axis?

12. Referring to the response shown in Fig. 8.60, determine the circuit time constant and the initial current through the inductor.

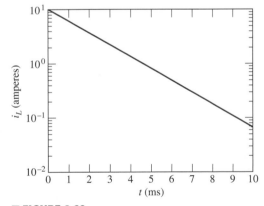

■ **FIGURE 8.59**

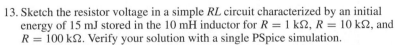

■ **FIGURE 8.60**

13. Sketch the resistor voltage in a simple RL circuit characterized by an initial energy of 15 mJ stored in the 10 mH inductor for $R = 1$ kΩ, $R = 10$ kΩ, and $R = 100$ kΩ. Verify your solution with a single PSpice simulation.

14. Take $R = 1$ MΩ and $L = 3.3\ \mu H$ in the circuit of Fig. 8.1. (a) Compute the circuit time constant. (b) If the inductor has an initial energy of 43 μJ at $t = 0$, determine i_L at $t = 5$ ps. (c) Verify your solution with a PSpice simulation.

15. A digital signal is sent through a loosely coiled wire having an inductance of 125.7 μH. Determine the maximum permitted value of the receiving equipment's Thévenin equivalent resistance if transients must last less than 100 ns.

16. The switch of Fig. 8.61 has been open for a long time before it closes at $t = 0$. For the time interval $-5 < t < 5 \ \mu s$, sketch: (a) $i_L(t)$; (b) $i_x(t)$.

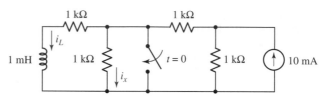

■ **FIGURE 8.61**

8.3 The Source-Free *RC* Circuit

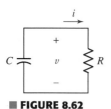

■ **FIGURE 8.62**

17. In the parallel *RC* circuit represented in Fig. 8.62, $C = 1 \ \mu F$ and $R = 100 \ M\Omega$ represents the losses in the dielectric of the capacitor. The capacitor is storing 1 mJ at $t = 0$. (a) Determine the circuit time constant. (b) Calculate i at 20 s. (c) Verify your solution with a PSpice simulation.

18. For the circuit of Fig. 8.62, assume $R = 1 \ \Omega$, $C = 2$ F, and $i(0) = 10$ V. Calculate v at (a) $t = 1$ s; (b) $t = 2$ s; (c) $t = 5$ s; (d) $t = 10$ s.

19. Use $R = 1 \ k\Omega$ and $C = 4$ mF in the circuit of Fig. 8.62. If $v(0) = 5$ V, calculate (a) v at $t = 1$ ms; (b) i at $t = 2$ ms; (c) the energy remaining in the capacitor at $t = 4$ ms.

20. For the *RC* circuit shown in Fig. 8.62, *C* is known to be 100 pF. (a) If $v(0) = 1.5$ V and $v(2 \ ns) = 100$ mV, determine the value of the resistor *R*. (b) Verify your solution with a PSpice simulation.

21. A stereo receiver has a power supply that includes two large parallel-connected 50 mF capacitors. When the power to the receiver is turned off, you notice that the amber LED used as a "power-on" indicator fades out slowly over a period of a few seconds. With nothing good on television, you decide to perform an experiment using a 35 mm camera with a variable shutter speed and some cheap film. Four shutter speeds are used: 150 ms, 1 s, 1.5 s, and 2.0 s. As the shutter speed is increased from 150 ms to 1.5 s, the image appearing on the developed film increases in brightness. No significant difference is discernable between images taken with shutter speeds of 1.5 s and 2.0 s, and for a shutter speed of 150 ms, the image appears to be about 14 percent of the intensity obtained at the slowest setting. Estimate the Thévenin equivalent resistance of the circuitry connected to the receiver's power supply.

22. (a) Find $v_C(t)$ for all time in the circuit of Fig. 8.63. (b) At what time is $v_C = 0.1v_C(0)$?

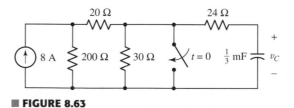

■ **FIGURE 8.63**

23. A 4 A current source, a 20 Ω resistor, and a 5 μF capacitor are all in parallel. The amplitude of the current source drops suddenly to zero (becoming a 0 A current source) at $t = 0$. At what time has (a) the capacitor voltage dropped to one-half of its initial value, and (b) the energy stored in the capacitor dropped to one-half of its initial value?

24. Determine $v_C(t)$ and $i_C(t)$ for the circuit of Fig. 8.64 and sketch both curves on the same time axis, $-0.1 < t < 0.1$ s.

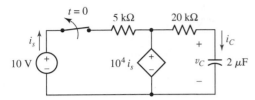

■ **FIGURE 8.64**

25. For the circuit of Fig. 8.65, determine the value of the current labeled i and the voltage labeled v at $t = 0^+$, $t = 1.5$ ms, and $t = 3.0$ ms.

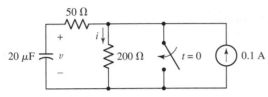

■ **FIGURE 8.65**

8.4 A More General Perspective

26. The switch in Fig. 8.66 opens at $t = 0$ after having been closed for an interminably long time. Find i_L and i_x at (a) $t = 0^-$; (b) $t = 0^+$; (c) $t = 300 \ \mu$s.

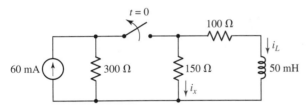

■ **FIGURE 8.66**

27. A 0.2 H inductor is in parallel with a 100 Ω resistor. The inductor current is 4 A at $t = 0$. (a) Find $i_L(t)$ at $t = 0.8$ ms. (b) If another 100 Ω resistor is connected in parallel with the inductor at $t = 1$ ms, calculate i_L at $t = 2$ ms.

28. A 20 mH inductor is in parallel with a 1 kΩ resistor. Let the value of the loop current be 40 mA at $t = 0$. (a) At what time will the current be 10 mA? (b) What series resistance should be switched into the circuit at $t = 10 \ \mu$s so that the current is 10 mA at $t = 15 \ \mu$s?

29. In the network of Fig. 8.67, initial values are $i_1(0) = 20$ mA and $i_2(0) = 15$ mA. (a) Determine $v(0)$. (b) Find $v(15 \ \mu$s). (c) At what time is $v(t) = 0.1v(0)$?

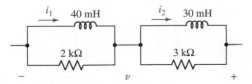

■ **FIGURE 8.67**

30. Select values for R_1 and R_2 in the circuit of Fig. 8.68 so that $v_R(0^+) = 10$ V and $v_R(1$ ms$) = 5$ V.

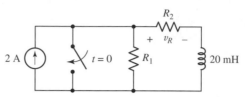

■ **FIGURE 8.68**

31. The switch in the circuit shown in Fig. 8.69 has been open for a long time before it closes at $t = 0$. (*a*) Find $i_L(t)$ for $t > 0$. (*b*) Sketch $v_x(t)$ for $-4 < t < 4$ ms.

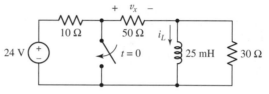

■ **FIGURE 8.69**

32. If $i_L(0) = 10$ A in the circuit of Fig. 8.70, find $i_L(t)$ for $t > 0$.

33. Refer to the circuit of Fig. 8.71 and determine i_1 at $t = -0.1, 0.03,$ and 0.1 s. Prepare a sketch of i_1 versus t, $-0.1 < t < 1$ s.

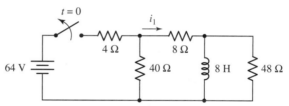

■ **FIGURE 8.71**

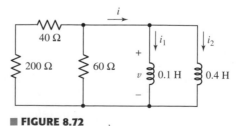

■ **FIGURE 8.70**

34. A circuit consists of a 0.5 H inductor, a 10 Ω resistor, and a 40 Ω resistor in series. The inductor current is 4 A at $t = 0$. (*a*) Calculate $i_L(15$ ms$)$. (*b*) The 40 Ω resistor is short-circuited at $t = 15$ ms. Calculate $i_L(30$ ms$)$.

35. The circuit shown in Fig. 8.72 contains two inductors in parallel, thus providing the opportunity of a trapped current circulating around the inductive loop. Let $i_1(0^-) = 10$ A and $i_2(0^-) = 20$ A. (*a*) Find $i_1(0^+)$, $i_2(0^+)$, and $i(0^+)$. (*b*) Determine the time constant τ for $i(t)$. (*c*) Find $i(t), t > 0$. (*d*) Find $v(t)$. (*e*) Find $i_1(t)$ and $i_2(t)$ from $v(t)$ and the initial values. (*f*) Show that the stored energy at $t = 0$ is equal to the sum of the energy dissipated in the resistive network between $t = 0$ and $t = \infty$, plus the energy stored in the inductors at $t = \infty$.

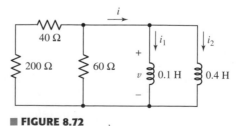

■ **FIGURE 8.72**

36. The circuit of Fig. 8.73 has been in the form shown since noon yesterday. The switch is opened at exactly 10:00 a.m. Find i_1 and v_C at (a) 9:59 a.m.; (b) 10:05 a.m. (c) Determine $i_1(t)$ at $t = 1.2\tau$. (d) Verify your solution with PSpice.

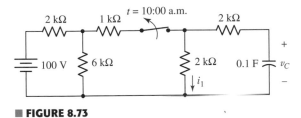

■ **FIGURE 8.73**

37. After being in the configuration shown for a long time, the switch in Fig. 8.74 is opened at $t = 0$. Determine values for (a) $i_s(0^-)$; (b) $i_x(0^-)$; (c) $i_x(0^+)$; (d) $i_s(0^+)$; (e) $i_x(0.4 \text{ s})$.

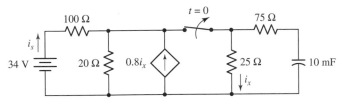

■ **FIGURE 8.74**

38. After being closed for a long time, the switch in the circuit of Fig. 8.75 is opened at $t = 0$. (a) Find $v_C(t)$ for $t > 0$. (b) Calculate values for $i_A(-100 \ \mu s)$ and $i_A(100 \ \mu s)$. (c) Verify your solution with PSpice.

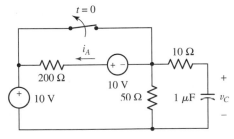

■ **FIGURE 8.75**

39. Many moons after the circuit of Fig. 8.76 was first assembled, its switch is closed at $t = 0$. (a) Find $i_1(t)$ for $t < 0$. (b) Find $i_1(t)$ for $t > 0$.

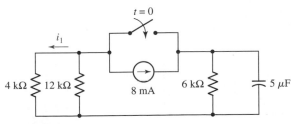

■ **FIGURE 8.76**

40. A long time after the circuit of Fig. 8.77 was assembled, both switches are opened simultaneously at $t = 0$, as indicated. (*a*) Obtain an expression for v_{out} for $t > 0$. (*b*) Obtain values for v_{out} at $t = 0^+$, 1 μs, and 5 μs.

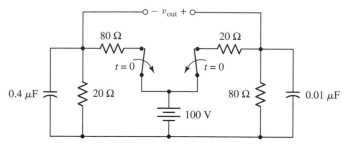

■ **FIGURE 8.77**

41. (*a*) Assume that the circuit shown in Fig. 8.78 has been in the form shown for a very long time. Find $v_C(t)$ for all t after the switch opens. (*b*) Compute $v_C(t)$ at $t = 3$ μs. (*c*) Verify your solution with PSpice.

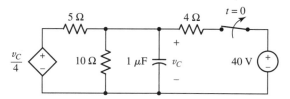

■ **FIGURE 8.78**

42. Determine values for R_0 and R_1 in the circuit of Fig. 8.79 so that $v_C = 50$ V at $t = 0.5$ ms and $v_C = 25$ V at $t = 2$ ms.

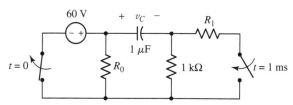

■ **FIGURE 8.79**

43. For the circuit shown in Fig. 8.80, determine $v_C(t)$ for (*a*) $t < 0$; (*b*) $t > 0$.

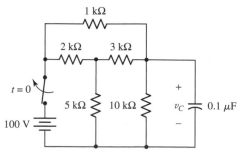

■ **FIGURE 8.80**

44. Find $i_1(t)$ for $t < 0$ and $t > 0$ in the circuit of Fig. 8.81.

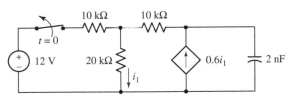

■ **FIGURE 8.81**

45. The switch in Fig. 8.82 is moved from A to B at $t = 0$ after being at A for a long time. This places the two capacitors in series, thus allowing equal and opposite dc voltages to be trapped on the capacitors. (a) Determine $v_1(0^-)$, $v_2(0^-)$, and $v_R(0^-)$. (b) Find $v_1(0^+)$, $v_2(0^+)$, and $v_R(0^+)$. (c) Determine the time constant of $v_R(t)$. (d) Find $v_R(t)$, $t > 0$. (e) Find $i(t)$. (f) Find $v_1(t)$ and $v_2(t)$ from $i(t)$ and the initial values. (g) Show that the stored energy at $t = \infty$ plus the total energy dissipated in the 20 kΩ resistor is equal to the energy stored in the capacitors at $t = 0$.

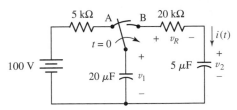

■ **FIGURE 8.82**

46. The value of i_s in the circuit of Fig. 8.83 is 1 mA for $t < 0$, and zero for $t > 0$. Find $v_x(t)$ for (a) $t < 0$; (b) $t > 0$.

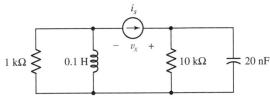

■ **FIGURE 8.83**

47. The value of v_s in the circuit of Fig. 8.84 is 20 V for $t < 0$, and zero for $t > 0$. Find $i_x(t)$ for (a) $t < 0$; (b) $t > 0$.

48. The switch in Fig. 8.85 has been closed for several hours, maybe longer. The fuse is a special type of resistor that overheats and melts if the current through it exceeds 1 A for more than 100 ms (other types of fuses are also available). The resistance of the fuse is 3 mΩ. If the switch is opened at $t = 0$, will the fuse blow? Verify your answer with PSpice.

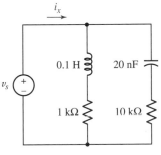

■ **FIGURE 8.84**

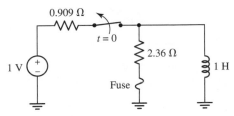

■ **FIGURE 8.85**

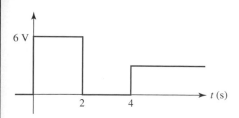

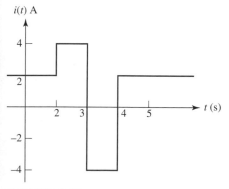

■ **FIGURE 8.86**

■ **FIGURE 8.87**

8.5 The Unit-Step Function

49. Using unit-step functions, create an expression that describes the waveform shown in Fig. 8.86.

50. Create an expression using unit-step functions that describes the waveform shown in Fig. 8.87.

51. Given the function $f(t) = 6u(-t) + 6u(t + 1) - 3u(t + 2)$, evaluate $f(t)$ at $t =$ (a) -1; (b) 0^-; (c) 0^+; (d) 1.5; (e) 3.

52. Given the function $g(t) = 9u(t) - 6u(t + 10) + 3u(t + 12)$, evaluate $g(t)$ at $t =$ (a) -1; (b) 0^+; (c) 5; (d) 11; (e) 30.

53. The source values in the circuit of Fig. 8.88 are $v_A = 300u(t - 1)$ V, $v_B = -120u(t + 1)$ V, and $i_C = 3u(-t)$ A. Find i_1 at $t = -1.5, -0.5, 0.5,$ and 1.5 s.

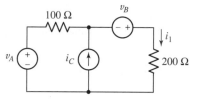

■ **FIGURE 8.88**

54. Source values for Fig. 8.88 are $v_A = 600tu(t + 1)$ V, $v_B = 600(t + 1)u(t)$ V, and $i_C = 6(t - 1)u(t - 1)$ A. (a) Find i_1 at $t = -1.5, -0.5, 0.5,$ and 1.5 s. (b) Sketch i_1 versus t, $-2.5 < t < 2.5$ s.

55. At $t = 2$, find the value of (a) $2u(1 - t) - 3u(t - 1) - 4u(t + 1)$; (b) $[5 - u(t)][2 + u(3 - t)][1 - u(1 - t)]$; (c) $4e^{-u(3-t)}u(3 - t)$.

56. Find i_x for $t < 0$ and for $t > 0$ in the circuit of Fig. 8.89 if the unknown branch contains: (a) a normally open switch that closes at $t = 0$, in series with a 60 V battery, + reference at top; (b) a voltage source, $60u(t)$ V, + reference at top.

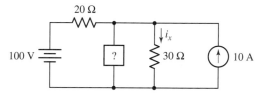

■ **FIGURE 8.89**

57. Find i_x in the circuit of Fig. 8.90 at 1 s intervals from $t = -0.5$ s to $t = 3.5$ s.

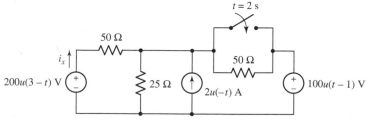

■ **FIGURE 8.90**

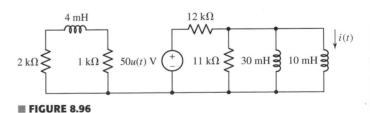

58. The switch in Fig. 8.91 is at position A for $t < 0$. At $t = 0$ it moves to B, and then it moves to C at $t = 4$ s and on to D at $t = 6$ s, where it remains. Sketch $v(t)$ as a function of time and express it as a sum of step forcing functions.

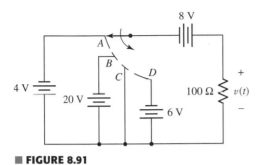

■ **FIGURE 8.91**

59. A voltage waveform appearing across an unknown element is given as $7u(t) - 0.2u(t) + 8u(t - 2) + 3$ V. (a) Determine the voltage at $t = 1$ s. (b) If the corresponding current through the element is $3.5u(t) - 0.1u(t) + 4u(t - 2) + 1.5$ A, what type of element is it, and what is its value?

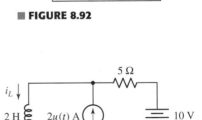

■ **FIGURE 8.92**

8.6 Driven *RL* Circuits

 60. For the circuit of Fig. 8.92, (a) find an expression for $v_R(t)$ valid for all time; (b) compute v_R at $t = 2$ ms; (c) verify your solution to part (b) using PSpice.

61. Refer to the circuit shown in Fig. 8.93 and (a) find $i_L(t)$; (b) use the expression for $i_L(t)$ to find $v_L(t)$.

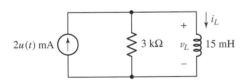

■ **FIGURE 8.93**

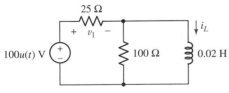

■ **FIGURE 8.94**

62. Find i_L in the circuit of Fig. 8.94 at t equal to (a) -0.5 s; (b) 0.5 s; (c) 1.5 s.

63. With reference to the circuit shown in Fig. 8.95, obtain an algebraic expression for and also sketch: (a) $i_L(t)$; (b) $v_1(t)$.

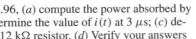

■ **FIGURE 8.95**

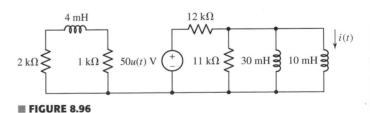

■ **FIGURE 8.96**

 64. With reference to the circuit of Fig. 8.96, (a) compute the power absorbed by the 2 kΩ resistor at $t = 1$ ms; (b) determine the value of $i(t)$ at 3 μs; (c) determine the peak current through the 12 kΩ resistor. (d) Verify your answers with PSpice.

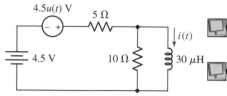

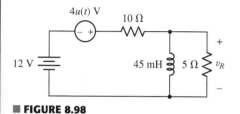

FIGURE 8.97

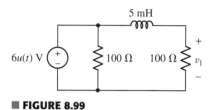

FIGURE 8.98

8.7 Natural and Forced Response

65. For the circuit depicted in Fig. 8.97, (*a*) find an expression for $i(t)$ valid for all time; (*b*) calculate $i(t)$ at $t = 1.5 \ \mu$s; (*c*) verify your result with an appropriate PSpice simulation.

66. For the *RL* circuit of Fig. 8.98, (*a*) find an expression for $v_R(t)$ valid for all time; (*b*) calculate $v_R(t)$ at $t = 2$ ms; (*c*) verify your result with an appropriate PSpice simulation.

67. Referring to the circuit of Fig. 8.99, compute $v_1(t)$ at $t = 27 \ \mu$s.

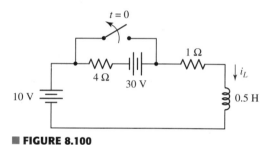

FIGURE 8.99

68. The switch shown in Fig. 8.100 has been closed for a long time. (*a*) Find i_L for $t < 0$. (*b*) Find $i_L(t)$ for all t after the switch opens at $t = 0$.

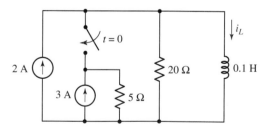

FIGURE 8.100

69. The switch in Fig. 8.101 has been open for a long time. (*a*) Find i_L for $t < 0$. (*b*) Find $i_L(t)$ for all t after the switch closes at $t = 0$.

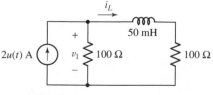

FIGURE 8.101

70. For the circuit shown in Fig. 8.102, find values for i_L and v_1 at t equal to (*a*) 0^-; (*b*) 0^+; (*c*) ∞; (*d*) 0.2 ms.

FIGURE 8.102

71. Equation [33] in Sec. 8.7 represents the general solution of the driven *RL* series circuit, where Q is a function of time in general and A and P are constants. Let $R = 125 \, \Omega$ and $L = 5$ H, and find $i(t)$ for $t > 0$ if the voltage forcing function $L Q(t)$ is (*a*) 10 V; (*b*) $10u(t)$ V; (*c*) $10 + 10u(t)$ V; (*d*) $10u(t) \cos 50t$ V.

72. The switch shown in Fig. 8.103 has been closed for a very long time. (*a*) Find i_L for $t < 0$. (*b*) Just after the switch is opened, find $i_L(0^+)$. (*c*) Find $i_L(\infty)$. (*d*) Derive an expression for $i_L(t)$ for $t > 0$.

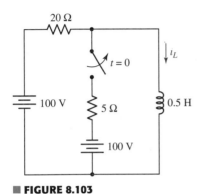

■ **FIGURE 8.103**

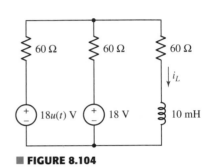

■ **FIGURE 8.104**

73. Find i_L for all t in the circuit of Fig. 8.104.

74. Assume that the switch in Fig. 8.105 has been closed for a long time and then opens at $t = 0$. Find i_x at t equal to (*a*) 0^-; (*b*) 0^+; (*c*) 40 ms.

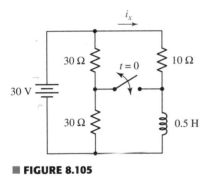

■ **FIGURE 8.105**

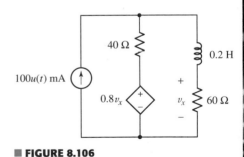

■ **FIGURE 8.106**

75. Assume that the switch in Fig. 8.105 has been open for a long time and then closes at $t = 0$. Find i_x at t equal to (*a*) 0^-; (*b*) 0^+; (*c*) 40 ms.

76. Find $v_x(t)$ for all t in the circuit of Fig. 8.106.

77. With reference to the circuit shown in Fig. 8.107, find (*a*) $i_L(t)$; (*b*) $i_1(t)$.

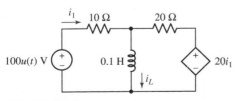

■ **FIGURE 8.107**

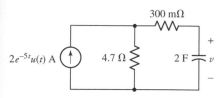

FIGURE 8.108

78. Find an expression for $v(t)$ in the circuit of Fig. 8.108 valid for all time.
79. Find an expression for $v(t)$ in the circuit of Fig. 8.109 valid for all time.

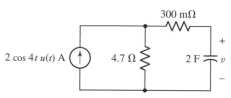

FIGURE 8.109

8.8 Driven *RC* Circuits

 80. (*a*) Find v_C in the circuit shown in Fig. 8.110 at $t = -2\ \mu s$ and $t = +2\ \mu s$. (*b*) Verify your solution with PSpice.

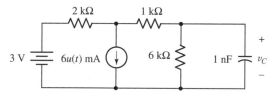

FIGURE 8.110

81. Referring to the *RC* circuit of Fig. 8.111, find an expression for $v_C(t)$ valid for all time.
82. After being closed for a long time, the switch shown in Fig. 8.112 opens at $t = 0$. Find i_A for all time.

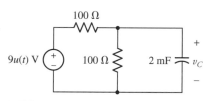

FIGURE 8.111

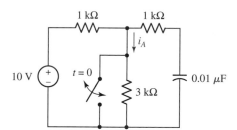

FIGURE 8.112

83. After being open for a long time, the switch shown in Fig. 8.112 closes at $t = 0$. Find i_A for all time.
84. The switch in the circuit of Fig. 8.113 has been open for a long time. It closes suddenly at $t = 0$. Find i_{in} at t equal to (*a*) -1.5 s; (*b*) 1.5 s.
85. Let $v_s = -12u(-t) + 24u(t)$ V in the circuit of Fig. 8.114. Over the time interval -5 ms $< t < 5$ ms, find an algebraic expression for and sketch the quantity (*a*) $v_C(t)$; (*b*) $i_{in}(t)$.

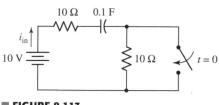

FIGURE 8.113

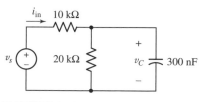

FIGURE 8.114

86. Find v_C for $t > 0$ in the circuit of Fig. 8.115.

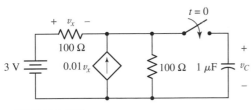

■ **FIGURE 8.115**

87. Find the value of $v_C(t)$ at $t = 0.4$ and 0.8 s in the circuit of Fig. 8.116.

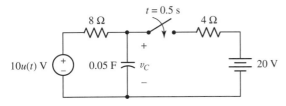

■ **FIGURE 8.116**

88. In the circuit of Fig. 8.117: (a) find $v_C(t)$ for all time, and (b) sketch $v_C(t)$ for $-1 < t < 2$ s. Verify your solution with PSpice.

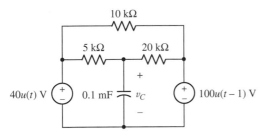

■ **FIGURE 8.117**

89. In the circuit of Fig. 8.118, find $v_R(t)$ for (a) $t < 0$; (b) $t > 0$. Now assume that the switch has been *closed* for a very long time and opens at $t = 0$. Find $v_R(t)$ for (c) $t < 0$; (d) $t > 0$.

90. The switch in Fig. 8.119 has been at A for a long time. It is moved to B at $t = 0$, and back to A at $t = 1$ ms. Find R_1 and R_2 so that $v_C(1\ \text{ms}) = 8$ V and $v_C(2\ \text{ms}) = 1$ V.

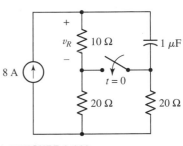

■ **FIGURE 8.118**

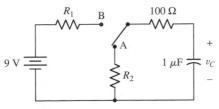

■ **FIGURE 8.119**

91. Find the first instant of time after $t = 0$ at which $v_x = 0$ in the circuit of Fig. 8.120.

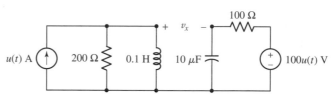

■ **FIGURE 8.120**

92. In the circuit of Fig. 8.121, one switch is opened at $t = 0$, while the other switch is simultaneously closed. Sketch the power absorbed by the 1 kΩ resistor over the interval -1 ms $\leq t \leq 7$ ms. At $t = 0$, the 1 mA source is also turned off.

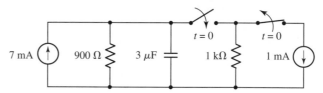

■ **FIGURE 8.121**

93. If the switch in Fig. 8.122 has been closed for several days, (*a*) determine v at $t = 5.45$ ms; (*b*) determine the power dissipated by the 4.7 kΩ resistor at $t = 1.7$ ms; (*c*) determine the total energy that will eventually be converted to heat by the 4.7 kΩ resistor after the switch is opened.

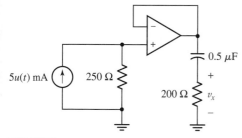

■ **FIGURE 8.123**

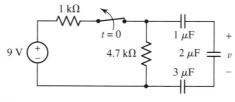

■ **FIGURE 8.122**

94. Assume that the op-amp shown in Fig. 8.123 is ideal, and find $v_x(t)$ for all t.

95. Assume that the op-amp shown in Fig. 8.124 is ideal, and (*a*) find $v_o(t)$ for all t. (*b*) Verify your solution with PSpice. *Hint:* You can plot functions in Probe by entering the expression into the **Trace Expression** box.

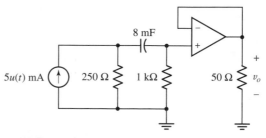

■ **FIGURE 8.124**

96. (*a*) Find $i_L(0)$ for the *RL* circuit of Fig. 8.125. (*b*) Using PSpice and the initial value found in part *a*, determine i_L at $t = 50$ ms.

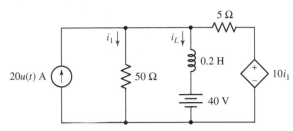

■ **FIGURE 8.125**

97. (*a*) Assume that the op-amp shown in Fig. 8.126 is ideal, and that $v_C(0) = 0$. Find $v_o(t)$ for all *t*. (*b*) Verify your solution with PSpice. Hint: You can plot functions in Probe by entering the expression into the **Trace Expression** box.

D 98. Design a circuit to allow a room light to remain on for 5 seconds after the switch has been turned off. Assume a 40 W light bulb and a 115 V ac supply.

D 99. A motion detector installed as part of a security system appears to be a little too sensitive to electrical power fluctuations. One solution is to insert a delay circuit between the sensor and the alarm circuit, so that false triggers are minimized. Assuming the Thévenin equivalent of the motion sensor is a 2.37 kΩ resistance in series with a 1.5 V source, and the Thévenin equivalent resistance of the alarm circuit is 1 MΩ, design a circuit that can be inserted between the sensor and the circuit that will require a signal from the sensor to last at least 1 full second. The motion sensor/alarm circuit works as follows: The sensor supplies a small current to the alarm circuit continuously unless motion is detected, in which case the current is interrupted.

8.9 Predicting the Response of Sequentially Switched Circuits

100. (*a*) Construct a pulse waveform in PSpice to model the voltage waveform v_B of Exer. 53 and plot it using Probe. (*Hint: Connect the source to a resistor to perform a simulation.*) (*b*) Construct a pulse waveform in PSpice to model the current waveform i_C of Exer. 53 and plot it using Probe.

101. (*a*) Sketch the resistor voltage v_R of the circuit in Fig. 8.127 in response to a pulsed waveform $v_S(t)$. The minimum value of $v_S(t)$ is 0 V, its maximum is 3 V, the pulse width is 2 s, and the period is 5 s. Confine your sketch to $0 \le t < 20$ s. (*b*) Verify your sketch by performing an appropriate PSpice simulation.

102. (*a*) Sketch the inductor current *i* of the circuit in Fig. 8.128 in response to a pulsed waveform $v_S(t)$. The minimum value of $v_S(t)$ is 0 V, its maximum is 5 V, the pulse width is 5 s, and the period is 5.5 s. Confine your sketch to $0 \le t < 20$ s. (*b*) Verify your sketch by performing an appropriate PSpice simulation.

103. The voltage source v_S of Fig. 8.129 is a pulsed source having a minimum value of 2 V, a maximum value of 10 V, and a pulse width of 4 *RC*. Sketch the capacitor voltage if the time between pulses of v_S is (*a*) 0.1 *RC*; (*b*) *RC*; (*c*) 10 *RC*.

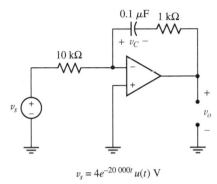

$v_s = 4e^{-20\,000t}\,u(t)$ V

■ **FIGURE 8.126**

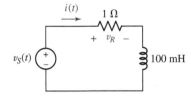

■ **FIGURE 8.127**

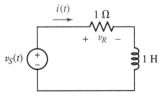

■ **FIGURE 8.128**

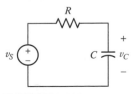

■ **FIGURE 8.129**

104. Referring to the circuit of Fig. 8.130, sketch $i_L(t)$ over the period $0 \le t \le t_4$ if $i(t)$ is as shown in Fig. 8.131.

(*a*) $t_1 = 4$ ns	(*b*) $t_1 = 150$ ns	(*c*) $t_1 = 150$ ns
$t_2 = 160$ ns	$t_2 = 300$ ns	$t_2 = 200$ ns
$t_3 = 164$ ns	$t_3 = 450$ ns	$t_3 = 350$ ns
$t_4 = 200$ ns	$t_4 = 500$ ns	$t_4 = 400$ ns

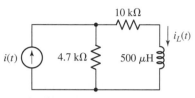

■ **FIGURE 8.130**

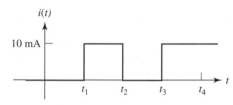

■ **FIGURE 8.131**

The *RLC* Circuit

Resonant Frequency and Damping Factor of Series and Parallel *RLC* Circuits

Overdamped Response

Critically Damped Response

Underdamped Response

Complete (Natural + Forced) Response of *RLC* Circuits

Representing Differential Equations Using Op Amp Circuits

INTRODUCTION

Our discussions in Chap. 8 focused exclusively on resistive circuits with *either* capacitors or inductors, but not both. The presence of inductance and capacitance in the same circuit produces at least a *second-order system,* which is characterized by one linear differential equation that includes a second-order derivative, or by two simultaneous linear first-order differential equations. This increase in order will make it necessary to evaluate two arbitrary constants. Furthermore, it becomes necessary to determine initial conditions for derivatives. We will find that such circuits, often referred to as *RLC* circuits, not only appear fairly often in practice, but also do very well as models for other types of systems. For example, an *RLC* circuit can be used to model the suspension system of an automobile, the behavior of a temperature controller used in growing semiconductor crystals, and even the response of an airplane to elevator and aileron controls.

9.1 • THE SOURCE-FREE PARALLEL CIRCUIT

Our first task is the determination of the natural response, which again is most conveniently done by considering the source-free circuit. We may then include dc sources, switches, or step sources in the circuit, representing the total response once again as the sum of the natural response and the forced response.

   We begin with the determination of the natural response of a simple circuit formed by connecting *R*, *L*, and *C* in parallel. This particular combination of ideal elements is a suitable model for portions of many communications networks. It represents, for example, an important part of some of the electronic amplifiers found in every radio receiver, and it enables the amplifiers to produce a large voltage amplification over a narrow band of signal frequencies with nearly zero amplification outside this band. Frequency selectivity of this kind enables us to listen to the transmission of one station while rejecting the transmission of any other station. Other applications include the use of parallel *RLC* circuits in frequency multiplexing and harmonic-suppression filters. However, even a simple discussion of these principles requires an understanding of such terms as *resonance, frequency response,* and *impedance,* which we have not yet discussed. Let it suffice to say, therefore, that an understanding of the natural behavior of the parallel *RLC* circuit is fundamentally important to future studies of communications networks and filter design, as well as many other applications.

When a *physical* capacitor is connected in parallel with an inductor and the capacitor has associated with it a finite resistance, the resulting network can be shown to have an equivalent circuit model like that shown in Fig. 9.1. The presence of this resistance can be used to model energy loss in the capacitor; over time, all real capacitors will eventually discharge, even if disconnected from a circuit. Energy losses in the physical inductor can also be taken into account by adding an ideal resistor (in series with the ideal inductor). For simplicity, however, we restrict our discussion to the case of an essentially ideal inductor in parallel with a "leaky" capacitor.

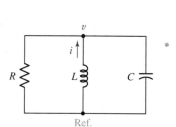

■ FIGURE 9.1 The source-free parallel *RLC* circuit.

Obtaining the Differential Equation for a Parallel *RLC* Circuit

In the following analysis we will assume that energy may be stored initially in both the inductor and the capacitor; in other words, nonzero initial values of both inductor current and capacitor voltage may be present. With reference to the circuit of Fig. 9.1, we may then write the single nodal equation

$$\frac{v}{R} + \frac{1}{L}\int_{t_0}^{t} v\, dt' - i(t_0) + C\frac{dv}{dt} = 0 \tag{1}$$

Note that the minus sign is a consequence of the assumed direction for *i*. We must solve Eq. [1] subject to the initial conditions

$$i(0^+) = I_0 \tag{2}$$

and

$$v(0^+) = V_0 \tag{3}$$

When both sides of Eq. [1] are differentiated once with respect to time, the result is the linear second-order homogeneous differential equation

$$C\frac{d^2v}{dt^2} + \frac{1}{R}\frac{dv}{dt} + \frac{1}{L}v = 0 \tag{4}$$

whose solution $v(t)$ is the desired natural response.

Solution of the Differential Equation

There are a number of interesting ways to solve Eq. [4]. Most of these methods we will leave to a course in differential equations, selecting only the quickest and simplest method to use now. We will assume a solution, relying upon our intuition and modest experience to select one of the several possible forms that are suitable. Our experience with first-order equations might suggest that we at least try the exponential form once more. Thus, we *assume*

$$v = Ae^{st} \tag{5}$$

being as general as possible by allowing A and s to be complex numbers if necessary. Substituting Eq. [5] in Eq. [4], we obtain

$$CAs^2e^{st} + \frac{1}{R}Ase^{st} + \frac{1}{L}Ae^{st} = 0$$

or

$$Ae^{st}\left(Cs^2 + \frac{1}{R}s + \frac{1}{L}\right) = 0$$

In order for this equation to be satisfied for all time, at least one of the three factors must be zero. If either of the first two factors is set equal to zero, then $v(t) = 0$. This is a trivial solution of the differential equation which cannot satisfy our given initial conditions. We therefore equate the remaining factor to zero:

$$Cs^2 + \frac{1}{R}s + \frac{1}{L} = 0 \tag{6}$$

This equation is usually called the *auxiliary equation* or the **characteristic equation,** as we discussed in Sec. 8.1. If it can be satisfied, then our assumed solution is correct. Since Eq. [6] is a quadratic equation, there are two solutions, identified as s_1 and s_2:

$$s_1 = -\frac{1}{2RC} + \sqrt{\left(\frac{1}{2RC}\right)^2 - \frac{1}{LC}} \tag{7}$$

and

$$s_2 = -\frac{1}{2RC} - \sqrt{\left(\frac{1}{2RC}\right)^2 - \frac{1}{LC}} \tag{8}$$

If *either* of these two values is used for s in the assumed solution, then that solution satisfies the given differential equation; it thus becomes a valid solution of the differential equation.

Let us assume that we replace s by s_1 in Eq. [5], obtaining

$$v_1 = A_1e^{s_1t}$$

and, similarly,

$$v_2 = A_2e^{s_2t}$$

The former satisfies the differential equation

$$C\frac{d^2v_1}{dt^2} + \frac{1}{R}\frac{dv_1}{dt} + \frac{1}{L}v_1 = 0$$

and the latter satisfies

$$C\frac{d^2v_2}{dt^2} + \frac{1}{R}\frac{dv_2}{dt} + \frac{1}{L}v_2 = 0$$

Adding these two differential equations and combining similar terms, we have

$$C\frac{d^2(v_1+v_2)}{dt^2} + \frac{1}{R}\frac{d(v_1+v_2)}{dt} + \frac{1}{L}(v_1+v_2) = 00$$

Linearity triumphs, and it is seen that the *sum* of the two solutions is also a solution. We thus have the general form of the natural response

$$v(t) = A_1 e^{s_1 t} + A_2 e^{s_2 t} \qquad [9]$$

where s_1 and s_2 are given by Eqs. [7] and [8]; A_1 and A_2 are two arbitrary constants which are to be selected to satisfy the two specified initial conditions.

Definition of Frequency Terms

The form of the natural response as given in Eq. [9] offers little insight into the nature of the curve we might obtain if $v(t)$ were plotted as a function of time. The relative amplitudes of A_1 and A_2, for example, will certainly be important in determining the shape of the response curve. Furthermore, the constants s_1 and s_2 can be real numbers or conjugate complex numbers, depending upon the values of R, L, and C in the given network. These two cases will produce fundamentally different response forms. Therefore, it will be helpful to make some simplifying substitutions in Eq. [9].

Since the exponents $s_1 t$ and $s_2 t$ must be dimensionless, s_1 and s_2 must have the unit of some dimensionless quantity "per second." From Eqs. [7] and [8] we therefore see that the units of $1/2RC$ and $1/\sqrt{LC}$ must also be s^{-1} (i.e., seconds^{-1}). Units of this type are called *frequencies.*

Let us define a new term, ω_0 (omega-sub-zero, or just omega-zero):

$$\omega_0 = \frac{1}{\sqrt{LC}} \qquad [10]$$

and reserve the term *resonant frequency* for it. On the other hand, we will call $1/2RC$ the **neper frequency,** or the **exponential damping coefficient,** and represent it by the symbol α (alpha):

$$\alpha = \frac{1}{2RC} \qquad [11]$$

This latter descriptive expression is used because α is a measure of how rapidly the natural response decays or damps out to its steady, final value (usually zero). Finally, s, s_1, and s_2, which are quantities that will form the basis for some of our later work, are called **complex frequencies.**

We should note that s_1, s_2, α, and ω_0 are merely symbols used to simplify the discussion of *RLC* circuits; they are not mysterious new properties of any kind. It is easier, for example, to say "*alpha*" than it is to say "*the reciprocal of 2RC.*"

Let us collect these results. The natural response of the parallel *RLC* circuit is

$$v(t) = A_1 e^{s_1 t} + A_2 e^{s_2 t} \qquad [9]$$

where

$$s_1 = -\alpha + \sqrt{\alpha^2 - \omega_0^2} \qquad [12]$$

$$s_2 = -\alpha - \sqrt{\alpha^2 - \omega_0^2} \qquad [13]$$

$$\alpha = \frac{1}{2RC} \tag{11}$$

$$\omega_0 = \frac{1}{\sqrt{LC}} \tag{10}$$

> The ratio of α to ω_0 is called the *damping ratio* by control system engineers and is designated by ζ (zeta).

and A_1 and A_2 must be found by applying the given initial conditions.

We note two basic scenarios possible with Eqs. [12] and [13] depending on the relative sizes of α and ω_0 (dictated by the values of R, L, and C). If $\alpha > \omega_0$, s_1 and s_2 will both be real numbers, leading to what is referred to as an ***overdamped response.*** In the opposite case, where $\alpha < \omega_0$, both s_1 and s_2 will have nonzero imaginary components, leading to what is known as an ***underdamped response.*** Both of these situations are considered separately in the following sections, along with the special case of $\alpha = \omega_0$, which leads to what is called a ***critically damped response.*** We should also note that the general response comprised by Eqs. [9] through [13] describes not only the voltage but all three branch currents in the parallel RLC circuit; the constants A_1 and A_2 will be different for each, of course.

EXAMPLE 9.1

Consider a parallel *RLC* circuit having an inductance of 10 mH and a capacitance of 100 μF. Determine the resistor values that would lead to overdamped and underdamped responses.

We first calculate the resonant frequency of the circuit:

$$\omega_0 = \sqrt{\frac{1}{LC}} = \sqrt{\frac{1}{(10 \times 10^{-3})(100 \times 10^{-6})}} = 10^3 \text{ rad/s}$$

An *overdamped* response will result if $\alpha > \omega_0$; an *underdamped* response will result if $\alpha < \omega_0$. Thus,

$$\frac{1}{2RC} > 10^3$$

and so

$$R < \frac{1}{(2000)(100 \times 10^{-6})}$$

or

$$R < 5 \, \Omega$$

leads to an overdamped response; $R > 5 \, \Omega$ leads to an underdamped response.

PRACTICE

9.1 A parallel *RLC* circuit contains a 100 Ω resistor and has the parameter values $\alpha = 1000$ s^{-1} and $\omega_0 = 800$ rad/s. Find: (*a*) C; (*b*) L; (*c*) s_1; (*d*) s_2.

Ans: 5 μF; 312.5 mH; -400 s^{-1}; -1600 s^{-1}.

9.2 THE OVERDAMPED PARALLEL *RLC* CIRCUIT

A comparison of Eqs. [10] and [11] shows that α will be greater than ω_0 if $LC > 4R^2C^2$. In this case the radical used in calculating s_1 and s_2 will be real, and both s_1 and s_2 will be real. Moreover, the following inequalities,

$$\sqrt{\alpha^2 - \omega_0^2} < \alpha$$

$$\left(-\alpha - \sqrt{\alpha^2 - \omega_0^2}\right) < \left(-\alpha + \sqrt{\alpha^2 - \omega_0^2}\right) < 0$$

may be applied to Eqs. [12] and [13] to show that both s_1 and s_2 are *negative* real numbers. Thus, the response $v(t)$ can be expressed as the (algebraic) sum of two decreasing exponential terms, both of which approach zero as time increases. In fact, since the absolute value of s_2 is larger than that of s_1, the term containing s_2 has the more rapid rate of decrease and, for large values of time, we may write the limiting expression

$$v(t) \to A_1 e^{s_1 t} \to 0 \qquad \text{as } t \to \infty$$

The next step is to determine the arbitrary constants A_1 and A_2 in conformance with the initial conditions. We select a parallel *RLC* circuit with $R = 6\,\Omega$, $L = 7$ H, and, for ease of computation, $C = \frac{1}{42}$ F. The initial energy storage is specified by choosing an initial voltage across the circuit $v(0) = 0$ and an initial inductor current $i(0) = 10$ A, where v and i are defined in Fig. 9.2.

We may easily determine the values of the several parameters

$$\begin{array}{cc} \alpha = 3.5 & \omega_0 = \sqrt{6} \\ s_1 = -1 & s_2 = -6 \end{array} \quad (\text{all } s^{-1})$$

and immediately write the general form of the natural response

$$v(t) = A_1 e^{-t} + A_2 e^{-6t} \qquad [14]$$

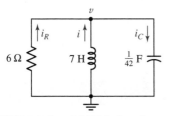

■ **FIGURE 9.2** A parallel *RLC* circuit used as a numerical example. The circuit is overdamped.

Finding Values for A_1 and A_2

Only the evaluation of the two constants A_1 and A_2 remains. If we knew the response $v(t)$ at two different values of time, these two values could be substituted in Eq. [14] and A_1 and A_2 easily found. However, we know only one instantaneous value of $v(t)$,

$$v(0) = 0$$

and, therefore,

$$0 = A_1 + A_2 \qquad [15]$$

We can obtain a second equation relating A_1 and A_2 by taking the derivative of $v(t)$ with respect to time in Eq. [14], determining the initial value of this derivative through the use of the remaining initial condition $i(0) = 10$, and equating the results. So, taking the derivative of both sides of Eq. [14],

$$\frac{dv}{dt} = -A_1 e^{-t} - 6A_2 e^{-6t}$$

and evaluating the derivative at $t = 0$,

$$\left.\frac{dv}{dt}\right|_{t=0} = -A_1 - 6A_2$$

we obtain a second equation. Although this may appear to be helpful, we do not have a numerical value for the initial value of the derivative, so we do not yet have two equations in two unknowns ... Or do we? The expression dv/dt suggests a capacitor current, since

$$i_C = C\frac{dv}{dt}$$

Kirchhoff's current law must hold at any instant in time, as it is based on conservation of electrons. Thus, we may write

$$-i_C(0) + i(0) + i_R(0) = 0$$

Substituting our expression for capacitor current and dividing by C,

$$\left.\frac{dv}{dt}\right|_{t=0} = \frac{i_C(0)}{C} = \frac{i(0) + i_R(0)}{C} = \frac{i(0)}{C} = 420 \text{ V/s}$$

since zero initial voltage across the resistor requires zero initial current through it. We thus have our second equation,

$$420 = -A_1 - 6A_2 \qquad [16]$$

and simultaneous solution of Eqs. [15] and [16] provides the two amplitudes $A_1 = 84$ and $A_2 = -84$. Therefore, the final numerical solution for the natural response of this circuit is

$$v(t) = 84(e^{-t} - e^{-6t}) \text{ V} \qquad [17]$$

For the remainder of our discussions concerning *RLC* circuits, we will always require two initial conditions in order to completely specify the response. One condition will usually be very easy to apply—either a voltage or current at $t = 0$. It is the second condition that usually gives us a little trouble. Although we will often have both an initial current and an initial voltage at our disposal, one of these will need to be applied indirectly through the derivative of our assumed solution.

EXAMPLE 9.2

Find an expression for $v_C(t)$ valid for $t > 0$ in the circuit of Fig. 9.3a.

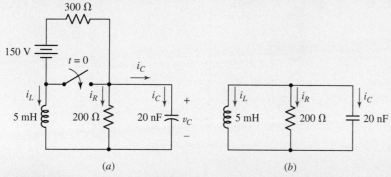

(a) (b)

■ **FIGURE 9.3** (a) An *RLC* circuit that becomes source-free at $t = 0$; (b) The circuit for $t > 0$, in which the 150 V source and the 300 Ω resistor have been shorted out by the switch, and so are of no further relevance to v_C.

▶ **Identify the goal of the problem.**

We are asked to find the capacitor voltage after the switch is thrown. This action leads to no sources remaining connected to either the inductor or the capacitor; we therefore expect v_C to decay with time.

(Continued on next page)

▶ **Collect the known information.**
After the switch is thrown, the capacitor is left in parallel with a 200 Ω resistor and a 5 mH inductor (Fig. 9.3*b*). Thus, $\alpha = 1/2RC = 125{,}000 \text{ s}^{-1}$, $\omega_0 = 1/\sqrt{LC} = 100{,}000 \text{ rad/s}$, $s_1 = -\alpha + \sqrt{\alpha^2 - \omega_0^2} = -50{,}000 \text{ s}^{-1}$ and $s_2 = -\alpha - \sqrt{\alpha^2 - \omega_0^2} = -200{,}000 \text{ s}^{-1}$.

▶ **Devise a plan.**
Since $\alpha > \omega_0$, the circuit is overdamped and so we expect a capacitor voltage of the form

$$v_C(t) = A_1 e^{s_1 t} + A_2 e^{s_2 t}$$

We know s_1 and s_2; we need to obtain and invoke two initial conditions to determine A_1 and A_2. To do this, we will analyze the circuit at $t = 0^-$ (Fig. 9.4*a*) to find $i_L(0^-)$ and $v_C(0^-)$. We will then analyze the circuit at $t = 0^+$ with the assumption that neither value changes.

▶ **Construct an appropriate set of equations.**
From Fig. 9.4*a*, in which the inductor has been replaced with a short circuit and the capacitor with an open circuit, we see that

$$i_L(0^-) = -\frac{150}{200 + 300} = -300 \text{ mA}$$

and

$$v_C(0^-) = 150\frac{200}{200 + 300} = 60 \text{ V}$$

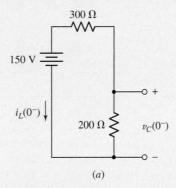

(*a*)

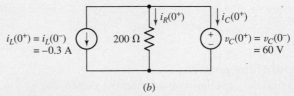

(*b*)

■ **FIGURE 9.4** (*a*) The equivalent circuit at $t = 0^-$; (*b*) Equivalent circuit at $t = 0^+$, drawn using ideal sources to represent the initial inductor current and initial capacitor voltage.

In Fig. 9.4*b*, we draw the circuit at $t = 0^+$, representing the inductor current and capacitor voltage by ideal sources for simplicity. Since neither can change in zero time, we know that $v_C(0^+) = 60$ V.

▶ **Determine if additional information is required.**
We have an equation for the capacitor voltage: $v_C(t) = A_1 e^{-50,000t} + A_2 e^{-200,000t}$. We now know $v_C(0) = 60$ V, but a third equation is still required. Differentiating our capacitor voltage equation,

$$\frac{dv_C}{dt} = -50,000 A_1 e^{-50,000t} - 200,000 A_2 e^{-200,000t}$$

which can be related to the capacitor current as $i_C = C(dv_C/dt)$. Returning to Fig. 9.4*b*, KCL yields that $i_C(0^+) = -i_L(0^+) - i_R(0^+) = 0.3 - \{v_C(0^+)/200\} = 0$.

▶ **Attempt a solution.**
Application of our first initial condition yields

$$v_C(0) = A_1 + A_2 = 60$$

and application of our second initial condition yields

$$i_C(0) = -20 \times 10^{-9}(50,000 A_1 + 200,000 A_2) = 0$$

Solving, $A_1 = 80$ V and $A_2 = -20$ V, so that $v_C(t) = 80 e^{-50,000t} - 20 e^{-200,000t}$ V, $t > 0$.

▶ **Verify the solution. Is it reasonable or expected?**
At the very least, we can check our solution at $t = 0$, verifying that $v_C(0) = 60$ V. Differentiating and multiplying by 20×10^{-9}, we can also verify that $i_C(0) = 0$.

PRACTICE

9.2 After being open for a long time, the switch in Fig. 9.5 closes at $t = 0$. Find (*a*) $i_L(0^-)$; (*b*) $v_C(0^-)$; (*c*) $i_R(0^+)$; (*d*) $i_C(0^+)$; (*e*) $v_C(0.2)$.

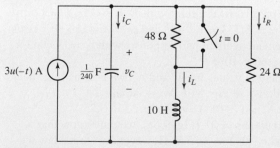

■ **FIGURE 9.5**

Ans: 1 A; 48 V; 2 A; −3 A; −17.54 V.

As noted previously, the form of the overdamped response applies to any voltage or current quantity, as we explore in the following example.

EXAMPLE 9.3

The circuit of Fig. 9.6*a* reduces to a simple parallel *RLC* circuit after $t = 0$. Determine an expression for the resistor current i_R valid for all time.

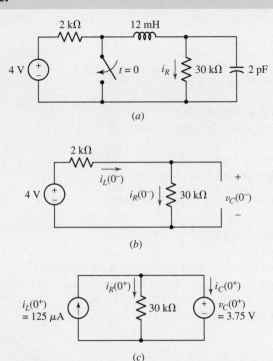

■ **FIGURE 9.6** (*a*) Circuit for which i_R is required. (*b*) Equivalent circuit for $t = 0^-$. (*c*) Equivalent circuit for $t = 0^+$.

If the circuit after $t > 0$ is overdamped, we expect a response of the form

$$i_R(t) = A_1 e^{s_1 t} + A_2 e^{s_2 t}, \qquad t > 0 \qquad [18]$$

For $t > 0$, we have a parallel *RLC* circuit with $R = 30$ kΩ, $L = 12$ mH, and $C = 2$ pF. Thus, $\alpha = 8.333 \times 10^6$ s^{-1} and $\omega_0 = 6.455 \times 10^6$ rad/s. We do therefore expect an overdamped response, with $s_1 = -3.063 \times 10^6$ s^{-1} and $s_2 = -13.60 \times 10^6$ s^{-1}.

To determine numerical values for A_1 and A_2, we first analyze the circuit at $t = 0^-$, as drawn in Fig. 9.6*b*. We see that $i_L(0^-) = i_R(0^-) = 4/32 \times 10^3 = 125$ μA, and $v_C(0^-) = 4 \times 30/32 = 3.75$ V.

In drawing the circuit at $t = 0^+$ (Fig. 9.6*c*), we only know that $i_L(0^+) = 125$ μA and $v_C(0^+) = 3.75$ V. However, by Ohm's law we can calculate that $i_R(0^+) = 3.75/30 \times 10^3 = 125$ μA, our first initial condition. Thus,

$$i_R(0) = A_1 + A_2 = 125 \times 10^{-6} \qquad [19]$$

How do we obtain a *second* initial condition? If we multiply Eq. [18] by 30×10^3, we obtain an expression for $v_C(t)$. Taking the derivative and multiplying by 2 pF yields an expression for $i_C(t)$:

$$i_C = C\frac{dv_C}{dt} = (2 \times 10^{-12})(30 \times 10^3)(A_1 s_1 e^{s_1 t} + A_2 s_2 e^{s_2 t})$$

By KCL,

$$i_C(0^+) = i_L(0^+) - i_R(0^+) = 0$$

Thus,

$$-(2 \times 10^{-12})(30 \times 10^3)(3.063 \times 10^6 A_1 + 13.60 \times 10^6 A_2) = 0 \quad [20]$$

Solving Eqs. [19] and [20], we find that $A_1 = 161.3 \ \mu A$ and $A_2 = -36.34 \ \mu A$. Thus,

$$i_R = \begin{cases} 125 \ \mu A, & t < 0 \\ 161.3 e^{-3.063 \times 10^6 t} - 36.34 e^{-13.6 \times 10^6 t} \ \mu A, & t > 0 \end{cases}$$

PRACTICE

9.3 Determine the current i_R through the resistor of Fig. 9.7 for $t > 0$ if $i_L(0^-) = 6$ A and $v_C(0^+) = 0$ V. The configuration of the circuit prior to $t = 0$ is not known.

Ans: $i_R(t) = 6.008(e^{-8.328 \times 10^{10} t} - e^{-6.003 \times 10^7 t})$ A, $t > 0$.

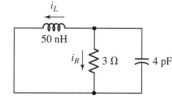

■ **FIGURE 9.7** Circuit for Practice Problem 9.3.

Graphical Representation of the Overdamped Response

Now let us return to Eq. [17] and see what additional information we can determine about this circuit. We may interpret the first exponential term as having a time constant of 1 s and the other exponential, a time constant of $\frac{1}{6}$ s. Each starts with unity amplitude, but the latter decays more rapidly; $v(t)$ is never negative. As time becomes infinite, each term approaches zero, and the response itself dies out as it should. We therefore have a response curve which is zero at $t = 0$, is zero at $t = \infty$, and is never negative; since it is not everywhere zero, it must possess at least one maximum, and this is not a difficult point to determine exactly. We differentiate the response

$$\frac{dv}{dt} = 84(-e^{-t} + 6e^{-6t})$$

set the derivative equal to zero to determine the time t_m at which the voltage becomes maximum,

$$0 = -e^{-t_m} + 6e^{-6t_m}$$

manipulate once,

$$e^{5t_m} = 6$$

and obtain

$$t_m = 0.358 \text{ s}$$

and

$$v(t_m) = 48.9 \text{ V}$$

A reasonable sketch of the response may be made by plotting the two exponential terms $84e^{-t}$ and $84e^{-6t}$ and then taking their difference. The usefulness of this technique is indicated by the curves of Fig. 9.8; the two exponentials are shown lightly, and their difference, the total response $v(t)$, is drawn as a colored line. The curves also verify our previous prediction that the functional behavior of $v(t)$ for very large t is $84e^{-t}$, the exponential term containing the smaller magnitude of s_1 and s_2.

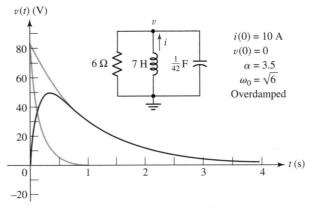

■ **FIGURE 9.8** The response $v(t) = 84(e^{-t} - e^{-6t})$ of the network shown in Fig. 9.2.

A frequently asked question is the length of time it actually takes for the transient part of the response to disappear (or "*damp out*"). In practice, it is often desirable to have this transient response approach zero as rapidly as possible, that is, to minimize the **settling time** t_s. Theoretically, of course, t_s is infinite, because $v(t)$ never settles to zero in a finite time. However, a negligible response is present after the magnitude of $v(t)$ has settled to values that remain less than 1 percent of its maximum absolute value $|v_m|$. The time that is required for this to occur we define as the settling time. Since $|v_m| = v_m = 48.9$ V for our example, the settling time is the time required for the response to drop to 0.489 V. Substituting this value for $v(t)$ in Eq. [17] and neglecting the second exponential term, known to be negligible here, the settling time is found to be 5.15 s.

EXAMPLE **9.4**

For $t > 0$, the capacitor current of a certain source-free parallel *RLC* circuit is given by $i_C(t) = 2e^{-2t} - 4e^{-t}$ A. Sketch the current in the range $0 < t < 5$ s, and determine the settling time.

We first sketch the two terms as shown in Fig. 9.9, then subtract them to find $i_C(t)$. The maximum value is clearly $|-2| = 2$ A. We therefore need to find the time at which $|i_C|$ has decreased to 20 mA, or

$$2e^{-2t_s} - 4e^{-t_s} = -0.02 \qquad [21]$$

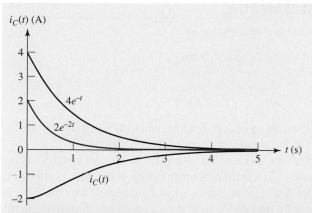

■ **FIGURE 9.9** The current response $i_C(t) = 2e^{-2t} - 4e^{-t}$ A, sketched alongside its two components.

This equation can be solved using an iterative solver routine on a scientific calculator, which returns the solution $t_s = 5.296$ s. If such an option is not available, however, we can approximate Eq. [21] for $t \geq t_s$ as

$$-4e^{-t_s} = -0.02 \qquad [22]$$

Solving,

$$t_s = -\ln\left(\frac{0.02}{4}\right) = 5.298 \text{ s} \qquad [23]$$

which is reasonably close (better than 0.1% accuracy) to the exact solution.

PRACTICE

9.4 (*a*) Sketch the voltage $v_R(t) = 2e^{-t} - 4e^{-3t}$ V in the range $0 < t < 5$ s; (*b*) Estimate the settling time; (*c*) Calculate the maximum positive value and the time at which it occurs.

Ans: (*a*) See Fig. 9.10; (*b*) 4.605 s; (*c*) 544 mV, 896 ms.

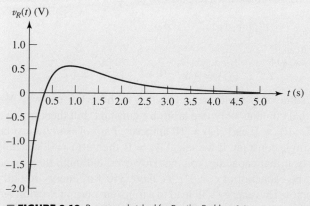

■ **FIGURE 9.10** Response sketched for Practice Problem 9.4*a*.

9.3 CRITICAL DAMPING

The overdamped case is characterized by

$$\alpha > \omega_0$$

or

$$LC > 4R^2C^2$$

and leads to negative real values for s_1 and s_2 and to a response expressed as the algebraic sum of two negative exponentials.

Now let us adjust the element values until α and ω_0 are equal. This is a very special case which is termed **critical damping.** If we were to attempt to construct a parallel *RLC* circuit that is critically damped, we would be attempting an essentially impossible task, for we could never make α exactly equal to ω_0. For completeness, however, we will discuss the critically damped circuit here, because it shows an interesting transition between overdamping and underdamping.

Critical damping is achieved when

or $\left. \begin{array}{l} \alpha = \omega_0 \\ LC = 4R^2C^2 \\ L = 4R^2C \end{array} \right\}$ critical damping

We can produce critical damping by changing the value of any of the three elements in the numerical example discussed at the end of Sec. 9.1. We will select R, increasing its value until critical damping is obtained, and thus leave ω_0 unchanged. The necessary value of R is $7\sqrt{6}/2$ Ω; L is still 7 H, and C remains $\frac{1}{42}$ F. We thus find

$$\alpha = \omega_0 = \sqrt{6} \text{ s}^{-1}$$
$$s_1 = s_2 = -\sqrt{6} \text{ s}^{-1}$$

and recall the initial conditions that were specified, $v(0) = 0$ and $i(0) = 10$ A.

Form of a Critically Damped Response

We proceed to attempt to construct a response as the sum of two exponentials,

$$v(t) \overset{?}{=} A_1 e^{-\sqrt{6}t} + A_2 e^{-\sqrt{6}t}$$

which may be written as

$$v(t) \overset{?}{=} A_3 e^{-\sqrt{6}t}$$

At this point, some of us might be feeling that we have lost our way. We have a response that contains only one arbitrary constant, but there are two initial conditions, $v(0) = 0$ and $i(0) = 10$ amperes, *both of which* must be satisfied by this single constant. If we select $A_3 = 0$, then $v(t) = 0$, which is consistent with our initial capacitor voltage. However, although there is no energy stored in the capacitor at $t = 0^+$, we have 350 J of energy initially stored in the inductor. This energy will lead to a transient current flowing out of the inductor, giving rise to a nonzero voltage across all three elements. This seems to be in direct conflict with our proposed solution.

"Impossible" is a pretty strong term. We make this statement because in practice it is unusual to obtain components that are closer than 1 percent of their specified values. Thus, obtaining *L* precisely equal to 4*R*²*C* is theoretically possible, but not very likely, even if we're willing to measure a drawer full of components until we find the right ones.

Our mathematics and our electricity have been unimpeachable; therefore, if a mistake has not led to our difficulties, we must have begun with an incorrect assumption, and only one assumption has been made. We originally hypothesized that the differential equation could be solved by assuming an exponential solution, and this turns out to be incorrect for this single special case of critical damping. When $\alpha = \omega_0$, the differential equation, Eq. [4], becomes

$$\frac{d^2v}{dt^2} + 2\alpha\frac{dv}{dt} + \alpha^2 v = 0$$

The solution of this equation is not a tremendously difficult process, but we will avoid developing it here, since the equation is a standard type found in the usual differential-equation texts. The solution is

$$v = e^{-\alpha t}(A_1 t + A_2) \qquad [24]$$

It should be noted that the solution is still expressed as the sum of two terms, where one term is the familiar negative exponential and the second is t times a negative exponential. We should also note that the solution contains the *two* expected arbitrary constants.

Finding Values for A_1 and A_2

Let us now complete our numerical example. After we substitute the known value of α in Eq. [24], obtaining

$$v = A_1 t e^{-\sqrt{6}t} + A_2 e^{-\sqrt{6}t}$$

we establish the values of A_1 and A_2 by first imposing the initial condition on $v(t)$ itself, $v(0) = 0$. Thus, $A_2 = 0$. This simple result occurs because the initial value of the response $v(t)$ was selected as zero; the more general case will require the solution of two equations simultaneously. The second initial condition must be applied to the derivative dv/dt just as in the overdamped case. We therefore differentiate, remembering that $A_2 = 0$:

$$\frac{dv}{dt} = A_1 t(-\sqrt{6})e^{-\sqrt{6}t} + A_1 e^{-\sqrt{6}t}$$

evaluate at $t = 0$:

$$\left.\frac{dv}{dt}\right|_{t=0} = A_1$$

and express the derivative in terms of the initial capacitor current:

$$\left.\frac{dv}{dt}\right|_{t=0} = \frac{i_C(0)}{C} = \frac{i_R(0)}{C} + \frac{i(0)}{C}$$

where reference directions for i_C, i_R, and i are defined in Fig. 9.2. Thus,

$$A_1 = 420 \text{ V}$$

The response is, therefore,

$$v(t) = 420te^{-2.45t} \text{ V} \qquad [25]$$

Graphical Representation of the Critically Damped Response

Before plotting this response in detail, let us again try to anticipate its form by qualitative reasoning. The specified initial value is zero, and Eq. [25] concurs. It is not immediately apparent that the response also approaches zero as t becomes infinitely large, because $te^{-2.45t}$ is an indeterminate form. However, this obstacle is easily overcome by use of L'Hôpital's rule, which yields

$$\lim_{t \to \infty} v(t) = 420 \lim_{t \to \infty} \frac{t}{e^{2.45t}} = 420 \lim_{t \to \infty} \frac{1}{2.45e^{2.45t}} = 0$$

and once again we have a response that begins and ends at zero and has positive values at all other times. A maximum value v_m again occurs at time t_m; for our example,

$$t_m = 0.408 \text{ s} \qquad \text{and} \qquad v_m = 63.1 \text{ V}$$

This maximum is larger than that obtained in the overdamped case, and is a result of the smaller losses that occur in the larger resistor; the time of the maximum response is slightly later than it was with overdamping. The settling time may also be determined by solving

$$\frac{v_m}{100} = 420 t_s e^{-2.45 t_s}$$

for t_s (by trial-and-error methods or a calculator's SOLVE routine):

$$t_s = 3.12 \text{ s}$$

which is a considerably smaller value than that which arose in the overdamped case (5.15 s). As a matter of fact, it can be shown that, for given values of L and C, the selection of that value of R which provides critical damping will always give a shorter settling time than any choice of R that produces an overdamped response. However, a slight improvement (reduction) in settling time may be obtained by a further slight increase in resistance; a slightly underdamped response that will undershoot the zero axis before it dies out will yield the shortest settling time.

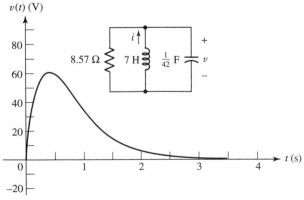

■ **FIGURE 9.11** The response $v(t) = 420te^{-2.45t}$ of the network shown in Fig. 9.2 with R changed to provide critical damping.

The response curve for critical damping is drawn in Fig. 9.11; it may be compared with the overdamped (and underdamped) case by reference to Fig. 9.16.

EXAMPLE **9.5**

Select a value for R_1 such that the circuit of Fig. 9.12 will be characterized by a critically damped response for $t > 0$, and a value for R_2 such that $v(0) = 2$ V.

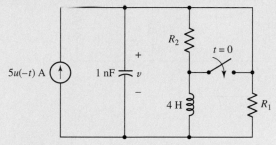

■ **FIGURE 9.12** A circuit that reduces to a parallel *RLC* circuit after the switch is thrown.

We note that at $t = 0^-$, the current source is on, and the inductor can be treated as a short circuit. Thus, $v(0^-)$ appears across R_2, and is given by

$$v(0^-) = 5R_2$$

and a value of 400 mΩ should be selected for R_2 to obtain $v(0) = 2$ V.

After the switch is thrown, the current source has turned itself off and R_2 is shorted. We are left with a parallel *RLC* circuit comprised of R_1, a 4 H inductor, and a 1 nF capacitor.

We may now calculate (for $t > 0$)

$$\alpha = \frac{1}{2RC}$$
$$= \frac{1}{2 \times 10^{-9} R_1}$$

and

$$\omega_0 = \frac{1}{\sqrt{LC}}$$
$$= \frac{1}{\sqrt{4 \times 10^{-9}}}$$
$$= 15,810 \text{ rad/s}$$

Therefore, to establish a critically damped response in the circuit for $t > 0$, we need to set $R_1 = 31.63$ kΩ. (Note: *since we have rounded to four significant figures, the pedantic can rightly argue that this is still not **exactly** a critically damped response—a difficult situation to create.*)

PRACTICE

9.5 (*a*) Choose R_1 in the circuit of Fig. 9.13 so that the response after $t = 0$ will be critically damped. (*b*) Now select R_2 to obtain $v(0) = 100$ V. (*c*) Find $v(t)$ at $t = 1$ ms.

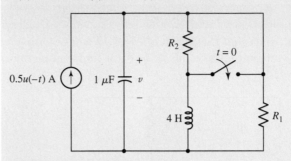

■ **FIGURE 9.13**

Ans: 1 kΩ; 250 Ω; −212 V.

9.4 THE UNDERDAMPED PARALLEL *RLC* CIRCUIT

Let us continue the process begun in Sec. 9.3 by increasing R once more to obtain what we will refer to as an ***underdamped*** response. Thus, the damping coefficient α decreases while ω_0 remains constant, α^2 becomes smaller than ω_0^2, and the radicand appearing in the expressions for s_1 and s_2 becomes negative. This causes the response to take on a much different character, but it is fortunately not necessary to return to the basic differential equation again. By using complex numbers, the exponential response turns into a *damped sinusoidal response;* this response is composed entirely of real quantities, the complex quantities being necessary only for the derivation.[1]

The Form of the Underdamped Response

We begin with the exponential form

$$v(t) = A_1 e^{s_1 t} + A_2 e^{s_2 t}$$

where

$$s_{1,2} = -\alpha \pm \sqrt{\alpha^2 - \omega_0^2}$$

and then let

$$\sqrt{\alpha^2 - \omega_0^2} = \sqrt{-1}\sqrt{\omega_0^2 - \alpha^2} = j\sqrt{\omega_0^2 - \alpha^2}$$

where $j \equiv \sqrt{-1}$.

We now take the new radical, which is real for the underdamped case, and call it ω_d, the ***natural resonant frequency:***

$$\omega_d = \sqrt{\omega_0^2 - \alpha^2}$$

The response may now be written as

$$v(t) = e^{-\alpha t}(A_1 e^{j\omega_d t} + A_2 e^{-j\omega_d t}) \qquad [26]$$

Electrical engineers use "*j*" instead of "*i*" to represent $\sqrt{-1}$ to avoid confusion with currents.

(1) A review of complex numbers is presented in Appendix 5.

or, in the longer but equivalent form,

$$v(t) = e^{-\alpha t} \left\{ (A_1 + A_2) \left[\frac{e^{j\omega_d t} + e^{-j\omega_d t}}{2} \right] + j(A_1 - A_2) \left[\frac{e^{j\omega_d t} - e^{-j\omega_d t}}{j2} \right] \right\}$$

Applying identities described in Appendix 5, the first square bracket in the preceding equation is identically equal to cos $\omega_d t$, and the second is identically sin $\omega_d t$. Hence,

$$v(t) = e^{-\alpha t} \left[(A_1 + A_2) \cos \omega_d t + j(A_1 - A_2) \sin \omega_d t \right]$$

and the multiplying factors may be assigned new symbols:

$$v(t) = e^{-\alpha t} (B_1 \cos \omega_d t + B_2 \sin \omega_d t) \qquad [27]$$

where Eqs. [26] and [27] are identical.

It may seem a little odd that our expression originally appeared to have a complex component, and now is purely real. However, we should remember that we originally allowed for A_1 and A_2 to be complex as well as s_1 and s_2. In any event, if we are dealing with the underdamped case, we have now left complex numbers behind. This must be true since α, ω_d, and t are real quantities, so that $v(t)$ itself must be a real quantity (which might be presented on an oscilloscope, a voltmeter, or a sheet of graph paper). Equation [27] is the desired functional form for the underdamped response, and its validity may be checked by direct substitution in the original differential equation; this exercise is left to the doubters. The two real constants B_1 and B_2 are again selected to fit the given initial conditions.

We return to our simple parallel *RLC* circuit of Fig. 9.2 with $R = 6\ \Omega$, $C = 1/42$ F, *and* $L = 7$ H, but now increase the resistance further to $10.5\ \Omega$. Thus,

$$\alpha = \frac{1}{2RC} = 2\ \text{s}^{-1}$$

$$\omega_0 = \frac{1}{\sqrt{LC}} = \sqrt{6}\ \text{s}^{-1}$$

and

$$\omega_d = \sqrt{\omega_0^2 - \alpha^2} = \sqrt{2}\ \text{rad/s}$$

Except for the evaluation of the arbitrary constants, the response is now known:

$$v(t) = e^{-2t} (B_1 \cos \sqrt{2}t + B_2 \sin \sqrt{2}t)$$

Finding Values for B_1 and B_2

The determination of the two constants proceeds as before. If we still assume that $v(0) = 0$ and $i(0) = 10$, then B_1 must be zero. Hence

$$v(t) = B_2 e^{-2t} \sin \sqrt{2}t$$

The derivative is

$$\frac{dv}{dt} = \sqrt{2} B_2 e^{-2t} \cos \sqrt{2}t - 2B_2 e^{-2t} \sin \sqrt{2}t$$

and at $t = 0$ it becomes

$$\left.\frac{dv}{dt}\right|_{t=0} = \sqrt{2}B_2 = \frac{i_C(0)}{C} = 420$$

where i_C is defined in Fig. 9.2. Therefore,

$$v(t) = 210\sqrt{2}e^{-2t}\sin\sqrt{2}t$$

Graphical Representation of the Underdamped Response

Notice that, as before, this response function has an initial value of zero because of the initial voltage condition we imposed, and a final value of zero because the exponential term vanishes for large values of t. As t increases from zero through small positive values, $v(t)$ increases as $210\sqrt{2}\sin\sqrt{2}t$, because the exponential term remains essentially equal to unity. But at some time t_m, the exponential function begins to decrease more rapidly than $\sin\sqrt{2}t$ is increasing; thus $v(t)$ reaches a maximum v_m and begins to decrease. We should note that t_m is not the value of t for which $\sin\sqrt{2}t$ is a maximum, but must occur somewhat before $\sin\sqrt{2}t$ reaches its maximum.

When $t = \pi/\sqrt{2}$, $v(t)$ is zero. Thus, in the interval $\pi/\sqrt{2} < t < \sqrt{2}\pi$, the response is negative, becoming zero again at $t = \sqrt{2}\pi$. Hence, $v(t)$ is an *oscillatory* function of time and crosses the time axis an infinite number of times at $t = n\pi/\sqrt{2}$, where n is any positive integer. In our example, however, the response is only slightly underdamped, and the exponential term causes the function to die out so rapidly that most of the zero crossings will not be evident in a sketch.

The oscillatory nature of the response becomes more noticeable as α decreases. If α is zero, which corresponds to an infinitely large resistance, then $v(t)$ is an undamped sinusoid that oscillates with constant amplitude. There is never a time at which $v(t)$ drops and stays below 1 percent of its maximum value; the settling time is therefore infinite. This is not perpetual motion; we have merely assumed an initial energy in the circuit and have not provided any means to dissipate this energy. It is transferred from its initial location in the inductor to the capacitor, then returns to the inductor, and so on, forever.

The Role of Finite Resistance

A finite R in the parallel RLC circuit acts as a kind of electrical transfer agent. Every time energy is transferred from L to C or from C to L, the agent exacts a commission. Before long, the agent has taken all the energy, wantonly dissipating every last joule. The L and C are left without a joule of their own, without voltage and without current. Actual parallel RLC circuits can be made to have effective values of R so large that a natural undamped sinusoidal response can be maintained for years without supplying any additional energy.

Returning to our specific numerical problem, differentiation locates the first maximum of $v(t)$,

$$v_{m_1} = 71.8 \text{ V} \qquad \text{at} \qquad t_{m_1} = 0.435 \text{ s}$$

the succeeding minimum,

$$v_{m_2} = -0.845 \text{ V} \qquad \text{at} \qquad t_{m_2} = 2.66 \text{ s}$$

and so on. The response curve is shown in Fig. 9.14. Additional response curves for increasingly more underdamped circuits are shown in Fig. 9.15.

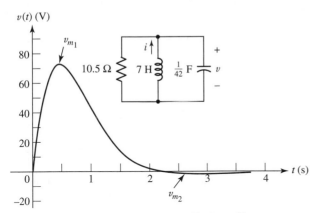

■ **FIGURE 9.14** The response $v(t) = 210\sqrt{2}e^{-2t} \sin \sqrt{2}t$ of the network shown in Fig. 9.2 with *R* increased to produce an underdamped response.

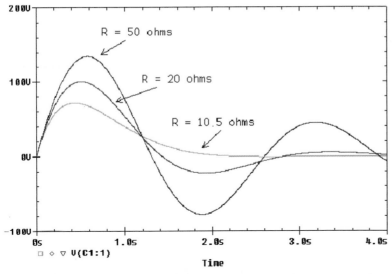

■ **FIGURE 9.15** Simulated underdamped voltage response of the network for three different resistance values, showing an increase in the oscillatory behavior as *R* is increased.

The settling time may be obtained by a trial-and-error solution, and for $R = 10.5\,\Omega$, it turns out to be 2.92 s, somewhat smaller than for critical damping. Note that t_s is *greater* than t_{m_2} because the magnitude of v_{m_2} is greater than 1 percent of the magnitude of v_{m_1}. This suggests that a slight decrease in *R* would reduce the magnitude of the undershoot and permit t_s to be less than t_{m_2}.

The overdamped, critically damped, and underdamped responses for this network as simulated by PSpice are shown on the same graph in Fig. 9.16. A comparison of the three curves makes the following general

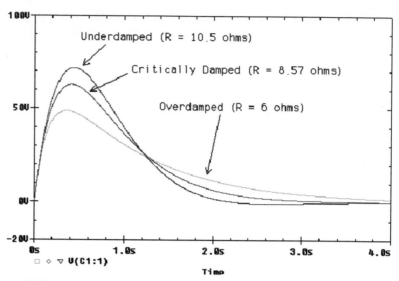

FIGURE 9.16 Simulated overdamped, critically damped, and underdamped voltage response for the example network, obtained by varying the value of the parallel resistance *R*.

conclusions plausible:

- When the damping is changed by increasing the size of the parallel resistance, the maximum magnitude of the response is greater and the amount of damping is smaller.

- The response becomes oscillatory when underdamping is present, and the minimum settling time is obtained for slight underdamping.

EXAMPLE 9.6

Determine $i_L(t)$ for the circuit of Fig. 9.17a, and plot the waveform.

At $t = 0$, both the 3 A source and the 48 Ω resistor are removed, leaving the circuit shown in Fig. 9.17b. Thus, $\alpha = 1.2$ s^{-1} and $\omega_0 = 4.899$ rad/s. Since $\alpha < \omega_0$, the circuit is *underdamped,* and we therefore expect a response of the form

$$i_L(t) = e^{-\alpha t}(B_1 \cos \omega_d t + B_2 \sin \omega_d t) \qquad [28]$$

where $\omega_d = \sqrt{\omega_0^2 - \alpha^2} = 4.750$ rad/s. The only remaining step is to find B_1 and B_2.

Fig. 9.17c shows the circuit as it exists at $t = 0^-$. We may replace the inductor with a short circuit and the capacitor with an open circuit; the result is $v_C(0^-) = 97.30$ V and $i_L(0^-) = 2.027$ A. Since neither quantity can change in zero time, $v_C(0^+) = 97.30$ V and $i_L(0^+) = 2.027$ A.

Substituting $i_L(0) = 2.027$ into Eq. [28] yields $B_1 = 2.027$ A. To determine the other constant, we first differentiate Eq. [28]:

$$\frac{di_L}{dt} = e^{-\alpha t}(-B_1 \omega_d \sin \omega_d t + B_2 \omega_d \cos \omega_d t)$$
$$- \alpha e^{-at}(B_1 \cos \omega_d t + B_2 \sin \omega_d t) \qquad [29]$$

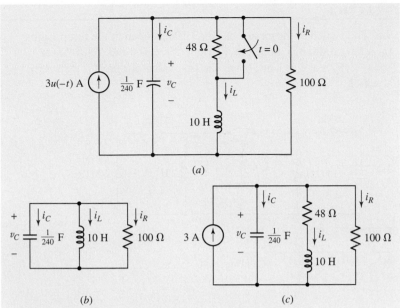

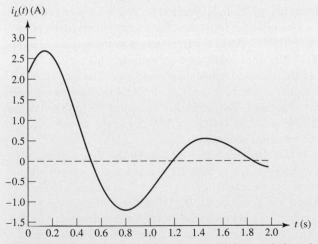

FIGURE 9.17 (*a*) A parallel *RLC* circuit for which the current $i_L(t)$ is desired. (*b*) Circuit for $t \geq 0$. (*c*) Circuit for determining the initial conditions.

and note that $v_L(t) = L(di_L/dt)$. Referring to the circuit of Fig. 9.17*b*, we see that $v_L(0^+) = v_C(0^+) = 97.3$ V. Thus, multiplying Eq. [29] by $L = 10$ H and setting $t = 0$, we find that

$$v_L(0) = 10(B_2\omega_d) - 10\alpha B_1 = 97.3$$

Solving, $B_2 = 2.561$ A, so that

$$i_L = e^{-1.2t}(2.027 \cos 4.75t + 2.561 \sin 4.75t) \text{ A}$$

which we have plotted in Fig. 9.18.

FIGURE 9.18 Plot of $i_L(t)$, showing obvious signs of being an underdamped response.

PRACTICE

9.6 The switch in the circuit of Fig. 9.19 has been in the left position for a long time; it is moved to the right at $t = 0$. Find (a) dv/dt at $t = 0^+$; (b) v at $t = 1$ ms; (c) t_0, the first value of t greater than zero at which $v = 0$.

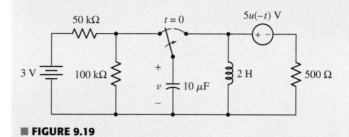

■ **FIGURE 9.19**

Ans: -1400 V/s; 0.695 V; 1.609 ms.

COMPUTER-AIDED ANALYSIS

One useful feature in Probe is the ability to perform mathematical operations on the voltages and currents that result from a simulation. In this example, we will make use of that ability to show the transfer of energy in a parallel *RLC* circuit from a capacitor that initially stores a specific amount of energy (1.25 μJ) to an inductor that initially stores no energy.

We choose a 100 nF capacitor and a 7 μH inductor, which immediately enables us to calculate $\omega_0 = 1.195 \times 10^6$ s^{-1}. In order to consider overdamped, critically damped, and underdamped cases, we need to select the parallel resistance in such a way as to obtain $\alpha > \omega_0$ (*overdamped*), $\alpha = \omega_0$ (*critically damped*), and $\alpha < \omega_0$ (*underdamped*). From our previous discussions, we know that for a parallel *RLC* circuit $\alpha = (2RC)^{-1}$. We select $R = 4.1833$ Ω as a close approximation to the critically damped case; obtaining α precisely equal to ω_0 is effectively impossible. If we increase the resistance, the energy stored in the other two elements is dissipated more slowly, resulting in an underdamped response. We select $R = 100$ Ω so that we are well into this regime, and use $R = 1$ Ω (a very small resistance) to obtain an overdamped response.

We therefore plan to run three separate simulations, varying only the resistance R between them. The 1.25 μJ of energy initially stored in the capacitor equates to an initial voltage of 5 V, and so we set the initial condition of our capacitor accordingly.

Once Probe is launched, we select **Add** under the **Trace** menu. We wish to plot the energy stored in both the inductor and the capacitor as a function of time. For the capacitor, $w = \frac{1}{2}Cv^2$, so we click in the **Trace Expression** window, type in "0.5*100E-9*" (without the quotes), click on V(C1:1), return to the **Trace Expression** window and enter "*", click on V(C1:1) once again and then select **Ok.** We repeat the sequence to obtain the energy stored in the inductor, using 7E-6 instead of 100E-9, and clicking on I(L1:1) instead of V(C1:1).

The Probe output plots for three separate simulations are provided in Fig. 9.20. In Fig. 9.20*a*, we see that the energy remaining in the circuit is continuously transferred back and forth between the capacitor and the inductor until it is (eventually) completely dissipated by the resistor. Decreasing the resistance to 4.1833 Ω yields a critically damped circuit, resulting in the energy plot of Fig. 9.20*b*. The oscillatory energy transfer between the capacitor and the inductor has been dramatically reduced. We see that the energy transferred to the inductor peaks at approximately 0.8 μs, and then drops to zero. The overdamped response is plotted in Fig. 9.20*c*. We note that the energy is dissipated much more quickly in the case of the overdamped response, and that very little energy is transferred to the inductor, since most of it is now quickly dissipated in the resistor.

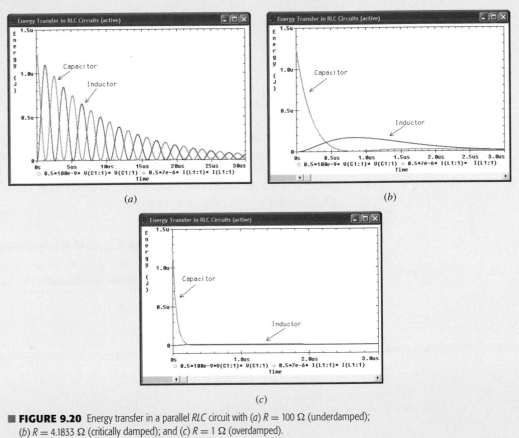

■ **FIGURE 9.20** Energy transfer in a parallel *RLC* circuit with (*a*) $R = 100 \ \Omega$ (underdamped); (*b*) $R = 4.1833 \ \Omega$ (critically damped); and (*c*) $R = 1 \ \Omega$ (overdamped).

9.5 • THE SOURCE-FREE SERIES *RLC* CIRCUIT

We now wish to determine the natural response of a circuit model composed of an ideal resistor, an ideal inductor, and an ideal capacitor connected in *series*. The ideal resistor may represent a physical resistor connected into a series *LC* or *RLC* circuit; it may represent the ohmic losses and the losses in the ferromagnetic core of the inductor; or it may be used to represent all these and other energy-absorbing devices.

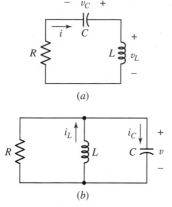

(a)

(b)

■ **FIGURE 9.21** (*a*) The series *RLC* circuit which is the dual of (*b*) a parallel *RLC* circuit. Element values are, of course, not identical in the two circuits.

The series *RLC* circuit is the *dual* of the parallel *RLC* circuit, and this single fact is sufficient to make its analysis a trivial affair. Figure 9.21*a* shows the series circuit. The fundamental integrodifferential equation is

$$L\frac{di}{dt} + Ri + \frac{1}{C}\int_{t_0}^{t} i\,dt' - v_C(t_0) = 0$$

and should be compared with the analogous equation for the parallel *RLC* circuit, drawn again in Fig. 9.21*b*,

$$C\frac{dv}{dt} + \frac{1}{R}v + \frac{1}{L}\int_{t_0}^{t} v\,dt' - i_L(t_0) = 0$$

The respective second-order equations obtained by differentiating these two equations with respect to time are also duals:

$$L\frac{d^2i}{dt^2} + R\frac{di}{dt} + \frac{1}{C}i = 0 \qquad [30]$$

$$C\frac{d^2v}{dt^2} + \frac{1}{R}\frac{dv}{dt} + \frac{1}{L}v = 0 \qquad [31]$$

Our complete discussion of the parallel *RLC* circuit is directly applicable to the series *RLC* circuit; the initial conditions on capacitor voltage and inductor current are equivalent to the initial conditions on inductor current and capacitor voltage; the *voltage* response becomes a *current* response. It is therefore possible to reread the previous four sections using dual language and thereby obtain a complete description of the series *RLC* circuit. This process, however, is apt to induce a mild neurosis after the first few paragraphs and does not really seem to be necessary.

A Brief Résumé of the Series Circuit Response

A brief résumé of the series circuit response is easily put together. In terms of the circuit shown in Fig. 9.21*a*, the *overdamped response* is

$$i(t) = A_1 e^{s_1 t} + A_2 e^{s_2 t}$$

where

$$s_{1,2} = -\frac{R}{2L} \pm \sqrt{\left(\frac{R}{2L}\right)^2 - \frac{1}{LC}} = -\alpha \pm \sqrt{\alpha^2 - \omega_0^2}$$

and thus

$$\alpha = \frac{R}{2L}$$

$$\omega_0 = \frac{1}{\sqrt{LC}}$$

The form of the *critically damped response* is

$$i(t) = e^{-\alpha t}(A_1 t + A_2)$$

and the *underdamped response* may be written

$$i(t) = e^{-\alpha t}(B_1 \cos \omega_d t + B_2 \sin \omega_d t)$$

TABLE ● **9.1** Summary of Relevant Equations for Source-Free *RLC* Circuits

Type	Condition	Criteria	α	ω_0	Response
Parallel	Overdamped	$\alpha > \omega_0$	$\dfrac{1}{2RC}$ $\dfrac{R}{2L}$	$\dfrac{1}{\sqrt{LC}}$	$A_1 e^{s_1 t} + A_2 e^{s_2 t}$, where $s_{1,2} = -\alpha \pm \sqrt{\alpha^2 - \omega^2}$
Series					
Parallel	Critically damped	$\alpha = \omega_0$	$\dfrac{1}{2RC}$ $\dfrac{R}{2L}$	$\dfrac{1}{\sqrt{LC}}$	$e^{-\alpha t}(A_1 t + A_2)$
Series					
Parallel	Underdamped	$\alpha < \omega_0$	$\dfrac{1}{2RC}$ $\dfrac{R}{2L}$	$\dfrac{1}{\sqrt{LC}}$	$e^{-\alpha t}(B_1 \cos \omega_d t + B_2 \sin \omega_d t)$, where $\omega_d = \sqrt{\omega_0^2 - \alpha^2}$
Series					

where

$$\omega_d = \sqrt{\omega_0^2 - \alpha^2}$$

It is evident that if we work in terms of the parameters α, ω_0, and ω_d, the mathematical forms of the responses for the dual situations are identical. An increase in α in either the series or parallel circuit, while keeping ω_0 constant, tends toward an overdamped response. The only caution that we need exert is in the computation of α, which is $1/2RC$ for the parallel circuit and $R/2L$ for the series circuit; thus, α is increased by increasing the series resistance or decreasing the parallel resistance. The key equations for parallel and series *RLC* circuits are summarized in Table 9.1 for convenience.

CAUTION

EXAMPLE **9.7**

Given the series *RLC* circuit of Fig. 9.22 in which $L = 1$ H, $R = 2$ kΩ, $C = 1/401$ μF, $i(0) = 2$ mA, and $v_C(0) = 2$ V, find and sketch $i(t)$, $t > 0$.

We find that $\alpha = R/2L = 1000$ s^{-1} and $\omega_0 = 1/\sqrt{LC} = 20{,}025$ rad/s. This indicates an *underdamped* response; we therefore calculate the value of ω_d and obtain 20,000 rad/s. Except for the evaluation of the two arbitrary constants, the response is now known:

$$i(t) = e^{-1000t}(B_1 \cos 20{,}000t + B_2 \sin 20{,}000t)$$

Since we know that $i(0) = 2$ mA, we may substitute this value into our equation for $i(t)$ to obtain

$$B_1 = 0.002 \text{ A}$$

and thus

$$i(t) = e^{-1000t}(0.002 \cos 20{,}000t + B_2 \sin 20{,}000t) \quad \text{A}$$

(Continued on next page)

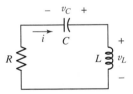

■ **FIGURE 9.22** A simple source-free *RLC* circuit with energy stored in both the inductor and the capacitor at $t = 0$.

The remaining initial condition must be applied to the derivative; thus,

$$\frac{di}{dt} = e^{-1000t}(-40\sin 20{,}000t + 20{,}000 B_2 \cos 20{,}000t$$

$$- 2\cos 20{,}000t - 1000 B_2 \sin 20{,}000t)$$

and

$$\left.\frac{di}{dt}\right|_{t=0} = 20{,}000 B_2 - 2 = \frac{v_L(0)}{L}$$

$$= \frac{v_C(0) - Ri(0)}{L}$$

$$= \frac{2 - 2000(0.002)}{1} = -2 \text{ A/s}$$

so that

$$B_2 = 0$$

The desired response is therefore

$$i(t) = 2e^{-1000t}\cos 20{,}000t \qquad \text{mA}$$

A good sketch may be made by first drawing in the two portions of the exponential *envelope*, $2e^{-1000t}$ and $-2e^{-1000t}$ mA, as shown by the broken lines in Fig. 9.23. The location of the quarter-cycle points of the sinusoidal wave at $20{,}000t = 0$, $\pi/2$, π, etc., or $t = 0.07854k$ ms, $k = 0, 1, 2, \ldots$, by light marks on the time axis then permits the oscillatory curve to be sketched in quickly.

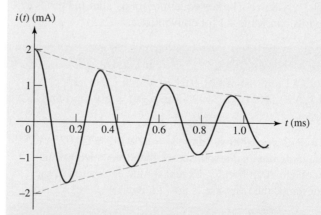

■ FIGURE 9.23 The current response in an underdamped series *RLC* circuit for which $\alpha = 1000 \text{ s}^{-1}$, $\omega_0 = 20{,}000 \text{ s}^{-1}$, $i(0) = 2$ mA, and $v_C(0) = 2$ V. The graphical construction is simplified by drawing in the envelope, shown as a pair of broken lines.

The settling time can be determined easily here by using the upper portion of the envelope. That is, we set $2e^{-1000t_s}$ mA equal to 1 percent of its maximum value, 2 mA. Thus, $e^{-1000t_s} = 0.01$, and $t_s = 4.61$ ms is the approximate value that is usually used.

PRACTICE

9.7 With reference to the circuit shown in Fig. 9.24, find (*a*) α; (*b*) ω_0; (*c*) $i(0^+)$; (*d*) $di/dt|_{t=0^+}$; (*e*) $i(12$ ms).

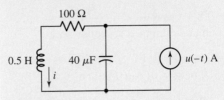

■ **FIGURE 9.24**

Ans: 100 s^{-1}; 224 rad/s; 1 A; 0; -0.1204 A.

As a final example, we pause to consider situations where the circuit includes a dependent source. If no controlling current or voltage associated with the dependent source is of interest, we may simply find the Thévenin equivalent connected to the inductor and capacitor. Otherwise, we are likely faced with having to write an appropriate integrodifferential equation, take the indicated derivative, and solve the resulting differential equation as best we can.

EXAMPLE 9.8

Find an expression for $v_C(t)$ in the circuit of Fig. 9.25a, valid for $t > 0$.

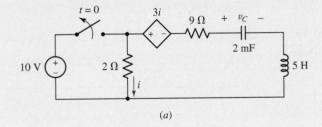

(*a*)

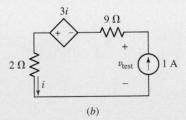

(*b*)

■ **FIGURE 9.25** (*a*) An *RLC* circuit containing a dependent source. (*b*) Circuit for finding R_{eq}.

As we are interested only in $v_C(t)$, it is perfectly acceptable to begin by finding the Thévenin equivalent resistance connected in series with the

(Continued on next page)

inductor and capacitor at $t = 0^+$. We do this by connecting a 1 A source as shown in Fig. 9.25*b*, from which we deduce that

$$v_{\text{test}} = 11i - 3i = 8i = 8(1) = 8 \text{ V}.$$

Thus, $R_{\text{eq}} = 8 \ \Omega$, so $\alpha = R/2L = 0.8 \ \text{s}^{-1}$ and $\omega_0 = 1/\sqrt{LC} = 10$ rad/s, meaning that we expect an underdamped response with $\omega_d = 9.968$ rad/s and the form:

$$v_C(t) = e^{-0.8t}(B_1 \cos 9.968t + B_2 \sin 9.968t) \qquad [32]$$

In considering the circuit at $t = 0^-$, we note that $i_L(0^-) = 0$ due to the presence of the capacitor. By Ohm's law, $i(0^-) = 5$ A, so

$$v_C(0^+) = v_C(0^-) = 10 - 3i = 10 - 15 = -5 \text{ V}$$

This last condition substituted into Eq. [32] yields $B_1 = -5$ V. Taking the derivative of Eq. [32] and evaluating at $t = 0$ yields

$$\frac{dv_C}{dt}\bigg|_{t=0} = -0.8B_1 + 9.968B_2 = 4 + 9.968B_2 \qquad [33]$$

We see from Fig. 9.25*a* that

$$i = -C\frac{dv_C}{dt}$$

Thus, making use of the fact that $i(0^+) = i_L(0^-) = 0$ in Eq. [33] yields $B_2 = -0.4013$ V, and we may write

$$v_C(t) = -e^{-0.8t}(5 \cos 9.968t + 0.4013 \sin 9.968t) \text{ V}, \quad t > 0$$

The PSpice simulation of this circuit, shown in Fig. 9.26, confirms our analysis.

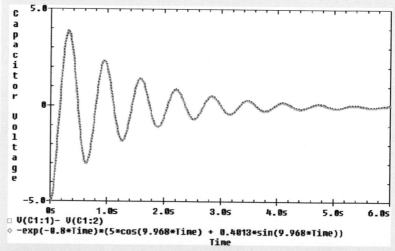

■ **FIGURE 9.26** PSpice simulation of the circuit shown in Fig. 9.25*a*. The analytical result is plotted using a dashed red line.

PRACTICE

9.8 Find an expression for $i_L(t)$ in the circuit of Fig. 9.27, valid for $t > 0$, if $v_C(0^-) = 10$ V and $i_L(0^-) = 0$. *Note that although it is not helpful to apply Thévenin techniques in this instance, the action of the dependent source links v_C and i_L such that a first-order linear differential equation results.*

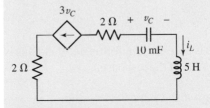

■ **FIGURE 9.27** Circuit for Practice Problem 9.8.

Ans: $i_L(t) = -30e^{-300t}$ A, $t > 0$.

9.6 THE COMPLETE RESPONSE OF THE *RLC* CIRCUIT

We now consider those *RLC* circuits in which dc sources are switched into the network and produce forced responses that do not necessarily vanish as time becomes infinite. The general solution is obtained by the same procedure that was followed for *RL* and *RC* circuits: the forced response is determined completely; the natural response is obtained as a suitable functional form containing the appropriate number of arbitrary constants; the complete response is written as the sum of the forced and the natural responses; and the initial conditions are then determined and applied to the complete response to find the values of the constants. *It is this last step which is quite frequently the most troublesome to students.* Consequently, although the determination of the initial conditions is basically no different for a circuit containing dc sources from what it is for the source-free circuits that we have already covered in some detail, this topic will receive particular emphasis in the examples that follow.

Most of the confusion in determining and applying the initial conditions arises for the simple reason that we do not have a rigorous set of rules laid down for us to follow. At some point in each analysis, a situation usually arises in which some thinking is involved that is more or less unique to that particular problem. This is almost always the source of the difficulty.

The Easy Part

The *complete* response (arbitrarily assumed to be a voltage response) of a second-order system consists of a *forced* response,

$$v_f(t) = V_f$$

which is a constant for dc excitation, and a *natural* response,

$$v_n(t) = Ae^{s_1t} + Be^{s_2t}$$

Thus,

$$v(t) = V_f + Ae^{s_1 t} + Be^{s_2 t}$$

We assume that s_1, s_2, and V_f have already been determined from the circuit and the given forcing functions; A and B remain to be found. The last equation shows the functional interdependence of A, B, v, and t, and substitution of the known value of v at $t = 0^+$ thus provides us with a single equation relating A and B, $v(0^+) = V_f + A + B$. *This is the easy part.*

The Other Part

Another relationship between A and B is necessary, unfortunately, and this is normally obtained by taking the derivative of the response,

$$\frac{dv}{dt} = 0 + s_1 Ae^{s_1 t} + s_2 Be^{s_2 t}$$

and inserting the known value of dv/dt at $t = 0^+$. We thus have two equations relating A and B, and these may be solved simultaneously to evaluate the two constants.

The only remaining problem is that of determining the values of v and dv/dt at $t = 0^+$. Let us suppose that v is a capacitor voltage, v_C. Since $i_C = C \, dv_C/dt$, we should recognize the relationship between the initial value of dv/dt and the initial value of some capacitor current. If we can establish a value for this initial capacitor current, then we will automatically establish the value of dv/dt. Students are usually able to get $v(0^+)$ very easily, but are inclined to stumble a bit in finding the initial value of dv/dt. If we had selected an inductor current i_L as our response, then the initial value of di_L/dt would be intimately related to the initial value of some inductor voltage. Variables other than capacitor voltages and inductor currents are determined by expressing their initial values and the initial values of their derivatives in terms of the corresponding values for v_C and i_L.

We will illustrate the procedure and find all these values by the careful analysis of the circuit shown in Fig. 9.28. To simplify the analysis, an unrealistically large capacitance is used again.

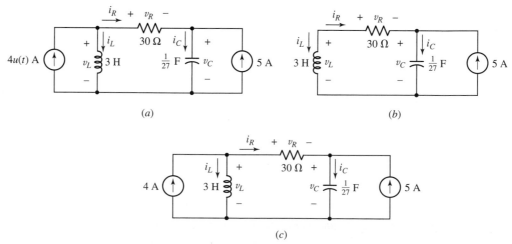

FIGURE 9.28 (*a*) An *RLC* circuit that is used to illustrate several procedures by which the initial conditions may be obtained. The desired response is nominally taken to be $v_C(t)$, (*b*) $t = 0^-$, (*c*) $t > 0$.

EXAMPLE **9.9**

There are three passive elements in the circuit shown in Fig. 9.28a, and a voltage and a current are defined for each. Find the values of these six quantities at both $t = 0^-$ and $t = 0^+$.

Our object is to find the value of each current and voltage at both $t = 0^-$ and $t = 0^+$. Once these quantities are known, the initial values of the derivatives may be found easily. We will employ a logical step-by-step method first.

1. $t = 0^-$ At $t = 0^-$, only the right-hand current source is active as depicted in Fig. 9.28b. The circuit is assumed to have been in this state forever, so all currents and voltages are constant. Thus, a dc current through the inductor requires zero voltage across it:

$$v_L(0^-) = 0$$

and a dc voltage across the capacitor $(-v_R)$ requires zero current through it:

$$i_C(0^-) = 0$$

We next apply Kirchhoff's current law to the right-hand node to obtain

$$i_R(0^-) = -5 \text{ A}$$

which also yields

$$v_R(0^-) = -150 \text{ V}$$

We may now use Kirchhoff's voltage law around the left-hand mesh, finding

$$v_C(0^-) = 150 \text{ V}$$

while KCL enables us to find the inductor current,

$$i_L(0^-) = 5 \text{ A}$$

2. $t = 0^+$ During the interval from $t = 0^-$ to $t = 0^+$, the left-hand current source becomes active and many of the voltage and current values at $t = 0^-$ will change abruptly. The corresponding circuit is shown in Fig. 9.28c. However, we should *begin by focusing our attention on those quantities which cannot change, namely, the inductor current and the capacitor voltage.* Both of these must remain constant during the switching interval. Thus,

$$i_L(0^+) = 5 \text{ A} \quad \text{and} \quad v_C(0^+) = 150 \text{ V}$$

Since two currents are now known at the left node, we next obtain

$$i_R(0^+) = -1 \text{ A} \quad \text{and} \quad v_R(0^+) = -30 \text{ V}$$

so that

$$i_C(0^+) = 4 \text{ A} \quad \text{and} \quad v_L(0^+) = 120 \text{ V}$$

and we have our six initial values at $t = 0^-$ and six more at $t = 0^+$. Among these last six values, only the capacitor voltage and the inductor current are unchanged from the $t - 0^-$ values.

We could have employed a slightly different method to evaluate these currents and voltages at $t = 0^-$ and $t = 0^+$. Prior to the switching operation, only direct currents and voltages exist in the circuit. The inductor may therefore be replaced by a short circuit, its dc equivalent, while the capacitor is replaced by an open circuit. Redrawn in this manner, the circuit of Fig. 9.28*a* appears as shown in Fig. 9.29*a*. Only the current source at the right is active, and its 5 A flow through the resistor and the inductor. We therefore have $i_R(0^-) = -5$ A and $v_R(0^-) = -150$ V, $i_L(0^-) = 5$ A and $v_L(0^-) = 0$, and $i_C(0^-) = 0$ and $v_C(0^-) = 150$ V, as before.

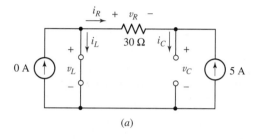

(a)

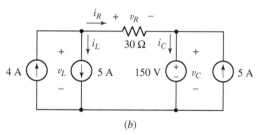

(b)

■ **FIGURE 9.29** (*a*) A simple circuit equivalent to the circuit of Fig. 9.28*a* for $t = 0^-$. (*b*) Equivalent circuit with labeled voltages and currents valid at the instant defined by $t = 0^+$.

We now turn to the problem of drawing an equivalent circuit that will assist us in determining the several voltages and currents at $t = 0^+$. *Each capacitor voltage and each inductor current must remain constant during the switching interval.* These conditions are ensured by replacing the inductor with a current source and the capacitor with a voltage source. Each source serves to maintain a constant response during the discontinuity. The equivalent circuit of Fig. 9.29*b* results. It should be noted that the circuit shown in Fig. 9.29*b* is valid *only for the interval between 0^- and 0^+.*

The voltages and currents at $t = 0^+$ are obtained by analyzing this dc circuit. The solution is not difficult, but the relatively large number of sources present in the network does produce a somewhat strange sight. However, problems of this type were solved in Chap. 3, and nothing new is involved. Attacking the currents first, we begin at the upper left node and see that $i_R(0^+) = 4 - 5 = -1$ A. Moving to the upper right node, we find that $i_C(0^+) = -1 + 5 = 4$ A. And, of course, $i_L(0^+) = 5$ A.

Next we consider the voltages. Using Ohm's law, we see that $v_R(0^+) = 30(-1) = -30$ V. For the inductor, KVL gives us $v_L(0^+) = -30 + 150 = 120$ V. Finally, including $v_C(0^+) = 150$ V, we have all the values at $t = 0^+$.

PRACTICE

9.9 Let $i_s = 10u(-t) - 20u(t)$ A in Fig. 9.30. Find (a) $i_L(0^-)$;
(b) $v_C(0^+)$; (c) $v_R(0^+)$; (d) $i_{L,(\infty)}$; (e) $i_L(0.1$ ms).

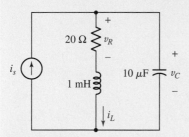

■ **FIGURE 9.30**

Ans: 10 A; 200 V; 200 V; -20 A; 2.07 A.

EXAMPLE **9.10**

Complete the determination of the initial conditions in the circuit of Fig. 9.28, repeated in Fig. 9.31, by finding values at $t = 0^+$ for the first derivatives of the three voltage and three current variables defined on the circuit diagram.

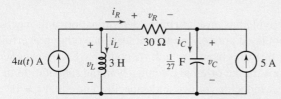

■ **FIGURE 9.31** Circuit of Fig. 9.28, repeated for Example 9.10.

We begin with the two energy-storage elements. For the inductor,

$$v_L = L \frac{di_L}{dt}$$

and, specifically,

$$v_L(0^+) = L \left. \frac{di_L}{dt} \right|_{t=0^+}$$

Thus,

$$\left. \frac{di_L}{dt} \right|_{t=0^+} = \frac{v_L(0^+)}{L} = \frac{120}{3} = 40 \text{ A/s}$$

Similarly,

$$\left. \frac{dv_C}{dt} \right|_{t=0^+} = \frac{i_C(0^+)}{C} = \frac{4}{1/27} = 108 \text{ V/s}$$

(Continued on next page)

The other four derivatives may be determined by realizing that KCL and KVL are both satisfied by the derivatives also. For example, at the left-hand node in Fig. 9.31,

$$4 - i_L - i_R = 0, \qquad t > 0$$

and thus,

$$0 - \frac{di_L}{dt} - \frac{di_R}{dt} = 0, \qquad t > 0$$

and therefore,

$$\left.\frac{di_R}{dt}\right|_{t=0^+} = -40 \text{ A/s}$$

The three remaining initial values of the derivatives are found to be

$$\left.\frac{dv_R}{dt}\right|_{t=0^+} = -1200 \text{ V/s}$$

$$\left.\frac{dv_L}{dt}\right|_{t=0^+} = -1092 \text{ V/s}$$

and

$$\left.\frac{di_C}{dt}\right|_{t=0^+} = -40 \text{ A/s}$$

Before leaving this problem of the determination of the necessary initial values, it should be pointed out that at least one other powerful method of determining them has been omitted: we could have written general nodal or loop equations for the original circuit. Then, the substitution of the known zero values of inductor voltage and capacitor current at $t = 0^-$ would uncover several other response values at $t = 0^-$ and enable the remainder to be found easily. A similar analysis at $t = 0^+$ must then be made. This is an important method, and it becomes a necessary one in more complicated circuits which cannot be analyzed by our simpler step-by-step procedures.

Now let us briefly complete the determination of the response $v_C(t)$ for the original circuit of Fig. 9.31. With both sources dead, the circuit appears as a series *RLC* circuit and s_1 and s_2 are easily found to be -1 and -9, respectively. The forced response may be found by inspection or, if necessary, by drawing the dc equivalent, which is similar to Fig. 9.29*a*, with the addition of a 4 A current source. The forced response is 150 V. Thus,

$$v_C(t) = 150 + Ae^{-t} + Be^{-9t}$$

and

$$v_C(0^+) = 150 = 150 + A + B$$

or

$$A + B = 0$$

Then,

$$\frac{dv_C}{dt} = -Ae^{-t} - 9Be^{-9t}$$

and

$$\left.\frac{dv_C}{dt}\right|_{t=0^+} = 108 = -A - 9B$$

Finally,

$$A = 13.5 \qquad B = -13.5$$

and

$$v_C(t) = 150 + 13.5(e^{-t} - e^{-9t}) \text{ V}$$

A Quick Summary of the Solution Process

In summary then, whenever we wish to determine the transient behavior of a simple three-element *RLC* circuit, we must first decide whether we are confronted with a series or a parallel circuit, so that we may use the correct relationship for α. The two equations are:

$$\alpha = \frac{1}{2RC} \qquad \text{(parallel RLC)}$$

$$\alpha = \frac{R}{2L} \qquad \text{(series RLC)}$$

Our second decision is made after comparing α with ω_0, which is given for either circuit by

$$\omega_0 = \frac{1}{\sqrt{LC}}$$

If $\alpha > \omega_0$, the circuit is *overdamped,* and the natural response has the form:

$$f_n(t) = A_1 e^{s_1 t} + A_2 e^{s_2 t}$$

where

$$s_{1,2} = -\alpha \pm \sqrt{\alpha^2 - \omega_0^2}$$

If $\alpha = \omega_0$, then the circuit is *critically damped* and

$$f_n(t) = e^{-\alpha t} (A_1 t + A_2)$$

And finally, if $\alpha < \omega_0$, then we are faced with the *underdamped* response,

$$f_n(t) = e^{-\alpha t} (A_1 \cos \omega_d t + A_2 \sin \omega_d t)$$

where

$$\omega_d = \sqrt{\omega_0^2 - \alpha^2}$$

Our last decision depends on the independent sources. If there are none acting in the circuit after the switching or discontinuity is completed, then the circuit is source-free and the natural response accounts for the complete response. If independent sources are still present, then the circuit is driven and a forced response must be determined. The complete response is then the sum

$$f(t) = f_f(t) + f_n(t)$$

Modeling Automotive Suspension Systems

In the introductory paragraph, we alluded to the fact that the concepts investigated in this chapter actually extend beyond the analysis of electric circuits. In fact, the general form of the differential equations we have been working with appear in many fields—we need only learn how to "translate" new parameter terminology. For example, consider a simple automotive suspension, as shown in Fig. 9.32. The piston is not attached to the cylinder, but *is* attached to both the spring and the wheel. The moving parts therefore are the spring, the piston, and the wheel.

We will model this physical system by first determining the forces in play. Defining a position function $p(t)$ which describes where the piston lies within the cylinder, we may write F_S, the force on the spring, as

$$F_S = Kp(t)$$

where K is known as the spring constant and has units of lb/ft. The force on the wheel F_W is equal to the mass of the wheel times its acceleration, or

$$F_W = m\frac{d^2 p(t)}{dt^2}$$

where m is measured in lb $\cdot$ s^2/ft. Last but not least is the force of friction F_f acting on the piston

$$F_f = \mu_f \frac{dp(t)}{dt}$$

■ FIGURE 9.32 Typical automotive suspension system.
© Transtock Inc./Alamy

where μ_f is the coefficient of friction, with units of lb $\cdot$ s/ft.

From our basic physics courses we know that all forces acting in our system must sum to zero, so that

$$m\frac{d^2 p(t)}{dt^2} + \mu_f \frac{dp(t)}{dt} + Kp(t) = 0 \qquad [34]$$

This equation most likely had the potential to give us nightmares at one point in our academic career, but no longer. We compare Eq. [32] to Eqs. [30] and [31] and immediately see a distinct resemblance, at least in the general form. Choosing Eq. [30], the differential equation describing the inductor current of a series-connected *RLC* circuit, we observe the following correspondences:

Mass	m	$\rightarrow$	inductance	L
Coefficient of friction	μ_f	$\rightarrow$	resistance	R
Spring constant	K	$\rightarrow$	inverse of the capacitance	C^{-1}
Position variable	$p(t)$	$\rightarrow$	current variable	$i(t)$

So, if we are willing to talk about feet instead of amperes, lb $\cdot$ s^2/ft instead of H, ft/lb instead of F, and lb $\cdot$ s/ft instead of Ω, we can apply our newly found skills at modeling *RLC* circuits to the task of evaluating automotive shock absorbers.

Take a typical car wheel of 70 lb. The mass is found by dividing the weight by the earth's gravitational acceleration (32.17 ft/s^2), resulting in $m = 2.176$ lb $\cdot$ s^2/ft. The curb weight of our car is 1985 lb, and the static displacement of the spring is 4 inches (no passengers). The spring constant is obtained by dividing the weight on each shock absorber by the static displacement, so that we have $K = (\frac{1}{4})(1985)(3\ \text{ft}^{-1}) = 1489$ lb/ft. We are also told that the coefficient of friction for our piston/cylinder assembly is 65 lb $\cdot$ s/ft. Thus, we can simulate our shock absorber by modeling it with a series *RLC* circuit having $R = 65\ \Omega$, $L = 2.176$ H, and $C = K^{-1} = 671.6\ \mu$F.

The resonant frequency of our shock absorber is $\omega_0 = (LC)^{-1/2} = 26.16$ rad/s, and the damping coefficient is $\alpha = R/2L = 14.94$ s^{-1}. Since $\alpha < \omega_0$, our shock absorber represents an underdamped system; this means that we expect a bounce or two after we run over a pothole. A stiffer shock (larger coefficient of friction, or a larger resistance in our circuit model) is typically desirable when curves are taken at high speeds—at some point this corresponds to an overdamped response. However, if most of our driving is over unpaved roads, a slightly underdamped response is preferable.

This is applicable to any current or voltage in the circuit. Our final step is to solve for unknown constants given the initial conditions.

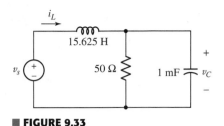

■ **FIGURE 9.33**

PRACTICE

9.10 Let $v_s = 10 + 20u(t)$ V in the circuit of Fig. 9.33. Find (a) $i_L(0)$; (b) $v_C(0)$; (c) $i_{L,f}$; (d) $i_L(0.1 \text{ s})$.

Ans: 0.2 A; 10 V; 0.6 A; 0.319 A.

9.7 THE LOSSLESS *LC* CIRCUIT

If the value of the resistance in a parallel *RLC* circuit becomes infinite, or zero in the case of a series *RLC* circuit, we have a simple *LC* loop in which an oscillatory response can be maintained forever. Let us look briefly at an example of such a circuit, and then discuss another means of obtaining an identical response without the need of supplying any inductance.

Consider the source-free circuit of Fig. 9.34, in which the large values $L = 4$ H and $C = \frac{1}{36}$ F are used so that the calculations will be simple. We let $i(0) = -\frac{1}{6}$ A and $v(0) = 0$. We find that $\alpha = 0$ and $\omega_0^2 = 9$ s^{-2}, so that $\omega_d = 3$ rad/s. In the absence of exponential damping, the voltage v is simply

$$v = A \cos 3t + B \sin 3t$$

Since $v(0) = 0$, we see that $A = 0$. Next,

$$\left.\frac{dv}{dt}\right|_{t=0} = 3B = -\frac{i(0)}{1/36}$$

But $i(0) = -\frac{1}{6}$ amperes, and therefore $dv/dt = 6$ V/s at $t = 0$. We must have $B = 2$ V and so

$$v = 2 \sin 3t \text{ V}$$

which is an undamped sinusoidal response; in other words, our voltage response does not decay.

Now let us see how we might obtain this voltage without using an *LC* circuit. Our intentions are to write the differential equation that v satisfies and then to develop a configuration of op amps that will yield the solution of the equation. Although we are working with a specific example, the technique is a general one that can be used to solve any linear homogeneous differential equation.

For the *LC* circuit of Fig. 9.34, we select v as our variable and set the sum of the downward inductor and capacitor currents equal to zero:

$$\frac{1}{4}\int_{t_0}^{t} v \, dt' - \frac{1}{6} + \frac{1}{36}\frac{dv}{dt} = 0$$

Differentiating once, we have

$$\frac{1}{4}v + \frac{1}{36}\frac{d^2v}{dt^2} = 0$$

or

$$\frac{d^2v}{dt^2} = -9v$$

■ **FIGURE 9.34** This circuit is lossless, and it provides the undamped response $v = 2 \sin 3t$ V, if $v(0) = 0$ and $i(0) = -\frac{1}{6}$ A.

In order to solve this equation, we plan to make use of the operational amplifier as an integrator. We assume that the highest-order derivative appearing in the differential equation here, d^2v/dt^2, is available in our configuration of op amps at an arbitrary point **A**. We now make use of the integrator, with $RC = 1$, as discussed in Sec. 7.5. The input is d^2v/dt^2, and the output must be $-dv/dt$, where the sign change results from using an inverting op-amp configuration for the integrator. The initial value of dv/dt is 6 V/s, as we showed when we first analyzed the circuit, and thus an initial value of -6 V must be set in the integrator. The negative of the first derivative now forms the input to a second integrator. Its output is therefore $v(t)$, and the initial value is $v(0) = 0$. Now it only remains to multiply v by -9 to obtain the second derivative we assumed at point **A**. This is amplification by 9 with a sign change, and it is easily accomplished by using the op amp as an inverting amplifier.

Figure 9.35 shows the circuit of an inverting amplifier. For an ideal op amp, both the input current and the input voltage are zero. Thus, the current going "east" through R_1 is v_s/R_1, while that traveling west through R_f is v_o/R_f. Since their sum is zero, we have

$$\frac{v_o}{v_s} = -\frac{R_f}{R_1}$$

Thus, we can design for a gain of -9 by setting $R_f = 90\,\text{k}\Omega$ and $R_1 = 10\,\text{k}\Omega$, for example.

If we let R be 1 MΩ and C be 1 μF in each of the integrators, then

$$v_o = -\int_0^t v_s \, dt' + v_o(0)$$

in each case. The output of the inverting amplifier now forms the assumed input at point **A**, leading to the configuration of op amps shown in Fig. 9.36. If the left switch is closed at $t = 0$ while the two initial-condition switches are opened at the same time, the output of the second integrator will be the undamped sine wave $v = 2 \sin 3t$ V.

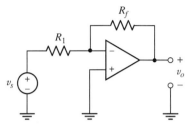

■ **FIGURE 9.35** The inverting operational amplifier provides a gain $v_o/v_s = -R_f/R_1$, assuming an ideal op amp.

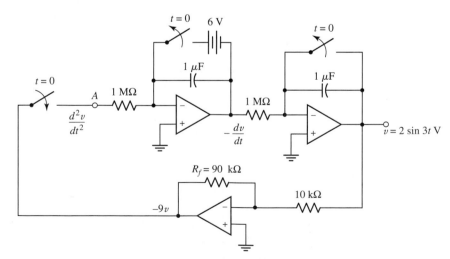

■ **FIGURE 9.36** Two integrators and an inverting amplifier are connected to provide the solution of the differential equation $d^2v/dt^2 = -9\,v$.

Note that both the *LC* circuit of Fig. 9.34 and the op-amp circuit of Fig. 9.36 have the same output, but the op-amp circuit does not contain a single inductor. It simply *acts* as though it contained an inductor, providing the appropriate sinusoidal voltage between its output terminal and ground. This can be a considerable practical or economic advantage in circuit design, as inductors are typically bulky, more costly than capacitors, and have more losses associated with them (and therefore are not as well approximated by the "ideal" model).

PRACTICE

9.11 Give new values for R_f and the two initial voltages in the circuit of Fig. 9.36 if the output represents the voltage $v(t)$ in the circuit of Fig. 9.37.

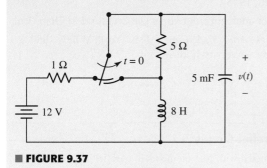

■ **FIGURE 9.37**

Ans: 250 kΩ; 400 V; 10 V.

SUMMARY AND REVIEW

❑ Circuits that contain two energy storage devices that cannot be combined using series/parallel combination techniques are described by a second-order differential equation.

❑ Series and parallel *RLC* circuits fall into one of three categories, depending on the relative values of R, L, and C:

$$\begin{array}{ll} \text{Overdamped} & (\alpha > \omega_0) \\ \text{Critically damped} & (\alpha = \omega_0) \\ \text{Underdamped} & (\alpha < \omega_0) \end{array}$$

❑ For series *RLC* circuits, $\alpha = R/2L$ and $\omega_0 = 1/\sqrt{LC}$.

❑ For parallel *RLC* circuits, $\alpha = 1/2RC$ and $\omega_0 = 1/\sqrt{LC}$.

❑ The typical form of an overdamped response is the sum of two exponential terms, one of which decays more quickly than the other: e.g., $A_1 e^{-t} + A_2 e^{-6t}$.

❑ The typical form of a critically damped response is $e^{-\alpha t}(A_1 t + A_2)$.

❑ The typical form of an underdamped response is an exponentially damped sinusoid: $e^{-\alpha t}(B_1 \cos \omega_d t + B_2 \sin \omega_d t)$.

❏ During the transient response of an *RLC* circuit, energy is transferred between energy storage elements to the extent allowed by the resistive component of the circuit, which acts to dissipate the energy initially stored.

❏ The complete response is the sum of the forced and natural responses. In this case the total response must be determined before solving for the constants.

READING FURTHER

An excellent discussion of employing PSpice in the modeling of automotive suspension systems can be found in

R.W. Goody, *MicroSim PSpice for Windows,* vol. I, 2nd ed. Englewood Cliffs, N.J.: Prentice-Hall, 1998.

Many detailed descriptions of analogous networks can be found in Chap. 3 of

E. Weber, *Linear Transient Analysis Volume I.* New York: Wiley, 1954. (Out of print, but in many university libraries.)

EXERCISES

9.1 The Source-Free Parallel Circuit

1. A certain circuit is constructed with four elements in parallel: a 4 Ω resistor, a 10 Ω resistor, a 1 μF capacitor, and a 2 mH inductor. (*a*) Compute α. (*b*) Compute ω_0. (*c*) Is the circuit underdamped, critically damped, or overdamped? Explain.

2. A parallel *RLC* circuit is constructed with a 2 H inductor and a 1 pF capacitor. What value of resistance should be added in parallel to ensure (*a*) an underdamped response; (*b*) a critically damped response?

3. A source-free *RLC* circuit has $R = 1\ \Omega$, $C = 1$ nF, and $L = 1$ pH. (*a*) Calculate α and ω_0. (*b*) Calculate s_1 and s_2. (*c*) What is the form of the inductor current response for $t > 0$?

4. A 22 aF capacitance is connected in parallel with a 1 fH inductance. What value of resistance connected in parallel will lead to (*a*) an underdamped response; (*b*) a critically damped response; (*c*) an overdamped response?

5. A source-free parallel *RLC* circuit contains an inductor for which the product $\omega_0 L$ is 10 Ω. If $s_1 = -6$ s^{-1} and $s_2 = -8$ s^{-1}, find R, L, and C.

6. The capacitor current in the circuit of Fig. 9.38 is $i_C = 40e^{-100t} - 30e^{-200t}$ mA. If $C = 1$ mF and $v(0) = -0.5$ V, find (*a*) $v(t)$; (*b*) $i_R(t)$; (*c*) $i(t)$.

7. A parallel *RLC* circuit is found to have a natural resonant frequency of $\omega_0 = 70.71 \times 10^{12}$ rad/s. It is known that the inductance $L = 2$ pH. (*a*) Compute C; (*b*) determine the value of resistance R that will lead to an exponential damping coefficient of 5 Gs^{-1}; (*c*) determine the neper frequency of the circuit; (*d*) compute s_1 and s_2; (*e*) calculate the damping ratio of the circuit.

8. Show that if $L = 4R^2 C$, the equation $v(t) = e^{-\alpha t}(A_1 t + A_2)$ is a solution to Eq. [4]. If $v(0) = 16$ V and $dv/dt|_{t=0} = 4$, find A_1 and A_2.

9. A 5 m length of 18 AWG solid copper wire is substituted for the resistor in Practice Problem 9.1. (*a*) Compute the resonant frequency of the new circuit. (*b*) Compute the new neper frequency of the circuit. (*c*) Calculate the percent change in the damping ratio.

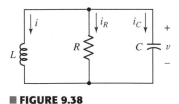

■ **FIGURE 9.38**

9.2 The Overdamped Parallel *RLC* Circuit

10. In the circuit of Fig. 9.39, let $L = 5$ H, $R = 8$ Ω, $C = 12.5$ mF, and $v(0^+) = 40$ V. Find (*a*) $v(t)$ if $i(0^+) = 8$ A; (*b*) $i(t)$ if $i_C(0^+) = 8$ A.

11. In the circuit of Fig. 9.39, $L = 1$ mH and $C = 100$ μF. (*a*) Choose $R = 0.1R_C$, where R_C is the value required to achieve critical damping. (*b*) If $i(0^-) = 4$ A and $v(0^-) = 10$ V, find $i(t)$ for $t > 0$.

12. The circuit shown schematically in Fig. 9.39 is constructed using $R = 20$ mΩ, $C = 50$ mF, and $L = 2$ mH. (*a*) Find an expression for $i_R(t)$ valid for $t > 0$ if $v(0^+) = 0$ and $i(0^-) = 2$ mA. (*b*) Sketch your solution over the range of $0 < t < 500$ ms. (*c*) Simulate the circuit using PSpice. Submit a properly labeled schematic with your plot. Does the simulation agree with your analytical result?

13. With reference to the circuit of Fig. 9.39, let $i(0) = 40$ A and $v(0) = 40$ V. If $L = 12.5$ mH, $R = 0.1$ Ω, and $C = 0.2$ F: (*a*) find $v(t)$, and (*b*) sketch i for $0 < t < 0.3$ s.

14. The values $R = 15$ $\mu\Omega$, $C = 50$ μF, and $L = 2$ μH are used in the circuit of Fig. 9.39. (*a*) Find an expression for $i_C(t)$ valid for $t > 0$ if $v(0^+) = 2$ and $i(0^-) = 0$. (*b*) Sketch your solution over the range of $0 < t < 5$ ns. (*c*) Simulate the circuit using PSpice. Submit a properly labeled schematic with your plot. Does the simulation agree with your analytical result?

15. For the circuit of Fig. 9.39, $R = 1$ Ω, $C = 4$ F, and $L = 20$ H. The initial conditions are $i(0) = 8$ A and $v(0) = 0$. (*a*) Find an expression for $v(t)$, $t > 0$. (*b*) Determine the peak value and the time at which it occurs. (*c*) Verify your analysis with a PSpice simulation. Be sure to submit a properly labeled schematic with your plot.

16. Obtain an expression for $i_L(t)$ in the circuit of Fig. 9.40 that is valid for all t.

17. Find $i_L(t)$ for $t \geq 0$ in the circuit shown in Fig. 9.41.

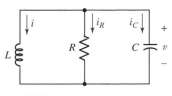

■ FIGURE 9.39

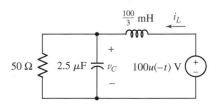

■ FIGURE 9.40

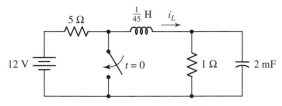

■ FIGURE 9.41

18. The circuit of Fig. 9.42 has been in the condition shown for a long time. After the switch closes at $t = 0$, find (*a*) $v(t)$; (*b*) $i(t)$; (*c*) the settling time for $v(t)$.

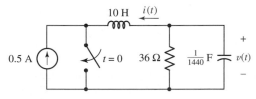

■ FIGURE 9.42

19. For the circuit of Fig. 9.42, the value of the inductance is 1250 mH. Determine $v(t)$ if it is known that the capacitor initially stores 390 J of energy and the inductor initially stores zero energy.

20. Referring to the circuit of Fig. 9.43, (*a*) what value of L will result in a transient response of the form $v = Ae^{-4t} + Be^{-6t}$? (*b*) Find A and B if $i_R(0^+) = 10$ A and $i_C(0^+) = 15$ A.

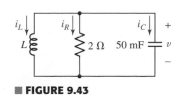

■ FIGURE 9.43

21. The switch in the circuit of Fig. 9.44 has been open since Alaska achieved statehood. Determine (*a*) $v_C(0^+)$; (*b*) $i_C(0^+)$; (*c*) $v_C(t)$. (*d*) Sketch $v_C(t)$. (*e*) Determine t when $v_C(t) = 0$. (*f*) Find the settling time.

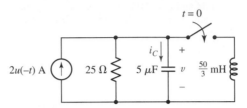

■ **FIGURE 9.44**

22. The switch in Fig. 9.45 was closed by the last crew aboard Mir before (at $t = 0$) it returned to earth. (*a*) Find $i_A(0^-)$. (*b*) Find $i_A(0^+)$. (*c*) Find $v_C(0^-)$. (*d*) Find the equivalent resistance in parallel with L and C for $t > 0$. (*e*) Find $i_A(t)$.

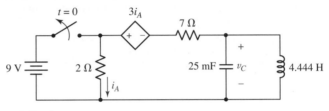

■ **FIGURE 9.45**

23. Two dimes are separated by a 1 mm-thick layer of ice at a temperature of 80 K. A coil of superconducting (and hence zero resistance) yttrium barium copper oxide wire having an inductance of 4 μH is carelessly blown off a nearby lab bench, falling so that each end is in contact with a different dime. The ice contains ionic impurities which cause it to be conducting. What resistance is needed for this bizarre structure to behave as an overdamped parallel *RLC* circuit?

9.3 Critical Damping

24. A parallel *RLC* circuit is constructed using a 1 mH inductor and a 12 μF capacitor. (*a*) Select R such that the circuit response is critically damped. (*b*) If $v_C(0^-) = 12$ V and $i_L(0^-) = 0$, find an expression for $v_C(t)$ valid for $t > 0$.

 25. A parallel *RLC* circuit is constructed using a 10 nH inductor and a 1 mF capacitor. (*a*) Select R such that the circuit response is critically damped. (*b*) If $v_C(0^-) = 0$ V and $i_L(0^-) = 10$ V, find an expression for $i_L(t)$ valid for $t > 0$. (*d*) Sketch your solution, and verify with a PSpice simulation. Include a properly labeled schematic with your plot. Do the two solutions agree?

26. Explain why it is unlikely one would encounter a circuit exhibiting a critically damped response in practice.

27. Change the inductance value in the circuit of Fig. 9.41 until the circuit is critically damped. (*a*) What is the new inductance? (*b*) Find i_L at $t = 5$ ms. (*c*) Find the settling time.

28. (*a*) What new value of resistance should be used in the circuit of Fig. 9.40 to achieve critical damping? (*b*) Using this value of resistance, find $v_C(t)$ for $t > 0$.

29. In the situation described in Exer. 23, what value of resistance must the ice have to result in a critically damped *RLC* circuit?

30. In the circuit of Fig. 9.39, let $v(0) = -400$ V and $i(0) = 0.1$ A. If $L = 5$ mH, $C = 10$ nF, and the circuit is critically damped: (*a*) find R; (*b*) find $|i|_{max}$; (*c*) find i_{max}.

31. A parallel *RLC* circuit has $\alpha = 1$ ms^{-1}, $R = 1$ MΩ, and is known to be critically damped. Assume the value of the inductor can be computed using the expression $L = \mu N^2 A / s$ where $\mu = 4\pi \times 10^{-7}$ H/m, $N =$ the number of complete turns of the coil, $A =$ the cross-sectional area of the coil, and $s =$ the axial length of the entire coil. The cross section of the inductor is 1 cm^2, there are 50 turns of wire per cm, and the coil is fabricated from a newly discovered element gluonium, which is superconducting up to temperatures of 100°F. How long is the coil?

9.4 The Underdamped Parallel *RLC* Circuit

32. For the circuit shown in Fig. 9.46, find (*a*) $i_L(0^+)$; (*b*) $v_C(0^+)$; (*c*) $di_L/dt|_{t=0^+}$; (*d*) $dv_C/dt|_{t=0^-}$; (*e*) $v_C(t)$. (*f*) Sketch $v_C(t)$, $-0.1 < t < 2$ s.

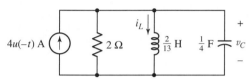

■ **FIGURE 9.46**

33. Find $i_C(t)$ for $t > 0$ in the circuit shown in Fig. 9.47.

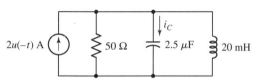

■ **FIGURE 9.47**

34. Let $\omega_d = 6$ rad/s in the circuit of Fig. 9.48. (*a*) Find L. (*b*) Obtain an expression for $i_L(t)$ valid for all t. (*c*) Sketch $i_L(t)$, $-0.1 < t < 0.6$ s.

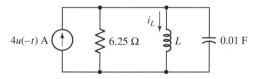

■ **FIGURE 9.48**

35. After being open for a long time, the switch in the circuit of Fig. 9.49 is closed at $t = 0$. For $t > 0$, find (*a*) $v_C(t)$; (*b*) $i_{SW}(t)$.

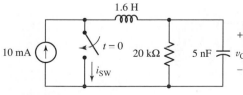

■ **FIGURE 9.49**

36. (*a*) Find $v(t)$ for $t > 0$ for the circuit shown in Fig. 9.50. (*b*) Make a quick sketch of $v(t)$ over the time interval $0 < t < 0.1$ s.

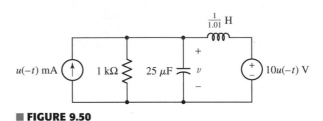

■ **FIGURE 9.50**

37. Find $i_1(t)$ for $t > 0$ in the circuit of Fig. 9.51.

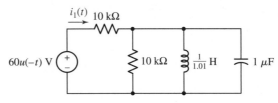

■ **FIGURE 9.51**

38. What minimum value of resistance should replace the 25 Ω resistor in the circuit of Fig. 9.44 if an underdamped response $v(t)$ is desired? Multiply your specified resistance by 1000 and plot the response. Use PSpice to determine the settling time, and include a properly labeled schematic with your plot.

39. Determine the value of R for the underdamped circuit of Fig. 9.14 [$L = 7$ H, $C = \frac{1}{42}$ F, $i(0) = 10$ A, $v(0) = 0$] that will lead to a minimum value of the settling time t_s. What is the value of t_s?

40. (*a*) Replace the 2 Ω resistor of Fig. 9.46 with a 5 Ω resistor. Obtain an expression for $i_L(t)$ and solve for $t = 2.5$ s. (*b*) Replace the 2 Ω resistor of Fig. 9.46 with a 0.5 Ω resistor. Obtain an expression for $i_L(t)$ and solve for $t = 250$ ms. (*c*) Simulate the circuits of parts (*a*) and (*b*), and plot the inductor current of each circuit on the same graph. Submit both the plot and a suitably labeled schematic.

41. (*a*) Model the circuit of Fig. 9.46 using PSpice. Instead of a $4u(-t)$ A current source, obtain an equivalent source-free circuit by specifying the appropriate initial conditions for the inductor and capacitor. Submit a properly labeled schematic. (*b*) Plot the current $i_L(t)$ using Probe, and compare to the solution obtained by hand. Use Probe to determine the settling time.

9.5 The Source-Free Series *RLC* Circuit

42. Find v_C, v_R, and v_L at $t = 40$ ms in the circuit shown in Fig. 9.52.

43. Find $i_L(t)$ for $t > 0$ in the circuit of Fig. 9.53.

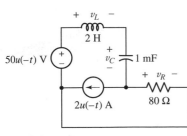

■ **FIGURE 9.52**

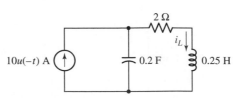

■ **FIGURE 9.53**

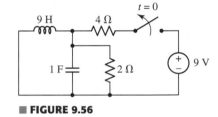

44. In the circuit of Fig. 9.21a, let $R = 300 \ \Omega$ and $C = 1 \ \mu F$ with the circuit critically damped. If $v_C(0) = -10$ V and $i(0) = -150$ mA, find (a) $v_C(t)$; (b) $|v_C|_{max}$; (c) $v_{C,max}$.

45. Write the dual of Exer. 16, including the dual of the circuit shown in Fig. 9.40. Solve the dual problem.

46. (a) Find $i_L(t)$ for $t > 0$ in the circuit shown in Fig. 9.54. (b) Find $|i_L|_{max}$ and $i_{L,max}$.

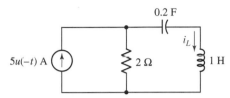

■ **FIGURE 9.54**

47. For the circuit of Fig. 9.55, $t > 0$, find (a) $i_L(t)$; (b) $v_C(t)$.

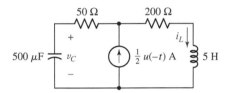

■ **FIGURE 9.55**

48. Determine the energy stored in the inductor of Fig. 9.56 at $t = 2$ s. Verify your answer with PSpice.

49. The switch in Fig. 9.57 has been closed an interminably long time. Determine the peak magnitude of the voltage that develops across the 500 mH inductor, and verify your answer with PSpice.

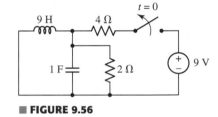

■ **FIGURE 9.56**

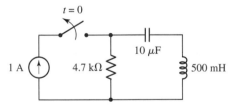

■ **FIGURE 9.57**

50. A very well built capacitor, once connected to a 12 V battery long enough to fully charge before the battery was put back in the snowmobile, is lying on the floor of a radio shack up in northern Canada. During a mild earthquake, an old coiled telephone cord falls off a shelf and onto the floor, with one end coming into contact with one terminal of the capacitor. The telephone cord has a resistance of 14 mΩ and an inductance of 5 μH; the capacitor is initially storing 144 mJ of energy. (a) What is the capacitor voltage just prior to the earthquake? (b) What is the capacitor voltage 1 s after the telephone cord hits the capacitor? (c) A soggy polar bear breaks into the shack looking for food and accidentally places one paw on the unconnected end of the telephone cord and another paw on the unconnected terminal of the capacitor. The polar bear's body jerks for 18 μs before it roars and runs out of the shack. If it takes 100 mA to make a bear twitch that violently, what was the resistance of the soggy fur coat?

51. Determine what resistance must replace the 2 Ω resistor in the circuit of Fig. 9.56 so that the circuit is critically damped. Calculate the energy stored in the inductor at $t = 100$ ms.

52. Find an expression for i_L as indicated in Fig. 9.58, valid for $t > 0$.

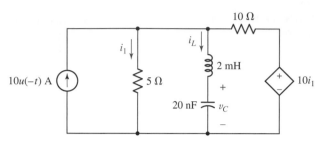

■ **FIGURE 9.58**

53. Find an expression for v_C as indicated in Fig. 9.58, valid for $t > 0$.
54. Referring to the circuit depicted in Fig. 9.59, obtain an expression for i_1 valid for all time if $C = 1$ F.

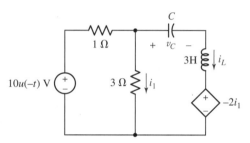

■ **FIGURE 9.59**

55. Referring to the circuit depicted in Fig. 9.59, obtain an expression for v_C valid for all time if $C = 1$ mF.

9.6 The Complete Response of the *RLC* Circuit

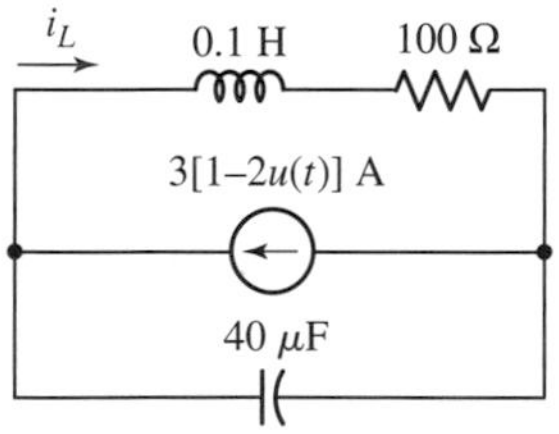

56. (*a*) Find $i_L(t)$ for all t in the circuit of Fig. 9.60. (*b*) At what instant of time after $t = 0$ is $i_L(t) = 0$?
57. The source in the circuit shown in Fig. 9.53 is changed to $10u(t)$ A. Find $i_L(t)$.
58. Replace the source in the circuit of Fig. 9.55 with $i_s = 0.5[1 - 2u(t)]$ A and find $i_L(t)$.
59. Replace the source shown in Fig. 9.47 with $i_s = 2[1 + u(t)]$ A and find $i_C(t)$ for $t > 0$.
60. (*a*) Find $v_C(t)$ for $t > 0$ in the circuit shown in Fig. 9.61. (*b*) Sketch $v_C(t)$ versus t, $-0.1 < t < 2$ ms.

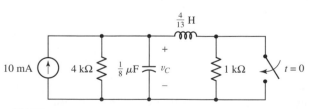

■ **FIGURE 9.61**

■ **FIGURE 9.60**

61. The switch in the circuit of Fig. 9.62 has been closed for a very long time. It opens at $t = 0$. Find $v_C(t)$ for $t > 0$.

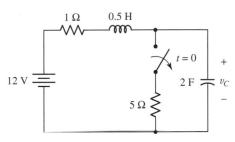

■ **FIGURE 9.62**

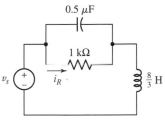

■ **FIGURE 9.63**

62. Find $i_R(t)$ for $t > 0$ in the circuit of Fig. 9.63 if $v_s(t)$ equals (a) $10u(-t)$ V; (b) $10u(t)$ V.

63. Find $i_s(t)$ for $t > 0$ in the circuit of Fig. 9.64 if $v_s(t)$ equals (a) $10u(-t)$ V; (b) $10u(t)$ V.

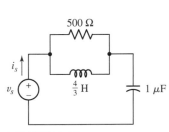

■ **FIGURE 9.64**   ■ **FIGURE 9.65**

64. Replace the 2 Ω resistor in the circuit of Fig. 9.65 with a 3 H inductor. Determine the energy stored in the capacitor at $t = 200$ ms if the current source increases from 15 A to 22 A at $t = 0$. Verify your answer with a PSpice simulation.

65. The current source in the circuit of Fig. 9.65 suddenly increases from 15 A to 22 A at $t = 0$. Find the voltage v_s at (a) $t = 0^-$; (b) $t = 0^+$; (c) $t = \infty$; (d) $t = 3.4$ s. Verify your answers with the appropriate PSpice simulations.

66. The current source in the circuit of Fig. 9.65 suddenly drops from 15 A to 0 A at $t = 0$, then increases to 3 A at $t = 1$ s. Plot the voltage $v_s(t)$. Verify your solution with a PSpice simulation.

67. A 5 mH inductor, a 25 μF capacitor, and a 20 Ω resistor are in series with a voltage source $v_x(t)$. The source voltage is zero prior to $t = 0$. At $t = 0$, it jumps to 75 V, at $t = 1$ ms it drops to zero, at $t = 2$ ms it again jumps to 75 V, and it continues in this periodic fashion thereafter. Find the source current at (a) $t = 0^-$; (b) $t = 0^+$; (c) $t = 1$ ms; (d) $t = 2$ ms.

68. Design a circuit that will produce a damped sinusoidal pulse with a peak voltage of 5 V, and at least three additional peaks with voltage magnitude greater than 1 V. Verify your design using PSpice.

69. A 12 V battery is sitting in a hut on a deserted island somewhere in the Pacific. The positive terminal of the battery is connected to one end of a 314.2 pF capacitor in series with a 869.1 μH inductor. An earthquake in the Bonin Islands of Japan triggers a tsunami that crashes into the hut, spilling salt water onto a rag that connects the other end of the inductor/capacitor combination to the negative terminal of the battery so that a series *RLC* circuit is formed. The resulting oscillation is detected by a nearby ship monitoring a radio beacon signal at 290.5 kHz (1.825 Mrad/s). What is the resistance of the damp rag?

70. Find the voltage $v_C(t)$ across the capacitor of Fig. 9.66 at $t = 1$ ms. Verify your answer with a PSpice simulation.

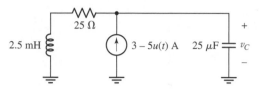

■ **FIGURE 9.66**

9.7 The Lossless *LC* Circuit

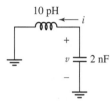

■ **FIGURE 9.67**

71. Design an op-amp circuit to model the voltage response of the *LC* circuit shown in Fig. 9.67. Verify your design by simulating both the circuit of Fig. 9.67 and your circuit using an LF 411 op amp, assuming $v(0) = 0$ and $i(0) = 1$ mA.

72. Refer to Fig. 9.68, and design an op-amp circuit whose output will be $i(t)$ for $t > 0$.

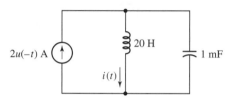

■ **FIGURE 9.68**

73. A source-free *RC* circuit is constructed using a 1 kΩ resistor and a 3.3 mF capacitor. The initial voltage across the capacitor is 1.2 V. (*a*) Write the differential equation for v, the voltage across the capacitor, for $t > 0$. (*b*) Design an op-amp circuit that provides $v(t)$ as the output.

74. Replace the capacitor in the circuit of Fig. 9.67 with a 20 H inductor in parallel with a 5 μF capacitor. Design an op-amp circuit whose output will be $i(t)$ for $t > 0$. Verify your design by simulating both the capacitor-inductor circuit and your op-amp circuit. Use an LM111 op amp in the PSpice simulation.

75. A source-free *RL* circuit contains a 20 Ω resistor and a 5 H inductor. If the initial value of the inductor current is 2 A: (*a*) write the differential equation for i for $t > 0$, and (*b*) design an op-amp integrator to provide $i(t)$ as the output, using $R_1 = 1$ MΩ and $C_f = 1$ μF.

Sinusoidal Steady-State Analysis

INTRODUCTION

The complete response of a linear electric circuit is composed of two parts, the *natural* response and the *forced* response. The natural response is the short-lived transient response of a circuit to a sudden change in its condition. The forced response is the long-term steady-state response of a circuit to any independent sources present. Up to this point, the only forced response we have considered is that due to dc sources. Another very common forcing function is the sinusoidal waveform. This function describes the voltage available at household electrical sockets as well as the voltage of power lines connected to residential and industrial areas.

In this chapter, we assume that the transient response is of little interest, and the steady-state response of a circuit (a television set, a toaster, or a power distribution network) to a sinusoidal voltage or current is needed. We will analyze such circuits using a powerful technique that transforms integrodifferential equations into algebraic equations.

Characteristics of Sinusoidal Functions

Phasor Representation of Sinusoids

Converting Between the Time and Frequency Domains

Impedance and Admittance

Reactance and Susceptance

Parallel and Series Combinations in the Frequency Domain

Determination of Forced Response Using Phasors

Application of Circuit Analysis Techniques in the Frequency Domain

10.1 CHARACTERISTICS OF SINUSOIDS

Consider a sinusoidally varying voltage

$$v(t) = V_m \sin \omega t$$

shown graphically in Figs. 10.1*a* and *b*. The *amplitude* of the sine wave is V_m, and the *argument* is ωt. The *radian frequency,* or *angular frequency,* is ω. In Fig. 10.1*a*, $V_m \sin \omega t$ is plotted as a function of the argument ωt, and the periodic nature of the sine wave is evident. The function repeats itself every 2π radians, and its **period** is therefore 2π radians. In Fig. 10.1*b*, $V_m \sin \omega t$ is plotted as a function of t and the *period* is now T. A sine wave having a period T must execute

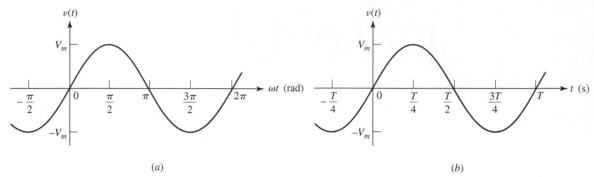

■ FIGURE 10.1 The sinusoidal function $v(t) = V_m \sin \omega t$ is plotted (a) versus ωt and (b) versus t.

$1/T$ periods each second; its **frequency** f is $1/T$ hertz, abbreviated Hz. Thus,

$$f = \frac{1}{T}$$

and since

$$\omega T = 2\pi$$

we obtain the common relationship between frequency and radian frequency,

$$\boxed{\omega = 2\pi f}$$

Lagging and Leading

A more general form of the sinusoid,

$$v(t) = V_m \sin(\omega t + \theta) \qquad [1]$$

includes a *phase angle* θ in its argument. Equation [1] is plotted in Fig. 10.2 as a function of ωt, and the phase angle appears as the number of radians by which the original sine wave (shown in green color in the sketch) is shifted to the left, or earlier in time. Since corresponding points on the sinusoid $V_m \sin(\omega t + \theta)$ occur θ rad, or θ/ω seconds, earlier, we say that $V_m \sin(\omega t + \theta)$ *leads* $V_m \sin \omega t$ by θ rad. Therefore, it is correct to describe $\sin \omega t$ as **lagging** $\sin(\omega t + \theta)$ by θ rad, as **leading** $\sin(\omega t + \theta)$ by $-\theta$ rad, or as leading $\sin(\omega t - \theta)$ by θ rad.

In either case, leading or lagging, we say that the sinusoids are *out of phase*. If the phase angles are equal, the sinusoids are said to be *in phase*.

In electrical engineering, the phase angle is commonly given in degrees, rather than radians; to avoid confusion we should be sure to always use the

Recall that to convert radians to degrees, we simply multiply the angle by $180/\pi$.

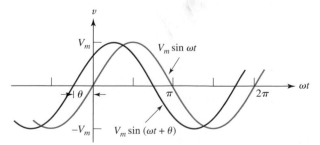

■ FIGURE 10.2 The sine wave $V_m \sin(\omega t + \theta)$ leads $V_m \sin \omega t$ by θ rad.

degree symbol. Thus, instead of writing

$$v = 100 \sin\left(2\pi \, 1000t - \frac{\pi}{6}\right)$$

we customarily use

$$v = 100 \sin(2\pi \, 1000t - 30°)$$

In evaluating this expression at a specific instant of time, e.g., $t = 10^{-4}$ s, $2\pi \, 1000t$ becomes 0.2π *radians,* and this should be expressed as $36°$ before $30°$ is subtracted from it. Don't confuse your apples with your oranges.

> *Two sinusoidal waves whose phases are to be compared must:*
>
> 1. Both be written as sine waves, or both as cosine waves.
> 2. Both be written with positive amplitudes.
> 3. Each have the same frequency.

Converting Sines to Cosines

The sine and cosine are essentially the same function, but with a $90°$ phase difference. Thus, $\sin \omega t = \cos(\omega t - 90°)$. Multiples of $360°$ may be added to or subtracted from the argument of any sinusoidal function without changing the value of the function. Hence, we may say that

$$v_1 = V_{m_1} \cos(5t + 10°)$$
$$= V_{m_1} \sin(5t + 90° + 10°)$$
$$= V_{m_1} \sin(5t + 100°)$$

Note that:

$-\sin \omega t = \sin(\omega t \pm 180°)$
$-\cos \omega t = \cos(\omega t \pm 180°)$
$\mp\sin \omega t = \cos(\omega t \pm 90°)$
$\pm\cos \omega t = \sin(\omega t \pm 90°)$

leads

$$v_2 = V_{m_2} \sin(5t - 30°)$$

by $130°$. It is also correct to say that v_1 *lags* v_2 by $230°$, since v_1 may be written as

$$v_1 = V_{m_1} \sin(5t - 260°)$$

We assume that V_{m_1} and V_{m_2} are both positive quantities. A graphical representation is provided in Fig. 10.3; note that the frequency of both sinusoids (5 rad/s in this case) must be the same, or the comparison is meaningless. Normally, the difference in phase between two sinusoids is expressed by that angle which is less than or equal to $180°$ in magnitude.

The concept of a leading or lagging relationship between two sinusoids will be used extensively, and the relationship should be recognizable both mathematically and graphically.

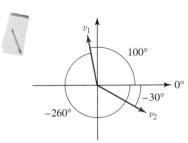

■ **FIGURE 10.3** A graphical representation of the two sinusoids v_1 and v_2. The magnitude of each sine function is represented by the length of the corresponding arrow, and the phase angle by the orientation with respect to the positive x axis. In this diagram, v_1 leads v_2 by $100° + 30° = 130°$, although it could also be argued that v_2 leads v_1 by $230°$. It is customary, however, to express the phase difference by an angle less than or equal to $180°$ in magnitude.

PRACTICE

10.1 Find the angle by which i_1 lags v_1 if $v_1 = 120 \cos(120\pi t - 40°)$ V and i_1 equals (a) $2.5 \cos(120\pi t + 20°)$ A; (b) $1.4 \sin(120\pi t - 70°)$ A; (c) $-0.8 \cos(120\pi t - 110°)$ A.

10.2 Find A, B, C, and ϕ if $40 \cos(100t - 40°) - 20 \sin(100t + 170°) = A \cos 100t + B \sin 100t = C \cos(100t + \phi)$.

Ans: 10.1: $-60°$; $120°$; $-110°$. 10.2: 27.2; 45.4; 52.9; $-59.1°$.

10.2 • FORCED RESPONSE TO SINUSOIDAL FUNCTIONS

Now that we are familiar with the mathematical characteristics of sinusoids, we are ready to apply a sinusoidal forcing function to a simple circuit and obtain the forced response. We will first write the differential equation that applies to the given circuit. The complete solution of this equation is composed of two parts, the complementary solution (which we call the *natural response*) and the particular integral (or *forced response*). The methods we plan to develop in this chapter assume that we are not interested in the short-lived transient or natural response of our circuit, but only in the long-term or "steady-state" response.

The Steady-State Response

The term *steady-state response* is used synonymously with *forced response,* and the circuits we are about to analyze are commonly said to be in the "sinusoidal steady state." Unfortunately, *steady state* carries the connotation of "not changing with time" in the minds of many students. This is true for dc forcing functions, but the sinusoidal steady-state response is definitely changing with time. The steady state simply refers to the condition that is reached after the transient or natural response has died out.

The forced response has the mathematical form of the forcing function, plus all its derivatives and its first integral. With this knowledge, one of the methods by which the forced response may be found is to assume a solution composed of a sum of such functions, where each function has an unknown amplitude to be determined by direct substitution in the differential equation. As we are about to see, this can be a lengthy process, so we will be sufficiently motivated to seek out a simpler alternative.

Consider the series *RL* circuit shown in Fig. 10.4. The sinusoidal source voltage $v_s = V_m \cos \omega t$ has been switched into the circuit at some remote time in the past, and the natural response has died out completely. We seek the forced (or "steady-state") response, which must satisfy the differential equation

$$L\frac{di}{dt} + Ri = V_m \cos \omega t$$

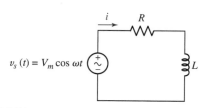

■ **FIGURE 10.4** A series *RL* circuit for which the forced response is desired.

obtained by applying KVL around the simple loop. At any instant where the derivative is equal to zero, we see that the current must have the form $i \propto \cos \omega t$. Similarly, at an instant where the current is equal to zero, the *derivative* must be proportional to cos ωt, implying a current of the form sin ωt. We might expect, therefore, that the forced response will have the general form

$$i(t) = I_1 \cos \omega t + I_2 \sin \omega t$$

where I_1 and I_2 are real constants whose values depend upon V_m, R, L, and ω. No constant or exponential function can be present. Substituting the assumed form for the solution in the differential equation yields

$$L(-I_1\omega \sin \omega t + I_2\omega \cos \omega t) + R(I_1 \cos \omega t + I_2 \sin \omega t) = V_m \cos \omega t$$

If we collect the cosine and sine terms, we obtain

$$(-L\,I_1\omega + RI_2) \sin \omega t + (L\,I_2\omega + RI_1 - V_m) \cos \omega t = 0$$

This equation must be true for all values of t, which can be achieved only if the factors multiplying $\cos \omega t$ and $\sin \omega t$ are each zero. Thus,

$$-\omega L\, I_1 + R I_2 = 0 \qquad \text{and} \qquad \omega L\, I_2 + R I_1 - V_m = 0$$

and simultaneous solution for I_1 and I_2 leads to

$$I_1 = \frac{R V_m}{R^2 + \omega^2 L^2} \qquad I_2 = \frac{\omega L V_m}{R^2 + \omega^2 L^2}$$

Thus, the forced response is obtained:

$$i(t) = \frac{R V_m}{R^2 + \omega^2 L^2} \cos \omega t + \frac{\omega L V_m}{R^2 + \omega^2 L^2} \sin \omega t \qquad [2]$$

A More Compact and User-Friendly Form

This expression is slightly cumbersome, however, and a clearer picture of the response can be obtained by expressing the response as a single sinusoid or cosinusoid with a phase angle. We choose to express the response as a cosine function:

$$i(t) = A \cos(\omega t - \theta) \qquad [3]$$

At least two methods of obtaining the values of A and θ suggest themselves. We might substitute Eq. [3] directly in the original differential equation, or we could simply equate the two solutions, Eqs. [2] and [3]. Selecting the latter method, and expanding the function $\cos(\omega t - \theta)$:

$$A \cos \theta \cos \omega t + A \sin \theta \sin \omega t = \frac{R V_m}{R^2 + \omega^2 L^2} \cos \omega t + \frac{\omega L V_m}{R^2 + \omega^2 L^2} \sin \omega t$$

Next, we collect the coefficients of $\cos \omega t$ and $\sin \omega t$, and find

$$A \cos \theta = \frac{R V_m}{R^2 + \omega^2 L^2} \qquad \text{and} \qquad A \sin \theta = \frac{\omega L V_m}{R^2 + \omega^2 L^2}$$

To determine A and θ, we divide one equation by the other:

$$\frac{A \sin \theta}{A \cos \theta} = \tan \theta = \frac{\omega L}{R}$$

and also square both equations and add the results:

$$A^2 \cos^2 \theta + A^2 \sin^2 \theta = A^2 = \frac{R^2 V_m^2}{(R^2 + \omega^2 L^2)^2} + \frac{\omega^2 L^2 V_m^2}{(R^2 + \omega^2 L^2)^2}$$

$$= \frac{V_m^2}{R^2 + \omega^2 L^2}$$

Hence,

$$\theta = \tan^{-1} \frac{\omega L}{R}$$

and

$$A = \frac{V_m}{\sqrt{R^2 + \omega^2 L^2}}$$

The *alternative form* of the forced response therefore becomes

$$i(t) = \frac{V_m}{\sqrt{R^2 + \omega^2 L^2}} \cos\left(\omega t - \tan^{-1} \frac{\omega L}{R}\right) \qquad [4]$$

Several useful trigonometric identities are provided on the inside cover of the book.

We see that the amplitude of the *response* is proportional to the amplitude of the *forcing function;* if not, the linearity concept would have to be discarded. The amplitude of the response also decreases as R, L, or ω is increased, but not proportionately. The current is seen to lag the applied voltage by $\tan^{-1}(\omega L/R)$, an angle between 0 and 90°. When $\omega = 0$ or $L = 0$, the current must be in phase with the voltage; since the former situation is direct current and the latter provides a resistive circuit, the result agrees with our previous experience. If $R = 0$, the current lags the voltage by 90°. In an inductor, then, if the passive sign convention is satisfied, the current lags the voltage by exactly 90°. In a similar manner[1] we can show that the current through a capacitor *leads* the voltage across it by 90°.

The phase difference between the current and voltage depends upon the ratio of the quantity ωL to R. We call ωL the *inductive reactance* of the inductor; it is measured in ohms, and it is a measure of the opposition that is offered by the inductor to the passage of a sinusoidal current.

Let us see how we can apply the results of this general analysis to a specific circuit that is not just a simple series loop. Note that we are implicitly ignoring the transient response now; the assumption is that we are concerned only with the steady-state or forced response of the circuit, so that any and all transients have long died out.

EXAMPLE 10.1

Find the current i_L in the circuit shown in Fig. 10.5a.

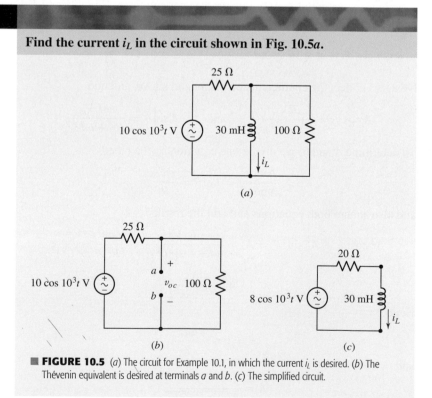

■ **FIGURE 10.5** (*a*) The circuit for Example 10.1, in which the current i_L is desired. (*b*) The Thévenin equivalent is desired at terminals *a* and *b*. (*c*) The simplified circuit.

(1) Once upon a time, the symbol E (for electromotive force) was used to designate voltages. Then every student learned the phase "ELI the ICE man" as a reminder that *voltage* leads *current* in an *inductive* circuit, while *current* leads *voltage* in a *capacitive* circuit. Now that we use V instead, it just isn't the same.

Although this circuit has a sinusoidal source and a single inductor, it contains two resistors and is not a single loop. In order to apply the results of the preceding analysis, we need to seek the Thévenin equivalent as viewed from terminals a and b in Fig. 10.5b.

The open-circuit voltage v_{oc} is

$$v_{oc} = (10\cos 10^3 t)\frac{100}{100 + 25} = 8\cos 10^3 t \qquad \text{V}$$

Since there are no dependent sources in sight, we find R_{th} by killing the independent source and calculating the resistance of the passive network, so $R_{th} = (25 \times 100)/(25 + 100) = 20\ \Omega$.

Now we do have a series RL circuit, with $L = 30$ mH, $R_{th} = 20\ \Omega$, and a source voltage of 8 cos $10^3 t$ V, as shown in Fig. 10.5c. Thus, applying Eq. [4], which was derived for a general RL series circuit,

$$i_L = \frac{8}{\sqrt{20^2 + (10^3 \times 30 \times 10^{-3})^2}}\cos\left(10^3 t - \tan^{-1}\frac{30}{20}\right)$$

$$= 222\cos(10^3 t - 56.3°)\ \text{mA}$$

The voltage and current waveforms are plotted in Fig. 10.6.

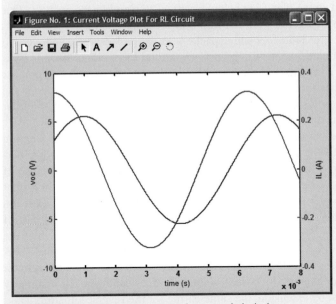

■ **FIGURE 10.6** Voltage and current waveforms on a dual axis plot, generated using MATLAB:
```
EDU» t = linspace(0,8e-3,1000);
EDU» v = 8*cos(1000*t);
EDU» i = 0.222*cos(1000*t − 56.3*pi/180);
EDU» plotyy(t,v,t,i);
EDU» xlabel('time (s)');
```

Note that there is not a 90° phase difference between the current and voltage waveforms of the plot. This is because we are not plotting the inductor voltage, which is left as an exercise for the reader.

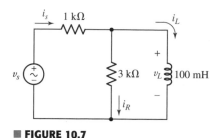

■ FIGURE 10.7

10.3 THE COMPLEX FORCING FUNCTION

The method by which we found the sinusoidal steady-state response for the
general series *RL* circuit was not a trivial problem. We might think of the
analytical complications as arising through the presence of the inductor; if
both the passive elements had been resistors, the analysis would have been
ridiculously easy, even with the sinusoidal forcing function present. The reason the analysis would be so easy results from the simple voltage-current
relationship specified by Ohm's law. The voltage-current relationship for an
inductor is not as simple, however; instead of solving an algebraic equation,
we were faced with a nonhomogeneous differential equation. It would be
rather impractical to analyze every circuit by the method described in the
example, and so we plan to develop a method to simplify the analysis. Our
result will be an algebraic relationship between sinusoidal current and sinusoidal voltage for inductors and capacitors as well as resistors, and we will
be able to produce a set of *algebraic* equations for a circuit of any complexity. The constants and the variables in the equations will be complex
numbers rather than real numbers, but the analysis of any circuit in the
sinusoidal steady state becomes almost as easy as the analysis of a similar
resistive circuit.

We are now ready to think about applying a complex forcing function
(that is, one that has both a *real* and an *imaginary* part) to an electrical network. This may seem like a strange idea, but we will find that the use of
complex quantities in sinusoidal steady-state analysis leads to methods that
are much simpler than those involving purely real quantities. We expect a
complex forcing function to produce a complex response; the real part of
the forcing function will produce the real part of the response, while the
imaginary portion of the forcing function will result in the imaginary portion of the response. Hopefully this seems reasonable: it would be difficult
to think of an example of a real voltage source leading to an imaginary
response, and, by extension, the same is true for the reverse situation.

In Fig. 10.8, a sinusoidal source

$$V_m \cos(\omega t + \theta) \qquad\qquad [5]$$

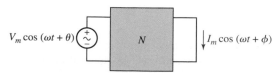

■ **FIGURE 10.8** The sinusoidal forcing function $V_m \cos(\omega t + \theta)$
produces the steady-state sinusoidal response $I_m \cos(\omega t + \phi)$.

is connected to a general network, which we will assume to contain only passive elements (i.e., no independent sources) in order to avoid having to invoke the superposition principle. A current response in some other branch of the network is to be determined, and the parameters appearing in Eq. [5] are all real quantities.

We have shown that we may represent the response by the general cosine function

$$I_m \cos(\omega t + \phi) \qquad [6]$$

A sinusoidal forcing function always produces a sinusoidal forced response of the same frequency in a linear circuit.

Now let us change our time reference by shifting the phase of the forcing function by 90°, or changing the instant that we call $t = 0$. Thus, the forcing function

$$V_m \cos(\omega t + \theta - 90°) = V_m \sin(\omega t + \theta) \qquad [7]$$

when applied to the same network will produce a corresponding response

$$I_m \cos(\omega t + \phi - 90°) = I_m \sin(\omega t + \phi) \qquad [8]$$

We next depart from physical reality by applying an imaginary forcing function, one that cannot be applied in the laboratory but can be applied mathematically.

Imaginary Sources Lead to . . . Imaginary Responses

We construct an imaginary source very simply; it is only necessary to multiply Eq. [7] by j, the imaginary operator. We thus apply

$$j V_m \sin(\omega t + \theta) \qquad [9]$$

Electrical engineers use "j" instead of "i" to represent $\sqrt{-1}$ to avoid confusion with currents.

What is the response? If we had doubled the source, then the principle of linearity would require that we double the response; multiplication of the forcing function by a constant k would result in the multiplication of the response by the same constant k. The fact that our constant is $\sqrt{-1}$ does not destroy this relationship. The response to the imaginary source of Eq. [9] is thus

$$j I_m \sin(\omega t + \phi) \qquad [10]$$

The imaginary source and response are indicated in Fig. 10.9.

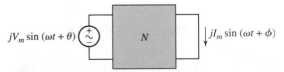

$jV_m \sin(\omega t + \theta)$ N $\downarrow jI_m \sin(\omega t + \phi)$

■ **FIGURE 10.9** The imaginary sinusoidal forcing function $jV_m \sin(\omega t + \theta)$ produces the imaginary sinusoidal response $jI_m \sin(\omega t + \phi)$ in the network of Fig. 10.8.

Applying a Complex Forcing Function

We have applied a *real source* and obtained a *real* response; we have also applied an *imaginary* source and obtained an *imaginary* response. Since we are dealing with a *linear* circuit, we may use the superposition theorem to find the response to a complex forcing function which is the sum of the real and imaginary forcing functions. Thus, the sum of the forcing functions of Eqs. [5] and [9],

$$V_m \cos(\omega t + \theta) + jV_m \sin(\omega t + \theta) \qquad [11]$$

must therefore produce a response that is the sum of Eqs. [6] and [10],

$$I_m \cos(\omega t + \phi) + jI_m \sin(\omega t + \phi) \qquad [12]$$

The complex source and response may be represented more simply by applying Euler's identity, which states that $\cos(\omega t + \theta) + j \sin(\omega t + \theta) = e^{j(\omega t + \theta)}$. Thus, the source of Eq. [11] may be written as

$$V_m e^{j(\omega t + \theta)} \qquad [13]$$

and the response of Eq. [12] is

$$I_m e^{j(\omega t + \phi)} \qquad [14]$$

The complex source and response are illustrated in Fig. 10.10.

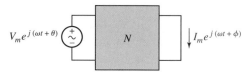

$V_m e^{j(\omega t + \theta)}$　N　$I_m e^{j(\omega t + \phi)}$

■ **FIGURE 10.10** The complex forcing function $V_m e^{j(\omega t + \theta)}$ produces the complex response $I_m e^{j(\omega t + \theta)}$ in the network of Fig. 10.8.

A real, an imaginary, or a complex forcing function will produce a real, an imaginary, or a complex response, respectively. Furthermore, through Euler's identity and the superposition theorem, a complex forcing function may be considered as the sum of a real and an imaginary forcing function; the *real* part of the complex response is produced by the *real* part of the complex forcing function, while the *imaginary* part of the response is caused by the *imaginary* part of the complex forcing function.

Our plan is that instead of applying a *real* forcing function to obtain the desired real response, we will substitute a *complex* forcing function whose real part is the given real forcing function; we expect to obtain a complex response whose real part is the desired real response. The advantage of this procedure is that the integrodifferential equations describing the steady-state response of a circuit will now become simple algebraic equations.

An Algebraic Alternative to Differential Equations

Let us try out this idea on the simple *RL* series circuit shown in Fig. 10.11. The real source $V_m \cos \omega t$ is applied; the real response $i(t)$ is desired. Since

$$\cos \omega t = \text{Re}\{e^{j\omega t}\}$$

the necessary complex source is

$$V_m e^{j\omega t}$$

We express the complex response that results in terms of an unknown amplitude I_m and an unknown phase angle ϕ:

$$I_m e^{j(\omega t + \phi)}$$

Appendix 5 defines the complex number and related terms, reviews complex arithmetic, and develops Euler's identity and the relationship between exponential and polar forms.

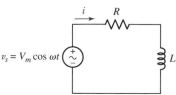

$v_s = V_m \cos \omega t$　i　R　L

■ **FIGURE 10.11** A simple circuit in the sinusoidal steady state is to be analyzed by the application of a complex forcing function.

Writing the differential equation for this particular circuit,

$$Ri + L\frac{di}{dt} = v_s$$

we insert our complex expressions for v_s and i:

$$RI_m e^{j(\omega t+\phi)} + L\frac{d}{dt}(I_m e^{j(\omega t+\phi)}) = V_m e^{j\omega t}$$

take the indicated derivative:

$$RI_m e^{j(\omega t+\phi)} + j\omega L I_m e^{j(\omega t+\phi)} = V_m e^{j\omega t}$$

and obtain an *algebraic* equation. In order to determine the value of I_m and ϕ, we divide throughout by the common factor $e^{j\omega t}$:

$$RI_m e^{j\phi} + j\omega L I_m e^{j\phi} = V_m$$

factor the left side:

$$I_m e^{j\phi}(R + j\omega L) = V_m$$

rearrange:

$$I_m e^{j\phi} = \frac{V_m}{R + j\omega L}$$

and identify I_m and ϕ by expressing the right side of the equation in exponential or polar form:

$$I_m e^{j\phi} = \frac{V_m}{\sqrt{R^2 + \omega^2 L^2}} e^{j(-\tan^{-1}(\omega L/R))} \qquad [15]$$

Thus,

$$I_m = \frac{V_m}{\sqrt{R^2 + \omega^2 L^2}}$$

and

$$\phi = -\tan^{-1}\frac{\omega L}{R}$$

In polar notation, this may be written as

$$I_m \underline{/\phi},$$

or

$$V_m/\sqrt{R^2 + \omega^2 L^2}\underline{/-\tan^{-1}\omega L/R}$$

The complex response is given by Eq. [15]. Since I_m and ϕ are readily identified, we can write the expression for $i(t)$ immediately. However, if we feel like using a more rigorous approach, we may obtain the real response $i(t)$ by reinserting the $e^{j\omega t}$ factor on both sides of Eq. [15] and taking the real part. Either way, we find that

$$i(t) = I_m \cos(\omega t + \phi) = \frac{V_m}{\sqrt{R^2 + \omega^2 L^2}}\cos\left(\omega t - \tan^{-1}\frac{\omega L}{R}\right)$$

which agrees with the response obtained in Eq. [4] for the same circuit.

EXAMPLE 10.2

Find the complex voltage across the series combination of a 500 Ω resistor and a 95 mH inductor if the complex current $8e^{j3000t}$ mA flows through the two elements in series.

The unknown complex voltage will have an amplitude V_m and phase angle ϕ, both of which must be determined. However, the voltage must have the same frequency as the current (3000 rad/s), so we may express this voltage as

$$V_m e^{j(3000t+\phi)}$$

Equating it to the sum of the resistor and inductor voltages

$$V_m e^{j(3000t+\phi)} = (500)0.008e^{j3000t} + (0.095)\frac{d(0.008e^{j3000t})}{dt}$$

and taking the indicated derivative, we find that

$$V_m e^{j(3000t+\phi)} = 4e^{j3000t} + j2.28e^{j3000t}$$

Factoring out the exponential term e^{j3000t}, we are left with

$$V_m e^{j\phi} = 4 + j2.28$$

Expressing the right-hand side in polar form yields

$$4 + j2.28 = 4.60e^{j29.7°}$$

from which we see that $V_m = 4.60$ V and $\phi = 29.7°$, so that the desired voltage is

$$4.60e^{j(3000t+29.7°)} \text{ V}$$

If anyone asks us to find the real response, we need only take the real part of the complex response:

$$\text{Re}\{4.60e^{j(3000t+29.7°)}\} = 4.60\cos(3000t + 29.7°) \qquad \text{V}$$

Thus, we are able to determine the forced response of a circuit containing an energy-storage element without resorting to solving differential equations!

PRACTICE

(If you have trouble working this practice problem, turn to Appendix 5.)

10.4 Evaluate and express the result in rectangular form:
(a) $[(2\underline{/30°})(5\underline{/-110°})](1 + j2)$; (b) $(5\underline{/-200°}) + 4\underline{/20°}$. Evaluate and express the result in polar form: (c) $(2 - j7)/(3 - j)$; (d) $8 - j4 + [(5\underline{/80°})/(2\underline{/20°})]$.

10.5 If the use of the passive sign convention is specified, find the (a) complex voltage that results when the complex current $4e^{j800t}$ A is applied to the series combination of a 1 mF capacitor and a 2 Ω resistor; (b) complex current that results when the complex voltage $100e^{j2000t}$ V is applied to the parallel combination of a 10 mH inductor and a 50 Ω resistor.

Ans: 10.4: $21.4 - j6.38$; $-0.940 + j3.08$; $2.30\underline{/-55.6°}$; $9.43\underline{/-11.22°}$. 10.5: $9.43e^{j(800t-32.0°)}$ V; $5.39e^{j(2000t-68.2°)}$ A.

10.4 THE PHASOR

A sinusoidal current or voltage *at a given frequency* is characterized by only two parameters, an amplitude and a phase angle. The complex representation of the voltage or current is also characterized by these same two parameters. For example, assume a sinusoidal current response given by

$$I_m \cos(\omega t + \phi)$$

where the corresponding representation of this current in complex form is

$$I_m e^{j(\omega t + \phi)}$$

Once I_m and ϕ are specified, the current is exactly defined. Throughout any linear circuit operating in the sinusoidal steady state at a single frequency ω, every current or voltage may be characterized completely by a knowledge of its amplitude and phase angle. Moreover, the complex representation of every voltage and current will contain the same factor $e^{j\omega t}$. Since it is the same for every quantity, it contains no useful information. Of course, the value of the frequency may be recognized by inspecting one of these factors, but it is a lot simpler to write down the value of the frequency near the circuit diagram once and for all and avoid carrying redundant information throughout the solution. Thus, we could simplify the voltage source and the current response of Example 10.1 by representing them concisely as

$$V_m \quad \text{or} \quad V_m e^{j0°}$$

and

$$I_m e^{j\phi}$$

These complex quantities are usually written in polar form rather than exponential form in order to achieve a slight additional saving of time and effort. Thus, the source voltage

$$v(t) = V_m \cos \omega t = V_m \cos(\omega t + 0°)$$

we now represent in complex form as

$$V_m \underline{/0°}$$

and the current response

$$i(t) = I_m \cos(\omega t + \phi)$$

becomes

$$I_m \underline{/\phi}$$

This abbreviated complex representation is called a ***phasor***.[2]

Let us review the steps by which a real sinusoidal voltage or current is transformed into a phasor, and then we will be able to define a phasor more meaningfully and to assign a symbol to represent it.

$e^{j0} = \cos 0 + j \sin 0 = 1$

Remember that none of the steady-state circuits we are considering will respond at a frequency other than that of the excitation source, so that the value of ω is always known.

(2) Not to be confused with the *phaser*, an interesting device featured in a popular television series. . . .

A real sinusoidal current

$$i(t) = I_m \cos(\omega t + \phi)$$

is expressed as the real part of a complex quantity by invoking Euler's identity

$$i(t) = \text{Re}\left\{I_m e^{j(\omega t + \phi)}\right\}$$

We then represent the current as a complex quantity by dropping the instruction Re{}, thus adding an imaginary component to the current without affecting the real component; further simplification is achieved by suppressing the factor $e^{j\omega t}$:

$$\mathbf{I} = I_m e^{j\phi}$$

and writing the result in polar form:

$$\mathbf{I} = I_m \underline{/\phi}$$

This abbreviated complex representation is the *phasor representation;* phasors are complex quantities and hence are printed in boldface type. Capital letters are used for the phasor representation of an electrical quantity because the phasor is not an instantaneous function of time; it contains only amplitude and phase information. We recognize this difference in viewpoint by referring to $i(t)$ as a *time-domain representation* and terming the phasor $\mathbf{I}$ a *frequency-domain representation.* It should be noted that the frequency-domain expression of a current or voltage does not explicitly include the frequency; however, we might think of the frequency as being so fundamental in the frequency domain that it is emphasized by its omission.

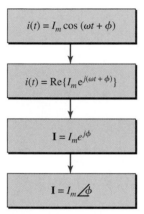

The process by which we change $i(t)$ into $\mathbf{I}$ is called a *phasor transformation* from the time domain to the frequency domain.

EXAMPLE **10.3**

Transform the time-domain voltage $v(t) = 100 \cos(400t - 30°)$ volts into the frequency domain.

The time-domain expression is already in the form of a cosine wave with a phase angle. Thus, suppressing $\omega = 400$ rad/s,

$$\mathbf{V} = 100\underline{/-30°} \text{ volts}$$

Note that we skipped several steps in writing this representation directly. Occasionally, this is a source of confusion for students, as they may forget that the phasor representation is *not* equal to the time-domain voltage $v(t)$. Rather, it is a simplified form of a complex function formed by adding an imaginary component to the real function $v(t)$.

PRACTICE

10.6 Transform each of the following functions of time into phasor form:
(*a*) $-5\sin(580t - 110°)$; (*b*) $3\cos 600t - 5\sin(600t + 110°)$;
(*c*) $8\cos(4t - 30°) + 4\sin(4t - 100°)$. Hint: First convert each into a single cosine function with a positive magnitude.

Ans: $5\underline{/-20°}$; $2.41\underline{/-134.8°}$; $4.46\underline{/-47.9°}$.

Several useful trigonometric identities are provided on the inside cover for convenience.

The process of returning to the time domain from the frequency domain is exactly the reverse of the previous sequence. Thus, given the phasor voltage

$$\mathbf{V} = 115\underline{/-45^\circ} \text{ volts}$$

and the knowledge that $\omega = 500$ rad/s, we can write the time-domain equivalent directly:

$$v(t) = 115\cos(500t - 45^\circ) \text{ volts}$$

If desired as a sine wave, $v(t)$ could also be written

$$v(t) = 115\sin(500t + 45^\circ) \text{ volts}$$

10.5 PHASOR RELATIONSHIPS FOR *R, L,* AND *C*

The real power of the phasor-based analysis technique lies in the fact that it is possible to define *algebraic* relationships between the voltage and current for inductors and capacitors, just as we have always been able to do in the case of resistors. Now that we are able to transform into and out of the frequency domain, we can proceed to our simplification of sinusoidal steady-state analysis by establishing the relationship between the phasor voltage and phasor current for each of the three passive elements.

The Resistor

The resistor provides the simplest case. In the time domain, as indicated by Fig. 10.12*a*, the defining equation is

$$v(t) = Ri(t)$$

Now let us apply the complex voltage

$$v(t) = V_m e^{j(\omega t + \theta)} = V_m \cos(\omega t + \theta) + jV_m \sin(\omega t + \theta) \qquad [16]$$

and assume the complex current response

$$i(t) = I_m e^{j(\omega t + \phi)} = I_m \cos(\omega t + \phi) + jI_m \sin(\omega t + \phi) \qquad [17]$$

so that

$$V_m e^{j(\omega t + \theta)} = Ri(t) = RI_m e^{j(\omega t + \phi)}$$

Dividing throughout by $e^{j\omega t}$, we find

$$V_m e^{j\theta} = RI_m e^{j\phi}$$

or, in polar form,

$$V_m\underline{/\theta} = RI_m\underline{/\phi}$$

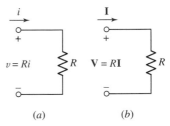

FIGURE 10.12 A resistor and its associated voltage and current in (*a*) the time domain, $v = Ri$; and (*b*) the frequency domain, $\mathbf{V} = R\mathbf{I}$.

But $V_m\underline{/\theta}$ and $I_m\underline{/\phi}$ merely represent the general voltage and current phasors **V** and **I**. Thus,

$$\mathbf{V} = R\mathbf{I} \tag{18}$$

The voltage-current relationship in phasor form for a resistor has the same form as the relationship between the time-domain voltage and current. The defining equation in phasor form is illustrated in Fig. 10.12b. The angles θ and ϕ are equal, so that the current and voltage are always in phase.

As an example of the use of both the time-domain and frequency-domain relationships, let us assume that a voltage of $8\cos(100t - 50°)$ V is across a 4 Ω resistor. Working in the time domain, we find that the current must be

$$i(t) = \frac{v(t)}{R} = 2\cos(100t - 50°) \quad \text{A}$$

The phasor form of the same voltage is $8\underline{/-50°}$ V, and therefore

$$\mathbf{I} = \frac{\mathbf{V}}{R} = 2\underline{/-50°} \quad \text{A}$$

If we transform this answer back to the time domain, it is evident that the same expression for the current is obtained. We conclude that there is no saving in time or effort when a *resistive* circuit is analyzed in the frequency domain.

> Ohm's law holds true both in the time domain and in the frequency domain. In other words, the voltage across a resistor is always given by the resistance times the current flowing through the element.

The Inductor

Let us now turn to the inductor. The time-domain representation is shown in Fig. 10.13a, and the defining equation, a time-domain expression, is

$$v(t) = L\frac{di(t)}{dt} \tag{19}$$

After substituting the complex voltage equation [16] and complex current equation [17] in Eq. [19], we have

$$V_m e^{j(\omega t+\theta)} = L\frac{d}{dt}I_m e^{j(\omega t+\phi)}$$

Taking the indicated derivative:

$$V_m e^{j(\omega t+\theta)} = j\omega L I_m e^{j(\omega t+\phi)}$$

and dividing through by $e^{j\omega t}$:

$$V_m e^{j\theta} = j\omega L I_m e^{j\phi}$$

we obtain the desired phasor relationship

$$\boxed{\mathbf{V} = j\omega L\mathbf{I}} \tag{20}$$

The time-domain differential equation [19] has become the algebraic equation [20] in the frequency domain. The phasor relationship is indicated in Fig. 10.13b. Note that the angle of the factor $j\omega L$ is exactly $+90°$ and that **I** must therefore lag **V** by $90°$ in an inductor.

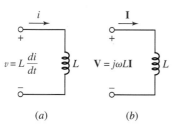

(a)　　　　　(b)

■ **FIGURE 10.13** An inductor and its associated voltage and current in (a) the time domain, $v = L\,di/dt$; and (b) the frequency domain, $\mathbf{V} = j\omega L\mathbf{I}$.

EXAMPLE **10.4**

Apply the voltage $8\underline{/-50^\circ}$ V at a frequency $\omega = 100$ rad/s to a 4 H inductor and determine the phasor current and the time-domain current.

We make use of the expression we just obtained for the inductor,

$$\mathbf{I} = \frac{\mathbf{V}}{j\omega L} = \frac{8\underline{/-50^\circ}}{j100(4)} = -j0.02\underline{/-50^\circ} = (1\underline{/-90^\circ})(0.02\underline{/-50^\circ})$$

or

$$\mathbf{I} = 0.02\underline{/-140^\circ} \text{ A}$$

If we express this current in the time domain, it becomes

$$i(t) = 0.02\cos(100t - 140^\circ) \text{ A} = 20\cos(100t - 140^\circ) \text{ mA}$$

The Capacitor

The final element to consider is the capacitor. The time-domain current-voltage relationship is

$$i(t) = C\frac{dv(t)}{dt}$$

The equivalent expression in the frequency domain is obtained once more by letting $v(t)$ and $i(t)$ be the complex quantities of Eqs. [16] and [17], taking the indicated derivative, suppressing $e^{j\omega t}$, and recognizing the phasors $\mathbf{V}$ and $\mathbf{I}$. Doing this, we find

$$\mathbf{I} = j\omega C\mathbf{V} \qquad [21]$$

Thus, $\mathbf{I}$ leads $\mathbf{V}$ by 90° in a capacitor. This, of course, does not mean that a current response is present one-quarter of a period earlier than the voltage that caused it! We are studying steady-state response, and we find that the current maximum is caused by the increasing voltage that occurs 90° earlier than the voltage maximum.

The time-domain and frequency-domain representations are compared in Fig. 10.14a and b. We have now obtained the **V-I** relationships for the three passive elements. These results are summarized in Table 10.1, where

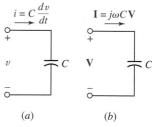

■ **FIGURE 10.14** (a) The time-domain and (b) the frequency-domain relationships between capacitor current and voltage.

TABLE   **10.1**   Comparison of Time-Domain and Frequency-Domain Voltage-Current Expressions

Time domain		Frequency domain	
$\xrightarrow{i} \overset{R}{\text{—}\wedge\wedge\vee\!\!-}$ $\underset{+\ \ v\ \ -}{}$	$v = Ri$	$\mathbf{V} = R\mathbf{I}$	$\xrightarrow{\mathbf{I}} \overset{R}{\text{—}\wedge\wedge\vee\!\!-}$ $\underset{+\ \ \mathbf{V}\ \ -}{}$
$\xrightarrow{i} \overset{L}{\text{—}\text{mmm}\text{—}}$ $\underset{+\ \ v\ \ -}{}$	$v = L\dfrac{di}{dt}$	$\mathbf{V} = j\omega L\mathbf{I}$	$\xrightarrow{\mathbf{I}} \overset{j\omega L}{\text{—}\text{mmm}\text{—}}$ $\underset{+\ \ \mathbf{V}\ \ -}{}$
$\xrightarrow{i} \overset{C}{\text{—}|\!\!(\text{—}}$ $\underset{+\ \ v\ \ -}{}$	$v = \dfrac{1}{C}\displaystyle\int i\,dt$	$\mathbf{V} = \dfrac{1}{j\omega C}\mathbf{I}$	$\xrightarrow{\mathbf{I}} \overset{1/j\omega C}{\text{—}|\!\!(\text{—}}$ $\underset{+\ \ \mathbf{V}\ \ -}{}$

the time-domain v-i expressions and the frequency-domain **V-I** relationships are shown in adjacent columns for the three circuit elements. All the phasor equations are algebraic. Each is also linear, and the equations relating to inductance and capacitance bear a great similarity to Ohm's law. In fact, we will indeed *use* them as we use Ohm's law.

Kirchhoff's Laws Using Phasors

Kirchhoff's voltage law in the time domain is

$$v_1(t) + v_2(t) + \cdots + v_N(t) = 0$$

We now use Euler's identity to replace each real voltage v_i by a complex voltage having the same real part, suppress $e^{j\omega t}$ throughout, and obtain

$$\mathbf{V}_1 + \mathbf{V}_2 + \cdots + \mathbf{V}_N = 0$$

Thus, we see that Kirchhoff's voltage law applies to phasor voltages just as it did in the time domain. Kirchhoff's current law can be shown to hold for phasor currents by a similar argument.

Now let us look briefly at the series RL circuit that we have considered several times before. The circuit is shown in Fig. 10.15, and a phasor current and several phasor voltages are indicated. We may obtain the desired response, a time-domain current, by first finding the phasor current. From Kirchhoff's voltage law,

$$\mathbf{V}_R + \mathbf{V}_L = \mathbf{V}_s$$

and using the recently obtained **V-I** relationships for the elements, we have

$$R\mathbf{I} + j\omega L\mathbf{I} = \mathbf{V}_s$$

The phasor current is then found in terms of the source voltage $\mathbf{V}_s$:

$$\mathbf{I} = \frac{\mathbf{V}_s}{R + j\omega L}$$

Let us select a source-voltage amplitude of V_m and phase angle of $0°$. Thus,

$$\mathbf{I} = \frac{V_m\underline{/0°}}{R + j\omega L}$$

The current may be transformed to the time domain by first writing it in polar form:

$$\mathbf{I} = \frac{V_m}{\sqrt{R^2 + \omega^2 L^2}}\underline{/(-\tan^{-1}(\omega L/R))}$$

and then following the familiar sequence of steps to obtain in a very simple manner the same result we obtained the "hard way" earlier in this chapter.

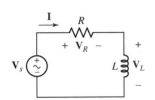

■ **FIGURE 10.15** The series RL circuit with a phasor voltage applied.

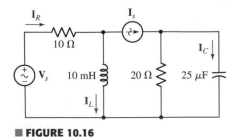

■ **FIGURE 10.16**

PRACTICE

10.8 In the circuit of Fig. 10.16, let $\omega = 1200$ rad/s, $\mathbf{I}_C = 1.2\underline{/28°}$ A, and $\mathbf{I}_L = 3\underline{/53°}$ A. Find (a) $\mathbf{I}_s$; (b) $\mathbf{V}_s$; (c) $i_R(t)$.

Ans: $2.33\underline{/-31.0°}$ A; $34.9\underline{/74.5°}$ V; $3.99\cos(1200t + 17.42°)$ A.

10.6 • IMPEDANCE

The current-voltage relationships for the three passive elements in the frequency domain are (assuming that the passive sign convention is satisfied)

$$\mathbf{V} = R\mathbf{I} \qquad \mathbf{V} = j\omega L\mathbf{I} \qquad \mathbf{V} = \frac{\mathbf{I}}{j\omega C}$$

If these equations are written as phasor voltage/phasor current ratios

$$\frac{\mathbf{V}}{\mathbf{I}} = R \qquad \frac{\mathbf{V}}{\mathbf{I}} = j\omega L \qquad \frac{\mathbf{V}}{\mathbf{I}} = \frac{1}{j\omega C}$$

we find that these ratios are simple quantities that depend on element values (and frequency also, in the case of inductance and capacitance). We treat these ratios in the same manner that we treat resistances, with the exception that they are complex quantities and all algebraic manipulations must be those appropriate for complex numbers.

Let us define the ratio of the phasor voltage to the phasor current as *impedance,* symbolized by the letter **Z**. The impedance is a complex quantity having the dimensions of ohms. Impedance is not a phasor and cannot be transformed to the time domain by multiplying by $e^{j\omega t}$ and taking the real part. Instead, we think of an inductor as being represented in the time domain by its inductance L and in the frequency domain by its impedance $j\omega L$. A capacitor in the time domain has a capacitance C; in the frequency domain, it has an impedance $1/j\omega C$. Impedance is a part of the frequency domain and not a concept that is a part of the time domain.

$$Z_R = R$$
$$Z_L = j\omega L$$
$$Z_C = \frac{1}{j\omega C}$$

Series Impedance Combinations

The validity of Kirchhoff's two laws in the frequency domain leads to the fact that impedances may be combined in series and parallel by the same rules we have already established for resistances. For example, at $\omega = 10 \times 10^3$ rad/s, a 5 mH inductor in series with a 100 μF capacitor may be replaced by the single impedance which is the sum of the individual impedances. The impedance of the inductor is

$$\mathbf{Z}_L = j\omega L = j50 \ \Omega$$

and the impedance of the capacitor is

$$\mathbf{Z}_C = \frac{1}{j\omega C} = \frac{-j}{\omega C} = -j1 \ \Omega$$

The impedance of the series combination is therefore

Note that $\frac{1}{j} = -j$.

$$\mathbf{Z}_{\text{eq}} = \mathbf{Z}_L + \mathbf{Z}_C = j50 - j1 = j49 \ \Omega$$

The impedance of inductors and capacitors is a function of frequency, and this equivalent impedance is thus applicable only at the single frequency at which it was calculated, $\omega = 10{,}000$ rad/s. If we change the frequency to $\omega = 5000$ rad/s, for example, $\mathbf{Z}_{\text{eq}} = j23 \ \Omega$.

Parallel Impedance Combinations

The *parallel* combination of the 5 mH inductor and the 100 μF capacitor at $\omega = 10{,}000$ rad/s is calculated in exactly the same fashion in which we calculated parallel resistances:

$$\mathbf{Z}_{eq} = \frac{(j50)(-j1)}{j50 - j1} = \frac{50}{j49} = -j1.020 \ \Omega$$

At $\omega = 5000$ rad/s, the parallel equivalent is $-j2.17 \ \Omega$.

The complex number or quantity representing impedance may be expressed in either polar or rectangular form. For example, the impedance $50 - j86.6 \ \Omega$ is said to have a *resistance* of $50 \ \Omega$ and a **reactance** of $-86.6 \ \Omega$. So, the real part of an impedance is referred to as the **resistance,** and the imaginary component (including the sign but not the j) is termed the **reactance,** often symbolized with X. Both have units of ohms. In rectangular form, $\mathbf{Z} = R + jX$, and in polar form, $\mathbf{Z} = |\mathbf{Z}|\underline{/\theta}$. Thus, a resistor has zero reactance, whereas (ideal) capacitors and inductors have zero resistance. This can also be noted directly from the polar form of an impedance. Consider again $\mathbf{Z} = 50 - j86.6 \ \Omega$, which may also be written as $100\underline{/-60^\circ} \ \Omega$. Since the phase angle is not zero, we know that the impedance is not purely resistive at frequency ω. Since it is not $+90^\circ$ we know it is not purely inductive, and likewise it is not purely capacitive or the phase angle would be -90°. Can a series or parallel combination include both a capacitor and an inductor, yet have zero reactance? *Absolutely.* Consider the simple situation where $\omega = 1$ rad/s, $L = 1$ H, and $C = 1$ F, all in series together with $R = 1 \ \Omega$. The equivalent impedance of this network is $\mathbf{Z} = 1 + j(1)(1) - j/(1)(1) = 1 \ \Omega$, as if (at a frequency of 1 rad/s) only a 1 Ω resistor were present.

EXAMPLE 10.5

Determine the equivalent impedance of the network shown in Fig. 10.17a, given an operating frequency of 5 rad/s.

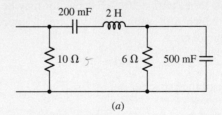

(a)

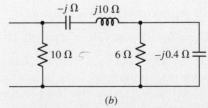

(b)

■ **FIGURE 10.17** (a) A network that is to be replaced by a single equivalent impedance. (b) The elements are replaced by their impedances at $\omega = 5$ rad/s.

We begin by converting the resistors, capacitors, and inductor into the corresponding impedances as shown in Fig. 10.17b.

Upon examining the resulting network, we observe that the 6 Ω impedance is in parallel with $-j0.4$ Ω. This combination is equivalent to

$$\frac{(6)(-j0.4)}{6 - j0.4} = 0.02655 - j0.3982 \ \Omega$$

which is in series with both the $-j$ Ω and $j10$ Ω impedances, so that we have

$$0.0265 - j0.3982 - j + j10 = 0.02655 + j8.602 \ \Omega$$

This new impedance is in parallel with 10 Ω, so that the equivalent impedance of the network is

$$10 \parallel (0.02655 + j8.602) = \frac{10(0.02655 + j8.602)}{10 + 0.02655 + j8.602}$$
$$= 4.255 + j4.929 \ \Omega$$

Alternatively, we can express the impedance in polar form as $6.511\underline{/49.20°}$ Ω.

PRACTICE

10.9 With reference to the network shown in Fig. 10.18, find the input impedance $\mathbf{Z}_{in}$ that would be measured between terminals: (a) a and g; (b) b and g; (c) a and b.

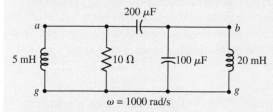

FIGURE 10.18

Ans: $2.81 + j4.49$ Ω; $1.798 - j1.124$ Ω; $0.1124 - j3.82$ Ω.

It is important to note that the resistive component of the impedance is not necessarily equal to the resistance of the resistor that is present in the network. For example, a 10 Ω resistor and a 5 H inductor in series at $\omega = 4$ rad/s have an equivalent impedance $\mathbf{Z} = 10 + j20$ Ω, or, in polar form, $22.4\underline{/63.4°}$ Ω. In this case, the resistive component of the impedance is equal to the resistance of the series resistor because the network is a simple series network. However, if these same two elements are placed in parallel, the equivalent impedance is $10(j20)/(10 + j20)$ Ω, or $8 + j4$ Ω. The resistive component of the impedance is now 8 Ω.

EXAMPLE 10.6

Find the current $i(t)$ in the circuit shown in Fig. 10.19a.

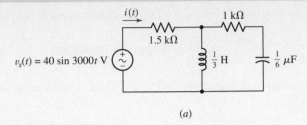

(a)

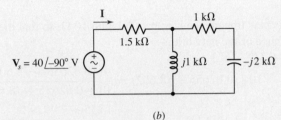

(b)

■ **FIGURE 10.19** (a) An *RLC* circuit for which the sinusoidal forced response $i(t)$ is desired. (b) The frequency-domain equivalent of the given circuit at $\omega = 3000$ rad/s.

▶ **Identify the goal of the problem.**

We need to find the sinusoidal steady-state current flowing through the 1.5 kΩ resistor due to the 3000 rad/s voltage source.

▶ **Collect the known information.**

We begin by drawing a frequency-domain circuit. The source is transformed to the frequency-domain representation $40/\underline{-90°}$ V, the frequency domain response is represented as **I**, and the impedances of the inductor and capacitor, determined at $\omega = 3000$ rad/s, are j kΩ and $-j2$ kΩ, respectively. The corresponding frequency-domain circuit is shown in Fig. 10.19b.

▶ **Devise a plan.**

We will analyze the circuit of Fig. 10.19b to obtain **I**; combining impedances and invoking Ohm's law is one possible approach. We will then make use of the fact that we know $\omega = 3000$ rad/s to convert **I** into a time-domain expression.

▶ **Construct an appropriate set of equations.**

$$\mathbf{Z}_{eq} = 1.5 + \frac{(j)(1-2j)}{j+1-2j} = 1.5 + \frac{2+j}{1-j}$$

$$= 1.5 + \frac{2+j}{1-j}\frac{1+j}{1+j} = 1.5 + \frac{1+j3}{2}$$

$$= 2 + j1.5 = 2.5/\underline{36.87°}\text{ k}\Omega$$

The phasor current is then simply

$$\mathbf{I} = \frac{\mathbf{V}_s}{\mathbf{Z}_{eq}}$$

▶ **Determine if additional information is required.**

Substituting known values, we find that

$$\mathbf{I} = \frac{40\underline{/-90°}}{2.5\underline{/36.87°}} \text{ mA}$$

which, along with the knowledge that $\omega = 3000$ rad/s, is sufficient to solve for $i(t)$.

▶ **Attempt a solution.**

This complex expression is easily simplified to a single complex number in polar form:

$$\mathbf{I} = \frac{40}{2.5}\underline{/-90° - 36.87°} \text{ mA} \quad = 16.00\underline{/-126.9°} \text{ mA}$$

Upon transforming the current to the time domain, the desired response is obtained:

$$i(t) = 16\cos(3000t - 126.9°) \text{ mA}$$

▶ **Verify the solution. Is it reasonable or expected?**

The effective impedance connected to the source has an angle of $+36.87°$, indicating that it has a net inductive character, or that the current will lag the voltage. Since the voltage source has a phase angle of $-90°$ (once converted to a cosine source), we see that our answer is consistent.

PRACTICE

10.10 In the frequency-domain circuit of Fig. 10.20, find (a) $\mathbf{I}_1$; (b) $\mathbf{I}_2$; (c) $\mathbf{I}_3$.

Ans: $28.3\underline{/45°}$ A; $20\underline{/90°}$ A; $20\underline{/0°}$ A.

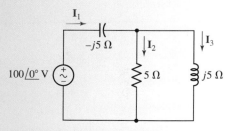

■ **FIGURE 10.20**

Before we begin to write great numbers of equations in the time domain or in the frequency domain, it is very important that we shun the construction of equations that are partly in the time domain, partly in the frequency domain, and wholly incorrect. One clue that a faux pas of this type has been committed is the sight of both a complex number and a t in the same equation, except in the factor $e^{j\omega t}$. And, since $e^{j\omega t}$ plays a much bigger role in derivations than in applications, it is pretty safe to say that students who find they have just created an equation containing j and t, or $\underline{/}$ and t, have created a monster that the world would be better off without.

For example, a few equations back we saw

$$\mathbf{I} = \frac{\mathbf{V}_s}{\mathbf{Z}_{eq}} = \frac{40\underline{/-90°}}{2.5\underline{/36.9°}} = 16\underline{/-126.9°} \text{ mA}$$

Please do not try anything like the following:

$$i(t) \not= \frac{40\sin 3000t}{2.5\underline{/36.9°}} \quad \text{or} \quad i(t) \not= \frac{40\sin 3000t}{2 + j1.5}$$

10.7 ADMITTANCE

Occasionally we find that the *reciprocal* of impedance is a more convenient quantity. In this spirit, we define the ***admittance*** **Y** of a circuit element as the ratio of phasor current to phasor voltage (assuming that the passive sign convention is satisfied):

$$\mathbf{Y} = \frac{\mathbf{I}}{\mathbf{V}}$$

and thus

$$\mathbf{Y} = \frac{1}{\mathbf{Z}}$$

$$Y_R = \frac{1}{R}$$
$$Y_L = \frac{1}{j\omega L}$$
$$Y_C = j\omega C$$

The real part of the admittance is the ***conductance*** G, and the imaginary part of the admittance is the ***susceptance*** B. Thus,

$$\mathbf{Y} = G + jB = \frac{1}{\mathbf{Z}} = \frac{1}{R + jX} \qquad [22]$$

Equation [22] should be scrutinized carefully; it does *not* state that the real part of the admittance is equal to the reciprocal of the real part of the impedance or that the imaginary part of the admittance is equal to the reciprocal of the imaginary part of the impedance!

Admittance, conductance, and susceptance are all measured in siemens. An impedance

$$\mathbf{Z} = 1 - j2 \ \Omega$$

which might be represented, for example, by a $1 \ \Omega$ resistor in series with a $0.1 \ \mu\text{F}$ capacitor at $\omega = 5$ Mrad/s, possesses an admittance

$$\mathbf{Y} = \frac{1}{\mathbf{Z}} = \frac{1}{1 - j2} = \frac{1}{1 - j2}\frac{1 + j2}{1 + j2} = 0.2 + j0.4 \ \text{S}$$

The equivalent admittance of a network consisting of a number of parallel branches is the sum of the admittances of the individual branches. Thus, the numerical value of the admittance just shown might be obtained from a conductance of 0.2 S in parallel with a positive susceptance of 0.4 S. The former could be represented by a $5 \ \Omega$ resistor and the latter by a $0.08 \ \mu\text{F}$ capacitor at $\omega = 5$ Mrad/s, since the admittance of a capacitor is $j\omega C$.

As a check on our analysis, let us compute the impedance of this latest network, a $5 \ \Omega$ resistor in parallel with a $0.08 \ \mu\text{F}$ capacitor at $\omega = 5$ Mrad/s. The equivalent impedance is

$$\mathbf{Z} = \frac{5(1/j\omega C)}{5 + 1/j\omega C} = \frac{5(-j2.5)}{5 - j2.5} = 1 - j2 \ \Omega$$

as before. These two networks represent only two of an infinite number of different networks that possess this same impedance and admittance at this frequency. They do, however, represent the only two-element networks, and thus might be considered to be the two simplest networks having an impedance of $1 - j2 \ \Omega$ and an admittance of $0.2 + j0.4 \ \text{S}$ at the frequency $\omega = 5 \times 10^6$ rad/s.

The term ***immittance,*** a combination of the words *impedance* and *admittance,* is sometimes used as a general term for both impedance and admittance. For example, it is evident that a knowledge of the phasor voltage across a known immittance enables the current through the immittance to be calculated.

PRACTICE

10.11 Determine the admittance (in rectangular form) of (*a*) an impedance $\mathbf{Z} = 1000 + j400\ \Omega$; (*b*) a network consisting of the parallel combination of an 800 Ω resistor, a 1 mH inductor, and a 2 nF capacitor, if $\omega = 1$ Mrad/s; (*c*) a network consisting of the series combination of an 800 Ω resistor, a 1 mH inductor, and a 2 nF capacitor, if $\omega = 1$ Mrad/s.

Ans: $0.862 - j0.345$ mS; $1.25 + j1$ mS; $0.899 - j0.562$ mS.

10.8 NODAL AND MESH ANALYSIS

We previously achieved a great deal with nodal and mesh analysis techniques, and it's reasonable to ask if a similar procedure might be valid in terms of phasors and impedances for the sinusoidal steady state. We already know that both of Kirchhoff's laws are valid for phasors; also, we have an Ohm-like law for the passive elements $\mathbf{V} = \mathbf{ZI}$. In other words, the laws upon which nodal analysis rests are true for phasors, and we may proceed, therefore, to analyze circuits by nodal techniques in the sinusoidal steady state. Using similar arguments, we can establish that mesh analysis methods are valid (and often useful) as well.

EXAMPLE 10.7

Find the time-domain node voltages $v_1(t)$ and $v_2(t)$ in the circuit shown in Fig. 10.21.

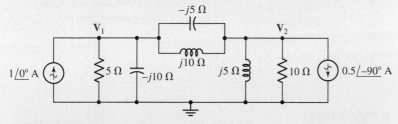

■ **FIGURE 10.21** A frequency-domain circuit for which node voltages V_1 and V_2 are identified.

Two current sources are given as phasors, and phasor node voltages $\mathbf{V}_1$ and $\mathbf{V}_2$ are indicated. At the left node we apply KCL, yielding:

$$\frac{\mathbf{V}_1}{5} + \frac{\mathbf{V}_1}{-j10} + \frac{\mathbf{V}_1 - \mathbf{V}_2}{-j5} + \frac{\mathbf{V}_1 - \mathbf{V}_2}{j10} = 1\underline{/0°} = 1 + j0$$

(Continued on next page)

At the right node,

$$\frac{\mathbf{V}_2 - \mathbf{V}_1}{-j5} + \frac{\mathbf{V}_2 - \mathbf{V}_1}{j10} + \frac{\mathbf{V}_2}{j5} + \frac{\mathbf{V}_2}{10} = -(0.5\underline{/-90^\circ}) = j0.5$$

Combining terms, we have

$$(0.2 + j0.2)\mathbf{V}_1 - j0.1\mathbf{V}_2 = 1$$

and

$$-j0.1\mathbf{V}_1 + (0.1 - j0.1)\mathbf{V}_2 = j0.5$$

These equations are easily solved on most scientific calculators, resulting in $\mathbf{V}_1 = 1 - j2$ V and $\mathbf{V}_2 = -2 + j4$ V.

The time-domain solutions are obtained by expressing $\mathbf{V}_1$ and $\mathbf{V}_2$ in polar form:

$$\mathbf{V}_1 = 2.24\underline{/-63.4^\circ}$$
$$\mathbf{V}_2 = 4.47\underline{/116.6^\circ}$$

and passing to the time domain:

$$v_1(t) = 2.24\cos(\omega t - 63.4^\circ) \text{ V}$$
$$v_2(t) = 4.47\cos(\omega t + 116.6^\circ) \text{ V}$$

Note that the value of ω would have to be known in order to compute the impedance values given on the circuit diagram. Also, *both sources must be operating at the same frequency.*

PRACTICE

10.12 Use nodal analysis on the circuit of Fig. 10.22 to find $\mathbf{V}_1$ and $\mathbf{V}_2$.

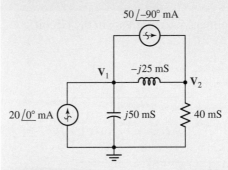

50 $\underline{/-90^\circ}$ mA

$\mathbf{V}_1$ $-j25$ mS

20 $\underline{/0^\circ}$ mA $\mathbf{V}_2$

$j50$ mS   40 mS

■ **FIGURE 10.22**

Ans: $1.062\underline{/23.3^\circ}$ V; $1.593\underline{/-50.0^\circ}$ V.

Now let us look at an example of mesh analysis, keeping in mind again that all sources must be operating at the same frequency. Otherwise, it is impossible to define a numerical value for any reactance in the circuit. As we see in the next section, the only way out of such a dilemma is to apply superposition.

EXAMPLE 10.8

Obtain expressions for the time-domain currents i_1 and i_2 in the circuit given as Fig. 10.23a.

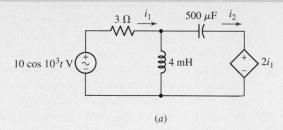

(a)

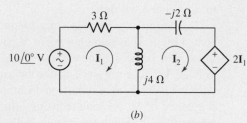

(b)

■ **FIGURE 10.23** (a) A time-domain circuit containing a dependent source. (b) The corresponding frequency-domain circuit.

Noting from the left source that $\omega = 10^3$ rad/s, we draw the frequency-domain circuit of Fig. 10.23b and assign mesh currents $\mathbf{I}_1$ and $\mathbf{I}_2$. Around mesh 1,

$$3\mathbf{I}_1 + j4(\mathbf{I}_1 - \mathbf{I}_2) = 10\underline{/0°}$$

or

$$(3 + j4)\mathbf{I}_1 - j4\mathbf{I}_2 = 10$$

while mesh 2 leads to

$$j4(\mathbf{I}_2 - \mathbf{I}_1) - j2\mathbf{I}_2 + 2\mathbf{I}_1 = 0$$

or

$$(2 - j4)\mathbf{I}_1 + j2\mathbf{I}_2 = 0$$

Solving,

$$\mathbf{I}_1 = \frac{14 + j8}{13} = 1.24\underline{/29.7°} \text{ A}$$

$$\mathbf{I}_2 = \frac{20 + j30}{13} = 2.77\underline{/56.3°} \text{ A}$$

Hence,

$$i_1(t) = 1.24 \cos(10^3 t + 29.7°) \text{ A}$$
$$i_2(t) = 2.77 \cos(10^3 t + 56.3°) \text{ A}$$

PRACTICE

10.13 Use mesh analysis on the circuit of Fig. 10.24 to find $\mathbf{I}_1$ and $\mathbf{I}_2$.

Ans: $4.87\underline{/-164.6°}$ A; $7.17\underline{/-144.9°}$ A.

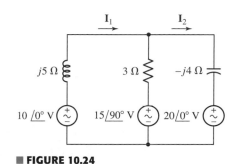

■ **FIGURE 10.24**

Cutoff Frequency of a Transistor Amplifier

Transistor-based amplifier circuits are an integral part of many modern electronic instruments. One common application is in mobile telephones (Fig. 10.25), where audio signals are superimposed on high-frequency carrier waves. Unfortunately, transistors have built-in capacitances that lead to limitations in the frequencies at which they can be used, and this fact must be considered when choosing a transistor for a particular application.

■ **FIGURE 10.25** Transistor amplifiers are used in many devices, including mobile phones. Linear circuit models are often used to analyze their performance as a function of frequency.
Courtesy of Nokia.

Figure 10.26a shows what is commonly referred to as a *high-frequency hybrid-π model* for a bipolar junction transistor. In practice, although transistors are *nonlinear* devices, we find that this simple *linear* circuit does a reasonably accurate job of modeling the actual device behavior. The two capacitors C_π and C_μ are used to represent internal capacitances that characterize the particular transistor being used; additional capacitors as well as resistors can be added to increase the accuracy of the model as needed. Figure 10.26b shows the transistor model inserted into an amplifier circuit known as a common emitter amplifier.

Assuming a sinusoidal steady-state signal represented by its Thévenin equivalent $\mathbf{V}_s$ and R_s, we are interested in the ratio of the output voltage $\mathbf{V}_{\text{out}}$ to the input voltage $\mathbf{V}_{\text{in}}$. The presence of the internal transistor capacitances leads to a reduction in amplification as the frequency of $\mathbf{V}_s$ is increased; this ultimately limits the frequencies at which the circuit will operate properly. Writing a single nodal equation at the output yields

$$-g_m\mathbf{V}_\pi = \frac{\mathbf{V}_{\text{out}} - \mathbf{V}_{\text{in}}}{(1/j\omega C_\mu)} + \frac{\mathbf{V}_{\text{out}}}{(R_C \parallel R_L)}$$

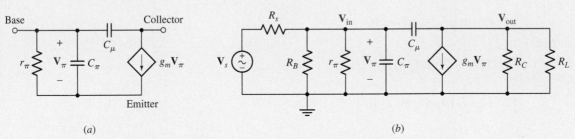

(a)	(b)

■ **FIGURE 10.26** (a) High-frequency hybrid-π transistor model. (b) Common-emitter amplifier circuit using the hybrid-π transistor model.

10.9 • SUPERPOSITION, SOURCE TRANSFORMATIONS, AND THÉVENIN'S THEOREM

After inductors and capacitors were introduced in Chap. 7, we found that circuits containing these elements were still linear, and that the benefits of linearity were again available. Included among these were the superposition principle, Thévenin's and Norton's theorems, and source transformations.

Solving for $\mathbf{V}_{out}$ in terms of $\mathbf{V}_{in}$, and noting that $\mathbf{V}_\pi = \mathbf{V}_{in}$, we obtain an expression for the amplifier gain

$$\frac{\mathbf{V}_{out}}{\mathbf{V}_{in}} = \frac{-g_m(R_C\|R_L)(1/j\omega C_\mu) + (R_C\|R_L)}{(R_C\|R_L) + (1/j\omega C_\mu)}$$
$$= \frac{-g_m(R_C\|R_L) + j\omega(R_C\|R_L)C_\mu}{1 + j\omega(R_C\|R_L)C_\mu}$$

Given the typical values $g_m = 30$ mS, $R_C = R_L = 2$ kΩ, and $C_\mu = 5$ pF, we can plot the magnitude of the gain as a function of frequency (recalling that $\omega = 2\pi f$). The semilogarithmic plot is shown in Fig. 10.27a, and the MATLAB script used to generate the figure is given in Fig. 10.27b. It is interesting, but maybe not totally surprising, to see that a characteristic such as the amplifier gain is dependent on frequency. In fact, we might be able to contemplate using such a circuit as a means of filtering out frequencies we aren't interested in. However, at least for relatively low frequencies, we see that the gain is essentially independent of the frequency of our input source.

When characterizing amplifiers, it is common to reference the frequency at which the gain is reduced to $1/\sqrt{2}$ times its maximum value. From Fig. 10.27a, we see that the maximum gain magnitude is 30, and the gain magnitude is reduced to $30/\sqrt{2} = 21$ at a frequency of approximately 30 MHz. This frequency is often called the *cutoff* or *corner* frequency of the amplifier. If operation at a higher frequency is required, either the internal capacitances must be reduced (i.e., a different transistor must be used) or the circuit must be redesigned in some way.

We should note at this point that defining the gain relative to $\mathbf{V}_{in}$ does not present a complete picture of the frequency-dependent behavior of the amplifier. This may be apparent if we briefly consider the capacitance C_π: as $\omega \to \infty$, $Z_{C_\pi} \to 0$, so $\mathbf{V}_{in} \to 0$. This effect does not manifest itself in the simple equation we derived. A more comprehensive approach is to develop an equation for $\mathbf{V}_{out}$ in terms of $\mathbf{V}_s$, in which case both capacitances will appear in the expression; this requires a little bit more algebra.

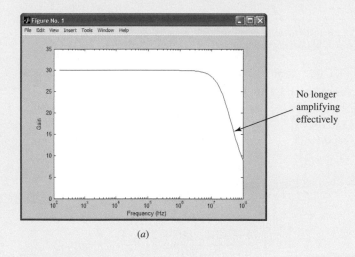

(a)

No longer amplifying effectively

```
EDU» frequency = logspace(3,9,100);
EDU» numerator = -30e-3*1000 + i*frequency*1000*5e-12;
EDU» denominator = 1 + i*frequency*1000*5e-12;
EDU» for k = 1:100
    gain(k) = abs(numerator(k)/denominator(k));
end
EDU» semilogx(frequency/2/pi,gain);
EDU» xlabel('Frequency (Hz)');
EDU» ylabel('Gain');
EDU» axis([100 1e8 0 35]);
```

(b)

■ **FIGURE 10.27** (a) Amplifier gain as a function of frequency. (b) MATLAB script used to create plot.

Thus, we know that these methods may be used on the circuits we are now considering; the fact that we happen to be applying sinusoidal sources and are seeking only the forced response is immaterial. The fact that we are analyzing the circuits in terms of phasors is also immaterial; they are still linear circuits. We might also remember that linearity and superposition were invoked when we combined real and imaginary sources to obtain a complex source.

EXAMPLE **10.9**

Use superposition to find V_1 for the circuit of Fig. 10.21, repeated as Fig. 10.28a for convenience.

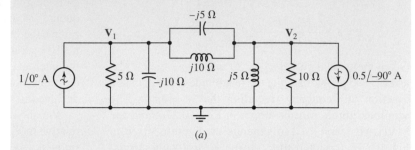

(a)

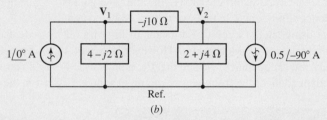

Ref.

(b)

■ **FIGURE 10.28** (a) Circuit of Fig. 10.21 for which V_1 is desired, (b) V_1 may be found by using superposition of the separate phasor responses.

First we redraw the circuit as Fig. 10.28b, where each pair of parallel impedances is replaced by a single equivalent impedance. That is, $5\| -j10\ \Omega$ is $4 - j2\ \Omega$; $j10\| -j5\ \Omega$ is $-j10\ \Omega$; and $10\| j5$ is equal to $2 + j4\ \Omega$. To find V_1, we first activate only the left source and find the partial response, V_{1L}. The $1\underline{/0°}$ source is in parallel with an impedance of

$$(4 - j2)\ \|\ (-j10 + 2 + j4)$$

so that

$$\mathbf{V}_{1L} = 1\underline{/0°}\cdot\frac{(4 - j2)(-j10 + 2 + j4)}{4 - j2 - j10 + 2 + j4}$$

$$= \frac{-4 - j28}{6 - j8} = 2 - j2\ \text{V}$$

With only the right source active, current division and Ohm's law yields

$$\mathbf{V}_{1R} = (-0.5\underline{/-90°})\left(\frac{2 + j4}{4 - j2 - j10 + 2 + j4}\right)(4 - j2) = -1\ \text{V}$$

Summing, then

$$\mathbf{V}_1 = \mathbf{V}_{1L} + \mathbf{V}_{1R} = 2 - j2 - 1 = 1 - j2 \qquad \text{V}$$

which agrees with our previous result from Example 10.7.

As we will see, superposition is also extremely useful when dealing with a circuit in which not all sources operate at the same frequency.

PRACTICE

10.14 If superposition is used on the circuit of Fig. 10.29, find $\mathbf{V}_1$ with (*a*) only the $20\underline{/0°}$-mA source operating; (*b*) only the $50\underline{/-90°}$-mA source operating.

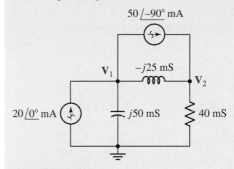

■ **FIGURE 10.29**

Ans: $0.1951 - j0.556$ V; $0.780 + j0.976$ V.

EXAMPLE **10.10**

Determine the Thévenin equivalent seen by the $-j10\ \Omega$ impedance of Fig. 10.30*a*, and use this to compute $\mathbf{V}_1$.

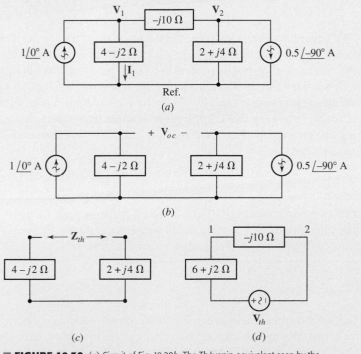

■ **FIGURE 10.30** (*a*) Circuit of Fig. 10.28*b*. The Thévenin equivalent seen by the $-j10\ \Omega$ impedance is desired. (*b*) $\mathbf{V}_{oc}$ is defined. (*c*) $\mathbf{Z}_{th}$ is defined. (*d*) The circuit is redrawn using the Thévenin equivalent.

(Continued on next page)

The open-circuit voltage, defined in Fig. 10.30b, is

$$\mathbf{V}_{oc} = (1\underline{/0^\circ})(4 - j2) - (-0.5\underline{/-90^\circ})(2 + j4)$$
$$= 4 - j2 + 2 - j1 = 6 - j3 \text{ V}$$

The impedance of the inactive circuit of Fig. 10.30c *as viewed from the load terminals* is simply the sum of the two remaining impedances. Hence,

$$\mathbf{Z}_{th} = 6 + j2 \ \Omega$$

Thus, when we reconnect the circuit as in Fig. 10.30d, the current directed from node 1 toward node 2 through the $-j10 \ \Omega$ load is

$$\mathbf{I}_{12} = \frac{6 - j3}{6 + j2 - j10} = 0.6 + j0.3 \text{ A}$$

We now know the current flowing through the $-j10 \ \Omega$ impedance of Fig. 10.30a. *Note that we are unable to compute* $\mathbf{V}_1$ *using the circuit of Fig. 10.30d as the reference node no longer exists.* Returning to the original circuit, then, and subtracting the $0.6 + j0.3$ A current from the left source current, the downward current through the $(4 - j2) \ \Omega$ branch is found:

$$\mathbf{I}_1 = 1 - 0.6 - j0.3 = 0.4 - j0.3 \quad \text{A}$$

and, thus,

$$\mathbf{V}_1 = (0.4 - j0.3)(4 - j2) = 1 - j2 \quad \text{V}$$

as before.

We might have been clever and used Norton's theorem on the three elements on the right of Fig. 10.30a, assuming that our chief interest is in $\mathbf{V}_1$. Source transformations can also be used repeatedly to simplify the circuit. Thus, all the shortcuts and tricks that arose in Chaps. 4 and 5 are available for circuit analysis in the frequency domain. The slight additional complexity that is apparent now arises from the necessity of using complex numbers and not from any more involved theoretical considerations.

PRACTICE

10.15 For the circuit of Fig. 10.31, find the (a) open-circuit voltage $\mathbf{V}_{ab}$; (b) downward current in a short circuit between a and b; (c) Thévenin-equivalent impedance $\mathbf{Z}_{ab}$ in parallel with the current source.

Ans: $16.77\underline{/-33.4^\circ}$ V; $2.60 + j1.500$ A; $2.5 - j5 \ \Omega$.

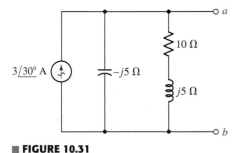

■ **FIGURE 10.31**

One final comment is in order. Up to this point, we have restricted ourselves to considering either single-source circuits or multiple-source circuits in which *every source operates at the exact same frequency*. This is necessary in order to define specific impedance values for inductive and capacitive elements. However, the concept of phasor analysis can be easily extended to circuits with multiple sources operating at different frequencies.

In such instances, we simply employ superposition to determine the voltages and currents due to each source, and then add the results *in the time domain*. If several sources are operating at the same frequency, superposition will also allow us to consider those sources at the same time, and add the resulting response to the response(s) of any other source(s) operating at a different frequency.

EXAMPLE 10.11

Determine the power dissipated by the 10 Ω resistor in the circuit of Fig. 10.32a.

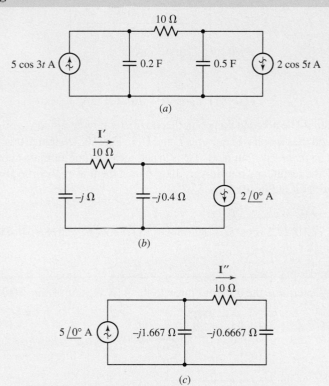

(a)

(b)

(c)

■ **FIGURE 10.32** (a) A simple circuit having sources operating at different frequencies. (b) Circuit with the left source killed. (c) Circuit with the right source killed.

After glancing at the circuit, we might be tempted to write two quick nodal equations, or perhaps perform two sets of source transformations and launch immediately into finding the voltage across the 10 Ω resistor.

Unfortunately, this is impossible, since we have *two* sources operating at *different* frequencies. In such a situation, there is no way to compute the impedance of any capacitor or inductor in the circuit—which ω would we use?

The only way out of this dilemma is to employ superposition, grouping all sources with the same frequency in the same subcircuit, as shown in Fig. 10.32b and c.

(Continued on next page)

In future studies of signal processing, we will also be introduced to the method of Jean-Baptiste Joseph Fourier, a French mathematician who developed a technique for representing almost any arbitrary function by a combination of sinusoids. When working with linear circuits, once we know the response of a particular circuit to a general sinusoidal forcing function, we can easily predict the response of the circuit to an arbitrary waveform represented by a Fourier series function, simply by using superposition.

In the subcircuit of Fig. 10.32*b*, we quickly compute the current $\mathbf{I}'$ using current division:

$$\mathbf{I}' = 2\underline{/0^\circ}\left[\frac{-j0.4}{10 - j - j0.4}\right]$$
$$= 79.23\underline{/-82.03^\circ}\text{ mA}$$

so that

$$i' = 79.23\cos(5t - 82.03^\circ)\text{ mA}$$

Likewise, we find that

$$\mathbf{I}'' = 5\underline{/0^\circ}\left[\frac{-j1.667}{10 - j0.6667 - j1.667}\right]$$
$$= 811.7\underline{/-76.86^\circ}\text{ mA}$$

so that

$$i'' = 811.7\cos(3t - 76.86^\circ)\text{ mA}$$

It should be noted at this point that no matter how tempted we might be to add the two phasor currents $\mathbf{I}'$ and $\mathbf{I}''$, in Fig. 10.32*b* and *c*, this *would be incorrect*. Our next step is to add the two time-domain currents, square the result, and multiply by 10 to obtain the power absorbed by the 10 Ω resistor in Fig. 10.32*a*:

$$p_{10} = (i' + i'')^2 \times 10$$
$$= 10[79.23\cos(5t - 82.03^\circ) + 811.7\cos(3t - 76.86^\circ)]^2\ \mu\text{W}$$

PRACTICE

10.16 Determine the current i through the 4 Ω resistor of Fig. 10.33.

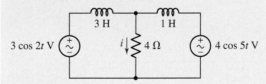

■ FIGURE 10.33

Ans: $i = 175.6\cos(2t - 20.55^\circ) + 547.1\cos(5t - 43.16^\circ)$ mA.

COMPUTER-AIDED ANALYSIS

We have several options in PSpice for the analysis of circuits in the sinusoidal steady state. Perhaps the most straightforward approach is to make use of two specially designed sources: VAC and IAC. The magnitude and phase of either source is selected by double-clicking on the part.

Let's simulate the circuit of Fig. 10.19*a*, shown redrawn in Fig. 10.34.

The frequency of the source is not selected through the Property Editor, but rather through the ac sweep analysis dialog box. This is

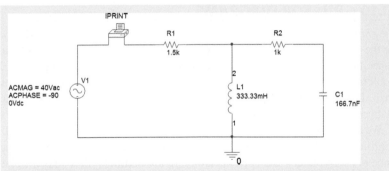

■ **FIGURE 10.34** The circuit of Fig. 10.19a, operating at $\omega = 3000$ rad/s. The current through the 1.5 kΩ resistor is desired.

accomplished by choosing **AC Sweep/Noise** for **Analysis** when presented with the Simulation Settings window. We select a **Linear** sweep and set **Total Points** to 1. Since we are only interested in the frequency of 3000 rad/s (477.5 Hz), we set both **Start Frequency** and **End Frequency** to 477.5 as shown in Fig. 10.35.

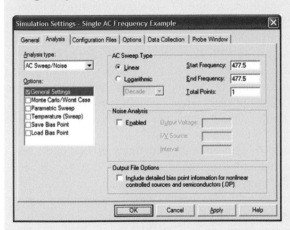

■ **FIGURE 10.35** Dialog box for setting source frequency.

Note that an additional "component" appears in the schematic. This component is called IPRINT, and allows a variety of current parameters to be printed. In this simulation, we are interested in the **AC, MAG,** and **PHASE** attributes. In order for PSpice to print these quantities, double-click on the IPRINT symbol in the schematic, and enter *yes* in each of the appropriate fields.

The simulation results are obtained by choosing **View Output File** under **PSpice** in the Capture CIS window.

```
FREQ        IM(V_PRINT1)  IP(V_PRINT1)

4.775E+02   1.600E-02     -1.269E+02
```

Thus, the current magnitude is 16 mA, and the phase angle is $-126.9°$, so that the current through the 1.5 kΩ resistor is

$$i = 16\cos(3000t - 126.9°) \text{ mA}$$
$$= 16\sin(3000t - 36.9°) \text{ mA}$$

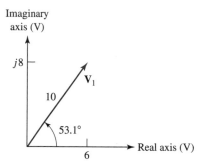

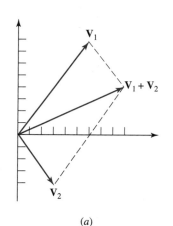

FIGURE 10.36 A simple phasor diagram shows the single voltage phasor $\mathbf{V}_1 = 6 + j8 = 10\underline{/53.1°}$ V.

10.10 PHASOR DIAGRAMS

The *phasor diagram* is a name given to a sketch in the complex plane showing the relationships of the phasor voltages and phasor currents throughout a specific circuit. It also provides a graphical method for solving certain problems which may be used to check more exact analytical methods. In Chap. 11 we encounter similar diagrams which display the complex power relationships in the sinusoidal steady state.

We are already familiar with the use of the complex plane in the graphical identification of complex numbers and in their addition and subtraction. Since phasor voltages and currents are complex numbers, they may also be identified as points in a complex plane. For example, the phasor voltage $\mathbf{V}_1 = 6 + j8 = 10\underline{/53.1°}$ V is identified on the complex voltage plane shown in Fig. 10.36. The x axis is the real voltage axis, and the y axis is the imaginary voltage axis; the voltage $\mathbf{V}_1$ is located by an arrow drawn from the origin. Since addition and subtraction are particularly easy to perform and display on a complex plane, phasors may be easily added and subtracted in a phasor diagram. Multiplication and division result in the addition and subtraction of angles and a change of amplitude. Figure 10.37a shows the sum of $\mathbf{V}_1$ and a second phasor voltage $\mathbf{V}_2 = 3 - j4 = 5\underline{/-53.1°}$ V, and Fig. 10.37b shows the current $\mathbf{I}_1$, which is the product of $\mathbf{V}_1$ and the admittance $\mathbf{Y} = 1 + j1$ S.

This last phasor diagram shows both current and voltage phasors on the same complex plane; it is understood that each will have its own amplitude scale, but a common angle scale. For example, a phasor voltage 1 cm long might represent 100 V, while a phasor current 1 cm long could indicate 3 mA. Plotting both phasors on the same diagram enables us to easily determine which waveform is leading and which is lagging.

The phasor diagram also offers an interesting interpretation of the time-domain to frequency-domain transformation, since the diagram may be interpreted from either the time- or the frequency-domain viewpoint. Up to this point, we have been using the frequency-domain interpretation, as we have been showing phasors directly on the phasor diagram. However, let us proceed to a time-domain viewpoint by first showing the phasor voltage $\mathbf{V} = V_m\underline{/\alpha}$ as sketched in Fig. 10.38a. In order to transform $\mathbf{V}$ to the time domain, the next necessary step is the multiplication of the phasor by $e^{j\omega t}$; thus we now have the complex voltage $V_m e^{j\alpha} e^{j\omega t} = V_m\underline{/\omega t + \alpha}$. This voltage may also be interpreted as a phasor, one which possesses a phase angle that increases linearly with time. On a phasor diagram it therefore

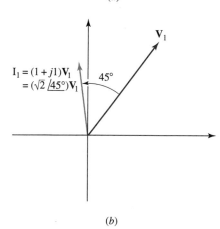

FIGURE 10.37 (*a*) A phasor diagram showing the sum of $\mathbf{V}_1 = 6 + j8$ V and $\mathbf{V}_2 = 3 - j4$ V, $\mathbf{V}_1 + \mathbf{V}_2 = 9 + j4$ V $= 9.85\underline{/24.0°}$ V. (*b*) The phasor diagram shows $\mathbf{V}_1$ and $\mathbf{I}_1$, where $\mathbf{I}_1 = \mathbf{Y}\mathbf{V}_1$ and $\mathbf{Y} = 1 + j$S $= \sqrt{2}\underline{/45°}$ S. The current and voltage amplitude scales are different.

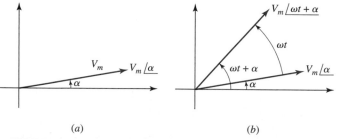

FIGURE 10.38 (*a*) The phasor voltage $V_m\underline{/\alpha}$. (*b*) The complex voltage $V_m\underline{/\omega t + \alpha}$ is shown as a phasor at a particular instant of time. This phasor leads $V_m\underline{/\alpha}$ by ωt radians.

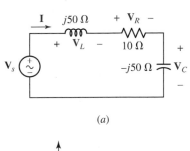

represents a rotating line segment, the instantaneous position being ωt radians ahead (counterclockwise) of $V_m/\underline{\alpha}$. Both $V_m/\underline{\alpha}$ and $V_m/\underline{\omega t + \alpha}$ are shown on the phasor diagram of Fig. 10.38b.

The passage to the time domain is now completed by taking the real part of $V_m/\underline{\omega t + \alpha}$. The real part of this complex quantity is the projection of $V_m/\underline{\omega t + \alpha}$ on the real axis: $V_m \cos(\omega t + \alpha)$.

In summary, then, the frequency-domain phasor appears on the phasor diagram, and the transformation to the time domain is accomplished by allowing the phasor to rotate in a counterclockwise direction at an angular velocity of ω rad/s and then visualizing the projection on the real axis. It is helpful to think of the arrow representing the phasor **V** on the phasor diagram as the photographic snapshot, taken at $\omega t = 0$, of a rotating arrow whose projection on the real axis is the instantaneous voltage $v(t)$.

Let us now construct the phasor diagrams for several simple circuits. The series *RLC* circuit shown in Fig. 10.39a has several different voltages associated with it, but only a single current. The phasor diagram is constructed most easily by employing the single current as the reference phasor. Let us arbitrarily select $\mathbf{I} = I_m/\underline{0°}$ and place it along the real axis of the phasor diagram, Fig. 10.39b. The resistor, capacitor, and inductor voltages may then be calculated and placed on the diagram, where the 90° phase relationships stand out clearly. The sum of these three voltages is the source voltage, and for this circuit, which is in what we will define in a subsequent chapter as the "resonant condition" since $\mathbf{Z}_C = -\mathbf{Z}_L$, the source voltage and resistor voltage are equal. The total voltage across the resistor and inductor or resistor and capacitor is obtained from the diagram by adding the appropriate phasors as shown.

Figure 10.40a is a simple parallel circuit in which it is logical to use the single voltage between the two nodes as a reference phasor. Suppose that $\mathbf{V} = 1/\underline{0°}$ V. The resistor current, $\mathbf{I}_R = 0.2/\underline{0°}$ A, is in phase with this voltage, and the capacitor current, $\mathbf{I}_C = j0.1$ A, leads the reference voltage by 90°. After these two currents are added to the phasor diagram, shown as Fig. 10.40b, they may be summed to obtain the source current. The result is $\mathbf{I}_s = 0.2 + j0.1$ A.

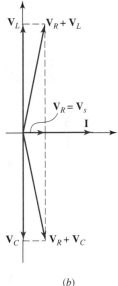

FIGURE 10.39 (*a*) A series *RLC* circuit. (*b*) The phasor diagram for this circuit; the current **I** is used as a convenient reference phasor.

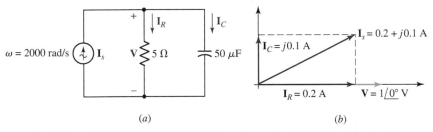

FIGURE 10.40 (*a*) A parallel *RC* circuit. (*b*) The phasor diagram for this circuit; the node voltage **V** is used as a convenient reference phasor.

If the source current is specified initially as the convenient value of $1/\underline{0°}$ A and the node voltage is not initially known, it is still convenient to begin construction of the phasor diagram by assuming a node voltage (for example, $\mathbf{V} = 1/\underline{0°}$ V once again) and using it as the reference phasor. The diagram is then completed as before, and the source current that flows as a result of the assumed node voltage is again found to be $0.2 + j0.1$ A. The true source current is $1/\underline{0°}$ A, however, and thus the true node voltage is obtained by multiplying the assumed node voltage by $1\underline{/0°}/(0.2 + j0.1)$; the true node voltage is therefore $4 - j2$ V $= \sqrt{20}\underline{/-26.6°}$ V. The assumed voltage leads to a phasor diagram which differs from the true phasor diagram by a change of scale (the assumed diagram is smaller by a factor of $1/\sqrt{20}$) and an angular rotation (the assumed diagram is rotated counterclockwise through $26.6°$).

Phasor diagrams are usually very simple to construct, and most sinusoidal steady-state analyses will be more meaningful if such diagrams are included. Additional examples of the use of phasor diagrams will appear frequently throughout the remainder of our study.

EXAMPLE 10.12

Construct a phasor diagram showing I_R, I_L, and I_C for the circuit of Fig. 10.41. Combining these currents, determine the angle by which I_s leads I_R, I_C, and I_x.

We begin by choosing a suitable reference phasor. Upon examining the circuit and the variables to be determined, we see that once $\mathbf{V}$ is known, $\mathbf{I}_R$, $\mathbf{I}_L$, and $\mathbf{I}_C$ can be computed by simple application of Ohm's law. Thus, we select $\mathbf{V} = 1\underline{/0°}$ V for simplicity's sake, and subsequently compute

$$\mathbf{I}_R = (0.2)1\underline{/0°} \quad = 0.2\underline{/0°} \text{ A}$$
$$\mathbf{I}_L = (-j0.1)1\underline{/0°} = 0.1\underline{/-90°} \text{ A}$$
$$\mathbf{I}_C = (j0.3)1\underline{/0°} \quad = 0.3\underline{/90°} \text{ A}$$

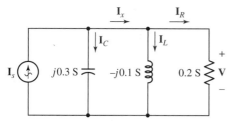

■ **FIGURE 10.41** A simple circuit for which several currents are required.

The corresponding phasor diagram is shown in Fig. 10.42*a*. We also need to find the phasor currents $\mathbf{I}_s$ and $\mathbf{I}_x$. Figure 10.42*b* shows the determination of $\mathbf{I}_x = \mathbf{I}_L + \mathbf{I}_R = 0.2 - j0.1 = 0.224\underline{/-26.6°}$ A, and Fig. 10.42*c* shows the determination of $\mathbf{I}_s = \mathbf{I}_C + \mathbf{I}_x = 0.283\underline{/45°}$ A. From Fig. 10.42*c*, we ascertain that $\mathbf{I}_s$ leads $\mathbf{I}_R$ by $45°$, $\mathbf{I}_C$ by $-45°$, and $\mathbf{I}_x$ by $45° + 26.6° = 71.6°$. These angles are only relative, however; the exact numerical values will depend on $\mathbf{I}_s$, upon which the actual value of $\mathbf{V}$ (assumed here to be $1\underline{/0°}$ V for convenience) also depends.

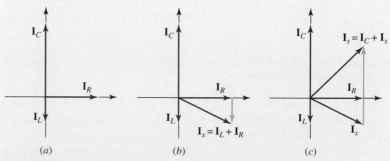

■ **FIGURE 10.42** (*a*) Phasor diagram constructed using a reference value of $\mathbf{V} = 1\underline{/0°}$. (*b*) Graphical determination of $\mathbf{I}_x = \mathbf{I}_L + \mathbf{I}_R$. (*c*) Graphical determination of $\mathbf{I}_s = \mathbf{I}_C + \mathbf{I}_x$.

FIGURE 10.43

PRACTICE

10.17 Select some convenient reference value for $\mathbf{I}_C$ in the circuit of Fig. 10.43, draw a phasor diagram showing $\mathbf{V}_R$, $\mathbf{V}_2$, $\mathbf{V}_1$, and $\mathbf{V}_s$, and measure the ratio of the lengths of (a) $\mathbf{V}_s$ to $\mathbf{V}_1$; (b) $\mathbf{V}_1$ to $\mathbf{V}_2$; (c) $\mathbf{V}_s$ to $\mathbf{V}_R$.

Ans: 1.90; 1.00; 2.12

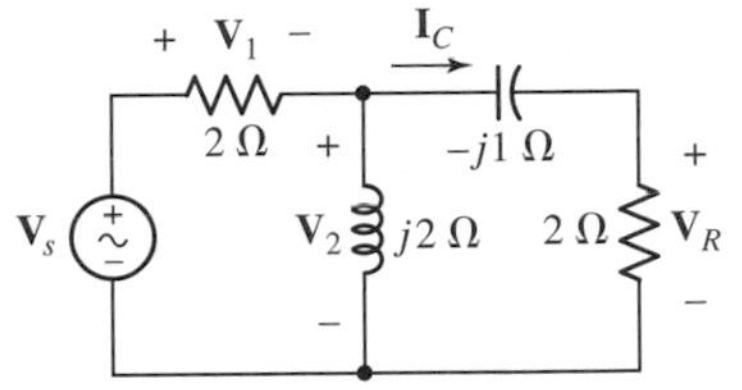

SUMMARY AND REVIEW

❑ If two sine waves (or two cosine waves) both have positive magnitudes and the same frequency, it is possible to determine which waveform is leading and which is lagging by comparing their phase angles.

❑ The forced response of a linear circuit to a sinusoidal voltage or current source can always be written as a single sinusoid having the same frequency as the sinusoidal source.

❑ A phasor transform may be performed on any sinusoidal function, and vice-versa: $V_m \cos(\omega t + \phi) \leftrightarrow V_m \underline{/\phi}$.

❑ A phasor has both a magnitude and a phase angle; the frequency is understood to be that of the sinusoidal source driving the circuit.

❑ When transforming a time-domain circuit into the corresponding frequency-domain circuit, resistors, capacitors, and inductors are replaced by impedances (or, occasionally, by admittances).

❑ The impedance of a resistor is simply its resistance.

❑ The impedance of a capacitor is $1/j\omega C$ Ω.

❑ The impedance of an inductor is $j\omega L$ Ω.

❑ Impedances combine both in series and in parallel combinations in the same manner as resistors.

❑ All analysis techniques previously used on resistive circuits apply to circuits with capacitors and/or inductors once all elements are replaced by their frequency-domain equivalents.

❑ Phasor analysis can only be performed on single-frequency circuits. Otherwise, superposition must be invoked, and the *time-domain* partial responses added to obtain the complete response.

❑ The power behind phasor diagrams is evident when a convenient forcing function is used initially, and the final result scaled appropriately.

READING FURTHER

A good reference to phasor-based analysis techniques can be found in:

R.A. DeCarlo and P.M. Lin, *Linear Circuit Analysis,* 2nd ed. New York: Oxford University Press, 2001.

Frequency-dependent transistor models are discussed from a phasor perspective in Chap. 7 of

W.H. Hayt, Jr. and G.W. Neudeck, *Electronic Circuit Analysis and Design,* 2nd ed. New York: Wiley, 1995.

EXERCISES

10.1 Characteristics of Sinusoids

1. A sine wave, $f(t)$, is zero and increasing at $t = 2.1$ ms, and the succeeding positive maximum of 8.5 occurs at $t = 7.5$ ms. Express the wave in the form $f(t)$ equals (a) $C_1 \sin(\omega t + \phi)$, where ϕ is positive, as small as possible, and in degrees; (b) $C_2 \cos(\omega t + \beta)$, where β has the smallest possible magnitude and is in degrees; (c) $C_3 \cos \omega t + C_4 \sin \omega t$.

2. (a) If $-10 \cos \omega t + 4 \sin \omega t = A \cos(\omega t + \phi)$, where $A > 0$ and $-180° < \phi \le 180°$, find A and ϕ. (b) If $200 \cos(5t + 130°) = F \cos 5t + G \sin 5t$, find F and G. (c) Find three values of t, $0 \le t \le 1$ s, at which $i(t) = 5 \cos 10t - 3 \sin 10t = 0$. (d) In what time interval between $t = 0$ and $t = 10$ ms is $10 \cos 100\pi t \ge 12 \sin 100\pi t$?

3. Given the two sinusoidal waveforms, $f(t) = -50 \cos \omega t - 30 \sin \omega t$ and $g(t) = 55 \cos \omega t - 15 \sin \omega t$, find (a) the amplitude of each, and (b) the phase angle by which $f(t)$ leads $g(t)$.

4. Substitute the assumed current response of Eq. [3], $i(t) = A \cos(\omega t - \theta)$, directly in the differential equation $L(di/dt) + Ri = V_m \cos \omega t$ to show that values for A and θ are obtained which agree with Eq. [4].

5. A certain power supply generates a voltage waveform that can be characterized as a cosine wave $V_m \cos(\omega t + \phi)$ having a frequency of 13.56 MHz. If the supply delivers a maximum power of 300 W to a 5 Ω load and the voltage reaches a minimum at $t = 21.15$ ms, what are V_m, ω, and ϕ?

6. Compare the following pairs of waveforms, and determine which one is leading: (a) $-33 \sin(8t - 9°)$ and $12 \cos(8t - 1°)$. (b) $15 \cos(1000t + 66°)$ and $-2 \cos(1000t + 450°)$. (c) $\sin(t - 13°)$ and $\cos(t - 90°)$. (d) $\sin t$ and $\cos(t - 90°)$.

7. Determine which waveform in each pair is lagging: (a) $6 \cos(2\pi 60t - 9°)$ and $-6 \cos(2\pi 60t + 9°)$. (b) $\cos(t - 100°)$ and $-\cos(t - 100°)$. (c) $-\sin t$ and $\sin t$. (d) $7000 \cos(t - \pi)$ and $9 \cos(t - 3.14°)$.

8. Show that the voltage $v(t) = V_1 \cos \omega t - V_2 \sin \omega t$ can be written as a single cosine function $V_m \cos(\omega t + \phi)$. Derive appropriate expressions for V_m and ϕ.

9. The Fourier theorem is a common tool in both science and engineering. It shows that the periodic waveform shown in Fig. 10.44 is equal to the infinite sum

$$v(t) = \frac{8}{\pi^2} \left(\sin \pi t - \frac{1}{3^2} \sin 3\pi t + \frac{1}{5^2} \sin 5\pi t - \frac{1}{7^2} \sin 7\pi t + \cdots \right)$$

(a) Compute the exact value of $v(t)$ at $t = 0.4$ s. Compute the approximate value of $v(t)$ using the Fourier series above using (b) the first term only; (c) the first four terms only; (d) the first five terms only.

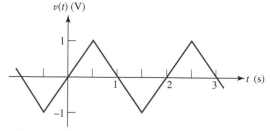

■ FIGURE 10.44

10. Household electrical voltages are typically quoted as either 110 V, 115 V, or 120 V. However, these values do not represent the peak ac voltage. Rather,

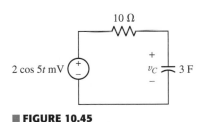

they represent what is known as the root mean square of the voltage, defined as

$$V_{rms} = \sqrt{\frac{1}{T} \int_0^T V_m^2 \cos^2(\omega t)\, dt}$$

where $T =$ the period of the waveform, V_m is the peak voltage, and $\omega =$ the waveform frequency ($f = 60$ Hz in North America).

(a) Perform the indicated integration, and show that for a sinusoidal voltage,

$$V_{rms} = \frac{V_m}{\sqrt{2}}$$

(b) Compute the peak voltages corresponding to the rms voltages of 110, 115, and 120 V.

10.2 Forced Response to Sinusoidal Functions

11. Find the steady-state voltage $v_C(t)$ as indicated in the circuit of Fig. 10.45.

12. Determine the inductor voltage $v_L(t)$ for the circuit of Fig. 10.46, assuming $R = 100\ \Omega$, $L = 2$ H, and all transients have long since died out.

13. Let $v_s = 20\cos 500t$ V in the circuit of Fig. 10.47. After simplifying the circuit a little, find $i_L(t)$.

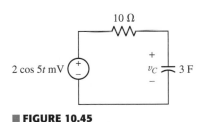

■ **FIGURE 10.45**

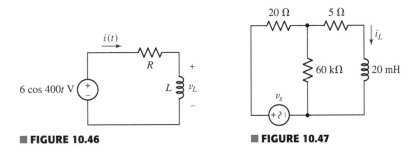

■ **FIGURE 10.46**   ■ **FIGURE 10.47**

14. If $i_s = 0.4\cos 500t$ A in the circuit shown in Fig. 10.48, simplify the circuit until it is in the form of Fig. 10.4 and then find (a) $i_L(t)$; (b) $i_x(t)$.

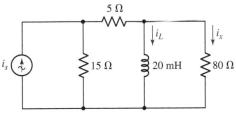

■ **FIGURE 10.48**

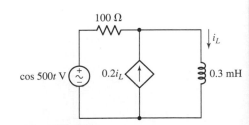

■ **FIGURE 10.49**

15. A sinusoidal voltage source $v_s = 100\cos 10^5 t$ V, a 500 Ω resistor, and an 8 mH inductor are in series. Determine those instants of time, $0 \le t < \frac{1}{2}T$, at which zero power is being: (a) delivered to the resistor; (b) delivered to the inductor; (c) generated by the source.

16. In the circuit of Fig. 10.49, let $v_s = 3\cos 10^5 t$ V and $i_s = 0.1\cos 10^5 t$. After making use of superposition and Thévenin's theorem, find the instantaneous values of i_L and v_L at $t = 10\ \mu s$.

17. Find $i_L(t)$ in the circuit shown in Fig. 10.50.

■ **FIGURE 10.50**

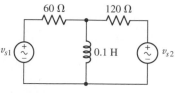

■ **FIGURE 10.51**

18. Both voltage sources in the circuit of Fig. 10.51 are given by $120\cos 120\pi t$ V. (a) Find an expression for the instantaneous energy stored in the inductor, and (b) use it to find the average value of the stored energy.

19. In the circuit of Fig. 10.51, the voltage sources are $v_{s1} = 120\cos 200t$ V and $v_{s2} = 180\cos 200t$ V. Find the downward inductor current.

20. Assume that the op-amp in Fig. 10.52 is ideal ($R_i = \infty$, $R_o = 0$, and $A = \infty$). Note also that the integrator input has two signals applied to it, $-V_m\cos\omega t$ and v_{out}. If the product $R_1 C_1$ is set equal to the ratio L/R in the circuit of Fig. 10.4, show that v_{out} equals the voltage across R (+ reference on the left) in Fig. 10.4.

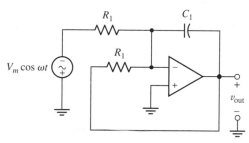

■ **FIGURE 10.52**

21. A voltage source $V_m\cos\omega t$, a resistor R, and a capacitor C are all in series. (a) Write an integrodifferential equation in terms of the loop current i and then differentiate it to obtain the differential equation for the circuit. (b) Assume a suitable general form for the forced response $i(t)$, substitute it in the differential equation, and determine the exact form of the forced response.

10.3 The Complex Forcing Function

22. Convert the following to rectangular form: (a) $7\underline{/-90°}$; (b) $3 + j + 7\underline{/-17°}$; (c) $14e^{j15°}$; (d) $1\underline{/0°}$. Convert the following to polar form: (e) $-2(1 + j9)$; (f) 3.

23. Perform the indicated operation(s), and express the answer as a single complex number in rectangular form: (a) $3 + 15\underline{/-23°}$; (b) $12j(17\underline{/180°})$; (c) $5 - 16\dfrac{(1+j)(2-j7)}{33\underline{/-9°}}$.

24. Perform the indicated operation(s), and express your answer as a single complex number in polar form: (a) $5\underline{/9°} - 9\underline{/-17°}$; (b) $(8 - j15)(4 + j16) - j$; (c) $\dfrac{(14 - j9)}{(2 - j8)} + 5\underline{/-30°}$; (d) $17\underline{/-33°} + 6\underline{/-21°} + j3$.

25. Express the following as a complex number in polar form: (a) $e^{j14°} + 9\underline{/3°} - \dfrac{8 - j6}{j^2}$; (b) $\dfrac{5\underline{/30°}}{2\underline{/-15°}} + \dfrac{2e^{j5}}{2 - j2}$.

26. Convert these complex numbers to rectangular form: (a) $5\underline{/-110°}$; (b) $6e^{j160°}$; (c) $(3 + j6)(2\underline{/50°})$. Convert to polar form: (d) $-100 - j40$; (e) $2\underline{/50°} + 3\underline{/-120°}$.

27. Carry out the indicated calculations, and express the result in polar form: (a) $40\underline{/-50°} - 18\underline{/25°}$; (b) $3 + \dfrac{2}{j} + \dfrac{2 - j5}{1 + j2}$. Express in rectangular form: (c) $(2.1\underline{/25°})^3$; (d) $0.7e^{j0.3}$.

28. In the circuit of Fig. 10.53, let i_C be expressed as the complex response $20e^{j(40t+30°)}$ A, and express v_s as a complex forcing function.

29. In the circuit of Fig. 10.54, let the current i_L be expressed as the complex response $20e^{j(10t+25°)}$ A, and express the source current $i_s(t)$ as a complex forcing function.

30. In a linear network, such as that shown in Fig. 10.8, a sinusoidal source voltage, $v_s = 80\cos(500t - 20°)$ V, produces the output current

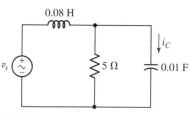

■ **FIGURE 10.53**

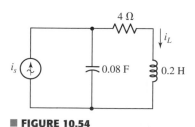

■ **FIGURE 10.54**

$i_{\text{out}} = 5\cos(500t + 12°)$ A. Find i_{out} if v_s equals (a) $40\cos(500t + 10°)$ V;
(b) $40\sin(500t + 10°)$ V; (c) $40e^{j(500t+10°)}$ V; (d) $(50 + j20)e^{j500t}$ V.

10.4 The Phasor

31. Express each of the following currents as a phasor: (a) $12\sin(400t + 110°)$ A;
(b) $-7\sin 800t - 3\cos 800t$ A; (c) $4\cos(200t - 30°) - 5\cos(200t + 20°)$ A.
If $\omega = 600$ rad/s, find the instantaneous value of each of these voltages at $t = 5$ ms: (d) $70\underline{/30°}$ V; (e) $-60 + j40$ V.

32. Let $\omega = 4$ krad/s, and determine the instantaneous value of i_x at $t = 1$ ms if $\mathbf{I}_x$
equals (a) $5\underline{/-80°}$ A; (b) $-4 + j1.5$ A. Express in polar form the phasor voltage $\mathbf{V}_x$ if $v_x(t)$ equals (c) $50\sin(250t - 40°)$ V; (d) $20\cos 108t - 30\sin 108t$ V;
(e) $33\cos(80t - 50°) + 41\cos(80t - 75°)$ V.

33. The phasor voltages $\mathbf{V}_1 = 10\underline{/90°}$ mV at $\omega = 500$ rad/s and $\mathbf{V}_2 = 8\underline{/90°}$ mV
at $\omega = 1200$ rad/s are added together in an op amp. If the op amp multiplies this input by a factor of -5, find the output at $t = 0.5$ ms.

34. If $\omega = 500$ rad/s and $\mathbf{I}_L = 2.5\underline{/40°}$ A in the circuit of Fig. 10.55, find $v_s(t)$.

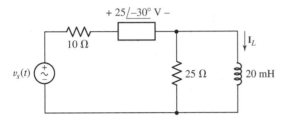

■ FIGURE 10.55

35. Let $\omega = 5$ krad/s in the circuit of Fig. 10.56. Find (a) $v_1(t)$; (b) $v_2(t)$; (c) $v_3(t)$.

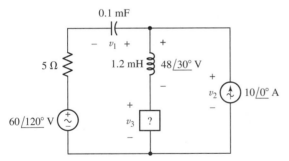

■ FIGURE 10.56

36. A phasor current of $1\underline{/0°}$ A is flowing through the series combination of 1 Ω,
1 H, and 1 F. At what frequency is the amplitude of the voltage across the network twice the amplitude of the voltage across the resistor?

37. Find v_x in the circuit shown in Fig. 10.57.

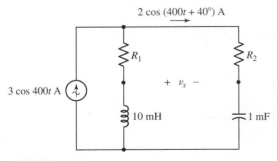

■ FIGURE 10.57

38. A black box with yellow stripes contains two current sources, $\mathbf{I}_{s1}$ and $\mathbf{I}_{s2}$. The output voltage is identified as $\mathbf{V}_{out}$. If $\mathbf{I}_{s1} = 2\underline{/20°}$ A and $\mathbf{I}_{s2} = 3\underline{/-30°}$ A, then $\mathbf{V}_{out} = 80\underline{/10°}$ V. However, if $\mathbf{I}_{s1} = \mathbf{I}_{s2} = 4\underline{/40°}$ A, then $\mathbf{V}_{out} = 90 - j30$ V. Find $\mathbf{V}_{out}$ if $\mathbf{I}_{s1} = 2.5\underline{/-60°}$ A and $\mathbf{I}_{s2} = 2.5\underline{/60°}$ A.

10.6 Impedance

39. Calculate the impedance of a series combination comprised of a 1 mF, 2 mF, and 3 mF capacitor if operated at a frequency of (a) 1 Hz; (b) 100 Hz; (c) 1 kHz; (d) 1 GHz.

40. Calculate the impedance of a 5 Ω resistor in parallel with a 1 nH inductor and a 5 nH inductor if the operating frequency is (a) 1 Hz; (b) 1 kHz; (c) 1 MHz; (d) 1 GHz; (e) 1 THz.

41. Find $\mathbf{Z}_{in}$ at terminals a and b in Fig. 10.58 if ω equals (a) 800 rad/s; (b) 1600 rad/s.

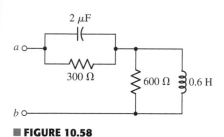

■ FIGURE 10.58

42. Let $\omega = 100$ rad/s in the circuit of Fig. 10.59. Find (a) $\mathbf{Z}_{in}$; (b) $\mathbf{Z}_{in}$ if a short circuit is connected from x to y.

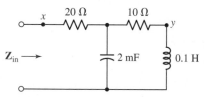

■ FIGURE 10.59

43. If a voltage source $v_s = 120 \cos 800t$ V is connected to terminals a and b in Fig. 10.58 (+ reference at the top), what current flows to the right in the 300 Ω resistance?

44. Find $\mathbf{V}$ in Fig. 10.60 if the box contains (a) 3 Ω in series with 2 mH; (b) 3 Ω in series with 125 μF; (c) 3 Ω, 2 mH, and 125 μF in series; (d) 3 Ω, 2 mH, and 125 μF in series, but $\omega = 4$ krad/s.

45. A 10 H inductor, a 200 Ω resistor, and a capacitor C are in parallel. (a) Find the impedance of the parallel combination at $\omega = 100$ rad/s if $C = 20$ μF. (b) If the magnitude of the impedance is 125 Ω at $\omega = 100$ rad/s, find C. (c) At what two values of ω is the magnitude of the impedance equal to 100 Ω if $C = 20$ μF?

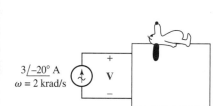

■ FIGURE 10.60

46. A 20 mH inductor and a 30 Ω resistor are in parallel. Find the frequency ω at which: (a) $|\mathbf{Z}_{in}| = 25$ Ω; (b) angle $(\mathbf{Z}_{in}) = 25°$; (c) Re$(\mathbf{Z}_{in}) = 25$ Ω; (d) Im $(\mathbf{Z}_{in}) = 10$ Ω.

47. Find R_1 and R_2 in the circuit of Fig. 10.57.

48. A two-element network has an input impedance of $200 + j80$ Ω at the frequency $\omega = 1200$ rad/s. What capacitance C should be placed in parallel with the network to provide an input impedance with (a) zero reactance? (b) a magnitude of 100 Ω?

49. For the network of Fig. 10.61, find $\mathbf{Z}_{in}$ at $\omega = 4$ rad/s if terminals a and b are (a) open-circuited; (b) short-circuited.

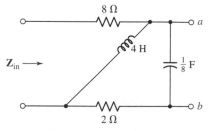

■ FIGURE 10.61

50. Find the equivalent impedance of the network shown in Fig. 10.62, assuming a frequency of $f = 1$ MHz.

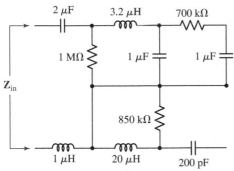

■ FIGURE 10.62

D 51. Design a combination of inductors, resistors, and capacitors that has (a) an impedance of $1 + j4\ \Omega$ at $\omega = 1$ rad/s; (b) an impedance of $5\ \Omega$ at $\omega = 1$ rad/s, constructed using at least one inductor; (c) an impedance of $7/80°\ \Omega$ at $\omega = 100$ rad/s; and (d) using the fewest components possible, an impedance of $5\ \Omega$ at $f = 3$ THz.

D 52. Design a combination of inductors, resistors, and capacitors that has (a) an impedance of $1 + j4\ \text{k}\Omega$ at $\omega = 230$ rad/s; (b) an impedance of $5\ \text{M}\Omega$ at $\omega = 10$ rad/s, constructed using at least one capacitor; (c) an impedance of $80/-22°\ \Omega$ at $\omega = 50$ rad/s; and (d) using the fewest components possible, an impedance of $300\ \Omega$ at $\omega = 3$ krad/s.

10.7 Admittance

53. Calculate the admittance of a parallel combination comprised of a 1 mF, 2 mF, and 4 mF capacitor if operated at a frequency of (a) 2 Hz; (b) 200 Hz; (c) 20 kHz; (d) 200 GHz.

54. What is the susceptance of the parallel combination of (a) two 100 Ω resistors; (b) a 1 Ω resistor in parallel with a 1 F capacitor if the operating frequency is 100 rad/s; (c) a 1 Ω resistor in series with a 2 H inductor if the operating frequency is 50 rad/s?

55. Find the input admittance Y_{ab} of the network shown in Fig. 10.63 and draw it as the parallel combination of a resistance R and an inductance L, giving values for R and L if $\omega = 1$ rad/s.

56. A 5 Ω resistance, a 20 mH inductance, and a 2 mF capacitance form a series network having terminals a and b. (a) Work with admittances to determine what size capacitance should be connected between a and b so that $Z_{in,ab} = R_{in,ab} + j0$ at $\omega = 500$ rad/s. (b) What is $R_{in, ab}$? (c) With your C in place, what is $Y_{in, ab}$ at $\omega = 100$ rad/s?

57. In the network shown in Fig. 10.64, find the frequency at which (a) $R_{in} = 550\ \Omega$; (b) $X_{in} = 50\ \Omega$; (c) $G_{in} = 1.8$ mS; (d) $B_{in} = -150\ \mu$S.

58. Two admittances, $Y_1 = 3 + j4$ mS and $Y_2 = 5 + j2$ mS, are in parallel, and a third admittance, $Y_3 = 2 - j4$ mS, is in series with the parallel combination. If a current $I_1 = 0.1/30°$ A is flowing through Y_1, find the magnitude of the voltage across (a) Y_1; (b) Y_2; (c) Y_3; (d) the entire network.

59. The admittance of the parallel combination of a 10 Ω resistance and a 50 μF capacitance at $\omega = 1$ krad/s is the same as the admittance of R_1 and C_1 in series at that frequency. (a) Find R_1 and C_1. (b) Repeat for $\omega = 2$ krad/s.

60. A cartesian coordinate plane contains a horizontal axis on which G_{in} is given in siemens, and a vertical axis along which B_{in} is measured, also in S. Let Y_{in} represent the series combination of a 1 Ω resistor and a 0.1 F capacitor. (a) Find Y_{in}, G_{in}, and B_{in} as functions of ω. (b) Locate the coordinate pairs

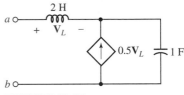

■ FIGURE 10.63

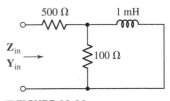

■ FIGURE 10.64

(G_{in}, B_{in}) on the plane at the frequency values $\omega = 0, 1, 2, 5, 10, 20,$ and 10^6 rad/s.

D 61. Design a combination of inductors, resistors, and capacitors that has (a) an admittance of $1 - j4$ S at $\omega = 1$ rad/s; (b) an admittance of 200 mS at $\omega = 1$ rad/s, constructed using at least one inductor; (c) an admittance of $7\underline{/80°}\ \mu$S at $\omega = 100$ rad/s; and (d) an admittance of 200 mΩ at $\omega = 3$ THz using the fewest components possible.

D 62. Design a combination of inductors, resistors, and capacitors that has (a) an admittance of $1 - j4$ pS at $\omega = 30$ rad/s; (b) an admittance of $5\ \mu$S at $\omega = 560$ rad/s, constructed using at least one capacitor; (c) an admittance of $4\underline{/-10°}$ nS at $\omega = 50$ rad/s; and (d) an admittance of 60 nS at $\omega = 300$ kHz using the fewest components possible.

10.8 Nodal and Mesh Analysis

63. Use phasors and nodal analysis on the circuit of Fig. 10.65 to find $\mathbf{V}_2$.

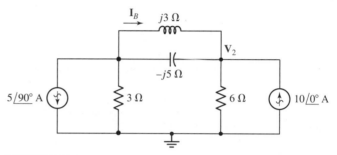

■ **FIGURE 10.65**

64. Use phasors and mesh analysis on the circuit of Fig. 10.65 to find $\mathbf{I}_B$.

65. Find $v_x(t)$ in the circuit of Fig. 10.66 if $v_{s1} = 20 \cos 1000t$ V and $v_{s2} = 20 \sin 1000t$ V.

66. (a) Find $\mathbf{V}_3$ in the circuit shown in Fig. 10.67. (b) To what identical values should the three capacitive impedances be changed so that $\mathbf{V}_3$ is 180° out of phase with the source voltage?

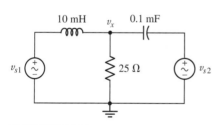

■ **FIGURE 10.66**

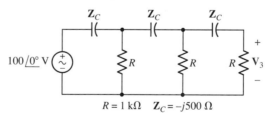

■ **FIGURE 10.67**

67. Use mesh analysis to find $i_x(t)$ in the circuit shown in Fig. 10.68.

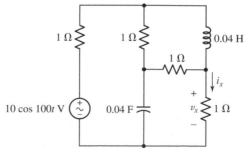

■ **FIGURE 10.68**

68. Find $v_x(t)$ for the circuit of Fig. 10.68 using phasors and nodal analysis.

69. The op amp shown in Fig. 10.69 has an infinite input impedance, zero output impedance, and a large but finite (positive, real) gain, $A = -\mathbf{V}_o/\mathbf{V}_i$ (a) Construct a basic differentiator by letting $\mathbf{Z}_f = R_f$, find $\mathbf{V}_o/\mathbf{V}_s$, and then show that $\mathbf{V}_o/\mathbf{V}_s \rightarrow -j\omega C_1 R_f$ as $A \rightarrow \infty$. (b) Let $\mathbf{Z}_f$ represent C_f and R_f in parallel, find $\mathbf{V}_o/\mathbf{V}_s$, and then show that $\mathbf{V}_o/\mathbf{V}_s \rightarrow -j\omega C_1 R_f/(1 + j\omega C_f R_f)$ as $A \rightarrow \infty$.

70. For the circuit of Fig. 10.70, determine the voltage v_2.

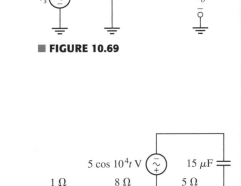

■ FIGURE 10.69

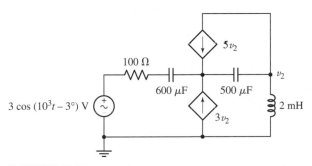

■ FIGURE 10.70

71. Compute the power dissipated by the 1 Ω resistor in Fig.10.71 at $t = 1$ ms.

72. Use phasor analysis to determine the three mesh currents $i_1(t)$, $i_2(t)$, and $i_3(t)$ in the circuit of Fig. 10.72.

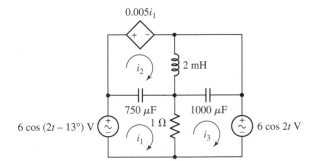

■ FIGURE 10.71

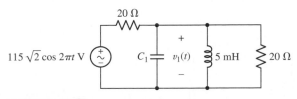

■ FIGURE 10.72

73. In the circuit of Fig. 10.73, the voltage $v_1(t) = 6.014 \cos(2\pi t + 85.76°)$ volts. What is the capacitance of C_1?

74. In the circuit of Fig. 10.74, the current $i_1(t) = 8.132 \cos 2\pi t$ A. What is the inductance of L_1?

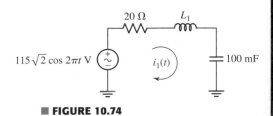

■ FIGURE 10.73   ■ FIGURE 10.74

75. Referring to the transistor amplifier circuit of Fig. 10.26b, (a) derive an equation for the phase angle of the output as a function of frequency, assuming an

input signal $\mathbf{V}_s = 1\underline{/0°}$ volts. (b) Plot your equation on a semilogarithmic scale for frequencies between 100 Hz and 10 GHz. Use $R_s = 300\,\Omega$, $R_B = 5\,k\Omega$, $r_\pi = 2.2\,k\Omega$, $C_\pi = 5\,pF$, $C_\mu = 2\,pF$, $g_m = 38\,mS$, $R_C = 4.7\,k\Omega$, and $R_L = 1.2\,k\Omega$. (c) Over what frequency range is the output exactly 180° out of phase with the input? At approximately what frequency does this phase relationship begin to change?

10.9 Superposition, Source Transformations, and Thévenin's Theorem

76. Find the frequency-domain Thévenin equivalent of the network shown in Fig. 10.75. Show the result as $\mathbf{V}_{th}$ in series with $\mathbf{Z}_{th}$.

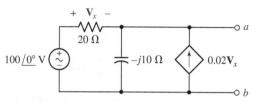

■ **FIGURE 10.75**

77. Find the input admittance of the circuit shown in Fig. 10.76, and represent it as the parallel combination of a resistance R and an inductance L, giving values for R and L if $\omega = 1$ rad/s.

78. With reference to the circuit of Fig. 10.77, think superposition and find that part of $v_1(t)$ due to (a) the voltage source acting alone; (b) the current source acting alone.

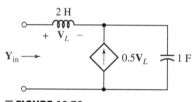

■ **FIGURE 10.76**

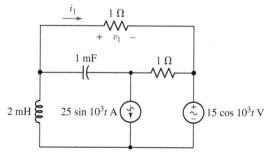

■ **FIGURE 10.77**

79. Use $\omega = 1$ rad/s, and find the Norton equivalent of the network shown in Fig. 10.78. Construct the Norton equivalent as a current source $\mathbf{I}_N$ in parallel with a resistance R_N and either an inductance L_N or a capacitance C_N.

80. In the circuit of Fig. 10.79, let $i_{s1} = 2\cos 200t$ A, $i_{s2} = 1\cos 100t$ A, and $v_{s3} = 2\sin 200t$ V. Find $v_L(t)$.

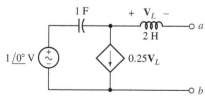

■ **FIGURE 10.78**

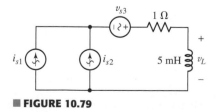

■ **FIGURE 10.79**

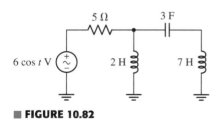

81. Find the Thévenin equivalent circuit for Fig. 10.80.

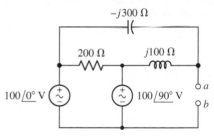

■ **FIGURE 10.80**

82. Determine the current $i(t)$ flowing from the voltage source of Fig. 10.81.

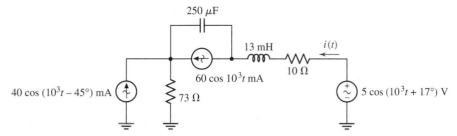

■ **FIGURE 10.81**

83. (a) Determine the voltage across the 3 F capacitor of Fig. 10.82. (b) Verify your answer with PSpice.

84. (a) Find the Thévenin equivalent seen by the $j5$ Ω inductor of Fig. 10.21. (b) Assuming a frequency of 100 rad/s, verify your answer with PSpice.

85. Using a single resistor, a single capacitor, a sinusoidal voltage source, and the principle of voltage division, design a circuit that will "filter out" high frequencies. (Hint: Define an output voltage across one of the two passive elements, and treat the sinusoidal source as the input. Interpret "filter out" to mean a reduced output voltage.)

■ **FIGURE 10.82**

86. Using a single resistor, a single capacitor, a sinusoidal voltage source, and the principle of voltage division, design a circuit that will "filter out" low frequencies. (Hint: Define an output voltage across one of the two passive elements, and treat the sinusoidal source as the input. Interpret "filter out" to mean a reduced output voltage.)

87. (a) Reduce the circuit in Fig. 10.83 to a simple series RC circuit. (b) Derive an equation for the magnitude of the voltage ratio $\mathbf{V}_{out}/\mathbf{V}_s$ as a function of frequency. (c) Plot your equation over the frequency range of 100 Hz to 1 MHz, and compare your result with an appropriate PSpice simulation of the original circuit.

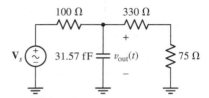

■ **FIGURE 10.83**

88. Refer to Fig. 10.26b. (a) Show that the maximum voltage gain of the amplifier circuit (defined as $\mathbf{V}_{out}/\mathbf{V}_\pi$) is $-g_m(R_C \| R_L)$. (b) If $R_S = 100$ Ω, $R_L = 8$ Ω, the maximum value of R_C is 10 kΩ, $r_\pi g_m = 300$, and all other parameters can

be varied, how would you modify the design to increase the maximum gain? (c) How does this design modification affect the cutoff frequency of the amplifier? How would you compensate for this?

89. Use superposition to find the voltages $v_1(t)$ and $v_2(t)$ in the circuit of Fig. 10.84.

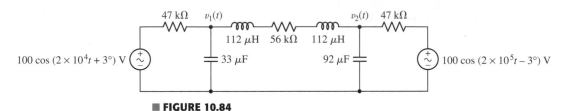

■ **FIGURE 10.84**

90. Use superposition to find the voltages $v_1(t)$ and $v_2(t)$ in the circuit of Fig. 10.85.

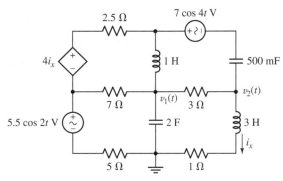

■ **FIGURE 10.85**

10.10 Phasor Diagrams

91. (a) Calculate values for $\mathbf{I}_L$, $\mathbf{I}_R$, $\mathbf{I}_C$, $\mathbf{V}_L$, $\mathbf{V}_R$, and $\mathbf{V}_C$ (plus $\mathbf{V}_s$) for the circuit shown in Fig. 10.86. (b) Using scales of 50 V to 1 in and 25 A to 1 in, show all seven quantities on a phasor diagram, and indicate that $\mathbf{I}_L = \mathbf{I}_R + \mathbf{I}_C$ and $\mathbf{V}_s = \mathbf{V}_L + \mathbf{V}_R$.

92. In the circuit of Fig. 10.87, find values for (a) $\mathbf{I}_1$, $\mathbf{I}_2$, and $\mathbf{I}_3$. (b) Show $\mathbf{V}_s$, $\mathbf{I}_1$, $\mathbf{I}_2$, and $\mathbf{I}_3$ on a phasor diagram (scales of 50 V/in and 2 A/in work fine). (c) Find $\mathbf{I}_s$ graphically and give its amplitude and phase angle.

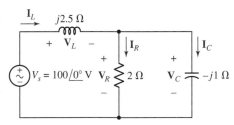

■ **FIGURE 10.86**

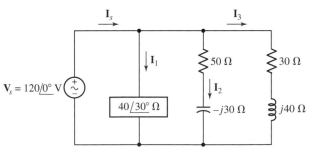

■ **FIGURE 10.87**

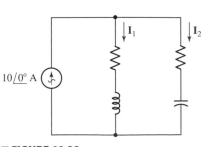

■ **FIGURE 10.88**

93. In the circuit sketched in Fig. 10.88, it is known that $|\mathbf{I}_1| = 5$ A and $|\mathbf{I}_2| = 7$ A. Find $\mathbf{I}_1$ and $\mathbf{I}_2$ with the help of compass, ruler, straightedge, protractor, and all that fun stuff.

94. Let $\mathbf{V}_1 = 100\underline{/0°}$ V, $|\mathbf{V}_2| = 140$ V, and $|\mathbf{V}_1 + \mathbf{V}_2| = 120$ V. Use graphical methods to find two possible values for the angle of $\mathbf{V}_2$.

AC Circuit Power Analysis

KEY CONCEPTS

Calculating Instantaneous Power

Average Power Supplied by a Sinusoidal Source

Root-Mean-Square (RMS) Values

Reactive Power

The Relationship Between Complex, Average, and Reactive Power

Power Factor of a Load

INTRODUCTION

Often an integral part of circuit analysis is the determination of either power delivered or power absorbed (or both). In the context of ac power, we find that the rather simple approach we have taken previously does not provide a convenient picture of how a particular system is operating, so we introduce several different power-related quantities in this chapter.

We will begin by considering *instantaneous* power, the product of the time-domain voltage and time-domain current associated with the element or network of interest. The instantaneous power is sometimes quite useful in its own right, because its maximum value might have to be limited in order to avoid exceeding the safe or useful operating range of a physical device. For example, transistor and vacuum-tube power amplifiers both produce a distorted output, and speakers give a distorted sound, when the peak power exceeds a certain limiting value. However, we are mainly interested in instantaneous power for the simple reason that it provides us with the means to calculate a more important quantity, the *average* power. In a similar way, the progress of a cross-country road trip is best described by the average velocity; our interest in the instantaneous velocity is limited to the avoidance of maximum velocities that will endanger our safety or arouse the highway patrol.

In practical problems we will deal with values of average power which range from the small fraction of a picowatt available in a telemetry signal from outer space, to the few watts of audio power supplied to the speakers in a high-fidelity stereo system, the several hundred watts required to run the morning coffeepot, or the 10 billion watts generated at the Grand Coulee Dam. Still, we will see that even the concept of average power has its limitations, especially when dealing with the energy exchange between reactive loads and power sources. This is easily handled by introducing the concepts of reactive power, complex power, and the power factor—all very common terms in the power industry.

11.1 • INSTANTANEOUS POWER

The *instantaneous power* delivered to any device is given by the product of the instantaneous voltage across the device and the instantaneous current through it (the passive sign convention is assumed). Thus,[1]

$$p(t) = v(t)i(t) \tag{1}$$

If the device in question is a resistor of resistance R, then the power may be expressed solely in terms of either the current or the voltage:

$$p(t) = v(t)i(t) = i^2(t)R = \frac{v^2(t)}{R} \tag{2}$$

If the voltage and current are associated with a device that is entirely inductive, then

$$p(t) = v(t)i(t) = Li(t)\frac{di(t)}{dt} = \frac{1}{L}v(t)\int_{-\infty}^{t} v(t')\,dt' \tag{3}$$

where we will arbitrarily assume that the voltage is zero at $t = -\infty$. In the case of a capacitor,

$$p(t) = v(t)i(t) = Cv(t)\frac{dv(t)}{dt} = \frac{1}{C}i(t)\int_{-\infty}^{t} i(t')\,dt' \tag{4}$$

where a similar assumption about the current is made. This listing of equations for power in terms of only a current or a voltage soon becomes unwieldy, however, as we begin to consider more general networks. The listing is also quite unnecessary, for we need only find both the current and voltage at the network terminals. As an example, we may consider the series RL circuit, as shown in Fig. 11.1, excited by a step-voltage source. The familiar current response is

$$i(t) = \frac{V_0}{R}(1 - e^{-Rt/L})u(t)$$

and thus the total power delivered by the source or absorbed by the passive network is

$$p(t) = v(t)i(t) = \frac{V_0^2}{R}(1 - e^{-Rt/L})u(t)$$

since the square of the unit-step function is simply the unit-step function itself.

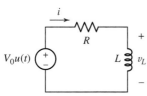

■ **FIGURE 11.1** The instantaneous power that is delivered to R is $p_R(t) = i^2(t)R = (V_0^2/R)(1 - e^{-Rt/L})^2u(t)$.

---

(1) Earlier, we agreed that lowercase variables in italics were understood to be functions of time, and we have carried on in this spirit up to now. However, in order to emphasize the fact that these quantities must be evaluated at a specific instant in time, we will explicitly denote the time dependence throughout this chapter.

The power delivered to the resistor is

$$p_R(t) = i^2(t)R = \frac{V_0^2}{R}(1 - e^{-Rt/L})^2 u(t)$$

In order to determine the power absorbed by the inductor, we first obtain the inductor voltage:

$$v_L(t) = L\frac{di(t)}{dt}$$

$$= V_0 e^{-Rt/L}u(t) + \frac{LV_0}{R}(1 - e^{-Rt/L})\frac{du(t)}{dt}$$

$$= V_0 e^{-Rt/L}u(t)$$

since $du(t)/dt$ is zero for $t > 0$ and $(1 - e^{-Rt/L})$ is zero at $t = 0$. The power absorbed by the inductor is therefore

$$p_L(t) = v_L(t)i(t) = \frac{V_0^2}{R}e^{-Rt/L}(1 - e^{-Rt/L})u(t)$$

Only a few algebraic manipulations are required to show that

$$p(t) = p_R(t) + p_L(t)$$

which serves to check the accuracy of our work; the results are sketched in Fig. 11.2.

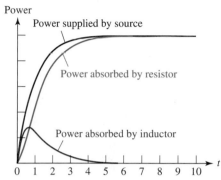

Power

Power supplied by source

Power absorbed by resistor

Power absorbed by inductor

0 1 2 3 4 5 6 7 8 9 10 → t

■ **FIGURE 11.2** Sketch of $p(t)$, $p_R(t)$, and $p_L(t)$. As the transient dies out, the circuit returns to steady-state operation. Since the only source remaining in the circuit is dc, the inductor eventually acts as a short circuit absorbing zero power.

Power Due to Sinusoidal Excitation

Let us change the voltage source in the circuit of Fig. 11.1 to the sinusoidal source $V_m \cos \omega t$. The familiar time-domain steady-state response is

$$i(t) = I_m \cos(\omega t + \phi)$$

where

$$I_m = \frac{V_m}{\sqrt{R^2 + \omega^2 L^2}} \quad \text{and} \quad \phi = -\tan^{-1}\frac{\omega L}{R}$$

The instantaneous power delivered to the entire circuit in the sinusoidal steady state is, therefore,

$$p(t) = v(t)i(t) = V_m I_m \cos(\omega t + \phi)\cos \omega t$$

which we will find convenient to rewrite in a form obtained by using the trigonometric identity for the product of two cosine functions. Thus,

$$p(t) = \frac{V_m I_m}{2}[\cos(2\omega t + \phi) + \cos \phi]$$

$$= \frac{V_m I_m}{2}\cos \phi + \frac{V_m I_m}{2}\cos(2\omega t + \phi)$$

The last equation possesses several characteristics that are true in general for circuits in the sinusoidal steady state. One term, the first, is not a function of time; and a second term is included which has a cyclic variation at *twice* the applied frequency. Since this term is a cosine wave, and since sine waves and cosine waves have average values which are zero (when averaged over an integral number of periods), this example suggests that the *average* power is $\frac{1}{2}V_m I_m \cos \phi$; as we will see shortly, this is indeed the case.

EXAMPLE 11.1

A voltage source, $40 + 60u(t)$ V, a 5 μF capacitor, and a 200 Ω resistor are in series. Find the power being absorbed by the capacitor and by the resistor at $t = 1.2$ ms.

At $t = 0^-$, no current is flowing and so 40 V appears across the capacitor. At $t = 0^+$, the voltage across the capacitor-resistor series combination jumps to 100 V. Since v_C cannot change in zero time, the resistor voltage at $t = 0^+$ is 60 V.

The current flowing through all three elements at $t = 0^+$ is therefore $60/200 = 300$ mA and for $t > 0$ is given by

$$i(t) = 300e^{-t/\tau} \text{ mA}$$

where $\tau = RC = 1$ ms. Thus, the current flowing at $t = 1.2$ ms is 90.36 mA, and the power being absorbed by the resistor *at that instant* is simply

$$i^2(t)R = 1.633 \text{ W}.$$

The instantaneous power absorbed by the capacitor is $i(t)v_C(t)$. Recognizing that the total voltage across both elements for $t > 0$ will always be 100 V, and that the resistor voltage is given by $60e^{-t/\tau}$,

$$v_C(t) = 100 - 60e^{-t/\tau}$$

and we find that $v_C(1.2 \text{ ms}) = 100 - 60e^{-1.2} = 81.93$ V. Thus, the power being absorbed by the capacitor at $t = 1.2$ ms is $(90.36 \text{ mA})(81.93 \text{ V}) = 7.403$ W.

PRACTICE

11.1 A current source of $12 \cos 2000t$ A, a 200 Ω resistor, and a 0.2 H inductor are in parallel. Assume steady-state conditions exist. At $t = 1$ ms, find the power being absorbed by the (a) resistor; (b) inductor; (c) sinusoidal source.

Ans: 13.98 kW; −5.63 kW; −8.35 kW.

11.2 AVERAGE POWER

When we speak of an average value for the instantaneous power, the time interval over which the averaging process takes place must be clearly defined. Let us first select a general interval of time from t_1 to t_2. We may then obtain the average value by integrating $p(t)$ from t_1 to t_2 and dividing the result by the time interval $t_2 - t_1$. Thus,

$$P = \frac{1}{t_2 - t_1} \int_{t_1}^{t_2} p(t)\, dt \qquad [5]$$

The average value is denoted by the capital letter P, since it is not a function of time, and it usually appears without any specific subscripts that identify it as an average value. Although P is not a function of time, it *is* a function of t_1 and t_2, the two instants of time which define the interval of integration. This dependence of P on a specific time interval may be expressed in a simpler manner if $p(t)$ is a periodic function. We consider this important case first.

Average Power for Periodic Waveforms

Let us assume that our forcing function and the circuit responses are all periodic; a steady-state condition has been reached, although not necessarily the sinusoidal steady state. We may define a *periodic* function $f(t)$ mathematically by requiring that

$$f(t) = f(t + T) \tag{6}$$

where T is the period. We now show that the average value of the instantaneous power as expressed by Eq. [5] may be computed over an interval of one period having an arbitrary beginning.

A general periodic waveform is shown in Fig. 11.3 and identified as $p(t)$. We first compute the average power by integrating from t_1 to a time t_2 which is one period later, $t_2 = t_1 + T$:

$$P_1 = \frac{1}{T} \int_{t_1}^{t_1+T} p(t)\, dt$$

and then by integrating from some other time t_x to $t_x + T$:

$$P_x = \frac{1}{T} \int_{t_x}^{t_x+T} p(t)\, dt$$

The equality of P_1 and P_x should be evident from the graphical interpretation of the integrals; the periodic nature of the curve requires the two areas to be equal. Thus, the **average power** may be computed by integrating the instantaneous power over any interval that is one period in length and then dividing by the period:

$$P = \frac{1}{T} \int_{t_x}^{t_x+T} p(t)\, dt \tag{7}$$

It is important to note that we may also integrate over any integral number of periods, provided that we divide by the same integral number of periods. Thus,

$$P = \frac{1}{nT} \int_{t_x}^{t_x+nT} p(t)\, dt \qquad n = 1, 2, 3, \ldots \tag{8}$$

If we carry this concept to the extreme by integrating over all time, another useful result is obtained. We first provide ourselves with symmetrical limits on the integral

$$P = \frac{1}{nT} \int_{-nT/2}^{nT/2} p(t)\, dt$$

and then take the limit as n becomes infinite,

$$P = \lim_{n \to \infty} \frac{1}{nT} \int_{-nT/2}^{nT/2} p(t)\, dt$$

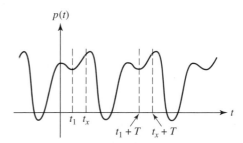

■ FIGURE 11.3 The average value P of a periodic function $p(t)$ is the same over any period T.

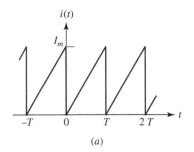

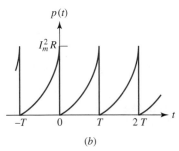

■ **FIGURE 11.4** (*a*) A sawtooth current waveform and (*b*) the instantaneous power waveform it produces in a resistor *R*.

As long as $p(t)$ is a mathematically well-behaved function, as all *physical* forcing functions and responses are, it is apparent that if a large integer n is replaced by a slightly larger number which is not an integer, then the value of the integral and of P is changed by a negligible amount; moreover, the error decreases as n increases. Without justifying this step rigorously, we therefore replace the discrete variable nT with the continuous variable τ:

$$P = \lim_{\tau \to \infty} \frac{1}{\tau} \int_{-\tau/2}^{\tau/2} p(t)\, dt \qquad [9]$$

We will find it convenient on several occasions to integrate periodic functions over this "infinite period." Examples of the use of Eqs. [7], [8], and [9] follow.

Let us illustrate the calculation of the average power of a periodic wave by finding the average power delivered to a resistor R by the (periodic) sawtooth current waveform shown in Fig. 11.4*a*. We have

$$i(t) = \frac{I_m}{T} t, \qquad 0 < t \le T$$

$$i(t) = \frac{I_m}{T}(t - T), \qquad T < t \le 2T$$

and so on; and

$$p(t) = \frac{1}{T^2} I_m^2 R t^2, \qquad 0 < t \le T$$

$$p(t) = \frac{1}{T^2} I_m^2 R (t - T)^2, \qquad T < t \le 2T$$

and so on, as sketched in Fig. 11.4*b*. Integrating over the simplest range of one period, from $t = 0$ to $t = T$, we have

$$P = \frac{1}{T} \int_0^T \frac{I_m^2 R}{T^2} t^2\, dt = \frac{1}{3} I_m^2 R$$

The selection of other ranges of one period, such as from $t = 0.1T$ to $t = 1.1T$, would produce the same answer. Integration from 0 to $2T$ and division by $2T$—that is, the application of Eq. [8] with $n = 2$ and $t_x = 0$—would also provide the same answer.

Average Power in the Sinusoidal Steady State

Now let us obtain the general result for the sinusoidal steady state. We will assume the general sinusoidal voltage

$$v(t) = V_m \cos(\omega t + \theta)$$

and current

$$i(t) = I_m \cos(\omega t + \phi)$$

associated with the device in question. The instantaneous power is

$$p(t) = V_m I_m \cos(\omega t + \theta)\cos(\omega t + \phi)$$

Again expressing the product of two cosine functions as one-half the sum of the cosine of the difference angle and the cosine of the sum angle,

$$p(t) = \tfrac{1}{2} V_m I_m \cos(\theta - \phi) + \tfrac{1}{2} V_m I_m \cos(2\omega t + \theta + \phi) \qquad [10]$$

we may save ourselves some integration by an inspection of the result. The first term is a constant, independent of t. The remaining term is a cosine function; $p(t)$ is therefore periodic, and its period is $\frac{1}{2}T$. Note that the period T is associated with the given current and voltage, and not with the power; the power function has a period $\frac{1}{2}T$. However, we may integrate over an interval of T to determine the average value if we wish; it is necessary only that we also divide by T. Our familiarity with cosine and sine waves, however, shows that the average value of either over a period is zero. There is thus no need to integrate Eq. [10] formally; by inspection, the average value of the second term is zero over a period T (or $\frac{1}{2}T$) and the average value of the first term, a constant, must be that constant itself. Thus,

> Recall that $T = \dfrac{1}{f} = \dfrac{2\pi}{\omega}$.

$$P = \tfrac{1}{2}V_m I_m \cos(\theta - \phi) \qquad [11]$$

This important result, introduced in the previous section for a specific circuit, is therefore quite general for the sinusoidal steady state. The average power is one-half the product of the crest amplitude of the voltage, the crest amplitude of the current, and the cosine of the phase-angle difference between the current and the voltage; the sense of the difference is immaterial.

EXAMPLE 11.2

Given the time-domain voltage $v = 4\cos(\pi t/6)$ V, find both the average power and an expression for the instantaneous power that result when the corresponding phasor voltage $V = 4\underline{/0°}$ V is applied across an impedance $Z = 2\underline{/60°}$ Ω.

The phasor current is $\mathbf{V}/\mathbf{Z} = 2\underline{/-60°}$ A, and the average power is

$$P = \tfrac{1}{2}(4)(2)\cos 60° = 2 \text{ W}$$

The time-domain voltage,

$$v(t) = 4\cos\frac{\pi t}{6} \text{ V}$$

time-domain current,

$$i(t) = 2\cos\left(\frac{\pi t}{6} - 60°\right) \text{ A}$$

and instantaneous power,

$$p(t) = 8\cos\frac{\pi t}{6}\cos\left(\frac{\pi t}{6} - 60°\right)$$

$$= 2 + 4\cos\left(\frac{\pi t}{3} - 60°\right) \text{ W}$$

(Continued on next page)

are all sketched on the same time axis in Fig. 11.5. Both the 2 W average value of the power and its period of 6 s, one-half the period of either the current or the voltage, are evident. The zero value of the instantaneous power at each instant when either the voltage or current is zero is also apparent.

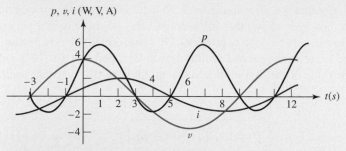

■ **FIGURE 11.5** Curves of $v(t)$, $i(t)$, and $p(t)$ are plotted as functions of time for a simple circuit in which the phasor voltage $\mathbf{V} = 4\underline{/0°}$ V is applied to the impedance $\mathbf{Z} = 2\underline{/60°}$ Ω at $\omega = \pi/6$ rad/s.

PRACTICE

11.2 Given the phasor voltage $\mathbf{V} = 115\sqrt{2}\underline{/45°}$ V across an impedance $\mathbf{Z} = 16.26\underline{/19.3°}$ Ω, obtain an expression for the instantaneous power, and compute the average power if $\omega = 50$ rad/s.

Ans: $767.5 + 813.2\cos(100t + 70.7°)$ W; 767.5 W.

Two special cases are worth isolating for consideration, the average power delivered to an ideal resistor, and that to an ideal reactor (any combination of only capacitors and inductors).

Average Power Absorbed by an Ideal Resistor

The phase-angle difference between the current through and the voltage across a pure resistor is zero. Thus,

$$P_R = \tfrac{1}{2}V_m I_m \cos 0 = \tfrac{1}{2}V_m I_m$$

or

$$P_R = \tfrac{1}{2}I_m^2 R \qquad [12]$$

or

$$P_R = \frac{V_m^2}{2R} \qquad [13]$$

Keep in mind that we are computing the *average* power delivered to a resistor by a sinusoidal source; take care not to confuse this quantity with the *instantaneous* power, which has a similar form.

The last two formulas, enabling us to determine the average power delivered to a pure resistance from a knowledge of either the sinusoidal current or voltage, are simple and important. Unfortunately, *they are often misused.* The most common error is made in trying to apply them in cases where the voltage included in Eq. [13] is *not the voltage across the resistor.* If care is taken to use the current through the resistor in Eq. [12] and the voltage across the resistor in Eq. [13], satisfactory operation is guaranteed. Also, do not forget the factor of $\tfrac{1}{2}$!

Average Power Absorbed by Purely Reactive Elements

The average power delivered to any device which is purely reactive (i.e., contains no resistors) must be zero. This is a direct result of the 90° phase difference which must exist between current and voltage; hence, $\cos(\theta - \phi) = \cos \pm 90° = 0$ and

$$P_X = 0$$

The *average* power delivered to any network composed entirely of ideal inductors and capacitors is zero; the *instantaneous* power is zero only at specific instants. Thus, power flows into the network for a part of the cycle and out of the network during another portion of the cycle, with *no* power lost.

EXAMPLE 11.3

Find the average power being delivered to an impedance $Z_L = 8 - j11\ \Omega$ by a current $I = 5\underline{/20°}$ A.

We may find the solution quite rapidly by using Eq. [12]. Only the 8 Ω resistance enters the average-power calculation, since the $j11\ \Omega$ component will not absorb any *average* power. Thus,

$$P = \tfrac{1}{2}(5^2)8 = 100 \text{ W}$$

PRACTICE

11.3 Calculate the average power delivered to the impedance $6\underline{/25°}\ \Omega$ by the current $I = 2 + j5$ A.

Ans: 78.85 W.

EXAMPLE 11.4

Find the average power absorbed by each of the three passive elements in Fig. 11.6, as well as the average power supplied by each source.

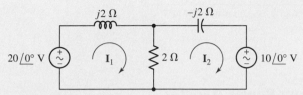

■ FIGURE 11.6 The average power delivered to each reactive element is zero in the sinusoidal steady state.

Without even analyzing the circuit, we already know that the average power absorbed by the two reactive elements is zero.

(Continued on next page)

The values of $\mathbf{I}_1$ and $\mathbf{I}_2$ are found by any of several methods, such as mesh analysis, nodal analysis, or superposition. They are

$$\mathbf{I}_1 = 5 - j10 = 11.18\underline{/-63.43°}\ \text{A}$$
$$\mathbf{I}_2 = 5 - j5 = 7.071\underline{/-45°}\ \text{A}$$

The downward current through the 2 Ω resistor is

$$\mathbf{I}_1 - \mathbf{I}_2 = -j5 = 5\underline{/-90°}\ \text{A}$$

so that $I_m = 5$ A, and the average power absorbed by the resistor is found most easily by Eq. [12]:

$$P_R = \tfrac{1}{2}I_m^2 R = \tfrac{1}{2}(5^2)2 = 25\ \text{W}$$

This result may be checked by using Eq. [11] or Eq. [13]. We next turn to the left source. The voltage $20\underline{/0°}$ V and associated current $\mathbf{I}_1 = 11.18\underline{/-63.43°}$ A satisfy the *active* sign convention, and thus the power *delivered* by this source is

$$P_{\text{left}} = \tfrac{1}{2}(20)(11.18)\cos[0° - (-63.43°)] = 50\ \text{W}$$

In a similar manner, we find the power *absorbed* by the right source using the *passive* sign convention,

$$P_{\text{right}} = \tfrac{1}{2}(10)(7.071)\cos(0° + 45°) = 25\ \text{W}$$

Since $50 = 25 + 25$, the power relations check.

PRACTICE

11.4 For the circuit of Fig. 11.7, compute the average power delivered to each of the passive elements. Verify your answer by computing the power delivered by the two sources.

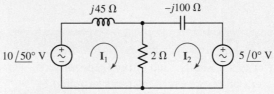

■ **FIGURE 11.7**

Ans: 0, 37.6 mW, 0, 42.0 mW, −4.4 mW.

Maximum Power Transfer

The notation $\mathbf{Z}^*$ denotes the **complex conjugate** of the complex number $\mathbf{Z}$. It is formed by replacing all "*j*" with "$-j$"s. See Appendix 5 for more details.

We previously considered the maximum power transfer theorem as it applied to resistive loads and resistive source impedances. For a Thévenin source $\mathbf{V}_{th}$ and impedance $\mathbf{Z}_{th} = R_{th} + jX_{th}$ connected to a load $\mathbf{Z}_L = R_L + jX_L$, it may be shown that the average power delivered to the load is a maximum when $R_L = R_{th}$ and $X_L = -X_{th}$, that is, when $\mathbf{Z}_L = \mathbf{Z}_{th}^*$. This result is often dignified by calling it the *maximum power transfer theorem for the* *sinusoidal steady state:*

> An independent voltage source in *series* with an impedance $\mathbf{Z}_{th}$ or an independent current source in *parallel* with an impedance $\mathbf{Z}_{th}$ delivers a ***maximum average power*** to that load impedance $\mathbf{Z}_L$ which is the conjugate of $\mathbf{Z}_{th}$, or $\mathbf{Z}_L = \mathbf{Z}_{th}^*$.

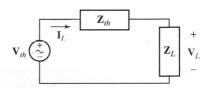

■ **FIGURE 11.8** A simple loop circuit used to illustrate the derivation of the maximum power transfer theorem as it applies to circuits operating in the sinusoidal steady state.

The details of the proof are left to Exercise 11, but the basic approach can be understood by considering the simple loop circuit of Fig. 11.8. The Thévenin equivalent impedance $\mathbf{Z}_{th}$ may be written as the sum of two components, $R_{th} + jX_{th}$, and in a similar fashion the load impedance $\mathbf{Z}_L$ may be written as $R_L + jX_L$. The current flowing through the loop is

$$
\begin{aligned}
\mathbf{I}_L &= \frac{\mathbf{V}_{th}}{\mathbf{Z}_{th} + \mathbf{Z}_L} \\
&= \frac{\mathbf{V}_{th}}{R_{th} + jX_{th} + R_L + jX_L} = \frac{\mathbf{V}_{th}}{R_{th} + R_L + j(X_{th} + X_L)}
\end{aligned}
$$

and

$$
\begin{aligned}
\mathbf{V}_L &= \mathbf{V}_{th}\frac{\mathbf{Z}_L}{\mathbf{Z}_{th} + \mathbf{Z}_L} \\
&= \mathbf{V}_{th}\frac{R_L + jX_L}{R_{th} + jX_{th} + R_L + jX_L} = \mathbf{V}_{th}\frac{R_L + jX_L}{R_{th} + R_L + j(X_{th} + X_L)}
\end{aligned}
$$

The magnitude of $\mathbf{I}_L$ is

$$
\frac{|\mathbf{V}_{th}|}{\sqrt{(R_{th} + R_L)^2 + (X_{th} + X_L)^2}}
$$

and the phase angle is

$$
\underline{/\mathbf{V}_{th}} - \tan^{-1}\left(\frac{X_{th} + X_L}{R_{th} + R_L}\right)
$$

Similarly, the magnitude of $\mathbf{V}_L$ is

$$
\frac{|\mathbf{V}_{th}|\sqrt{R_L^2 + X_L^2}}{\sqrt{(R_{th} + R_L)^2 + (X_{th} + X_L)^2}}
$$

and its phase angle is

$$
\underline{/\mathbf{V}_{th}} + \tan^{-1}\left(\frac{X_L}{R_L}\right) - \tan^{-1}\left(\frac{X_{th} + X_L}{R_{th} + R_L}\right)
$$

Referring to Eq. [11], then, we find an expression for the average power P delivered to the load impedance $\mathbf{Z}_L$:

$$
P = \frac{\frac{1}{2}|\mathbf{V}_{th}|^2\sqrt{R_L^2 + X_L^2}}{(R_{th} + R_L)^2 + (X_{th} + X_L)^2}\cos\left(\tan^{-1}\left(\frac{X_L}{R_L}\right)\right) \qquad [14]
$$

In order to prove that maximum average power is indeed delivered to the load when $\mathbf{Z}_L = \mathbf{Z}_{th}^*$, we must perform two separate steps. First, the derivative of Eq. [14] with respect to R_L must be set to zero. Second, the derivative of Eq. [14] with respect to X_L must be set to zero. The remaining details are left as an exercise for the avid reader.

EXAMPLE 11.5

A particular circuit is composed of the series combination of a sinusoidal voltage source $3 \cos(100t - 3°)$ V, a 500 Ω resistor, a 30 mH inductor, and an unknown impedance. If we are assured that the voltage source is delivering maximum average power to the unknown impedance, what is its value?

The phasor representation of the circuit is sketched in Fig. 11.9. The circuit is easily seen as an unknown impedance $\mathbf{Z}_?$ in series with a Thévenin equivalent consisting of the $3\underline{/-3°}$ V source and a Thévenin impedance $500 + j3$ Ω.

Since the circuit of Fig. 11.9 is already in the form required to employ the maximum average power transfer theorem, we know that maximum average power will be transferred to an impedance equal to the complex conjugate of $\mathbf{Z}_{th}$, or

$$\mathbf{Z}_? = \mathbf{Z}_{th}^* = 500 - j3 \text{ Ω}$$

This impedance can be constructed in several ways, the simplest being a 500 Ω resistor in series with a capacitor having impedance $-j3$ Ω. Since the operating frequency of the circuit is 100 rad/s, this corresponds to a capacitance of 3.333 mF.

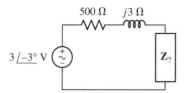

■ **FIGURE 11.9** The phasor representation of a simple series circuit composed of a sinusoidal voltage source, a resistor, an inductor, and an unknown impedance.

PRACTICE

11.5 If the 30 mH inductor of Example 11.5 is replaced with a 10 μF capacitor, what is the value of the inductive component of the unknown impedance $\mathbf{Z}_?$ if it is known that $\mathbf{Z}_?$ is absorbing maximum power?

Ans: 10 H.

Average Power for Nonperiodic Functions

We should pay some attention to *nonperiodic* functions. One practical example of a nonperiodic power function for which an average power value is desired is the power output of a radio telescope directed toward a "radio star." Another is the sum of a number of periodic functions, each function having a different period, such that no greater common period can be found for the combination. For example, the current

$$i(t) = \sin t + \sin \pi t \qquad [15]$$

is nonperiodic because the ratio of the periods of the two sine waves is an irrational number. At $t = 0$, both terms are zero and increasing. But the first term is zero and increasing only when $t = 2\pi n$, where n is an integer, and thus periodicity demands that πt or $\pi(2\pi n)$ must equal $2\pi m$, where m is also an integer. No solution (integral values for both m and n) for this equation is possible. It may be illuminating to compare the nonperiodic expression in Eq. [15] with the *periodic* function

$$i(t) = \sin t + \sin 3.14t \qquad [16]$$

where 3.14 is an exact decimal expression and is *not* to be interpreted as 3.141592. . . . With a little effort,[2] it can be shown that the period of this current wave is 100π seconds.

The average value of the power delivered to a 1 Ω resistor by either a periodic current such as Eq. [16] or a nonperiodic current such as Eq. [15] may be found by integrating over an infinite interval. Much of the actual integration can be avoided because of our thorough knowledge of the average values of simple functions. We therefore obtain the average power delivered by the current in Eq. [15] by applying Eq. [9]:

$$P = \lim_{\tau \to \infty} \frac{1}{\tau} \int_{-\tau/2}^{\tau/2} (\sin^2 t + \sin^2 \pi t + 2 \sin t \sin \pi t) \, dt$$

We now consider P as the sum of three average values. The average value of $\sin^2 t$ over an infinite interval is found by replacing $\sin^2 t$ with $(\frac{1}{2} - \frac{1}{2} \cos 2t)$; the average is simply $\frac{1}{2}$. Similarly, the average value of $\sin^2 \pi t$ is also $\frac{1}{2}$. And the last term can be expressed as the sum of two cosine functions, each of which must certainly have an average value of zero. Thus,

$$P = \tfrac{1}{2} + \tfrac{1}{2} = 1 \text{ W}$$

An identical result is obtained for the periodic current of Eq. [16]. Applying this same method to a current function which is the sum of several sinusoids of *different periods* and arbitrary amplitudes,

$$i(t) = I_{m1} \cos \omega_1 t + I_{m2} \cos \omega_2 t + \cdots + I_{mN} \cos \omega_N t \qquad [17]$$

we find the average power delivered to a resistance R,

$$P = \tfrac{1}{2} \left(I_{m1}^2 + I_{m2}^2 + \cdots + I_{mN}^2 \right) R \qquad [18]$$

The result is unchanged if an arbitrary phase angle is assigned to each component of the current. This important result is surprisingly simple when we think of the steps required for its derivation: squaring the current function, integrating, and taking the limit. The result is also just plain surprising, because it shows that, *in this special case of a current such as Eq. [17], where each term has a unique frequency, superposition is applicable to power.* Superposition is *not* applicable for a current which is the sum of two direct currents, nor is it applicable for a current which is the sum of two sinusoids of the same frequency.

EXAMPLE 11.6

Find the average power delivered to a 4 Ω resistor by the current $i_1 = 2 \cos 10t - 3 \cos 20t$ A.

Since the two cosine terms are at *different* frequencies, the two average-power values may be calculated separately and added. Thus, this current delivers $\frac{1}{2}(2^2)4 + \frac{1}{2}(3^2)4 = 8 + 18 = 26$ W to a 4 Ω resistor.

---

(2) $T_1 = 2\pi$ and $T_2 = 2\pi/3.14$. Therefore, we seek integral values of m and n such that $2\pi n = 2\pi m/3.14$, or $3.14n = m$, or $\frac{314}{100}n = m$ or $157n = 50 m$. Thus, the smallest integral values for n and m are $n = 50$ and $m = 157$. The period is therefore $T = 2\pi n = 100\pi$, or $T = 2\pi(157/3.14) = 100\pi$ s.

EXAMPLE 11.7

Find the average power delivered to a 4 Ω resistor by the current $i_2 = 2 \cos 10t - 3 \cos 10t$ A.

Here, the two components of the current are at the *same* frequency, and they must therefore be combined into a single sinusoid at that frequency. Thus, $i_2 = 2 \cos 10t - 3 \cos 10t = -\cos 10t$ delivers only $\frac{1}{2}(1^2)4 = 2$ W of average power to a 4 Ω resistor.

PRACTICE

11.6 A voltage source v_s is connected across a 4 Ω resistor. Find the average power absorbed by the resistor if v_s equals (*a*) 8 sin 200t V; (*b*) 8 sin 200t − 6 cos(200t − 45°) V; (*c*) 8 sin 200t − 4 sin 100t V; (*d*) 8 sin 200t − 6 cos(200t − 45°) − 5 sin 100t + 4 V.

Ans: 8.00 W; 4.01 W; 10.00 W; 11.14 W.

11.3 EFFECTIVE VALUES OF CURRENT AND VOLTAGE

In North America, most power outlets deliver a sinusoidal voltage having a frequency of 60 Hz and a "voltage" of 115 V (elsewhere, 50 Hz and 240 V is typically encountered). But what is meant by "115 volts"? This is certainly not the instantaneous value of the voltage, for the voltage is not a constant. The value of 115 V is also not the amplitude which we have been symbolizing as V_m; if we displayed the voltage waveform on a calibrated oscilloscope, we would find that the amplitude of this voltage at one of our ac outlets is $115\sqrt{2}$, or 162.6, volts. We also cannot fit the concept of an average value to the 115 V, because the average value of the sine wave is zero. We might come a little closer by trying the magnitude of the average over a positive or negative half cycle; by using a rectifier-type voltmeter at the outlet, we should measure 103.5 V. As it turns out, however, the 115 V is the ***effective value*** of this sinusoidal voltage. This value is a measure of the effectiveness of a voltage source in delivering power to a resistive load.

Effective Value of a Periodic Waveform

Let us arbitrarily define effective value in terms of a current waveform, although a voltage could equally well be selected. The *effective value* of any periodic current is equal to the value of the direct current which, flowing through an R-ohm resistor, delivers the same average power to the resistor as does the periodic current.

In other words, we allow the given periodic current to flow through the resistor, determine the instantaneous power i^2R, and then find the average value of i^2R over a period; this is the average power. We then cause a direct current to flow through this same resistor and adjust the value of the direct current until the same value of average power is obtained. The resulting magnitude of the direct current is equal to the effective value of the given periodic current. These ideas are illustrated in Fig. 11.10.

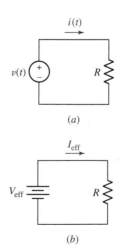

■ **FIGURE 11.10** If the resistor receives the same average power in parts *a* and *b*, then the effective value of $i(t)$ is equal to I_{eff}, and the effective value of $v(t)$ is equal to V_{eff}.

The general mathematical expression for the effective value of $i(t)$ is now easily obtained. The average power delivered to the resistor by the periodic current $i(t)$ is

$$P = \frac{1}{T} \int_0^T i^2 R \, dt = \frac{R}{T} \int_0^T i^2 \, dt$$

where the period of $i(t)$ is T. The power delivered by the direct current is

$$P = I_{\text{eff}}^2 R$$

Equating the power expressions and solving for I_{eff}, we get

$$I_{\text{eff}} = \sqrt{\frac{1}{T} \int_0^T i^2 \, dt} \qquad [19]$$

The result is independent of the resistance R, as it must be to provide us with a worthwhile concept. A similar expression is obtained for the effective value of a periodic voltage by replacing i and I_{eff} by v and V_{eff}, respectively.

Notice that the effective value is obtained by first squaring the time function, then taking the average value of the squared function over a period, and finally taking the square root of the average of the squared function. In abbreviated language, the operation involved in finding an effective value is the (square) *root* of the *mean* of the *square;* for this reason, the effective value is often called the ***root-mean-square*** value, or simply the ***rms*** value.

Effective (RMS) Value of a Sinusoidal Waveform

The most important special case is that of the sinusoidal waveform. Let us select the sinusoidal current

$$i(t) = I_m \cos(\omega t + \phi)$$

which has a period

$$T = \frac{2\pi}{\omega}$$

and substitute in Eq. [19] to obtain the effective value

$$I_{\text{eff}} = \sqrt{\frac{1}{T} \int_0^T I_m^2 \cos^2(\omega t + \phi) \, dt}$$

$$= I_m \sqrt{\frac{\omega}{2\pi} \int_0^{2\pi/\omega} \left[\frac{1}{2} + \frac{1}{2} \cos(2\omega t + 2\phi) \right] dt}$$

$$= I_m \sqrt{\frac{\omega}{4\pi} [t]_0^{2\pi/\omega}}$$

$$= \frac{I_m}{\sqrt{2}}$$

Thus the effective value of a sinusoidal current is a real quantity which is independent of the phase angle and numerically equal to $1/\sqrt{2} = 0.707$ times the amplitude of the current. A current $\sqrt{2} \cos(\omega t + \phi)$ A, therefore, has an effective value of 1 A and will deliver the **same** average power to any resistor as will a **direct** current of 1 A.

It should be noted carefully that the $\sqrt{2}$ factor that we obtained as the ratio of the amplitude of the periodic current to the effective value is applicable only when the periodic function is *sinusoidal*. For the sawtooth waveform of Fig. 11.4, for example, the effective value is equal to the maximum value divided by $\sqrt{3}$. The factor by which the maximum value must be divided to obtain the effective value depends on the mathematical form of the given periodic function; it may be either rational or irrational, depending on the nature of the function.

Use of RMS Values to Compute Average Power

The use of the effective value also simplifies slightly the expression for the average power delivered by a sinusoidal current or voltage by avoiding use of the factor $\frac{1}{2}$. For example, the average power delivered to an R-ohm resistor by a sinusoidal current is

$$P = \tfrac{1}{2}I_m^2 R$$

Since $I_{\text{eff}} = I_m/\sqrt{2}$, the average power may be written as

$$P = I_{\text{eff}}^2 R \tag{20}$$

> The fact that the effective value is defined in terms of an equivalent dc quantity provides us with average power formulas for resistive circuits which are identical with those used in dc analysis.

The other power expressions may also be written in terms of effective values:

$$P = V_{\text{eff}} I_{\text{eff}} \cos(\theta - \phi) \tag{21}$$

$$P = \frac{V_{\text{eff}}^2}{R} \tag{22}$$

Although we have succeeded in eliminating the factor $\frac{1}{2}$ from our average-power relationships, we must now take care to determine whether a sinusoidal quantity is expressed in terms of its amplitude or its effective value. In practice, the effective value is usually used in the fields of power transmission or distribution and of rotating machinery; in the areas of electronics and communications, the amplitude is more often used. We will assume that the amplitude is specified unless the term "rms" is explicitly used, or we are otherwise instructed.

In the sinusoidal steady state, phasor voltages and currents may be given either as effective values or as amplitudes; the two expressions differ only by a factor of $\sqrt{2}$. The voltage $50\underline{/30^\circ}$ V is expressed in terms of an amplitude; as an rms voltage, we should describe the same voltage as $35.4\underline{/30^\circ}$ V rms.

Effective Value with Multiple-Frequency Circuits

In order to determine the effective value of a periodic or nonperiodic waveform which is composed of the sum of a number of sinusoids of different frequencies, we may use the appropriate average-power relationship of Eq. [18], developed in Sec. 11.2, rewritten in terms of the effective values of the several components:

$$P = \left(I_{1\text{eff}}^2 + I_{2\text{eff}}^2 + \cdots + I_{N\text{eff}}^2\right)R \tag{23}$$

From this we see that the effective value of a current which is composed of any number of sinusoidal currents of *different* frequencies can be expressed as

$$I_{\text{eff}} = \sqrt{I_{1\text{eff}}^2 + I_{2\text{eff}}^2 + \cdots + I_{N\text{eff}}^2} \qquad [24]$$

These results indicate that if a sinusoidal current of 5 A rms at 60 Hz flows through a 2 Ω resistor, an average power of $5^2(2) = 50$ W is absorbed by the resistor; if a second current—perhaps 3 A rms at 120 Hz, for example—is also present, the absorbed power is $3^2(2) + 50 = 68$ W. Using Eq. [24] instead, we find that the effective value of the sum of the 60 and 120 Hz currents is 5.831 A. Thus, $P = 5.831^2(2) = 68$ W as before. However, if the second current is also at 60 Hz, the effective value of the sum of the two 60 Hz currents may have any value between 2 and 8 A. Thus, the absorbed power may have *any* value between 8 W and 128 W, depending on the relative phase of the two current components.

> Note that the effective value of a dc quantity K is simply K, not $\dfrac{K}{2}$.

PRACTICE

11.7 Calculate the effective value of each of the periodic voltages:
(a) $6 \cos 25t$; (b) $6 \cos 25t + 4 \sin(25t + 30°)$; (c) $6 \cos 25t + 5 \cos^2(25t)$; (d) $6 \cos 25t + 5 \sin 30t + 4$ V.

Ans: 4.24 V; 6.16 V; 5.23 V; 6.82 V.

COMPUTER-AIDED ANALYSIS

Several useful techniques are available through PSpice for calculation of power quantities. In particular, the built-in functions of Probe allow us to both plot the instantaneous power and compute the average power. For example, consider the simple voltage divider circuit of Fig. 11.11, which is being driven by a 60 Hz sine wave with an

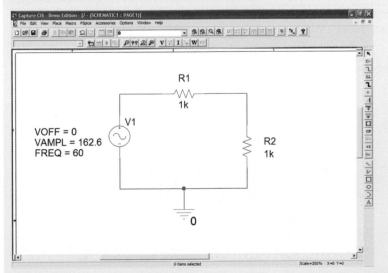

■ **FIGURE 11.11** A simple voltage divider circuit driven by a 115 V rms source operating at 60 Hz.

(Continued on next page)

amplitude of $115\sqrt{2}$ V. We begin by performing a transient simulation over one period of the voltage waveform, $\frac{1}{60}$ s.

The current along with the instantaneous power dissipated in resistor R1 is plotted in Fig. 11.12 by employing the **Add** **P**lot **to Window** option under **P**lot. The instantaneous power is periodic, with a nonzero average value and a peak of 6.61 W.

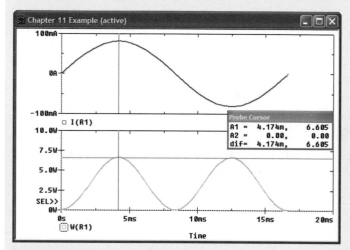

■ **FIGURE 11.12** Current and instantaneous power associated with resistor R1 of Fig. 11.11.

The easiest means of using Probe to obtain the average power, which we expect to be $\frac{1}{2}\left(162.6\frac{1000}{1000+1000}\right)(81.3 \times 10^{-3}) = 3.305$ W, is to make use of the built-in "running average" function. Once the **Add Traces** dialog box appears (**Trace**, ⤢ **Add Trace . . .**), type

$$\text{AVG(I(R1)}^*\text{I(R1)}^*\text{1000)}$$

in the **Trace Expression** window.

As can be seen in Fig. 11.13, the average value of the power over either one or two periods is 3.305 W, in agreement with the hand calculation.

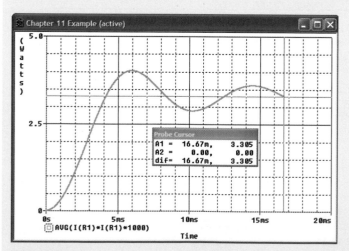

■ **FIGURE 11.13** Calculated running average of the power dissipated by resistor R1.

Probe also allows us to compute the average over a specific interval using the built-in function **avgx.** For example, to use this function to compute the average power over a single period, which in this case is $1/120 = 8.33$ ms, we would enter

$$\text{AVGX(I(R1)} * \text{I(R1)} * 1000, 8.33 \text{ m)}$$

Either approach will result in a value of 3.305 W at the endpoint of the plot.

11.4 APPARENT POWER AND POWER FACTOR

Historically, the introduction of the concepts of apparent power and power factor can be traced to the electric power industry, where large amounts of electric energy must be transferred from one point to another; the efficiency with which this transfer is effected is related directly to the cost of the electric energy, which is eventually paid by the consumer. Customers who provide loads which result in a relatively poor transmission efficiency must pay a greater price for each **kilowatthour** (kWh) of electric energy they actually receive and use. In a similar way, customers who require a costlier investment in transmission and distribution equipment by the power company will also pay more for each kilowatthour unless the company is benevolent and enjoys losing money.

Let us first define **apparent power** and **power factor** and then show briefly how these terms are related to the aforementioned economic situations. We assume that the sinusoidal voltage

$$v = V_m \cos(\omega t + \theta)$$

is applied to a network and the resultant sinusoidal current is

$$i = I_m \cos(\omega t + \phi)$$

The phase angle by which the voltage leads the current is therefore $(\theta - \phi)$. The average power delivered to the network, assuming a passive sign convention at its input terminals, may be expressed either in terms of the maximum values:

$$P = \tfrac{1}{2} V_m I_m \cos(\theta - \phi)$$

or in terms of the effective values:

$$P = V_{\text{eff}} I_{\text{eff}} \cos(\theta - \phi)$$

If our applied voltage and current responses had been dc quantities, the average power delivered to the network would have been given simply by the product of the voltage and the current. Applying this dc technique to the sinusoidal problem, we should obtain a value for the absorbed power which is "apparently" given by the familiar product $V_{\text{eff}} I_{\text{eff}}$. However, this product of the *effective* values of the voltage and current is not the average power; we define it as the **apparent power.** Dimensionally, apparent power must be measured in the same units as real power, since $\cos(\theta - \phi)$ is

Apparent power is not a concept which is limited to sinusoidal forcing functions and responses. It may be determined for any current and voltage waveshapes by simply taking the product of the effective values of the current and voltage.

dimensionless; but in order to avoid confusion, the term **volt-amperes,** or VA, is applied to the apparent power. Since $\cos(\theta - \phi)$ cannot have a magnitude greater than unity, it is evident that the magnitude of the real power can never be greater than the magnitude of the apparent power.

The ratio of the real or average power to the apparent power is called the **power factor,** symbolized by PF. Hence,

$$PF = \frac{\text{average power}}{\text{apparent power}} = \frac{P}{V_{\text{eff}} I_{\text{eff}}}$$

In the sinusoidal case, the power factor is simply $\cos(\theta - \phi)$, where $(\theta - \phi)$ is the angle by which the voltage leads the current. This relationship is the reason why the angle $(\theta - \phi)$ is often referred to as the **PF angle.**

For a purely resistive load, the voltage and current are in phase, $(\theta - \phi)$ is zero, and the PF is unity. In other words, the apparent power and the average power are equal. Unity PF, however, may also be achieved for loads that contain both inductance and capacitance if the element values and the operating frequency are carefully selected to provide an input impedance having a zero phase angle. A purely reactive load, that is, one containing no resistance, will cause a phase difference between the voltage and current of either plus or minus 90°, and the PF is therefore zero.

Between these two extreme cases there are the general networks for which the PF can range from zero to unity. A PF of 0.5, for example, indicates a load having an input impedance with a phase angle of either 60° or −60°; the former describes an inductive load, since the voltage leads the current by 60°, while the latter refers to a capacitive load. The ambiguity in the exact nature of the load is resolved by referring to a leading PF or a lagging PF, the terms *leading* or *lagging* referring to the *phase of the current with respect to the voltage.* Thus, an inductive load will have a lagging PF and a capacitive load a leading PF.

EXAMPLE 11.8

Calculate values for the average power delivered to each of the two loads shown in Fig. 11.14, the apparent power supplied by the source, and the power factor of the combined loads.

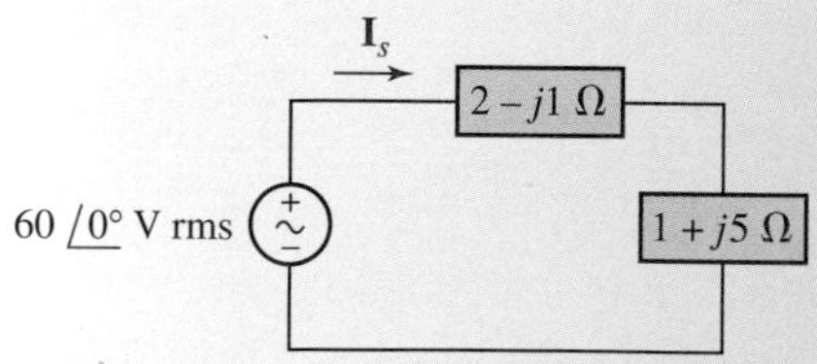

▸ **Identify the goal of the problem.**
The average power refers to the power drawn by the resistive components of the load elements; the apparent power is the product of the effective voltage and the effective current of the load combination.

▸ **Collect the known information.**
The effective voltage is 60 V rms, which appears across a combined load of $2 - j + 1 + j5 = 3 + j4 \ \Omega$.

▸ **Devise a plan.**
Simple phasor analysis will provide the current. Knowing voltage and current will enable us to calculate average power and apparent power; these two quantities can be used to obtain the power factor.

■ FIGURE 11.14 A circuit in which we seek the average power delivered to each element, the apparent power supplied by the source, and the power factor of the combined load.

In the circuit: $\mathbf{I}_s$, $60 \underline{/0°}$ V rms, $2 - j1 \ \Omega$, $1 + j5 \ \Omega$.

▶ **Construct an appropriate set of equations.**

The average power is given by

$$P = V_{\text{eff}} I_{\text{eff}} \cos(ang\ \mathbf{V} - ang\ \mathbf{I})$$

The apparent power is simply $V_{\text{eff}} I_{\text{eff}}$.

The power factor is calculated as the ratio of these two quantities:

$$\text{PF} = \frac{\text{average power}}{\text{apparent power}} = \frac{P}{V_{\text{eff}} I_{\text{eff}}}$$

▶ **Determine if additional information is required.**

We require I_{eff}:

$$\mathbf{I}_s = \frac{60\underline{/0°}}{3 + j4} = 12\underline{/-53.13°} \text{ A rms}$$

so $I_{\text{eff}} = 12$ A rms, and $ang\ \mathbf{I}_s = -53.13°$.

▶ **Attempt a solution.**

The average power delivered to the top load is given by

$$P_{\text{upper}} = I_{\text{eff}}^2 R_{\text{top}} = (12)^2(2) = 288 \text{ W}$$

and the average power delivered to the right load is given by

$$P_{\text{lower}} = I_{\text{eff}}^2 R_{\text{right}} = (12)^2(1) = 144 \text{ W}$$

The source itself supplies an apparent power of $V_{\text{eff}} I_{\text{eff}} = (60)(12) = 720$ VA.

Finally, the power factor of the combined loads is found by considering the voltage and current associated with the combined loads. This power factor is, of course, identical to the power factor for the source. Thus

$$\text{PF} = \frac{P}{V_{\text{eff}} I_{\text{eff}}} = \frac{432}{60(12)} = 0.6 \text{ lagging}$$

since the combined load is *inductive*.

▶ **Verify the solution. Is it reasonable or expected?**

The total average power delivered to the source is $288 + 144 = 432$ W. The average power supplied by the source is

$$P = V_{\text{eff}} I_{\text{eff}} \cos(ang\ \mathbf{V} - ang\ \mathbf{I}) = (60)(12) \cos(0 + 53.13°) = 432 \text{ W}$$

so we see the power balance is correct.

We might also write the combined load impedance as $5\underline{/53.1°}\ \Omega$, identify 53.1° as the PF angle, and thus have a PF of $\cos 53.1° = 0.6$ *lagging*.

PRACTICE
●

11.8 For the circuit of Fig. 11.15, determine the power factor of the combined loads if $Z_L = 10\ \Omega$.

Ans: 0.9966 leading.

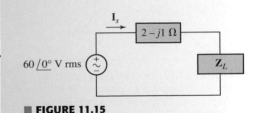

■ **FIGURE 11.15**

11.5 • COMPLEX POWER

Some simplification in power calculations is achieved if power is considered to be a complex quantity. The magnitude of the complex power will be found to be the apparent power, and the real part of the complex power will be shown to be the (real) average power. The new quantity, the imaginary part of the complex power, we will call *reactive power.*

We define complex power with reference to a general sinusoidal voltage $\mathbf{V}_{\text{eff}} = V_{\text{eff}}\underline{/\theta}$ across a pair of terminals and a general sinusoidal current $\mathbf{I}_{\text{eff}} = I_{\text{eff}}\underline{/\phi}$ flowing into one of the terminals in such a way as to satisfy the passive sign convention. The average power P absorbed by the two-terminal network is thus

$$P = V_{\text{eff}}I_{\text{eff}}\cos(\theta - \phi)$$

Complex nomenclature is next introduced by making use of Euler's formula in the same way as we did in introducing phasors. We express P as

$$P = V_{\text{eff}}I_{\text{eff}}\operatorname{Re}\{e^{j(\theta-\phi)}\}$$

or

$$P = \operatorname{Re}\{V_{\text{eff}}e^{j\theta}I_{\text{eff}}e^{-j\phi}\}$$

The phasor voltage may now be recognized as the first two factors within the brackets in the preceding equation, but the second two factors do not quite correspond to the phasor current, because the angle includes a minus sign, which is not present in the expression for the phasor current. That is, the phasor current is

$$\mathbf{I}_{\text{eff}} = I_{\text{eff}}\,e^{j\phi}$$

and we therefore must make use of conjugate notation:

$$\mathbf{I}^*_{\text{eff}} = I_{\text{eff}}\,e^{-j\phi}$$

Hence

$$P = \operatorname{Re}\{\mathbf{V}_{\text{eff}}\mathbf{I}^*_{\text{eff}}\}$$

and we may now let power become complex by defining the **complex power S** as

$$\mathbf{S} = \mathbf{V}_{\text{eff}}\mathbf{I}^*_{\text{eff}} \tag{25}$$

If we first inspect the polar or exponential form of the complex power,

$$\mathbf{S} = V_{\text{eff}}I_{\text{eff}}\,e^{j(\theta-\phi)}$$

it is evident that the magnitude of **S**, $V_{\text{eff}}I_{\text{eff}}$, is the apparent power and the angle of **S**, $(\theta - \phi)$, is the PF angle (i.e., the angle by which the voltage leads the current).

In rectangular form, we have

$$\mathbf{S} = P + jQ \tag{26}$$

where P is the average power, as before. The imaginary part of the complex power is symbolized as Q and is termed the *reactive power*. The dimensions of Q are the same as those of the real power P, the complex power $\mathbf{S}$, and the apparent power $|\mathbf{S}|$. In order to avoid confusion with these other quantities, the unit of Q is defined as the *volt-ampere-reactive* (abbreviated VAR). From Eqs. [25] and [26], it is seen that

$$Q = V_{\text{eff}}I_{\text{eff}}\sin(\theta - \phi) \tag{27}$$

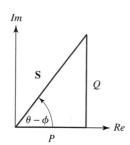

■ FIGURE 11.16 The power triangle representation of complex power.

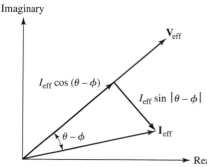

■ FIGURE 11.17 The current phasor $\mathbf{I}_{\text{eff}}$ is resolved into two components, one in phase with the voltage phasor $\mathbf{V}_{\text{eff}}$ and the other 90° out of phase with the voltage phasor. This latter component is called a *quadrature component.*

TABLE ● 11.1 Summary of Quantities Related to Complex Power

Quantity	Symbol	Formula	Units
Average power	P	$V_{\text{eff}} I_{\text{eff}} \cos(\theta - \phi)$	watt (W)
Reactive power	Q	$V_{\text{eff}} I_{\text{eff}} \sin(\theta - \phi)$	volt-ampere-reactive (VAR)
Complex power	$\mathbf{S}$	$P + jQ$	
		$V_{\text{eff}} I_{\text{eff}} \underline{/\theta - \phi}$	volt-ampere (VA)
		$\mathbf{V}_{\text{eff}} \mathbf{I}_{\text{eff}}^*$	
Apparent power	$|\mathbf{S}|$	$V_{\text{eff}} I_{\text{eff}}$	volt-ampere (VA)

The physical interpretation of reactive power is the time rate of energy flow back and forth between the source (i.e., the utility company) and the reactive components of the load (i.e., inductances and capacitances). These components alternately charge and discharge, which leads to current flow from and to the source, respectively.

The relevant quantities are summarized in Table 11.1 for convenience.

The sign of the reactive power characterizes the nature of a passive load at which $\mathbf{V}_{\text{eff}}$ and $\mathbf{I}_{\text{eff}}$ are specified. If the load is inductive, then $(\theta - \phi)$ is an angle between 0 and 90°, the sine of this angle is positive, and the reactive power is *positive*. A capacitive load results in a *negative* reactive power.

The Power Triangle

A commonly employed graphical representation of complex power is known as the ***power triangle,*** and is illustrated in Fig. 11.16. The diagram shows that only two of the three power quantities are required, as the third may be obtained by trigonometric relationships. If the power triangle lies in the first quadrant ($\theta - \phi > 0$), the power factor is *lagging* (corresponding to an inductive load), and if the power triangle lies in the fourth quadrant ($\theta - \phi < 0$), the power factor is *leading* (corresponding to a capacitive load). A great deal of qualitative information concerning our load is therefore available at a glance.

Another interpretation of reactive power may be seen by constructing a phasor diagram containing $\mathbf{V}_{\text{eff}}$ and $\mathbf{I}_{\text{eff}}$ as shown in Fig. 11.17. If the phasor current is resolved into two components, one in phase with the voltage, having a magnitude $\mathbf{I}_{\text{eff}} \cos(\theta - \phi)$, and one 90° out of phase with the voltage, with magnitude equal to $\mathbf{I}_{\text{eff}} \sin |\theta - \phi|$, then it is clear that the real power is given by the product of the magnitude of the voltage phasor and the component of the phasor current which is in phase with the voltage. Moreover, the product of the magnitude of the voltage phasor and the component of the phasor current which is 90° out of phase with the voltage is the reactive power Q. It is common to speak of the component of a phasor which is 90° out of phase with some other phasor as a *quadrature component*. Thus Q is simply $\mathbf{V}_{\text{eff}}$ times the quadrature component of $\mathbf{I}_{\text{eff}}$. Q is also known as the *quadrature power*.

Power Measurement

Strictly speaking, a wattmeter measures average real power P drawn by a load, and a varmeter reads the average reactive power Q drawn by a load. However, it is common to find both features in the same meter, which is often also capable of measuring apparent power and power factor as well (Fig. 11.18).

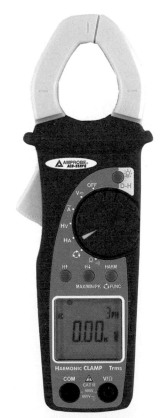

■ **FIGURE 11.18** A clamp-on digital powermeter manufactured by Amprobe, capable of measuring ac currents up to 400 A and voltages up to 600 V. Copyright AMPROBE

Power Factor Correction

When electric power is being supplied to large industrial consumers by a power company, the company will frequently include a PF clause in its rate schedules. Under this clause, an additional charge is made to the consumer whenever the PF drops below a certain specified value, usually about 0.85 lagging. Very little industrial power is consumed at leading PFs, because of the nature of typical industrial loads. There are several reasons that force the power company to make this additional charge for low PFs. In the first place, it is evident that larger current-carrying capacity must be built into its generators in order to provide the larger currents that go with lower-PF operation at constant power and constant voltage. Another reason is found in the increased losses in its transmission and distribution system.

In an effort to recoup losses and encourage its customers to operate at high PF, a certain utility charges a penalty of $0.22/kVAR for each kVAR above a benchmark value computed as 0.62 times the average power demand:

$$\mathbf{S} = P + jQ = P + j0.62P = P(1 + j0.62)$$
$$= P(1.177\underline{/31.8°})$$

This benchmark targets a PF of 0.85 lagging, as $\cos 31.8° = 0.85$ and Q is positive; this is represented graphically in Fig. 11.19. Customers with a PF smaller than the benchmark value are subject to financial penalties.

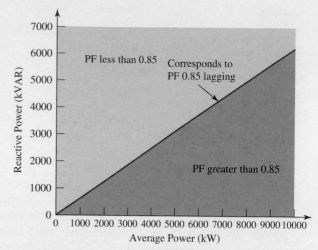

■ **FIGURE 11.19** Plot showing acceptable ratio of reactive power to average power for power factor benchmark of 0.85 lagging.

The reactive power requirement is commonly adjusted through the installation of compensation capacitors placed in parallel with the load (typically at the substation outside the customer's facility). The value of the required capacitance can be shown to be

$$C = \frac{P(\tan\theta_{\text{old}} - \tan\theta_{\text{new}})}{\omega V_{\text{rms}}^2} \qquad [28]$$

where ω is the frequency, θ_{old} is the present PF angle,

It is easy to show that the complex power delivered to several interconnected loads is the sum of the complex powers delivered to each of the individual loads, no matter how the loads are interconnected. For example, consider the two loads shown connected in parallel in Fig. 11.21. If rms values are assumed, the complex power drawn by the combined load is

$$\mathbf{S} = \mathbf{VI}^* = \mathbf{V}(\mathbf{I}_1 + \mathbf{I}_2)^* = \mathbf{V}(\mathbf{I}_1^* + \mathbf{I}_2^*)$$

and thus

$$\mathbf{S} = \mathbf{VI}_1^* + \mathbf{VI}_2^*$$

as stated.

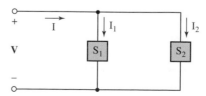

■ **FIGURE 11.21** A circuit used to show that the complex power drawn by two parallel loads is the sum of the complex powers drawn by the individual loads.

and θ_{new} is the target PF angle. For convenience, however, compensation capacitor banks are manufactured in specific increments rated in units of kVAR capacity. An example of such an installation is shown in Fig. 11.20.

Now let us consider a specific example. A particular industrial machine plant has a monthly peak demand of 5000 kW and a monthly reactive requirement of 6000 kVAR. Using the rate schedule above, what is the annual cost to this utility customer associated with PF penalties? If compensation is available through the utility company at a cost of $2390 per 1000 kVAR increment and $3130 per 2000 kVAR increment, what is the most cost-effective solution for the customer?

The PF of the installation is the angle of the complex power **S**, which in this case is $5000 + j6000$ kVA. Thus, the angle is $\tan^{-1}(6000/5000) = 50.19°$ and the PF is 0.64 lagging. The benchmark reactive power value, computed as 0.62 times the peak demand, is $0.62(5000) = 3100$ kVAR. So, the plant is drawing $6000 - 3100 = 2900$ kVAR more reactive power than the utility company is willing to allow without penalty. This represents an annual assessment of $12(2900)(0.22) = \$7656$ in addition to regular electricity costs.

If the customer chooses to have a single 1000 kVAR increment installed (at a cost of $2390), the excess reactive power draw is reduced to $2900 - 1000 = 1900$ kVAR, so that the annual penalty is now $12(1900)(0.22) = \$5016$. The total cost this year is then $\$5016 + \$2390 = \$7406$, for a savings of $250. If the customer chooses to have a single 2000 kVAR increment installed (at a cost of $3130), the excess reactive power draw is reduced to $2900 - 2000 = 900$ kVAR, so that the annual penalty is now $12(900)(0.22) = \$2376$. The total cost this year is then $\$2376 + \$3130 = \$5506$, for a first-year savings of $2150. If, however, the customer goes overboard and installs 3000 kVAR of compensation capacitors so that no penalty is assessed, it will actually cost $14 more in the first year than if only 2000 kVAR were installed.

■ **FIGURE 11.20** A compensation capacitor installation. (Courtesy of Nokian Capacitors Ltd.)

EXAMPLE **11.9**

An industrial consumer is operating a 50 kW (67.1 hp) induction motor at a lagging PF of 0.8. The source voltage is 230 V rms. In order to obtain lower electrical rates, the customer wishes to raise the PF to 0.95 lagging. Specify a suitable solution.

Although the PF might be raised by increasing the real power and maintaining the reactive power constant, this would not result in a lower bill and is not a cure that interests the consumer. A purely reactive load must be added to the system, and it is clear that it must be added in parallel, since the supply voltage to the induction motor must not change. The circuit of Fig. 11.22 is thus applicable if we interpret $\mathbf{S}_1$ as the induction motor's complex power and $\mathbf{S}_2$ as the complex power drawn by the corrective device.

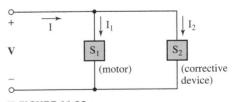

■ **FIGURE 11.22**

(Continued on next page)

The complex power supplied to the induction motor must have a real part of 50 kW and an angle of $\cos^{-1}(0.8)$, or $36.9°$. Hence,

$$\mathbf{S}_1 = \frac{50\underline{/36.9°}}{0.8} = 50 + j37.5 \text{ kVA}$$

In order to achieve a PF of 0.95, the total complex power must become

$$\mathbf{S} = \mathbf{S}_1 + \mathbf{S}_2 = \frac{50}{0.95}\underline{/\cos^{-1}(0.95)} = 50 + j16.43 \text{ kVA}$$

Thus, the complex power drawn by the corrective load is

$$\mathbf{S}_2 = -j21.07 \text{ kVA}$$

The necessary load impedance $\mathbf{Z}_2$ may be found in several simple steps. We select a phase angle of $0°$ for the voltage source, and therefore the current drawn by $\mathbf{Z}_2$ is

$$\mathbf{I}_2^* = \frac{\mathbf{S}_2}{\mathbf{V}} = \frac{-j21{,}070}{230} = -j91.6 \text{ A}$$

or

$$\mathbf{I}_2 = j91.6 \text{ A}$$

Therefore,

$$\mathbf{Z}_2 = \frac{\mathbf{V}}{\mathbf{I}_2} = \frac{230}{j91.6} = -j2.51 \text{ } \Omega$$

If the operating frequency is 60 Hz, this load can be provided by a 1056 μF capacitor connected in parallel with the motor. However, its initial cost, maintenance, and depreciation must be covered by the reduction in the electric bill.

PRACTICE

11.9 For the circuit shown in Fig. 11.23, find the complex power absorbed by the (a) 1 Ω resistor; (b) $-j10$ Ω capacitor; (c) $5 + j10$ Ω impedance; (d) source.

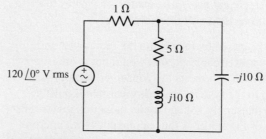

■ FIGURE 11.23

Ans: $26.6 + j0$ VA; $0 - j1331$ VA; $532 + j1065$ VA; $-559 + j266$ VA.

11.6 COMPARISON OF POWER TERMINOLOGY

We have been introduced to a possibly daunting array of power terminologies in this chapter, and it may be worthwhile to pause and consider them all together for a moment. A summary along with a brief description of each is provided in Table 11.2.

The practical importance of these new terms can be shown by considering the following practical situation. Let us first assume that we have a sinusoidal ac generator, which is a rotating machine driven by some other device whose output is a mechanical torque, such as a steam turbine, an electric motor, or an internal-combustion engine. We will let our generator produce an output voltage of 200 V rms at 60 Hz. Suppose that in addition, the rating of the generator is stated as a maximum power output of 1 kW. The generator would therefore be capable of delivering an rms current of 5 A to a resistive load. If, however, a load requiring 1 kW at a lagging power factor of 0.5 is connected to the generator, then an rms current of 10 A is necessary. As the PF decreases, greater and greater currents must be delivered to the load if operation at 200 V and 1 kW is to be maintained. If our generator were correctly and economically designed to furnish safely a maximum current of 5 A, then these greater currents would cause unsatisfactory operation, such as causing the insulation to overheat and begin smoking, which could be injurious to its health.

The rating of the generator is more informatively given in terms of apparent power in volt-amperes. Thus a 1000 VA rating at 200 V indicates that the generator can deliver a maximum current of 5 A at rated voltage; the power it delivers depends on the load, and in an extreme case might be zero. An apparent power rating is equivalent to a current rating when operation is at a constant voltage.

TABLE 11.2 A Summary of Relevant Terms

Term	Symbol	Unit	Description
Instantaneous power	$p(t)$	W	$p(t) = v(t)i(t)$. It is the value of the power at a specific instant in time. It is *not* the product of the voltage and current phasors!
Average power	P	W	In the sinusoidal steady state, $P = \frac{1}{2}V_m I_m \cos(\theta - \phi)$, where θ is the angle of the voltage and ϕ is the angle of the current. Reactances do not contribute to P.
Effective or rms value	V_{rms} or I_{rms}	V or A	Defined, e.g., as $I_{eff} = \sqrt{\dfrac{1}{T}\displaystyle\int_0^T i^2\,dt}$; if $i(t)$ is sinusoidal, then $I_{eff} = I_m/\sqrt{2}$.
Apparent power	$|\mathbf{S}|$	VA	$|\mathbf{S}| = V_{eff}I_{eff}$, and is the maximum value the average power can be; $P = |\mathbf{S}|$ only for purely resistive loads.
Power factor	PF	None	Ratio of the average power to the apparent power. The PF is unity for a purely resistive load, and zero for a purely reactive load.
Reactive power	Q	VAR	A means of measuring the energy flow rate to and from reactive loads.
Complex power	$\mathbf{S}$	VA	A convenient complex quantity that contains both the average power P and the reactive power Q: $\mathbf{S} = P + jQ$.

PRACTICE

11.10 A 440 V rms source supplies power to a load $Z_L = 10 + j2\ \Omega$ through a transmission line having a total resistance of 1.5 Ω. Find (a) the average and apparent power supplied to the load; (b) the average and apparent power lost in the transmission line; (c) the average and apparent power supplied by the source; (d) the power factor at which the source operates.

Ans: 14.21 kW, 14.49 kVA; 2.131 kW, 2.131 kVA; 16.34 kW, 16.59 kVA; 0.985 lag.

SUMMARY AND REVIEW

❑ The instantaneous power absorbed by an element is given by the expression $p(t) = v(t)i(t)$.

❑ The average power delivered to an impedance by a sinusoidal source is $\frac{1}{2}V_m I_m \cos(\theta - \phi)$, where θ = the voltage phase angle, and ϕ = the phase angle of the current.

❑ Only the *resistive* component of a load draws nonzero average power. The average power delivered to the *reactive* component of a load is zero.

❑ Maximum average power transfer occurs when the condition $\mathbf{Z}_L = \mathbf{Z}_{th}^*$ is satisfied.

❑ The effective or rms value of a sinusoidal waveform is obtained by dividing its amplitude by $\sqrt{2}$.

❑ The power factor (PF) of a load is the ratio of its average dissipated power to the apparent power.

❑ A purely resistive load will have a unity power factor. A purely reactive load will have a zero power factor.

❑ Complex power is defined as $\mathbf{S} = P + jQ$, or $\mathbf{S} = \mathbf{V}_{eff}\mathbf{I}_{eff}^*$. It is measured in units of volt-amperes (VA).

❑ Reactive power Q is the imaginary component of the complex power, and is a measure of the energy flow rate into or out of the reactive components of a load. Its unit is the volt-ampere-reactive (VAR).

❑ Capacitors are commonly used to improve the PF of industrial loads to minimize the reactive power required from the utility company.

READING FURTHER

A good overview of ac power concepts can be found in Chap. 2 of:

B.M. Weedy, *Electric Power Systems,* 3rd ed. Chichester, England: Wiley, 1984.

Contemporary issues pertaining to ac power systems can be found in:

International Journal of Electrical Power & Energy Systems. Guildford, England: IPC Science and Technology Press, 1979–. ISSN: 0142-0615.

EXERCISES

11.1 Instantaneous Power

1. A current source, $i_s(t) = 2 \cos 500t$ A, a 50 Ω resistor, and a 25 μF capacitor are in parallel. Find the power being supplied by the source, being absorbed by the resistor, and being absorbed by the capacitor, all at $t = \pi/2$ ms.

2. The current $i = 2t^2 - 1$ A, $1 \le t \le 3$ s, is flowing through a certain circuit element. (a) If the element is a 4 H inductor, what energy is delivered to it in the given time interval? (b) If the element is a 0.2 F capacitor with $v(1) = 2$ V, what power is being delivered to it at $t = 2$ s?

3. If $v_C(0) = -2$ V and $i(0) = 4$ A in the circuit of Fig. 11.24, find the power being absorbed by the capacitor at t equal to (a) 0^+; (b) 0.2 s; (c) 0.4 s.

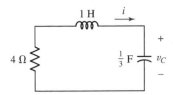

■ **FIGURE 11.24**

4. Find the power being absorbed by each passive element in the circuit of Fig. 11.25 at $t = 0$ if $v_s = 20 \cos(1000t + 30°)$ V. Verify your answer with PSpice.

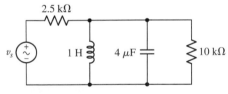

■ **FIGURE 11.25**

5. The circuit shown in Fig. 11.26 has reached steady-state conditions. Find the power being absorbed by each of the four circuit elements at $t = 0.1$ s.

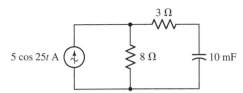

■ **FIGURE 11.26**

6. Consider the *RL* circuit depicted in Fig. 11.27. Determine the instantaneous power absorbed by the resistor at $t = $ (a) 0^+; (b) 1 s; (c) 2 s.

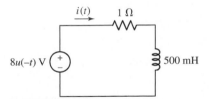

■ **FIGURE 11.27**

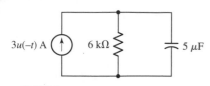

■ **FIGURE 11.28**

7. Consider the *RC* circuit depicted in Fig. 11.28. Determine the instantaneous power absorbed by the resistor at $t = $ (a) 0^+; (b) 30 ms; (c) 90 ms.

8. If we take a typical cloud-to-ground lightning *stroke* to represent a current of 30 kA over an interval of 150 μs, calculate (a) the instantaneous power delivered to a copper rod having resistance 1.2 mΩ during the stroke; and (b) the total energy delivered to the rod.

9. A 100 mF capacitor is storing 100 mJ of energy up until the point when a conductor of resistance 1.2 Ω falls across its terminals. What is the instantaneous power dissipated in the conductor at $t = $ 120 ms? If the specific heat capacity[3] of the conductor is 0.9 kJ/kg · K and its mass is 1 g, estimate the increase in temperature of the conductor in the first second of the capacitor discharge assuming both elements are initially at 23°C.

10. A semiconductor light-emitting diode runs at a voltage of 2.76 V, and draws a current of 130 mA. Neglecting any internal capacitance, what is the instantaneous power drawn by the LED 2 s after being switched on? If instead it is connected to a sinusoidal signal source described by $v(t) = 2.76 \cos(1000t)$ V, what other information is required in order to compute the instantaneous power at $t = $ 500 ms, assuming all transients have died out by that time?

11.2 Average Power

11. Find the average power being absorbed by each of the five circuit elements shown in Fig. 11.29.

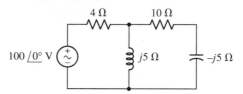

■ **FIGURE 11.29**

12. Calculate the average power generated by each source and the average power delivered to each impedance in the circuit of Fig. 11.30.

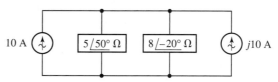

■ **FIGURE 11.30**

13. In the circuit shown in Fig. 11.31, find the average power being (a) dissipated in the 3 Ω resistor; (b) generated by the source.

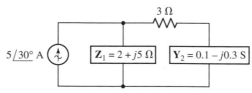

■ **FIGURE 11.31**

(3) Assume the specific heat capacity c is given by $c = Q/m \cdot \Delta T$, where $Q = $ the energy delivered to the conductor, m is its mass, and ΔT is the increase in temperature.

14. Find the average power absorbed by each of the five circuit elements shown in Fig. 11.32.

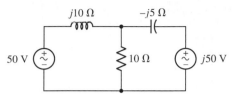

■ **FIGURE 11.32**

15. Determine the average power supplied by the dependent source in the circuit of Fig. 11.33.

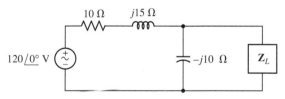

■ **FIGURE 11.33**

16. A frequency-domain Thévenin equivalent circuit consists of a sinusoidal source $\mathbf{V}_{th}$ in series with an impedance $\mathbf{Z}_{th} = R_{th} + jX_{th}$. Specify the conditions on a load $\mathbf{Z}_L = R_L + jX_L$ if it is to receive a maximum average power subject to the constraint that (a) $X_{th} = 0$; (b) R_L and X_L may be selected independently; (c) R_L is fixed (not equal to R_{th}); (d) X_L is fixed (independent of X_{th}); (e) $X_L = 0$.

17. For the circuit of Fig. 11.34; (a) what value of $\mathbf{Z}_L$ will absorb a maximum average power? (b) What is the value of this maximum power?

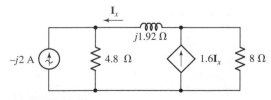

■ **FIGURE 11.34**

18. For the circuit of Fig. 11.34, it is required that the load be a pure resistance R_L. What value of R_L will absorb a maximum average power, and what is the value of this power?

19. Find the average power supplied by the dependent source of Fig. 11.35.

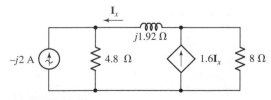

■ **FIGURE 11.35**

20. For the network of Fig. 11.36; (*a*) what impedance $\mathbf{Z}_L$ should be connected between *a* and *b* so that a maximum average power will be absorbed by it? (*b*) What is this maximum average power?

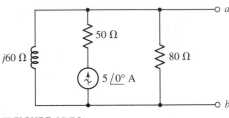

■ **FIGURE 11.36**

21. Determine the average power delivered to each of the boxed networks in the circuit of Fig. 11.37 if the $10\underline{/0°}$ A source is replaced with a $5\underline{/-30°}$ A source operating at a frequency of 50 Hz.

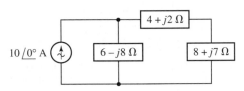

■ **FIGURE 11.37**

22. Find the value of R_L in the circuit of Fig. 11.38 that will absorb a maximum power, and specify the value of that power.

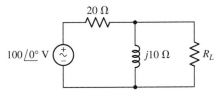

■ **FIGURE 11.38**

23. Determine the average power delivered to each resistor in the network shown in Fig. 11.39 if (*a*) $\lambda = 0$; (*b*) $\lambda = 1$. (*c*) Verify your answer with PSpice assuming the physical circuit operates at 60 Hz.

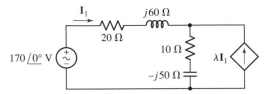

■ **FIGURE 11.39**

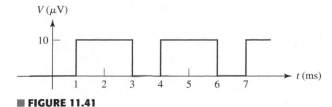

24. (a) Calculate the average value of each of the waveforms shown in Fig. 11.40.
   (b) If each of these waveforms is now squared, find the average value of each
   of the new periodic waveforms (in A^2).

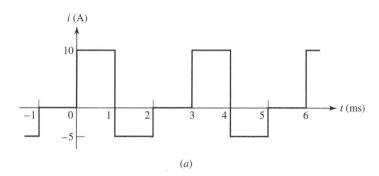

(a)

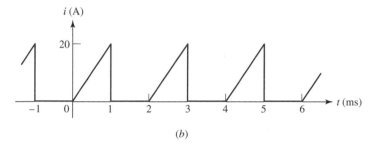

(b)

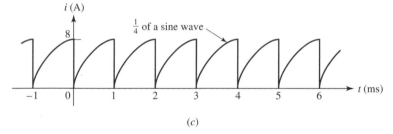

(c)

■ **FIGURE 11.40**

25. Find the average power delivered to each element of the circuit shown in
   Fig. 11.25 if $v_S = 400\sqrt{2} \cos(120\pi t - 9°)$ V. Verify your answer with PSpice.

11.3 Effective Values of Current and Voltage

26. Compute the effective value of the following: (a) 12 cos(1000t) V;
   (b) 12 sin(1000t) V; (c) 12 cos(500t) V; (d) 12 cos(500t − 88°) V.

27. Compute the effective value of the following: (a) 2 cos(10t) A;
   (b) 2 sin(10t) A; (c) 2 cos(5t) A; (d) 2 cos(5t − 32°) A.

28. Determine the effective value of the waveform depicted in Fig. 11.41.

V (μV)

■ **FIGURE 11.41**

29. Determine the effective value of the waveform depicted in Fig. 11.42.

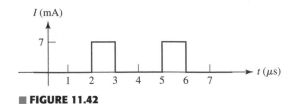

■ **FIGURE 11.42**

30. What is the effective value of (a) 1 V; (b) $1 + \cos 10t$ V; (c) $1 + \cos(10t + 10°)$ V?

31. Find the effective value of (a) $v(t) = 10 + 9 \cos 100t + 6 \sin 100t$; (b) the waveform appearing as Fig. 11.43. (c) Also find the average value of this waveform.

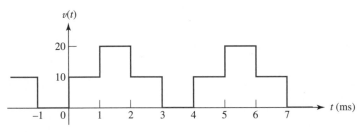

■ **FIGURE 11.43**

32. Find the effective value of (a) $g(t) = 2 + 3 \cos 100t + 4 \cos(100t − 120°)$; (b) $h(t) = 2 + 3 \cos 100t + 4 \cos(101t − 120°)$; (c) the waveform of Fig. 11.44.

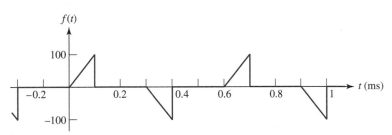

■ **FIGURE 11.44**

33. Given the periodic waveform $f(t) = (2 − 3 \cos 100t)^2$, find (a) its average value; (b) its rms value.

34. Calculate the effective value of each of the three periodic waveforms shown in Fig. 11.40.

35. Four ideal independent voltage sources, $A \cos 10t$, $B \sin(10t + 45°)$, $C \cos 40t$, and the constant D, are connected in series with a 4 Ω resistor. Find the average power dissipated in the resistor if (a) $A = B = 10$ V, $C = D = 0$; (b) $A = C = 10$ V, $B = D = 0$; (c) $A = 10$ V, $B = −10$ V, $C = D = 0$; (d) $A = B = C = 10$ V, $D = 0$; (e) $A = B = C = D = 10$ V.

36. (a) What value of R will cause the rms voltages across the inductors in Fig. 11.45 to be equal? (b) What is the value of the rms voltage? (c) Verify your answers with PSpice.

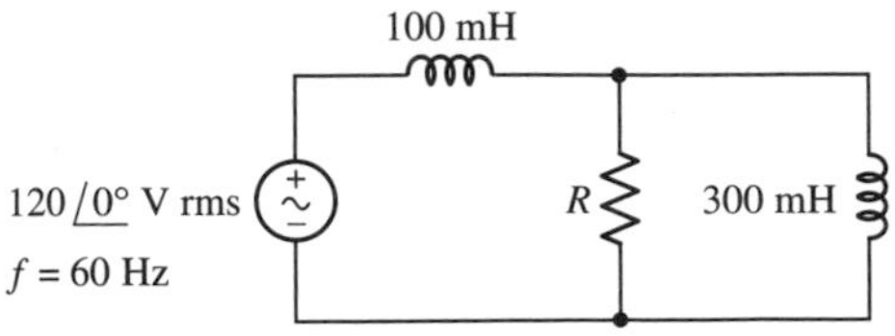

100 mH

$120 \underline{/0°}$ V rms
$f = 60$ Hz

R 300 mH

■ **FIGURE 11.45**

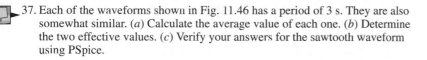

37. Each of the waveforms shown in Fig. 11.46 has a period of 3 s. They are also somewhat similar. (a) Calculate the average value of each one. (b) Determine the two effective values. (c) Verify your answers for the sawtooth waveform using PSpice.

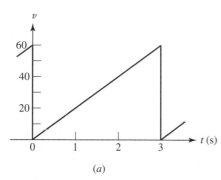

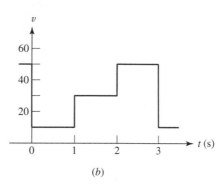

(a)   (b)

■ **FIGURE 11.46**

38. Replace the 100 mH inductor of Fig. 11.45 with a 1 μF capacitor, and the 300 mH inductor with a 3 μF capacitor. (a) What value of R will cause the rms currents through the capacitors to be equal? (b) What is the value of the rms current? (c) Verify your answers with PSpice.

39. A voltage waveform has a period of 5 s, and it is expressed as $v(t) = 10t[u(t) - u(t-2)] + 16e^{-0.5(t-3)}[u(t-3) - u(t-5)]$ V in the interval $0 < t < 5$ s. Find the effective value of the waveform.

40. The series combination of a 1 kΩ resistor and a 2 H inductor must not dissipate more than 250 mW of power at any instant. Assuming a sinusoidal current with $\omega = 500$ rad/s, what is the largest rms current that can be tolerated?

11.4 Apparent Power and Power Factor

41. In Fig. 11.47, let $\mathbf{I} = 4/35°$ A rms, and find the average power being supplied: (a) by the source; (b) to the 20 Ω resistor; (c) to the load. Find the apparent power being supplied: (d) by the source; (e) to the 20 Ω resistor; (f) to the load. (g) What is the load PF?

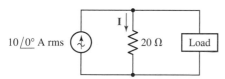

■ **FIGURE 11.47**

42. (a) Find the power factor at which the source in the circuit of Fig. 11.48 is operating. (b) Find the average power being supplied by the source. (c) What size capacitor should be placed in parallel with the source to cause its power factor to be unity? (d) Verify your answers with PSpice.

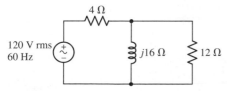

■ **FIGURE 11.48**

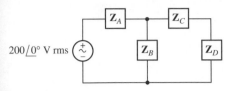

■ **FIGURE 11.49**

43. In the circuit shown in Fig. 11.49, let $\mathbf{Z}_A = 5 + j2\ \Omega$, $\mathbf{Z}_B = 20 - j10\ \Omega$, $\mathbf{Z}_C = 10\underline{/30°}\ \Omega$, and $\mathbf{Z}_D = 10\underline{/-60°}\ \Omega$. Find the apparent power delivered to each load and the apparent power generated by the source.

44. Let us visualize a network operating at $f = 50$ Hz that utilizes loads connected in series and carrying a common current of $10\underline{/0°}$ A rms. Such a system is the dual of one operating with parallel loads and a common voltage. In the series system, a load would be turned off by short-circuiting it; open circuits would cause all kinds of fireworks. Two loads are on this particular system: $\mathbf{Z}_1 = 30\underline{/15°}\ \Omega$ and $\mathbf{Z}_2 = 40\underline{/40°}\ \Omega$. (*a*) At what PF is the source operating? (*b*) What is the apparent power drawn by the combination of the two loads? (*c*) Is the combined load inductive or capacitive in character?

11.5 Complex Power

45. A composite load consists of three loads connected in parallel. One draws 100 W at a PF of 0.92 lagging, another takes 250 W at a PF of 0.8 lagging, and the third requires 150 W at a unity PF. The parallel load is supplied by a source $\mathbf{V}_s$ in series with a 10 Ω resistor. The loads must all operate at 115 V rms. Determine (*a*) the rms current through the source; (*b*) the PF of the composite load.

46. The load in Fig. 11.50 draws 10 kVA at PF = 0.8 lagging. If $|\mathbf{I}_L| = 40$ A rms, what must be the value of C to cause the source to operate at PF = 0.9 lagging?

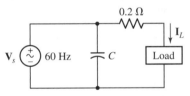

■ **FIGURE 11.50**

47. Consider the circuit of Fig. 11.51. Specify the value of capacitance required to raise the PF of the total load connected to the source to 0.92 lagging if the capacitance is added (*a*) in *series* with the 100 mH inductor; (*b*) in *parallel* with the 100 mH inductor. Verify your answers to parts (*a*) and (*b*) using PSpice.

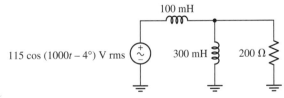

■ **FIGURE 11.51**

48. Analyze the circuit of Fig. 11.52 to find the complex power absorbed by each of the five circuit elements.

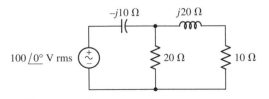

■ **FIGURE 11.52**

49. Both sources shown in Fig. 11.53 are operating at the same frequency. Find the complex power generated by each source and the complex power absorbed by each passive circuit element.

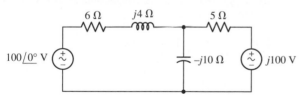

■ **FIGURE 11.53**

50. Find the complex power being delivered to a load that (a) draws 500 VA at a leading PF of 0.75; (b) draws 500 W at a leading PF of 0.75; (c) draws −500 VAR at a PF of 0.75.

51. A capacitive impedance, $\mathbf{Z}_C = -j120\ \Omega$, is in parallel with a load $\mathbf{Z}_L$. The parallel combination is supplied by a source, $\mathbf{V}_s = 400\underline{/0°}$ V rms, that generates a complex power of $1.6 + j0.5$ kVA. Find the (a) complex power delivered to $\mathbf{Z}_L$; (b) PF of $\mathbf{Z}_L$; (c) PF of the source.

52. A source of 230 V rms is supplying three loads in parallel: 1.2 kVA at a lagging PF of 0.8, 1.6 kVA at a lagging PF of 0.9, and 900 W at unity PF. Find (a) the amplitude of the source current; (b) the PF at which the source is operating; (c) the complex power being furnished by the source.

53. A 250 V rms system is supplying three parallel loads. One draws 20 kW at unity power factor, a second uses 25 kVA at PF = 0.8 lagging, and the third requires a power of 30 kW at a lagging PF of 0.75. (a) Find the total power supplied by the source. (b) Find the total apparent power supplied by the source. (c) At what PF does the source operate?

54. A cookie-baking operation has a monthly average demand of 200 kW and a monthly average reactive requirement of 280 kVAR. In an effort to recoup losses and encourage its customers to operate at high PF, a certain local utility charges a penalty of $0.22/kVAR for each kVAR above a benchmark value computed as 0.65 times the peak average power demand. (a) Using the rate schedule above, what is the annual cost to this utility customer associated with PF penalties? (b) Calculate the target PF on which the utility policy is based. (c) If compensation is available through the utility company at a cost of $200 per 100 kVAR increment and $395 per 200 kVAR increment, what is the most cost-effective solution for the customer?

55. Derive Eq. [28].

11.6 Comparison of Power Terminology

56. A voltage source $339\cos(100\pi t - 66°)$ V is connected to a purely resistive load of 1 kΩ. (a) What is the effective voltage of the source? (b) What is the peak instantaneous power absorbed by the load? (c) What is the minimum instantaneous power absorbed by the load? (d) Compute the apparent power delivered by the source. (e) Calculate the reactive power delivered by the source. (f) What is the complex power delivered to the load?

57. A voltage source $339\cos(100\pi t - 66°)$ V is connected to a purely inductive load of 150 mH. (a) What is the effective current through the circuit? (b) What is the peak instantaneous power absorbed by the load? (c) What is the minimum instantaneous power absorbed by the load? (d) Compute the apparent power delivered by the source. (e) Calculate the reactive power delivered by the source. (f) What is the complex power delivered to the load?

58. For the circuit of Fig. 11.25, $v_S = 5\cos 1000t$ V. (a) What is the peak instantaneous power delivered to the 10 kΩ resistor? (b) Calculate the reactive power delivered to the 10 kΩ resistor. (c) Find the apparent power delivered to the 10 kΩ resistor. (d) What is the complex power delivered by the source?

59. (*a*) Find the complex power delivered to each passive element in the circuit of Fig. 11.54 and (*b*) show that the *sum* of those values is equal to the complex power generated by the source. (*c*) Is this result true for the values of apparent power? (*d*) What is the average power delivered by the source? (*e*) What is the reactive power delivered by the source?

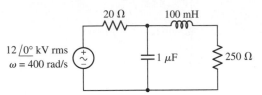

■ **FIGURE 11.54**

60. A load operating at 2300 V rms draws 28 A rms at a power factor of 0.812 lagging. Find (*a*) the peak current in amperes; (*b*) the instantaneous power at $t = 2.5$ ms assuming an operating frequency of 60 Hz; (*c*) the real power taken by the load; (*d*) the complex power; (*e*) the apparent power; (*f*) the impedance of the load; and (*g*) the reactive power.

12 Polyphase Circuits

KEY CONCEPTS

Single-Phase Power Systems

Three-Phase Power Systems

Three-Phase Sources

Line Versus Phase Voltage

Line Versus Phase Current

Y-Connected Networks

Δ-Connected Networks

Balanced Loads

Per-Phase Analysis

Power Measurement in Three-Phase Systems

INTRODUCTION

Utility companies supply electricity to residential and industrial customers in the form of sinusoidal voltages and currents, typically referred to as alternating current (ac). Most residential electricity in North America is supplied in the form of a sinusoidal waveform having a frequency of 60 Hz and an rms voltage of approximately 120 V. In other parts of the world, electricity is provided at a frequency of 50 Hz and an rms voltage of approximately 240 V. It was originally proposed by Thomas Edison that utility companies should distribute power through dc networks, but Nikola Tesla and George Westinghouse, two other pioneers in the field of electricity, were strong advocates of using ac. Ultimately, their arguments were more persuasive.

The transient response of ac power systems is of interest when determining the peak power demand, since most equipment requires more current to start up than it does to run continuously. In most instances, however, it is the steady-state operation that is of primary interest, so our experience with phasor-based analysis will prove to be handy. We will be introduced to a new type of voltage source, the three-phase source, which can be connected in either a three- or four-wire Y configuration or a three-wire Δ configuration. Similarly, we will find that loads can also be either Y- or Δ-connected, depending on the application.

12.1 • POLYPHASE SYSTEMS

So far, whenever we have used the term "sinusoidal source" we pictured a single sinusoidal voltage or current having a particular amplitude, frequency, and phase. In this chapter, we introduce the concept of *polyphase* sources, focusing on three-phase systems in particular. There are distinct advantages in using rotating machinery to generate three-phase power rather than single-phase power, and there are economical advantages in favor of the transmission of power in a three-phase system. Although most of the electrical equipment we have encountered so far is single-phase, three-phase equipment is not uncommon, especially in manufacturing environments. In particular, motors used in large refrigeration systems and in machining facilities are often wired for three-phase power. For the remaining applications, once we have become familiar with the basics of polyphase systems, we will find that it is simple to obtain single-phase power by just connecting to a single "leg" of a polyphase system.

Let us look briefly at the most common polyphase system, a balanced three-phase system. The source has three terminals (not counting a *neutral* or *ground* connection), and voltmeter measurements will show that sinusoidal voltages of equal amplitude are present between any two terminals. However, these voltages are not in phase; each of the three voltages is 120° out of phase with each of the other two, the sign of the phase angle depending on the sense of the voltages. One possible set of voltages is shown in Fig. 12.1. A *balanced load* draws power equally from all three phases. *At no instant does the instantaneous power drawn by the total load reach zero; in fact, the total instantaneous power is constant.* This is an advantage in rotating machinery, for it keeps the torque on the rotor much more constant than it would be if a single-phase source were used. As a result, there is less vibration.

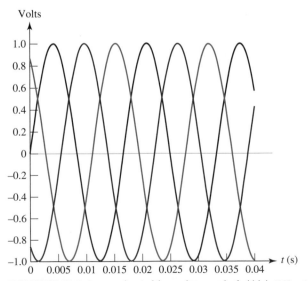

■ **FIGURE 12.1** An example set of three voltages, each of which is 120° out of phase with the other two. As can be seen, only one of the voltages is zero at any particular instant.

The use of a higher number of phases, such as 6- and 12-phase systems, is limited almost entirely to the supply of power to large **rectifiers.** Rectifiers convert alternating current to direct current by only allowing current to flow to the load in one direction, so that the sign of the voltage across the load remains the same. The rectifier output is a direct current plus a smaller pulsating component, or ripple, which decreases as the number of phases increases.

Almost without exception, polyphase systems in practice contain sources which may be closely approximated by ideal voltage sources or by ideal voltage sources in series with small internal impedances. Three-phase current sources are extremely rare.

Double-Subscript Notation

It is convenient to describe polyphase voltages and currents using **double-subscript notation.** With this notation, a voltage or current, such as $\mathbf{V}_{ab}$ or $\mathbf{I}_{aA}$, has more meaning than if it were indicated simply as $\mathbf{V}_3$ or $\mathbf{I}_x$. By definition, the voltage of point a with respect to point b is $\mathbf{V}_{ab}$. Thus, the plus sign is located at a, as indicated in Fig. 12.2a. We therefore consider the double subscripts to be *equivalent* to a plus-minus sign pair; the use of both would be redundant. With reference to Fig. 12.2b, for example, we see that $\mathbf{V}_{ad} = \mathbf{V}_{ab} + \mathbf{V}_{cd}$. The advantage of the double-subscript notation lies in the fact that Kirchhoff's voltage law requires the voltage between two points to be the same, regardless of the path chosen between the points; thus $\mathbf{V}_{ad} = \mathbf{V}_{ab} + \mathbf{V}_{bd} = \mathbf{V}_{ac} + \mathbf{V}_{cd} = \mathbf{V}_{ab} + \mathbf{V}_{bc} + \mathbf{V}_{cd}$, and so forth. The benefit of this is that KVL may be satisfied without reference to the circuit diagram; correct equations may be written even though a point, or subscript letter, is included which is not marked on the diagram. For example, we might have written $\mathbf{V}_{ad} = \mathbf{V}_{ax} + \mathbf{V}_{xd}$, where x identifies the location of any interesting point of our choice.

One possible representation of a three-phase system of voltages[1] is shown in Fig. 12.3. Let us assume that the voltages $\mathbf{V}_{an}$, $\mathbf{V}_{bn}$, and $\mathbf{V}_{cn}$ are known:

$$\mathbf{V}_{an} = 100\underline{/0°} \text{ V}$$
$$\mathbf{V}_{bn} = 100\underline{/-120°} \text{ V}$$
$$\mathbf{V}_{cn} = 100\underline{/-240°} \text{ V}$$

and thus the voltage $\mathbf{V}_{ab}$ may be found, with an eye on the subscripts:

$$\begin{aligned}
\mathbf{V}_{ab} &= \mathbf{V}_{an} + \mathbf{V}_{nb} = \mathbf{V}_{an} - \mathbf{V}_{bn} \\
&= 100\underline{/0°} - 100\underline{/-120°} \text{ V} \\
&= 100 - (-50 - j86.6) \text{ V} \\
&= 173.2\underline{/30°} \text{ V}
\end{aligned}$$

The three given voltages and the construction of the phasor $\mathbf{V}_{ab}$ are shown on the phasor diagram of Fig. 12.4.

A double-subscript notation may also be applied to currents. We define the current $\mathbf{I}_{ab}$ as the current flowing from a to b by the most direct path. In

(1) In keeping with power industry convention, rms values for currents and voltages will be used *implicitly* throughout this chapter.

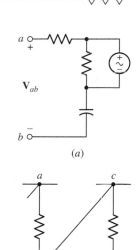

(a)

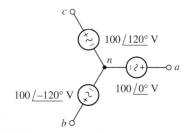

(b)

■ **FIGURE 12.2** (a) The definition of the voltage V_{ab}. (b) $\mathsf{V}_{ad} = \mathsf{V}_{ab} + \mathsf{V}_{bc} + \mathsf{V}_{cd} = \mathsf{V}_{ab} + \mathsf{V}_{cd}$.

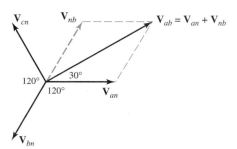

■ **FIGURE 12.3** A network used as a numerical example of double-subscript voltage notation.

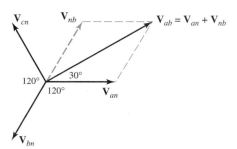

■ **FIGURE 12.4** This phasor diagram illustrates the graphical use of the double-subscript voltage convention to obtain V_{ab} for the network of Fig. 12.3.

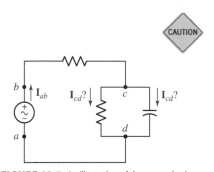

■ **FIGURE 12.5** An illustration of the use and *misuse* of the double-subscript convention for current notation.

every complete circuit we consider, there must of course be at least two possible paths between the points *a* and *b*, and we agree that we will not use double-subscript notation unless it is obvious that one path is much shorter, or much more direct. Usually this path is through a single element. Thus, the current $\mathbf{I}_{ab}$ is correctly indicated in Fig. 12.5. In fact, we do not even need the direction arrow when talking about this current; the subscripts *tell* us the direction. However, the identification of a current as $\mathbf{I}_{cd}$ for the circuit of Fig. 12.5 would cause confusion.

PRACTICE

12.1 Let $\mathbf{V}_{ab} = 100\underline{/0°}$ V, $\mathbf{V}_{bd} = 40\underline{/80°}$ V, and $\mathbf{V}_{ca} = 70\underline{/200°}$ V. Find (a) $\mathbf{V}_{ad}$; (b) $\mathbf{V}_{bc}$; (c) $\mathbf{V}_{cd}$.
12.2 Refer to the circuit of Fig. 12.6 and let $\mathbf{I}_{fj} = 3$ A, $\mathbf{I}_{de} = 2$ A, and $\mathbf{I}_{hd} = -6$ A. Find (a) $\mathbf{I}_{cd}$; (b) $\mathbf{I}_{ef}$; (c) $\mathbf{I}_{ij}$.

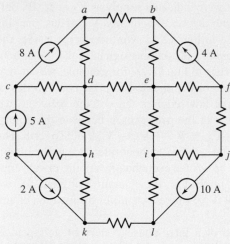

■ **FIGURE 12.6**

Ans: 12.1: $114.0\underline{/20.2°}$ V; $41.8\underline{/145.0°}$ V; $44.0\underline{/20.6°}$ V. 12.2: -3 A; 7 A; 7 A.

12.2 • SINGLE-PHASE THREE-WIRE SYSTEMS

A *single-phase three-wire source* is defined as a source having three output terminals, such as *a*, *n*, and *b* in Fig. 12.7*a*, at which the phasor voltages $\mathbf{V}_{an}$ and $\mathbf{V}_{nb}$ are equal. The source may therefore be represented by the combination of two identical voltage sources; in Fig. 12.7*b*, $\mathbf{V}_{an} = \mathbf{V}_{nb} = \mathbf{V}_1$. It is apparent that $\mathbf{V}_{ab} = 2\mathbf{V}_{an} = 2\mathbf{V}_{nb}$, and we therefore have a source to which loads operating at either of two voltages may be connected. The normal North American household system is single-phase three-wire, permitting the operation of both 110 V and 220 V appliances. The higher-voltage appliances are normally those drawing larger amounts of power; operation at higher voltage results in a smaller current draw for the same power. Smaller-diameter wire may consequently be used safely in the appliance, the household distribution system, and the distribution system of the utility

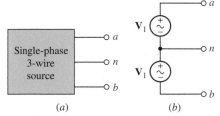

■ **FIGURE 12.7** (*a*) A single-phase three-wire source. (*b*) The representation of a single-phase three-wire source by two identical voltage sources.

company, as larger-diameter wire must be used with higher currents to reduce the heat produced due to the resistance of the wire.

The name ***single-phase*** arises because the voltages $\mathbf{V}_{an}$ and $\mathbf{V}_{nb}$, being equal, must have the same phase angle. From another viewpoint, however, the voltages between the outer wires and the central wire, which is usually referred to as the *neutral,* are exactly $180°$ out of phase. That is, $\mathbf{V}_{an} = -\mathbf{V}_{bn}$, and $\mathbf{V}_{an} + \mathbf{V}_{bn} = 0$. Later, we will see that balanced polyphase systems are characterized by a set of voltages of equal *amplitude* whose (phasor) sum is zero. From this viewpoint, then, the single-phase three-wire system is really a balanced two-phase system. *Two-phase,* however, is a term that is traditionally reserved for a relatively unimportant unbalanced system utilizing two voltage sources $90°$ out of phase.

Let us now consider a single-phase three-wire system that contains identical loads $\mathbf{Z}_p$ between each outer wire and the neutral (Fig. 12.8). We will first assume that the wires connecting the source to the load are perfect conductors. Since

$$\mathbf{V}_{an} = \mathbf{V}_{nb}$$

then,

$$\mathbf{I}_{aA} = \frac{\mathbf{V}_{an}}{\mathbf{Z}_p} = \mathbf{I}_{Bb} = \frac{\mathbf{V}_{nb}}{\mathbf{Z}_p}$$

and therefore

$$\mathbf{I}_{nN} = \mathbf{I}_{Bb} + \mathbf{I}_{Aa} = \mathbf{I}_{Bb} - \mathbf{I}_{aA} = 0$$

Thus there is no current in the neutral wire, and it could be removed without changing any current or voltage in the system. This result is achieved through the equality of the two loads and of the two sources.

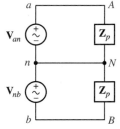

■ FIGURE 12.8 A simple single-phase three-wire system. The two loads are identical, and the neutral current is zero.

Effect of Finite Wire Impedance

We next consider the effect of a finite impedance in each of the wires. If lines aA and bB each have the same impedance, this impedance may be added to $\mathbf{Z}_p$, resulting in two equal loads once more, and zero neutral current. Now let us allow the neutral wire to possess some impedance $\mathbf{Z}_n$. Without carrying out any detailed analysis, superposition should show us that the symmetry of the circuit will still cause zero neutral current. Moreover, the addition of any impedance connected directly from one of the outer lines to the other outer line also yields a symmetrical circuit and zero neutral current. Thus, zero neutral current is a consequence of a balanced, or symmetrical, load; nonzero impedance in the neutral wire does not destroy the symmetry.

The most general single-phase three-wire system will contain unequal loads between each outside line and the neutral and another load directly between the two outer lines; the impedances of the two outer lines may be expected to be approximately equal, but the neutral impedance is often slightly larger. Let us consider an example of such a system, with particular interest in the current that may flow now through the neutral wire, as well as the overall efficiency with which our system is transmitting power to the unbalanced load.

EXAMPLE 12.1

Analyze the system shown in Fig. 12.9 and determine the power delivered to each of the three loads as well as the power lost in the neutral wire and each of the two lines.

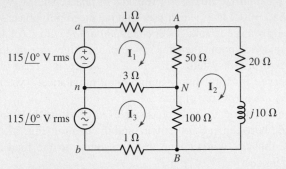

■ **FIGURE 12.9** A typical single-phase three-wire system.

▶ *Identify the goal of the problem.*
The three loads in the circuit are: the 50 Ω resistor, the 100 Ω resistor, and a $20 + j10$ Ω impedance. Each of the two lines has a resistance of 1 Ω, and the neutral wire has a resistance of 3 Ω. We need the current through each of these in order to determine power.

▶ *Collect the known information.*
We have a single-phase three-wire system; the circuit diagram of Fig. 12.9 is completely labeled. The computed currents will be in rms units.

▶ *Devise a plan.*
The circuit is conducive to mesh analysis, having three clearly defined meshes. The result of the analysis will be a set of mesh currents, which can then be used to compute absorbed power.

▶ *Construct an appropriate set of equations.*
The three mesh equations are:

$$-115\underline{/0^\circ} + \mathbf{I}_1 + 50(\mathbf{I}_1 - \mathbf{I}_2) + 3(\mathbf{I}_1 - \mathbf{I}_3) = 0$$
$$(20 + j10)\mathbf{I}_2 + 100(\mathbf{I}_2 - \mathbf{I}_3) + 50(\mathbf{I}_2 - \mathbf{I}_1) = 0$$
$$-115\underline{/0^\circ} + 3(\mathbf{I}_3 - \mathbf{I}_1) + 100(\mathbf{I}_3 - \mathbf{I}_2) + \mathbf{I}_3 = 0$$

which can be rearranged to obtain the following three equations

$$\begin{aligned} 54\mathbf{I}_1 \quad & \quad -50\mathbf{I}_2 \quad & -3\mathbf{I}_3 \quad & = 115\underline{/0^\circ} \\ -50\mathbf{I}_1 \quad & +(170 + j10)\mathbf{I}_2 \quad & -100\mathbf{I}_3 \quad & = 0 \\ -3\mathbf{I}_1 \quad & \quad -100\mathbf{I}_2 \quad & +104\mathbf{I}_3 \quad & = 115\underline{/0^\circ} \end{aligned}$$

▶ *Determine if additional information is required.*
We have a set of three equations in three unknowns, so it is possible to attempt a solution at this point.

▶ *Attempt a solution.*

Solving for the phasor currents $\mathbf{I}_1$, $\mathbf{I}_2$, and $\mathbf{I}_3$ using a scientific calculator, we find

$$\mathbf{I}_1 = 11.24\underline{/-19.83°}\ \text{A}$$
$$\mathbf{I}_2 = 9.389\underline{/-24.47°}\ \text{A}$$
$$\mathbf{I}_3 = 10.37\underline{/-21.80°}\ \text{A}$$

The currents in the outer lines are thus

$$\mathbf{I}_{aA} = \mathbf{I}_1 = 11.24\underline{/-19.83°}\ \text{A}$$

and

$$\mathbf{I}_{bB} = -\mathbf{I}_3 = 10.37\underline{/158.20°}\ \text{A}$$

and the smaller neutral current is

$$\mathbf{I}_{nN} = \mathbf{I}_3 - \mathbf{I}_1 = 0.9459\underline{/-177.7°}\ \text{A}$$

The average power drawn by each load may thus be determined:

$$P_{50} = |\mathbf{I}_1 - \mathbf{I}_2|^2 (50) = 206\ \text{W}$$
$$P_{100} = |\mathbf{I}_3 - \mathbf{I}_2|^2 (100) = 117\ \text{W}$$
$$P_{20+j10} = |\mathbf{I}_2|^2 (20) = 1763\ \text{W}$$

> Note that we do not need to include a factor of $\frac{1}{2}$ since we are working with rms current values.

The total load power is 2086 W. The loss in each of the wires is next found:

$$P_{aA} = |\mathbf{I}_1|^2 (1) = 126\ \text{W}$$
$$P_{bB} = |\mathbf{I}_3|^2 (1) = 108\ \text{W}$$
$$P_{nN} = |\mathbf{I}_{nN}|^2 (3) = 3\ \text{W}$$

giving a total line loss of 237 W. The wires are evidently quite long; otherwise, the relatively high power loss in the two outer lines would cause a dangerous temperature rise.

> Imagine the heat produced by two 100 W light bulbs! These outer wires must dissipate the same amount of energy. In order to keep their temperature down, a large surface area is required.

▶ *Verify the solution. Is it reasonable or expected?*

The total absorbed power is $206 + 117 + 1763 + 237$, or 2323 W, which may be checked by finding the power delivered by each voltage source:

$$P_{an} = 115(11.24)\cos 19.83° = 1216\ \text{W}$$
$$P_{bn} = 115(10.37)\cos 21.80° = 1107\ \text{W}$$

or a total of 2323 W. The **transmission efficiency** for the system is

$$\eta = \frac{\text{total power delivered to load}}{\text{total power generated}} = \frac{2086}{2086 + 237} = 89.8\%$$

This value would be unbelievable for a steam engine or an internal combustion engine, but it is too low for a well-designed distribution system. Larger-diameter wires should be used if the source and the load cannot be placed closer to each other.

(Continued on next page)

A phasor diagram showing the two source voltages, the currents in the outer lines, and the current in the neutral is constructed in Fig. 12.10. The fact that $\mathbf{I}_{aA} + \mathbf{I}_{bB} + \mathbf{I}_{nN} = 0$ is indicated on the diagram.

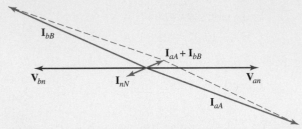

■ **FIGURE 12.10** The source voltages and three of the currents in the circuit of Fig. 12.9 are shown on a phasor diagram. Note that $\mathbf{I}_{aA} + \mathbf{I}_{bB} + \mathbf{I}_{nN} = 0$.

PRACTICE

12.3 Modify Fig. 12.9 by adding a 1.5 Ω resistance to each of the two outer lines, and a 2.5 Ω resistance to the neutral wire. Find the average power delivered to each of the three loads.

Ans: 153.1 W; 95.8 W; 1374 W.

12.3 • THREE-PHASE Y-Y CONNECTION

Three-phase sources have three terminals, called the *line* terminals, and they may or may not have a fourth terminal, the *neutral* connection. We will begin by discussing a three-phase source that does have a neutral connection. It may be represented by three ideal voltage sources connected in a Y, as shown in Fig. 12.11; terminals *a*, *b*, *c*, and *n* are available. We will consider only balanced three-phase sources, which may be defined as having

$$|\mathbf{V}_{an}| = |\mathbf{V}_{bn}| = |\mathbf{V}_{cn}|$$

and

$$\mathbf{V}_{an} + \mathbf{V}_{bn} + \mathbf{V}_{cn} = 0$$

These three voltages, each existing between one line and the neutral, are called *phase voltages*. If we arbitrarily choose $\mathbf{V}_{an}$ as the reference, or define

$$\mathbf{V}_{an} = V_p \underline{/0^\circ}$$

where we will consistently use V_p to represent the rms *amplitude* of any of the phase voltages, then the definition of the three-phase source indicates that either

$$\mathbf{V}_{bn} = V_p \underline{/-120^\circ} \quad \text{and} \quad \mathbf{V}_{cn} = V_p \underline{/-240^\circ}$$

or

$$\mathbf{V}_{bn} = V_p \underline{/120^\circ} \quad \text{and} \quad \mathbf{V}_{cn} = V_p \underline{/240^\circ}$$

The former is called *positive phase sequence,* or *abc phase sequence,* and is shown in Fig. 12.12*a*; the latter is termed *negative phase sequence,* or *cba phase sequence,* and is indicated by the phasor diagram of Fig. 12.12*b*.

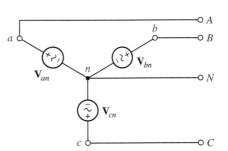

■ **FIGURE 12.11** A Y-connected three-phase four-wire source.

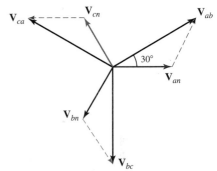

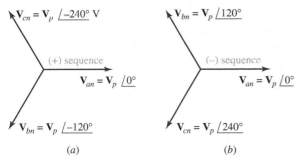

$$\mathbf{V}_{cn} = \mathbf{V}_p \angle{-240°} \text{ V}$$
$$\mathbf{V}_{bn} = \mathbf{V}_p \angle{120°}$$

(+) sequence   (−) sequence

$$\mathbf{V}_{an} = \mathbf{V}_p \angle{0°}$$ $$\mathbf{V}_{an} = \mathbf{V}_p \angle{0°}$$

$$\mathbf{V}_{bn} = \mathbf{V}_p \angle{-120°}$$ $$\mathbf{V}_{cn} = \mathbf{V}_p \angle{240°}$$

(a)   (b)

■ **FIGURE 12.12** (a) Positive, or abc, phase sequence. (b) Negative, or cba, phase sequence.

The actual phase sequence of a physical three-phase source depends on the arbitrary choice of the three terminals to be lettered *a*, *b*, and *c*. They may always be chosen to provide positive phase sequence, and we will assume that this has been done in most of the systems we consider.

Line-to-Line Voltages

Let us next find the line-to-line voltages (often simply called the **line voltages**) which are present when the phase voltages are those of Fig. 12.12*a*. It is easiest to do this with the help of a phasor diagram, since the angles are all multiples of 30°. The necessary construction is shown in Fig. 12.13; the results are

$$\mathbf{V}_{ab} = \sqrt{3}V_p \angle{30°} \qquad [1]$$

$$\mathbf{V}_{bc} = \sqrt{3}V_p \angle{-90°} \qquad [2]$$

$$\mathbf{V}_{ca} = \sqrt{3}V_p \angle{-210°} \qquad [3]$$

Kirchhoff's voltage law requires the sum of these three voltages to be zero; the reader is encouraged to verify this as an exercise.

If the rms amplitude of any of the line voltages is denoted by V_L, then one of the important characteristics of the Y-connected three-phase source may be expressed as

$$\boxed{V_L = \sqrt{3}V_p}$$

Note that with positive phase sequence, $\mathbf{V}_{an}$ leads $\mathbf{V}_{bn}$ and $\mathbf{V}_{bn}$ leads $\mathbf{V}_{cn}$, in each case by 120°, and also that $\mathbf{V}_{ab}$ leads $\mathbf{V}_{bc}$ and $\mathbf{V}_{bc}$ leads $\mathbf{V}_{ca}$, again by 120°. The statement is true for negative phase sequence if "lags" is substituted for "leads."

Now let us connect a balanced Y-connected three-phase load to our source, using three lines and a neutral, as drawn in Fig. 12.14. The load is

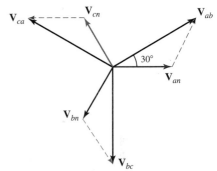

■ **FIGURE 12.13** A phasor diagram which is used to determine the line voltages from the given phase voltages. Or, algebraically, $\mathbf{V}_{ab} = \mathbf{V}_{an} - \mathbf{V}_{bn} = V_p \angle{0°} - V_p \angle{-120°} = V_p - V_p \cos(-120°) - j V_p \sin(-120°) = V_p(1 + \frac{1}{2} + j\sqrt{3}/2) = \sqrt{3}V_p \angle{30°}$.

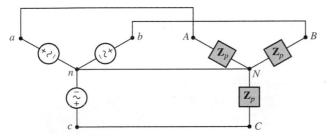

■ **FIGURE 12.14** A balanced three-phase system, connected Y-Y and including a neutral.

represented by an impedance $\mathbf{Z}_p$ between each line and the neutral. The three line currents are found very easily, since we really have three single-phase circuits that possess one common lead:[2]

$$\mathbf{I}_{aA} = \frac{\mathbf{V}_{an}}{\mathbf{Z}_p}$$

$$\mathbf{I}_{bB} = \frac{\mathbf{V}_{bn}}{\mathbf{Z}_p} = \frac{\mathbf{V}_{an}\underline{/-120^\circ}}{\mathbf{Z}_p} = \mathbf{I}_{aA}\underline{/-120^\circ}$$

$$\mathbf{I}_{cC} = \mathbf{I}_{aA}\underline{/-240^\circ}$$

and therefore

$$\mathbf{I}_{Nn} = \mathbf{I}_{aA} + \mathbf{I}_{bB} + \mathbf{I}_{cC} = 0$$

Thus, the neutral carries no current if the source and load are both balanced and if the four wires have zero impedance. How will this change if an impedance $\mathbf{Z}_L$ is inserted in series with each of the three lines and an impedance $\mathbf{Z}_n$ is inserted in the neutral? The line impedances may be combined with the three load impedances; this effective load is still balanced, and a perfectly conducting neutral wire could be removed. Thus, if no change is produced in the system with a short circuit or an open circuit between n and N, any impedance may be inserted in the neutral and the neutral current will remain zero.

It follows that, if we have balanced sources, balanced loads, and balanced line impedances, a neutral wire of any impedance may be replaced by any other impedance, including a short circuit or an open circuit; the replacement will not affect the system's voltages or currents. It is often helpful to *visualize* a short circuit between the two neutral points, whether a neutral wire is actually present or not; the problem is then reduced to three single-phase problems, all identical except for the consistent difference in phase angle. We say that we thus work the problem on a "per-phase" basis.

EXAMPLE 12.2

For the circuit of Fig. 12.15, find both the phase and line currents and voltages throughout the circuit, and calculate the total power dissipated in the load.

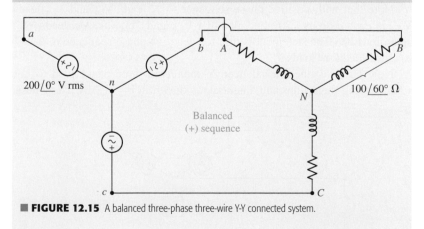

■ **FIGURE 12.15** A balanced three-phase three-wire Y-Y connected system.

---

(2) This can be seen to be true by applying superposition and looking at each phase one at a time.

Since one of the source phase voltages is given and we are told to use the positive phase sequence, the three phase voltages are:

$$\mathbf{V}_{an} = 200\underline{/0^\circ}\ \text{V} \qquad \mathbf{V}_{bn} = 200\underline{/-120^\circ}\ \text{V} \qquad \mathbf{V}_{cn} = 200\underline{/-240^\circ}\ \text{V}$$

The line voltage is $200\sqrt{3} = 346$ V; the phase angle of each line voltage can be determined by constructing a phasor diagram, as we did in Fig. 12.13 (as a matter of fact, the phasor diagram of Fig. 12.13 is applicable), subtracting the phase voltages using a scientific calculator, or by invoking Eqs. [1] to [3]. We find that $\mathbf{V}_{ab}$ is $346\underline{/30^\circ}$ V, $\mathbf{V}_{bc} = 346\underline{/-90^\circ}$ V, and $\mathbf{V}_{ca} = 346\underline{/-210^\circ}$ V.

Let us work with phase A. The line current is

$$\mathbf{I}_{aA} = \frac{\mathbf{V}_{an}}{\mathbf{Z}_p} = \frac{200\underline{/0^\circ}}{100\underline{/60^\circ}} = 2\underline{/-60^\circ}\ \text{A}$$

Since we know this is a balanced three-phase system, we may easily write the remaining line currents based on $\mathbf{I}_{aA}$:

$$\mathbf{I}_{bB} = 2\underline{/(-60^\circ - 120^\circ)} = 2\underline{/-180^\circ}\ \text{A}$$
$$\mathbf{I}_{cC} = 2\underline{/(-60^\circ - 240^\circ)} = 2\underline{/-300^\circ}\ \text{A}$$

The power absorbed by phase A is

$$P_{AN} = 200(2)\cos(0^\circ + 60^\circ) = 200\ \text{W}$$

Thus, the total average power drawn by the three-phase load is 600 W.

The phasor diagram for this circuit is shown in Fig. 12.16. Once we knew any of the line voltage magnitudes and any of the line current magnitudes, the angles for all three voltages and all three currents could have been easily obtained by reading the diagram.

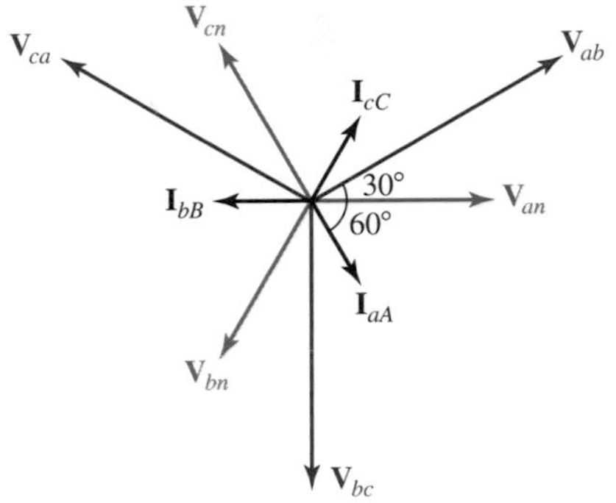

■ **FIGURE 12.16** The phasor diagram that applies to the circuit of Fig. 12.15.

PRACTICE

12.4 A balanced three-phase three-wire system has a Y-connected load. Each phase contains three loads in parallel: $-j100\ \Omega$, $100\ \Omega$, and $50 + j50\ \Omega$. Assume positive phase sequence with $\mathbf{V}_{ab} = 400\underline{/0^\circ}$ V. Find (a) $\mathbf{V}_{an}$; (b) $\mathbf{I}_{aA}$; (c) the total power drawn by the load.

Ans: $231\underline{/-30^\circ}$ V; $4.62\underline{/-30^\circ}$ A; 3200 W.

Before working another example, this would be a good opportunity to quickly explore a statement made in Sec. 12.1, i.e., that even though phase voltages and currents have zero value at specific instants in time (every 1/120 s in North America), the instantaneous power delivered to the *total* load is never zero. Consider phase A of Example 12.2 once more, with the phase voltage and current written in the time domain:

$$v_{AN} = 200\sqrt{2}\cos(120\pi t + 0^\circ)\ \text{V}$$

and

$$i_{AN} = 2\sqrt{2}\cos(120\pi t - 60^\circ)\ \text{A}$$

The factor of $\sqrt{2}$ is required to convert from rms units.

Thus, the instantaneous power absorbed by phase A is

$$
\begin{aligned}
p_A(t) = v_{AN}i_{AN} &= 800\cos(120\pi t)\cos(120\pi t - 60°) \\
&= 400[\cos(-60°) + \cos(240\pi t - 60°)] \\
&= 200 + 400\cos(240\pi t - 60°) \text{ W}
\end{aligned}
$$

in a similar fashion,

$$
p_B(t) = 200 + 400\cos(240\pi t - 300°) \text{ W}
$$

and

$$
p_C(t) = 200 + 400\cos(240\pi t - 180°) \text{ W}
$$

The instantaneous power absorbed by the *total* load is therefore

$$
p(t) = p_A(t) + p_B(t) + p_C(t) = 600 \text{ W}
$$

independent of time, and the same value as the average power computed in Example 12.2.

EXAMPLE 12.3

A balanced three-phase system with a line voltage of 300 V is supplying a balanced Y-connected load with 1200 W at a leading PF of 0.8. Find the line current and the per-phase load impedance.

The phase voltage is $300/\sqrt{3}$ V and the per-phase power is $1200/3 = 400$ W. Thus the line current may be found from the power relationship

$$
400 = \frac{300}{\sqrt{3}}(I_L)(0.8)
$$

and the line current is therefore 2.89 A. The phase impedance is given by

$$
|\mathbf{Z}_p| = \frac{V_p}{I_L} = \frac{300/\sqrt{3}}{2.89} = 60 \ \Omega
$$

Since the PF is 0.8 leading, the impedance phase angle is $-36.9°$; thus $\mathbf{Z}_p = 60\underline{/-36.9°} \ \Omega$.

More complicated loads can be handled easily, since the problems reduce to simpler single-phase problems.

PRACTICE

12.5 A balanced three-phase three-wire system has a line voltage of 500 V. Two balanced Y-connected loads are present. One is a capacitive load with $7 - j2 \ \Omega$ per phase, and the other is an inductive load with $4 + j2 \ \Omega$ per phase. Find (*a*) the phase voltage; (*b*) the line current; (*c*) the total power drawn by the load; (*d*) the power factor at which the source is operating.

Ans: 289 V; 97.5 A; 83.0 kW; 0.983 lagging.

EXAMPLE **12.4**

A balanced 600 W lighting load is added (in parallel) to the system of Example 12.3. Determine the new line current.

We first sketch a suitable per-phase circuit, as shown in Fig. 12.17. The 600 W load is assumed to be a balanced load evenly distributed among the three phases, resulting in an additional 200 W consumed by each phase.

The amplitude of the lighting current is determined by

$$200 = \frac{300}{\sqrt{3}} |\mathbf{I}_1| \cos 0°$$

so that

$$|\mathbf{I}_1| = 1.155 \text{ A}$$

In a similar way, the amplitude of the capacitive load current is found to be unchanged from its previous value, since the voltage across it has remained the same:

$$|\mathbf{I}_2| = 2.89 \text{ A}$$

If we assume that the phase with which we are working has a phase voltage with an angle of 0°, then

$$\mathbf{I}_1 = 1.155\underline{/0°} \text{ A} \qquad \mathbf{I}_2 = 2.89\underline{/+36.9°} \text{ A}$$

and the line current is

$$\mathbf{I}_L = \mathbf{I}_1 + \mathbf{I}_2 = 3.87\underline{/+26.6°} \text{ A}$$

Furthermore, the power generated by this phase of the source is

$$P_p = \frac{300}{\sqrt{3}} 3.87 \cos(+26.6°) = 600 \text{ W}$$

which agrees with the fact that the individual phase is known to be supplying 200 W to the new lighting load, as well as 400 W to the original load.

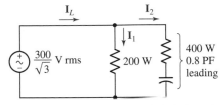

■ **FIGURE 12.17** The per-phase circuit that is used to analyze a *balanced* three-phase example.

PRACTICE

12.6 Three balanced Y-connected loads are installed on a balanced three-phase four-wire system. Load 1 draws a total power of 6 kW at unity PF, load 2 requires 10 kVA at PF = 0.96 lagging, and load 3 needs 7 kW at 0.85 lagging. If the phase voltage at the loads is 135 V, if each line has a resistance of 0.1 Ω, and if the neutral has a resistance of 1 Ω, find (*a*) the total power drawn by the loads; (*b*) the combined PF of the loads; (*c*) the total power lost in the four lines; (*d*) the phase voltage at the source; (*e*) the power factor at which the source is operating.

Ans: 22.6 kW; 0.954 lag; 1027 W; 140.6 V; 0.957 lagging.

If an *unbalanced* Y-connected load is present in an otherwise balanced three-phase system, the circuit may still be analyzed on a per-phase basis *if* the neutral wire is present and *if* it has zero impedance. If either of these conditions is not met, other methods must be used, such as mesh or nodal analysis. However, engineers who spend most of their time with unbalanced three-phase systems will find the use of *symmetrical components* a great timesaver. We will not discuss this method here.

12.4 THE DELTA (Δ) CONNECTION

An alternative configuration to the Y-connected load is the Δ-connected load, as shown in Fig. 12.18. This type of configuration is very common, and does not possess a neutral connection.

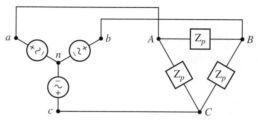

■ **FIGURE 12.18** A balanced Δ-connected load is present on a three-wire three-phase system. The source happens to be Y-connected.

Let us consider a balanced Δ-connected load which consists of an impedance $\mathbf{Z}_p$ inserted between each pair of lines. With reference to Fig. 12.18, let us assume known line voltages

$$V_L = |\mathbf{V}_{ab}| = |\mathbf{V}_{bc}| = |\mathbf{V}_{ca}|$$

or known phase voltages

$$V_p = |\mathbf{V}_{an}| = |\mathbf{V}_{bn}| = |\mathbf{V}_{cn}|$$

where

$$V_L = \sqrt{3}V_p \qquad \text{and} \qquad \mathbf{V}_{ab} = \sqrt{3}V_p\underline{/30°}$$

as we found previously. Because the voltage across each branch of the Δ is known, the *phase currents* are easily found:

$$\mathbf{I}_{AB} = \frac{\mathbf{V}_{ab}}{\mathbf{Z}_p} \qquad \mathbf{I}_{BC} = \frac{\mathbf{V}_{bc}}{\mathbf{Z}_p} \qquad \mathbf{I}_{CA} = \frac{\mathbf{V}_{ca}}{\mathbf{Z}_p}$$

and their differences provide us with the line currents, such as

$$\mathbf{I}_{aA} = \mathbf{I}_{AB} - \mathbf{I}_{CA}$$

Since we are working with a balanced system, the three phase currents are of equal amplitude:

$$I_p = |\mathbf{I}_{AB}| = |\mathbf{I}_{BC}| = |\mathbf{I}_{CA}|$$

The line currents are also equal in amplitude; the symmetry is apparent from the phasor diagram of Fig. 12.19. We thus have

$$I_L = |\mathbf{I}_{aA}| = |\mathbf{I}_{bB}| = |\mathbf{I}_{cC}|$$

and

$$I_L = \sqrt{3}I_p$$

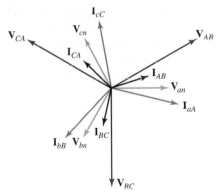

FIGURE 12.19 A phasor diagram that could apply to the circuit of Fig. 12.18 if $\mathbf{Z}_p$ were an inductive impedance.

Let us disregard the source for the moment and consider only the balanced load. If the load is Δ-connected, then the phase voltage and the line voltage are indistinguishable, but the line current is larger than the phase current by a factor of $\sqrt{3}$; with a Y-connected load, however, the phase current and the line current refer to the same current, and the line voltage is greater than the phase voltage by a factor of $\sqrt{3}$.

EXAMPLE 12.5

Determine the amplitude of the line current in a three-phase system with a line voltage of 300 V that supplies 1200 W to a Δ-connected load at a lagging PF of 0.8, then find the phase impedance.

Let us again consider a single phase. It draws 400 W, 0.8 lagging PF, at a 300 V line voltage. Thus,

$$400 = 300(I_p)(0.8)$$

and

$$I_p = 1.667 \text{ A}$$

and the relationship between phase currents and line currents yields

$$I_L = \sqrt{3}(1.667) = 2.89 \text{ A}$$

Next, the phase angle of the load is $\cos^{-1}(0.8) = 36.9°$, and therefore the impedance in each phase must be

$$\mathbf{Z}_p = \frac{300}{1.667} \underline{/36.9°} = 180\underline{/36.9°} \ \Omega$$

> Again, keep in mind that we are assuming all voltages and currents are quoted as rms values.

PRACTICE

12.7 Each phase of a balanced three-phase Δ-connected load consists of a 200 mH inductor in series with the parallel combination of a 5 μF capacitor and a 200 Ω resistance. Assume zero line resistance and a phase voltage of 200 V at $\omega = 400$ rad/s. Find (a) the phase current; (b) the line current; (c) the total power absorbed by the load.

Ans: 1.158 A; 2.01 A; 693 W.

EXAMPLE 12.6

Determine the amplitude of the line current in a three-phase system with a 300 V line voltage that supplies 1200 W to a Y-connected load at a lagging PF of 0.8. (*This is the same circuit as in Example 12.5, but with a Y-connected load instead.*)

On a per-phase basis, we now have a phase voltage of $300/\sqrt{3}$ V, a power of 400 W, and a lagging PF of 0.8. Thus,

$$400 = \frac{300}{\sqrt{3}}(I_p)(0.8)$$

and

$$I_p = 2.89 \qquad \text{(and so } I_L = 2.89 \text{ A)}$$

The phase angle of the load is again $36.9°$, and thus the impedance in each phase of the Y is

$$\mathbf{Z}_p = \frac{300/\sqrt{3}}{2.89}\underline{/36.9°} = 60\underline{/36.9°} \; \Omega$$

The $\sqrt{3}$ factor not only relates phase and line quantities but also appears in a useful expression for the total power drawn by any balanced three-phase load. If we assume a Y-connected load with a power-factor angle θ, the power taken by any phase is

$$P_p = V_p I_p \cos\theta = V_p I_L \cos\theta = \frac{V_L}{\sqrt{3}} I_L \cos\theta$$

and the total power is

$$P = 3P_p = \sqrt{3} V_L I_L \cos\theta$$

In a similar way, the power delivered to each phase of a Δ-connected load is

$$P_p = V_p I_p \cos\theta = V_L I_p \cos\theta = V_L \frac{I_L}{\sqrt{3}} \cos\theta$$

giving a total power

$$P = 3P_p \qquad\qquad\qquad [4]$$
$$P = \sqrt{3} V_L I_L \cos\theta$$

Thus Eq. [4] enables us to calculate the total power delivered to a balanced load from a knowledge of the magnitude of the line voltage, of the line current, and of the phase angle of the load impedance (or admittance), regardless of whether the load is Y-connected or Δ-connected. The line current in Examples 12.5 and 12.6 can now be obtained in two simple steps:

$$1200 = \sqrt{3}(300)(I_L)(0.8)$$

Therefore,

$$I_L = \frac{5}{\sqrt{3}} = 2.89 \text{ A}$$

TABLE 12.1 Comparison of Y- and Δ-Connected Three-Phase Loads. V_p Is the Voltage Magnitude of Each Y-Connected *Source* Phase

Load	Phase Voltage	Line Voltage	Phase Current	Line Current	Power per Phase
Y	$\mathbf{V}_{AN} = V_p\underline{/0°}$ $\mathbf{V}_{BN} = V_p\underline{/-120°}$ $\mathbf{V}_{CN} = V_p\underline{/-240°}$	$\mathbf{V}_{AB} = \mathbf{V}_{ab}$ $= (\sqrt{3}\underline{/30°})\mathbf{V}_{AN}$ $= \sqrt{3}V_p\underline{/30°}$ $\mathbf{V}_{BC} = \mathbf{V}_{bc}$ $= (\sqrt{3}\underline{/30°})\mathbf{V}_{BN}$ $= \sqrt{3}V_p\underline{/-90°}$ $\mathbf{V}_{CA} = \mathbf{V}_{ca}$ $= (\sqrt{3}\underline{/30°})\mathbf{V}_{CN}$ $= \sqrt{3}V_p\underline{/-210°}$	$\mathbf{I}_{aA} = \mathbf{I}_{AN} = \dfrac{\mathbf{V}_{AN}}{\mathbf{Z}_p}$ $\mathbf{I}_{bB} = \mathbf{I}_{BN} = \dfrac{\mathbf{V}_{BN}}{\mathbf{Z}_p}$ $\mathbf{I}_{cC} = \mathbf{I}_{CN} = \dfrac{\mathbf{V}_{CN}}{\mathbf{Z}_p}$	$\mathbf{I}_{aA} = \mathbf{I}_{AN} = \dfrac{\mathbf{V}_{AN}}{\mathbf{Z}_p}$ $\mathbf{I}_{bB} = \mathbf{I}_{BN} = \dfrac{\mathbf{V}_{BN}}{\mathbf{Z}_p}$ $\mathbf{I}_{cC} = \mathbf{I}_{CN} = \dfrac{\mathbf{V}_{CN}}{\mathbf{Z}_p}$	$\sqrt{3}V_L I_L \cos\theta$ where $\cos\theta =$ power factor of the load
Δ	$\mathbf{V}_{AB} = \mathbf{V}_{ab}$ $= \sqrt{3}V_p\underline{/30°}$ $\mathbf{V}_{BC} = \mathbf{V}_{bc}$ $= \sqrt{3}V_p\underline{/-90°}$ $\mathbf{V}_{CA} = \mathbf{V}_{ca}$ $= \sqrt{3}V_p\underline{/-210°}$	$\mathbf{V}_{AB} = \mathbf{V}_{ab}$ $= \sqrt{3}V_p\underline{/30°}$ $\mathbf{V}_{BC} = \mathbf{V}_{bc}$ $= \sqrt{3}V_p\underline{/-90°}$ $\mathbf{V}_{CA} = \mathbf{V}_{ca}$ $= \sqrt{3}V_p\underline{/-210°}$	$\mathbf{I}_{AB} = \dfrac{\mathbf{V}_{AB}}{\mathbf{Z}_p}$ $\mathbf{I}_{BC} = \dfrac{\mathbf{V}_{BC}}{\mathbf{Z}_p}$ $\mathbf{I}_{CA} = \dfrac{\mathbf{V}_{CA}}{\mathbf{Z}_p}$	$\mathbf{I}_{aA} = (\sqrt{3}\underline{/-30°})\dfrac{\mathbf{V}_{AB}}{\mathbf{Z}_p}$ $\mathbf{I}_{bB} = (\sqrt{3}\underline{/-30°})\dfrac{\mathbf{V}_{BC}}{\mathbf{Z}_p}$ $\mathbf{I}_{cC} = (\sqrt{3}\underline{/-30°})\dfrac{\mathbf{V}_{CA}}{\mathbf{Z}_p}$	$\sqrt{3}V_L I_L \cos\theta$ where $\cos\theta =$ power factor of the load

A brief comparison of phase and line voltages as well as phase and line currents is presented in Table 12.1 for both Y- and Δ-connected loads powered by a Y-connected three-phase source.

PRACTICE

12.8 A balanced three-phase three-wire system is terminated with two Δ-connected loads in parallel. Load 1 draws 40 kVA at a lagging PF of 0.8, while load 2 absorbs 24 kW at a leading PF of 0.9. Assume no line resistance, and let $\mathbf{V}_{ab} = 440\underline{/30°}$ V. Find (a) the total power drawn by the loads; (b) the phase current $\mathbf{I}_{AB1}$ for the lagging load; (c) $\mathbf{I}_{AB2}$; (d) $\mathbf{I}_{aA}$.

Ans: 56.0 kW; $30.3\underline{/-6.87°}$ A; $20.2\underline{/55.8°}$ A; $75.3\underline{/-12.46°}$ A.

Δ-Connected Sources

The source may also be connected in a Δ configuration. This is not typical, however, for a slight unbalance in the source phases can lead to large currents circulating in the Δ loop. For example, let us call the three single-phase sources $\mathbf{V}_{ab}$, $\mathbf{V}_{bc}$, and $\mathbf{V}_{cd}$. Before closing the Δ by connecting d to a, let us determine the unbalance by measuring the sum $\mathbf{V}_{ab} + \mathbf{V}_{bc} + \mathbf{V}_{ca}$. Suppose that the amplitude of the result is only 1 percent of the line voltage. The circulating current is thus approximately $\frac{1}{3}$ percent of the line voltage divided by the internal impedance of any source. How large is this impedance

Power-Generating Systems

A rather wide variety of techniques can be used to generate electrical power. For example, direct conversion of solar energy into electricity using photovoltaic (solar cell) technology results in the production of dc power. Despite representing a very environmentally friendly technology, however, photovoltaic-based installations are presently more expensive than other means of producing electricity, and require the use of inverters to convert the dc power into ac. Other technologies, such as wind turbine, geothermal, hydrodynamic, nuclear, and fossil fuel–based generators are much more economical by comparison. In these systems, a shaft is rotated through the action of a *prime mover,* such as wind on a propeller, or water or steam on turbine blades (Fig. 12.20).

Once a prime mover has been harnessed to generate rotational movement of a shaft, there are several means of converting this mechanical energy into electrical energy. One example is a *synchronous generator* (Fig. 12.21). These machines are composed of two main sections: a stationary part, called the *stator,* and a rotating part, termed the *rotor.* DC current is supplied to coils of wire wound about the rotor to generate a magnetic field, which is rotated through the action of the prime mover. A set of three-phase voltages is then induced at a second set of windings around the stator. Synchronous generators get their name from the fact that the frequency of the ac voltage produced is synchronized with the mechanical rotation of the rotor.

The actual demand on a stand-alone generator can vary greatly as various loads are added or removed, such as when air conditioning units kick on, lighting is turned on or off, etc. The voltage output of a generator should ideally be independent of the load, but this is not the case in practice. The voltage $\mathbf{E}_A$ induced in any given stator phase, often referred to as the *internal generated voltage,* has a magnitude given by

$$E_A = K\phi\omega$$

where K is a constant dependent on the way the machine is constructed, ϕ is the magnetic flux produced by the field windings on the rotor (and hence is independent of the load), and ω is the speed of rotation, which depends only on the prime mover and not the load attached to the generator. Thus, *changing the load does not affect the magnitude of* $\mathbf{E}_A$. The internal generated voltage can be related to the phase voltage $\mathbf{V}_\phi$ and the phase current $\mathbf{I}_A$ by

$$\mathbf{E}_A = \mathbf{V}_\phi + jX_S\mathbf{I}_A$$

where X_S is the *synchronous reactance* of the generator.

■ **FIGURE 12.20** Wind-energy harvesting installation at Altamont Pass, California, which consists of over 7000 individual windmills. (© Digital Vision/PunchStock)

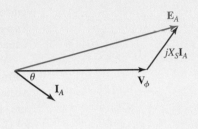

(a)

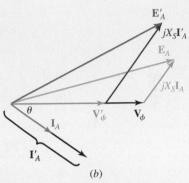

(b)

■ **FIGURE 12.22** Phasor diagrams describing the effect of loading on a stand-alone synchronous generator. (a) Generator connected to a load having a lagging power factor of $\cos\theta$. (b) An additional load is added without changing the power factor. The magnitude of the internal generated voltage E_A remains the same while the output current increases. Consequently, the output voltage V_ϕ is reduced.

■ **FIGURE 12.21** The 24-pole rotor of a synchronous generator as it is being lowered into position.
Photo courtesy Dr. Wade Enright, Te Kura Pukaha Vira O Te Whare Wananga O Waitaha, Aotearoa.

If the load is increased, then a larger current $\mathbf{I}'_A$ will be drawn from the generator. If the power factor is not changed (i.e., the angle between $\mathbf{V}_\phi$ and $\mathbf{I}_A$ remains constant), $\mathbf{V}_\phi$ will be reduced since E_A cannot change.

For example, consider the phasor diagram of Fig. 12.22a, which depicts the voltage-current output of a single phase of a generator connected to a load with a lagging power factor of $\cos\theta$. The internal generated voltage $\mathbf{E}_A$ is also shown. If an additional load is added without changing the power factor, as represented in Fig. 12.22b, the supplied current $\mathbf{I}_A$ increases to $\mathbf{I}'_A$. However, the magnitude of the internal generated voltage, formed by the sum of the phasors $jX_S\mathbf{I}'_A$ and $\mathbf{V}'_\phi$, must remain unchanged. Thus, $E'_A = E_A$, and so the voltage *output* ($\mathbf{V}'_\phi$) of the generator will be slightly reduced, as depicted in Fig. 12.22b.

The *voltage regulation* of a generator is defined as

$$\% \text{ regulation} = \frac{V_{\text{no load}} - V_{\text{full load}}}{V_{\text{full load}}} \times 100$$

Ideally, the regulation should be as close to zero as possible, but this can only be accomplished if the dc current used to control the flux ϕ around the field winding is varied in order to compensate for changing load conditions; this can quickly become rather cumbersome. For this reason, when designing a power generation facility several smaller generators connected in parallel are usually preferable to one large generator capable of handling the peak load. Each generator can be operated at or near full load, so that the voltage output is essentially constant; individual generators can be added or removed from the system depending on the demand.

apt to be? It must depend on the current that the source is expected to deliver with a negligible drop in terminal voltage. If we assume that this maximum current causes a 1 percent drop in the terminal voltage, then *the circulating current is one-third of the maximum current!* This reduces the useful current capacity of the source and also increases the losses in the system.

We should also note that balanced three-phase sources may be transformed from Y to Δ, or vice versa, without affecting the load currents or voltages. The necessary relationships between the line and phase voltages are shown in Fig. 12.13 for the case where $\mathbf{V}_{an}$ has a reference phase angle of 0°. This transformation enables us to use whichever source connection we prefer, and all the load relationships will be correct. Of course, we cannot specify any currents or voltages within the source until we know how it is actually connected. Balanced three-phase loads may be transformed between Y- and Δ-connected configurations using the relation

$$Z_Y = \frac{Z_\Delta}{3}$$

which is probably worth remembering.

12.5 · POWER MEASUREMENT IN THREE-PHASE SYSTEMS

Use of the Wattmeter

Before embarking on a discussion of the specialized techniques used to measure power in three-phase systems, it is to our advantage to briefly consider how a ***wattmeter*** is used in a single-phase circuit.

Power measurement is most often accomplished at frequencies below a few hundred Hz through the use of a wattmeter that contains two separate coils. One of these coils is made of heavy wire, having a very low resistance, and is called the *current coil;* the second coil is composed of a much greater number of turns of fine wire, with relatively high resistance, and is termed the *potential coil,* or *voltage coil.* Additional resistance may also be inserted internally or externally in series with the potential coil. The torque applied to the moving system and the pointer is proportional to the instantaneous product of the currents flowing in the two coils. The mechanical inertia of the moving system, however, causes a deflection that is proportional to the *average* value of this torque.

The wattmeter is used by connecting it into a network in such a way that the current flowing in the current coil is the current flowing into the network and the voltage across the potential coil is the voltage across the two terminals of the network. The current in the potential coil is thus the input voltage divided by the resistance of the potential coil.

It is apparent that the wattmeter has four available terminals, and correct connections must be made to these terminals in order to obtain an upscale reading on the meter. To be specific, let us assume that we are measuring the power absorbed by a passive network. The current coil is inserted in series with one of the two conductors connected to the load, and the potential coil is installed between the two conductors, usually on the "load side" of the current coil. The potential coil terminals are often indicated by arrows, as

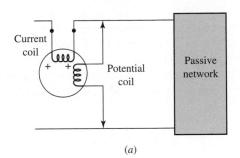

(a)

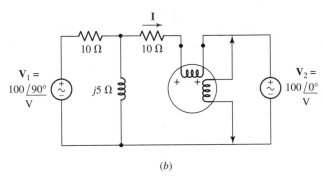

(b)

■ **FIGURE 12.23** (a) A wattmeter connection that will ensure an upscale reading for the power absorbed by the passive network. (b) An example in which the wattmeter is installed to give an upscale indication of the power absorbed by the right source.

shown in Fig. 12.23a. Each coil has two terminals, and the proper relationship between the sense of the current and voltage must be observed. One end of each coil is usually marked (+), and an upscale reading is obtained if a positive current is flowing into the (+) end of the current coil while the (+) terminal of the potential coil is positive with respect to the unmarked end. The wattmeter shown in the network of Fig. 12.23a therefore gives an upscale deflection when the network to the right is absorbing power. A reversal of either coil, but not both, will cause the meter to try to deflect downscale; a reversal of both coils will never affect the reading.

As an example of the use of such a wattmeter in measuring average power, let us consider the circuit shown in Fig. 12.23b. The connection of the wattmeter is such that an upscale reading corresponds to a positive absorbed power for the network to the right of the meter, that is, the right source. The power absorbed by this source is given by

$$P = |\mathbf{V}_2| \, |\mathbf{I}| \cos(\text{ang } \mathbf{V}_2 - \text{ang } \mathbf{I})$$

Using superposition or mesh analysis, we find the current is

$$\mathbf{I} = 11.18\underline{/153.4°} \text{ A}$$

and thus the absorbed power is

$$P = (100)(11.18) \cos(0° - 153.4°) = -1000 \text{ W}$$

The pointer therefore rests against the downscale stop. In practice, the potential coil can be reversed more quickly than the current coil, and this reversal provides an upscale reading of 1000 W.

PRACTICE

12.9 Determine the wattmeter reading in Fig. 12.24, state whether or not the potential coil had to be reversed in order to obtain an upscale reading, and identify the device or devices absorbing or generating this power. The (+) terminal of the wattmeter is connected to: (a) x; (b) y; (c) z.

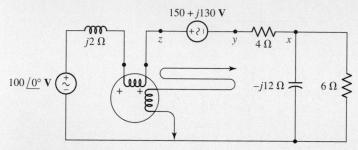

■ **FIGURE 12.24**

Ans: 1200 W, as is, $P_{6\Omega}$ (absorbed); 2200 W, as is, $P_{4\Omega} + P_{6\Omega}$ (absorbed); 500 W, reversed, absorbed by 100 V.

The Wattmeter in a Three-Phase System

At first glance, measurement of the power drawn by a three-phase load seems to be a simple problem. We need place only one wattmeter in each of the three phases and add the results. For example, the proper connections for a Y-connected load are shown in Fig. 12.25a. Each wattmeter has its current coil inserted in one phase of the load and its potential coil connected between the line side of that load and the neutral. In a similar way, three

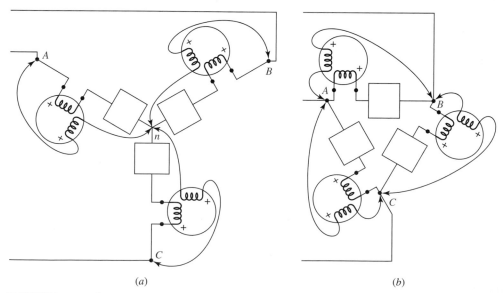

(a)   (b)

■ **FIGURE 12.25** Three wattmeters are connected in such a way that each reads the power taken by one phase of a three-phase load, and the sum of the readings is the total power. (a) A Y-connected load. (b) A Δ-connected load. Neither the loads nor the source need be balanced.

wattmeters may be connected as shown in Fig. 12.25*b* to measure the total power taken by a Δ-connected load. The methods are theoretically correct, but they may be useless in practice because the neutral of the Y is not always accessible and the phases of the Δ are not available. A three-phase rotating machine, for example, has only three accessible terminals, those which we have been calling *A*, *B*, and *C*.

Clearly, we have a need for a method of measuring the total power drawn by a three-phase load having only three accessible terminals; measurements may be made on the "line" side of these terminals, but not on the "load" side. Such a method is available, and is capable of measuring the power taken by an *unbalanced* load from an *unbalanced* source. Let us connect three wattmeters in such a way that each has its current coil in one line and its voltage coil between that line and some common point *x*, as shown in Fig. 12.26. Although a system with a Y-connected load is illustrated, the arguments we present are equally valid for a Δ-connected load. The point *x* may be some unspecified point in the three-phase system, or it may be merely a point in space at which the three potential coils have a common node. The average power indicated by wattmeter *A* must be

$$P_A = \frac{1}{T} \int_0^T v_{Ax} i_{aA} \, dt$$

where *T* is the period of all the source voltages. The readings of the other two wattmeters are given by similar expressions, and the total average power drawn by the load is therefore

$$P = P_A + P_B + P_C = \frac{1}{T} \int_0^T (v_{Ax} i_{aA} + v_{Bx} i_{bB} + v_{Cx} i_{cC}) \, dt$$

Each of the three voltages in the preceding expression may be written in terms of a phase voltage and the voltage between point *x* and the neutral,

$$v_{Ax} = v_{AN} + v_{Nx}$$
$$v_{Bx} = v_{BN} + v_{Nx}$$
$$v_{Cx} = v_{CN} + v_{Nx}$$

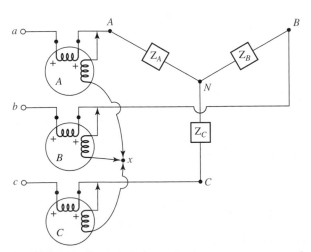

■ **FIGURE 12.26** A method of connecting three wattmeters to measure the total power taken by a three-phase load. Only the three terminals of the load are accessible.

and, therefore,

$$P = \frac{1}{T} \int_0^T (v_{AN} i_{aA} + v_{BN} i_{bB} + v_{CN} i_{cC}) \, dt$$
$$+ \frac{1}{T} \int_0^T v_{Nx} (i_{aA} + i_{bB} + i_{cC}) \, dt$$

However, the entire three-phase load may be considered to be a supernode, and Kirchhoff's current law requires

$$i_{aA} + i_{bB} + i_{cC} = 0$$

Thus

$$P = \frac{1}{T} \int_0^T (v_{AN} i_{aA} + v_{BN} i_{bB} + v_{CN} i_{cC}) \, dt$$

Reference to the circuit diagram shows that this sum is indeed the sum of the average powers taken by each phase of the load, and the sum of the readings of the three wattmeters therefore represents the total average power drawn by the entire load!

Let us illustrate this procedure with a numerical example before we discover that one of these three wattmeters is really superfluous. We will assume a balanced source,

$$\mathbf{V}_{ab} = 100\underline{/0°} \quad \text{V}$$
$$\mathbf{V}_{bc} = 100\underline{/-120°} \quad \text{V}$$
$$\mathbf{V}_{ca} = 100\underline{/-240°} \quad \text{V}$$

or

$$\mathbf{V}_{an} = \frac{100}{\sqrt{3}}\underline{/-30°} \quad \text{V}$$
$$\mathbf{V}_{bn} = \frac{100}{\sqrt{3}}\underline{/-150°} \quad \text{V}$$
$$\mathbf{V}_{cn} = \frac{100}{\sqrt{3}}\underline{/-270°} \quad \text{V}$$

and an unbalanced load,

$$\mathbf{Z}_A = -j10 \ \Omega$$
$$\mathbf{Z}_B = j10 \ \Omega$$
$$\mathbf{Z}_C = 10 \ \Omega$$

Let us assume ideal wattmeters, connected as illustrated in Fig. 12.26, with point x located on the neutral of the source n. The three line currents may be obtained by mesh analysis,

$$\mathbf{I}_{aA} = 19.32\underline{/15°} \quad \text{A}$$
$$\mathbf{I}_{bB} = 19.32\underline{/165°} \quad \text{A}$$
$$\mathbf{I}_{cC} = 10\underline{/-90°} \quad \text{A}$$

The voltage between the neutrals is

$$\mathbf{V}_{nN} = \mathbf{V}_{nb} + \mathbf{V}_{BN} = \mathbf{V}_{nb} + \mathbf{I}_{bB}(j10) = 157.7\underline{/-90°}$$

The average power indicated by each wattmeter may be calculated,

$$P_A = V_p I_{aA} \cos(\text{ang} \mathbf{V}_{an} - \text{ang}\, \mathbf{I}_{aA})$$

$$= \frac{100}{\sqrt{3}} 19.32 \cos(-30° - 15°) = 788.7 \text{ W}$$

$$P_B = \frac{100}{\sqrt{3}} 19.32 \cos(-150° - 165°) = 788.7 \text{ W}$$

$$P_C = \frac{100}{\sqrt{3}} 10 \cos(-270° + 90°) = -577.4 \text{ W}$$

or a total power of 1 kW. Since an rms current of 10 A flows through the *resistive* load, the total power drawn by the load is

$$P = 10^2(10) = 1 \text{ kW}$$

and the two methods agree.

The Two-Wattmeter Method

We have proved that point x, the common connection of the three potential coils, may be located any place we wish without affecting the algebraic sum of the three wattmeter readings. Let us now consider the effect of placing point x, this common connection of the three wattmeters, directly on one of the lines. If, for example, one end of each potential coil is returned to B, then there is no voltage across the potential coil of wattmeter B and *this meter must read zero*. It may therefore be removed, and the algebraic sum of the remaining two wattmeter readings is still the total power drawn by the load. When the location of x is selected in this way, we describe the method of power measurement as the **two-wattmeter** method. The sum of the readings indicates the total power, regardless of (1) load unbalance, (2) source unbalance, (3) differences in the two wattmeters, and (4) the waveform of the periodic source. The only assumption we have made is that wattmeter corrections are sufficiently small so that we can ignore them. In Fig. 12.26, for example, the current coil of each meter has passing through it the line current drawn by the load plus the current taken by the potential coil. Since the latter current is usually quite small, its effect may be estimated from a knowledge of the resistance of the potential coil and voltage across it. These two quantities enable a close estimate to be made of the power dissipated in the potential coil.

In the numerical example described previously, let us now assume that two wattmeters are used, one with current coil in line A and potential coil between lines A and B, the other with current coil in line C and potential coil between C and B. The first meter reads

$$P_1 = V_{AB} I_{aA} \cos(\text{ang } V_{AB} - \text{ang } I_{aA})$$

$$= 100(19.32) \cos(0° - 15°)$$

$$= 1866 \text{ W}$$

and the second

$$P_2 = V_{CB} I_{cC} \cos(\text{ang } V_{CB} - \text{ang } I_{cC})$$

$$= 100(10) \cos(60° + 90°)$$

$$= -866 \text{ W}$$

Note that the reading of one of the wattmeters is negative. Our previous discussion on the basic use of a wattmeter indicates that an upscale reading on that meter can only be obtained after either the potential coil or the current coil is reversed.

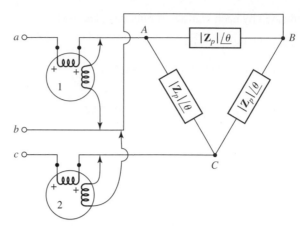

■ **FIGURE 12.27** Two wattmeters connected to read the total power drawn by a balanced three-phase load.

and, therefore,

$$P = P_1 + P_2 = 1866 - 866 = 1000 \text{ W}$$

as we expect from recent experience with the circuit.

In the case of a balanced load, the two-wattmeter method enables the PF angle to be determined, as well as the total power drawn by the load. Let us assume a load impedance with a phase angle θ; either a Y or Δ connection may be used and we will assume the Δ connection shown in Fig. 12.27. The construction of a standard phasor diagram, such as that of Fig. 12.19, enables us to determine the proper phase angle between the several line voltages and line currents. We therefore determine the readings

$$P_1 = |\mathbf{V}_{AB}| \, |\mathbf{I}_{aA}| \cos(\text{ang } \mathbf{V}_{AB} - \text{ang } \mathbf{I}_{aA})$$
$$= V_L I_L \cos(30° + \theta)$$

and

$$P_2 = |\mathbf{V}_{CB}| \, |\mathbf{I}_{cC}| \cos(\text{ang } \mathbf{V}_{CB} - \text{ang } \mathbf{I}_{cC})$$
$$= V_L I_L \cos(30° - \theta)$$

The ratio of the two readings is

$$\frac{P_1}{P_2} = \frac{\cos(30° + \theta)}{\cos(30° - \theta)} \tag{5}$$

If we expand the cosine terms, this equation is easily solved for $\tan \theta$,

$$\tan \theta = \sqrt{3} \frac{P_2 - P_1}{P_2 + P_1} \tag{6}$$

Thus, equal wattmeter readings indicate a unity PF load, equal and opposite readings indicate a purely reactive load, a reading of P_2 which is (algebraically) greater than P_1 indicates an inductive impedance, and a reading of P_2 which is less than P_1 signifies a capacitive load. How can we tell which wattmeter reads P_1 and which reads P_2? It is true that P_1 is in line A, and P_2 is in line C, and our positive phase-sequence system forces V_{an} to lag V_{cn}. This is enough information to differentiate between two wattmeters, but it is confusing to apply in practice. Even if we were unable to distinguish between the two, we know the magnitude of the phase angle, but not its sign. This is often sufficient information; if the load is an induction motor, the angle must be positive and we do not need to make any tests to determine which reading is which. If no previous knowledge of the load is assumed, then there are several methods of resolving the ambiguity. Perhaps the simplest method is that which involves adding a high-impedance reactive load, say a three-phase capacitor, across the unknown load. The load must become more capacitive. Thus, if the magnitude of $\tan \theta$ (or the magnitude of θ) decreases, then the load was inductive, whereas an increase in the magnitude of $\tan \theta$ signifies an original capacitive impedance.

EXAMPLE **12.7**

The balanced load in Fig. 12.28 is fed by a balanced three-phase system having $V_{ab} = 230\underline{/0°}$ V rms and positive phase sequence. Find the reading of each wattmeter and the total power drawn by the load.

The potential coil of wattmeter #1 is connected to measure the voltage $\mathbf{V}_{ac}$, and its current coil is measuring the phase current $\mathbf{I}_{aA}$. Since we know to use the positive phase sequence, the line voltages are

$$\mathbf{V}_{ab} = 230\underline{/0°} \quad \text{V}$$
$$\mathbf{V}_{bc} = 230\underline{/-120°} \quad \text{V}$$
$$\mathbf{V}_{ca} = 230\underline{/120°} \quad \text{V}$$

Note that $\mathbf{V}_{ac} = -\mathbf{V}_{ca} = 230\underline{/-60°}$ V.

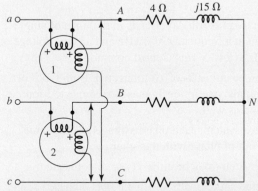

■ **FIGURE 12.28** A balanced three-phase system connected to a balanced three-phase load, the power of which is being measured using the two-wattmeter technique.

The phase current $\mathbf{I}_{aA}$ is given by the phase voltage $\mathbf{V}_{an}$ divided by the phase impedance $4 + j15 \ \Omega$,

$$\mathbf{I}_{aA} = \frac{\mathbf{V}_{an}}{4 + j15} = \frac{(230/\sqrt{3})\underline{/-30°}}{4 + j15} \text{ A}$$
$$= 8.554\underline{/-105.1°} \text{ A}$$

(Continued on next page)

We may now compute the power measured by wattmeter #1 as

$$P_1 = |\mathbf{V}_{ac}| \, |\mathbf{I}_{aA}| \cos(\text{ang } \mathbf{V}_{ac} - \text{ang } \mathbf{I}_{aA})$$
$$= (230)(8.554) \cos(-60° + 105.1°) \text{ W}$$
$$= 1389 \text{ W}$$

In a similar fashion, we determine that

$$P_2 = |\mathbf{V}_{bc}| \, |\mathbf{I}_{bB}| \cos(\text{ang } \mathbf{V}_{bc} - \text{ang } \mathbf{I}_{bB})$$
$$= (230)(8.554) \cos(-120° - 134.9°) \text{ W}$$
$$= -512.5 \text{ W}$$

Thus, the total average power absorbed by the load is

$$P = P_1 + P_2 = 876.5 \text{ W}$$

Since this measurement would result in the meter pegged at downscale, one of the coils would need to be reversed in order to take the reading.

PRACTICE

12.10 For the circuit of Fig. 12.26, let the loads be $\mathbf{Z}_A = 25\underline{/60°}$ Ω, $\mathbf{Z}_B = 50\underline{/-60°}$ Ω, $\mathbf{Z}_C = 50\underline{/60°}$ Ω, $\mathbf{V}_{AB} = 600\underline{/0°}$ V rms with (+) phase sequence, and locate point x at C. Find (a) P_A; (b) P_B; (c) P_C.

Ans: 0; 7200 W; 0.

SUMMARY AND REVIEW

❏ The majority of electricity production is in the form of three-phase power.

❏ Most residential electricity in North America is in the form of single-phase alternating current at a frequency of 60 Hz and an rms voltage of 115 V. Elsewhere, 50 Hz at 240 V rms is most common.

❏ Three-phase sources can be either Y- or Δ-connected. Both types of sources have three terminals, one for each phase; Y-connected sources have a neutral connection as well.

❏ In a balanced three-phase system, each phase voltage has the same magnitude, but is 120° out of phase with the other two.

❏ Loads in a three-phase system may be either Y- or Δ-connected.

❏ In a balanced Y-connected source with positive ("*abc*") phase sequence, the line voltages are

$$\mathbf{V}_{ab} = \sqrt{3}V_p\underline{/30°} \qquad \mathbf{V}_{bc} = \sqrt{3}V_p\underline{/-90°}$$
$$\mathbf{V}_{ca} = \sqrt{3}V_p\underline{/-210°}$$

where the phase voltages are

$$\mathbf{V}_{an} = V_p\underline{/0°} \qquad \mathbf{V}_{bn} = V_p\underline{/-120°} \qquad \mathbf{V}_{cn} = V_p\underline{/-240°}$$

❏ In a system with a Y-connected load, the line currents are equal to the phase currents.

❑ In a Δ-connected load, the line voltages are equal to the phase voltages.

❑ In a balanced system with positive phase sequence and a balanced Δ-connected load, the line currents are

$$\mathbf{I}_a = \mathbf{I}_{AB}\sqrt{3}\underline{/-30^\circ} \qquad \mathbf{I}_b = \mathbf{I}_{BC}\sqrt{3}\underline{/-150^\circ} \qquad \mathbf{I}_c = \mathbf{I}_{CA}\sqrt{3}\underline{/+90^\circ}$$

where the phase currents are

$$\mathbf{I}_{AB} = \frac{\mathbf{V}_{AB}}{\mathbf{Z}_\Delta} = \frac{\mathbf{V}_{ab}}{\mathbf{Z}_\Delta} \qquad \mathbf{I}_{BC} = \frac{\mathbf{V}_{BC}}{\mathbf{Z}_\Delta} = \frac{\mathbf{V}_{bc}}{\mathbf{Z}_\Delta} \qquad \mathbf{I}_{CA} = \frac{\mathbf{V}_{CA}}{\mathbf{Z}_\Delta} = \frac{\mathbf{V}_{ca}}{\mathbf{Z}_\Delta}$$

❑ Most power calculations are performed on a per-phase basis, assuming a balanced system; otherwise, nodal/mesh analysis is always a valid approach.

❑ The power in a three-phase system (balanced or unbalanced) can be measured with only two wattmeters.

❑ The instantaneous power in any balanced three-phase system is constant.

READING FURTHER

A good overview of ac power concepts can be found in Chap. 2 of:

B.M. Weedy, *Electric Power Systems,* 3rd ed. Chichester, England: Wiley, 1984.

A comprehensive book on generation of electrical power from wind is:

T. Burton, D. Sharpe, N. Jenkins, and E. Bossanyi, *Wind Energy Handbook.* Chichester, England: Wiley, 2001.

EXERCISES

12.1 Polyphase Systems

1. A circuit built in the laboratory is measured with a voltmeter, with the results $V_{be} = 0.7$ V and $V_{ce} = 10$ V. Find V_{bc}, V_{eb}, and V_{cb}.

2. A common-source field effect transistor amplifier is simulated using SPICE. (a) If $V_{gs} = -1$ V and $V_{ds} = 5$ V, find V_{gd}. (b) If $V_{ds} = 4$ V and $V_{gd} = 2.5$ V, find V_{sg}.

3. A six-phase power system is developed as part of a high-current dc magnet power supply. Write the phase voltages for (a) positive phase sequence; (b) negative phase sequence.

4. If $\mathbf{V}_{xy} = 110\underline{/20^\circ}$ V, $\mathbf{V}_{xz} = 160\underline{/-50^\circ}$ V, and $\mathbf{V}_{ay} = 80\underline{/130^\circ}$ V, find: (a) $\mathbf{V}_{yz}$; (b) $\mathbf{V}_{az}$; (c) $\mathbf{V}_{zx}/\mathbf{V}_{xy}$.

5. For a particular circuit, it is known that $\mathbf{V}_{12} = 100\underline{/0^\circ}$, $\mathbf{V}_{45} = 60\underline{/75^\circ}$, $\mathbf{V}_{42} = 80\underline{/120^\circ}$, and $\mathbf{V}_{35} = -j120$, all in volts. Find (a) $\mathbf{V}_{25}$; (b) $\mathbf{V}_{13}$.

6. In a particular ac system, $\mathbf{V}_{12} = 9\underline{/87^\circ}$ V and $\mathbf{V}_{23} = 8\underline{/45^\circ}$ V. Calculate (a) $\mathbf{V}_{21}$; (b) $\mathbf{V}_{32}$; (c) $\mathbf{V}_{12} - \mathbf{V}_{32}$.

7. A power system is known to have $\mathbf{V}_{an} = 400\underline{/-45^\circ}$ V and $\mathbf{V}_{bn} = 400\underline{/75^\circ}$ V. (a) Draw a phasor diagram, including $\mathbf{V}_{cn}$. (b) Is the phase sequence of the system positive or negative? Explain.

8. Given the ac currents $\mathbf{I}_{12} = 33\underline{/12^\circ}$ A and $\mathbf{I}_{23} = 40\underline{/12^\circ}$ A, what is $\mathbf{I}_{31}$?

9. If an ac circuit is known to have currents $\mathbf{I}_{12} = 5\underline{/55^\circ}$ A and $\mathbf{I}_{23} = 4\underline{/33^\circ}$ A, what is $\mathbf{I}_{31}$ if the operating frequency is 50 Hz?

12.2 Single-Phase Three-Wire Systems

10. The 230/460 V rms 60 Hz three-wire system shown in Fig. 12.29 supplies power to three loads: load AN draws a complex power of $10\underline{/40^\circ}$ kVA, load NB uses $8\underline{/10^\circ}$ kVA, and load AB requires $4\underline{/-80^\circ}$ kVA. Find the two line currents and the neutral current.

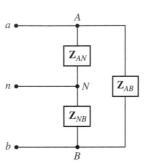

■ **FIGURE 12.29**

11. A balanced three-wire single-phase system has loads $\mathbf{Z}_{AN} = \mathbf{Z}_{NB} = 10\ \Omega$, and a load $\mathbf{Z}_{AB} = 16 + j12\ \Omega$. The three lines may be assumed to be resistance-less. Let $\mathbf{V}_{an} = \mathbf{V}_{nb} = 120\underline{/0^\circ}$ V. (a) Find I_{aA} and I_{nN}. (b) The system is unbalanced by connecting another $10\ \Omega$ resistance in parallel with $\mathbf{Z}_{AN}$. Find I_{aA}, I_{bB}, and I_{nN}.

12. An inefficient three-wire single-phase system has source voltages of $\mathbf{V}_{an} = \mathbf{V}_{nb} = 720\underline{/0^\circ}$ V, line resistances $R_{aA} = R_{bB} = 1\ \Omega$ with $R_{nN} = 10\ \Omega$, and loads $\mathbf{Z}_{AN} = 10 + j3\ \Omega$, $\mathbf{Z}_{NB} = 8 + j2\ \Omega$, and $\mathbf{Z}_{AB} = 18 + j0\ \Omega$. Find (a) $\mathbf{I}_{aA}$; (b) $\mathbf{I}_{nN}$; (c) $P_{\text{wiring,total}}$; (d) $P_{\text{gen.total}}$.

13. In the balanced three-wire single-phase system of Fig. 12.30, let $V_{AN} = 220$ V at 60 Hz. (a) What size should C be to provide a unity-power-factor load? (b) How many kVA does C handle?

14. The balanced three-wire single-phase system of Fig. 12.29 has source voltages $\mathbf{V}_{an} = \mathbf{V}_{nb} = 200\underline{/0^\circ}$ V, zero line and neutral resistance, and loads $\mathbf{Z}_{AN} = \mathbf{Z}_{NB} = 12 + j3\ \Omega$. Find $\mathbf{Z}_{AB}$ so that: (a) $X_{AB} = 0$ and $I_{aA} = 30$ A rms; (b) $R_{AB} = 0$ and ang $\mathbf{I}_{aA} = 0^\circ$.

12.3 Three-Phase Y-Y Connection

15. Figure 12.31 shows a balanced three-phase three-wire system with positive phase sequence. Let $\mathbf{V}_{BC} = 120\underline{/60^\circ}$ V and $R_w = 0.6\ \Omega$. If the total load (including wire resistance) draws 5 kVA at PF = 0.8 lagging, find (a) the total power lost in the line resistance, and (b) $\mathbf{V}_{an}$.

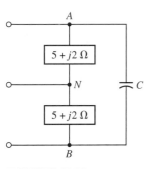

■ **FIGURE 12.30**

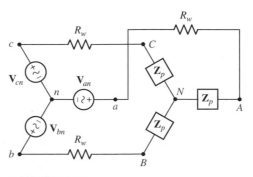

■ **FIGURE 12.31**

16. Let $\mathbf{V}_{an} = 2300\underline{/0^\circ}$ V in the balanced system shown in Fig. 12.31, and set $R_w = 2\ \Omega$. Assume positive phase sequence with the source supplying a total complex power of $\mathbf{S} = 100 + j30$ kVA. Find (a) $\mathbf{I}_{aA}$; (b) $\mathbf{V}_{AN}$; (c) $\mathbf{Z}_p$; (d) the transmission efficiency.

17. In the balanced three-phase system of Fig. 12.31, let $\mathbf{Z}_p = 12 + j5\ \Omega$ and $\mathbf{I}_{bB} = 20\underline{/0^\circ}$ A with (+) phase sequence. If the source is operating with a power factor of 0.935, find (a) R_w; (b) $\mathbf{V}_{bn}$; (c) $\mathbf{V}_{AB}$; (d) the complex power supplied by the source.

18. A three-phase three-wire system has a balanced Y-connected load with a 75 Ω resistance, 125 mH inductance, and 55 μF capacitance in series from each line to the neutral point. Assuming positive phase sequence with $V_p = 125$ V at 60 Hz, find the line current, the total power taken by the load, and the power factor of the load.

19. A lossless neutral conductor is installed between nodes n and N in the three-phase system shown in Fig. 12.31. Assume a balanced system with (+) phase sequence, but connect unbalanced loads: $\mathbf{Z}_{AN} = 8 + j6\ \Omega$, $\mathbf{Z}_{BN} = 12 - j16\ \Omega$, and $\mathbf{Z}_{CN} = 5\ \Omega$. If $\mathbf{V}_{an} = 120\underline{/0^\circ}$ V rms and $R_w = 0.5\ \Omega$, find $\mathbf{I}_{nN}$.

20. For the balanced circuit of Fig. 12.31, $\mathbf{V}_{an} = 40\underline{/0^\circ}$ V (positive phase sequence). Determine the line current and the total power delivered to the load if the phase impedance $\mathbf{Z}_p = 5 + j10\ \Omega$ and $R_w = (a)\ 0\ \Omega$; $(b)\ 3\ \Omega$.

21. The phase impedance $\mathbf{Z}_p$ in the system shown in Fig. 12.31 consists of an impedance of $75\underline{/25^\circ}\ \Omega$ in parallel with a 25 μF capacitance. Let $\mathbf{V}_{an} = 240\underline{/0^\circ}$ V at 60 Hz, and $R_w = 2\ \Omega$. Find (a) $\mathbf{I}_{aA}$; (b) P_{wires}; (c) P_{load}; (d) the source power factor.

22. Each load in the circuit of Fig. 12.31 is composed of an inductive impedance $100\underline{/28^\circ}\ \Omega$ in parallel with a 500 nF capacitor. The resistance labeled $R_w = 1\ \Omega$. Using positive phase sequence with $\mathbf{V}_{ab} = 240\underline{/0^\circ}$ V at $f = 50$ Hz, determine the rms line current, the total power delivered to the load, and the power lost in the wiring. Verify your answers with an appropriate PSpice simulation.

23. The balanced three-phase system shown in Fig. 12.31 has $R_w = 0$ and $\mathbf{Z}_p = 10 + j5\ \Omega$ per phase. (a) At what power factor is the source operating? (b) Assuming $f = 60$ Hz, what size capacitor must be placed in parallel with each phase impedance to raise the PF to 0.93 lagging? (c) How much reactive power is drawn by each capacitor if the line voltage at the load is 440 V?

24. Each load in the circuit of Fig. 12.31 is composed of a 1.5 H inductor in parallel with a 100 μF capacitor and a 1 kΩ resistor. The resistance labeled $R_w = 0\ \Omega$. Using positive phase sequence with $\mathbf{V}_{ab} = 115\underline{/0^\circ}$ V at $f = 60$ Hz, determine the rms line current and the total power delivered to the load. Verify your answers with an appropriate PSpice simulation.

12.4 The Delta (Δ) Connection

25. Figure 12.32 shows a balanced three-wire three-phase circuit. Let $R_w = 0$ and $\mathbf{V}_{an} = 200\underline{/60^\circ}$ V. Each phase of the load absorbs a complex power, $\mathbf{S}_p = 2 - j1$ kVA. If (+) phase sequence is assumed, find: (a) $\mathbf{V}_{bc}$; (b) $\mathbf{Z}_p$; (c) $\mathbf{I}_{aA}$.

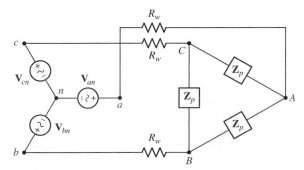

■ **FIGURE 12.32**

26. The balanced Δ load of Fig. 12.32 requires 15 kVA at a lagging PF of 0.8. Assume $(+)$ phase sequence with $\mathbf{V}_{BC} = 180\underline{/30°}$ V. If $R_w = 0.75$ Ω, find (a) $\mathbf{V}_{bc}$; (b) the total complex power generated by the source.

27. The load in the balanced system of Fig. 12.32 draws a total complex power of $3 + j1.8$ kVA, while the source generates $3.45 + j1.8$ kVA. If $R_w = 5$ Ω, find (a) I_{aA}; (b) I_{AB}; (c) V_{an}.

28. The Δ-connected load in the circuit of Fig. 12.32 draws 1800 W at a lagging PF of $\sqrt{2}/2$, and 240 W is lost in the wire resistance, $R_w = 2.3$ Ω. Find the rms phase voltage of the source and the rms phase current of the load.

29. The source in Fig. 12.33 is balanced and exhibits $(+)$ phase sequence. Find (a) $\mathbf{I}_{aA}$; (b) $\mathbf{I}_{bB}$; (c) $\mathbf{I}_{cC}$; (d) the total complex power supplied by the source.

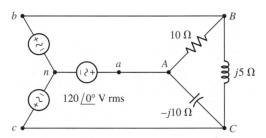

■ **FIGURE 12.33**

30. For the circuit depicted in Fig. 12.32, $\mathbf{V}_{AB} = 200\underline{/0°}$ V rms with $(+)$ phase sequence, $R_w = 200$ mΩ, and the phase impedance $\mathbf{Z}_p$ consists of a 10 Ω resistance in parallel with an inductive reactance of 30 Ω. Determine the total power supplied by the source, the power factor at which it operates, and the transmission efficiency.

31. The balanced three-phase Y-connected source in Fig. 12.32 has $\mathbf{V}_{an} = 140\underline{/0°}$ V rms with $(+)$ phase sequence. Let $R_w = 0$. The balanced three-phase load draws 15 kW and $+9$ kVAR. Find (a) $\mathbf{V}_{AB}$; (b) $\mathbf{I}_{AB}$, (c) $\mathbf{I}_{aA}$.

32. For the three-phase system of Fig. 12.34, assume a balanced source with a positive phase sequence. If the operating frequency is 60 Hz, find the magnitude of (a) $\mathbf{V}_{AN}$; (b) $\mathbf{V}_{BN}$; (c) $\mathbf{V}_{CN}$. Verify your answers with an appropriate PSpice simulation.

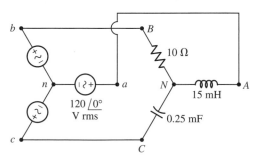

■ **FIGURE 12.34**

33. (a) Insert 1 Ω of resistance in each of the lines of Fig. 12.33 and work Exer. 29 again. (b) Verify your solution with an appropriate PSpice simulation.

34. A balanced three-phase system having line voltage 240 V rms contains a Δ-connected load of $12 + j$ kΩ per phase and also a Y-connected load of $5 + j3$ kΩ per phase. Find the line current, the power taken by the combined load, and the power factor of the load.

12.5 Power Measurement in Three-Phase Systems

35. Determine the wattmeter reading (stating whether or not the leads had to be reversed to obtain it) in the circuit of Fig. 12.35 if terminals A and B, respectively, are connected to (a) x and y; (b) x and z; (c) y and z.

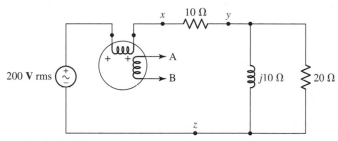

■ **FIGURE 12.35**

 36. A wattmeter is connected into the circuit of Fig. 12.36 so that $\mathbf{I}_1$ enters the (+) terminal of the current coil, while $\mathbf{V}_2$ is the voltage across the potential coil. Find the wattmeter reading, and verify your solution with an appropriate PSpice simulation.

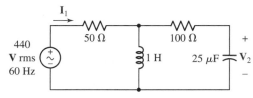

■ **FIGURE 12.36**

37. Find the reading of the wattmeter connected in the circuit of Fig. 12.37.

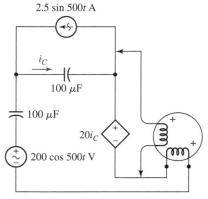

■ **FIGURE 12.37**

38. (a) Find both wattmeter readings in Fig. 12.38 if $\mathbf{V}_A = 100\underline{/0°}$ V rms, $\mathbf{V}_B = 50\underline{/90°}$ V rms, $\mathbf{Z}_A = 10 - j10\ \Omega$, $\mathbf{Z}_B = 8 + j6\ \Omega$, and $\mathbf{Z}_C = 30 + j10\ \Omega$. (b) Is the sum of these readings equal to the total power taken by the three loads? Verify your answer with an appropriate PSpice simulation.

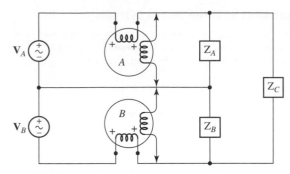

■ **FIGURE 12.38**

39. Circuit values for Fig. 12.39 are $\mathbf{V}_{ab} = 200\underline{/0°}$, $\mathbf{V}_{bc} = 200\underline{/120°}$, $\mathbf{V}_{ca} = 200\underline{/240°}$ V rms, $\mathbf{Z}_4 = \mathbf{Z}_5 = \mathbf{Z}_6 = 25\underline{/30°}$ Ω, $\mathbf{Z}_1 = \mathbf{Z}_2 = \mathbf{Z}_3 = 50\underline{/-60°}$ Ω. Find the reading for each wattmeter.

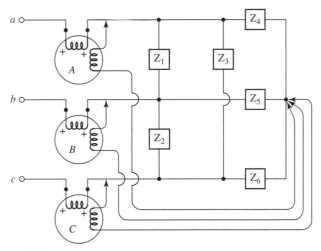

■ **FIGURE 12.39**

40. For the circuit of Fig. 12.32, show how the power absorbed by the load could be measured using (a) three wattmeters; (b) the two-wattmeter method.

41. For the circuit of Fig. 12.31, show how the power absorbed by the load could be measured using (a) three wattmeters; (b) the two-wattmeter method.

Magnetically Coupled Circuits

KEY CONCEPTS

Mutual Inductance

Self-Inductance

The Dot Convention

Reflected Impedance

T and Π Equivalent
Networks

The Ideal Transformer

Turns Ratio of an Ideal
Transformer

Impedance Matching

Voltage Level Adjustment

PSpice Analysis of Circuits
with Transformers

INTRODUCTION

Whenever current flows through a conductor, whether as ac or dc, a magnetic field is generated about that conductor. In the context of circuits, we often refer to the *magnetic flux* through a loop of wire, which is simply the average normal component of the magnetic field density emanating from the loop multiplied by the surface area of the loop. When a time-varying magnetic field generated by one loop penetrates a second loop, a voltage is induced between the ends of the second wire. In order to distinguish this phenomenon from the "inductance" we defined earlier, more properly termed "self-inductance," we will define a new term, *mutual inductance*.

There is no such device as a "mutual inductor," but the principle forms the basis for an extremely important device—the *transformer*. A transformer consists of two coils of wire separated by a small distance, and is commonly used to convert ac voltages to higher or lower values depending on the application. Every electrical appliance that requires dc current to operate but plugs into an ac wall outlet makes use of a transformer to adjust voltage levels prior to *rectification,* a function typically performed by diodes and described in every introductory electronics text.

13.1 · MUTUAL INDUCTANCE

When we defined inductance in Chap. 7, we did so by specifying the relationship between the terminal voltage and current,

$$v(t) = L \frac{di(t)}{dt}$$

where the passive sign convention is assumed. The physical basis for such a current-voltage characteristic rests upon two things:

1. The production of a ***magnetic flux*** by a current, the flux being proportional to the current in linear inductors.
2. The production of a voltage by the time-varying magnetic field, the voltage being proportional to the time rate of change of the magnetic field or the magnetic flux.

Coefficient of Mutual Inductance

Mutual inductance results from a slight extension of this same argument. A current flowing in one coil establishes a magnetic flux about that coil and also about a second coil nearby. The time-varying flux surrounding the second coil produces a voltage across the terminals of the second coil; this voltage is proportional to the time rate of change of the current flowing through the first coil. Figure 13.1a shows a simple model of two coils L_1 and L_2, sufficiently close together that the flux produced by a current $i_1(t)$ flowing through L_1 establishes an open-circuit voltage $v_2(t)$ across the terminals of L_2. Without considering the proper algebraic sign for the relationship at this point, we define the *coefficient of mutual inductance,* or simply ***mutual inductance, M_{21},***

$$v_2(t) = M_{21} \frac{di_1(t)}{dt} \qquad [1]$$

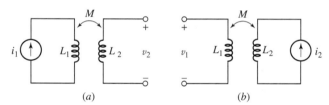

■ **FIGURE 13.1** (a) A current i_1 through L_1 produces an open-circuit voltage v_2 across L_2. (b) A current i_2 through L_2 produces an open-circuit voltage v_1 across L_1.

The order of the subscripts on M_{21} indicates that a voltage response is produced at L_2 by a current source at L_1. If the system is reversed, as indicated in Fig. 13.1b, then we have

$$v_1(t) = M_{12} \frac{di_2(t)}{dt} \qquad [2]$$

Two coefficients of mutual inductance are not necessary, however; we will use energy relationships a little later to prove that M_{12} and M_{21} are equal. Thus, $M_{12} = M_{21} = M$. The existence of mutual coupling between two coils is indicated by a double-headed arrow, as shown in Fig. 13.1a and b.

Mutual inductance is measured in henrys and, like resistance, inductance, and capacitance, is a positive quantity.[1] The voltage $M \, di/dt$, however, may appear as either a positive or a negative quantity depending on whether the current is increasing or decreasing at a particular instant in time.

---

(1) Mutual inductance is not universally assumed to be positive. It is particularly convenient to allow it to "carry its own sign" when three or more coils are involved and each coil interacts with each other coil. We will restrict our attention to the more important simple case of two coils.

Dot Convention

The inductor is a two-terminal element, and we are able to use the passive sign convention in order to select the correct sign for the voltage $L\, di/dt$ or $j\omega L\mathbf{I}$. If the current enters the terminal at which the positive voltage reference is located, then the positive sign is used. Mutual inductance, however, cannot be treated in exactly the same way because four terminals are involved. The choice of a correct sign is established by use of one of several possibilities that include the ***"dot convention,"*** or by an examination of the particular way in which each coil is wound. We will use the dot convention and merely look briefly at the physical construction of the coils; the use of other special symbols is not necessary when only two coils are coupled.

The dot convention makes use of a large dot placed at one end of each of the two coils which are mutually coupled. We determine the sign of the mutual voltage as follows:

> A current entering the *dotted* terminal of one coil produces an open-circuit voltage with a *positive* voltage reference at the *dotted* terminal of the second coil.

Thus, in Fig. 13.2*a*, i_1 enters the dotted terminal of L_1, v_2 is sensed positively at the dotted terminal of L_2, and $v_2 = M\, di_1/dt$. We have found previously that it is often not possible to select voltages or currents throughout a circuit so that the passive sign convention is everywhere satisfied; the same situation arises with mutual coupling. For example, it may be more convenient to represent v_2 by a positive voltage reference at the undotted terminal, as shown in Fig. 13.2*b*; then $v_2 = -M\, di_1/dt$. Currents that enter the dotted terminal are also not always available, as indicated by Fig. 13.2*c* and *d*. We note then that:

> A current entering the *undotted* terminal of one coil provides a voltage that is *positively* sensed at the *undotted* terminal of the second coil.

Note that the preceding discussion does not include any contribution to the voltage from self-induction, which would occur if i_2 were nonzero. We will consider this important situation in detail, but a quick example first is appropriate.

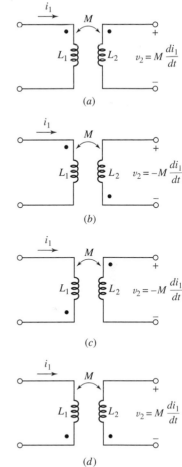

■ **FIGURE 13.2** Current entering the dotted terminal of one coil produces a voltage that is sensed positively at the dotted terminal of the second coil. Current entering the undotted terminal of one coil produces a voltage that is sensed positively at the undotted terminal of the second coil.

EXAMPLE 13.1

For the circuit shown in Fig. 13.3, (*a*) determine v_1 if $i_2 = 5 \sin 45t$ A and $i_1 = 0$; (*b*) determine v_2 if $i_1 = -8e^{-t}$ A and $i_2 = 0$.

(*a*) Since the current i_2 is entering the *undotted* terminal of the right coil, the positive reference for the voltage induced across the left coil is the undotted terminal. Thus, we have an open-circuit voltage

$$v_1 = -(2)(45)(5 \cos 45t) = -450 \cos 45t \text{ V}$$

appearing across the terminals of the left coil as a result of the time-varying magnetic flux generated by i_2 flowing into the right coil.

(Continued on next page)

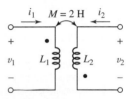

■ **FIGURE 13.3** The dot convention provides a relationship between the terminal at which a current enters one coil, and the positive voltage reference for the other coil.

Since no current flows through the coil on the left, there is no contribution to v_1 from self-induction.

(b) We now have a current entering a *dotted* terminal, but v_2 has its positive reference at the *undotted* terminal. Thus,

$$v_2 = -(2)(-1)(-8e^{-t}) = -16e^{-t} \text{ V}$$

PRACTICE

13.1 Assuming $M = 10$ H, coil L_2 is open circuited, and $i_1 = -2e^{-5t}$ A, find the voltage v_2 for (a) Fig. 13.2a; (b) Fig. 13.2b.

Ans: $100e^{-5t}$ V; $-100e^{-5t}$ V.

Combined Mutual and Self-Induction Voltage

So far, we have considered only a mutual voltage present across an *open-circuited* coil. In general, a nonzero current will be flowing in each of the two coils, and a mutual voltage will be produced in each coil because of the current flowing in the other coil. *This mutual voltage is present independently of and in addition to any voltage of self-induction.* In other words, the voltage across the terminals of L_1 will be composed of two terms, $L_1 \, di_1/dt$ and $M \, di_2/dt$, each carrying a sign depending on the current directions, the assumed voltage sense, and the placement of the two dots. In the portion of a circuit drawn in Fig. 13.4, currents i_1 and i_2 are shown, each arbitrarily assumed entering the dotted terminal. The voltage across L_1 is thus composed of two parts,

$$v_1 = L_1 \frac{di_1}{dt} + M \frac{di_2}{dt}$$

as is the voltage across L_2,

$$v_2 = L_2 \frac{di_2}{dt} + M \frac{di_1}{dt}$$

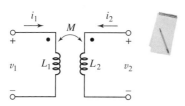

■ **FIGURE 13.4** Since the pairs v_1, i_1 and v_2, i_2 each satisfy the passive sign convention, the voltages of self-induction are both positive; since i_1 and i_2 each enter dotted terminals, and since v_1 and v_2 are both positively sensed at the dotted terminals, the voltages of mutual induction are also both positive.

In Fig. 13.5 the currents and voltages are not selected with the object of obtaining all positive terms for v_1 and v_2. By inspecting only the reference symbols for i_1 and v_1, it is apparent that the passive sign convention is not satisfied and the sign of $L_1 \, di_1/dt$ must therefore be negative. An identical conclusion is reached for the term $L_2 \, di_2/dt$. The mutual term of v_2 is signed by inspecting the direction of i_1 and v_2; since i_1 enters the dotted terminal and v_2 is sensed positive at the dotted terminal, the sign of $M \, di_1/dt$ must be positive. Finally, i_2 enters the undotted terminal of L_2, and v_1 is sensed positive at the undotted terminal of L_1; hence, the mutual portion of v_1, $M \, di_2/dt$, must also be positive. Thus, we have

$$v_1 = -L_1 \frac{di_1}{dt} + M \frac{di_2}{dt} \qquad v_2 = -L_2 \frac{di_2}{dt} + M \frac{di_1}{dt}$$

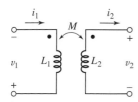

■ **FIGURE 13.5** Since the pairs v_1, i_1 and v_2, i_2 are not sensed according to the passive sign convention, the voltages of self-induction are both negative; since i_1 enters the dotted terminal and v_2 is positively sensed at the dotted terminal, the mutual term of v_2 is positive; and since i_2 enters the undotted terminal and v_1 is positively sensed at the undotted terminal, the mutual term of v_1 is also positive.

The same considerations lead to identical choices of signs for excitation by a sinusoidal source operating at frequency ω

$$\mathbf{V}_1 = -j\omega L_1 \mathbf{I}_1 + j\omega M \mathbf{I}_2 \qquad \mathbf{V}_2 = -j\omega L_2 \mathbf{I}_2 + j\omega M \mathbf{I}_1$$

Physical Basis of the Dot Convention

We can gain a more complete understanding of the dot symbolism by looking at the physical basis for the convention; the meaning of the dots is now interpreted in terms of *magnetic flux*. Two coils are shown wound on a cylindrical form in Fig. 13.6, and the direction of each winding is evident. Let us assume that the current i_1 is positive and increasing with time. The magnetic flux that i_1 produces within the form has a direction which may be found by the right-hand rule: when the right hand is wrapped around the coil with the fingers pointing in the direction of current flow, the thumb indicates the direction of the flux within the coil. Thus i_1 produces a flux which is directed downward; since i_1 is increasing with time, the flux, which is proportional to i_1, is also increasing with time. Turning now to the second coil, let us also think of i_2 as positive and increasing; the application of the right-hand rule shows that i_2 also produces a magnetic flux which is directed downward and is increasing. In other words, the assumed currents i_1 and i_2 produce *additive* fluxes.

The voltage across the terminals of any coil results from the time rate of change of the flux within that coil. The voltage across the terminals of the first coil is therefore greater with i_2 flowing than it would be if i_2 were zero. Thus i_2 induces a voltage in the first coil which has the same sense as the self-induced voltage in that coil. The sign of the self-induced voltage is known from the passive sign convention, and the sign of the mutual voltage is thus obtained.

The dot convention merely enables us to suppress the physical construction of the coils by placing a dot at one terminal of each coil such that currents entering dot-marked terminals produce additive fluxes. It is apparent that there are always two possible locations for the dots, because both dots may always be moved to the other ends of the coils and additive fluxes will still result.

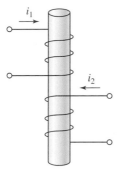

■ FIGURE 13.6 The physical construction of two mutually coupled coils. From a consideration of the direction of magnetic flux produced by each coil, it is shown that dots may be placed either on the upper terminal of each coil or on the lower terminal of each coil.

EXAMPLE 13.2

For the circuit shown in Fig. 13.7a, find the ratio of the output voltage across the 400 Ω resistor to the source voltage, expressed using phasor notation.

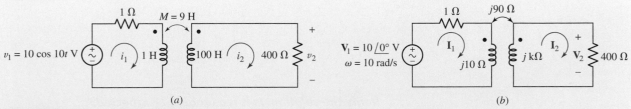

■ FIGURE 13.7 (a) A circuit containing mutual inductance in which the voltage ratio V_2/V_1 is desired. (b) Self and mutual inductances are replaced by the corresponding impedances.

▶ *Identify the goal of the problem.*
We need a numerical value for $\mathbf{V}_2$. We will then divide by $10\underline{/0°}$ V. *(Continued on next page)*

▶ **Collect the known information.**
We begin by replacing the 1 H and 100 H by their corresponding impedances, $j10\ \Omega$ and $j\ k\Omega$, respectively (Fig. 13.7b). We also replace the 9 H mutual inductance by $j\omega M = j90\ \Omega$.

▶ **Devise a plan.**
Mesh analysis is likely to be a good approach, as we have a circuit with two clearly defined meshes. Once we find $\mathbf{I}_2$, $\mathbf{V}_2$ is simply $400\ \mathbf{I}_2$.

▶ **Construct an appropriate set of equations.**
In the left mesh, the sign of the mutual term is determined by applying the dot convention. Since $\mathbf{I}_2$ enters the undotted terminal of L_2, the mutual voltage across L_1 must have the positive reference at the undotted terminal. Thus,

$$(1 + j10)\mathbf{I}_1 - j90\mathbf{I}_2 = 10\underline{/0^\circ}$$

Since $\mathbf{I}_1$ enters the dot-marked terminal, the mutual term in the right mesh has its (+) reference at the dotted terminal of the 100 H inductor. Therefore, we may write

$$(400 + j1000)\mathbf{I}_2 - j90\mathbf{I}_1 = 0$$

▶ **Determine if additional information is required.**
We have two equations in two unknowns, $\mathbf{I}_1$ and $\mathbf{I}_2$. Once we solve for the two currents, the output voltage $\mathbf{V}_2$ may be obtained by multiplying $\mathbf{I}_2$ by $400\ \Omega$.

▶ **Attempt a solution.**
Upon solving these two equations with a scientific calculator, we find that

$$\mathbf{I}_2 = 0.172\underline{/-16.70^\circ}\ \text{A}$$

Thus,

$$\frac{\mathbf{V}_2}{\mathbf{V}_1} = \frac{400(0.172\underline{/-16.70^\circ})}{10\underline{/0^\circ}}$$

$$= 6.880\underline{/-16.70^\circ}$$

▶ **Verify the solution. Is it reasonable or expected?**
We note that the output voltage $\mathbf{V}_2$ is actually larger in magnitude than the input voltage $\mathbf{V}_1$. Should we always expect this result? The answer is no. As we will see in later sections, the transformer can be constructed to achieve *either* a reduction *or* an increase in the voltage. We can perform a quick estimate, however, and at least find an upper

and lower bound for our answer. If the 400 Ω resistor is replaced with a short circuit, $\mathbf{V}_2 = 0$. If instead we replace the 400 Ω resistor with an open circuit, $\mathbf{I}_2 = 0$ and hence

$$\mathbf{V}_1 = (1 + j\omega L_1)\mathbf{I}_1$$

and

$$\mathbf{V}_2 = j\omega M\mathbf{I}_1$$

Solving, we find that the maximum value we could expect for $\mathbf{V}_2/\mathbf{V}_1$ is $8.955\underline{/5.711°}$. Thus, our answer at least appears reasonable.

The output voltage of the circuit in Fig. 13.7a is greater in magnitude than the input voltage, so that a voltage gain is possible with this type of circuit. It is also interesting to consider this voltage ratio as a function of ω.

To find $\mathbf{I}_2(j\omega)$ for this particular circuit, we write the mesh equations in terms of an unspecified angular frequency ω:

$$(1 + j\omega)\,\mathbf{I}_1 \qquad -j\omega9\,\mathbf{I}_2 = 10\underline{/0°}$$

and

$$-j\omega9\,\mathbf{I}_1 + (400 + j\omega100)\,\mathbf{I}_2 = 0$$

Solving by substitution, we find that

$$\mathbf{I}_2 = \frac{j90\omega}{400 + j500\omega - 19\omega^2}$$

Thus, we obtain the ratio of output voltage to input voltage as a function of frequency ω

$$\frac{\mathbf{V}_2}{\mathbf{V}_1} = \frac{400\mathbf{I}_2}{10}$$

$$= \frac{j\omega3600}{400 + j500\omega - 19\omega^2}$$

The magnitude of this ratio, sometimes referred to as the ***circuit transfer function,*** is plotted in Fig. 13.8 and has a peak magnitude of approximately 7 near a frequency of 4.6 rad/s. However, for very small or very large frequencies, the magnitude of the transfer function is less than unity.

The circuit is still passive, except for the voltage source, and the *voltage gain* must not be mistakenly interpreted as a *power gain.* At $\omega = 10$ rad/s, the voltage gain is 6.88, but the ideal voltage source, having a terminal voltage of 10 V, delivers a total power of 8.07 W, of which only 5.94 W reaches the 400 Ω resistor. The ratio of the output power to the source power, which we may define as the ***power gain,*** is thus 0.736.

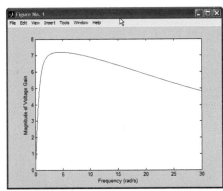

■ **FIGURE 13.8** The voltage gain $|\mathbf{V}_2/\mathbf{V}_1|$ of the circuit shown in Fig. 13.7a is plotted as a function of ω using the following MATLAB script:

```
>> w = linspace(0,30,1000);
>> num = j*w*3600;
>> for indx = 1:1000
   den = 400 + j*500*w(indx) - 19*w(indx)*w(indx);
   gain(indx) = num(indx)/den;
   end
>> plot(w, abs(gain));
>> xlabel('Frequency (rad/s)');
>> ylabel('Magnitude of Voltage Gain');
```

PRACTICE

13.2 For the circuit of Fig. 13.9, write appropriate mesh equations for the left mesh and the right mesh if $v_s = 20e^{-1000t}$ V.

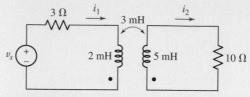

■ **FIGURE 13.9**

Ans: $20e^{-1000t} = 3i_1 + 0.002 \, di_1/dt - 0.003 \, di_2/dt$; $10i_2 + 0.005 di_2/dt - 0.003 \, di_1/dt = 0$.

EXAMPLE 13.3

Write a complete set of phasor equations for the circuit of Fig. 13.10a.

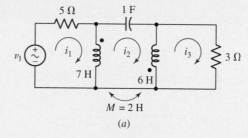

(a)

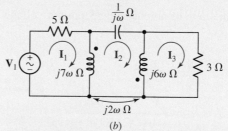

(b)

■ **FIGURE 13.10** (a) A three-mesh circuit with mutual coupling. (b) The 1 F capacitance as well as the self- and mutual inductances are replaced by their corresponding impedances.

The circuit contains three meshes, and the three mesh currents have already been assigned. Once again, our first step is to replace both the mutual inductance and the two self-inductances with their corresponding impedances as shown in Fig. 13.10b. Applying Kirchhoff's voltage law to the first mesh, a positive sign for the mutual term is assured by selecting $(\mathbf{I}_3 - \mathbf{I}_2)$ as the current through the second coil. Thus,

$$5\mathbf{I}_1 + 7j\omega(\mathbf{I}_1 - \mathbf{I}_2) + 2j\omega(\mathbf{I}_3 - \mathbf{I}_2) = \mathbf{V}_1$$

or

$$(5 + 7j\omega)\mathbf{I}_1 - 9j\omega\mathbf{I}_2 + 2j\omega\mathbf{I}_3 = \mathbf{V}_1 \qquad [3]$$

The second mesh requires two self-inductance terms and two mutual-inductance terms; the equation cannot be written carelessly. We obtain

$$7j\omega(\mathbf{I}_2 - \mathbf{I}_1) + 2j\omega(\mathbf{I}_2 - \mathbf{I}_3) + \frac{1}{j\omega}\mathbf{I}_2 + 6j\omega(\mathbf{I}_2 - \mathbf{I}_3)$$
$$+ 2j\omega(\mathbf{I}_2 - \mathbf{I}_1) = 0$$

or

$$-9j\omega\mathbf{I}_1 + \left(17j\omega + \frac{1}{j\omega}\right)\mathbf{I}_2 - 8j\omega\mathbf{I}_3 = 0 \qquad [4]$$

Finally, for the third mesh,

$$6j\omega(\mathbf{I}_3 - \mathbf{I}_2) + 2j\omega(\mathbf{I}_1 - \mathbf{I}_2) + 3\mathbf{I}_3 = 0$$

or

$$2j\omega\mathbf{I}_1 - 8j\omega\mathbf{I}_2 + (3 + 6j\omega)\mathbf{I}_2 = 0 \qquad [5]$$

Equations [3] to [5] may be solved by any of the conventional methods.

PRACTICE

13.3 For the circuit of Fig. 13.11, write an appropriate mesh equation in terms of the phasor currents $\mathbf{I}_1$ and $\mathbf{I}_2$ for the (a) left mesh; (b) right mesh.

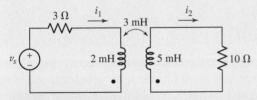

■ FIGURE 13.11

Ans: $\mathbf{V}_s = (3 + j10)\mathbf{I}_1 - j15\mathbf{I}_2$; $0 = -j15\mathbf{I}_1 + (10 + j25)\mathbf{I}_2$.

13.2 • ENERGY CONSIDERATIONS

Let us now consider the energy stored in a pair of mutually coupled inductors. The results will be useful in several different ways. We will first justify our assumption that $M_{12} = M_{21}$, and we may then determine the maximum possible value of the mutual inductance between two given inductors.

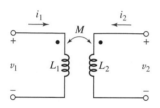

FIGURE 13.12 A pair of coupled coils with a mutual inductance of $M_{12} = M_{21} = M$.

Equality of M_{12} and M_{21}

The pair of coupled coils shown in Fig. 13.12 has currents, voltages, and polarity dots indicated. In order to show that $M_{12} = M_{21}$ we begin by letting all currents and voltages be zero, thus establishing zero initial energy storage in the network. We then open-circuit the right-hand terminal pair and increase i_1 from zero to some constant (dc) value I_1 at time $t = t_1$. The power entering the network from the left at any instant is

$$v_1 i_1 = L_1 \frac{di_1}{dt} i_1$$

and the power entering from the right is

$$v_2 i_2 = 0$$

since $i_2 = 0$.

The energy stored within the network when $i_1 = I_1$ is thus

$$\int_0^{t_1} v_1 i_1 \, dt = \int_0^{I_1} L_1 i_1 \, di_1 = \frac{1}{2} L_1 I_1^2$$

We now hold i_1 constant, $(i_1 = I_1)$, and let i_2 change from zero at $t = t_1$ to some constant value I_2 at $t = t_2$. The energy delivered from the right-hand source is thus

$$\int_{t_1}^{t_2} v_2 i_2 \, dt = \int_0^{I_2} L_2 i_2 \, di_2 = \frac{1}{2} L_2 I_2^2$$

However, even though the value of i_1 remains constant, the left-hand source also delivers energy to the network during this time interval:

$$\int_{t_1}^{t_2} v_1 i_1 \, dt = \int_{t_1}^{t_2} M_{12} \frac{di_2}{dt} i_1 \, dt = M_{12} I_1 \int_0^{I_2} di_2 = M_{12} I_1 I_2$$

The total energy stored in the network when both i_1 and i_2 have reached constant values is

$$W_{\text{total}} = \tfrac{1}{2} L_1 I_1^2 + \tfrac{1}{2} L_2 I_2^2 + M_{12} I_1 I_2$$

Now, we may establish the same final currents in this network by allowing the currents to reach their final values in the reverse order, that is, first increasing i_2 from zero to I_2 and then holding i_2 constant while i_1 increases from zero to I_1. If the total energy stored is calculated for this experiment, the result is found to be

$$W_{\text{total}} = \tfrac{1}{2} L_1 I_1^2 + \tfrac{1}{2} L_2 I_2^2 + M_{21} I_1 I_2$$

The only difference is the interchange of the mutual inductances M_{21} and M_{12}. The initial and final conditions in the network are the same, however, and so the two values of the stored energy must be identical. Thus,

$$M_{12} = M_{21} = M$$

and

$$W = \tfrac{1}{2} L_1 I_1^2 + \tfrac{1}{2} L_2 I_2^2 + M I_1 I_2 \qquad [6]$$

If one current enters a dot-marked terminal while the other leaves a dot-marked terminal, the sign of the mutual energy term is reversed:

$$W = \tfrac{1}{2}L_1 I_1^2 + \tfrac{1}{2}L_2 I_2^2 - M I_1 I_2 \qquad [7]$$

Although Eqs. [6] and [7] were derived by treating the final values of the two currents as constants, these "constants" can have any value, and the energy expressions correctly represent the energy stored when the *instantaneous* values of i_1 and i_2 are I_1 and I_2, respectively. In other words, lower-case symbols might just as well be used:

$$w(t) = \tfrac{1}{2}L_1 \left[i_1(t)\right]^2 + \tfrac{1}{2}L_2 \left[i_2(t)\right]^2 \pm M \left[i_1(t)\right]\left[i_2(t)\right] \qquad [8]$$

The only assumption upon which Eq. [8] is based is the logical establishment of a zero-energy reference level when both currents are zero.

Establishing an Upper Limit for *M*

Equation [8] may now be used to establish an upper limit for the value of M. Since $w(t)$ represents the energy stored within a *passive* network, it cannot be negative for any values of i_1, i_2, L_1, L_2, or M. Let us assume first that i_1 and i_2 are either both positive or both negative; their product is therefore positive. From Eq. [8], the only case in which the energy could possibly be negative is

$$w = \tfrac{1}{2}L_1 i_1^2 + \tfrac{1}{2}L_2 i_2^2 - M i_1 i_2$$

which we may write, by completing the square, as

$$w = \tfrac{1}{2}\left(\sqrt{L_1}\,i_1 - \sqrt{L_2}\,i_2\right)^2 + \sqrt{L_1 L_2}\,i_1 i_2 - M i_1 i_2$$

Since in reality the energy cannot be negative, the right-hand side of this equation cannot be negative. The first term, however, may be as small as zero, so we have the restriction that the sum of the last two terms cannot be negative. Hence,

$$\sqrt{L_1 L_2} \geq M$$

or

$$M \leq \sqrt{L_1 L_2} \qquad [9]$$

There is, therefore, an upper limit to the possible magnitude of the mutual inductance; it can be no larger than the geometric mean of the inductances of the two coils between which the mutual inductance exists. Although we have derived this inequality on the assumption that i_1 and i_2 carried the same algebraic sign, a similar development is possible if the signs are opposite; it is necessary only to select the positive sign in Eq. [8].

We might also have demonstrated the truth of inequality [9] from a physical consideration of the magnetic coupling; if we think of i_2 as being zero and the current i_1 as establishing the magnetic flux linking both L_1 and L_2, it is apparent that the flux within L_2 cannot be greater than the flux within L_1, which represents the total flux. Qualitatively, then, there is an upper limit to the magnitude of the mutual inductance possible between two given inductors.

The Coupling Coefficient

The degree to which M approaches its maximum value is described by the ***coupling coefficient,*** defined as

$$k = \frac{M}{\sqrt{L_1 L_2}} \qquad [10]$$

Since $M \leq \sqrt{L_1 L_2}$,

$$0 \leq k \leq 1$$

The larger values of the coefficient of coupling are obtained with coils which are physically closer, which are wound or oriented to provide a larger common magnetic flux, or which are provided with a common path through a material which serves to concentrate and localize the magnetic flux (a high-permeability material). Coils having a coefficient of coupling close to unity are said to be *tightly coupled.*

EXAMPLE 13.4

In Fig. 13.13, let $L_1 = 0.4$ H, $L_2 = 2.5$ H, $k = 0.6$, and $i_1 = 4i_2 = 20\cos(500t - 20°)$ mA. Evaluate the following quantities at $t = 0$: (a) i_2; (b) v_1; (c) the total energy stored in the system.

(a) $i_2(t) = 5\cos(500t - 20°)$ mA, so $i_2(0) = 5\cos(-20°) = 4.698$ mA.

(b) In order to determine the value of v_1, we need to include the contributions from both the self-inductance of coil 1 and the mutual inductance. Thus, paying attention to the dot convention,

$$v_1(t) = L_1 \frac{di_1}{dt} + M \frac{di_2}{dt}$$

To evaluate this quantity, we require a value for M. This is obtained from Eq. [10],

$$M = k\sqrt{L_1 L_2} = 0.6\sqrt{(0.4)(2.5)} = 0.6 \text{ H}$$

Thus, $v_1(0) = 0.4[-10\sin(-20°)] + 0.6[-2.5\sin(-20°)] = 1.881$ V.

(c) The total energy is found by summing the energy stored in each inductor, which has three separate components since the two coils are known to be magnetically coupled. Since both currents enter a "dotted" terminal,

$$w(t) = \tfrac{1}{2}L_1[i_1(t)]^2 + \tfrac{1}{2}L_2[i_2(t)]^2 + M[i_1(t)][i_2(t)]$$

With the knowledge from part (a) that $i_2(0) = 4.698$ mA and $i_1(0) = 4i_2(0) = 18.79$ mA, we find that the total energy stored in the two coils at $t = 0$ is 151.2 μJ.

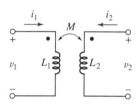

■ **FIGURE 13.13** Two coils with a coupling coefficient of 0.6, $L_1 = 0.4$ H and $L_2 = 2.5$ H.

PRACTICE

13.4 Let $i_s = 2\cos 10t$ A in the circuit of Fig. 13.14, and find the total energy stored in the passive network at $t = 0$ if $k = 0.6$ and terminals x and y are (a) left open-circuited; (b) short-circuited.

Ans: 0.8 J; 0.512 J.

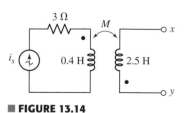

■ **FIGURE 13.14**

13.3 • THE LINEAR TRANSFORMER

We are now ready to apply our knowledge of magnetic coupling to the description of two specific practical devices, each of which may be represented by a model containing mutual inductance. Both of the devices are transformers, a term which we define as a network containing two or more coils which are deliberately coupled magnetically (Fig. 13.15). In this section we will consider the linear transformer, which happens to be an excellent model for the practical linear transformer used at radio frequencies, or higher frequencies. In Sec. 13.4 we will consider the ideal transformer, which is an idealized unity-coupled model of a physical transformer that has a core made of some magnetic material, usually an iron alloy.

■ **FIGURE 13.15** A selection of small transformers for use in electronic applications; the AA battery is shown for scale only.

In Fig. 13.16 a transformer is shown with two mesh currents identified. The first mesh, usually containing the source, is called the ***primary,*** while the second mesh, usually containing the load, is known as the ***secondary.*** The inductors labeled L_1 and L_2 are also referred to as the primary and secondary, respectively, of the transformer. We will assume that the transformer is *linear.* This implies that no magnetic material (which may cause a *nonlinear* flux-versus-current relationship) is employed. Without such material, however, it is difficult to achieve a coupling coefficient greater than a few tenths. The two resistors serve to account for the resistance of the wire out of which the primary and secondary coils are wound, and any other losses.

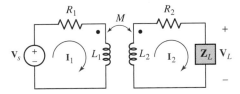

■ **FIGURE 13.16** A linear transformer containing a source in the primary circuit and a load in the secondary circuit. Resistance is also included in both the primary and the secondary.

Reflected Impedance

Consider the input impedance offered at the terminals of the primary circuit. The two mesh equations are

$$\mathbf{V}_s = (R_1 + j\omega L_1)\mathbf{I}_1 - j\omega M\mathbf{I}_2 \qquad [11]$$

and

$$0 = -j\omega M \mathbf{I}_1 + (R_2 + j\omega L_2 + \mathbf{Z}_L)\mathbf{I}_2 \qquad [12]$$

We may simplify by defining

$$\mathbf{Z}_{11} = R_1 + j\omega L_1 \qquad \text{and} \qquad \mathbf{Z}_{22} = R_2 + j\omega L_2 + \mathbf{Z}_L$$

so that

$$\mathbf{V}_s = \mathbf{Z}_{11}\mathbf{I}_1 - j\omega M \mathbf{I}_2 \qquad [13]$$

$$0 = -j\omega M \mathbf{I}_1 + \mathbf{Z}_{22}\mathbf{I}_2 \qquad [14]$$

Solving the second equation for $\mathbf{I}_2$ and inserting the result in the first equation enables us to find the input impedance,

$$\mathbf{Z}_{\text{in}} = \frac{\mathbf{V}_s}{\mathbf{I}_1} = \mathbf{Z}_{11} - \frac{(j\omega)^2 M^2}{\mathbf{Z}_{22}} \qquad [15]$$

$\mathbf{Z}_{\text{in}}$ is the impedance seen looking into the primary coil of the transformer.

Before manipulating this expression any further, we can draw several exciting conclusions. In the first place, this result is independent of the location of the dots on either winding, for if either dot is moved to the other end of the coil, the result is a change in sign of each term involving M in Eqs. [11] to [14]. This same effect could be obtained by replacing M by $(-M)$, and such a change cannot affect the input impedance, as Eq. [15] demonstrates. We also may note in Eq. [15] that the input impedance is simply $\mathbf{Z}_{11}$ if the coupling is reduced to zero. As the coupling is increased from zero, the input impedance differs from $\mathbf{Z}_{11}$ by an amount $\omega^2 M^2/\mathbf{Z}_{22}$, termed the **reflected impedance.** The nature of this change is more evident if we expand this expression

$$\mathbf{Z}_{\text{in}} = \mathbf{Z}_{11} + \frac{\omega^2 M^2}{R_{22} + j X_{22}}$$

and rationalize the reflected impedance,

$$\mathbf{Z}_{\text{in}} = \mathbf{Z}_{11} + \frac{\omega^2 M^2 R_{22}}{R_{22}^2 + X_{22}^2} + \frac{-j\omega^2 M^2 X_{22}}{R_{22}^2 + X_{22}^2}$$

Since $\omega^2 M^2 R_{22}/(R_{22}^2 + X_{22}^2)$ must be positive, it is evident that the presence of the secondary increases the losses in the primary circuit. In other words, the presence of the secondary might be accounted for in the primary circuit by increasing the value of R_1. Moreover, the reactance which the secondary reflects into the primary circuit has a sign which is opposite to that of X_{22}, the net reactance around the secondary loop. This reactance X_{22} is the sum of ωL_2 and X_L; it is necessarily positive for inductive loads and either positive or negative for capacitive loads, depending on the magnitude of the load reactance.

PRACTICE

13.5 Element values for a certain linear transformer are $R_1 = 3\ \Omega$, $R_2 = 6\ \Omega$, $L_1 = 2\ \text{mH}$, $L_2 = 10\ \text{mH}$, and $M = 4\ \text{mH}$. If $\omega = 5000\ \text{rad/s}$, find $\mathbf{Z}_{\text{in}}$ for $\mathbf{Z}_L$ equal to (a) $10\ \Omega$; (b) $j20\ \Omega$; (c) $10 + j20\ \Omega$; (d) $-j20\ \Omega$.

Ans: $5.32 + j2.74\ \Omega$; $3.49 + j4.33\ \Omega$; $4.24 + j4.57\ \Omega$; $5.56 - j2.82\ \Omega$.

T and Π Equivalent Networks

It is often convenient to replace a transformer with an equivalent network in the form of a T or Π. If we separate the primary and secondary resistances from the transformer, only the pair of mutually coupled inductors remains, as shown in Fig. 13.17. Note that the two lower terminals of the transformer are connected together to form a three-terminal network. We do this because both of our equivalent networks are also three-terminal networks. The differential equations describing this circuit are, once again,

$$v_1 = L_1 \frac{di_1}{dt} + M \frac{di_2}{dt} \qquad [16]$$

and

$$v_2 = M \frac{di_1}{dt} + L_2 \frac{di_2}{dt} \qquad [17]$$

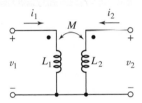

■ **FIGURE 13.17** A given transformer which is to be replaced by an equivalent Π or T network.

The form of these two equations is familiar and may be easily interpreted in terms of mesh analysis. Let us select a clockwise i_1 and a counterclockwise i_2 so that i_1 and i_2 are exactly identifiable with the currents in Fig. 13.17. The terms $M\, di_2/dt$ in Eq. [16] and $M\, di_1/dt$ in Eq. [17] indicate that the two meshes must then have a common *self*-inductance M. Since the total inductance around the left-hand mesh is L_1, a self-inductance of $L_1 - M$ must be inserted in the first mesh, but not in the second mesh. Similarly, a self-inductance of $L_2 - M$ is required in the second mesh, but not in the first mesh. The resultant equivalent network is shown in Fig. 13.18. The equivalence is guaranteed by the identical pairs of equations relating v_1, i_1, v_2, and i_2 for the two networks.

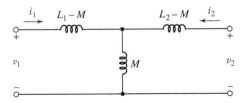

■ **FIGURE 13.18** The T equivalent of the transformer shown in Fig. 13.17.

If either of the dots on the windings of the given transformer is placed on the opposite end of its coil, the sign of the mutual terms in Eqs. [16] and [17] will be negative. This is analogous to replacing M with $-M$, and such a replacement in the network of Fig. 13.18 leads to the correct equivalent for this case. The three self-inductance values would then be $L_1 + M$, $-M$, and $L_2 + M$.

The inductances in the T equivalent are all self-inductances; no mutual inductance is present. It is possible that negative values of inductance may be obtained for the equivalent circuit, but this is immaterial if our only desire is a mathematical analysis; the actual construction of the equivalent network is, of course, impossible in any form involving a negative inductance. However, there are times when procedures for synthesizing networks to provide a desired transfer function lead to circuits containing a T network having a negative inductance; this network may then be realized by use of an appropriate linear transformer.

EXAMPLE 13.5

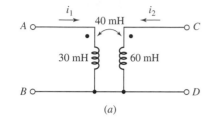

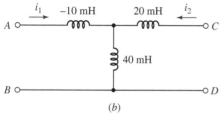

(a)

(b)

FIGURE 13.19 (a) A linear transformer used as an example. (b) The T-equivalent network of the transformer.

Find the T equivalent of the linear transformer shown in Fig. 13.19a.

We identify $L_1 = 30$ mH, $L_2 = 60$ mH, and $M = 40$ mH, and note that the dots are both at the upper terminals, as they are in the basic circuit of Fig. 13.17.

Hence, $L_1 - M = -10$ mH is in the upper left arm, $L_2 - M = 20$ mH is at the upper right, and the center stem contains $M = 40$ mH. The complete equivalent T is shown in Fig. 13.19b.

To demonstrate the equivalence, let us leave terminals C and D open-circuited and apply $v_{AB} = 10 \cos 100t$ V to the input in Fig. 13.19a. Thus,

$$i_1 = \frac{1}{30 \times 10^{-3}} \int 10 \cos(100t) \, dt = 3.33 \sin 100t \text{ A}$$

and

$$v_{CD} = M \frac{di_1}{dt} = 40 \times 10^{-3} \times 3.33 \times 100 \cos 100t$$
$$= 13.33 \cos 100t \quad \text{V}$$

Applying the same voltage in the T equivalent, we find that

$$i_1 = \frac{1}{(-10 + 40) \times 10^{-3}} \int 10 \cos(100t) \, dt = 3.33 \sin 100t \quad \text{A}$$

once again. Also, the voltage at C and D is equal to the voltage across the 40 mH inductor. Thus,

$$v_{CD} = 40 \times 10^{-3} \times 3.33 \times 100 \cos 100t = 13.33 \cos 100t \quad \text{V}$$

and the two networks yield equal results.

PRACTICE

13.6 (a) If the two networks shown in Fig. 13.20 are equivalent, specify values for L_x, L_y, and L_z. (b) Repeat if the dot on the secondary in Fig. 13.20b is located at the bottom of the coil.

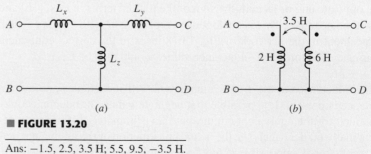

(a)   (b)

FIGURE 13.20

Ans: -1.5, 2.5, 3.5 H; 5.5, 9.5, -3.5 H.

The equivalent Π network is not obtained as easily. It is more complicated, and it is not used as much. We develop it by solving Eq. [17] for di_2/dt and substituting the result in Eq. [16]:

$$v_1 = L_1 \frac{di_1}{dt} + \frac{M}{L_2} v_2 - \frac{M^2}{L_2} \frac{di_1}{dt}$$

or

$$\frac{di_1}{dt} = \frac{L_2}{L_1 L_2 - M^2} v_1 - \frac{M}{L_1 L_2 - M^2} v_2$$

If we now integrate from 0 to t, we obtain

$$i_1 - i_1(0)u(t) = \frac{L_2}{L_1 L_2 - M^2} \int_0^t v_1 \, dt' - \frac{M}{L_1 L_2 - M^2} \int_0^t v_2 \, dt' \qquad [18]$$

In a similar fashion, we also have

$$i_2 - i_2(0)u(t) = \frac{-M}{L_1 L_2 - M^2} \int_0^t v_1 \, dt' + \frac{L_1}{L_1 L_2 - M^2} \int_0^t v_2 \, dt' \qquad [19]$$

Equations [18] and [19] may be interpreted as a pair of nodal equations; a step-current source must be installed at each node in order to provide the proper initial conditions. The factors multiplying each integral have the general form of inverses of certain equivalent inductances. Thus, the second coefficient in Eq. [18], $M/(L_1 L_2 - M^2)$, is $1/L_B$, or the reciprocal of the inductance extending between nodes 1 and 2, as shown on the equivalent Π network, Fig. 13.21. So

$$L_B = \frac{L_1 L_2 - M^2}{M}$$

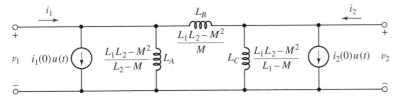

■ FIGURE 13.21 The Π network which is equivalent to the transformer shown in Fig. 13.17.

The first coefficient in Eq. [18], $L_2/(L_1 L_2 - M^2)$, is $1/L_A + 1/L_B$. Thus,

$$\frac{1}{L_A} = \frac{L_2}{L_1 L_2 - M^2} - \frac{M}{L_1 L_2 - M^2}$$

or

$$L_A = \frac{L_1 L_2 - M^2}{L_2 - M}$$

Finally,

$$L_C = \frac{L_1 L_2 - M^2}{L_1 - M}$$

No magnetic coupling is present among the inductors in the equivalent Π, and the initial currents in the three *self-inductances* are zero.

We may compensate for a reversal of either dot in the given transformer by merely changing the sign of M in the equivalent network. Also, just as we found in the equivalent T, negative self-inductances may appear in the equivalent Π network.

EXAMPLE 13.6

Find the equivalent Π network of the transformer in Fig. 13.19a, assuming zero initial currents.

Since the term $L_1 L_2 - M^2$ is common to L_A, L_B, and L_C, we begin by evaluating this quantity, obtaining

$$30 \times 10^{-3} \times 60 \times 10^{-3} - (40 \times 10^{-3})^2 = 2 \times 10^{-4}\ \text{H}^2$$

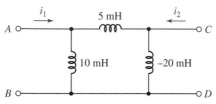

Thus,

$$L_A = \frac{(L_1 L_2 - M^2)}{(L_2 - M)} = \frac{2 \times 10^{-4}}{(20 \times 10^{-3})} = 10\ \text{mH}$$

$$L_C = \frac{(L_1 L_2 - M^2)}{(L_1 - M)} = -20\ \text{mH}$$

■ **FIGURE 13.22** The Π equivalent of the linear transformer shown in Fig. 13.19a. It is assumed that $i_1(0) = 0$ and $i_2(0) = 0$.

and

$$L_B = \frac{(L_1 L_2 - M^2)}{M} = 5\ \text{mH}$$

The equivalent Π network is shown in Fig. 13.22.

If we again check our result by letting $v_{AB} = 10\cos 100t$ V with terminals C-D open-circuited, the output voltage is quickly obtained by voltage division:

$$v_{CD} = \frac{-20 \times 10^{-3}}{5 \times 10^{-3} - 20 \times 10^{-3}} 10\cos 100t = 13.33\cos 100t \qquad \text{V}$$

as before. Thus, the network in Fig. 13.22 is electrically equivalent to the networks in Fig. 13.19a and b.

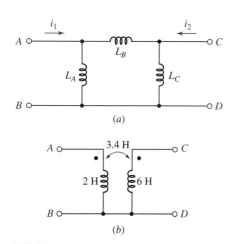

(a)

(b)

■ **FIGURE 13.23**

PRACTICE

13.7 If the networks in Fig. 13.23 are equivalent, specify values (in mH) for L_A, L_B, and L_C.

Ans: $L_A = 169.2$ mH, $L_B = 129.4$ mH, $L_C = -314.3$ mH.

COMPUTER-AIDED ANALYSIS

The ability to simulate circuits that contain magnetically coupled inductances is a useful skill, especially with modern circuit dimensions continuing to decrease. As various loops and partial loops of conductors are brought closer in new designs, various circuits and subcircuits that are intended to be isolated from one another inadvertently become coupled through stray magnetic fields, and interact with one another. PSpice allows us to incorporate this effect through the component **K_Linear,** which links a pair of inductors in the schematic by a coupling coefficient k in the range of $0 \le k \le 1$.

For example, let's simulate the circuit of Fig. 13.19*a*, which consists of two coils whose coupling is described by a mutual inductance of $M = 40$ mH, corresponding to a coupling coefficient of $k = 0.9428$. The basic circuit schematic is shown in Fig. 13.24. Note that no "dot" appears next to the inductor symbols. When first placed horizontally in the schematic, the dotted terminal is on the left, and this is the pin about which the symbol is rotated. Also note that the **K_Linear** component is not "wired" into the schematic anywhere; its location is arbitrary. The two coupled inductors, **L1** and **L2**, are specified along with the coupling coefficient through the component dialog box.

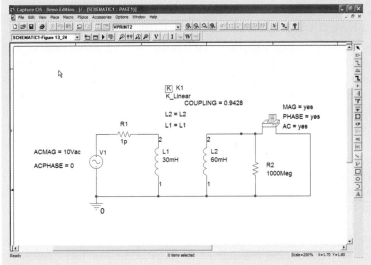

■ **FIGURE 13.24** Schematic of circuit based on Fig. 13.19*a*.

The circuit is connected to a 100 rad/s (15.92 Hz) sinusoidal voltage source, a fact which is incorporated by performing a single-frequency ac sweep. It is also necessary to add two resistors to the schematic in order for PSpice to perform the simulation without generating an error message. First, a small series resistance has been inserted between the voltage source and L1; a value of 1 pΩ was selected to minimize its effects. Second, a 1000 MΩ resistor (essentially infinite) was connected to L2. The output of the simulation is a voltage magnitude of 13.33 V and a phase angle of -3.819×10^{-8} degrees (essentially zero), in agreement with the values calculated by hand in Example 13.5.

PSpice also provides two different transformer models, a linear transformer, **XFRM_LINEAR,** and an ideal transformer **XFRM_NONLINEAR,** a circuit element which is the subject of the following section. The linear transformer requires that values be specified for the coupling coefficient and both coil inductances. The ideal transformer also requires a coupling coefficient, but, as we shall see, an *ideal* transformer has infinite or nearly infinite inductance values. Thus, the remaining parameters required for the part **XFRM_NONLINEAR** are the number of turns of wire that form each coil.

13.4 THE IDEAL TRANSFORMER

An *ideal transformer* is a useful approximation of a very tightly coupled transformer in which the coupling coefficient is essentially unity and both the primary and secondary inductive reactances are extremely large in comparison with the terminating impedances. These characteristics are closely approached by most well-designed iron-core transformers over a reasonable range of frequencies for a reasonable range of terminal impedances. The approximate analysis of a circuit containing an iron-core transformer may be achieved very simply by replacing that transformer with an ideal transformer; the ideal transformer may be thought of as a first-order model of an iron-core transformer.

Turns Ratio of an Ideal Transformer

One new concept arises with the ideal transformer: the *turns ratio* a. The self-inductance of a coil is proportional to the square of the number of turns of wire forming the coil. This relationship is valid only if all the flux established by the current flowing in the coil links all the turns. In order to develop this result quantitatively it is necessary to utilize magnetic field concepts, a subject that is not included in our discussion of circuit analysis. However, a qualitative argument may suffice. If a current i flows through a coil of N turns, then N times the magnetic flux of a single-turn coil will be produced. If we think of the N turns as being coincident, then all the flux certainly links all the turns. As the current and flux change with time, a voltage is then induced *in each turn* which is N times larger than that caused by a single-turn coil. Thus, the voltage induced *in the N-turn coil* must be N^2 times the single-turn voltage. From this, the proportionality between inductance and the square of the numbers of turns arises. It follows that

$$\frac{L_2}{L_1} = \frac{N_2^2}{N_1^2} = a^2 \qquad [20]$$

or

$$\boxed{a = \frac{N_2}{N_1}} \qquad [21]$$

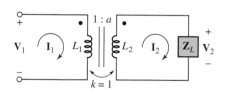

■ FIGURE 13.25 An ideal transformer is connected to a general load impedance.

Figure 13.25 shows an ideal transformer to which a secondary load is connected. The ideal nature of the transformer is established by several conventions: the use of the vertical lines between the two coils to indicate the iron laminations present in many iron-core transformers, the unity value of the coupling coefficient, and the presence of the symbol 1:a, suggesting a turns ratio of N_1 to N_2.

Let us analyze this transformer in the sinusoidal steady state in order to interpret our assumptions in the simplest context. The two mesh equations are

$$\mathbf{V}_1 = j\omega L_1 \mathbf{I}_1 - j\omega M \mathbf{I}_2 \qquad [22]$$

and

$$0 = -j\omega M \mathbf{I}_1 + (\mathbf{Z}_L + j\omega L_2)\mathbf{I}_2 \qquad [23]$$

We first determine the input impedance of an ideal transformer. By solving Eq. [23] for $\mathbf{I}_2$ and substituting in Eq. [22], we obtain

$$\mathbf{V}_1 = \mathbf{I}_1 j\omega L_1 + \mathbf{I}_1 \frac{\omega^2 M^2}{\mathbf{Z}_L + j\omega L_2}$$

and

$$\mathbf{Z}_{\text{in}} = \frac{\mathbf{V}_1}{\mathbf{I}_1} = j\omega L_1 + \frac{\omega^2 M^2}{\mathbf{Z}_L + j\omega L_2}$$

Since $k = 1$, $M^2 = L_1 L_2$ so

$$\mathbf{Z}_{\text{in}} = j\omega L_1 + \frac{\omega^2 L_1 L_2}{\mathbf{Z}_L + j\omega L_2}$$

Besides a unity coupling coefficient, another characteristic of an ideal transformer is an extremely large impedance for both the primary and secondary coils, regardless of the operating frequency. This suggests that the ideal case would be for both L_1 and L_2 to tend to infinity. Their ratio, however, must remain finite, as specified by the turns ratio. Thus,

$$L_2 = a^2 L_1$$

leads to

$$\mathbf{Z}_{\text{in}} = j\omega L_1 + \frac{\omega^2 a^2 L_1^2}{\mathbf{Z}_L + j\omega a^2 L_1}$$

Now if we let L_1 become infinite, both of the terms on the right-hand side of the preceding expression become infinite, and the result is indeterminate. Thus, it is necessary to first combine these two terms:

$$\mathbf{Z}_{\text{in}} = \frac{j\omega L_1 \mathbf{Z}_L - \omega^2 a^2 L_1^2 + \omega^2 a^2 L_1^2}{\mathbf{Z}_L + j\omega a^2 L_1} \qquad [24]$$

or

$$\mathbf{Z}_{\text{in}} = \frac{j\omega L_1 \mathbf{Z}_L}{\mathbf{Z}_L + j\omega a^2 L_1} = \frac{\mathbf{Z}_L}{\mathbf{Z}_L/j\omega L_1 + a^2} \qquad [25]$$

Now as $L_1 \to \infty$, we see that $\mathbf{Z}_{\text{in}}$ becomes

$$\mathbf{Z}_{\text{in}} = \frac{\mathbf{Z}_L}{a^2} \qquad [26]$$

for finite $\mathbf{Z}_L$.

This result has some interesting implications, and at least one of them appears to contradict one of the characteristics of the linear transformer. The input impedance of an ideal transformer is proportional to the load impedance, the proportionality constant being the reciprocal of the square of the turns ratio. In other words, if the *load* impedance is a capacitive impedance, then the *input* impedance is a capacitive impedance. In the linear transformer, however, the reflected impedance suffered a sign change in its reactive part; a capacitive load led to an inductive contribution to the input impedance. The explanation of this occurrence is achieved by first realizing that $\mathbf{Z}_L/a^2$ is *not* the reflected impedance, although it is often loosely called by that name. The true reflected impedance is infinite in the ideal transformer; otherwise it could not "cancel" the infinite impedance of the primary inductance. This cancellation occurs in the numerator of Eq. [24]. The impedance $\mathbf{Z}_L/a^2$ represents a small term which is the amount by which an exact cancellation does not occur. The true reflected impedance in the ideal transformer does change sign in its reactive part; as the primary and secondary inductances become infinite, however, the effect of the infinite primary-coil reactance and the infinite, but negative, reflected reactance of the secondary coil is one of cancellation.

The first important characteristic of the ideal transformer is therefore its ability to change the magnitude of an impedance, or to change impedance level. An ideal transformer having 100 primary turns and 10,000 secondary turns has a turns ratio of 10,000/100, or 100. Any impedance placed across the secondary then appears at the primary terminals reduced in magnitude by a factor of 100^2, or 10,000. A 20,000 Ω resistor looks like 2 Ω, a 200 mH inductor looks like 20 μH, and a 100 pF capacitor looks like 1 μF. If the primary and secondary windings are interchanged, then $a = 0.01$ and the load impedance is apparently increased in magnitude. In practice, this exact change in magnitude does not always occur, for we must remember that as we took the last step in our derivation and allowed L_1 to become infinite in Eq. [25], it was necessary to neglect $\mathbf{Z}_L$ in comparison with $j\omega L_2$. Since L_2 can never be infinite, it is evident that the ideal transformer model will become invalid if the load impedances are very large.

Use of Transformers for Impedance Matching

A practical example of the use of an iron-core transformer as a device for changing impedance level is in the coupling of a vacuum tube audio power amplifier to a speaker system. In order to achieve maximum power transfer, we know that the resistance of the load should be equal to the internal resistance of the source; the speaker usually has an impedance magnitude (often assumed to be a resistance) of only a few ohms, while the power amplifier typically possesses an internal resistance of several thousand ohms. Thus, an ideal transformer is required in which $N_2 < N_1$. For example, if the amplifier (or generator) internal impedance is 4000 Ω and the speaker impedance is 8 Ω, then we desire that

$$\mathbf{Z}_g = 4000 = \frac{\mathbf{Z}_L}{a^2} = \frac{8}{a^2}$$

or

$$a = \frac{1}{22.4}$$

and thus

$$\frac{N_1}{N_2} = 22.4$$

There is a simple relationship between the primary and secondary currents $\mathbf{I}_1$ and $\mathbf{I}_2$ in an ideal transformer. From Eq. [23],

$$\frac{\mathbf{I}_2}{\mathbf{I}_1} = \frac{j\omega M}{\mathbf{Z}_L + j\omega L_2}$$

Once again we allow L_2 to become infinite, and it follows that

$$\frac{\mathbf{I}_2}{\mathbf{I}_1} = \frac{j\omega M}{j\omega L_2} = \sqrt{\frac{L_1}{L_2}}$$

or

$$\boxed{\frac{\mathbf{I}_2}{\mathbf{I}_1} = \frac{1}{a}} \qquad [27]$$

Thus, the ratio of the primary and secondary currents is the turns ratio. If we have $N_2 > N_1$, then $a > 1$, and it is apparent that the larger current flows in the winding with the smaller number of turns. In other words,

$$N_1 \mathbf{I}_1 = N_2 \mathbf{I}_2$$

It should also be noted that the current ratio is the *negative* of the turns ratio if either current is reversed or if either dot location is changed.

In our example in which an ideal transformer was used to change the impedance level to efficiently match a speaker to a power amplifier, an rms current of 50 mA at 1000 Hz in the primary causes an rms current of 1.12 A at 1000 Hz in the secondary. The power delivered to the speaker is $(1.12)^2(8)$, or 10 W, and the power delivered to the transformer by the power amplifier is $(0.05)^2 4000$, or 10 W. The result is comforting, since the ideal transformer contains neither an active device which can generate power nor any resistor which can absorb power.

Use of Transformers for Voltage Level Adjustment

Since the power delivered to the ideal transformer is identical with that delivered to the load, whereas the primary and secondary currents are related by the turns ratio, it should seem reasonable that the primary and secondary voltages must also be related to the turns ratio. If we define the secondary voltage, or load voltage, as

$$\mathbf{V}_2 = \mathbf{I}_2 \mathbf{Z}_L$$

and the primary voltage as the voltage across L_1, then

$$\mathbf{V}_1 = \mathbf{I}_1 \mathbf{Z}_{\text{in}} = \mathbf{I}_1 \frac{\mathbf{Z}_L}{a^2}$$

The ratio of the two voltages then becomes

$$\frac{\mathbf{V}_2}{\mathbf{V}_1} = a^2 \frac{\mathbf{I}_2}{\mathbf{I}_1}$$

or

$$\boxed{\frac{\mathbf{V}_2}{\mathbf{V}_1} = a = \frac{N_2}{N_1}} \qquad [28]$$

(a)

(b)

(c)

■ **FIGURE 13.26** (a) A step-up transformer used to increase the generator output voltage for transmission. (b) Substation transformer used to reduce the voltage from the 220 kV transmission level to several tens of kilovolts for local distribution. (c) Step-down transformer used to reduce the distribution voltage level to 240 V for power consumption.
Photos courtesy of Dr. Wade Enright, Te Kura Pukaha Vira O Te Whare Wananga O Waitaha, Aotearoa.

The ratio of the secondary to the primary voltage is equal to the turns ratio. We should take care to note that this equation is opposite that of Eq. [27], and this is a common source of error for students. This ratio may also be negative if either voltage is reversed or either dot location is changed.

Simply by choosing the turns ratio, therefore, we now have the ability to change any ac voltage to any other ac voltage. If $a > 1$, the secondary voltage will be greater than the primary voltage, and we have what is commonly referred to as a ***step-up transformer.*** If $a < 1$, the secondary voltage will be less than the primary voltage, and we would have a ***step-down transformer.*** Utility companies typically generate power at a voltage in the range of 12 to 25 kV. Although this is a rather large voltage, transmission losses over long distances can be reduced by increasing the level to several hundred thousand volts using a step-up transformer (Fig. 13.26a). This voltage is then reduced to several tens of kilovolts at substations for local power distribution using step-down transformers (Fig. 13.26b). Additional step-down transformers are located outside buildings to reduce the voltage from the transmission voltage to the 110 or 220 V level required to operate machinery (Fig. 13.26c).

Combining the voltage and current ratios, Eqs. [27] and [28],

$$\mathbf{V}_2\mathbf{I}_2 = \mathbf{V}_1\mathbf{I}_1$$

and we see that the primary and secondary complex voltamperes are equal. The magnitude of this product is usually specified as a maximum allowable value on power transformers. If the load has a phase angle θ, or

$$\mathbf{Z}_L = |\mathbf{Z}_L|\underline{/\theta}$$

then $\mathbf{V}_2$ leads $\mathbf{I}_2$ by an angle θ. Moreover, the input impedance is $\mathbf{Z}_L/a^2$, and thus $\mathbf{V}_1$ also leads $\mathbf{I}_1$ by the same angle θ. If we let the voltage and current represent rms values, then $|\mathbf{V}_2|\,|\mathbf{I}_2|\cos\theta$ must equal $|\mathbf{V}_1|\,|\mathbf{I}_1|\cos\theta$, and all the power delivered to the primary terminals reaches the load; none is absorbed by or delivered to the ideal transformer.

The characteristics of the ideal transformer that we have obtained have all been determined by phasor analysis. They are certainly true in the sinusoidal steady state, but we have no reason to believe that they are correct for the *complete* response. Actually, they are applicable in general, and the demonstration that this statement is true is much simpler than the phasor-based analysis we have just completed. Our analysis, however, has served to point out the specific approximations that must be made on a more exact model of an actual transformer in order to obtain an ideal transformer. For example, we have seen that the reactance of the secondary winding must be much greater in magnitude than the impedance of any load that is connected to the secondary. Some feeling for those operating conditions under which a transformer ceases to behave as an ideal transformer is thus achieved.

EXAMPLE 13.7

For the circuit given in Fig. 13.27, determine the average power dissipated in the 10 kΩ resistor.

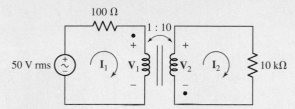

■ **FIGURE 13.27** A simple ideal transformer circuit.

The average power dissipated by the 10 kΩ resistor is simply

$$P = 10,000|\mathbf{I}_2|^2$$

The 50 V rms source "sees" a transformer input impedance of $\mathbf{Z}_L/a^2$ or 100 Ω. Thus, we obtain

$$\mathbf{I}_1 = \frac{50}{100 + 100} = 250 \text{ mA rms}$$

From Eq. [27], $\mathbf{I}_2 = (1/a)\mathbf{I}_1 = 25$ mA rms, so we find that the 10 kΩ resistor dissipates 6.25 W.

> The phase angles can be ignored in this example as they do not impact the calculation of average power dissipated by a purely resistive load.

PRACTICE
●

13.8 Repeat Example 13.7 using voltages to compute the dissipated power.

Ans: 6.25 W.

Voltage Relationship in the Time Domain

Let us now determine how the time-domain quantities v_1 and v_2 are related in the ideal transformer. Returning to the circuit shown in Fig. 13.17 and the two equations, [16] and [17], describing it, we may solve the second equation for di_2/dt and substitute in the first equation:

$$v_1 = L_1\frac{di_1}{dt} + \frac{M}{L_2}v_2 - \frac{M^2}{L_2}\frac{di_1}{dt}$$

However, for unity coupling, $M^2 = L_1L_2$, and thus

$$v_1 = \frac{M}{L_2}v_2 = \sqrt{\frac{L_1}{L_2}}v_2 = \frac{1}{a}v_2$$

The relationship between primary and secondary voltage is therefore found to apply to the complete time-domain response.

PRACTICAL APPLICATION

Superconducting Transformers

For the most part, we have neglected the various types of losses that may be present in a particular transformer. When dealing with large power transformers, however, close attention must be paid to such nonidealities, despite overall efficiencies of typically 97 percent or more. Although such a high efficiency may seem nearly ideal, it can represent a great deal of wasted energy when the transformer is handling several thousand amperes. So-called i^2R (pronounced "eye-squared-R") losses represent power dissipated as heat, which can increase the temperature of the transformer coils. Wire resistance increases with temperature, so heating only leads to greater losses. High temperatures can also lead to degradation of the wire insulation, resulting in shorter transformer life. As a result, many modern power transformers employ a liquid oil bath to remove excess heat from the transformer coils. Such an approach has its drawbacks, however, including environmental impact and fire danger from leaking oil as a result of corrosion over time (Fig. 13.28).

One possible means of improving the performance of such transformers is to make use of superconducting wire to replace the resistive coils of a standard transformer design. Superconductors are materials that are resistive at high temperature, but suddenly show no resistance to the flow of current below a critical temperature. Most elements are superconducting only near

■ **FIGURE 13.28** Fire that broke out in 2004 at the 340,000 V American Electric Power Substation near Mishawaka, Indiana. © AP/Wide World Photos

An expression relating primary and secondary current in the time domain is most quickly obtained by dividing Eq. [16] throughout by L_1,

$$\frac{v_1}{L_1} = \frac{di_1}{dt} + \frac{M}{L_1}\frac{di_2}{dt} = \frac{di_1}{dt} + a\frac{di_2}{dt}$$

and then invoking one of the hypotheses underlying the ideal transformer: L_1 must be infinite. If we assume that v_1 is not infinite, then

$$\frac{di_1}{dt} = -a\frac{di_2}{dt}$$

Integrating,

$$i_1 = -ai_2 + A$$

absolute zero, requiring expensive liquid helium–based cryogenic cooling. With the discovery in the 1980s of ceramic superconductors having critical temperatures of 90 K (−183°C) and higher, it became possible to replace helium-based equipment with significantly cheaper liquid nitrogen systems.

Figure 13.29 shows a prototype partial-core superconducting transformer being developed at the University of Canterbury. This design employs environmentally benign liquid nitrogen in place of an oil bath, and is also significantly smaller than a comparably rated conventional transformer. The result is a measurable improvement in overall transformer efficiency, which translates into operational cost savings for the owner.

Still, all designs have disadvantages that must be weighed against their potential advantages, and superconducting transformers are no exception. The most significant obstacle at present is the relatively high cost of fabricating superconducting wire several kilometers in length compared to copper wire. Part of this is due to the challenge of fabricating long wires from ceramic materials, but part of it is also due to the silver tubing used to surround the superconductor to provide a low-resistance current path in the event of a cooling system failure (although less expensive than silver, copper reacts with the ceramic and is therefore not a viable alternative). The net result is that although a superconducting transformer is likely to save a utility money over a long period of time—many transformers see over 30 years of service—

■ **FIGURE 13.29** Prototype 15 kVA partial core superconducting power transformer.
Photo courtesy of Department of Electrical and Computer Engineering, University of Canterbury.

the initial cost is much higher than for a traditional resistive transformer. At present, many companies (including utilities) are driven by short-term cost considerations and are not always eager to make large capital investments with only long-term cost benefits.

where A is a constant of integration that does not vary with time. Thus, if we neglect any direct currents in the two windings and fix our attention only on the time-varying portion of the response, then

$$i_1 = -ai_2$$

The minus sign arises from the placement of the dots and selection of the current directions in Fig. 13.17.

The same current and voltage relationships are therefore obtained in the time domain as were obtained previously in the frequency domain, provided that dc components are ignored. The time-domain results are more general, but they have been obtained by a less informative process.

The characteristics of the ideal transformer which we have established may be utilized to simplify circuits in which ideal transformers appear. Let us assume, for purposes of illustration, that everything to the left of the primary terminals has been replaced by its Thévenin equivalent, as has the network to the right of the secondary terminals. We thus consider the circuit shown in Fig. 13.30. Excitation at any frequency ω is assumed.

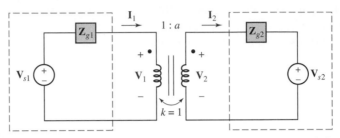

■ **FIGURE 13.30** The networks connected to the primary and secondary terminals of an ideal transformer are represented by their Thévenin equivalents.

Equivalent Circuits

Thévenin's or Norton's theorem may now be used to achieve an equivalent circuit that does not contain a transformer. For example, let us determine the Thévenin equivalent of the network to the left of the secondary terminals. Open-circuiting the secondary, $\mathbf{I}_2 = 0$ and therefore $\mathbf{I}_1 = 0$ (remember that L_1 is infinite). No voltage appears across $\mathbf{Z}_{g1}$, and thus $\mathbf{V}_1 = \mathbf{V}_{s1}$ and $\mathbf{V}_{2oc} = a\mathbf{V}_{s1}$. The Thévenin impedance is obtained by killing $\mathbf{V}_{s1}$ and utilizing the square of the turns ratio, being careful to use the reciprocal turns ratio, since we are looking in at the secondary terminals. Thus, $\mathbf{Z}_{th2} = \mathbf{Z}_{g1}a^2$.

As a check on our equivalent, let us also determine the short-circuit secondary current $\mathbf{I}_{2sc}$. With the secondary short-circuited, the primary generator faces an impedance of $\mathbf{Z}_{g1}$, and, thus, $\mathbf{I}_1 = \mathbf{V}_{s1}/\mathbf{Z}_{g1}$. Therefore, $\mathbf{I}_{2sc} = \mathbf{V}_{s1}/a\mathbf{Z}_{g1}$. The ratio of the open-circuit voltage to the short-circuit current is $a^2\mathbf{Z}_{g1}$, as it should be. The Thévenin equivalent of the transformer and the primary circuit is shown in the circuit of Fig. 13.31.

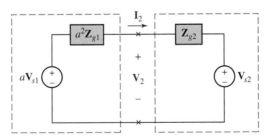

■ **FIGURE 13.31** The Thévenin equivalent of the network to the left of the secondary terminals in Fig. 13.30 is used to simplify that circuit.

Each primary voltage may therefore be multiplied by the turns ratio, each primary current divided by the turns ratio, and each primary impedance multiplied by the square of the turns ratio; and then these modified voltages, currents, and impedances replace the given voltages, currents, and impedances plus the transformer. If either dot is interchanged, the equivalent may be obtained by using the negative of the turns ratio.

Note that this equivalence, as illustrated by Fig. 13.31, is possible only if the network connected to the two primary terminals, and that connected to the two secondary terminals, can be replaced by their Thévenin equivalents. That is, each must be a two-terminal network. For example, if we cut the two primary leads at the transformer, the circuit must be divided into two separate networks; there can be no element or network bridging across the transformer between primary and secondary.

A similar analysis of the transformer and the secondary network shows that everything to the right of the primary terminals may be replaced by an identical network without the transformer, each voltage being divided by a, each current being multiplied by a, and each impedance being divided by a^2. A reversal of either winding requires the use of a turns ratio of $-a$.

EXAMPLE 13.8

For the circuit given in Fig. 13.32, determine the equivalent circuit in which the transformer and the secondary circuit are replaced, and also that in which the transformer and the primary circuit are replaced.

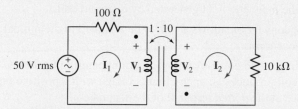

■ **FIGURE 13.32** A simple circuit in which a resistive load is matched to the source impedance by means of an ideal transformer.

This is the same circuit we analyzed in Example 13.7. As before, the input impedance is $10,000/(10)^2$, or $100\ \Omega$ and so $|\mathbf{I}_1| = 250$ mA rms. We can also compute the voltage across the primary coil

$$|\mathbf{V}_1| = |50 - 100\mathbf{I}_1| = 25 \text{ V rms}$$

and thus find that the source delivers $(25 \times 10^{-3})(50) = 12.5$ W, of which $(25 \times 10^{-3})^2(100) = 6.25$ W is dissipated in the internal resistance of the source and $12.5 - 6.25 = 6.25$ W is delivered to the load. This is the condition for maximum power transfer to the load.

If the secondary circuit and the ideal transformer are removed by the use of the Thévenin equivalent, the 50 V source and $100\ \Omega$ resistor simply see a $100\ \Omega$ impedance, and the simplified circuit of Fig. 13.33a is obtained. The primary current and voltage are now immediately evident.

If, instead, the network to the left of the secondary terminals is replaced by its Thévenin equivalent, we find (keeping in mind the location of the dots) $\mathbf{V}_{th} = -10(50) = -500$ V rms, and $\mathbf{Z}_{th} = (-10)^2(100) = 10$ kΩ; the resulting circuit is shown in Fig. 13.33b.

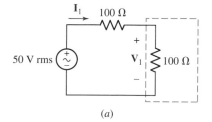

(a)

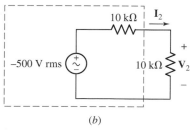

(b)

■ **FIGURE 13.33** The circuit of Fig. 13.32 is simplified by replacing (a) the transformer and secondary circuit by the Thévenin equivalent or (b) the transformer and primary circuit by the Thévenin equivalent.

PRACTICE

13.9 Let $N_1 = 1000$ turns and $N_2 = 5000$ turns in the ideal transformer shown in Fig. 13.34. If $\mathbf{Z}_L = 500 - j400\ \Omega$, find the average power delivered to $\mathbf{Z}_L$ for (a) $\mathbf{I}_2 = 1.4\underline{/20°}$ A rms; (b) $\mathbf{V}_2 = 900\underline{/40°}$ V rms; (c) $\mathbf{V}_1 = 80\underline{/100°}$ V rms; (d) $\mathbf{I}_1 = 6\underline{/45°}$ A rms; (e) $\mathbf{V}_s = 200\underline{/0°}$ V rms.

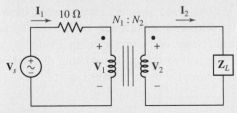

■ **FIGURE 13.34**

Ans: 980 W; 988 W; 195.1 W; 720 W; 692 W.

SUMMARY AND REVIEW

❏ Mutual inductance describes the voltage induced at the ends of a coil due to the magnetic field generated by a second coil.

❏ The dot convention allows a sign to be assigned to the mutual inductance term.

❏ According to the dot convention, a current entering the *dotted* terminal of one coil produces an open-circuit voltage with a positive voltage reference at the *dotted* terminal of the second coil.

❏ The total energy stored in a pair of coupled coils has three separate terms: the energy stored in each self-inductance ($\frac{1}{2}Li^2$), and the energy stored in the mutual inductance (Mi_1i_2).

❏ The coupling coefficient is given by $k = M/\sqrt{L_1L_2}$, and is restricted to values between 0 and 1.

❏ A linear transformer consists of two coupled coils: the primary winding and the secondary winding.

❏ An ideal transformer is a useful approximation for practical iron-core transformers. The coupling coefficient is taken to be unity, and the inductance values are assumed to be infinite.

❏ The turns ratio $a = N_2/N_1$ of an ideal transformer relates the primary and secondary coil voltages: $\mathbf{V}_2 = a\mathbf{V}_1$.

❏ The turns ratio a also relates the currents in the primary and secondary coils: $\mathbf{I}_1 = a\mathbf{I}_2$.

READING FURTHER

Almost everything you ever wanted to know about transformers can be found in

M. Heathcote, *J&P Transformer Book,* 12th ed. Oxford: Reed Educational and Professional Publishing Ltd., 1998.

Another comprehensive transformer title is

> W.T. McLyman, *Transformer and Inductor Design Handbook,* 3rd ed. New York: Marcel Dekker, 2004.

A good transformer book with a strong economic focus is

> B.K. Kennedy, *Energy Efficient Transformers.* New York: McGraw-Hill, 1998.

EXERCISES

13.1 Mutual Inductance

1. Consider the circuit of Fig. 13.35. If $i_1(t) = 400\cos 120\pi t$ A and the maximum value of $v_2(t)$ is 100 V, what is the value of the mutual inductance linking L_1 and L_2?

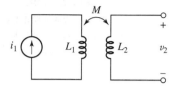

■ **FIGURE 13.35**

2. In the circuit of Fig. 13.36, the voltage $v_1(t)$ is known to be $115\sqrt{2}\cos(120\pi t - 16°)$ V. If the peak value of the current i_2 is measured as 45 A, what is the value of the mutual inductance linking the two inductors L_1 and L_2?

3. The physical construction of three pairs of coupled coils is shown in Fig. 13.37. Show the two different possible locations for the two dots on each pair of coils.

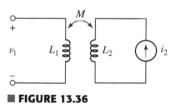

■ **FIGURE 13.36**

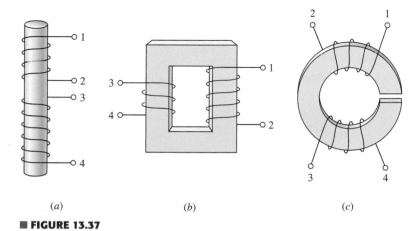

| (a) | (b) | (c) |

■ **FIGURE 13.37**

4. The two coupled inductors of Fig. 13.38 are connected in a circuit with voltages and currents as shown. $L_1 = 1$ H, $L_2 = 3$ H, and $M = 0.5$ H. If $i_1 = 30\sin 80t$ A and $i_2 = 30\cos 80t$ A, compute (*a*) v_1; (*b*) v_2.

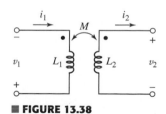

■ **FIGURE 13.38**

5. The two coupled inductors of Fig. 13.39 are connected in a circuit with voltages and currents as shown. $L_1 = 22\ \mu$H, $L_2 = 15\ \mu$H, and $M = 5\ \mu$H. If $i_1 = 3 \cos 800t$ nA and $i_2 = 2 \cos 800t$ nA, compute (a) v_1; (b) v_2.

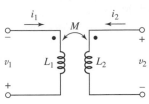

■ **FIGURE 13.39**

6. Referring to Fig. 13.40, assume that $v_1 = 5e^{-t}$ V and $v_2 = 3e^{-2t}$ V. If $L_1 = L_2 = 8$ H and $M = 0.4$ H, determine (a) di_1/dt; (b) di_2/dt; (c) $i_1(t)$ if there is no energy stored at $t = 0$.

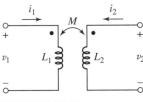

■ **FIGURE 13.40**

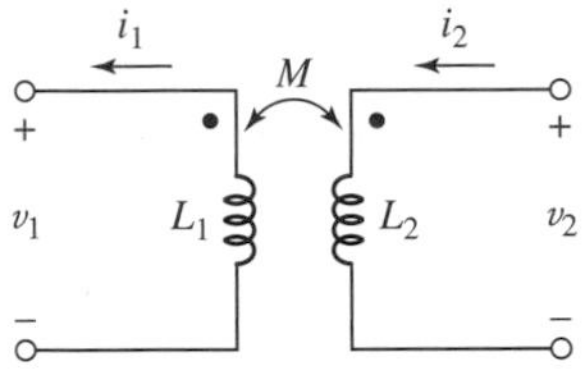

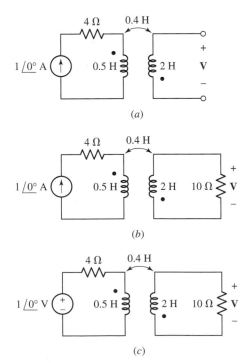

7. In Fig. 13.41, assume that $v_1 = 2e^{-t}$ V and $v_2 = 4e^{-3t}$ V. If $L_1 = L_2 = 2$ mH and $M = 1.5$ mH, determine (a) di_1/dt; (b) di_2/dt; (c) $i_2(t)$ if there is no energy stored at $t = 0$.

8. Find $v(t)$ for each network shown in Fig. 13.42 if $f = 50$ Hz.

■ **FIGURE 13.41**

(a)

(b)

(c)

■ **FIGURE 13.42**

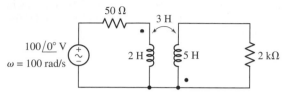

9. In the circuit shown in Fig. 13.43, find the average power absorbed by (a) the source; (b) each of the two resistors; (c) each of the two inductances; (d) the mutual inductance.

■ FIGURE 13.43

10. Let $i_{s1}(t) = 4t$ A and $i_{s2}(t) = 10t$ A in the circuit shown in Fig. 13.44. Find (a) v_{AG}; (b) v_{CG}; (c) v_{DG}.

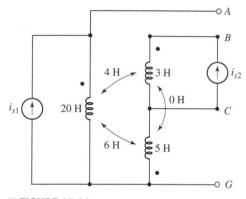

■ FIGURE 13.44

11. (a) Find the Thévenin equivalent network faced by the 2 kΩ resistor in the circuit of Exer. 9. (b) What is the maximum average power that can be drawn from the network by an optimum value of $\mathbf{Z}_L$ (instead of 2 kΩ)?

12. For the circuit of Fig. 13.45, find the currents $i_1(t)$, $i_2(t)$, and $i_3(t)$ if $f = 50$ Hz.

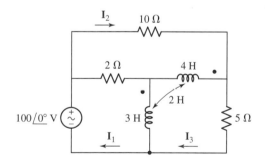

■ FIGURE 13.45

13. Determine an expression for $i_C(t)$ valid for $t > 0$ in the circuit of Fig. 13.46, if $v_s(t) = 10t^2 u(t)/(t^2 + 0.01)$ V.

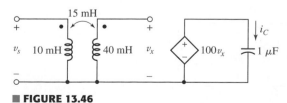

■ FIGURE 13.46

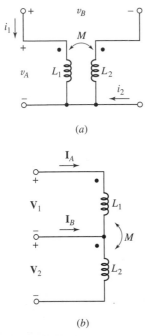

(a)

(b)

■ **FIGURE 13.47**

14. (a) For the network of Fig. 13.47a, write two equations giving $v_A(t)$ and $v_B(t)$ as functions of $i_1(t)$ and $i_2(t)$. (b) Write two equations giving $\mathbf{V}_1(j\omega)$ and $\mathbf{V}_2(j\omega)$ as functions of $\mathbf{I}_A(j\omega)$ and $\mathbf{I}_B(j\omega)$ for the network of Fig. 13.47b.

15. Note that there is no mutual coupling between the 5 H and 6 H inductors in the circuit of Fig. 13.48. (a) Write a set of equations in terms of $\mathbf{I}_1(j\omega)$, $\mathbf{I}_2(j\omega)$, and $\mathbf{I}_3(j\omega)$. (b) Find $\mathbf{I}_3(j\omega)$ if $\omega = 2$ rad/s.

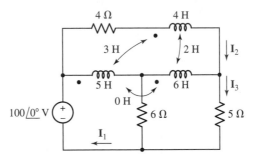

■ **FIGURE 13.48**

16. Find $\mathbf{V}_1(j\omega)$ and $\mathbf{V}_2(j\omega)$ in terms of $\mathbf{I}_1(j\omega)$ and $\mathbf{I}_2(j\omega)$ for each circuit of Fig. 13.49.

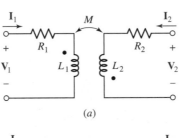

(a)

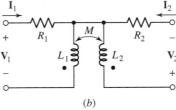

(b)

■ **FIGURE 13.49**

17. (a) Find $\mathbf{Z}_{in}(j\omega)$ for the network of Fig. 13.50. (b) Plot $\mathbf{Z}_{in}$ over the frequency range of $0 \le \omega \le 1000$ rad/s. (c) Find $\mathbf{Z}_{in}(j\omega)$ for $\omega = 50$ rad/s.

18. With reference to the circuit of Fig. 13.51, what value of M will cause exactly 3.2 W average power to be delivered to the 8 Ω bass speaker at an audio frequency of 160 Hz?

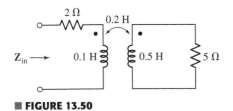

■ **FIGURE 13.50**

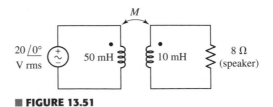

■ **FIGURE 13.51**

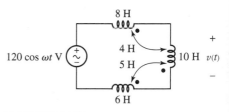

19. Let $i_{s1} = 2 \cos 10t$ A and $i_{s2} = 1.2 \cos 10t$ A in Fig. 13.52. Find (a) $v_1(t)$; (b) $v_2(t)$; (c) the average power being supplied by each source.

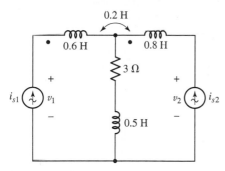

■ **FIGURE 13.52**

20. It is possible to arrange three coils physically in such a way that there is mutual coupling between coils A and B and between B and C, but not between A and C. Such an arrangement is shown in Fig. 13.53. Find $v(t)$.

■ **FIGURE 13.53**

21. Find $\mathbf{I}_L$ in the circuit shown in Fig. 13.54.

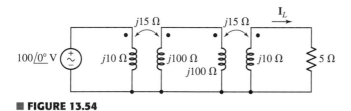

■ **FIGURE 13.54**

13.2 Energy Considerations

22. Let $i_s = 2 \cos 10t$ A in the circuit of Fig. 13.55. Find the total energy stored at $t = 0$ if (a) a-b is open-circuited as shown; (b) a-b is short-circuited.

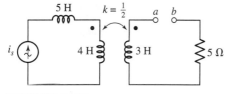

■ **FIGURE 13.55**

23. Let $\mathbf{V}_s = 12\underline{/0°}$ V rms in the linear transformer of Fig. 13.56. With $\omega = 100$ rad/s, find the average power supplied to the 24 Ω resistor as a function of k.

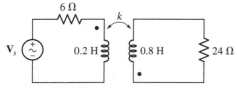

■ **FIGURE 13.56**

24. Two mutually coupled coils for which $L_1 = 2$ μH, $L_2 = 80$ μH, and $k = 1$ have a load $\mathbf{Z}_L = 2 + j10$ Ω connected across L_2. Find $\mathbf{Z}_{in}$ at the terminals of L_1 if $\omega = 250$ krad/s.

25. Let $\omega = 100$ rad/s in the circuit of Fig. 13.57 and find the average power:
(a) delivered to the 10 Ω load; (b) delivered to the 20 Ω load; (c) generated by
the source.

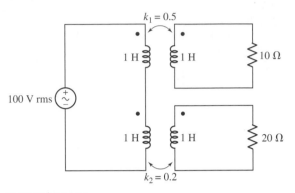

■ **FIGURE 13.57**

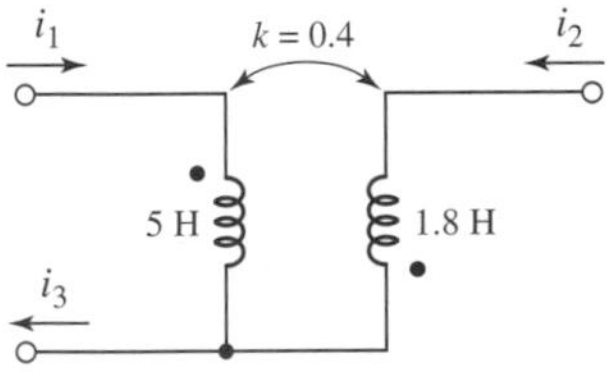

26. For the coupled coils shown in Fig. 13.58, let $i_1(t) = 4e^{-t/10}$ A and
$i_3(t) = 5e^{-t/5}$ A. Find (a) M; (b) $i_2(t)$; (c) the total energy stored in the system
at $t = 0$.

27. Let $\omega = 1000$ rad/s for the circuit of Fig. 13.59 and determine the value of the
ratio $\mathbf{V}_2/\mathbf{V}_s$ if (a) $L_1 = 1$ mH, $L_2 = 25$ mH, and $k = 1$; (b) $L_1 = 1$ H,
$L_2 = 25$ H, and $k = 0.99$; (c) $L_1 = 1$ H, $L_2 = 25$ H, and $k = 1$.

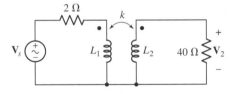

■ **FIGURE 13.59**

28. (a) An inductance bridge used on the coupled coils of Fig. 13.60 measures
the following values under short-circuit or open-circuit conditions:
$L_{AB,CD=OC} = 10$ mH, $L_{CD,AB=OC} = 5$ mH, $L_{AB,CD=SC} = 8$ mH. Find k.
(b) Assuming dots at A and D, and with $i_1 = 5$ A, what value should i_2 have in
order for 100 mJ to be stored in the system?

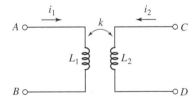

■ **FIGURE 13.60**

29. For the circuit shown in Fig. 13.61, $f = 60$ Hz. Calculate $\mathbf{V}_2$ as a function of k
and plot $|\mathbf{V}_2|$ versus k.

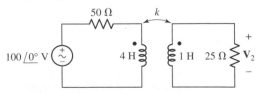

■ **FIGURE 13.61**

■ **FIGURE 13.58**

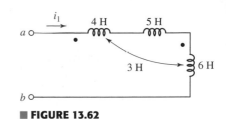

30. If $i_1 = 2 \cos 500t$ A in the network of Fig. 13.62, find the value of the maximum energy stored in the network.

13.3 The Linear Transformer

31. The load impedance $\mathbf{Z}_L$ of the circuit in Fig. 13.63 is known to be $7\underline{/32°}$ Ω at the operating frequency of 50 Hz. The mutual inductance linking the primary and secondary coils has a value of 800 nH. Calculate the (a) reflected impedance and (b) the input impedance seen by $\mathbf{V}_s$.

■ **FIGURE 13.62**

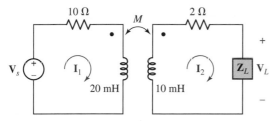

■ **FIGURE 13.63**

32. If the circuit of Fig. 13.64 is operating at 60 Hz and $\text{Re}\{\mathbf{Z}_L\} = 2$ Ω, what reactance is required of $\mathbf{Z}_L$ for the reflected impedance to equal $\mathbf{Z}_{11}$ when $M = 1$ mH? $(\mathbf{Z}_{11} \overset{\Delta}{=} R_1 + j\omega L_1)$.

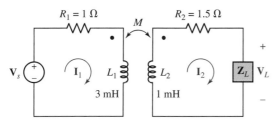

■ **FIGURE 13.64**

33. The networks of Fig. 13.65 are equivalent. Calculate values for L_1, L_2, and M.

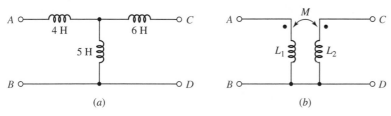

(a)   (b)

■ **FIGURE 13.65**

34. What values for L_z, L_y, and L_3 are required if the two networks of Fig. 13.66 are to be equivalent?

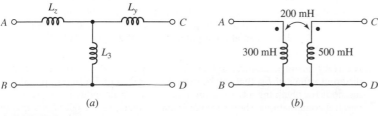

(a)   (b)

■ **FIGURE 13.66**

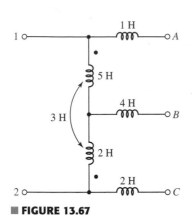

■ FIGURE 13.67

35. Find the equivalent inductance seen at terminals 1 and 2 in the network of Fig. 13.67 if the following terminals are connected together: (a) none; (b) A to B; (c) B to C; (d) A to C.

36. Refer to Fig. 13.68 and (a) use the equivalent T to help find the ratio $I_L(j\omega)/V_s(j\omega)$. (b) Set $v_s(t) = 100u(t)$ V and find $i_L(t)$. [Hint: You may want to write the two differential equations for the circuit to help find di_L/dt at $t = 0^+$.]

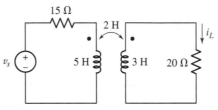

■ FIGURE 13.68

37. Determine equivalent T's for both dot locations in a lossless linear transformer for which $L_1 = 4$ mH, $L_2 = 18$ mH, and $M = 8$ mH. Use the T's to find the three equivalent input inductances obtained when the secondary is (a) open-circuited; (b) short-circuited; (c) connected in parallel with the primary.

38. Find $H(j\omega) = V_o/V_s$ for the circuit shown in Fig. 13.69.

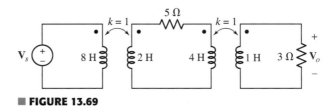

■ FIGURE 13.69

39. Use the equivalent T to help determine the input impedance $Z(j\omega)$ for the network shown in Fig. 13.70.

40. Let $V_s = 100\underline{/0°}$ V rms and $\omega = 100$ rad/s in the circuit of Fig. 13.71. Find the Thévenin equivalent of the network: (a) to the right of terminals a and b; (b) to the left of terminals c and d.

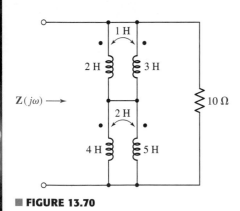

■ FIGURE 13.70

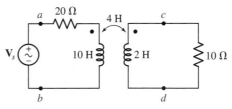

■ FIGURE 13.71

41. A load Z_L is connected to the secondary of a linear transformer that is characterized by inductances $L_1 = 1$ H and $L_2 = 4$ H and a unity coefficient of coupling. If $\omega = 1000$ rad/s, find the equivalent series network (R, L, and C values) seen at the input terminals if Z_L is (a) 100 Ω; (b) 0.1 H; (c) 10 μF.

42. A linear transformer has $L_1 = 6$ H, $L_2 = 12$ H, and $M = 5$ H. Find the eight different values of L_{in} that can be obtained for the eight different possible methods for obtaining a two-terminal network (single inductances, series and parallel combinations, short-circuited transformers, various dot combinations). Show each network and give its L_{in}.

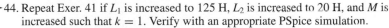

43. In the circuit of Fig. 13.72, let $\mathbf{Z}_L$ be a 100 μF capacitor having impedance $-j31.83$ Ω. Calculate $\mathbf{Z}_{in}$ when $k = $ (a) 0; (b) 0.5; (c) 0.9; (d) 1. Verify with appropriate PSpice simulations.

44. Repeat Exer. 41 if L_1 is increased to 125 H, L_2 is increased to 20 H, and M is increased such that $k = 1$. Verify with an appropriate PSpice simulation.

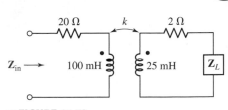

■ **FIGURE 13.72**

13.4 The Ideal Transformer

45. Find the average power delivered to each of the four resistors in the circuit of Fig. 13.73. Verify with an appropriate PSpice simulation.

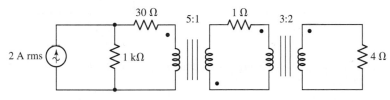

■ **FIGURE 13.73**

46. (a) What is the maximum value of average power that can be delivered to R_L in the circuit shown in Fig. 13.74? (b) Let $R_L = 100$ Ω and connect a 40 Ω resistor between the upper terminals of the primary and secondary. Find P_L.

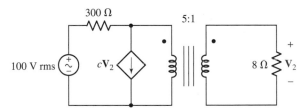

■ **FIGURE 13.74**

47. Find the average power delivered to the 8 Ω load in the circuit of Fig. 13.75 if c equals (a) 0; (b) 0.04 S; (c) −0.04 S.

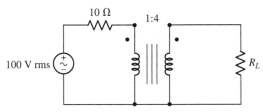

■ **FIGURE 13.75**

48. Find the Thévenin equivalent at terminals a and b for the network shown in Fig. 13.76.

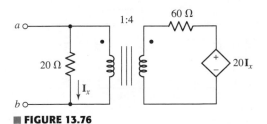

■ **FIGURE 13.76**

49. Select values for a and b in the circuit of Fig. 13.77 so that the ideal source supplies 1000 W, half of which is delivered to the 100 Ω load.

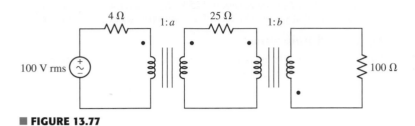

■ **FIGURE 13.77**

50. For the circuit shown in Fig. 13.78, find (a) $\mathbf{I}_1$; (b) $\mathbf{I}_2$; (c) $\mathbf{I}_3$; (d) $P_{25\Omega}$; (e) $P_{2\Omega}$; (f) $P_{3\Omega}$.

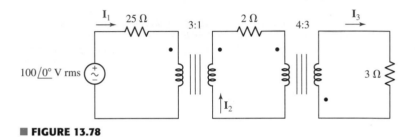

■ **FIGURE 13.78**

51. Find $\mathbf{V}_2$ in the circuit of Fig. 13.79.

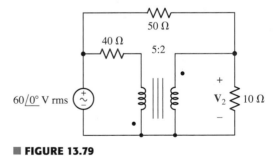

■ **FIGURE 13.79**

52. Find the power being dissipated in each resistor in the circuit of Fig. 13.80.

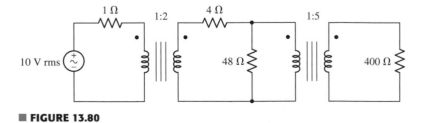

■ **FIGURE 13.80**

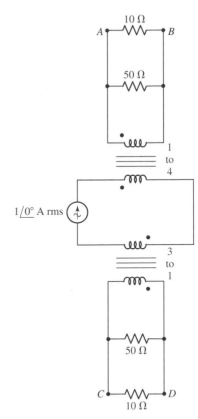

53. Find $\mathbf{I}_x$ in the circuit of Fig. 13.81.

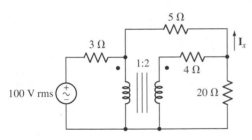

■ **FIGURE 13.81**

54. (*a*) Find the average power delivered to each 10 Ω resistor in the circuit shown in Fig. 13.82. (*b*) Repeat after connecting *A* to *C* and *B* to *D*.

55. Show how two ideal transformers may be used to match a generator having an output impedance of $4 + j0$ kΩ to a load consisting of one 8 W and one 10 W speaker so that the 8 W speaker receives *twice* the average power that the 10 W speaker does. Draw a suitable circuit diagram and specify the required turns ratios.

56. A transformer whose nameplate reads $\boxed{2300/230\ \text{V},\ 25\ \text{kVA}}$ operates with primary and secondary voltages of 2300 V and 230 V rms, respectively, and can supply 25 kVA from its secondary winding. If this transformer is supplied with 2300 V rms and is connected to secondary loads requiring 8 kW at unity PF and 15 kVA at 0.8 PF lagging, (*a*) what is the primary current? (*b*) How many kilowatts can the transformer still supply to a load operating at 0.95 PF lagging? (*c*) Verify your answers with PSpice.

57. Late at night, an advertisement on TV is selling a device for $19.95 that will measure your IQ. In a moment of weakness, you reach for the phone and place an order; 4 to 6 weeks later, your purchase arrives. You are instructed to turn a dial marked R_H to your height (cm), a dial marked R_M to your mass (kg), and a third dial R_A to your age (years). Disgusted with the number on the display, you toss it across the room where the back panel falls off, revealing the schematic shown in Fig. 13.83. You note that cm, kg, and years all correspond to ohms, and the power measured by the wattmeter in mW is displayed as IQ. (*a*) What IQ would it predict for your roommate? (*b*) What are the characteristics of the individual who would show the greatest IQ? (*c*) How much money are you out?

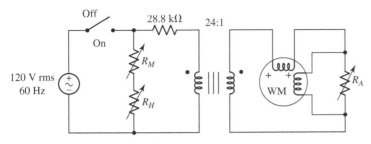

■ **FIGURE 13.83**

58. The company for which you work asks you to travel from Fresno, California, where power is supplied as 120 V rms, 60 Hz to Rostock, Germany (where power is supplied as 240 V rms, 50 Hz) for 6 weeks to help set up a new semi-conductor fabrication facility. Fortunately your laptop power supply can be plugged into outlets in either country provided you have a plug adapter.

■ **FIGURE 13.82**

However, your external CD writer runs only on 120 V ac. Design a circuit to allow you to use your CD writer in Germany, assuming that it can run on 50 Hz. (Transformers designed only for use at 60 Hz typically have a lighter-weight iron core than those designed for 50 Hz, so they will likely overheat at 50 Hz. Many transformers, however, are marked 50/60 Hz.)

D 59. As your first assignment in a new job, you are asked to design a circuit to allow a helium cryocompressor designed in the United States to be used in Australia. The cryocompressor consists of a three-phase motor which draws 10 A rms per phase at a line voltage of 208 V. The only three-phase power available at the site in Australia is 400 V rms. Design the necessary circuit.

D 60. The network shown in Fig. 13.84 has the unusual property of allowing only positive voltages $v(t)$ to pass through to the output; negative values result in $v_o(t) = 0$. (a) If an output voltage $v_o(t)$ is required having a peak voltage of 5 V, design an appropriate circuit using a 115 V rms supply and the network of Fig. 13.84. Sketch the output of your design. (b) If a "smoother" output is desired (i.e., one with less "ripple"), suggest a modification to your design.

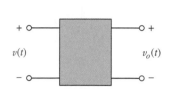

■ FIGURE 13.84

Complex Frequency and the Laplace Transform

KEY CONCEPTS

Neper Frequency

Complex Frequency

Laplace Transform

Use of Transform Tables

Method of Residuals

Using MATLAB to Manipulate Polynomials

Using MATLAB to Determine Residues of Rational Fractions

Initial Value Theorem

Final Value Theorem

INTRODUCTION

We are now about to begin the fourth major portion of our study of circuit analysis, a discussion of the concept of complex frequency. This, we will see, is a remarkably unifying concept that will enable us to tie together all our previously developed analytical techniques into one neat package. Resistive circuit analysis, steady-state sinusoidal analysis, transient analysis, the forced response, the complete response, and the analysis of circuits excited by exponential forcing functions and exponentially damped sinusoidal forcing functions will all become special cases of the general techniques of circuit analysis which are associated with the complex-frequency concept.

A common approach to this topic is to launch immediately into the Laplace transform integral, but this does not convey any real sense of understanding or intuition. Therefore, we will first examine the basic concept of complex frequency and its relevance to circuit analysis. From this, we introduce the Laplace transform as a means of handling circuits with more general time-dependent sources, learn how to perform inverse transforms to obtain time-domain responses, and consider a few special theorems that can exploit key properties of functions in the frequency domain. These techniques are then expanded in Chap. 15 to cover a broad range of analysis situations.

14.1 COMPLEX FREQUENCY

We introduce the notion of "complex frequency" by considering an exponentially damped sinusoidal function, such as the voltage

$$v(t) = V_m e^{\sigma t} \cos(\omega t + \theta) \qquad [1]$$

where σ (sigma) is a real quantity and is usually negative. Although we often refer to this function as being "damped," it is possible that the sinusoidal amplitude may increase. This occurs if $\sigma > 0$; the more practical case, however, is that of the damped function. Our work with the natural response of the RLC circuit also indicates that σ is the negative of the exponential damping coefficient.

We may construct a constant voltage from Eq. [1] by letting $\sigma = \omega = 0$:

$$v(t) = V_m \cos\theta = V_0 \qquad [2]$$

If we set only σ equal to zero, then we obtain a general sinusoidal voltage

$$v(t) = V_m \cos(\omega t + \theta) \qquad [3]$$

and if $\omega = 0$, we have the exponential voltage

$$v(t) = V_m \cos\theta \; e^{\sigma t} = V_0 e^{\sigma t} \qquad [4]$$

Thus, the damped sinusoid of Eq. [1] includes as special cases the dc Eq. [2], sinusoidal Eq. [3], and exponential Eq. [4] functions.

Some additional insight into the significance of σ can be obtained by comparing the exponential function of Eq. [4] with the complex representation of a sinusoidal function with a zero-degree phase angle,

$$v(t) = V_0 e^{j\omega t} \qquad [5]$$

It is apparent that the two functions, Eqs. [4] and [5], have much in common. The only difference is that the exponent in Eq. [4] is real and the one in Eq. [5] is imaginary. The similarity between the two functions is emphasized by describing σ as a "frequency." This choice of terminology will be discussed in detail in the following sections, but for now we need merely note that σ is specifically termed the *real part* of the complex frequency. It should not be called the "real frequency," however, for this is a term that is more suitable for f (or, loosely, for ω). We will also refer to σ as the **neper frequency,** the name arising from the dimensionless unit of the exponent of e. Thus, given e^{7t}, the dimensions of $7t$ are **nepers** (Np), and 7 is the neper frequency in nepers per second.

The neper itself was named after the Scottish philosopher and mathematician John Napier (1550–1617) and his napierian logarithm system; the spelling of his name is, interestingly enough, historically uncertain (see, for example, H. A. Wheeler, *IRE Transactions on Circuit Theory* **2**, 1955, p. 219.)

The General Form

The forced response of a network to a general forcing function of the form of Eq. [1] can be found very simply by using a method almost identical with that used in phasor-based analysis. Once we are able to find the forced response to this damped sinusoid, we should realize that we will also have found the forced response to a dc voltage, an exponential voltage, and a sinusoidal voltage. Now we will see how we may consider σ and ω as the real and imaginary parts of a complex frequency.

Let us first provide ourselves with a purely mathematical definition of complex frequency and then gradually develop a physical interpretation as the chapter progresses. We suggest that any function that may be written in the form

$$f(t) = \mathbf{K}e^{st} \qquad [6]$$

where $\mathbf{K}$ and $\mathbf{s}$ are complex constants (independent of time) is characterized by the **complex frequency s.** The complex frequency $\mathbf{s}$ is therefore simply the factor that multiplies t in this complex exponential representation. Until we are able to determine the complex frequency of a given function by inspection, it is necessary to write the function in the form of Eq. [6].

The DC Case

We may apply this definition first to the more familiar forcing functions. For example, a constant voltage

$$v(t) = V_0$$

may be written in the form

$$v(t) = V_0 e^{(0)t}$$

Therefore, we conclude that the complex frequency of a dc voltage or current is zero (i.e., $\mathbf{s} = 0$).

The Exponential Case

The next simple case is the exponential function

$$v(t) = V_0 e^{\sigma t}$$

which is already in the required form. The complex frequency of this voltage is therefore σ (i.e., $\mathbf{s} = \sigma + j0$).

The Sinusoidal Case

Now let us consider a sinusoidal voltage, one that may provide a slight surprise. Given

$$v(t) = V_m \cos(\omega t + \theta)$$

we desire to find an equivalent expression in terms of the complex exponential. From our past experience, we therefore use the formula we derived from Euler's identity,

$$\cos(\omega t + \theta) = \tfrac{1}{2}[e^{j(\omega t + \theta)} + e^{-j(\omega t + \theta)}]$$

and obtain

$$v(t) = \tfrac{1}{2} V_m [e^{j(\omega t + \theta)} + e^{-j(\omega t + \theta)}]$$
$$= \left(\tfrac{1}{2} V_m e^{j\theta}\right) e^{j\omega t} + \left(\tfrac{1}{2} V_m e^{-j\theta}\right) e^{-j\omega t}$$

or

$$v(t) = \mathbf{K}_1 e^{\mathbf{s}_1 t} + \mathbf{K}_2 e^{\mathbf{s}_2 t}$$

We have the *sum* of two complex exponentials, and *two* complex frequencies are therefore present, one for each term. The complex frequency of the first term is $\mathbf{s} = \mathbf{s}_1 = j\omega$, and that of the second term is $\mathbf{s} = \mathbf{s}_2 = -j\omega$. These two values of $\mathbf{s}$ are *conjugates*, or $\mathbf{s}_2 = \mathbf{s}_1^*$ and the two values of $\mathbf{K}$ are also conjugates: $\mathbf{K}_1 = \tfrac{1}{2} V_m e^{j\theta}$ and $\mathbf{K}_2 = \mathbf{K}_1^* = \tfrac{1}{2} V_m e^{-j\theta}$. The entire first term and the entire second term are therefore conjugates, which we might have expected inasmuch as their sum must be a real quantity, $v(t)$.

The complex conjugate of any number can be obtained by simply replacing all occurrences of "j" with "$-j$." The concept arises from our arbitrary choice of $j = +\sqrt{-1}$. However, the negative root is just as valid, which leads us to the definition of a complex conjugate.

The Exponentially Damped Sinusoidal Case

Finally, let us determine the complex frequency or frequencies associated with the exponentially damped sinusoidal function, Eq. [1]. We again use Euler's formula to obtain a complex exponential representation:

$$v(t) = V_m e^{\sigma t} \cos(\omega t + \theta)$$
$$= \tfrac{1}{2} V_m e^{\sigma t} [e^{j(\omega t + \theta)} + e^{-j(\omega t + \theta)}]$$

and thus

$$v(t) = \tfrac{1}{2} V_m e^{j\theta} e^{j(\sigma + j\omega)t} + \tfrac{1}{2} V_m e^{-j\theta} e^{j(\sigma - j\omega)t}$$

We find that a conjugate complex pair of frequencies, $\mathbf{s}_1 = \sigma + j\omega$ and $\mathbf{s}_2 = \mathbf{s}_1^* = \sigma - j\omega$, is also required to describe the exponentially damped sinusoid. In general, neither σ nor ω is zero, and the exponentially varying sinusoidal waveform is the general case; the constant, sinusoidal, and exponential waveforms are special cases.

The Relationship of s to Reality

A positive real value of $\mathbf{s}$, e.g., $\mathbf{s} = 5 + j0$, identifies an exponentially increasing function $\mathbf{K}e^{+5t}$, where $\mathbf{K}$ must be real if the function is to be a physical one. A negative real value for $\mathbf{s}$, such as $\mathbf{s} = -5 + j0$, refers to an exponentially decreasing function $\mathbf{K}e^{-5t}$.

A purely imaginary value of $\mathbf{s}$, such as $j10$, can never be associated with a purely real quantity. The functional form is $\mathbf{K}e^{j10t}$, which can also be written as $\mathbf{K}(\cos 10t + j \sin 10t)$; it obviously possesses both a real and an imaginary part, each of which is sinusoidal. In order to construct a real function, it is necessary to consider conjugate values of $\mathbf{s}$, such as $\mathbf{s}_{1,2} = \pm j10$, with which must be associated conjugate values of $\mathbf{K}$. Loosely speaking, however, we may identify either of the complex frequencies $\mathbf{s}_1 = +j10$ or $\mathbf{s}_2 = -j10$ with a sinusoidal voltage at the radian frequency of 10 rad/s; the presence of the conjugate complex frequency is understood. The amplitude and phase angle of the sinusoidal voltage will depend on the choice of $\mathbf{K}$ for each of the two frequencies. Thus, selecting $\mathbf{s}_1 = j10$ and $\mathbf{K}_1 = 6 - j8$, where

$$v(t) = \mathbf{K}_1 e^{\mathbf{s}_1 t} + \mathbf{K}_2 e^{\mathbf{s}_2 t}, \qquad \mathbf{s}_2 = \mathbf{s}_1^* \qquad \text{and} \qquad \mathbf{K}_2 = \mathbf{K}_1^*$$

we obtain the real sinusoid $20 \cos(10t - 53.1°)$.

In a similar manner, a general value for $\mathbf{s}$, such as $3 - j5$, can be associated with a real quantity only if it is accompanied by its conjugate, $3 + j5$. Speaking loosely again, we may think of either of these two conjugate frequencies as describing an exponentially increasing sinusoidal function, $e^{3t} \cos 5t$; the specific amplitude and phase angle will again depend on the specific values of the conjugate complex $\mathbf{K}$'s.

By now we should have achieved some appreciation of the physical nature of the complex frequency $\mathbf{s}$; in general, it describes an exponentially varying sinusoid. The real part of $\mathbf{s}$ is associated with the exponential variation; if it is negative, the function decays as t increases; if it is positive, the function increases; and if it is zero, the sinusoidal amplitude is constant. The larger the *magnitude* of the real part of $\mathbf{s}$, the greater is the rate of exponential increase or decrease. The imaginary part of $\mathbf{s}$ describes the sinusoidal variation; it is specifically the radian frequency. A large magnitude for the imaginary part of $\mathbf{s}$ indicates a more rapidly changing function of time.

Note that $|6 - j8| = 10$, so that $V_m = 2|\mathbf{K}| = 20$. Also, $ang(6 - j8) = -53.13°$.

Large magnitudes for the real part of $\mathbf{s}$, the imaginary part of $\mathbf{s}$, or the magnitude of $\mathbf{s}$ indicate a rapidly varying function.

It is customary to use the letter σ to designate the real part of **s**, and ω (*not* $j\omega$) to designate the imaginary part:

$$\boxed{\mathbf{s} = \sigma + j\omega} \qquad\qquad [7]$$

The radian frequency is sometimes referred to as the "real frequency," but this terminology can be very confusing when we find that we must then say that "the real frequency is the imaginary part of the complex frequency"! When we need to be specific, we will call **s** the complex frequency, σ the neper frequency, ω the radian frequency, and $f = \omega/2\pi$ the cyclic frequency; when no confusion seems likely, it is permissible to use "frequency" to refer to any of these four quantities. The *neper frequency* is measured in nepers per second, *radian frequency* is measured in radians per second, and *complex frequency* **s** is measured in units which are variously termed complex nepers per second or complex radians per second.

PRACTICE

14.1 Identify all the complex frequencies present in the real time functions: (*a*) $(2e^{-100t} + e^{-200t})\sin 2000t$; (*b*) $(2 - e^{-10t})\cos(4t + \phi)$; (*c*) $e^{-10t}\cos 10t \sin 40t$.

14.2 Use real constants A, B, C, ϕ, and so forth, to construct the general form of the real function of time for a current having components at these frequencies: (*a*) $0, 10, -10$ s^{-1}; (*b*) $-5, j8$, $-5 - j8$ s^{-1}; (*c*) $-20, 20, -20 + j20, 20 - j20$ s^{-1}.

Ans: 14.1: $-100 + j2000, -100 - j2000, -200 + j2000, -200 - j2000$ s^{-1}; $j4, -j4, -10 + j4, -10 - j4$ s^{-1}; $-10 + j30, -10 - j30, -10 + j50$, $-10 - j50$ s^{-1}; 14.2: $A + Be^{10t} + Ce^{-10t}$; $Ae^{-5t} + B\cos(8t + \phi_1) + Ce^{-5t}\cos(8t + \phi_2)$; $Ae^{-20t} + Be^{20t} + Ce^{-20t}\cos(20t + \phi_1) + De^{20t}\cos(20t + \phi_2)$.

14.2 THE DAMPED SINUSOIDAL FORCING FUNCTION

We have devoted enough time to the definition and introductory interpretation of complex frequency; now it is time to put this concept to work and become familiar with it by seeing what it will do and how it is used.

The general exponentially varying sinusoid, which we may represent with the voltage function

$$v(t) = V_m e^{\sigma t} \cos(\omega t + \theta) \qquad\qquad [8]$$

can be expressed in terms of the complex frequency **s** by making use of Euler's identity as before:

$$v(t) = \text{Re}\{V_m e^{\sigma t} e^{j(\omega t + \theta)}\} \qquad\qquad [9]$$

or

$$v(t) = \text{Re}\{V_m e^{\sigma t} e^{j(-\omega t - \theta)}\} \qquad\qquad [10]$$

Either representation is suitable, and the two expressions should remind us that a pair of conjugate complex frequencies is associated with a sinusoid or an exponentially damped sinusoid. Equation [9] is more directly related to the given damped sinusoid, and we will concern ourselves principally with it. Collecting factors, we now substitute $\mathbf{s} = \sigma + j\omega$ into

$$v(t) = \text{Re}\{V_m e^{j\theta} e^{(\sigma + j\omega)t}\}$$

and obtain

$$v(t) = \text{Re}\{V_m e^{j\theta} e^{\mathbf{s}t}\} \qquad [11]$$

Before we apply a forcing function of this form to any circuit, we should note the resemblance of this last representation of the damped sinusoid to the corresponding representation of the *undamped* sinusoid that we studied back in Chap. 10,

$$\text{Re}\{V_m e^{j\theta} e^{j\omega t}\}$$

The only difference is that we now have $\mathbf{s}$ where we previously had $j\omega$. Instead of restricting ourselves to sinusoidal forcing functions and their radian frequencies, we have now extended our notation to include the damped sinusoidal forcing function at a complex frequency. It should be no surprise at all to see later in this section that we will develop a *frequency-domain* description of the exponentially damped sinusoid in exactly the same way that we did for the sinusoid; we will simply omit the Re{ } notation and suppress $e^{\mathbf{s}t}$.

We are now ready to apply the exponentially damped sinusoid, as given by Eq. [8], [9], [10], or [11], to an electrical network, where the forced response—perhaps a current in some branch of the network—is the desired response. Since the forced response has the form of the forcing function, its integral and its derivatives, the response may be assumed to be

$$i(t) = I_m e^{\sigma t} \cos(\omega t + \phi)$$

or

$$i(t) = \text{Re}\{I_m e^{j\phi} e^{\mathbf{s}t}\}$$

where the complex frequency of both the source and the response must be identical.

If we now recall that the *real* part of a complex forcing function produces the *real* part of the response while the *imaginary* part of the complex forcing function causes the *imaginary* part of the response, then we are again led to the application of a *complex* forcing function to our network. We will obtain a complex response whose real part is the desired real response. Actually, we will work with the Re{ } notation omitted, but we should realize that it may be reinserted at any time and that it *must* be reinserted whenever we desire the time-domain response. Thus, given the real forcing function

$$v(t) = \text{Re}\{V_m e^{j\theta} e^{\mathbf{s}t}\}$$

we apply the complex forcing function $V_m e^{j\theta} e^{\mathbf{s}t}$; the resultant forced response $I_m e^{j\phi} e^{\mathbf{s}t}$ is complex, and it must have as its real part the desired time-domain forced response

$$i(t) = \text{Re}\{I_m e^{j\phi} e^{\mathbf{s}t}\}$$

The solution of our circuit analysis problem consists of the determination of the unknown response amplitude I_m and phase angle ϕ.

Before we actually carry out the details of an analysis problem and see how the procedure resembles what we used in sinusoidal analysis, it is worthwhile to outline the steps of the basic method.

- We first characterize the circuit with a set of loop or nodal integrodifferential equations.
- The given forcing functions, in complex form, and the assumed forced responses, also in complex form, are then substituted in the equations and the indicated integrations and differentiations are performed.
- Each term in every equation will then contain the same factor e^{st}. We therefore divide throughout by this factor, or "suppress e^{st}," understanding that it must be reinserted if a time-domain description of any response function is desired.

With the Re{ } notation and the e^{st} factor gone, we have converted all the voltages and currents from the *time domain* to the *frequency domain*. The integrodifferential equations have become algebraic equations, and their solution is obtained just as easily as in the sinusoidal steady state. Let us illustrate the basic method by a numerical example.

EXAMPLE 14.1

Apply the forcing function $v(t) = 60e^{-2t}\cos(4t + 10°)$ V to the series *RLC* circuit shown in Fig. 14.1, and specify the forced response by finding values for I_m and ϕ in the time-domain expression $i(t) = I_m e^{-2t}\cos(4t + \phi)$.

We first express the forcing function in Re{ } notation:

$$v(t) = 60e^{-2t}\cos(4t + 10°) = \text{Re}\{60e^{-2t}e^{j(4t+10°)}\}$$

$$= \text{Re}\{60e^{j10°}e^{(-2+j4)t}\}$$

or

$$v(t) = \text{Re}\{\mathbf{V}e^{st}\}$$

where

$$\mathbf{V} = 60\underline{/10°} \qquad \text{and} \qquad s = -2 + j4$$

After dropping Re{ }, we are left with the complex forcing function

$$60\underline{/10°}e^{st}$$

In a similar manner, we represent the unknown response by the complex quantity $\mathbf{I}e^{st}$, where $\mathbf{I} = I_m\underline{/\phi}$.

Our next step must be the integrodifferential equation for the circuit. From Kirchhoff's voltage law, we obtain

$$v(t) = Ri + L\frac{di}{dt} + \frac{1}{C}\int i\,dt = 2i + 3\frac{di}{dt} + 10\int i\,dt$$

(Continued on next page)

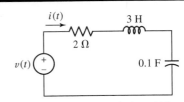

FIGURE 14.1 A series *RLC* circuit to which a damped sinusoidal forcing function is applied. A frequency-domain solution for $i(t)$ is desired.

and we substitute the given complex forcing function and the assumed complex forced response in this equation:

$$60\underline{/10°}e^{st} = 2\mathbf{I}e^{st} + 3s\mathbf{I}e^{st} + \frac{10}{\mathbf{s}}\mathbf{I}e^{st}$$

The common factor e^{st} is next suppressed:

$$60\underline{/10°} = 2\mathbf{I} + 3s\mathbf{I} + \frac{10}{\mathbf{s}}\mathbf{I}$$

and thus

$$\mathbf{I} = \frac{60\underline{/10°}}{2 + 3\mathbf{s} + 10/\mathbf{s}}$$

We now let $\mathbf{s} = -2 + j4$ and solve for the complex current $\mathbf{I}$:

$$\mathbf{I} = \frac{60\underline{/10°}}{2 + 3(-2 + j4) + 10/(-2 + j4)}$$

After manipulating the complex numbers, we find

$$\mathbf{I} = 5.37\underline{/-106.6°}$$

Thus, I_m is 5.37 A, ϕ is $-106.6°$, and the forced response is

$$i(t) = 5.37e^{-2t}\cos(4t - 106.6°) \text{ A}$$

We have thus solved the problem by reducing a *calculus*-based expression to an *algebraic* expression. This is only a small indication of the power of the technique we are about to study.

PRACTICE

14.3 Give the phasor current that is equivalent to the time-domain current: (*a*) $24\sin(90t + 60°)$ A; (*b*) $24e^{-10t}\cos(90t + 60°)$ A; (*c*) $24e^{-10t}\cos 60° \times \cos 90t$ A. If $\mathbf{V} = 12\underline{/35°}$ V, find $v(t)$ for $\mathbf{s}$ equal to (*d*) 0; (*e*) -20 s^{-1}; (*f*) $-20 + j5$ s^{-1}.

Ans: $24\underline{/-30°}$ A; $24\underline{/60°}$ A; $12\underline{/0°}$ A; 9.83 V; $9.83e^{-20t}$ V; $12e^{-20t}\cos(5t + 35°)$ V.

14.3 • DEFINITION OF THE LAPLACE TRANSFORM

Our constant goal has been one of analysis: given some forcing function at one point in a linear circuit, determine the response at some other point. For the first several chapters, we played only with dc forcing functions and responses of the form V_0e^0. However, after the introduction of inductance and capacitance, the sudden dc excitation of simple *RL* and *RC* circuits produced responses varying exponentially with time: $V_0e^{\sigma t}$. When we considered the *RLC* circuit, the responses took on the form of the exponentially varying sinusoid, $V_0e^{\sigma t}\cos(\omega t + \theta)$. All this work was accomplished in the time domain, and the dc forcing function was the only one we considered.

As we advanced to the use of the sinusoidal forcing function, the tedium and complexity of solving the integrodifferential equations caused us to begin casting about for an easier way to work problems. The phasor transform was the result, and we might remember that we were led to it through consideration of a complex forcing function of the form $V_0 e^{j\theta} e^{j\omega t}$. As soon as we concluded that we did not need the factor containing t, we were left with the phasor $V_0 e^{j\theta}$; we had arrived at the *frequency domain.*

Now a little flexing of our cerebral cortex has caused us to apply a forcing function of the form $V_0 e^{j\theta} e^{(\sigma + j\omega)t}$, leading to the invention of the complex frequency $\mathbf{s}$, and thereby relegating all our previous functional forms to special cases: dc ($\mathbf{s} = 0$), exponential ($\mathbf{s} = \sigma$), sinusoidal ($\mathbf{s} = j\omega$), and exponential sinusoid ($\mathbf{s} = \sigma + j\omega$). By analogy to our previous experience with phasors, we saw that in these cases we may omit the factor containing t, and once again obtain a solution by working in the frequency domain.

The Two-Sided Laplace Transform

We know that sinusoidal forcing functions lead to sinusoidal responses, and also that exponential forcing functions lead to exponential responses. However, as practicing engineers we will encounter many waveforms that are neither sinusoidal nor exponential, such as square waves, sawtooth waveforms, and pulses beginning at arbitrary instants of time. When such forcing functions are applied to a linear circuit, we will see that the response is neither similar to the form of the excitation waveform nor exponential. As a result, we are not able to eliminate the terms containing t to form a frequency-domain response. This is rather unfortunate, as working in the frequency domain has proved to be much more pleasant.

There is a solution, however, which makes use of a technique that allows us to expand any function into a *sum* of exponential waveforms, each with its own complex frequency. Since we are considering linear circuits, we know that the total response of our circuit can be obtained by simply adding the individual response to each exponential waveform. And, in dealing with each exponential waveform, we may once again neglect any terms containing t, and work instead in the *frequency* domain. It unfortunately takes an infinite number of exponential terms to accurately represent a general time function, so that taking a brute-force approach and applying superposition to the exponential series might be somewhat insane. Instead, we will sum these terms by performing an integration, leading to a frequency-domain function.

We formalize this approach using what is known as a ***Laplace transform,*** defined for a general function $f(t)$ as

$$\mathbf{F(s)} = \int_{-\infty}^{\infty} e^{-\mathbf{s}t} f(t) \, dt \qquad [12]$$

The mathematical derivation of this integral operation requires an understanding of Fourier series and the Fourier transform, which are discussed in Chap. 18. The fundamental concept behind the Laplace transform, however, can be understood based on our discussion of complex frequency and our prior experience with phasors and converting back and forth between the time domain and the frequency domain. In fact, that is precisely what the Laplace transform does: it converts the general time-domain function $f(t)$ into a corresponding frequency-domain representation, $\mathbf{F(s)}$.

The Two-Sided Inverse Laplace Transform

Equation [12] defines the *two-sided,* or *bilateral,* Laplace transform of $f(t)$. The term *two-sided* or *bilateral* is used to emphasize the fact that both positive and negative values of t are included in the range of integration. The inverse operation, often referred to as the ***inverse Laplace transform,*** is also defined as an integral expression[1]

$$f(t) = \frac{1}{2\pi j} \int_{\sigma_0-j\infty}^{\sigma_0+j\infty} e^{st} \mathbf{F}(\mathbf{s}) \, d\mathbf{s} \qquad [13]$$

where the real constant σ_0 is included in the limits to ensure convergence of this improper integral; the two equations [12] and [13] constitute the two-sided Laplace transform pair. The good news is that Eq. [13] need never be invoked in the study of circuit analysis: there is a quick and easy alternative to look forward to learning.

The One-Sided Laplace Transform

In many of our circuit analysis problems, the forcing and response functions do not exist forever in time, but rather they are initiated at some specific instant that we usually select as $t = 0$. Thus, for time functions that do not exist for $t < 0$, or for those time functions whose behavior for $t < 0$ is of no interest, the time-domain description can be thought of as $v(t)u(t)$. The defining integral for the Laplace transform is taken with the lower limit at $t = 0^-$ in order to include the effect of any discontinuity at $t = 0$, such as an impulse or a higher-order singularity. The corresponding Laplace transform is then

$$\mathbf{F}(\mathbf{s}) = \int_{-\infty}^{\infty} e^{-st} f(t) u(t) \, dt = \int_{0^-}^{\infty} e^{-st} f(t) \, dt$$

This defines the *one-sided* Laplace transform of $f(t)$, or simply the *Laplace transform* of $f(t)$, one-sided being understood. The inverse transform expression remains unchanged, but when evaluated, it is understood to be valid only for $t > 0$. Here then is the definition of the Laplace transform pair that we will use from now on:

$$\boxed{\mathbf{F}(\mathbf{s}) = \int_{0^-}^{\infty} e^{-st} f(t) \, dt} \qquad [14]$$

$$\boxed{\begin{array}{l} f(t) = \dfrac{1}{2\pi j} \displaystyle\int_{\sigma_0-j\infty}^{\sigma_0+j\infty} e^{st} \mathbf{F}(\mathbf{s}) \, d\mathbf{s} \\[4mm] f(t) \Leftrightarrow \mathbf{F}(\mathbf{s}) \end{array}} \qquad [15]$$

The script $\mathcal{L}$ may also be used to indicate the direct or inverse Laplace transform operation:

$$\mathbf{F}(\mathbf{s}) = \mathcal{L}\{f(t)\} \qquad \text{and} \qquad f(t) = \mathcal{L}^{-1}\{\mathbf{F}(\mathbf{s})\}$$

---

(1) If we ignore the distracting factor of $1/2\pi j$ and view the integral as a summation over all frequencies such that $f(t) \propto \Sigma[\mathbf{F}(\mathbf{s}) \, d\mathbf{s}] e^{st}$, this reinforces the notion that $f(t)$ is indeed a sum of complex frequency terms having a magnitude proportional to $\mathbf{F}(\mathbf{s})$.

EXAMPLE **14.2**

Find the Laplace transform of the function $f(t) + 2u(t - 3)$**.**

In order to find the one-sided Laplace transform of $f(t) = 2u(t - 3)$, we must evaluate the integral

$$\mathbf{F(s)} = \int_{0^-}^{\infty} e^{-st} f(t) \, dt$$

$$= \int_{0^-}^{\infty} e^{-st} 2u(t - 3) \, dt$$

$$= 2 \int_{3}^{\infty} e^{-st} \, dt$$

Simplifying, we find

$$\mathbf{F(s)} = \frac{-2}{\mathbf{s}} e^{-st} \bigg|_{3}^{\infty} = \frac{-2}{\mathbf{s}}(0 - e^{-3\mathbf{s}}) = \frac{2}{\mathbf{s}} e^{-3\mathbf{s}}$$

PRACTICE
●

14.4 Let $f(t) = -6e^{-2t}[u(t + 3) - u(t - 2)]$. Find the (*a*) two-sided $\mathbf{F(s)}$; (*b*) one-sided $\mathbf{F(s)}$.

Ans: $\frac{6}{2+s}[e^{-4-2s} - e^{6+3s}]$; $\frac{6}{2+s}[e^{-4-2s} - 1]$.

14.4 • LAPLACE TRANSFORMS OF SIMPLE
TIME FUNCTIONS

In this section we will begin to build up a catalog of Laplace transforms for those time functions most frequently encountered in circuit analysis; we will assume for now that the function of interest is a voltage, although such a choice is strictly arbitrary. We will create this catalog, at least initially, by utilizing the definition,

$$\mathbf{V(s)} = \int_{0^-}^{\infty} e^{-st} v(t) \, dt = \mathcal{L}\{v(t)\}$$

which, along with the expression for the inverse transform,

$$v(t) = \frac{1}{2\pi j} \int_{\sigma_0 - j\infty}^{\sigma_0 + j\infty} e^{st} \mathbf{V(s)} \, d\mathbf{s} = \mathcal{L}^{-1}\{\mathbf{V(s)}\}$$

establishes a one-to-one correspondence between $v(t)$ and $\mathbf{V(s)}$. That is, for every $v(t)$ for which $\mathbf{V(s)}$ exists, there is a unique $\mathbf{V(s)}$. At this point, we may be looking with some trepidation at the rather ominous form given for the inverse transform. Fear not! As we will see shortly, *an introductory study of Laplace transform theory does not require actual evaluation of this integral*. By going from the time domain to the frequency domain and taking advantage of the uniqueness just mentioned, we will be able to generate a catalog of transform pairs that will already contain the corresponding time function for nearly every transform that we wish to invert.

Before we continue, however, we should pause to consider whether there is any chance that the transform may not even exist for some $v(t)$ that concerns us. A set of conditions sufficient to ensure the absolute convergence of the Laplace integral for $\text{Re}\{s\} > \sigma_0$ is

1. The function $v(t)$ is integrable in every finite interval $t_1 < t < t_2$, where $0 \leq t_1 < t_2 < \infty$.

2. $\lim_{t \to \infty} e^{-\sigma_0 t} |v(t)|$ exists for some value of σ_0.

Time functions that do not satisfy these conditions are seldom encountered by the circuit analyst.[2]

The Unit-Step Function $u(t)$

Now let us look at some specific transforms. We first examine the Laplace transform of the unit-step function $u(t)$. From the defining equation, we may write

$$\mathcal{L}\{u(t)\} = \int_{0^-}^{\infty} e^{-st} u(t)\, dt = \int_0^{\infty} e^{-st}\, dt$$

$$= -\frac{1}{s} e^{-st} \Big|_0^{\infty} = \frac{1}{s}$$

for $\text{Re}\{s\} > 0$, to satisfy condition 2. Thus,

$$u(t) \Leftrightarrow \frac{1}{s} \qquad\qquad [16]$$

and our first Laplace transform pair has been established with great ease.

> The double arrow notation is commonly used to indicate Laplace transform pairs.

The Unit-Impulse Function $\delta(t - t_0)$

A singularity function whose transform is of considerable interest is the unit-impulse function $\delta(t - t_0)$. This function, plotted in Fig. 14.2, seems rather strange at first but is enormously useful in practice. The unit-impulse function is defined to have an area of unity, so that

$$\delta(t - t_0) = 0 \qquad t \neq t_0$$

$$\int_{t_0 - \varepsilon}^{t_0 + \varepsilon} \delta(t - t_0)\, dt = 1$$

where ε is a small constant. Thus, this "function" (a naming that makes many purist mathematicians cringe), has a nonzero value only at the point t_0. For $t_0 > 0^-$, we therefore find the Laplace transform to be

$$\mathcal{L}\{\delta(t - t_0)\} = \int_{0^-}^{\infty} e^{-st} \delta(t - t_0)\, dt = e^{-st_0}$$

$$\delta(t - t_0) \Leftrightarrow e^{-st_0} \qquad\qquad [17]$$

In particular, note that we obtain

$$\delta(t) \Leftrightarrow 1 \qquad\qquad [18]$$

for $t_0 = 0$.

An interesting feature of the unit-impulse function is known as the *sifting property.* Consider the integral of the impulse function multiplied by

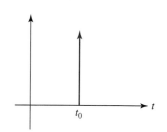

■ FIGURE 14.2 The unit-impulse function $\delta(t - t_0)$. This function is often used to approximate a signal pulse whose duration is very short compared to circuit time constants.

---

(2) Examples of such functions are e^{t^2} and e^{e^t}, but not t^n or n^t. For a somewhat more detailed discussion of the Laplace transform and its applications, refer to Clare D. McGillem and George R. Cooper, *Continuous and Discrete Signal and System Analysis*, 3d ed. Oxford University Press, North Carolina: 1991, Chap. 5.

an arbitrary function $f(t)$:

$$\int_{-\infty}^{\infty} f(t)\delta(t - t_0)\, dt$$

Since the function $\delta(t - t_0)$ is zero everywhere except at $t = t_0$, the value of this integral is simply $f(t_0)$. The property turns out to be *very* useful in simplifying integral expressions containing the unit-impulse function.

The Exponential Function $e^{-\alpha t}$

Recalling our past interest in the exponential function, we examine its transform,

$$\mathcal{L}\{e^{-\alpha t}u(t)\} = \int_{0^-}^{\infty} e^{-\alpha t}e^{-st}\, dt$$

$$= -\frac{1}{s + \alpha}e^{-(s+\alpha)t}\Big|_0^{\infty} = \frac{1}{s + \alpha}$$

and therefore,

$$e^{-\alpha t}u(t) \iff \frac{1}{s + \alpha} \qquad [19]$$

It is understood that $\text{Re}\{s\} > -\alpha$.

The Ramp Function $t\, u(t)$

As a final example, for the moment, let us consider the ramp function $tu(t)$. We obtain

$$\mathcal{L}\{tu(t)\} = \int_{0^-}^{\infty} te^{-st}\, dt = \frac{1}{s^2}$$

$$tu(t) \Leftrightarrow \frac{1}{s^2} \qquad [20]$$

either by a straightforward integration by parts or from a table of integrals.

So what of the function $te^{-\alpha t}u(t)$? We leave it to the reader to show that

$$te^{-\alpha t}u(t) \Leftrightarrow \frac{1}{(s + \alpha)^2} \qquad [21]$$

There are, of course, quite a few additional time functions worth considering, but it may be best if we pause for the moment to consider the reverse of the process—the inverse Laplace transform—before returning to add to our list.

PRACTICE

14.5 Determine $V(s)$ if $v(t)$ equals (*a*) $4\delta(t) - 3u(t)$; (*b*) $4\delta(t - 2) - 3tu(t)$; (*c*) $[u(t)][u(t - 2)]$.
14.6 Determine $v(t)$ if $V(s)$ equals (*a*) 10; (*b*) $10/s$; (*c*) $10/s^2$; (*d*) $10/[s(s + 10)]$; (*e*) $10s/(s + 10)$.

Ans: 14.5: $(4s - 3)/s$; $4e^{-2s} - (3/s^2)$; e^{-2s}/s. 14.6: $10\delta(t)$; $10u(t)$; $10tu(t)$; $u(t) - e^{-10t}u(t)$; $10\delta(t) - 100e^{-10t}u(t)$.

14.5 · INVERSE TRANSFORM TECHNIQUES

The Linearity Theorem

We did mention that an integral expression (Eq. [13]) could be applied to convert an **s**-domain expression back into a time-domain expression. We also alluded to the fact that such an approach could be avoided by exploiting the uniqueness of any Laplace transform pair. In order to fully capitalize on this fact, we must first introduce one of several helpful and well-known Laplace transform theorems—the *linearity theorem*. This theorem states that the Laplace transform of the sum of two or more time functions is equal to the sum of the transforms of the individual time functions. For two time functions we have

$$\mathcal{L}\{f_1(t) + f_2(t)\} = \int_{0^-}^{\infty} e^{-st}[f_1(t) + f_2(t)]\,dt$$

$$= \int_{0^-}^{\infty} e^{-st} f_1(t)\,dt + \int_{0^-}^{\infty} e^{-st} f_2(t)\,dt$$

$$= \mathbf{F}_1(\mathbf{s}) + \mathbf{F}_2(\mathbf{s})$$

This is known as the "additive property" of the Laplace transform.

As an example of the use of this theorem, suppose that we have a Laplace transform $\mathbf{V}(\mathbf{s})$ and want to know the corresponding time function $v(t)$. It will often be possible to decompose $\mathbf{V}(\mathbf{s})$ into the sum of two or more functions, say, $\mathbf{V}_1(\mathbf{s})$ and $\mathbf{V}_2(\mathbf{s})$, whose inverse transforms, $v_1(t)$ and $v_2(t)$, are already tabulated. It then becomes a simple matter to apply the linearity theorem and write

$$v(t) = \mathcal{L}^{-1}\{\mathbf{V}(\mathbf{s})\} = \mathcal{L}^{-1}\{\mathbf{V}_1(\mathbf{s}) + \mathbf{V}_2(\mathbf{s})\}$$

$$= \mathcal{L}^{-1}\{\mathbf{V}_1(\mathbf{s})\} + \mathcal{L}^{-1}\{\mathbf{V}_2(\mathbf{s})\} = v_1(t) + v_2(t)$$

Another important consequence of the linearity theorem is evident by studying the definition of the Laplace transform. Since we are simply working with an integral, *the Laplace transform of a constant times a function is equal to the constant times the Laplace transform of the function*. In other words,

$$\mathcal{L}\{kv(t)\} = k\mathcal{L}\{v(t)\}$$

This is known as the "homogeneity property" of the Laplace transform.

or

$$kv(t) \Leftrightarrow k\mathbf{V}(\mathbf{s}) \tag{22}$$

where k is a constant of proportionality. This result is extremely handy in many situations that arise from circuit analysis, as we are about to see.

EXAMPLE 14.3

Given the function $G(s) = 7/s - 31/(s + 17)$, find $g(t)$.

This **s**-domain function is composed of the sum of two terms, $7/\mathbf{s}$ and $-31/(\mathbf{s} + 17)$. Through the linearity theorem we know that $g(t)$ will be composed of two terms as well, each the inverse Laplace transform of one of the two **s**-domain terms:

$$g(t) = \mathcal{L}^{-1}\left\{\frac{7}{\mathbf{s}}\right\} - \mathcal{L}^{-1}\left\{\frac{31}{\mathbf{s}+17}\right\}.$$

Let's begin with the first term. The homogeneity property of the Laplace transform allows us to write that

$$\mathcal{L}^{-1}\left\{\frac{7}{\mathbf{s}}\right\} = 7\mathcal{L}^{-1}\left\{\frac{1}{\mathbf{s}}\right\} = 7u(t).$$

Thus, we have made use of the known transform pair $u(t) \Leftrightarrow 1/\mathbf{s}$ and the homogeneity property to find this first component of $g(t)$. In a similar fashion, we find that $\mathcal{L}^{-1}\left\{\dfrac{31}{\mathbf{s}+17}\right\} = 31e^{-17t}u(t)$. Putting these two terms together,

$$g(t) = [7 - 31e^{-17t}]u(t).$$

PRACTICE

14.7 Given the function $\mathbf{H(s)} = \dfrac{7}{\mathbf{s}^2} + \dfrac{31}{(\mathbf{s}+17)^2}$, find $h(t)$.

Ans: $h(t) = [7 + 31e^{-17t}]tu(t)$.

Inverse Transform Techniques for Rational Functions

In analyzing circuits with multiple energy-storage elements, we will often encounter **s**-domain expressions that are ratios of **s**-polynomials. We thus expect to routinely encounter expressions of the form

$$\mathbf{V(s)} = \frac{\mathbf{N(s)}}{\mathbf{D(s)}}$$

where $\mathbf{N(s)}$ and $\mathbf{D(s)}$ are polynomials in **s**. The values of **s** which lead to $\mathbf{N(s)} = 0$ are referred to as *zeros* of $\mathbf{V(s)}$, and those values of **s** which lead to $\mathbf{D(s)} = 0$ are referred to as *poles* of $\mathbf{V(s)}$.

Rather than rolling up our sleeves and invoking Eq. [13] each time we need to find an inverse transform, it is often possible to decompose these expressions using the method of residues into simpler terms whose inverse transforms are already known. The criterion for this is that $\mathbf{V(s)}$ must be a *rational function* for which the degree of the numerator $\mathbf{N(s)}$ must be less than that of the denominator $\mathbf{D(s)}$. If it is not, we must first perform a simple division step, as shown in the following example. The result will include an impulse function (assuming the degree of the numerator is the same as that of the denominator) and a rational function. The inverse transform of the first is simple; the straightforward method of residues applies to the rational function if its inverse transform is not already known.

In practice, it is seldom necessary to ever invoke Eq. [13] for functions encountered in circuit analysis, provided that we are clever in using the various techniques presented in this chapter.

EXAMPLE 14.4

Find the inverse transform of $F(s) = 2\dfrac{s+2}{s}$.

$F(s)$ is not a rational function, so we begin by performing long division:

$$F(s) = s\overline{)\begin{array}{l} 2 \\ 2s + 4 \\ \underline{2s} \\ 4 \end{array}}$$

so that $F(s) = 2 + (4/s)$. By the linearity theorem,

$$\mathcal{L}^{-1}\{F(s)\} = \mathcal{L}^{-1}\{2\} + \mathcal{L}^{-1}\left\{\frac{4}{s}\right\} = 2\delta(t) + 4u(t).$$

(It should be noted that this particular function can be simplified without the process of long division; such a route was chosen to provide an example of the basic process.)

PRACTICE

14.8 Given the function $Q(s) = \dfrac{3s^2 - 4}{s^2}$, find $q(t)$.

Ans: $q(t) = 3\delta(t) - 4tu(t)$.

In employing the method of residues, essentially performing a partial fraction expansion of $V(s)$, we focus our attention on the roots of the denominator. Thus, it is first necessary to factor the s-polynomial that comprises $D(s)$ into a product of binomial terms. The roots of $D(s)$ may be any combination of distinct or repeated roots, and may be real or complex. It is worth noting, however, that complex roots always occur as conjugate pairs, provided that the coefficients of $D(s)$ are real.

Distinct Poles and the Method of Residues

As a specific example, let us determine the inverse Laplace transform of

$$V(s) = \frac{1}{(s+\alpha)(s+\beta)}$$

The denominator has been factored into two distinct roots, $-\alpha$ and $-\beta$. Although it is possible to substitute this expression in the defining equation for the inverse transform, it is much easier to utilize the linearity theorem. Using partial-fraction expansion, we can split the given transform into the sum of two simpler transforms,

$$V(s) = \frac{A}{(s+\alpha)} + \frac{B}{(s+\beta)}$$

where A and B may be found by any of several methods. Perhaps the quickest solution is obtained by recognizing that

$$A = \lim_{s \to -\alpha}\left[(s+\alpha)V(s) - \frac{(s+\alpha)}{(s+\beta)}B\right]$$

$$= \lim_{s \to -\alpha}\left[\frac{1}{(s+\beta)} - 0\right] = \frac{1}{\beta - \alpha}$$

In this equation, we use the single-fraction (i.e., nonexpanded) version of $V(s)$.

Recognizing that the second term is always zero, in practice we would simply write

$$A = (s + \alpha)\mathbf{V}(s)|_{s=-\alpha}$$

Similarly,

$$B = (s + \beta)\mathbf{V}(s)|_{s=-\beta} = \frac{1}{\alpha - \beta}$$

and therefore,

$$\mathbf{V}(s) = \frac{1/(\beta - \alpha)}{(s + \alpha)} + \frac{1/(\alpha - \beta)}{(s + \beta)}$$

We have already evaluated inverse transforms of this form, and so

$$v(t) = \frac{1}{\beta - \alpha}e^{-\alpha t}u(t) + \frac{1}{\alpha - \beta}e^{-\beta t}u(t)$$

$$= \frac{1}{\beta - \alpha}(e^{-\alpha t} - e^{-\beta t})u(t)$$

If we wished, we could now include this as a new entry in our catalog of Laplace pairs,

$$\frac{1}{\beta - \alpha}(e^{-\alpha t} - e^{-\beta t})u(t) \Leftrightarrow \frac{1}{(s + \alpha)(s + \beta)}$$

This approach is easily extended to functions whose denominators are higher-order **s**-polynomials, although the operations can become somewhat tedious. It should also be noted that we did not specify that the constants A and B must be real. However, in situations where α and β are complex, we will find that α and β are also complex conjugates (this is not required mathematically, but is required for physical circuits). In such instances, we will also find that $A = B^*$; in other words, the coefficients will be complex conjugates as well.

EXAMPLE 14.5

Find the inverse transform of

$$\mathbf{P}(s) = \frac{7s + 5}{s^2 + s}.$$

We see that $\mathbf{P}(s)$ is a rational function (the degree of the numerator is *one*, whereas the degree of the denominator is *two*), so we begin by factoring the denominator and write:

$$\mathbf{P}(s) = \frac{7s + 5}{s(s + 1)} = \frac{a}{s} + \frac{b}{s + 1}.$$

where our next step is to determine values for a and b. Applying the method of residues,

$$a = \frac{7s + 5}{s + 1}\bigg|_{s=0} = 5, \quad \text{and} \quad b = \frac{7s + 5}{s}\bigg|_{s=-1} = 2.$$

(Continued on next page)

We may then write $\mathbf{P(s)}$ as

$$\mathbf{P(s)} = \frac{5}{\mathbf{s}} + \frac{2}{\mathbf{s}+1}$$

the inverse transform of which is simply $p(t) = [5 + 2e^{-t}]u(t)$.

PRACTICE

14.9   Given the function $\mathbf{Q(s)} = \dfrac{11s + 30}{s^2 + 3s}$, find $q(t)$.

Ans: $q(t) = [10 + e^{-3t}]u(t)$.

Repeated Poles

The remaining situation is that of repeated poles. Consider the function

$$\mathbf{V(s)} = \frac{\mathbf{N(s)}}{(\mathbf{s} - p)^n}$$

which we will expand into

$$\mathbf{V(s)} = \frac{a_n}{(\mathbf{s} - p)^n} + \frac{a_{n-1}}{(\mathbf{s} - p)^{n-1}} + \cdots + \frac{a_1}{(\mathbf{s} - p)}$$

To determine each constant, we first multiply the nonexpanded version of $\mathbf{V(s)}$ by $(\mathbf{s} - p)^n$. The constant a_n is found by simply evaluating the resulting expression at $\mathbf{s} = p$. The remaining constants are found by differentiating the expression $(\mathbf{s} - p)^n \mathbf{V(s)}$ the appropriate number of times prior to evaluating at $\mathbf{s} = p$, and dividing by a factorial term. The differentiation procedure removes the constants previously found, and evaluating at $\mathbf{s} = p$ removes the remaining constants. For example, a_{n-2} is found by evaluating

$$\frac{1}{2!} \frac{d^2}{d\mathbf{s}^2} [(\mathbf{s} - p)^n \mathbf{V(s)}]_{\mathbf{s}=p}$$

and the term a_{n-k} is found by evaluating

$$\frac{1}{k!} \frac{d^k}{d\mathbf{s}^k} [(\mathbf{s} - p)^n \mathbf{V(s)}]_{\mathbf{s}=p}$$

To illustrate the basic procedure, let's find the inverse Laplace transform of a function having a combination of both situations: one pole at $\mathbf{s} = 0$ and two poles at $\mathbf{s} = -6$.

EXAMPLE 14.6

Find the inverse transform of the function
$$\mathbf{V(s)} = \frac{2}{s^3 + 12s^2 + 36s}$$

We note that the denominator can be easily factored, leading to

$$\mathbf{V(s)} = \frac{2}{\mathbf{s(s + 6)(s + 6)}} = \frac{2}{\mathbf{s(s + 6)^2}}$$

As promised, we see that there are indeed three poles, one at $s = 0$, and two at $s = -6$. Next, we expand the function into

$$\mathbf{V(s)} = \frac{a_1}{(s+6)^2} + \frac{a_2}{(s+6)} + \frac{a_3}{s}$$

and apply our new procedure to obtain the unknown constants a_1 and a_2; we will find a_3 using the previous procedure. Thus,

$$a_1 = \left[(s+6)^2 \frac{2}{s(s+6)^2} \right]_{s=-6} = \frac{2}{s} \bigg|_{s=-6} = \frac{-1}{3}$$

and

$$a_2 = \frac{d}{ds} \left[(s+6)^2 \frac{2}{s(s+6)^2} \right]_{s=-6} = \frac{d}{ds} \left(\frac{2}{s} \right) \bigg|_{s=-6} = \frac{-2}{s^2} \bigg|_{s=-6} = \frac{-1}{18}$$

The remaining constant a_3 is found using the procedure for distinct poles

$$a_3 = s \frac{2}{s(s+6)^2} \bigg|_{s=0} = \frac{2}{6^2} = \frac{1}{18}$$

Thus, we may now write $\mathbf{V(s)}$ as

$$\mathbf{V(s)} = \frac{-\frac{1}{3}}{(s+6)^2} + \frac{-\frac{1}{18}}{(s+6)} + \frac{\frac{1}{18}}{s}$$

Using the linearity theorem, the inverse transform of $\mathbf{V(s)}$ can now be found by simply determining the inverse transform of each of these three terms. We see that the first term on the right is of the form

$$\frac{1}{(s+\alpha)^2}$$

and making use of Eq. [21] we find that its inverse transform is $-\frac{1}{3}te^{-6t}u(t)$. In a similar fashion, we find that the inverse transform of the second term is $-\frac{1}{18}e^{-6t}u(t)$, and that of the third term is simply $\frac{1}{18}u(t)$. Thus,

$$v(t) = -\tfrac{1}{3}te^{-6t}u(t) - \tfrac{1}{18}e^{-6t}u(t) + \tfrac{1}{18}u(t)$$

or, more compactly,

$$v(t) = \tfrac{1}{18}[1 - (1+6t)e^{-6t}]u(t)$$

PRACTICE

14.10 Find $v(t)$ if $\mathbf{V(s)} = 2s/(s^2+4)^2$.

Ans: $\frac{1}{2}t \sin 2t u(t)$.

COMPUTER-AIDED ANALYSIS

MATLAB, a very powerful numerical analysis package, can be used to assist in the solution of equations arising from the analysis of circuits with time-varying excitation in several different ways. The most straightforward technique makes use of ordinary differential equation

(ODE) solver routines *ode*23() and *ode*45(). These two routines are based on numerical methods of solving differential equations, with *ode*45() having greater accuracy. The solution is determined only at discrete points, however, and therefore is not known for all values of time. For many applications this is adequate, provided a sufficient density of points is used.

The Laplace transform technique provides us with the means of obtaining an exact expression for the solution of differential equations, and as such has many advantages over the use of numerical ODE solution techniques. Another significant advantage to the Laplace transform technique will become apparent in subsequent chapters when we study the significance of the form of *s*-domain expressions, particularly once we factor the denominator polynomials.

As we have already seen, lookup tables are extremely handy when working with Laplace transforms, although the method of residues can become somewhat tedious for functions with high-order polynomials in their denominators. In these instances MATLAB can also be of assistance, as it contains several useful functions for the manipulation of polynomial expressions.

In MATLAB, the polynomial

$$p(x) = a_n x^n + a_{n-1} x^{n-1} + \cdots + a_1 x + a_0$$

is stored as the vector $[a_n\, a_{n-1}\, \ldots\, a_1\, a_0]$.

Thus, to define the polynomials $\mathbf{N(s)} = 2$ and $\mathbf{D(s)} = \mathbf{s}^3 + 12\mathbf{s}^2 + 36\mathbf{s}$ we write

EDU» N = [2];

EDU» D = [1 12 36 0];

The roots of either polynomial can be obtained by invoking the function *roots*(**p**), where **p** is a vector containing the coefficients of the polynomial. For example,

EDU» q = [1 8 16];

EDU» roots(q)

yields

ans =

−4

−4

MATLAB also enables us to determine the residues of the rational function $\mathbf{N(s)}/\mathbf{D(s)}$ using the function *residue*(). For example,

EDU» [r p y] = residue(N, D);

returns three vectors **r**, **p**, and **y**, such that

$$\frac{\mathbf{N(s)}}{\mathbf{D(s)}} = \frac{r_1}{x - p_1} + \frac{r_2}{x - p_2} + \cdots + \frac{r_n}{x - p_n} + \mathbf{y(s)}$$

in the case of no multiple poles, and in the case of *n* multiple poles

$$\frac{\mathbf{N(s)}}{\mathbf{D(s)}} = \frac{r_1}{(x - p)} + \frac{r_2}{(x - p)^2} + \cdots + \frac{r_n}{(x - p)^n} + \mathbf{y(s)}$$

Note that as long as the order of the numerator polynomial is less than the order of the denominator polynomial, the vector $\mathbf{y(s)}$ will always be empty.

Executing the command without the semicolon results in the output

$$r =$$
$$-0.0556$$
$$-0.3333$$
$$0.0556$$

$$p =$$
$$-6$$
$$-6$$
$$0$$

$$y =$$
$$[\,]$$

which agrees with the answer found in Example 14.6.

14.6 • BASIC THEOREMS FOR THE LAPLACE TRANSFORM

We are now able to consider two theorems that might be considered collectively the *raison d'être* for Laplace transforms in circuit analysis—the time differentiation and integration theorems. These will help us transform the derivatives and integrals appearing in the time-domain circuit equations.

Time Differentiation Theorem

Let us look at time differentiation first by considering a time function $v(t)$ whose Laplace transform $\mathbf{V(s)}$ is known to exist. We want the transform of the first derivative of $v(t)$,

$$\mathcal{L}\left\{\frac{dv}{dt}\right\} = \int_{0^-}^{\infty} e^{-st}\frac{dv}{dt}\,dt$$

This can be integrated by parts:

$$U = e^{-st} \qquad dV = \frac{dv}{dt}dt$$

with the result

$$\mathcal{L}\left\{\frac{dv}{dt}\right\} = v(t)e^{-st}\Big|_{0^-}^{\infty} + \mathbf{s}\int_{0^-}^{\infty} e^{-st}v(t)\,dt$$

The first term on the right must approach zero as t increases without limit; otherwise $\mathbf{V(s)}$ would not exist. Hence,

$$\mathcal{L}\left\{\frac{dv}{dt}\right\} = 0 - v(0^-) + \mathbf{s}\mathbf{V(s)}$$

and

$$\frac{dv}{dt} \Leftrightarrow \mathbf{s}\mathbf{V(s)} - v(0^-) \qquad\qquad [23]$$

Similar relationships may be developed for higher-order derivatives:

$$\frac{d^2v}{dt^2} \Leftrightarrow s^2\mathbf{V}(s) - sv(0^-) - v'(0^-) \qquad [24]$$

$$\frac{d^3v}{dt^3} \Leftrightarrow s^3\mathbf{V}(s) - s^2v(0^-) - sv'(0^-) - v''(0^-) \qquad [25]$$

where $v'(0^-)$ is the value of the first derivative of $v(t)$ evaluated at $t = 0^-$, $v''(0^-)$ is the initial value of the second derivative of $v(t)$, and so forth. When all initial conditions are zero, we see that differentiating once with respect to t in the time domain corresponds to multiplication by $\mathbf{s}$ in the frequency domain; differentiating twice in the time domain corresponds to multiplication by $\mathbf{s}^2$ in the frequency domain, and so on. Thus, *differentiation in the time domain is equivalent to multiplication in the frequency domain*. This is a substantial simplification! We should also begin to see that, when the initial conditions are not zero, their presence is still accounted for. A simple example will serve to demonstrate this.

EXAMPLE 14.7

Given the series *RL* circuit shown in Fig. 14.3, find the current through the 4 Ω resistor.

▶ **Identify the goal of the problem.**
We need to find an expression for the current labeled $i(t)$.

▶ **Collect the known information.**
The network is driven by a step voltage, and we have an initial value of the current (at $t = 0^-$) of 5 A.

▶ **Devise a plan.**
Applying KVL to the circuit will result in a differential equation with $i(t)$ as the unknown. Taking the Laplace transform of both sides of this equation will convert it to the **s**-domain. Solving the resulting algebraic equation for $\mathbf{I}(s)$, the only remaining task will be to take the inverse Laplace transform to find $i(t)$.

▶ **Construct an appropriate set of equations.**
Using KVL to write the single-loop equation in the time domain, we find

$$2\frac{di}{dt} + 4i = 3u(t)$$

Now, we take the Laplace transform of each term, so that

$$2[s\mathbf{I}(s) - i(0^-)] + 4\mathbf{I}(s) = \frac{3}{s}$$

▶ **Determine if additional information is required.**
We have an equation that may be solved for the frequency-domain representation $\mathbf{I}(s)$ of our goal, $i(t)$.

▶ **Attempt a solution.**
We next solve for $\mathbf{I}(s)$, substituting $i(0^-) = 5$:

$$(2s + 4)\mathbf{I}(s) = \frac{3}{s} + 10$$

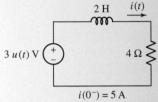

2 H $i(t)$

3 $u(t)$ V   4 Ω

$i(0^-) = 5$ A

■ **FIGURE 14.3** A circuit that is analyzed by transforming the differential equation $2\,di/dt + 4i = 3u(t)$ into $2[s\mathbf{I}(s) - i(0^-)] + 4\mathbf{I}(s) = 3/s$.

and

$$I(s) = \frac{1.5}{s(s+2)} + \frac{5}{s+2}$$

Applying the method of residues to the first term,

$$\frac{1.5}{s+2}\Big|_{s=0} = 0.75 \quad \text{and} \quad \frac{1.5}{s}\Big|_{s=-2} = -0.75$$

so that

$$I(s) = \frac{0.75}{s} + \frac{4.25}{s+2}$$

We then use our known transform pairs to invert:

$$i(t) = 0.75u(t) + 4.25e^{-2t}u(t)$$
$$= (0.75 + 4.25e^{-2t})u(t) \quad \text{A}$$

▶ **Verify the solution. Is it reasonable or expected?**
Based on our previous experience with this type of circuit, we
expected a dc forced response plus an exponentially decaying natural
response. At $t = 0$, we obtain $i(0) = 5$ A, as required, and as $t \to \infty$,
$i(t) \to \frac{3}{4}$ A as we would expect.

Our solution for $i(t)$ is therefore complete. Both the forced response $0.75u(t)$ and the natural response $4.25e^{-2t}u(t)$ are present,
and the initial condition was automatically incorporated into the
solution. The method illustrates a very painless way of obtaining the
complete solution of many differential equations.

PRACTICE
●

14.11 Use Laplace transform methods to find $i(t)$ in the circuit of
Fig. 14.4.

Ans: $(0.25 + 4.75e^{-20t})u(t)$ A.

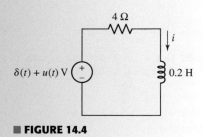

■ **FIGURE 14.4**

Time-Integration Theorem

The same kind of simplification can be accomplished when we meet the
operation of integration with respect to time in our circuit equations. Let
us determine the Laplace transform of the time function described by
$\int_{0^-}^{t} v(x)\,dx$,

$$\mathcal{L}\left\{\int_{0^-}^{t} v(x)\,dx\right\} = \int_{0^-}^{\infty} e^{-st} \left[\int_{0^-}^{t} v(x)\,dx\right] dt$$

Integrating by parts, we let

$$u = \int_{0^-}^{t} v(x)\,dx \qquad dv = e^{-st}\,dt$$

$$du = v(t)\,dt \qquad v = -\frac{1}{s}e^{-st}$$

Then

$$\mathcal{L}\left\{\int_{0^-}^t v(x)\,dx\right\} = \left\{\left[\int_{0^-}^t v(x)\,dx\right]\left[-\frac{1}{s}e^{-st}\right]\right\}_{t=0^-}^{t=\infty} - \int_{0^-}^\infty -\frac{1}{s}e^{-st}v(t)\,dt$$

$$= \left[-\frac{1}{s}e^{-st}\int_{0^-}^t v(x)\,dx\right]_{0^-}^\infty + \frac{1}{s}\mathbf{V}(s)$$

But, since $e^{-st} \to 0$ as $t \to \infty$, the first term on the right vanishes at the upper limit, and when $t \to 0^-$, the integral in this term likewise vanishes. This leaves only the $\mathbf{V}(s)/s$ term, so that

$$\int_{0^-}^t v(x)\,dx \Leftrightarrow \frac{\mathbf{V}(s)}{s} \qquad [26]$$

and thus *integration in the time domain corresponds to division by* s *in the frequency domain*. Once more, a relatively complicated calculus operation in the time domain simplifies to an algebraic operation in the frequency domain.

EXAMPLE 14.8

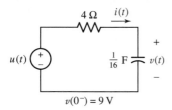

$v(0^-) = 9$ V

■ **FIGURE 14.5** A circuit illustrating the use of the Laplace transform pair $\int_{0^-}^t i(t')\,dt' \Leftrightarrow \frac{1}{s}\mathbf{I}(s)$.

Determine $i(t)$ for $t > 0$ in the series RC circuit shown in Fig. 14.5.

We first write the single-loop equation,

$$u(t) = 4i(t) + 16\int_{-\infty}^t i(t')\,dt'$$

In order to apply the time-integration theorem, we must arrange for the lower limit of integration to be 0^-. Thus, we set

$$16\int_{-\infty}^t i(t')\,dt' = 16\int_{-\infty}^{0^-} i(t')\,dt' + 16\int_{0^-}^t i(t')\,dt'$$

$$= v(0^-) + 16\int_{0^-}^t i(t')\,dt'$$

Therefore,

$$u(t) = 4i(t) + v(0^-) + 16\int_{0^-}^t i(t')\,dt'$$

We next take the Laplace transform of both sides of this equation. Since we are utilizing the one-sided transform, $\mathcal{L}\{v(0^-)\}$ is simply $\mathcal{L}\{v(0^-)u(t)\}$, and thus

$$\frac{1}{s} = 4\mathbf{I}(s) + \frac{9}{s} + \frac{16}{s}\mathbf{I}(s)$$

and solving for $\mathbf{I}(s)$,

$$\mathbf{I}(s) = \frac{-2}{s+4}$$

the desired result is immediately obtained,

$$i(t) = -2e^{-4t}u(t) \qquad \text{A}$$

EXAMPLE **14.9**

Find $v(t)$ for the same circuit, repeated as Fig. 14.6 for convenience.

This time we simply write a single nodal equation,

$$\frac{v(t) - u(t)}{4} + \frac{1}{16}\frac{dv}{dt} = 0$$

Taking the Laplace transform, we obtain

$$\frac{\mathbf{V(s)}}{4} - \frac{1}{4\mathbf{s}} + \frac{1}{16}\mathbf{s}\mathbf{V(s)} - \frac{v(0^-)}{16} = 0$$

or

$$\mathbf{V(s)}\left(1 + \frac{\mathbf{s}}{4}\right) = \frac{1}{\mathbf{s}} + \frac{9}{4}$$

Thus,

$$\mathbf{V(s)} = \frac{4}{\mathbf{s}(\mathbf{s}+4)} + \frac{9}{\mathbf{s}+4}$$

$$= \frac{1}{\mathbf{s}} - \frac{1}{\mathbf{s}+4} + \frac{9}{\mathbf{s}+4}$$

$$= \frac{1}{\mathbf{s}} + \frac{8}{\mathbf{s}+4}$$

and taking the inverse transform,

$$v(t) = (1 + 8e^{-4t})u(t)$$

we quickly obtain the desired capacitor voltage without recourse to the usual differential equation solution.

To check this result, we note that $(\frac{1}{16})dv/dt$ should yield the previous expression for $i(t)$. For $t > 0$,

$$\frac{1}{16}\frac{dv}{dt} = \frac{1}{16}(-32)e^{-4t} = -2e^{-4t}$$

which is in agreement with what was found in Example 14.8.

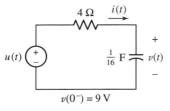

■ **FIGURE 14.6** The circuit of Fig. 14.5 repeated, in which the voltage $v(t)$ is sought.

PRACTICE

14.12 Find $v(t)$ at $t = 800$ ms for the circuit of Fig. 14.7.

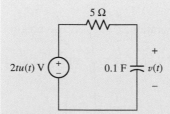

■ **FIGURE 14.7**

Ans: 802 mV.

Laplace Transforms of Sinusoids

To illustrate the use of both the linearity theorem and the time-differentiation theorem, not to mention the addition of a most important pair to our forthcoming Laplace transform table, let us establish the Laplace transform of $\sin \omega t \, u(t)$. We could use the defining integral expression with integration by parts, but this is needlessly difficult. Instead, we use the relationship

$$\sin \omega t = \frac{1}{2j}(e^{j\omega t} - e^{-j\omega t})$$

The transform of the sum of these two terms is just the sum of the transforms, and each term is an exponential function for which we already have the transform. We may immediately write

$$\mathcal{L}\{\sin \omega t \, u(t)\} = \frac{1}{2j}\left(\frac{1}{\mathbf{s} - j\omega} - \frac{1}{\mathbf{s} + j\omega}\right) = \frac{\omega}{\mathbf{s}^2 + \omega^2}$$

$$\sin \omega t \, u(t) \Leftrightarrow \frac{\omega}{\mathbf{s}^2 + \omega^2} \qquad [27]$$

We next use the time-differentiation theorem to determine the transform of $\cos \omega t \, u(t)$, which is proportional to the derivative of $\sin \omega t$. That is,

$$\mathcal{L}\{\cos \omega t \, u(t)\} = \mathcal{L}\left\{\frac{1}{\omega}\frac{d}{dt}[\sin \omega t \, u(t)]\right\} = \frac{1}{\omega}\mathbf{s}\frac{\omega}{\mathbf{s}^2 + \omega^2}$$

$$\cos \omega t \, u(t) \Leftrightarrow \frac{\mathbf{s}}{\mathbf{s}^2 + \omega^2} \qquad [28]$$

Note that we have made use of the fact that $\sin \omega t \big|_{t=0} = 0$.

The Time-Shift Theorem

As we have seen in some of our earlier transient problems, not all forcing functions begin at $t = 0$. What happens to the transform of a time function if that function is simply shifted in time by some known amount? In particular, if the transform of $f(t)u(t)$ is the known function $\mathbf{F}(\mathbf{s})$, then what is the transform of $f(t-a)u(t-a)$, the original time function delayed by a seconds (and not existing for $t < a$)? Working directly from the definition of the Laplace transform, we get

$$\mathcal{L}\{f(t-a)u(t-a)\} = \int_{0^-}^{\infty} e^{-\mathbf{s}t} f(t-a)u(t-a)\, dt$$

$$= \int_{a^-}^{\infty} e^{-\mathbf{s}t} f(t-a)\, dt$$

for $t \geq a^-$. Choosing a new variable of integration, $\tau = t - a$, we obtain

$$\mathcal{L}\{f(t-a)u(t-a)\} = \int_{0^-}^{\infty} e^{-\mathbf{s}(\tau+a)} f(\tau)\, d\tau = e^{-a\mathbf{s}}\mathbf{F}(\mathbf{s})$$

Therefore,

$$f(t-a)u(t-a) \Leftrightarrow e^{-a\mathbf{s}}\mathbf{F}(\mathbf{s}) \qquad (a \geq 0) \qquad [29]$$

This result is known as the *time-shift theorem,* and it simply states that if a time function is delayed by a time a in the time domain, the result in the frequency domain is a multiplication by $e^{-a\mathbf{s}}$.

EXAMPLE 14.10

Determine the transform of the rectangular pulse $v(t) = u(t - 2) - u(t - 5)$.

This pulse, shown plotted in Fig. 14.8, has unit value for the time interval $2 < t < 5$, and zero value elsewhere. We know that the transform of $u(t)$ is just $1/s$, and since $u(t - 2)$ is simply $u(t)$ delayed by 2 s, the transform of this delayed function is e^{-2s}/s. Similarly, the transform of $u(t - 5)$ is e^{-5s}/s. It follows, then, that the desired transform is

$$\mathbf{V(s)} = \frac{e^{-2s}}{s} - \frac{e^{-5s}}{s} = \frac{e^{-2s} - e^{-5s}}{s}$$

It was not necessary to revert to the definition of the Laplace transform in order to determine $\mathbf{V(s)}$.

PRACTICE
●
14.13 Find the Laplace transform of the time function shown in Fig. 14.9.

Ans: $(5/s)(2e^{-2s} - e^{-4s} - e^{-5s})$.

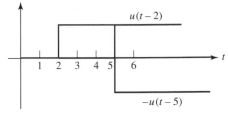

■ **FIGURE 14.8** Plot of $u(t - 2) - u(t - 5)$.

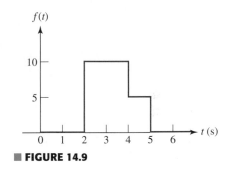

■ **FIGURE 14.9**

At this point we have obtained a number of entries for the catalog of Laplace transform pairs that we agreed to construct earlier. Included are the transforms of the *impulse function,* the *step function,* the *exponential function,* the *ramp function,* the *sine* and *cosine functions,* and the sum of two exponentials. In addition, we have noted the consequences in the **s** domain of the time-domain operations of addition, multiplication by a constant, differentiation, and integration. These results are collected together in Tables 14.1 and 14.2; several others which are derived in Appendix 7 are also included.

TABLE ● 14.1 Laplace Transform Pairs

$f(t) = L^{-1}\{F(s)\}$	$F(s) = L\{f(t)\}$	$f(t) = L^{-1}\{F(s)\}$	$F(s) = L\{f(t)\}$
$\delta(t)$	1	$\dfrac{1}{\beta - \alpha}(e^{-\alpha t} - e^{-\beta t})u(t)$	$\dfrac{1}{(s + \alpha)(s + \beta)}$
$u(t)$	$\dfrac{1}{s}$	$\sin \omega t\, u(t)$	$\dfrac{\omega}{s^2 + \omega^2}$
$tu(t)$	$\dfrac{1}{s^2}$	$\cos \omega t\, u(t)$	$\dfrac{s}{s^2 + \omega^2}$
$\dfrac{t^{n-1}}{(n-1)!}u(t), n = 1, 2, \ldots$	$\dfrac{1}{s^n}$	$\sin(\omega t + \theta)\, u(t)$	$\dfrac{s \sin \theta + \omega \cos \theta}{s^2 + \omega^2}$
$e^{-\alpha t}u(t)$	$\dfrac{1}{s + \alpha}$	$\cos(\omega t + \theta)\, u(t)$	$\dfrac{s \cos \theta - \omega \sin \theta}{s^2 + \omega^2}$
$te^{-\alpha t}u(t)$	$\dfrac{1}{(s + \alpha)^2}$	$e^{-\alpha t} \sin \omega t\, u(t)$	$\dfrac{\omega}{(s + \alpha)^2 + \omega^2}$
$\dfrac{t^{n-1}}{(n-1)!}e^{-\alpha t}u(t), n = 1, 2, \ldots$	$\dfrac{1}{(s + \alpha)^n}$	$e^{-\alpha t} \cos \omega t\, u(t)$	$\dfrac{s + \alpha}{(s + \alpha)^2 + \omega^2}$

PRACTICAL APPLICATION

Stability of a System

Many years ago (or so it seems), one of the authors was driving along a country road and attempting to make use of the electronic speed control ("cruise control") feature of the automobile. After turning the system on and manually setting the vehicular speed at precisely the posted speed limit,[3] the "set" button was depressed and the accelerator pedal released; at this point the system was expected to maintain the set speed by regulating the fuel flow as needed.

© Donovan Reese/Getty Images

Unfortunately, something else happened instead. The vehicle speed immediately dropped by about 10 percent, to which the cruise control electronics responded by increasing the fuel flow. The two events were not well matched, so that a few moments later the vehicle speed overshot the set point—resulting in a sudden (and significant) drop in fuel flow—which led to a reduction in the vehicle speed. The cycle continued to the consternation of the driver, who eventually gave up and turned off the system.

Clearly the system response was not optimized—in fact, as built the system was *unstable*. System stability is a major engineering concern across a wide variety of problems (cruise controls, temperature regulators, and tracking systems, to name but a few), and the techniques developed in this chapter are invaluable in allowing the stability of a particular system to be examined.

One of the powerful aspects of working in the s-domain as made possible by the Laplace transform is that instead of describing the response of a particular system through an integrodifferential equation, we can obtain a system transfer function represented by the ratio of two s-polynomials. The issue of stability is then easily

addressed by studying the denominator of the transfer function: *no pole should have a positive real component.*

There are quite a few techniques that can be applied to the problem of determining the stability of a particular system. One simple technique is known as the **Routh test**. Consider the s-domain system function (a concept developed further in Chap. 15)

$$\mathbf{H(s)} = \frac{\mathbf{N(s)}}{\mathbf{D(s)}}$$

The s-polynomial represented by $\mathbf{D(s)}$ can be written as $a_n\mathbf{s}^n + a_{n-1}\mathbf{s}^{n-1} + \cdots + a_1\mathbf{s} + a_0$. Without factoring the polynomial, not much about the poles can be determined at a glance. If all the coefficients $a_n \ldots a_0$ are positive and nonzero, the Routh procedure has us arrange them in the following pattern:

a_n	a_{n-2}	a_{n-4}	$\ldots$
a_{n-1}	a_{n-3}	a_{n-5}	$\ldots$

We next create a third row by cross-multiplying the two rows:

$$\frac{a_{n-1}a_{n-2} - a_na_{n-3}}{a_{n-1}} \qquad \frac{a_{n-1}a_{n-4} - a_na_{n-5}}{a_{n-1}}$$

and a fourth row by cross-multiplying the second and third rows. This process continues until we have $n + 1$ rows of numerical values. All that remains is to scan down the leftmost column for changes in sign. The number of sign changes indicates the number of poles with a positive real component; any sign changes indicate an unstable system.

For example, let's assume the automobile cruise control system behind the author's vexation has a system transfer function with denominator

$$\mathbf{D(s)} = 7\mathbf{s}^4 + 4\mathbf{s}^3 + \mathbf{s}^2 + 13\mathbf{s} + 2$$

All of the coefficients of this fourth-order s-polynomial are positive and nonzero, so we construct the corresponding Routh table:

7	1	2
4	13	0
−21.75	2	
13.37		
2		

from which we see two sign changes in the leftmost column. Thus, the system is in fact unstable (which explains its failure to perform) as two of its poles have positive real components.

TABLE 14.2 Laplace Transform Operations

Operation	$f(t)$	$F(s)$
Addition	$f_1(t) \pm f_2(t)$	$\mathbf{F}_1(\mathbf{s}) \pm \mathbf{F}_2(\mathbf{s})$
Scalar multiplication	$kf(t)$	$k\mathbf{F}(\mathbf{s})$
Time differentiation	$\dfrac{df}{dt}$	$\mathbf{s}\mathbf{F}(\mathbf{s}) - f(0^-)$
	$\dfrac{d^2 f}{dt^2}$	$\mathbf{s}^2\mathbf{F}(\mathbf{s}) - \mathbf{s}f(0^-) - f'(0^-)$
	$\dfrac{d^3 f}{dt^3}$	$\mathbf{s}^3\mathbf{F}(\mathbf{s}) - \mathbf{s}^2 f(0^-) - \mathbf{s}f'(0^-) - f''(0^-)$
Time integration	$\displaystyle\int_{0^-}^{t} f(t)\, dt$	$\dfrac{1}{\mathbf{s}}\mathbf{F}(\mathbf{s})$
	$\displaystyle\int_{-\infty}^{t} f(t)\, dt$	$\dfrac{1}{\mathbf{s}}\mathbf{F}(\mathbf{s}) + \dfrac{1}{\mathbf{s}}\displaystyle\int_{-\infty}^{0^-} f(t)\, dt$
Convolution	$f_1(t) * f_2(t)$	$\mathbf{F}_1(\mathbf{s})\mathbf{F}_2(\mathbf{s})$
Time shift	$f(t-a)u(t-a),\, a \geq 0$	$e^{-a\mathbf{s}}\mathbf{F}(\mathbf{s})$
Frequency shift	$f(t)e^{-at}$	$\mathbf{F}(\mathbf{s}+a)$
Frequency differentiation	$-tf(t)$	$\dfrac{d\mathbf{F}(\mathbf{s})}{d\mathbf{s}}$
Frequency integration	$\dfrac{f(t)}{t}$	$\displaystyle\int_{\mathbf{s}}^{\infty} \mathbf{F}(\mathbf{s})\, d\mathbf{s}$
Scaling	$f(at),\, a \geq 0$	$\dfrac{1}{a}\mathbf{F}\left(\dfrac{\mathbf{s}}{a}\right)$
Initial value	$f(0^+)$	$\displaystyle\lim_{\mathbf{s}\to\infty} \mathbf{s}\mathbf{F}(\mathbf{s})$
Final value	$f(\infty)$	$\displaystyle\lim_{\mathbf{s}\to 0} \mathbf{s}\mathbf{F}(\mathbf{s})$, all poles of $\mathbf{s}\mathbf{F}(\mathbf{s})$ in LHP
Time periodicity	$f(t) = f(t+nT),$ $n = 1, 2, \ldots$	$\dfrac{1}{1-e^{-T\mathbf{s}}}\mathbf{F}_1(\mathbf{s}),$ where $\mathbf{F}_1(\mathbf{s}) = \displaystyle\int_{0^-}^{T} f(t)e^{-\mathbf{s}t}\, dt$

14.7 THE INITIAL-VALUE AND FINAL-VALUE THEOREMS

The last two fundamental theorems that we will discuss are known as the initial-value and final-value theorems. They will enable us to evaluate $f(0^+)$ and $f(\infty)$ by examining the limiting values of $\mathbf{s}\mathbf{F}(\mathbf{s})$. Such an ability can be invaluable; if only the initial and final values are needed for a particular function of interest, there is no need to take the time to perform an inverse transform operation.

The Initial-Value Theorem

To derive the initial-value theorem, we consider the Laplace transform of the derivative once again,

$$\mathcal{L}\left\{\dfrac{df}{dt}\right\} = \mathbf{s}\mathbf{F}(\mathbf{s}) - f(0^-) = \int_{0^-}^{\infty} e^{-\mathbf{s}t}\dfrac{df}{dt}\, dt$$

We now let **s** approach infinity. By breaking the integral into two parts,

$$\lim_{s\to\infty}[sF(s) - f(0^-)] = \lim_{s\to\infty}\left(\int_{0^-}^{0^+} e^0\frac{df}{dt}\,dt + \int_{0^+}^{\infty} e^{-st}\frac{df}{dt}\,dt\right)$$

we see that the second integral must approach zero in the limit, since the integrand itself approaches zero. Also, $f(0^-)$ is not a function of **s**, and it may be removed from the left limit:

$$-f(0^-) + \lim_{s\to\infty}[sF(s)] = \lim_{s\to\infty}\int_{0^-}^{0^+} df = \lim_{s\to\infty}[f(0^+) - f(0^-)]$$
$$= f(0^+) - f(0^-)$$

and finally,

$$f(0^+) = \lim_{s\to\infty}[sF(s)]$$

or

$$\lim_{t\to0^+} f(t) = \lim_{s\to\infty}[sF(s)] \qquad [30]$$

This is the mathematical statement of the ***initial-value theorem.*** It states that the initial value of the time function $f(t)$ can be obtained from its Laplace transform $F(s)$ by first multiplying the transform by **s** and then letting **s** approach infinity. Note that the initial value of $f(t)$ we obtain is the limit from the right.

The initial-value theorem, along with the final-value theorem that we will consider in a moment, is useful in checking the results of a transformation or an inverse transformation. For example, when we first calculated the transform of $\cos(\omega_0 t)u(t)$, we obtained $s/(s^2 + \omega_0^2)$. After noting that $f(0^+) = 1$, we can make a partial check on the validity of this result by applying the initial-value theorem:

$$\lim_{s\to\infty}\left(s\frac{s}{s^2 + \omega_0^2}\right) = 1$$

and the check is accomplished.

The Final-Value Theorem

The final-value theorem is not quite as useful as the initial-value theorem, for it can be used only with a certain class of transforms. In order to determine whether a transform fits into this class, the denominator of $F(s)$ must be evaluated to find all values of **s** for which it is zero; i.e., the ***poles*** of $F(s)$. Only those transforms $F(s)$ whose poles lie entirely within the left half of the **s** plane (i.e., $\sigma < 0$), except for a simple pole at $s = 0$, are suitable for use with the final-value theorem. We again consider the Laplace transform of df/dt,

$$\int_{0^-}^{\infty} e^{-st}\frac{df}{dt}\,dt = sF(s) - f(0^-)$$

this time in the limit as **s** approaches zero,

$$\lim_{s\to0}\int_{0^-}^{\infty} e^{-st}\frac{df}{dt}\,dt = \lim_{s\to0}[sF(s) - f(0^-)] = \int_{0^-}^{\infty}\frac{df}{dt}\,dt$$

We assume that both $f(t)$ and its first derivative are transformable. Now, the last term of this equation is readily expressed as a limit,

$$\int_{0^-}^{\infty} \frac{df}{dt}\, dt = \lim_{t \to \infty} \int_{0^-}^{t} \frac{df}{dt}\, dt$$

$$= \lim_{t \to \infty} [f(t) - f(0^-)]$$

By recognizing that $f(0^-)$ is a constant, a comparison of the last two equations shows us that

$$\lim_{t \to \infty} f(t) = \lim_{s \to 0}[sF(s)] \qquad [31]$$

which is the **final-value theorem.** In applying this theorem, it is necessary to know that $f(\infty)$, the limit of $f(t)$ as t becomes infinite, exists, or—what amounts to the same thing—that the poles of $F(s)$ all lie *within* the left half of the **s** plane except for (possibly) a simple pole at the origin. The product $sF(s)$ thus has all of its poles lying *within* the left half plane.

EXAMPLE **14.11**

Use the final-value theorem to determine $f(\infty)$ for the function $(1 - e^{-at})u(t)$, where $a > 0$.

Without even using the final-value theorem, we see immediately that $f(\infty) = 1$. The transform of $f(t)$ is

$$F(s) = \frac{1}{s} - \frac{1}{s+a}$$

$$= \frac{a}{s(s+a)}$$

The poles of $F(s)$ are $s = 0$ and $s = -a$. Thus, the nonzero pole of $F(s)$ lies in the left-hand **s**-plane, as we were assured that $a > 0$; we find that we may indeed apply the final-value theorem to this function. Multiplying by **s** and letting **s** approach zero, we obtain

$$\lim_{s \to 0}[sF(s)] = \lim_{s \to 0} \frac{a}{s+a} = 1$$

which agrees with $f(\infty)$.

If $f(t)$ is a sinusoid, however, so that $F(s)$ has poles on the $j\omega$ axis, then a blind use of the final-value theorem might lead us to conclude that the final value is zero. We know, however, that the final value of either $\sin \omega_0 t$ or $\cos \omega_0 t$ is indeterminate. So, beware of $j\omega$-axis poles!

CAUTION

PRACTICE

14.14 Without finding $f(t)$ first, determine $f(0^+)$ and $f(\infty)$ for each of the following transforms: (*a*) $4e^{-2s}(s + 50)/s$; (*b*) $(s^2 + 6)/(s^2 + 7)$; (*c*) $(5s^2 + 10)/[2s(s^2 + 3s + 5)]$.

Ans: 0, 200; ∞, indeterminate (poles lie on the $j\omega$ axis); 2.5, 1.

SUMMARY AND REVIEW

❑ The concept of complex frequency allows us to consider the exponentially damped and oscillatory components of a function simultaneously.

❑ The complex frequency $\mathbf{s} = \sigma + j\omega$ is the general case; dc ($\mathbf{s} = 0$), exponential ($\omega = 0$), and sinusoidal ($\sigma = 0$) functions are special cases.

❑ Analyzing circuits in the $\mathbf{s}$-domain results in the conversion of time-domain *integrodifferential* equations into frequency-domain algebraic equations.

❑ In circuit-analysis problems, we convert time-domain functions into the frequency domain using the one-sided Laplace transform: $\mathbf{F}(\mathbf{s}) = \int_{0^-}^{\infty} e^{-st} f(t)\, dt$.

❑ The inverse Laplace transform converts frequency-domain expressions into the time domain. However, it is seldom needed due to the existence of tables listing Laplace transform pairs.

❑ The unit-impulse function is a common approximation to pulses with very narrow widths compared to circuit time constants. It is nonzero only at a single point, and has unity area.

❑ $\mathcal{L}\{f_1(t) + f_2(t)\} = \mathcal{L}\{f_1(t)\} + \mathcal{L}\{f_2(t)\}$ (*additive property*)

❑ $\mathcal{L}\{kf(t)\} = k\mathcal{L}\{f(t)\},\, k = \text{constant}$ (*homogeneity property*)

❑ The differentiation and integration theorems allow us to convert integrodifferential equations in the time domain into simple algebraic equations in the frequency domain.

❑ Inverse transforms are typically found using a combination of partial-fraction expansion techniques and various operations (Table 14.2) to simplify $\mathbf{s}$-domain quantities into expressions that can be found in transform tables (such as Table 14.1).

❑ The initial-value and final-value theorems are useful when only the specific values $f(t = 0^+)$ or $f(t \to \infty)$ are desired.

READING FURTHER

An easily readable development of the Laplace transform and some of its key properties can be found in Chap. 4 of

A. Pinkus and S. Zafrany, *Fourier Series and Integral Transforms,* Cambridge, United Kingdom: Cambridge University Press, 1997.

A much more detailed treatment of integral transforms and their application to science and engineering problems can be found in

B. Davies, *Integral Transforms and Their Applications,* 3rd ed. New York: Springer-Verlag, 2002.

Stability and the Routh test are discussed in Chap. 5 of

K. Ogata, *Modern Control Engineering,* 4th ed. Englewood Cliffs, N.J.: Prentice-Hall, 2002.

EXERCISES

14.1 Complex Frequency

1. Determine the complex frequency of each term: (a) $v(t) = 5$ V; (b) $i(t) = 3\cos 9t$ μA; (c) $i(t) = 2.5e^{-8t}$ mA; (d) $v(t) = 65e^{-1000t}\cos 1000t$ V; (e) $v(t) = 8 + 2\cos t$ mV.

2. State the complex frequency **s** of each of the following:
(a) $v(t) = 33.3$ V; (b) $i(t) = 3\cos 77t$ A; (c) $q(t) = 7e^{-5t}$ C; (d) $q(t) = 7e^{-5t} - 19e^{-5t}\sin(8t - 42°)$ C.

3. Obtain the complex conjugate of each term, expressing your answers in polar form: (a) $8e^{-t}$; (b) 19; (c) $9 - j7$; (d) e^{jwt}; (e) $\cos 4t$; (f) $\sin 4t$; (g) $88\underline{/-9°}$.

4. Determine the complex conjugate of each term: (a) $6 - j$; (b) 9; (c) $-j30$; (d) $5e^{-j6}$; (e) $24\underline{/-45°}$; (f) $\dfrac{4 - j18}{3.33 + j}$; (g) $\dfrac{5\underline{/0.1°}}{4 - j7}$; (h) $4 - 22\underline{/92.5°}$.

5. The charge emitted from a particular field emission array is represented for convenience as $\mathbf{Q} = 9\underline{/43°}$ μC at a complex frequency $\mathbf{s} = j20\pi$ s^{-1}. (a) How much charge is being emitted at $t = 1$ s? (b) What is the maximum amount of charge that will ever be emitted by the array at any one time? (c) Is the array showing any sign of deterioration? What might be an indication, based on the complex frequency of **Q**?

6. Your new assistant has measured the signal coming from a piece of test equipment, writing $v(t) = \mathbf{V}_x e^{(-2+j60)t}$, where $\mathbf{V}_x = 8 - j100$ V. (a) There is a missing term. What is it, and how can you tell it's missing? (b) What is the complex frequency of the signal? (c) What is the significance of the fact that Im$\{\mathbf{V}_x\}$ > Re$\{\mathbf{V}_x\}$? (d) What is the significance of the fact that $|\text{Re}\{\mathbf{s}\}| < |\text{Im}\{\mathbf{s}\}|$?

7. Let the real part of the complex time-varying current $\mathbf{i}(t)$ be $i(t)$. Find (a) $i_x(t)$ if $\mathbf{i}_x(t) = (4 - j7)e^{(-3+j15)t}$; (b) $i_y(t)$ if $\mathbf{i}_y(t) = (4 + j7)e^{-3t}(\cos 15t - j\sin 15t)$; (c) $i_A(0.4)$ if $\mathbf{i}_A(t) = \mathbf{K}_A e^{\mathbf{s}_A t}$, where $\mathbf{K}_A = 5 - j8$ and $\mathbf{s}_A = -1.5 + j12$; (d) $i_B(0.4)$ if $\mathbf{i}_B(t) = \mathbf{K}_B e^{\mathbf{s}_B t}$, where $\mathbf{K}_B$ is the conjugate of $\mathbf{K}_A$ and $\mathbf{s}_B$ is the conjugate of $\mathbf{s}_A$.

8. A periodic signal current $i(t) = 2.33\cos(279 \times 10^6 t)$ fA is detected by a radio telescope trained on the Orion nebula. (a) What is the frequency (in Hz) of the signal? (b) If the signal is detected by measuring the voltage that results when the current flows through a precision 1 TΩ resistor, write the voltage signal as a sum of two complex exponentials.

9. If a complex time-varying voltage is given as $\mathbf{v}_s(t) = (20 - j30)e^{(-2+j50)t}$ V, find (a) $\mathbf{v}_s(0.1)$ in polar form; (b) Re $\{\mathbf{v}_s(t)\}$; (c) Re $[\mathbf{v}_s(0.1)]$; (d) **s**; (e) $\mathbf{s}^*$.

14.2 The Damped Sinusoidal Forcing Function

10. The circuit of Fig. 14.10 is driven by a source having magnitude 10 V, phase angle 3°, and complex frequency $-2 + j10$ s^{-1}. (a) Determine $i(t)$. (b) Determine $v_1(t)$ and $v_2(t)$.

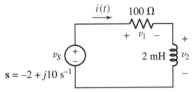

$i(t)$ 100 Ω

v_S $+ \; v_1 \; -$ 2 mH v_2

$\mathbf{s} = -2 + j10$ s^{-1}

■ **FIGURE 14.10**

11. (a) Extend the concept of phasors introduced in Chap. 10 to derive impedance expressions for inductors, capacitors, and resistors driven by a complex

frequency **s**. (*b*) What is the impedance of the resistor and the inductor of Fig. 14.10, respectively? (*c*) When Re{**s**} = 0, do your expressions reduce to those of Chap. 10?

12. A simple series *RL* circuit is connected to a North American wall outlet having voltage $v(t) = 179\cos(120\pi t)$ V. If $R = 100\ \Omega$ and $L = 500\ \mu$H, (*a*) determine the complex frequency of the corresponding frequency-domain voltage **V**(**s**). (*b*) Work in the frequency domain to determine the current **I**(**s**) flowing through the circuit. (*c*) Find $i(t)$.

13. (*a*) Let $v_s = 10e^{-2t}\cos(10t + 30°)$ V in the circuit of Fig. 14.11, and work in the frequency domain to find $\mathbf{I}_x$. (*b*) Find $i_x(t)$.

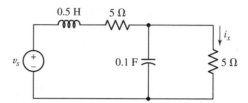

■ **FIGURE 14.11**

14. A simple series *RC* circuit is connected to a wall outlet in Japan with $v(t) = 339\cos(100\pi t)$ V. If $R = 2\ \text{k}\Omega$ and $C = 100\ \mu$F, (*a*) determine the complex frequency of the corresponding frequency-domain voltage **V**(**s**). (*b*) Work in the frequency domain to determine the current **I**(**s**) flowing through the circuit. (*c*) Find $i(t)$.

15. Let $i_{s1} = 20e^{-3t}\cos 4t$ A and $i_{s2} = 30e^{-3t}\sin 4t$ A in the circuit of Fig. 14.12. (*a*) Work in the frequency domain to find $\mathbf{V}_x$. (*b*) Find $v_x(t)$.

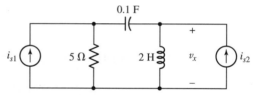

■ **FIGURE 14.12**

16. The Thévenin equivalent resistance of a large piece of industrial electronic equipment, as seen from the output terminals of its dc power supply, is approximately 3 mΩ (it is slightly larger when the equipment is idle). If the power supply can be modeled as an exponential voltage source $v(t) = 240\sqrt{2}e^{-2t} \times \cos(120\pi t)$ V from the moment the power coming from the wall outlet is cut off, (*a*) find the current $i(t)$ flowing through the resistor by working in the frequency domain. (*b*) Check your answer to (*a*) by working in the time domain. (*c*) Rework part (*a*) if a 1000 mF capacitor is added across the output terminals of the power supply (i.e., in parallel with R_{TH}).

14.3 Definition of the Laplace Transform

17. Find the one-sided Laplace transform of $Ku(t)$, where K is an unknown real constant.

18. Use Eq. [14] to find the Laplace transform of the following: (*a*) $3u(t)$; (*b*) $3u(t-3)$; (*c*) $3u(t-3) - 3$; (*d*) $3u(3-t)$.

19. Use Eq. [14] to find the Laplace transform of the following: (*a*) $2 + 3u(t)$; (*b*) $3e^{-8t}$; (*c*) $u(-t)$; (*d*) K, where K is an unknown real constant.

20. A current source provides a current of $4e^{-t}u(t)$ mA through a 1 Ω resistor. (*a*) Find the frequency-domain representation of the voltage across the resistor. (*b*) Recalling that $\mathbf{s} = \sigma + j\omega$, plot the magnitude of the frequency-domain current as a function of σ if $\omega = 0$.

21. A voltage source $v(t) = 5u(t) - 5u(t-2)$ V is connected across a 1 Ω resistor. (a) Find the frequency-domain representation of the voltage. (b) Find the frequency-domain representation of the current flowing through the resistor.

14.4 Laplace Transforms of Simple Time Functions

22. Specify the range of σ over which the Laplace transform exists if $f(t)$ equals (a) $t+1$; (b) $(t+1)u(t)$; (c) $e^{50t}u(t)$; (d) $e^{50t}u(t-5)$; (e) $e^{-50t}u(t-5)$.

23. For each of the following functions, determine the one-sided Laplace transform: (a) $8e^{-2t}[u(t+3) - u(t-3)]$; (b) $8e^{2t}[u(t+3) - u(t-3)]$; (c) $8e^{-2|t|}[u(t+3) - u(t-3)]$.

24. Determine the one-sided Laplace transform of each of the following: (a) $\mathcal{L}^{-1}\{s^{-1}\}$; (b) $1 + u(t) + [u(t)]^2$; (c) $tu(t) - 3$; (d) $1 - \delta(t) + \delta(t-1) - \delta(t-2)$.

25. Without resorting to Eq. [15], find the inverse transform of each of the following: (a) $\dfrac{1}{s+3}$; (b) 1; (c) s^{-2}; (d) 275; (e) $\dfrac{s^2}{s^3}$.

26. Show that as long as individual Laplace transforms of $f_1(t)$ and $f_2(t)$ exist, $\mathcal{L}\{f_1(t) + f_2(t)\} = \mathcal{L}\{f_1(t)\} + \mathcal{L}\{f_2(t)\}$.

27. Use the definition of the Laplace transform to calculate the value of $\mathbf{F}(1 + j2)$ if $f(t)$ equals (a) $2u(t-2)$; (b) $2\delta(t-2)$; (c) $e^{-t}u(t-2)$.

28. Evaluate the following: (a) $\int_{-\infty}^{\infty} 8\sin 5t\, \delta(t-1)\, dt$; (b) $\int_{-\infty}^{\infty} (t-5)^2\, \delta(t-2)\, dt$; (c) $\int_{-\infty}^{\infty} 5e^{-3000t}\delta(t-3.333 \times 10^{-4})\, dt$; (d) $\int_{-\infty}^{\infty} K\delta(t-2)\, dt$, where K is a real constant.

29. Use the definition of the (one-sided) Laplace transform to find $\mathbf{F}(\mathbf{s})$ if $f(t)$ equals (a) $[u(5-t)][u(t-2)]u(t)$; (b) $4u(t-2)$; (c) $4e^{-3t}u(t-2)$; (d) $4\delta(t-2)$; (e) $5\delta(t)\sin(10t + 0.2\pi)$.

30. Evaluate the following: (a) $\int_{-\infty}^{\infty} \cos 500t\, \delta(t)\, dt$; (b) $\int_{-\infty}^{\infty} t^5\delta(t-2)\, dt$; (c) $\int_{-\infty}^{\infty} 2.5e^{-0.001t}\delta(t-1000)\, dt$; (d) $\int_{-\infty}^{\infty} -K^2\delta(t-c)\, dt$, where K and c are real constants.

31. Using the one-sided Laplace transform, find $\mathbf{F}(\mathbf{s})$ if $f(t)$ equals (a) $[2u(t-1)][u(3-t)]u(t^3)$; (b) $2u(t-4)$; (c) $3e^{-2t}u(t-4)$; (d) $3\delta(t-5)$; (e) $4\delta(t-1)[\cos \pi t - \sin \pi t]$.

14.5 Inverse Transform Techniques

32. Determine $f(t)$ if $\mathbf{F}(\mathbf{s})$ is (a) $3 + 1/s$; (b) $3 + 1/s^2$; (c) $\dfrac{1}{(s+3)(s+4)}$; (d) $\dfrac{1}{(s+3)(s+4)(s+5)}$.

33. Determine $g(t)$ if $\mathbf{G}(\mathbf{s})$ is (a) $90 - 4.5/s$; (b) $11 + 2s/s^2$; (c) $\dfrac{1}{(s+1)(s+1)}$; (d) $\dfrac{1}{(s+1)(s+2)(s+3)}$.

34. Find the inverse transform of each of the following without performing any integrations and without resorting to the use of MATLAB: (a) $5s^{-1} - 16 + (s+4.4)^{-1}$; (b) $1 - s^{-1} + s^{-2}$; (c) $5(s+7)^{-1} + 88s^{-1} + \dfrac{17}{(s+6)(s+1)}$.

35. The frequency-domain voltage across a 2 kΩ resistor is given by $\mathbf{V}(\mathbf{s}) = 5s^{-1}$ V. What is the current through the resistor at $t = 1$ ms?

36. The frequency-domain current through a 100 MΩ resistor is $5(s+10)^{-1}$ pA. (a) Plot the voltage $v(t)$ across the resistor as a function of time. (b) What is the power absorbed by the resistor at $t = 100$ ms? (c) At what time has the voltage across the resistor dropped to 1 percent of its maximum value?

37. Determine $f(t)$ if $\mathbf{F}(\mathbf{s})$ equals (a) $[(s+1)/s] + [2/(s+1)]$; (b) $(e^{-s} + 1)^2$; (c) $2e^{-(s+1)}$; (d) $2e^{-3s}\cosh 2s$.

38. If $\mathbf{N(s)} = 5s$ find $\mathcal{L}^{-1}\{\mathbf{N(s)}/\mathbf{D(s)}\}$ for $\mathbf{D(s)} = $ (a) $s^2 - 9$; (b) $(s+3)(s^2 + 19s + 90)$; (c) $(4s + 12)(8s^2 + 6s + 1)$. (d) Check your answers to (a)-(c) with MATLAB.

39. Given the following expressions for $\mathbf{F(s)}$, find $f(t)$: (a) $5/(s+1)$; (b) $5/(s+1) - 2/(s+4)$; (c) $18/[(s+1)(s+4)]$; (d) $18s/[(s+1)(s+4)]$; (e) $18s^2/[(s+1)(s+4)]$.

40. If $\mathbf{N(s)} = 2s^2$ find $\mathcal{L}^{-1}\{\mathbf{N(s)}/\mathbf{D(s)}\}$ for $\mathbf{D(s)} = $ (a) $s^2 - 1$; (b) $(s+3)(s^2 + 19s + 90)$; (c) $(8s + 12)(16s^2 + 12s + 2)$. (d) Check your answers to (a)-(c) with MATLAB.

41. Find $f(t)$ if $\mathbf{F(s)}$ equals

   (a) $\dfrac{2}{s} - \dfrac{3}{s+1}$; (b) $\dfrac{2s + 10}{s + 3}$; (c) $3e^{-0.8s}$; (d) $\dfrac{12}{(s+2)(s+6)}$; (e) $\dfrac{12}{(s+2)^2(s+6)}$.

42. Find $\mathcal{L}^{-1}\{\mathbf{F(s)}\}$ if $\mathbf{F(s)} = 2 - s^{-1} + \dfrac{\pi}{(s^3 + 4s^2 + 5s + 2)}$.

43. Obtain partial-fraction expansions for each of the following functions, and then determine the corresponding time functions: (a) $\mathbf{F(s)} = [(s+1)(s+2)]/[s(s+3)]$; (b) $\mathbf{F(s)} = (s+2)/[s^2(s^2+4)]$.

44. Find $\mathcal{L}^{-1}\{\mathbf{G(s)}\}$ if $\mathbf{G(s)}$ is (a) $\dfrac{12s^3}{(s+1)(s+2)}$; (b) $\dfrac{12s^3}{(s^2 + 2s + 1)(s+2)}$;

   (c) $3s - \dfrac{12s^3}{(s+1)(s+2)(s+3)}$.

45. Find $\mathcal{L}^{-1}\{\mathbf{H(s)}\}$ if $\mathbf{H(s)}$ is (a) $\dfrac{(s+1)^2}{(s+1)(s+2)}$; (b) $\dfrac{s+3}{(s+1)(s+2)}$;

   (c) $3s - \dfrac{s^4}{(s^2 + 2s + 1)(s+3)} + 1$.

14.6 Basic Theorems for the Laplace Transform

46. Take the Laplace transform of the following equations:

   (a) $5\,di/dt - 7\,d^2i/dt^2 + 9i = 4$; (b) $m\dfrac{d^2p}{dt^2} + \mu_f\dfrac{dp}{dt} + kp(t) = 0$,

   the equation that describes the "force-free" response of a simple shock absorber system; (c) $\dfrac{d\Delta n_p}{dt} = -\dfrac{\Delta n_p}{\tau} + G_L$, with $\tau = $ constant, which describes the recombination rate of excess electrons (Δn_p) in p-type silicon under optical illumination (G_L is a constant proportional to the light intensity).

47. If $f(0^-) = -3$ and $15u(t) - 4\delta(t) = 8f(t) + 6f'(t)$, find $f(t)$ by taking the Laplace transform of the differential equation, solving for $\mathbf{F(s)}$, and inverting to find $f(t)$.

48. Referring to the RL circuit of Fig. 14.13, (a) write a differential equation for the inductor current $i_L(t)$. (b) Find $\mathbf{I}_L(s)$, the Laplace transform of $i_L(t)$. (c) Solve for $i_L(t)$ by taking the inverse Laplace transform of $\mathbf{I}_L(s)$.

49. (a) Find $v_C(0^-)$ and $v_C(0^+)$ for the circuit shown in Fig. 14.14. (b) Obtain an equation for $v_C(t)$ that holds for $t > 0$. (c) Use Laplace transform techniques to solve for $\mathbf{V}_C(s)$ and then find $v_C(t)$.

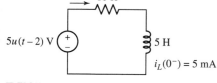

$i_L(t)$ $10\,\Omega$

$5u(t-2)$ V 5 H

$i_L(0^-) = 5$ mA

■ **FIGURE 14.13**

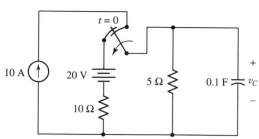

$t = 0$

10 A 20 V $5\,\Omega$ 0.1 F v_C

$10\,\Omega$

■ **FIGURE 14.14**

50. (*a*) Add a voltage source $v_S(t) = -5u(t)$ V in series with the $5u(t-2)$ V source in Fig. 14.13, and repeat Exer. 48. (*b*) Sketch the inductor current, and compare with an appropriate PSpice simulation.

51. Given the differential equation $12u(t) = 20f_2'(t) + 3f_2(t)$, where $f_2(0^-) = 2$, take its Laplace transform, solve for $\mathbf{F}_2(\mathbf{s})$, and then find $f_2(t)$.

52. Find the inverse Laplace transform of each of the following: (*a*) $2/\mathbf{s} - 4$; (*b*) $\mathbf{s}/(\mathbf{s}^2 + 99)$; (*c*) $1/(\mathbf{s}^2 + 5\mathbf{s} + 6) - 5$; (*d*) $\mathbf{s}$; (*e*) $\mathbf{s}^2$.

53. Given the two differential equations $x' + y = 2u(t)$ and $y' - 2x + 3y = 8u(t)$, where $x(0^-) = 5$ and $y(0^-) = 8$, find $x(t)$ and $y(t)$.

54. Find $f(t)$ if $\mathbf{F}(\mathbf{s})$ is given by: (a) $8\mathbf{s} + 8 + 8\mathbf{s}^{-1}$, $f(0^-) = 0$; (b) $\mathbf{s}^2/(\mathbf{s}+2) - \mathbf{s} + 2$.

55. (*a*) Determine $i_C(0^-)$ and $i_C(0^+)$ for the circuit of Fig. 14.15. (*b*) Write an equation for $i_C(t)$ in the time domain that is valid for $t > 0$. (*c*) Use Laplace transform methods to solve for $\mathbf{I}_C(\mathbf{s})$ and then find the inverse transform.

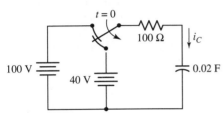

■ FIGURE 14.15

56. Find $\mathbf{V}(\mathbf{s})$ if $v(t) = $ (*a*) $4\cos(100t)$ V; (*b*) $2\sin(10^3 t) - 3\cos(100t)$ V; (*c*) $14\cos(8t) - 2\sin(8°)$ V; (*d*) $\delta(t) + \sin(6t)u(6t)$; (*e*) $\cos(5t)\sin(3t)$ V.

57. A resistor R, a capacitor C, an inductor L, and an ideal current source $i_s = 100e^{-5t}u(t)$ A are in parallel. Let the voltage v be across the source with the positive reference at the terminal at which $i_s(t)$ leaves the source. Then $i_s = v' + 4v + 3\int_{0^-}^{t} v\, dx$. (*a*) Find R, L, and C. (*b*) Use Laplace transform techniques to find $v(t)$.

58. Find $\mathcal{L}\{v(t)\}$ if $v(t) = $ (*a*) $7 + (t-2)u(t-2)$ V; (*b*) $e^{-t+2}u(t-2)$ V; (*c*) $48\delta(t-1)u(t-1)$ V.

59. Obtain a single integrodifferential equation in terms of i_C for the circuit of Fig. 14.16, take the Laplace transform, solve for $\mathbf{I}_C(\mathbf{s})$, and then find $i_C(t)$ by making use of the inverse transform.

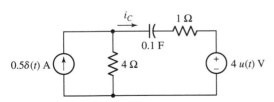

■ FIGURE 14.16

60. Given the differential equation $v' + 6v + 9\int_{0^-}^{t} v(z)\, dz = 24(t-2)u(t-2)$, let $v(0^-) = 0$ and find $v(t)$.

61. Apply the Routh test to the following system functions, and state whether the system is *stable* or *unstable*:
(*a*) $\mathbf{H}(\mathbf{s}) = \dfrac{\mathbf{s} - 500}{\mathbf{s}^3 + 13\mathbf{s}^2 + 47\mathbf{s} + 35}$; (*b*) $\mathbf{H}(\mathbf{s}) = \dfrac{\mathbf{s} - 500}{\mathbf{s}^3 + 13\mathbf{s}^2 + \mathbf{s} + 35}$.

62. Apply the Routh test to the following system functions, and state whether the system is *stable* or *unstable*, then factor each denominator to identify the poles of $\mathbf{H}(s)$ and verify the accuracy of the Routh test for these functions:

(*a*) $\mathbf{H}(s) = \dfrac{4s}{s^2 + 3s + 8}$; (*b*) $\mathbf{H}(s) = \dfrac{s - 9}{s^2 + 2s + 1}$.

63. Apply the Routh test to the following system functions, and state whether the system is *stable* or *unstable*:

(*a*) $\mathbf{H}(s) = \dfrac{s^2}{s^4 + 3s^3 + 3s^2 + 3s + 1}$; (*b*) $\mathbf{H}(s) = \dfrac{2}{s + 3}$.

14.7 The Initial-Value and Final-Value Theorems

64. Given the function $v(t) = 7u(t) + 8e^{-3t}u(t)$ V, (*a*) apply the initial value theorem to $\mathbf{V}(s)$. (*b*) Verify your answer by evaluating $v(t)$ at $t = 0$.

65. Given the function $v(t) = 7u(t) + 8e^{-3t}u(t)$ V, (*a*) apply the final value theorem to $\mathbf{V}(s)$. (*b*) Verify your answer by evaluating $v(t)$ at $t = \infty$.

66. Find $f(0^+)$ and $f(\infty)$ for a time function whose Laplace transform is (*a*) $5(s^2 + 1)/(s^3 + 1)$; (*b*) $5(s^2 + 1)/(s^4 + 16)$; (*c*) $(s + 1)(1 + e^{-4s})/(s^2 + 2)$.

67. Without finding $f(t)$ first, determine $f(0^+)$ and $f(\infty)$ for each of the following transforms: (*a*) $(2s^2 + 6)/[s(s^2 + 5s + 2)]$; (*b*) $2e^{-s}/(s + 3)$; (*c*) $(s^2 + 1)/(s^2 + 5)$.

68. Find $f(\infty)$ and $f(0^+)$ for a time function whose Laplace transform is (*a*) $5(s^2 + 1)/(s + 1)^3$; (*b*) $5(s^2 + 1)/[s(s + 1)^3]$; (*c*) $(1 - e^{-3s})/s^2$.

69. Let $f(t) = (1/t)(e^{-at} - e^{-bt})u(t)$. (*a*) Find $\mathbf{F}(s)$. (*b*) Evaluate both sides of the equation $\lim\limits_{t \to 0^+} f(t) = \lim\limits_{s \to \infty} [s\mathbf{F}(s)]$.

70. Find both the initial and final values (or show that they do not exist) of the time functions corresponding to

(*a*) $\dfrac{8s - 2}{s^2 + 6s + 10}$; (*b*) $\dfrac{2s^3 - s^2 - 3s - 5}{s^3 + 6s^2 + 10s}$; (*c*) $\dfrac{8s - 2}{s^2 - 6s + 10}$;

(*d*) $\dfrac{8s^2 - 2}{(s + 2)^2(s + 1)(s^2 + 6s + 10)}$.

Circuit Analysis in the *s*-Domain

KEY CONCEPTS

Extending the Concept of Impedance to the s-Domain

Modeling Initial Conditions with Ideal Sources

Applying Nodal, Mesh, Superposition, and Source Transformation in the s-Domain

Thévenin and Norton's Theorems Applied to s-Domain Circuits

Manipulating s-Domain Algebraic Expressions with MATLAB

Identifying Poles and Zeros in Circuit Transfer Functions

Impulse Response of a Circuit

Use of Convolution to Determine System Response

Response as a Function of σ and ω

Using Pole-Zero Plots to Predict the Natural Response of Circuits

Synthesizing Specific Voltage Transfer Functions Using Op Amps

INTRODUCTION

In Chap. 14 we developed the concept of complex frequency and were introduced to using Laplace transforms as a means of solving the types of differential equations we encounter in circuit analysis. After a little practice, we became adept at moving back and forth between the time domain and the frequency domain as necessary. Now we are ready to bring these formidable techniques to bear on the analysis of circuits in a structured fashion. The resulting set of skills will enable us to efficiently analyze any linear circuit to obtain the complete response—transient plus steady-state— regardless of the nature of the excitation sources.

15.1 • Z(s) AND Y(s)

The key concept that makes phasors so useful in the analysis of sinusoidal steady-state circuits is the transformation of resistors, capacitors, and inductors into *impedances*. Circuit analysis then proceeds using the basic techniques of nodal or mesh analysis, superposition, source transformation, as well as Thévenin or Norton equivalents. As we may already suspect, this concept can be extended to the *s*-domain, since the sinusoidal steady state is included in **s**-domain analysis as a special case (where $\sigma = 0$).

Resistors in the Frequency Domain

Let's begin with the simplest situation, that of a resistor connected to a voltage source $v(t)$. Ohm's law specifies that

$$v(t) = Ri(t)$$

Taking the Laplace transform of both sides,

$$\mathbf{V}(s) = R\mathbf{I}(s)$$

we find that the ratio of the frequency-domain representation of the voltage to the frequency-domain representation of the current is simply the resistance, R. Thus,

$$\mathbf{Z(s)} \equiv \frac{\mathbf{V(s)}}{\mathbf{I(s)}} = R \qquad [1]$$

Since we are working in the frequency domain, we refer to this quantity as an *impedance* for the sake of clarity, but still assign it the unit ohms (Ω). Just as we found in working with phasors in the sinusoidal steady state, the impedance of a resistor does not depend on frequency. The *admittance* $\mathbf{Y(s)}$ of a resistor, defined as the ratio of $\mathbf{I(s)}$ to $\mathbf{V(s)}$, is simply $1/R$; the unit of admittance is the siemen (S).

Inductors in the Frequency Domain

Next, we consider an inductor connected to some time-varying voltage source $v(t)$, as shown in Fig. 15.1a. We know that

$$v(t) = L\frac{di}{dt}$$

Taking the Laplace transform of both sides of this equation, we find

$$\mathbf{V(s)} = L[\mathbf{sI(s)} - i(0^-)] \qquad [2]$$

We now have two terms: $sL\mathbf{I(s)}$ and $Li(0^-)$. In situations where we have zero initial energy stored in the inductor (i.e., $i(0^-) = 0$), then

$$\mathbf{V(s)} = sL\mathbf{I(s)}$$

so that

$$\mathbf{Z(s)} \equiv \frac{\mathbf{V(s)}}{\mathbf{I(s)}} = sL \qquad [3]$$

Equation [3] may be further simplified if we are only interested in the sinusoidal steady-state response. It is permissible to neglect the initial conditions in such instances as they only affect the nature of the transient response. Thus, we substitute $\mathbf{s} = j\omega$ and find

$$\mathbf{Z}(j\omega) = j\omega L$$

as was obtained previously in Chap. 10.

Modeling Inductors in the s-Domain

Although we refer to the quantity in Eq. [3] as the impedance of an inductor, we must remember that it was obtained by assuming zero initial current. In the more general situation where energy is stored in the element at $t = 0^-$, this quantity is not sufficient to represent the inductor in the frequency domain. Fortunately, it is possible to include the initial condition by modeling an inductor as an impedance in combination with either a voltage or current source. To do this, we begin by rearranging Eq. [2] as

$$\mathbf{V(s)} = sL\mathbf{I(s)} - Li(0^-) \qquad [4]$$

The second term on the right will be a constant: the inductance L in henrys multiplied by the initial current $i(0^-)$ in amperes. The result is a constant

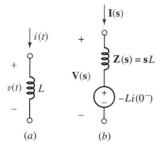

FIGURE 15.1 (a) Inductor in the time domain. (b) The complete model for an inductor in the frequency domain, consisting of an impedance sL and a voltage source $-Li(0^-)$ that incorporates the effects of nonzero initial conditions on the element.

voltage term that is subtracted from the frequency-dependent term $s L\mathbf{I}(\mathbf{s})$. A small leap of intuition at this point leads us to the realization that we can model a single inductor L as a two-component frequency-domain element, as shown in Fig. 15.1b.

The frequency-domain inductor model shown in Fig. 15.1b consists of an impedance sL and a voltage source $Li(0^-)$. The voltage across the impedance sL is given by Ohm's law as $s L\mathbf{I}(\mathbf{s})$. Since the two-element combination in Fig. 15.1b is linear, every circuit analysis technique previously explored can be brought to bear in the s-domain as well. For example, it is possible to perform a source transformation on the model in order to obtain an impedance sL in parallel with a current source $[-Li(0^-)]/sL = -i(0^-)/\mathbf{s}$. This can be verified by taking Eq. [4] and solving for $\mathbf{I}(\mathbf{s})$:

$$\mathbf{I}(\mathbf{s}) = \frac{\mathbf{V}(\mathbf{s}) + Li(0^-)}{sL}$$

$$= \frac{\mathbf{V}(\mathbf{s})}{sL} + \frac{i(0^-)}{\mathbf{s}}$$

[5]

We are once again left with two terms. The first term on the right is simply an admittance $1/sL$ times the voltage $\mathbf{V}(\mathbf{s})$. The second term on the right is a current, although it has units of ampere · seconds. Thus, we can model this equation with two separate components: an admittance $1/sL$ in parallel with a current source $i(0^-)/\mathbf{s}$; the resulting model is shown in Fig. 15.2. The choice of whether to use the model of Fig. 15.1b or that shown in Fig. 15.2 is usually made depending on which one will result in simpler equations when analyzing a complete circuit containing the inductor. Note that although Fig. 15.2 shows the inductor symbol labeled with an admittance $\mathbf{Y}(\mathbf{s}) = 1/sL$, it can also be viewed as an impedance $\mathbf{Z}(\mathbf{s}) = sL$; again, the choice of which to use is generally based on personal preference and convenience.

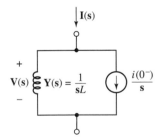

■ **FIGURE 15.2** An alternative frequency-domain model for the inductor, consisting of an admittance $1/sL$ and a current source $i(0^-)/\mathbf{s}$.

A brief comment on units is in order. When we take the Laplace transform of a current $i(t)$, we are integrating over time. Thus, the units of $\mathbf{I}(\mathbf{s})$ are technically ampere·seconds; in a similar fashion, the units of $\mathbf{V}(\mathbf{s})$ are volt·seconds. However, it is the convention to drop the seconds and assign $\mathbf{I}(\mathbf{s})$ the units of amperes, and to measure $\mathbf{V}(\mathbf{s})$ in volts. This convention does not present any problems until we scrutinize an equation such as Eq. [5], and see a term like $i(0^-)/\mathbf{s}$ seemingly in conflict with the units of $\mathbf{I}(\mathbf{s})$ on the left-hand side. Although we will continue to measure these phasor quantities in "amperes" and "volts," when checking the units of an equation to verify algebra, we must remember the seconds!

EXAMPLE 15.1

Calculate the voltage $v(t)$ shown in Fig. 15.3a, given an initial current $i(0^-) = 1$ A.

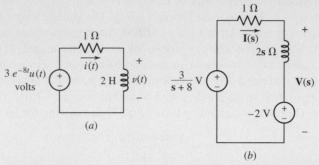

(a)

(b)

■ **FIGURE 15.3** (a) A simple resistor-inductor circuit for which the voltage $v(t)$ is desired. (b) The equivalent frequency-domain circuit, including the initial current in the inductor through the use of a series voltage source $-Li(0^-)$.

We begin by first converting the circuit in Fig. 15.3a to its frequency-domain equivalent, shown in Fig. 15.3b; the inductor has been replaced with a two-component model: an impedance $\mathbf{s}L = 2\mathbf{s}$ Ω, and an independent voltage source $-Li(0^-) = -2$ V.

We seek the quantity labeled $\mathbf{V(s)}$, as its inverse transform will result in $v(t)$. Note that $\mathbf{V(s)}$ appears across the *entire* inductor model, and not just the impedance component.

Taking a straightforward route, we write

$$\mathbf{I(s)} = \frac{\left[\dfrac{3}{\mathbf{s}+8} + 2\right]}{1 + 2\mathbf{s}} = \frac{\mathbf{s} + 9.5}{(\mathbf{s}+8)(\mathbf{s}+0.5)}$$

and

$$\mathbf{V(s)} = 2\mathbf{s}\,\mathbf{I(s)} - 2$$

so that

$$\mathbf{V(s)} = \frac{2\mathbf{s}(\mathbf{s}+9.5)}{(\mathbf{s}+8)(\mathbf{s}+0.5)} - 2$$

Before attempting to take the inverse Laplace transform of this expression, it is well worth a little time and effort to simplify it first. Thus,

$$\mathbf{V(s)} = \frac{2\mathbf{s} - 8}{(\mathbf{s}+8)(\mathbf{s}+0.5)}$$

Employing the technique of partial fraction expansion (on paper or with the assistance of MATLAB), we find that

$$\mathbf{V(s)} = \frac{3.2}{\mathbf{s}+8} - \frac{1.2}{\mathbf{s}+0.5}$$

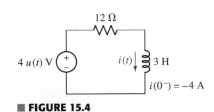

FIGURE 15.4

Referring to Table 14.1, then, the inverse transform is found to be

$$v(t) = [3.2e^{-8t} - 1.2e^{-0.5t}]u(t) \qquad \text{volts}$$

PRACTICE

15.1 Determine the current $i(t)$ in the circuit of Fig. 15.4.

Ans: $\frac{1}{3}[1 - 13e^{-4t}]u(t)$ A.

Modeling Capacitors in the *s*-Domain

The same concepts apply to capacitors in the *s*-domain as well. Following the passive sign convention as illustrated in Fig. 15.5*a*, the governing equation for capacitors is

$$i = C\frac{dv}{dt}$$

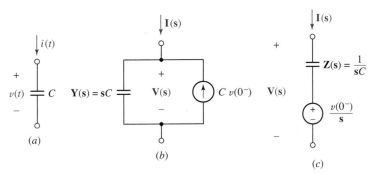

■ **FIGURE 15.5** (*a*) Capacitor in the time domain, with $v(t)$ and $i(t)$ labeled. (*b*) Frequency-domain model of a capacitor with initial voltage $v(0^-)$. (*c*) An equivalent model obtained by performing a source transformation.

Taking the Laplace transform of both sides results in

$$\mathbf{I}(s) = C[s\mathbf{V}(s) - v(0^-)]$$

or

$$\mathbf{I}(s) = sC\mathbf{V}(s) - Cv(0^-) \qquad [6]$$

which can be modeled as an admittance sC in parallel with a current source $Cv(0^-)$ as shown in Fig. 15.5*b*. Performing a source transformation on this circuit (taking care to follow the passive sign convention) results in an equivalent model for the capacitor consisting of an impedance $1/sC$ in series with a voltage source $v(0^-)/s$, as shown in Fig. 15.5*c*.

In working with these *s*-domain equivalents, we should be careful not to be confused with the independent sources being used to include initial conditions. The initial condition for an inductor is given as $i(0^-)$; this term may appear as part of either a voltage source or a current source, depending on which model is chosen. The initial condition for a capacitor is given as $v(0^-)$; this term may thus appear as part of either a voltage source or a

CAUTION

current source. A very common mistake for students working with *s*-domain analysis for the first time is to always use $v(0^-)$ for the voltage source component of the model, even when dealing with an inductor.

EXAMPLE 15.2

Find $v_C(t)$ in the circuit of Fig. 15.6*a*, assuming an initial voltage $v_C(0^-) = -2$ V.

▶ **Identify the goal of the problem.**
We are in need of an expression for the capacitor voltage, $v_C(t)$.

▶ **Collect the known information.**
The problem specifies an initial capacitor voltage of -2 V.

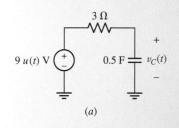
(a)

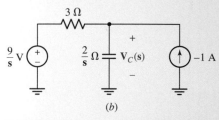
(b)

■ **FIGURE 15.6** (*a*) A circuit for which the current $v_C(t)$ is required. (*b*) Frequency-domain equivalent circuit, employing the current-source based model to account for the initial condition of the capacitor.

▶ **Devise a plan.**
Our first step is to draw the equivalent **s**-domain circuit; in doing so, we must choose between the two possible capacitor models. With no clear benefit in choosing one over the other, we select the current-source-based model, as in Fig. 15.6*b*.

▶ **Construct an appropriate set of equations.**
We proceed with the analysis by writing a single nodal equation:

$$-1 = \frac{\mathbf{V}_C}{2/s} + \frac{\mathbf{V}_C - 9/s}{3}$$

▶ **Determine if additional information is required.**
We have one equation in one unknown, the frequency-domain representation of the desired capacitor voltage.

▶ **Attempt a solution.**
Solving for $\mathbf{V}_C$, we find that

$$\mathbf{V}_C = \frac{18/s - 6}{3s + 2} = -2\frac{(s - 3)}{s(s + 2/3)}$$

Partial fraction expansion yields

$$\mathbf{V}_C = \frac{9}{s} - \frac{11}{s + 2/3}$$

We obtain $v_C(t)$ by taking the inverse Laplace transform of this expression, resulting in

$$v_C(t) = 9u(t) - 11e^{-2t/3}u(t) \qquad \text{V}$$

or, more compactly,

$$v_C(t) = [9 - 11e^{-2t/3}]u(t) \qquad \text{V}$$

▶ **Verify the solution. Is it reasonable or expected?**
A quick check for $t = 0$ yields $v_C(t) = -2$ V, as it should based on our knowledge of the initial condition. Also, as $t \to \infty$, $v_C(t) \to 9$ V, as we would expect from Fig. 15.6*a* once the transient has died out.

PRACTICE

15.2 Repeat Example 15.2 using the voltage-source-based capacitor model.

Ans: $[9 - 11e^{-2t/3}]u(t)$ V.

The results of this section are summarized in Table 15.1. Note that in each case, we have assumed the passive sign convention.

TABLE **15.1** Summary of Element Representations in the Time and Frequency Domains

Time Domain	Frequency Domain	
Resistor		

Resistor

$v(t) = R\,i(t)$

$V(s) = R\,I(s)$

$I(s) = \dfrac{1}{R}V(s)$

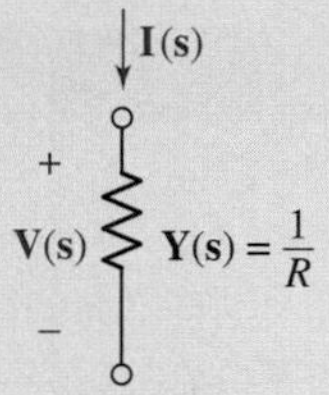

$Z(s) = R$ $\qquad Y(s) = \dfrac{1}{R}$

Inductor

$v(t) = L\dfrac{di}{dt}$

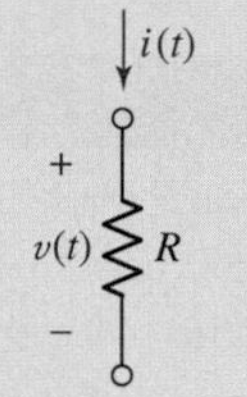

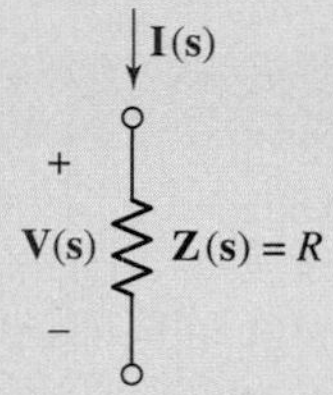

$V(s) = sLI(s) - Li(0^-)$

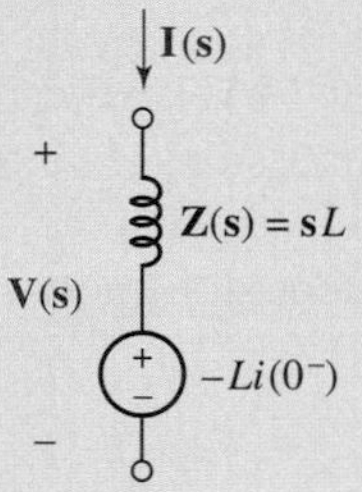

$I(s) = \dfrac{V(s)}{sL} + \dfrac{i(0^-)}{s}$

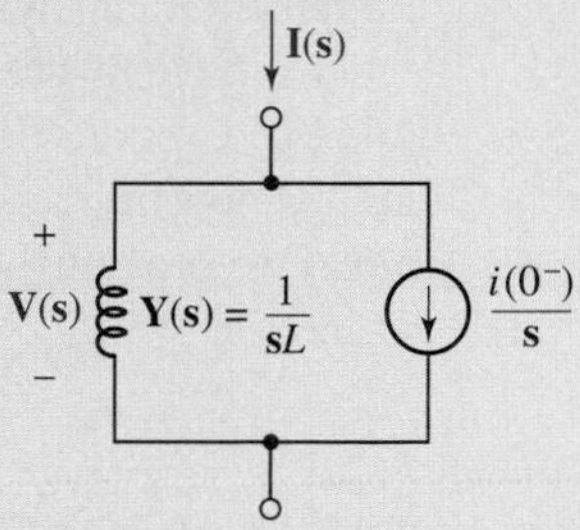

$Z(s) = sL$ $\qquad Y(s) = \dfrac{1}{sL}$ $\qquad \dfrac{i(0^-)}{s}$

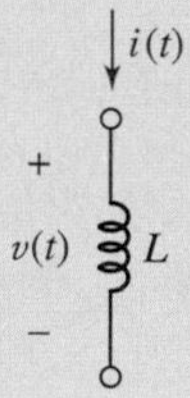

$-Li(0^-)$

Capacitor

$i(t) = C\dfrac{dv}{dt}$

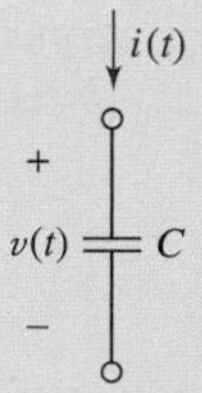

$V(s) = \dfrac{I(s)}{sC} + \dfrac{v(0^-)}{s}$

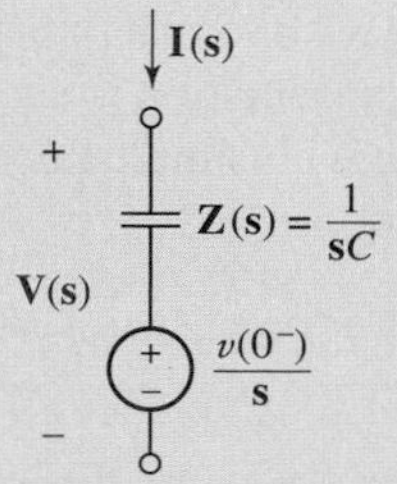

$I(s) = sCV(s) - Cv(0^-)$

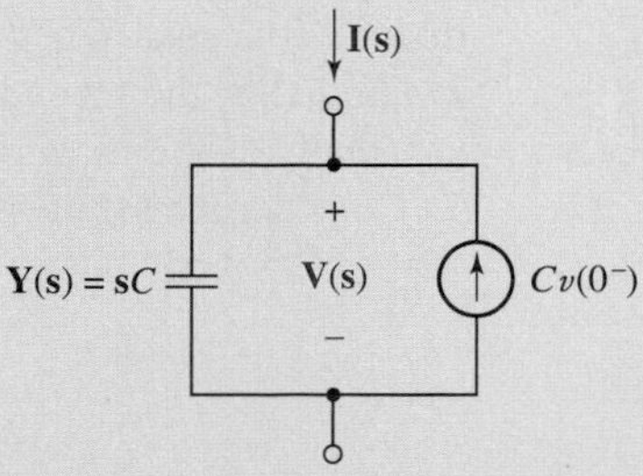

$Z(s) = \dfrac{1}{sC}$ $\qquad Y(s) = sC$ $\qquad Cv(0^-)$

$\dfrac{v(0^-)}{s}$

15.2 • NODAL AND MESH ANALYSIS IN THE *s*-DOMAIN

In Chap. 10, we learned how to transform time-domain circuits driven by sinusoidal sources into their frequency-domain equivalents. The benefits of this transformation were immediately evident, as we were no longer required to solve integrodifferential equations. Nodal and mesh analysis of such circuits (restricted to determining only the steady-state response) resulted in algebraic expressions in terms of $j\omega$, ω being the frequency of the sources.

We have now seen that the concept of impedance can be extended to the more general case of complex frequency ($\mathbf{s} = \sigma + j\omega$). Once we transform circuits from the time domain into the frequency domain, performing nodal or mesh analysis will once again result in purely algebraic expressions, this time in terms of the complex frequency $\mathbf{s}$. Solution of the resulting equations requires the use of variable substitution, Cramer's rule, or software capable of symbolic algebra manipulation (e.g., MATLAB). In this section, we present two examples of reasonable complexity so that we may examine these issues in more detail. First, however, we consider how MATLAB can be used to assist us in such endeavors.

COMPUTER-AIDED ANALYSIS

In Chap. 14, we saw that MATLAB can be used to determine the residues of rational functions in the *s*-domain, making the inverse Laplace transform process significantly easier. However, the software package is actually much more powerful, having numerous built-in routines for manipulation of algebraic expressions. In fact, as we will see in this example, MATLAB is even capable of performing inverse Laplace transforms directly from the rational functions we obtain through circuit analysis.

Let's begin by seeing how MATLAB can be used to work with algebraic expressions. These expressions are stored as character strings, with the apostrophe symbol (') used in the defining expression. For example, we previously represented the polynomial $\mathbf{p(s)} = \mathbf{s}^3 - 12\mathbf{s} + 6$ as a vector:

EDU» p = [1   0   −12   6];

However, we can also represent it symbolically:

EDU» p = 's^3 − 12*s + 6';

These two representations are not equal in MATLAB; they are two distinct concepts. When we wish to manipulate an algebraic expression *symbolically,* the second representation is necessary. This ability is especially useful in working with simultaneous equations.

Consider the set of equations

$$(3\mathbf{s} + 10)\mathbf{I}_1 - 10\mathbf{I}_2 = \frac{4}{\mathbf{s} + 2}$$

$$-10\mathbf{I}_1 + (4\mathbf{s} + 10)\mathbf{I}_2 = \frac{-2}{\mathbf{s} + 1}$$

Using MATLAB's symbolic notation, we define two string variables:

> EDU» eqn1 = '(3*s+10)*I1 − 10*I2 = 4/(s+2)';
>
> EDU» eqn2 = '−10*I1 + (4*s+10)*I2 = −2/(s+1)';

Note that the entire equation has been included in each string; our goal is to solve the two equations for the variables I1 and I2. MATLAB provides a special routine, *solve*(), that can manipulate the equations for us. It is invoked by listing the separate equations (defined as strings), followed by a list of the unknowns (also defined as strings):

> EDU» solution = solve(eqn1, eqn2, 'I1', 'I2');

The answer is stored in the variable *solution,* although in a somewhat unexpected form. MATLAB returns the answer in what is termed a structure, a construct that will be familiar to C programmers. At this stage, however, all we need to know is how to extract our answer. If we type

> EDU» I1 = solution.I1

we obtain the response

> I1 =
>
> 2*(4*s+9)/(s+1)/(6*s^2+47*s+70)

indicating that an **s**-polynomial expression has been assigned to the variable I1; a similar operation is used for the variable I2.

We can now proceed directly to determining the inverse Laplace transform using the function *ilaplace*():

> EDU» i1 = ilaplace(I1)
>
> i1 =
>
> 10/29*exp(−t)−172/667*exp(−35/6*t)−2/23*exp(−2*t)

In this manner, we can quickly obtain the solution to simultaneous equations resulting from nodal or mesh analysis, and also obtain the inverse Laplace transforms. The command *ezplot*(i1) allows us to see what the solution looks like, if we're so inclined. It should be noted that complicated expressions sometimes may confuse MATLAB; in such situations, *ilaplace*() may not return a useful answer.

It is worth mentioning a few related functions, as they can also be used to quickly check answers obtained by hand. The function *numden*() converts a rational function into two separate variables: one containing the numerator, and the other containing the denominator. For example,

> EDU» [N,D] = numden(I1)

returns two algebraic expressions stored in N and D, respectively:

> N =
>
> 8*s+18
>
> D =
>
> (s+1)*(6*s^2+47*s+70)

(Continued on next page)

In order to apply our previous experience with the function *residue()*, we need to convert each symbolic (string) expression into a vector containing the coefficients of the polynomial. This is achieved using the command *sym2poly()*:

$$EDU\text{» } n = sym2poly(N);$$

and

$$EDU\text{» } d = sym2poly(D)$$
$$d =$$
$$6 \quad 53 \quad 117 \quad 70$$

after which we can determine the residues:

$$EDU\text{» } [r\ p\ y] = residue(n,d)$$

r =	p =	y =
−0.2579	−5.8333	[ ]
−0.0870	−2.0000	
0.3448	−1.0000	

which is in agreement with what we obtained using *ilaplace()*.

With these new MATLAB skills, (or a deep-seated desire to try an alternative approach such as Cramer's rule or direct substitution), we are ready to proceed to analyze a few circuits.

EXAMPLE 15.3

Determine the two mesh currents i_1 and i_2 in the circuit of Fig. 15.7a. There is no energy initially stored in the circuit.

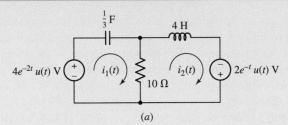

(a)

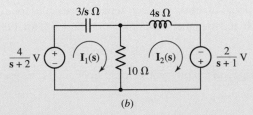

(b)

■ **FIGURE 15.7** (*a*) A two-mesh circuit for which the individual mesh currents are desired. (*b*) The frequency-domain equivalent circuit.

As always, our first step is to draw the appropriate frequency-domain equivalent circuit. Since we have zero energy stored in the circuit at

$t = 0^-$, we replace the $\frac{1}{3}$ F capacitor with a $3/s$ Ω impedance, and the 4 H inductor with a $4s$ Ω impedance, as shown in Fig. 15.7*b*.

Next, we write two mesh equations just as we have before:

$$-\frac{4}{s+2} + \frac{3}{s}\mathbf{I}_1 + 10\mathbf{I}_1 - 10\mathbf{I}_2 = 0$$

or

$$\left(\frac{3}{s} + 10\right)\mathbf{I}_1 - 10\mathbf{I}_2 = \frac{4}{s+2} \qquad \text{(mesh 1)}$$

and

$$-\frac{2}{s+1} + 10\mathbf{I}_2 - 10\mathbf{I}_1 + 4s\mathbf{I}_2 = 0$$

or

$$-10\mathbf{I}_1 + (4s+10)\mathbf{I}_2 = \frac{2}{s+1} \qquad \text{(mesh 2)}$$

Solving for $\mathbf{I}_1$ and $\mathbf{I}_2$, we find that

$$\mathbf{I}_1 = \frac{2s(4s^2 + 19s + 20)}{(20s^4 + 66s^3 + 73s^2 + 57s + 30)} \quad \text{A}$$

and

$$\mathbf{I}_2 = \frac{30s^2 + 43s + 6}{(s+2)(20s^3 + 26s^2 + 21s + 15)} \quad \text{A}$$

All that remains is for us to take the inverse Laplace transform of each function, after which we find that

$$i_1(t) = -96.39e^{-2t} - 344.8e^{-t} + 841.2e^{-0.15t}\cos 0.8529t$$
$$+ 197.7e^{-0.15t}\sin 0.8529t \qquad \text{mA}$$

and

$$i_2(t) = -481.9e^{-2t} - 241.4e^{-t} + 723.3e^{-0.15t}\cos 0.8529t$$
$$+ 472.8e^{-0.15t}\sin 0.8529t \qquad \text{mA}$$

> We were (indirectly) told that no current flows through the inductor at $t = 0^-$. Therefore, $i_2(0^-) = 0$, and consequently $i_2(0^+)$ must be zero as well. Does this result hold true for our answer?

PRACTICE

15.3 Find the mesh currents i_1 and i_2 in the circuit of Fig. 15.8. You may assume no energy is stored in the circuit at $t = 0^-$.

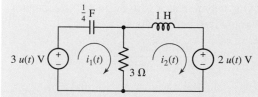

■ **FIGURE 15.8**

Ans: $i_1 = e^{-2t/3}\cos\left(\frac{4}{3}\sqrt{2}t\right) + \left(\sqrt{2}/8\right)e^{-2t/3}\sin\left(\frac{4}{3}\sqrt{2}t\right)$ A;
$i_2 = -\frac{2}{3} + \frac{2}{3}e^{-2t/3}\cos\left(\frac{4}{3}\sqrt{2}t\right) + \left(13\sqrt{2}/24\right)e^{-2t/3}\sin\left(\frac{4}{3}\sqrt{2}t\right)$ A.

EXAMPLE 15.4

Calculate the voltage v_x in the circuit of Fig. 15.9 using nodal analysis techniques.

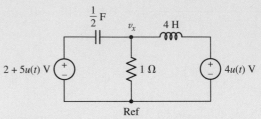

■ **FIGURE 15.9** A simple four-node circuit containing two energy storage elements.

The first step is to draw the corresponding **s**-domain circuit. We see that the $\frac{1}{2}$ F capacitor has an initial voltage of 2 V across it at $t = 0^-$, requiring that we employ one of the two models of Fig. 15.5. Since we are to use nodal analysis, perhaps the model of Fig. 15.5*b* is the better route. The resulting circuit is shown in Fig. 15.10.

With two of the three nodal voltages specified, we have only one nodal equation to write:

$$-1 = \frac{\mathbf{V}_x - \dfrac{7}{\mathbf{s}}}{\dfrac{2}{\mathbf{s}}} + \mathbf{V}_x + \frac{\mathbf{V}_x - \dfrac{4}{\mathbf{s}}}{4\mathbf{s}}$$

so that

$$\mathbf{V}_x = \frac{10\mathbf{s}^2 + 4}{\mathbf{s}(2\mathbf{s}^2 + 4\mathbf{s} + 1)} = \frac{5\mathbf{s}^2 + 2}{\mathbf{s}\left(\mathbf{s} + 1 + \dfrac{\sqrt{2}}{2}\right)\left(\mathbf{s} + 1 - \dfrac{\sqrt{2}}{2}\right)}$$

The nodal voltage v_x is found by taking the inverse Laplace transform, whereby we find that

$$v_x = [4 + 6.864e^{-1.707t} - 5.864e^{-0.2929t}]u(t)$$

or

$$v_x = \left[4 - e^{-t}\left(9\sqrt{2}\sinh\frac{\sqrt{2}}{2}t - \cosh\frac{\sqrt{2}}{2}t\right)\right]u(t)$$

Is our answer correct? One way to check is to evaluate the capacitor voltage at $t = 0$, since we know it to be 2 V. Thus,

$$\mathbf{V}_C = \frac{7}{\mathbf{s}} - \mathbf{V}_x = \frac{4\mathbf{s}^2 + 28\mathbf{s} + 3}{\mathbf{s}(2\mathbf{s}^2 + 4\mathbf{s} + 1)}$$

Multiplying $\mathbf{V}_C$ by $\mathbf{s}$ and taking the limit as $\mathbf{s} \to \infty$, we find that

$$v_c(0^+) = \lim_{\mathbf{s} \to \infty}\left[\frac{4\mathbf{s}^2 + 28\mathbf{s} + 3}{2\mathbf{s}^2 + 4\mathbf{s} + 1}\right] = 2 \text{ V}$$

as expected.

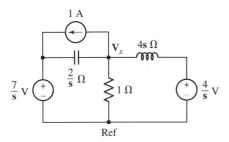

■ **FIGURE 15.10** The **s**-domain equivalent circuit of Fig. 15.9.

PRACTICE

15.4 Employ nodal analysis to calculate $v_x(t)$ for the circuit of Fig. 15.11.

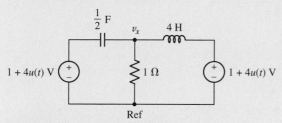

■ **FIGURE 15.11** For Practice Problem 15.4.

Ans: $[5 + 5.657(e^{-1.707t} - e^{-0.2929t})]u(t)$.

EXAMPLE 15.5

Use nodal analysis to determine the voltages v_1, v_2, and v_3 in the circuit of Fig. 15.12a. No energy is stored in the circuit at $t = 0^-$.

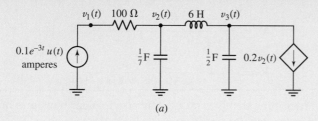

(a)

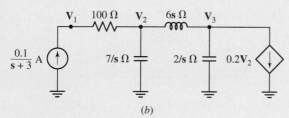

(b)

■ **FIGURE 15.12** (a) A four-node circuit containing two capacitors and one inductor, none of which are storing energy at $t = 0^-$. (b) The frequency-domain equivalent circuit.

This circuit consists of three separate energy storage elements, none of which is storing any energy at $t = 0^-$. Thus, each may be replaced by their corresponding impedance as shown in Fig. 15.12b. We also note the presence of a dependent current source controlled by the nodal voltage $v_2(t)$.

Beginning at node 1, we can write the following equation:

$$\frac{0.1}{s+3} = \frac{\mathbf{V}_1 - \mathbf{V}_2}{100}$$

(Continued on next page)

or

$$\frac{10}{s+3} = V_1 - V_2 \qquad \text{(node 1)}$$

and at node 2,

$$0 = \frac{V_2 - V_1}{100} + \frac{V_2}{7/s} + \frac{V_2 - V_3}{6s}$$

or

$$-42sV_1 + (600s^2 + 42s + 700)V_2 - 700V_3 = 0 \qquad \text{(node 2)}$$

and finally, at node 3,

$$-0.2V_2 = \frac{V_3 - V_2}{6s} + \frac{V_3}{2/s}$$

or

$$(1.2s - 1)V_2 + (3s^2 + 1)V_3 = 0$$

Solving this set of equations for the nodal voltages, we obtain

$$V_1 = 3\frac{100s^3 + 7s^2 + 150s + 49}{(s+3)(30s^3 + 45s + 14)}$$

$$V_2 = 7\frac{3s^2 + 1}{(s+3)(30s^3 + 45s + 14)}$$

$$V_3 = -1.4\frac{6s - 5}{(s+3)(30s^3 + 45s + 14)}$$

The only remaining step is to take the inverse Laplace transform of each voltage, so that, for $t > 0$,

$$v_1(t) = 9.789e^{-3t} + 0.06173e^{-0.2941t} + 0.1488e^{0.1471t}\cos(1.251t)$$
$$+ 0.05172e^{0.1471t}\sin(1.251t)\ \text{V}$$

$$v_2(t) = -0.2105e^{-3t} + 0.06173e^{-0.2941t} + 0.1488e^{0.1471t}\cos(1.251t)$$
$$+ 0.05172e^{0.1471t}\sin(1.251t)\ \text{V}$$

$$v_3(t) = -0.03459e^{-3t} + 0.06631e^{-0.2941t} - 0.03172e^{0.1471t}\cos(1.251t)$$
$$- 0.06362e^{0.1471t}\sin(1.251t)\ \text{V}$$

Note that the response grows exponentially as a result of the action of the dependent current source. In essence, the circuit is "running away," indicating that at some point a component will melt, explode, or fail in some related fashion. Although analyzing such circuits can evidently entail a great deal of work, the advantages to *s*-domain techniques are clear once we contemplate performing the analysis in the time domain!

PRACTICE

15.5 Use nodal analysis to determine the voltages v_1, v_2, and v_3 in the circuit of Fig. 15.13. Assume there is zero energy stored in the inductors at $t = 0^-$.

Ans: $v_1(t) = -30\delta(t) - 14u(t)\ \text{V}$; $v_2(t) = -14u(t)\ \text{V}$; $v_3(t) = 24\delta(t) - 14u(t)\ \text{V}$.

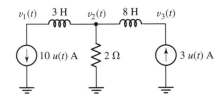

■ **FIGURE 15.13**

15.3 ADDITIONAL CIRCUIT ANALYSIS TECHNIQUES

Depending on the specific goal in analyzing a particular circuit, we often find that we can simplify our task by carefully choosing our analysis technique. For example, it is seldom desirable to apply superposition to a circuit containing 215 independent sources, as such an approach requires analysis of 215 separate circuits! By treating passive elements such as capacitors and inductors as impedances, however, we are free to employ any of the circuit analysis techniques studied in Chaps. 3, 4, and 5 to circuits that have been transformed to their s-domain equivalents.

Thus, superposition, source transformations, Thévenin's theorem, and Norton's theorem all apply in the s-domain.

EXAMPLE 15.6

Simplify the circuit of Fig. 15.14a using source transformations, and determine an expression for the voltage $v(t)$.

With no initial currents or voltages specified, and a $u(t)$ multiplying the voltage source, we conclude that there is no energy initially stored in the circuit. Thus, we draw the frequency-domain circuit as shown in Fig. 15.14b.

Our strategy will be to perform several source transformations in succession in order to combine the two $2/\mathbf{s}$ Ω impedances and the 10 Ω resistor; we must leave the $9\mathbf{s}$ Ω impedance alone as the desired quantity $\mathbf{V}(\mathbf{s})$ appears across its terminals. We may now transform the voltage source and the leftmost $2/\mathbf{s}$ Ω impedance into a current source

$$\mathbf{I}(\mathbf{s}) = \left(\frac{2\mathbf{s}}{\mathbf{s}^2 + 9}\right)\left(\frac{\mathbf{s}}{2}\right) = \frac{\mathbf{s}^2}{\mathbf{s}^2 + 9} \quad \mathrm{A}$$

in parallel with a $2/\mathbf{s}$ Ω impedance.

As depicted in Fig. 15.15a, after this transformation, we have $\mathbf{Z}_1 \equiv (2/\mathbf{s}) \| 10 = 20/(10\mathbf{s} + 2)$ Ω facing the current source. Performing another source transformation, we obtain a voltage source $\mathbf{V}_2(\mathbf{s})$ such that

$$\mathbf{V}_2(\mathbf{s}) = \left(\frac{\mathbf{s}^2}{\mathbf{s}^2 + 9}\right)\left(\frac{20}{10\mathbf{s} + 2}\right)$$

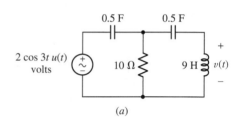

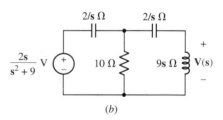

■ **FIGURE 15.14** (a) Circuit to be simplified using source transformations. (b) Frequency-domain representation.

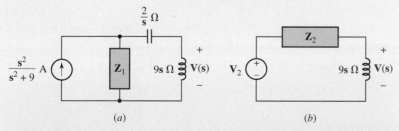

(a)   (b)

■ **FIGURE 15.15** (a) Circuit after first source transformation. (b) Final circuit to be analyzed for $\mathbf{V}(\mathbf{s})$.

This voltage source is in series with $\mathbf{Z}_1$ and also with the remaining $2/\mathbf{s}$ impedance; combining $\mathbf{Z}_1$ and $2/\mathbf{s}$ into a new impedance $\mathbf{Z}_2$ yields

$$\mathbf{Z}_2 = \frac{20}{10\mathbf{s} + 2} + \frac{2}{\mathbf{s}} = \frac{40\mathbf{s} + 4}{\mathbf{s}(10\mathbf{s} + 2)} \quad \Omega$$

(Continued on next page)

The resulting circuit is shown in Fig. 15.15*b*. At this stage, we are now ready to obtain an expression for the voltage $\mathbf{V(s)}$ using simple voltage division:

$$\mathbf{V(s)} = \left(\frac{s^2}{s^2+9}\right)\left(\frac{20}{10s+2}\right)\frac{9s}{9s+\left[\dfrac{40s+4}{s(10s+2)}\right]}$$

$$= \frac{180s^4}{(s^2+9)(90s^3+18s^2+40s+4)}$$

Both terms in the denominator possess complex roots. Employing MATLAB to expand the denominator and then determine the residues,

> EDU» d1 = 's^2 + 9';
> EDU» d2 = '90*s^3 + 18*s^2 + 40*s + 4';
> EDU» d = symmul(d1,d2);
> EDU» denominator = expand(d);
> EDU» den = sym2poly(denominator);
> EDU» num = [180 0 0 0 0];
> EDU» [r p y] = residue(num,den);

we find

$$\mathbf{V(s)} = \frac{1.047 + j0.0716}{s - j3} + \frac{1.047 - j0.0716}{s + j3} - \frac{0.0471 + j0.0191}{s + 0.04885 - j0.6573}$$

$$- \frac{0.0471 - j0.0191}{s + 0.04885 + j0.6573} + \frac{5.590 \times 10^{-5}}{s + 0.1023}$$

> Note that each term having a complex pole has a companion term that is its complex conjugate. For any physical system, complex poles will always occur in conjugate pairs.

Taking the inverse transform of each term, writing $1.047 + j0.0716$ as $1.049e^{j3.912°}$ and $0.0471 + j0.0191$ as $0.05083e^{j157.9°}$ results in

$$v(t) = 1.049e^{j3.912°}e^{j3t}u(t) + 1.049e^{-j3.912°}e^{-j3t}u(t)$$

$$+ 0.05083e^{-j157.9°}e^{-0.04885t}e^{-j0.6573t}u(t)$$

$$+ 0.05083e^{+j157.9°}e^{-0.04885t}e^{+j0.6573t}u(t)$$

$$+ 5.590 \times 10^{-5}e^{-0.1023t}u(t)$$

Converting the complex exponentials to sinusoids then allows us to write a slightly simplified expression for our voltage

$$v(t) = [5.590 \times 10^{-5}e^{-0.1023t} + 2.098\cos(3t + 3.912°)$$

$$+ 0.1017e^{-0.04885t}\cos(0.6573t + 157.9°)]u(t) \qquad \text{V}$$

PRACTICE

15.6 Using the method of source transformation, reduce the circuit of Fig. 15.16 to a single **s**-domain current source in parallel with a single impedance.

Ans: $\mathbf{I}_s = \dfrac{35}{s^2(18s+63)}$ A, $\mathbf{Z}_s = \dfrac{72s^2 + 252s}{18s^3 + 63s^2 + 12s + 28}$ Ω.

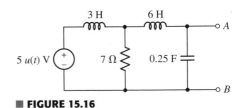

■ **FIGURE 15.16**

EXAMPLE **15.7**

Find the frequency-domain Thévenin equivalent of the highlighted network shown in Fig. 15.17a.

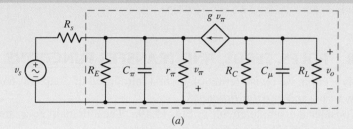

(a)

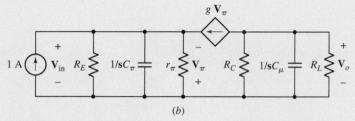

(b)

■ **FIGURE 15.17** (a) An equivalent circuit for the "common base" transistor amplifier. (b) The frequency-domain equivalent circuit with a 1 A test source substituted for the input source represented by v_s and R_s.

We are being asked to determine the Thévenin equivalent of the circuit connected to the input device; this quantity is often referred to as the *input impedance* of the amplifier circuit. After converting the circuit to its frequency-domain equivalent, we replace the input device (v_s and R_s) with a 1 A "test" source, as shown in Fig. 15.17b. The input impedance $\mathbf{Z}_{\text{in}}$ is then

$$\mathbf{Z}_{\text{in}} = \frac{\mathbf{V}_{\text{in}}}{1}$$

or simply $\mathbf{V}_{\text{in}}$. We must find an expression for this quantity in terms of the 1 A source, resistors and capacitors, and/or the dependent source parameter g.

Writing a single nodal equation at the input, then, we find that

$$1 + g\mathbf{V}_{\pi} = \frac{\mathbf{V}_{\text{in}}}{\mathbf{Z}_{\text{eq}}}$$

where

$$\mathbf{Z}_{\text{eq}} \equiv R_E \left\| \frac{1}{sC_{\pi}} \right\| r_{\pi} = \frac{R_E r_{\pi}}{r_{\pi} + R_E + sR_E r_{\pi} C_{\pi}}$$

Since $\mathbf{V}_{\pi} = -\mathbf{V}_{\text{in}}$, we find that

$$\mathbf{Z}_{\text{in}} = \mathbf{V}_{\text{in}} = \frac{R_E r_{\pi}}{r_{\pi} + R_E + sR_E r_{\pi} C_{\pi} + gR_E r_{\pi}} \quad \Omega$$

This particular circuit is known as the "hybrid π" model for a special type of single-transistor circuit known as the common base amplifier. The two capacitors, C_{π} and C_{μ}, represent capacitances internal to the transistor, and are typically on the order of a few pF. The resistor R_L in the circuit represents the Thévenin equivalent resistance of the output device, which could be a speaker or even a semiconductor laser. The voltage source v_s and the resistor R_s together represent the Thévenin equivalent of the input device, which may be a microphone, a light-sensitive resistor, or possibly a radio antenna.

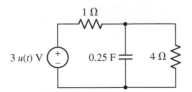

■ FIGURE 15.18

15.4 ● POLES, ZEROS, AND TRANSFER FUNCTIONS

In this section, we revisit terminology first introduced in Chap. 14, namely *poles, zeros,* and *transfer functions.*

Consider the simple circuit in Fig. 15.19*a*. The *s*-domain equivalent is given in Fig. 15.19*b*, and nodal analysis yields

$$0 = \frac{\mathbf{V}_{out}}{1/sC} + \frac{\mathbf{V}_{out} - \mathbf{V}_{in}}{R}$$

Rearranging and solving for $\mathbf{V}_{out}$, we find

$$\mathbf{V}_{out} = \frac{\mathbf{V}_{in}}{1 + \mathbf{s}RC}$$

or

$$\mathbf{H}(\mathbf{s}) \equiv \frac{\mathbf{V}_{out}}{\mathbf{V}_{in}} = \frac{1}{1 + \mathbf{s}RC} \qquad [7]$$

where $\mathbf{H}(\mathbf{s})$ is the *transfer function* of the circuit, defined as the ratio of the output to the input. We could just as easily specify a particular current as either the input or output quantity, leading to a different transfer function for the same circuit. Circuit schematics are typically read from left to right, so designers often place the input of a circuit on the left of the schematic and the output terminals on the right, at least to the extent where it is possible.

The concept of a transfer function is very important, both in terms of circuit analysis as well as other areas of engineering. There are two reasons for this. First, once we know the transfer function of a particular circuit, we can easily find the output that results from *any* input. All we need to do is multiply $\mathbf{H}(\mathbf{s})$ by the input quantity, and take the inverse transform of the resulting expression. Second, the form of the transfer function contains a great deal of information about the behavior we might expect from a particular circuit (or system).

As alluded to in the Practical Application of Chap. 14, in order to evaluate the stability of a system it is necessary to determine the poles and zeros of the transfer function $\mathbf{H}(\mathbf{s})$; we will explore this issue in detail shortly. Equation [7] may be written as

$$\mathbf{H}(\mathbf{s}) = \frac{1/RC}{\mathbf{s} + 1/RC} \qquad [8]$$

The magnitude of this function approaches zero as $\mathbf{s} \to \infty$. Thus, we say that $\mathbf{H}(\mathbf{s})$ has a *zero* at $\mathbf{s} = \infty$. The function approaches infinity at $\mathbf{s} = -1/RC$; we therefore say that $\mathbf{H}(\mathbf{s})$ has a *pole* at $\mathbf{s} = -1/RC$. These frequencies are termed *critical frequencies,* and their early identification simplifies the construction of the response curves we will develop in Section 15.7.

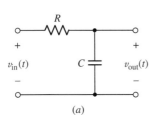

(*a*)

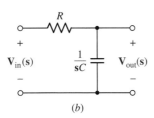

(*b*)

■ FIGURE 15.19 (*a*) A simple resistor-capacitor circuit, with an input voltage and output voltage specified. (*b*) The s-domain equivalent circuit.

When computing magnitude, it is customary to consider $+\infty$ and $-\infty$ as being the same frequency. The phase angle of the response at very large positive and negative values of ω need not be the same, however.

15.5 • CONVOLUTION

The s-domain techniques we have developed up to this point are very use-ful in determining the current and voltage response of a particular circuit. However, in practice we are often faced with circuits to which arbitrary sources can be connected, and require an efficient means of determining the new output each time. This is easily accomplished if we can characterize the basic circuit by a transfer function called the **system function.** Interestingly enough, we are about to see that this system function is the Laplace trans-form of the unit impulse response of the circuit.

The analysis can proceed in either the time domain or the frequency do-main, although it is generally more useful to work in the frequency domain. In such situations, we have a simple four-step process:

---

1. Determine the circuit system function (if not already known);
2. Obtain the Laplace transform of the forcing function to be applied;
3. Multiply this transform and the system function; and finally
4. Perform an inverse transform operation on the product to find the output.

By these means some relatively complicated integral expressions will be reduced to simple functions of **s**, and the mathematical operations of inte-gration and differentiation will be replaced by the simpler operations of al-gebraic multiplication and division. With these remarks in mind, let us now proceed to examine the unit-impulse response of a circuit and establish its relation to the system function. Then we can look at some specific analysis problems.

The Impulse Response

Consider a linear electrical network N, without initial stored energy, to which a forcing function $x(t)$ is applied. At some point in this circuit, a response function $y(t)$ is present. We show this in block diagram form in Fig. 15.20a along with sketches of generic time functions. The forcing function is shown to exist only in the interval $a < t < b$. Thus, $y(t)$ exists only for $t > a$.

The question that we now wish to answer is this: *"If we know the form of $x(t)$, then how is $y(t)$ described?"* To answer this question, we need to know something about N. Suppose, therefore, that our knowledge of N consists of knowing its response when the forcing function is a unit impulse $\delta(t)$. That is, we are assuming that we know $h(t)$, the response function re-sulting when a unit impulse is supplied as the forcing function at $t = 0$, as shown in Fig. 15.20b. The function $h(t)$ is commonly called the unit-impulse response function, or the **impulse response.** This is a very important descriptive property for an electric circuit.

Based on our knowledge of Laplace transforms, we can view this from a slightly different perspective. Transforming $x(t)$ into $\mathbf{X}(\mathbf{s})$ and $y(t)$ into $\mathbf{Y}(\mathbf{s})$, we define the system transfer function $\mathbf{H}(\mathbf{s})$ as

$$\mathbf{H}(\mathbf{s}) \equiv \frac{\mathbf{Y}(\mathbf{s})}{\mathbf{X}(\mathbf{s})}$$

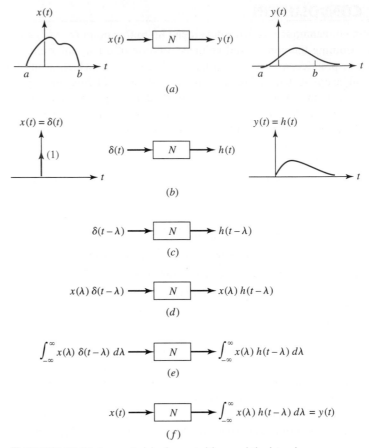

■ **FIGURE 15.20** A conceptual development of the convolution integral.

If $x(t) = \delta(t)$, then according to Table 14.1, $\mathbf{X(s)} = 1$. Thus, $\mathbf{H(s)} = \mathbf{Y(s)}$ and so *in this instance* $h(t) = y(t)$.

Instead of applying the unit impulse at time $t = 0$, let us now suppose that it is applied at time $t = \lambda$ (lambda). We see that the only change in the output is a time delay. Thus, the output becomes $h(t - \lambda)$ when the input is $\delta(t - \lambda)$, as shown in Fig. 15.20c. Next, suppose that the input impulse has some strength other than unity. Specifically, let the strength of the impulse be numerically equal to the value of $x(t)$ when $t = \lambda$. This value $x(\lambda)$ is a constant; we know that the multiplication of a single forcing function in a *linear* circuit by a constant simply causes the response to change proportionately. Thus, if the input is changed to $x(\lambda)\delta(t - \lambda)$, then the response becomes $x(\lambda)h(t - \lambda)$, as shown in Fig. 15.20d.

Now let us sum this latest input over all possible values of λ and use the result as a forcing function for N. Linearity decrees that the output must be equal to the sum of the responses resulting from the use of all possible values of λ. Loosely speaking, the integral of the input produces the integral of the output, as shown in Fig. 15.20e. But what is the input now? Given the sifting property[1] of the unit impulse, we see that the input is simply $x(t)$, the original input. Thus, Fig. 15.20e may be represented as in Fig. 15.20f.

(1) The sifting property of the impulse function, described in Section 14.5, states that $\int_{-\infty}^{\infty} f(t)\delta(t - t_0)\, dt = f(t_0)$.

The Convolution Integral

If the input to our system N is the forcing function $x(t)$, we know the output must be the function $y(t)$ as depicted in Fig. 15.20a. Thus, from Fig. 15.20f we conclude that

$$y(t) = \int_{-\infty}^{\infty} x(\lambda)h(t-\lambda)\,d\lambda \qquad [9]$$

where $h(t)$ is the impulse response of N. This important relationship is known far and wide as the ***convolution integral.*** In words, this last equation states that *the output is equal to the input convolved with the impulse response.* It is often abbreviated by means of

$$y(t) = x(t) * h(t)$$

where the asterisk is read "convolved with."

Equation [9] sometimes appears in a slightly different but equivalent form. If we let $z = t - \lambda$, then $d\lambda = -dz$, and the expression for $y(t)$ becomes

$$y(t) = \int_{\infty}^{-\infty} -x(t-z)h(z)\,dz = \int_{-\infty}^{\infty} x(t-z)h(z)\,dz$$

and since the symbol that we use for the variable of integration is unimportant, we can modify Eq. [9] to write

$$\begin{aligned} y(t) = x(t) * h(t) &= \int_{-\infty}^{\infty} x(z)h(t-z)\,dz \\ &= \int_{-\infty}^{\infty} x(t-z)h(z)\,dz \end{aligned} \qquad [10]$$

> Be careful not to confuse this new notation with multiplication!

Convolution and Realizable Systems

The result that we have in Eq. [10] is very general; it applies to any linear system. However, we are usually interested in ***physically realizable systems,*** those that *do* exist or *could* exist, and such systems have a property that modifies the convolution integral slightly. That is, *the response of the system cannot begin before the forcing function is applied.* In particular, $h(t)$ is the response of the system resulting from the application of a unit impulse at $t = 0$. Therefore, $h(t)$ cannot exist for $t < 0$. It follows that, in the second integral of Eq. [10], the integrand is zero when $z < 0$; in the first integral, the integrand is zero when $(t - z)$ is negative, or when $z > t$. Therefore, for *realizable* systems the limits of integration change in the convolution integrals:

$$\begin{aligned} y(t) = x(t) * h(t) &= \int_{-\infty}^{t} x(z)h(t-z)\,dz \\ &= \int_{0}^{\infty} x(t-z)h(z)\,dz \end{aligned} \qquad [11]$$

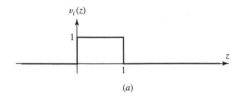

(a)

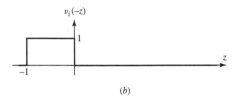

(b)

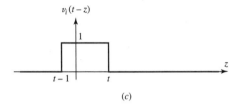

(c)

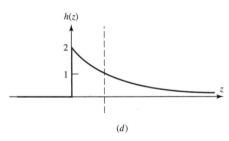

(d)

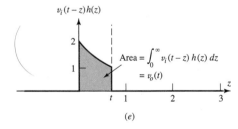

(e)

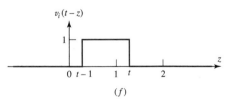

(f)

■ **FIGURE 15.21** Graphical concepts in evaluating a convolution integral.

Equations [10] and [11] are both valid, but the latter is more specific when we are speaking of *realizable* linear systems, and well worth memorizing.

Graphical Method of Convolution

Before discussing the significance of the impulse response of a circuit any further, let us consider a numerical example that will give us some insight into just how the convolution integral can be evaluated. Although the expression itself is simple enough, the evaluation is sometimes troublesome, especially with regard to the values used as the limits of integration.

Suppose that the input is a rectangular voltage pulse that starts at $t = 0$, has a duration of 1 second, and is 1 V in amplitude:

$$x(t) = v_i(t) = u(t) - u(t - 1)$$

Suppose also that this voltage pulse is applied to a circuit whose impulse response is known to be an exponential function of the form:

$$h(t) = 2e^{-t}u(t)$$

We wish to evaluate the output voltage $v_o(t)$, and we can write the answer immediately in integral form,

$$y(t) = v_o(t) = v_i(t) * h(t) = \int_0^\infty v_i(t - z)h(z)\, dz$$
$$= \int_0^\infty [u(t - z) - u(t - z - 1)][2e^{-z}u(z)]\, dz$$

Obtaining this expression for $v_o(t)$ is simple enough, but the presence of the many unit-step functions tends to make its evaluation confusing and possibly even a little obnoxious. Careful attention must be paid to the determination of those portions of the range of integration in which the integrand is zero.

Let us use some graphical assistance to help us understand what the convolution integral says. We begin by drawing several z axes lined up one above the other, as shown in Fig. 15.21. We know what $v_i(t)$ looks like, and so we know what $v_i(z)$ looks like also; this is plotted as Fig. 15.21*a*. The function $v_i(-z)$ is simply $v_i(z)$ run backward with respect to z, or rotated about the ordinate axis; it is shown in Fig. 15.21*b*. Next we wish to represent $v_i(t - z)$, which is $v_i(-z)$ after it is shifted to the right by an amount $z = t$ as shown in Fig. 15.21*c*. On the next z axis, in Fig. 15.21*d*, our impulse response $h(z) = 2e^{-z}u(z)$ is plotted.

The next step is to multiply the two functions $v_i(t - z)$ and $h(z)$; the result for an arbitrary value of $t < 1$ is shown in Fig. 15.21*e*. We are after a value for the output $v_o(t)$, which is given by the *area* under the product curve (shown shaded in the figure).

Let's first consider $t < 0$. In this case, there is no overlap between $v_i(t - z)$ and $h(z)$, so $v_o = 0$. As we increase t, we slide the pulse shown in Fig. 15.21*c* to the right, leading to an overlap with $h(z)$ once $t > 0$. The area under the corresponding curve of Fig. 15.21*e* continues to increase as we increase the value of t until we reach $t = 1$. As t increases above this value, a gap opens up between $z = 0$ and the leading edge of the pulse, as shown in Fig. 15.21*f*. As a result, the overlap with $h(z)$ decreases.

In other words, for values of t that lie between zero and unity, we must integrate from $z = 0$ to $z = t$; for values of t that exceed unity, the range of integration is $t - 1 < z < t$. Thus, we may write

$$v_o(t) = \begin{cases} 0 & t < 0 \\ \displaystyle\int_0^t 2e^{-z}\,dz = 2(1 - e^{-t}) & 0 \le t \le 1 \\ \displaystyle\int_{t-1}^t 2e^{-z}\,dz = 2(e - 1)e^{-t} & t > 1 \end{cases}$$

This function is shown plotted versus the time variable t in Fig. 15.22, and our solution is completed.

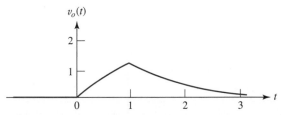

■ FIGURE 15.22 The output function v_o obtained by graphical convolution.

Apply a unit-step function, $x(t) = u(t)$, as the input to a system whose impulse response is $h(t) = u(t) - 2u(t - 1) + u(t - 2)$, and determine the corresponding output $y(t) = x(t) * h(t)$.

Our first step is to plot both $x(t)$ and $h(t)$, as shown in Fig. 15.23.

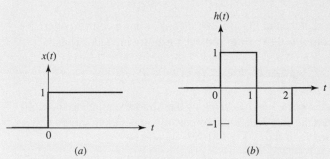

■ FIGURE 15.23 Sketches of (*a*) the input signal $x(t) = u(t)$ and (*b*) the unit-impulse response $h(t) = u(t) - 2u(t - 1) + u(t - 2)$, for a linear system.

We arbitrarily choose to evaluate the first integral of Eq. [11],

$$y(t) = \int_{-\infty}^t x(z)h(t - z)\,dz$$

and prepare a sequence of sketches to help select the correct limits of integration. Figure 15.24 shows these functions in order: the input $x(z)$ as a function of z; the impulse response $h(z)$; the curve of $h(-z)$,

(Continued on next page)

which is just $h(z)$ rotated about the vertical axis; and $h(t-z)$, obtained by sliding $h(-z)$ to the right t units. For this sketch, we have selected t in the range $0 < t < 1$.

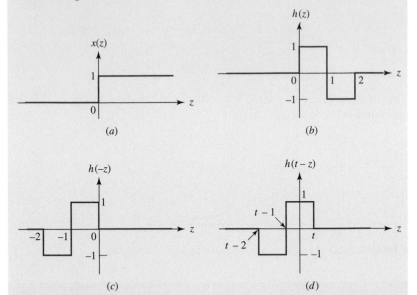

■ **FIGURE 15.24** (*a*) The input signal and (*b*) the impulse response are plotted as functions of *z*. (*c*) $h(-z)$ is obtained by flipping $h(z)$ about the vertical axis, and (*d*) $h(t-z)$ results when $h(-z)$ is slid *t* units to the right.

It is now easy to visualize the product of the first graph, $x(z)$, and the last, $h(t-z)$, for the various ranges of t. When t is less than zero, there is no overlap, and

$$y(t) = 0 \qquad t < 0$$

For the case sketched in Fig. 15.24*d*, $h(t-z)$ has a nonzero overlap with $x(z)$ from $z = 0$ to $z = t$, and each is unity in value. Thus,

$$y(t) = \int_0^t (1 \times 1) \, dz = t \qquad 0 < t < 1$$

When t lies between 1 and 2, $h(t-z)$ has slid far enough to the right to bring under the step function that part of the negative square wave extending from $z = 0$ to $z = t - 1$. We then have

$$y(t) = \int_0^{t-1} [1 \times (-1)] \, dz + \int_{t-1}^t (1 \times 1) \, dz = -z \Big|_{z=0}^{z=t-1} + z \Big|_{z=t-1}^{z=t}$$

Therefore,

$$y(t) = -(t-1) + t - (t-1) = 2 - t, \qquad 1 < t < 2$$

Finally, when t is greater than 2, $h(t-z)$ has slid far enough to the right so that it lies entirely to the right of $z = 0$. The intersection with the unit step is complete, and

$$y(t) = \int_{t-2}^{t-1} [1 \times (-1)] \, dz + \int_{t-1}^t (1 \times 1) \, dz = -z \Big|_{z=t-2}^{z=t-1} + z \Big|_{z=t-1}^{z=t}$$

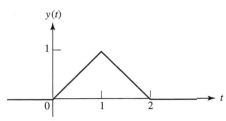

or

$$y(t) = -(t - 1) + (t - 2) + t - (t - 1) = 0, \qquad t > 2$$

These four segments of $y(t)$ are collected as a continuous curve in Fig. 15.25.

PRACTICE

15.8 Repeat Example 15.8 using the *second* integral of Eq. [11].

15.9 The impulse response of a network is given by $h(t) = 5u(t - 1)$. If an input signal $x(t) = 2[u(t) - u(t - 3)]$ is applied, determine the output $y(t)$ at t equal to (a) -0.5; (b) 0.5; (c) 2.5; (d) 3.5.

Ans: 15.9: 0, 0, 15, 25.

■ FIGURE 15.25 The result of convolving the $x(t)$ and $h(t)$ shown in Fig. 15.23.

Convolution and the Laplace Transform

Convolution has applications in a wide variety of disciplines beyond linear circuit analysis, including image processing, communications, and semiconductor transport theory. It is often helpful therefore to have a graphical intuition of the basic process, even if the integral expressions of Eqs. [10] and [11] are not always the best solution route. One very powerful alternative approach makes use of properties of the Laplace transform—hence our introduction to convolution in this chapter.

Let $\mathbf{F}_1(\mathbf{s})$ and $\mathbf{F}_2(\mathbf{s})$ be the Laplace transforms of $f_1(t)$ and $f_2(t)$, respectively, and consider the Laplace transform of $f_1(t) * f_2(t)$,

$$\mathcal{L}\{f_1(t) * f_2(t)\} = \mathcal{L}\left\{ \int_{-\infty}^{\infty} f_1(\lambda) f_2(t - \lambda) \, d\lambda \right\}$$

One of these time functions will typically be the forcing function that is applied at the input terminals of a linear circuit, and the other will be the unit-impulse response of the circuit.

Since we are now dealing with time functions that do not exist prior to $t = 0^-$ (the definition of the Laplace transform forces us to assume this), the lower limit of integration can be changed to 0^-. Then, using the definition of the Laplace transform, we get

$$\mathcal{L}\{f_1(t) * f_2(t)\} = \int_{0^-}^{\infty} e^{-st} \left[\int_{0^-}^{\infty} f_1(\lambda) f_2(t - \lambda) \, d\lambda \right] dt$$

Since e^{-st} does not depend upon λ, we can move this factor inside the inner integral. If we do this and also reverse the order of integration, the result is

$$\mathcal{L}\{f_1(t) * f_2(t)\} = \int_{0^-}^{\infty} \left[\int_{0^-}^{\infty} e^{-st} f_1(\lambda) f_2(t - \lambda) \, dt \right] d\lambda$$

Continuing with the same type of trickery, we note that $f_1(\lambda)$ does not depend upon t, and so it can be moved outside the inner integral:

$$\mathcal{L}\{f_1(t) * f_2(t)\} = \int_{0^-}^{\infty} f_1(\lambda) \left[\int_{0^-}^{\infty} e^{-st} f_2(t - \lambda) \, dt \right] d\lambda$$

We then make the substitution $x = t - \lambda$ in the bracketed integral (where we may treat λ as a constant):

$$\mathcal{L}\{f_1(t) * f_2(t)\} = \int_{0^-}^{\infty} f_1(\lambda) \left[\int_{-\lambda}^{\infty} e^{-s(x+\lambda)} f_2(x) \, dx \right] d\lambda$$

$$= \int_{0^-}^{\infty} f_1(\lambda) e^{-s\lambda} \left[\int_{-\lambda}^{\infty} e^{-sx} f_2(x) \, dx \right] d\lambda$$

$$= \int_{0^-}^{\infty} f_1(\lambda) e^{-s\lambda} [\mathbf{F}_2(\mathbf{s})] \, d\lambda$$

$$= \mathbf{F}_2(\mathbf{s}) \int_{0^-}^{\infty} f_1(\lambda) e^{-s\lambda} d\lambda$$

Since the remaining integral is simply $\mathbf{F}_1(\mathbf{s})$, we find that

$$\boxed{\mathcal{L}\{f_1(t) * f_2(t)\} = \mathbf{F}_1(\mathbf{s}) \cdot \mathbf{F}_2(\mathbf{s})} \qquad [12]$$

Stated slightly differently, we may conclude that the inverse transform of the product of two transforms is the convolution of the individual inverse transforms, a result that is sometimes useful in obtaining inverse transforms.

EXAMPLE **15.9**

Find $v(t)$ by applying convolution techniques, given that $V(s) = 1/[(s + \alpha)(s + \beta)]$.

We obtained the inverse transform of this particular function in Sec. 14.5 using a partial-fraction expansion. We now identify $\mathbf{V}(\mathbf{s})$ as the product of two transforms,

$$\mathbf{V}_1(\mathbf{s}) = \frac{1}{(\mathbf{s} + \alpha)}$$

and

$$\mathbf{V}_2(\mathbf{s}) = \frac{1}{(\mathbf{s} + \beta)}$$

where

$$v_1(t) = e^{-\alpha t} u(t)$$

and

$$v_2(t) = e^{-\beta t} u(t)$$

The desired $v(t)$ can be immediately expressed as

$$v(t) = \mathcal{L}^{-1}\{\mathbf{V}_1(\mathbf{s})\mathbf{V}_2(\mathbf{s})\} = v_1(t) * v_2(t) = \int_{0^-}^{\infty} v_1(\lambda) v_2(t - \lambda) \, d\lambda$$

$$= \int_{0^-}^{\infty} e^{-\alpha \lambda} u(\lambda) e^{-\beta(t-\lambda)} u(t - \lambda) \, d\lambda = \int_{0^-}^{t} e^{-\alpha \lambda} e^{-\beta t} e^{\beta \lambda} \, d\lambda$$

$$= e^{-\beta t} \int_{0^-}^{t} e^{(\beta-\alpha)\lambda} \, d\lambda = e^{-\beta t} \frac{e^{(\beta-\alpha)t} - 1}{\beta - \alpha} u(t)$$

or, more compactly,

$$v(t) = \frac{1}{\beta - \alpha}(e^{-\alpha t} - e^{-\beta t})u(t)$$

which is the same result that we obtained before using partial-fraction expansion. Note that it is necessary to insert the unit step $u(t)$ in the result because all (one-sided) Laplace transforms are valid only for nonnegative time.

Was the result easier to obtain by this method? Not unless one is in love with convolution integrals! The partial-fraction-expansion method is usually simpler, assuming that the expansion itself is not too cumbersome. However, the operation of convolution is easier to perform in the s-domain, since it only requires multiplication.

PRACTICE

15.10 Repeat Example 15.8, performing the convolution in the **s**-domain.

Further Comments on Transfer Functions

As we have noted several times before, the output $v_o(t)$ at some point in a linear circuit can be obtained by convolving the input $v_i(t)$ with the unit-impulse response $h(t)$. However, we must remember that the impulse response results from the application of a unit impulse at $t = 0$ *with all initial conditions zero.* Under these conditions, the Laplace transform of $v_o(t)$ is

$$\mathcal{L}\{v_o(t)\} = \mathbf{V}_o(\mathbf{s}) = \mathcal{L}\{v_i(t) * h(t)\} = \mathbf{V}_i(\mathbf{s})[\mathcal{L}\{h(t)\}]$$

Thus, the ratio $\mathbf{V}_o(\mathbf{s})/\mathbf{V}_i(\mathbf{s})$ is equal to the transform of the impulse response, which we shall denote by $\mathbf{H}(\mathbf{s})$,

$$\mathcal{L}\{h(t)\} = \mathbf{H}(\mathbf{s}) = \frac{\mathbf{V}_o(\mathbf{s})}{\mathbf{V}_i(\mathbf{s})} \qquad [13]$$

From Eq. [13] we see that the impulse response and the transfer function make up a Laplace transform pair,

$$h(t) \Leftrightarrow \mathbf{H}(\mathbf{s})$$

This is an important fact that we shall explore further in Sec. 15.7, after becoming familiar with the concept of pole-zero plots and the complex-frequency plane. At this point, however, we are already able to exploit this new concept of convolution for circuit analysis.

EXAMPLE 15.10

Determine the impulse response of the circuit in Fig. 15.26*a*, and use this to compute the forced response $v_o(t)$ if the input $v_{in}(t) = 6e^{-t}u(t)$ V.

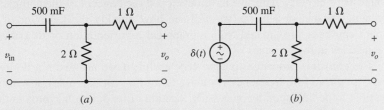

■ **FIGURE 15.26** (*a*) A simple circuit to which an exponential input is applied at $t = 0$. (*b*) Circuit used to determine $h(t)$.

(Continued on next page)

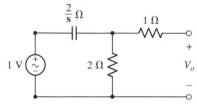

■ **FIGURE 15.27** Circuit used to find H(s).

We first connect an impulse voltage pulse $\delta(t)$ V to the circuit as shown in Fig. 15.26*b*. Although we may work in either the time domain with $h(t)$ or the **s**-domain with $\mathbf{H}(\mathbf{s})$, we choose the latter, so we next consider the **s**-domain representation of Fig. 15.26*b* as depicted in Fig. 15.27.

The impulse response $\mathbf{H}(\mathbf{s})$ is given by

$$\mathbf{H}(\mathbf{s}) = \frac{\mathbf{V}_o}{1}$$

so our immediate goal is to find $\mathbf{V}_o$—a task easily performed by simple voltage division:

$$\mathbf{V}_o\bigg|_{v_{\text{in}}=\partial(t)} = \frac{2}{\dfrac{2}{\mathbf{s}}+2} = \frac{\mathbf{s}}{\mathbf{s}+1} = \mathbf{H}(\mathbf{s})$$

We may now find $v_o(t)$ when $v_{\text{in}} = 6e^{-t}u(t)$ using convolution, as

$$v_{\text{in}} = \mathcal{L}^{-1}\{\mathbf{V}_{\text{in}}(\mathbf{s}) \cdot \mathbf{H}(\mathbf{s})\}$$

Since $\mathbf{V}_{\text{in}}(\mathbf{s}) = 6/(\mathbf{s}+1)$,

$$\mathbf{V}_o = \frac{6\mathbf{s}}{(\mathbf{s}+1)^2} = \frac{6}{\mathbf{s}+1} - \frac{6}{(\mathbf{s}+1)^2}$$

Taking the inverse Laplace transform, we find that

$$v_o(t) = 6e^{-t}(1-t)u(t) \text{ V.}$$

PRACTICE

15.11 Referring to the circuit of Fig. 15.26*a*, use convolution to obtain $v_o(t)$ if $v_{\text{in}} = tu(t)$ V.

Ans: $v_o(t) = (1 - e^{-t})u(t)$ V.

15.6 THE COMPLEX-FREQUENCY PLANE

We now plan to develop a more general graphical presentation by plotting quantities as functions of **s**; that is, we wish to show the response simultaneously as functions of both σ and ω. Such a graphical portrayal of the forced response as a function of the complex frequency **s** is a useful, enlightening technique in the analysis of circuits, as well as in the design or synthesis of circuits. After we have developed the concept of the complex-frequency plane, or *s*-plane, we will see how quickly the behavior of a circuit can be approximated from a graphical representation of its critical frequencies in this *s*-plane.

The converse procedure is also very useful: if we are given a desired response curve (the frequency response of a filter, for example), it will be possible to decide upon the necessary location of its poles and zeros in the **s**-plane and then to synthesize the filter. The **s**-plane is also the basic tool with which the possible presence of undesired oscillations is investigated in feedback amplifiers and automatic control systems.

Response as a Function of σ

Let us develop a method of obtaining circuit response as a function of **s** by first considering the response as a function of either σ or ω. Consider, for example, the input or "driving-point" impedance of a network composed of a 3 Ω resistor in series with a 4 H inductor. As a function of **s**, we have

$$\mathbf{Z(s)} = 3 + 4\mathbf{s} \ \Omega$$

If we wish to obtain a graphical interpretation of the impedance variation with σ, we let $\mathbf{s} = \sigma + j0$:

$$\mathbf{Z}(\sigma) = 3 + 4\sigma \ \Omega$$

and recognize a zero at $\sigma = -\frac{3}{4}$ and a pole at infinity. These critical frequencies are marked on a σ axis, and after identifying the value of $\mathbf{Z}(\sigma)$ at some convenient noncritical frequency [perhaps $\mathbf{Z}(0) = 3$], it is easy to sketch $|\mathbf{Z}(\sigma)|$ versus σ as shown in Fig. 15.28. This provides us with information regarding the impedance when connected to a simple exponential forcing function $e^{\sigma t}$. In particular, note that the dc case ($\sigma = \omega = 0$) results in an impedance of 3 Ω, as we would expect.

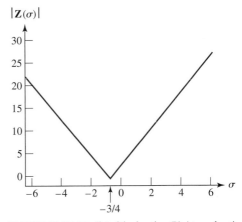

■ **FIGURE 15.28** Plot of the function $|\mathbf{Z}(\sigma)|$ as a function of frequency σ.

Response as a Function of ω

In order to plot the response as a function of the radian frequency ω, we let $\mathbf{s} = 0 + j\omega$:

$$\mathbf{Z}(j\omega) = 3 + j4\omega$$

and then obtain both the magnitude and phase angle of $\mathbf{Z}(j\omega)$ as functions of ω:

$$|\mathbf{Z}(j\omega)| = \sqrt{9 + 16\omega^2} \qquad [14]$$

$$\text{ang } \mathbf{Z}(j\omega) = \tan^{-1}\frac{4\omega}{3} \qquad [15]$$

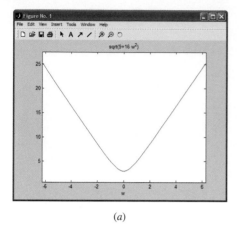

(a)

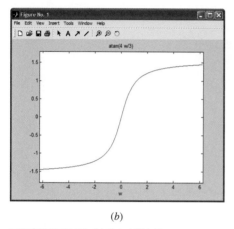

(b)

■ **FIGURE 15.29** (a) Plot of $|\mathbf{Z}(j\omega)|$ as a function of frequency. The plot was generated using the MATLAB command line
EDU >> ezplot('sqrt(9 + 16*w^2)')
(b) Plot of the angle of $\mathbf{Z}(j\omega)$ as a function of frequency.

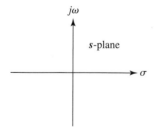

■ **FIGURE 15.30** The complex-frequency plane, or s-plane.

The magnitude function shows a single pole at infinity and a minimum at $\omega = 0$; it can be sketched readily as a curve of $|\mathbf{Z}(j\omega)|$ versus ω. As the frequency is increased, the magnitude of the impedance also increases, which is precisely the behavior we expect from the inductor. The phase angle is an inverse tangent function, zero at $\omega = 0$ and $\pm 90°$ at $\omega = \pm\infty$; it is also easily presented as a plot of ang $\mathbf{Z}(j\omega)$ versus ω. Equations [14] and [15] are plotted in Fig. 15.29.

In graphing the response $\mathbf{Z}(j\omega)$ as a function of ω, two two-dimensional plots are required: both the magnitude and phase angle as functions of ω. When exponential excitation is assumed, we can present all the information on a single two-dimensional graph by allowing both positive and negative values of $\mathbf{Z}(\sigma)$ versus σ. However, we chose to plot the *magnitude* of $\mathbf{Z}(\sigma)$ in order that our sketches would compare more closely with those depicting the magnitude of $\mathbf{Z}(j\omega)$. The phase angle ($0°$, $\pm 180°$ only) of $\mathbf{Z}(\sigma)$ was largely ignored. The important point to note is that there is only one independent variable, σ in the case of exponential excitation and ω in the sinusoidal case. Now let us consider what alternatives are available to us if we wish to plot a response as a function of **s**.

Graphing on the Complex Frequency Plane

The complex frequency **s** requires two parameters, σ and ω, for its complete specification. The response is also a complex function, and we must therefore consider sketching both the magnitude and phase angle as functions of **s**. Either of these quantities—for example, the magnitude—is a function of the two parameters σ and ω, and we can plot it in two dimensions only as a family of curves, such as magnitude versus ω, with σ as the parameter.

A better method of representing the magnitude of some complex response graphically involves using a *three*-dimensional model. Although such a model is difficult to draw on a two-dimensional sheet of paper, we will find that the model is not difficult to visualize; most of the drawing will be done mentally, since in one's head few supplies are needed and construction, correction, and erasures are quickly accomplished. Let us think of a σ axis and a $j\omega$ axis, perpendicular to each other, laid out on a horizontal surface such as the floor. The floor now represents a *complex-frequency plane,* or s-plane, as sketched in Fig. 15.30. To each point in this plane there corresponds exactly one value of **s**, and with each value of **s** we may associate a single point in this complex plane.

Since we are already quite familiar with the type of time-domain function associated with a particular value of the complex frequency **s**, it is now possible to associate the functional form of a forcing function or forced response with a specific region in the s-plane. The origin, for example, represents a dc quantity. Points lying on the σ axis represent exponential functions, decaying for $\sigma < 0$, increasing for $\sigma > 0$. Pure sinusoids are associated with points on the positive or negative $j\omega$ axis. The right half of the s-plane, usually referred to simply as the RHP, contains points describing frequencies with positive real parts and thus corresponds to time-domain quantities that are exponentially *increasing* sinusoids, except on the σ axis. Correspondingly, points in the left half of the s-plane (LHP) describe the frequencies of exponentially *decreasing* sinusoids, again with the exception of the negative σ axis. Figure 15.31 summarizes the relationship between the time domain and the various regions of the s-plane.

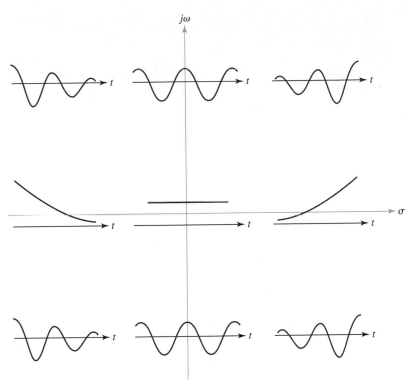

■ **FIGURE 15.31** The nature of the time-domain function is sketched in the region of the complex-frequency plane to which it corresponds.

Let us now return to our search for an appropriate method of representing a response graphically as a function of the complex frequency **s**. The magnitude of the response may be represented by constructing a model out of clay whose height above the floor at every point corresponds to the magnitude of the response at that value of **s**. In other words, we have added a third axis, perpendicular to both the σ axis and the $j\omega$ axis and passing through the origin; this axis is labeled $|\mathbf{Z}|$, $|\mathbf{Y}|$, $|\mathbf{V}_2/\mathbf{V}_1|$, or with another appropriate symbol. The response magnitude is determined for every value of **s**, and the resultant plot is a surface lying above (or just touching) the **s**-plane.

EXAMPLE 15.11

Sketch the admittance of the series combination of a 1 H inductor and a 3 Ω resistor as a function of both $j\omega$ and σ.

The admittance of these two series elements is given by

$$\mathbf{Y}(\mathbf{s}) = \frac{1}{\mathbf{s} + 3}$$

Substituting $\mathbf{s} = \sigma + j\omega$, we find the magnitude of the function is

$$|\mathbf{Y}(\mathbf{s})| = \frac{1}{\sqrt{(\sigma + 3)^2 + \omega^2}}$$

(Continued on next page)

When $\mathbf{s} = -3 + j0$, the response magnitude is infinite; when $\mathbf{s}$ is infinite, the magnitude of $\mathbf{Y}(\mathbf{s})$ is zero. Thus our model must have infinite height over the point $(-3 + j0)$, and it must have zero height at all points infinitely far away from the origin. A cutaway view of such a model is shown in Fig. 15.32a.

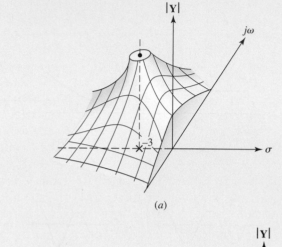

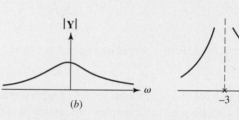

■ **FIGURE 15.32** (*a*) A cutaway view of a clay model whose top surface represents $|\mathbf{Y}(\mathbf{s})|$ for the series combination of a 1 H inductor and a 3 Ω resistor. (*b*) $|\mathbf{Y}(\mathbf{s})|$ as a function of ω. (*c*) $|\mathbf{Y}(\mathbf{s})|$ as a function of σ.

Once the model is constructed, it is simple to visualize the variation of $|\mathbf{Y}|$ as a function of ω (with $\sigma = 0$) by cutting the model with a perpendicular plane containing the $j\omega$ axis. The model shown in Fig. 15.32a happens to be cut along this plane, and the desired plot of $|\mathbf{Y}|$ versus ω can be seen; the curve is also drawn in Fig. 15.32b. In a similar manner, a vertical plane containing the σ axis enables us to obtain $|\mathbf{Y}|$ versus σ (with $\omega = 0$), as shown in Fig. 15.32c.

PRACTICE

15.12 Sketch the magnitude of the impedance $\mathbf{Z}(\mathbf{s}) = 2 + 5\mathbf{s}$ as a function of σ and $j\omega$.

Ans: See Fig. 15.33.

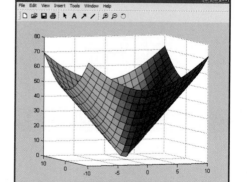

■ **FIGURE 15.33** Solution for Practice Problem 15.12, generated with the following code:

```
EDU» sigma = linspace(−10,10,21);
EDU» omega = linspace(−10,10,21);
EDU» [X, Y] = meshgrid(sigma,omega);
EDU» Z = abs(2 + 5*X + j*5*Y);
EDU» colormap(hsv);
EDU» s = [−5 3 8];
EDU» surfl(X,Y,Z,s);
EDU» view (−20,5)
```

Pole-Zero Constellations

This approach works well for relatively simple functions, but a more practical method is needed in general. Let us visualize the **s**-plane once again as the floor and then imagine a large elastic sheet laid on it. We now fix our attention on all the poles and zeros of the response. At each zero, the response is zero, the height of the sheet must be zero, and we therefore tack the sheet to the floor. At the value of **s** corresponding to each pole, we may prop up the sheet with a thin vertical rod. Zeros and poles at infinity must be treated by using a large-radius clamping ring or a high circular fence, respectively. If we have used an infinitely large, weightless, perfectly elastic sheet, tacked down with vanishingly small tacks, and propped up with infinitely long, zero-diameter rods, then the elastic sheet assumes a height that is exactly proportional to the magnitude of the response.

These comments may be illustrated by considering the configuration of the poles and zeros, sometimes called a ***pole-zero constellation,*** that locates all the critical frequencies of a frequency-domain quantity—for example, an impedance $\mathbf{Z}(\mathbf{s})$. A pole-zero constellation for an example impedance is shown in Fig. 15.34; in such a diagram, poles are denoted by crosses and zeros by circles. If we visualize an elastic-sheet model, tacked down at $\mathbf{s} = -2 + j0$ and propped up at $\mathbf{s} = -1 + j5$ and at $\mathbf{s} = -1 - j5$, we should see a terrain whose distinguishing features are two mountains and one conical crater or depression. The portion of the model for the upper LHP is shown in Fig. 15.34*b*.

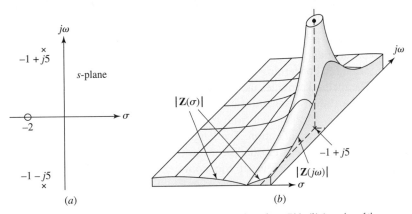

■ **FIGURE 15.34** (*a*) The pole-zero constellation of some impedance $\mathbf{Z}(\mathbf{s})$. (*b*) A portion of the elastic-sheet model of the magnitude of $\mathbf{Z}(\mathbf{s})$.

Let us now build up the expression for $\mathbf{Z}(\mathbf{s})$ that leads to this pole-zero configuration. The zero requires a factor of $(\mathbf{s} + 2)$ in the numerator, and the two poles require the factors $(\mathbf{s} + 1 - j5)$ and $(\mathbf{s} + 1 + j5)$ in the denominator. Except for a multiplying constant k, we now know the form of $\mathbf{Z}(\mathbf{s})$:

$$\mathbf{Z}(\mathbf{s}) = k\frac{\mathbf{s} + 2}{(\mathbf{s} + 1 - j5)(\mathbf{s} + 1 + j5)}$$

or

$$\mathbf{Z}(\mathbf{s}) = k\frac{\mathbf{s} + 2}{\mathbf{s}^2 + 2\mathbf{s} + 26} \qquad [16]$$

In order to determine k, we require a value for $\mathbf{Z}(\mathbf{s})$ at some $\mathbf{s}$ other than a critical frequency. For this function, let us suppose we are told $\mathbf{Z}(0) = 1$.

By direct substitution in Eq. [16], we find that k is 13, and therefore

$$\mathbf{Z}(\mathbf{s}) = 13\frac{\mathbf{s}+2}{\mathbf{s}^2 + 2\mathbf{s} + 26} \tag{17}$$

The plots $|\mathbf{Z}(\sigma)|$ versus σ and $|\mathbf{Z}(j\omega)|$ versus ω may be obtained exactly from Eq. [17], but the general form of the function is apparent from the pole-zero configuration and the elastic-sheet analogy. Portions of these two curves appear at the sides of the model shown in Fig. 15.34b.

PRACTICE
●

15.13 The parallel combination of 0.25 mH and 5 Ω is in series with the parallel combination of 40 μF and 5 Ω. (a) Find $\mathbf{Z}_{in}(\mathbf{s})$, the input impedance of the series combination. (b) Specify all the zeros of $\mathbf{Z}_{in}(\mathbf{s})$. (c) Specify all the poles of $\mathbf{Z}_{in}(\mathbf{s})$. (d) Draw the pole-zero configuration.

Ans: $5(\mathbf{s}^2 + 10{,}000\mathbf{s} + 10^8)/(\mathbf{s}^2 + 25{,}000\mathbf{s} + 10^8)$ Ω; $-5 \pm j8.66$ krad/s; $-5, -20$ krad/s.

Frequency Dependence of Magnitude and Phase Angle

Thus far, we have been using the **s**-plane and the elastic-sheet model to obtain *qualitative* information about the variation of the magnitude of the **s**-domain function with frequency. It is possible, however, to get *quantitative* information concerning the variation of both the *magnitude* and *phase angle*. The method provides us with a powerful new tool.

Consider the representation of a complex frequency in polar form, as suggested by an arrow drawn from the origin of the **s**-plane to the complex frequency under consideration. The length of the arrow is the magnitude of the frequency, and the angle that the arrow makes with the positive direction of the σ axis is the angle of the complex frequency. The frequency $\mathbf{s}_1 = -3 + j4 = 5\underline{/126.9°}$ is indicated in Fig. 15.35a.

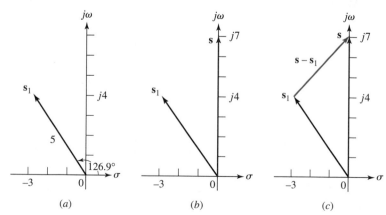

■ **FIGURE 15.35** (a) The complex frequency $\mathbf{s}_1 = -3 + j4$ is indicated by drawing an arrow from the origin to $\mathbf{s}_1$. (b) The frequency $\mathbf{s} = j7$ is also represented vectorially. (c) The difference $\mathbf{s} - \mathbf{s}_1$ is represented by the vector drawn from $\mathbf{s}_1$ to $\mathbf{s}$.

We can also represent the difference between two values of **s** as an arrow or vector on the complex plane. Let us select a value of **s** that corresponds to a sinusoid $\mathbf{s} = j7$ and indicate it also as a vector, as shown in Fig. 15.35b. The *difference* $\mathbf{s} - \mathbf{s}_1$ is seen to be the vector drawn from the last-named point $\mathbf{s}_1$ to the first-named point **s**; the vector $\mathbf{s} - \mathbf{s}_1$ is drawn in Fig. 15.35c. Note that $\mathbf{s}_1 + (\mathbf{s} - \mathbf{s}_1) = \mathbf{s}$. Numerically, $\mathbf{s} - \mathbf{s}_1 = j7 - (-3 + j4) = 3 + j3 = 4.24\underline{/45°}$, and this value agrees with the graphical difference.

Let us see how this graphical interpretation of the difference $(\mathbf{s} - \mathbf{s}_1)$ enables us to determine frequency response. Consider the admittance

$$\mathbf{Y(s)} = \mathbf{s} + 2$$

This expression has a zero at $\mathbf{s}_2 = -2 + j0$. The factor $\mathbf{s} + 2$, which may be written as $\mathbf{s} - \mathbf{s}_2$, is represented by the vector drawn from the zero location $\mathbf{s}_2$ to the frequency **s** at which the response is desired. If the sinusoidal response is desired, **s** must lie on the $j\omega$ axis, as illustrated in Fig. 15.36a. The magnitude of $\mathbf{s} + 2$ may now be visualized as ω varies from zero to infinity. When **s** is zero, the vector has a magnitude of 2 and an angle of $0°$. Thus $\mathbf{Y}(0) = 2$. As ω increases, the magnitude increases, slowly at first, and then almost linearly with ω; the phase angle increases almost linearly at first, and then gradually approaches $90°$ as ω becomes infinite. The magnitude and phase of $\mathbf{Y(s)}$ are sketched as functions of ω in Fig. 15.36b.

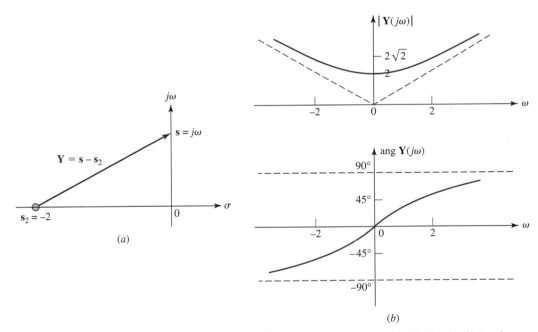

■ **FIGURE 15.36** (a) The vector representing the admittance $Y(s) = s + 2$ is shown for $s = j\omega$. (b) Sketches of $|Y(j\omega)|$ and ang $Y(j\omega)$ as they might be obtained from the performance of the vector as **s** moves up or down the $j\omega$ axis from the origin.

Let us now construct a more realistic example by considering a frequency-domain function given by the quotient of two factors,

$$\mathbf{V(s)} = \frac{\mathbf{s} + 2}{\mathbf{s} + 3}$$

We again select a value of **s** that corresponds to sinusoidal excitation and draw the vectors **s** + 2 and **s** + 3, the first from the zero to the chosen point on the $j\omega$ axis and the second from the pole to the chosen point. The two vectors are sketched in Fig. 15.37a. *The quotient of these two vectors has a magnitude equal to the quotient of the magnitudes, and a phase angle equal to the difference of the numerator and denominator phase angles.* An investigation of the variation of $|\mathbf{V}(\mathbf{s})|$ versus ω is made by allowing **s** to move from the origin up the $j\omega$ axis and considering the ratio of the distance from the zero to $\mathbf{s} = j\omega$ and the distance from the pole to the same point on the $j\omega$ axis. The ratio is $\frac{2}{3}$ at $\omega = 0$ and approaches unity as ω becomes infinite, as shown in Fig. 15.37b.

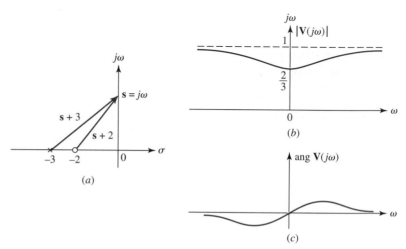

■ **FIGURE 15.37** (a) Vectors are drawn from the two critical frequencies of the voltage response $\mathbf{V}(\mathbf{s}) = (\mathbf{s} + 2)/(\mathbf{s} + 3)$. (b, c) Sketches of the magnitude and the phase angle, respectively, of $\mathbf{V}(j\omega)$ as obtained from the quotient of the two vectors shown in part a.

A consideration of the difference of the two phase angles shows that ang $\mathbf{V}(j\omega)$ is $0°$ at $\omega = 0$. It increases at first as ω increases, since the angle of the vector $\mathbf{s} + 2$ is greater than that of $\mathbf{s} + 3$. It then decreases with a further increase in ω, finally approaching $0°$ at infinite frequency, where both vectors possess $90°$ angles. These results are sketched in Fig. 15.37c. Although no quantitative markings are present on these sketches, it is important to note that they could be obtained easily. For example, the complex response at $\mathbf{s} = j4$ must be given by the ratio

$$\mathbf{V}(j4) = \frac{\sqrt{4 + 16}\,\underline{/\tan^{-1}\left(\frac{4}{2}\right)}}{\sqrt{9 + 16}\,\underline{/\tan^{-1}\left(\frac{4}{3}\right)}}$$

$$= \sqrt{\tfrac{20}{25}}\,\underline{/\left(\tan^{-1}2 - \tan^{-1}\left(\tfrac{4}{3}\right)\right)}$$

$$= 0.894\underline{/10.3°}$$

In designing circuits to produce some desired response, the behavior of the vectors drawn from the respective critical frequencies to a general point on the $j\omega$ axis is an important aid. For example, if it were necessary to

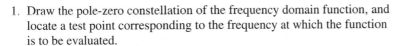

increase the hump in the phase response of Fig. 15.37c, we can see that we would have to provide a greater difference in the angles of the two vectors. This may be achieved in Fig. 15.37a either by moving the zero closer to the origin or by locating the pole farther from the origin, or both.

The ideas we have been discussing to help in the graphical determination of the magnitude and angular variation of some frequency-domain function with frequency will be needed in Chap. 16 when we investigate the frequency performance of highly selective filters, or resonant circuits. These concepts are fundamental in obtaining a quick, clear understanding of the behavior of electrical networks and other engineering systems. The procedure is briefly summarized as follows:

1. Draw the pole-zero constellation of the frequency domain function, and locate a test point corresponding to the frequency at which the function is to be evaluated.

2. Draw an arrow from each pole and zero to the test point.

3. Determine the length of each arrow, and the value of each pole-arrow angle and each zero-arrow angle.

4. Divide the product of the zero-arrow lengths by the product of the pole-arrow lengths. This quotient is the magnitude of the frequency-domain function for the test frequency (*within a multiplying constant, since* $\mathbf{F(s)}$ *and* $k\mathbf{F(s)}$ *have the same pole-zero constellations*).

5. Subtract the sum of the pole-arrow angles from the sum of the zero-arrow angles. The result is the angle of the frequency-domain function, evaluated at the frequency of the test point. The angle does not depend upon the value of the real multiplying constant k.

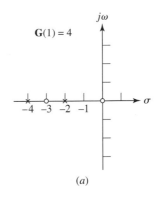

(a)

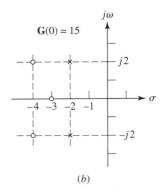

(b)

PRACTICE

15.14 Three pole-zero constellations are shown in Fig. 15.38. Each applies to a voltage gain $\mathbf{G}$. Obtain an expression for each gain that is a ratio of polynomials in $\mathbf{s}$.

15.15 The pole-zero configuration for an admittance $\mathbf{Y(s)}$ has one pole at $\mathbf{s} = -10 + j0 \ s^{-1}$ and one zero at $\mathbf{s} = z_1 + j0$, where $z_1 < 0$. Let $\mathbf{Y}(0) = 0.1$ S. Find the value of z_1 if (a) ang $\mathbf{Y}(j5) = 20°$; (b) $|\mathbf{Y}(j5)| = 0.2$ S.

Ans: 15.14 $(15s^2 + 45s)/(s^2 + 6s + 8)$; $(2s^3 + 22s^2 + 88s + 120)/(s^2 + 4s + 8)$; $(3s^2 + 27)/(s^2 + 2s)$. 15.15: -4.73 Np/s; -2.50 Np/s.

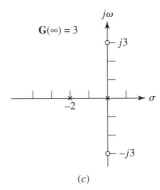

(c)

■ **FIGURE 15.38**

15.7 NATURAL RESPONSE AND THE s-PLANE

At the beginning of this chapter, we explored how working in the frequency domain through the Laplace transform allows us to consider a broad range of time-varying circuits, discarding integrodifferential equations and working algebraically instead. This approach is very powerful, but it does suffer from not being a very visual process. In contrast, there is

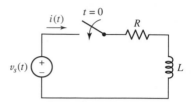

■ **FIGURE 15.39** An example that illustrates the determination of the complete response through a knowledge of the critical frequencies of the impedance faced by the source.

a *tremendous* amount of information contained in the pole-zero plot of a forced response. In this section, we consider how such plots can be used to obtain the *complete* response of a circuit—natural plus forced—provided the initial conditions are known. The advantage of such an approach is a more *intuitive* linkage between the location of the critical frequencies, easily visualized through the pole-zero plot, and the desired response.

Let us introduce the method by considering the simplest example, a series *RL* circuit as shown in Fig. 15.39. A general voltage source $v_s(t)$ causes the current $i(t)$ to flow after closure of the switch at $t = 0$. The complete response $i(t)$ for $t > 0$ is composed of a natural response and a forced response:

$$i(t) = i_n(t) + i_f(t)$$

We may find the forced response by working in the frequency domain, assuming, of course, that $v_s(t)$ has a functional form that we can transform to the frequency domain; if $v_s(t) = 1/(1 + t^2)$, for example, we must proceed as best we can from the basic differential equation for the circuit. For the circuit of Fig. 15.39, we have

$$\mathbf{I}_f(\mathbf{s}) = \frac{\mathbf{V}_s}{R + \mathbf{s}L}$$

or

$$\mathbf{I}_f(\mathbf{s}) = \frac{1}{L}\frac{\mathbf{V}_s}{\mathbf{s} + R/L} \qquad [18]$$

Next we consider the natural response. From previous experience, we know that the form will be a decaying exponential with the time constant L/R, but let's pretend that we are finding it for the first time. The *form* of the natural (source-free) response is, by definition, independent of the forcing function; the forcing function contributes only to the *magnitude* of the natural response. To find the proper form, we kill all independent sources; here, $v_s(t)$ is replaced by a short circuit. Next, we try to obtain the natural response as a limiting case of the forced response. Returning to the frequency-domain expression of Eq. [18], we obediently set $\mathbf{V}_s = 0$. On the surface, it appears that $\mathbf{I}(\mathbf{s})$ must also be zero, but this is not necessarily true if we are operating at a complex frequency that is a simple pole of $\mathbf{I}(\mathbf{s})$. That is, the denominator and the numerator may *both* be zero so that $\mathbf{I}(\mathbf{s})$ need not be zero.

Let us inspect this new idea from a slightly different vantage point. We fix our attention on the ratio of the desired forced response to the forcing function. We will designate this ratio $\mathbf{H}(\mathbf{s})$ and define it to be the circuit transfer function. Then,

$$\frac{\mathbf{I}_f(\mathbf{s})}{\mathbf{V}_s} = \mathbf{H}(\mathbf{s}) = \frac{1}{L(\mathbf{s} + R/L)}$$

In this example, the transfer function is the input admittance faced by $\mathbf{V}_s$. We seek the natural (source-free) response by setting $\mathbf{V}_s = 0$. However, $\mathbf{I}_f(\mathbf{s}) = \mathbf{V}_s\mathbf{H}(\mathbf{s})$, and if $\mathbf{V}_s = 0$, a nonzero value for the current can be obtained only by operating at a pole of $\mathbf{H}(\mathbf{s})$. The poles of the transfer function therefore assume a special significance.

What does it mean to "operate" at a complex frequency? How could we possibly accomplish such a thing in a real laboratory? In this instance, it is important to remember how we invented complex frequency to begin with: it is a means of describing a sinusoidal function of frequency ω multiplied by an exponential function $e^{\sigma t}$. Such types of signals are very easy to generate with real (i.e., nonimaginary) laboratory equipment. Thus, we need only set the value for σ and the value for ω in order to "operate" at $\mathbf{s} = \sigma + j\omega$.

In this particular example, we see that the pole of the transfer function occurs at $\mathbf{s} = -R/L + j0$, as shown in Fig. 15.40. If we choose to operate at this particular complex frequency, the only *finite* current that could result must be a constant in the **s**-domain (i.e., frequency independent). We thus obtain the natural response

$$\mathbf{I}\left(\mathbf{s} = -\frac{R}{L} + j0\right) = A$$

where A is an unknown constant. We next desire to transform this natural response to the time domain. Our knee-jerk reaction might be to attempt to apply inverse Laplace transform techniques in this situation. However, we have already specified a value of **s**, so that such an approach is not valid. Instead, we look to the real part of our general function $e^{\mathbf{s}t}$, such that

$$i_n(t) = \text{Re}\{Ae^{\mathbf{s}t}\} = \text{Re}\{Ae^{-Rt/L}\}$$

In this case we find

$$i_n(t) = Ae^{-Rt/L}$$

so that the total response is then

$$i(t) = Ae^{-Rt/L} + i_f(t)$$

and A may be determined once the initial conditions are specified for this circuit. The forced response $i_f(t)$ is obtained by finding the inverse Laplace transform of $\mathbf{I}_f(\mathbf{s})$.

A More General Perspective

Figure 15.41*a* and *b* shows single sources connected to networks containing no independent sources. The desired response, which might be some current $\mathbf{I}_1(\mathbf{s})$ or some voltage $\mathbf{V}_2(\mathbf{s})$, may be expressed by a transfer function that displays all the critical frequencies. To be specific, we select the response $\mathbf{V}_2(\mathbf{s})$ in Fig. 15.41*a*:

$$\frac{\mathbf{V}_2(\mathbf{s})}{\mathbf{V}_s} = \mathbf{H}(\mathbf{s}) = k\frac{(\mathbf{s} - \mathbf{s}_1)(\mathbf{s} - \mathbf{s}_3)\cdots}{(\mathbf{s} - \mathbf{s}_2)(\mathbf{s} - \mathbf{s}_4)\cdots} \qquad [19]$$

The poles of $\mathbf{H}(\mathbf{s})$ occur at $\mathbf{s} = \mathbf{s}_2, \mathbf{s}_4, \ldots$, and so a finite voltage $\mathbf{V}_2(\mathbf{s})$ at each of these frequencies must be a possible functional form for the natural

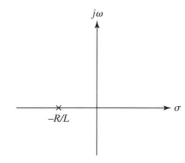

■ FIGURE 15.40 Pole-zero constellation of the transfer function $\mathbf{H}(\mathbf{s})$ showing the single pole at $\mathbf{s} = -R/L$.

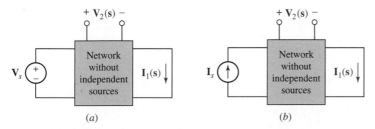

■ FIGURE 15.41 The poles of the response, $\mathbf{I}_1(\mathbf{s})$ or $\mathbf{V}_2(\mathbf{s})$, produced by (*a*) a voltage source $\mathbf{V}_s$ or (*b*) a current source $\mathbf{I}_s$. The poles determine the form of the natural response, $i_{1n}(t)$ or $v_{2n}(t)$, that occurs when $\mathbf{V}_s$ is replaced by a short circuit or $\mathbf{I}_s$ by an open circuit and some initial energy is available.

response. Thus, we think of a zero-volt source (which is just a short-circuit) applied to the input terminals; the natural response that occurs when the input terminals are short-circuited must therefore have the form

$$v_{2n}(t) = \mathbf{A}_2 e^{\mathbf{s}_2 t} + \mathbf{A}_4 e^{\mathbf{s}_4 t} + \cdots$$

where each $\mathbf{A}$ must be evaluated in terms of the initial conditions (including the initial value of any voltage source applied at the input terminals).

To find the form of the natural response $i_{1n}(t)$ in Fig. 15.41a, we should determine the poles of the transfer function, $\mathbf{H}(\mathbf{s}) = \mathbf{I}_1(\mathbf{s})/\mathbf{V}_s$. The transfer functions applying to the situations depicted in Fig. 15.41b would be $\mathbf{I}_1(\mathbf{s})/\mathbf{I}_s$ and $\mathbf{V}_2(\mathbf{s})/\mathbf{I}_s$, and their poles then determine the natural responses $i_{1n}(t)$ and $v_{2n}(t)$, respectively.

If the natural response is desired for a network that does not contain any independent sources, then a source $\mathbf{V}_s$ or $\mathbf{I}_s$ may be inserted at any convenient point, restricted only by the condition that the original network is obtained when the source is killed. The corresponding transfer function is then determined and its poles specify the natural frequencies. Note that the same frequencies must be obtained for any of the many source locations possible. If the network already contains a source, that source may be set equal to zero and another source inserted at a more convenient point.

A Special Case

Before we illustrate this method with an example, completeness requires us to acknowledge a special case that might arise. This occurs when the network in Fig. 15.41a or b contains two or more parts that are isolated from each other. For example, we might have the parallel combination of three networks: R_1 in series with C, R_2 in series with L, and a short circuit. Clearly, a voltage source in series with R_1 and C cannot produce any current in R_2 and L; that transfer function would be zero. To find the form of the natural response of the inductor voltage, for example, the voltage source must be installed in the $R_2 L$ network. A case of this type can often be recognized by an inspection of the network before a source is installed; but if it is not, then a transfer function equal to zero will be obtained. When $\mathbf{H}(\mathbf{s}) = 0$, we obtain no information about the frequencies characterizing the natural response, and a more suitable location for the source must be used.

EXAMPLE 15.12

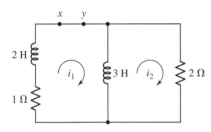

■ **FIGURE 15.42** A circuit for which the natural responses i_1 and i_2 are desired.

For the source-free circuit of Fig. 15.42, determine expressions for i_1 and i_2 for $t > 0$, given the initial conditions $i_1(0) = i_2(0) = $ **11 amperes.**

Let us install a voltage source $\mathbf{V}_s$ between points x and y and find the transfer function $\mathbf{H}(\mathbf{s}) = \mathbf{I}_1(\mathbf{s})/\mathbf{V}_s$, which also happens to be the input admittance seen by the voltage source. We have

$$\mathbf{I}_1(\mathbf{s}) = \frac{\mathbf{V}_s}{2\mathbf{s} + 1 + 6\mathbf{s}/(3\mathbf{s} + 2)} = \frac{(3\mathbf{s} + 2)\mathbf{V}_s}{6\mathbf{s}^2 + 13\mathbf{s} + 2}$$

or

$$H(s) = \frac{I_1(s)}{V_s} = \frac{\frac{1}{2}\left(s + \frac{2}{3}\right)}{(s+2)\left(s + \frac{1}{6}\right)}$$

From recent experience, we know at a glance that i_1 must be of the form

$$i_1(t) = Ae^{-2t} + Be^{-t/6}$$

The solution is completed by using the given initial conditions to establish the values of A and B. Since $i_1(0)$ is given as 11 amperes,

$$11 = A + B$$

The necessary additional equation is obtained by writing the KVL equation around the perimeter of our circuit:

$$1i_1 + 2\frac{di_1}{dt} + 2i_2 = 0$$

and solving for the derivative:

$$\left.\frac{di_1}{dt}\right|_{t=0} = -\frac{1}{2}[2i_2(0) + 1i_1(0)] = -\frac{22 + 11}{2} = -2A - \frac{1}{6}B$$

Thus, $A = 8$ and $B = 3$, and so the desired solution is

$$i_1(t) = 8e^{-2t} + 3e^{-t/6} \qquad \text{amperes}$$

The natural frequencies constituting i_2 are the same as those of i_1, and a similar procedure used to evaluate the arbitrary constants leads to

$$i_2(t) = 12e^{-2t} - e^{-t/6} \qquad \text{amperes}$$

PRACTICE

15.16 If a current source $i_1(t) = u(t)$ A is present at a-b in Fig. 15.43 with the arrow entering a, find $H(s) = V_{cd}/I_1$, and specify the natural frequencies present in $v_{cd}(t)$.

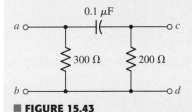

■ **FIGURE 15.43**

Ans: $120s/(s + 20,000)$ Ω, $-20,000$ s^{-1}.

The process that we must pursue to evaluate the amplitude coefficients of the natural response is a detailed one, except in those cases where the initial values of the desired response and its derivatives are obvious. However, we should not lose sight of the ease and rapidity with which the *form* of the natural response can be obtained.

Design of Oscillator Circuits

At several points throughout this book, we have investigated the behavior of various circuits responding to sinusoidal excitation. The creation of sinusoidal waveforms, however, is an interesting topic in itself. Generation of large sinusoidal voltages and currents is straightforward using magnets and rotating coils of wire, for example, but such an approach is not easily scaled down for creation of small signals. Instead, low-current applications typically make use of what is known as an *oscillator*, which exploits the concept of *positive feedback* using an appropriate amplifier circuit. Oscillator circuits are an integral component of many consumer products, such as the global positioning satellite (GPS) receiver of Fig. 15.44.

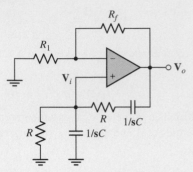

■ **FIGURE 15.45** A Wien-bridge oscillator circuit.

■ **FIGURE 15.44** Many consumer electronic products, such as this GPS receiver, rely on oscillator circuits to provide a reference frequency. (© Royalty-Free/CORBIS)

One straightforward but useful oscillator circuit is known as the **Wien-bridge oscillator,** shown in Fig. 15.45.

The circuit resembles a noninverting op amp circuit, with a resistor R_1 connected between the inverting input pin and ground, and a resistor R_f connected between the output and the inverting input pin. The resistor R_f supplies what is referred to as a ***negative feedback path,*** since it connects the output of the amplifier to the inverting input. Any increase ΔV_o in the output then leads to a reduction of the input, which in turn leads to a smaller output; this process increases the stability of the output voltage V_o. The *gain* of the op amp, defined as the ratio of V_o to V_i, is determined by the relative sizes of R_1 and R_f.

The *positive* feedback loop consists of two separate resistor-capacitor combinations, defined as $Z_s = R + 1/sC$ and $Z_p = R \| (1/sC)$. The values we choose for R and C allow us to design an oscillator having a specific frequency (*the internal capacitances of the op amp itself will limit the maximum frequency that can be obtained*). In order to determine the relationship between R, C, and the oscillation frequency, we seek an expression for the amplifier gain, V_o/V_i.

Recalling the two ideal op amp rules as discussed in Chap. 6 and examining the circuit in Fig. 15.45 closely, we recognize that Z_p and Z_s form a voltage divider such that

$$V_i = V_o \frac{Z_p}{Z_p + Z_s} \qquad [20]$$

Simplifying the expressions for $Z_p = R \| (1/sC) = R/(1 + sRC)$, and $Z_s = R + 1/sC = (1 + sRC)/sC$,

15.8 A TECHNIQUE FOR SYNTHESIZING THE VOLTAGE RATIO H(s) = V$_{out}$/V$_{in}$

Much of the discussion in this chapter has been related to the poles and zeros of a transfer function. We have located them on the complex-frequency plane, we have used them to express transfer functions as ratios of factors or polynomials in **s**, we have calculated forced responses from

we find that

$$\frac{\mathbf{V}_i}{\mathbf{V}_o} = \frac{\dfrac{R}{1+sRC}}{\dfrac{1+sRC}{sC} + \dfrac{R}{1+sRC}}$$

$$= \frac{sRC}{1+3sRC+s^2R^2C^2} \qquad [21]$$

Since we are interested in the sinusoidal steady-state operation of the amplifier, we replace $\mathbf{s}$ with $j\omega$, so that

$$\frac{\mathbf{V}_i}{\mathbf{V}_o} = \frac{j\omega RC}{1+3j\omega RC+(j\omega)^2 R^2 C^2}$$

$$= \frac{j\omega RC}{1-\omega^2 R^2 C^2 + 3j\omega RC} \qquad [22]$$

This expression for the gain is real only when $\omega = 1/RC$. Thus, we can design an amplifier to operate at a particular frequency $f = \omega/2\pi = 1/2\pi RC$ by selecting values for R and C.

As an example, let's design a Wien-bridge oscillator to generate a sinusoidal signal at a frequency of 20 Hz, the commonly accepted lower frequency of the audio range. We require a frequency $\omega = 2\pi f = (6.28)(20) = 125.6$ rad/s. Once we specify a value for R, the necessary value for C is known (and vice versa). Assuming that we happen to have a 1 μF capacitor handy, we thus compute a required resistance of $R = 7962\ \Omega$. Since this is not a standard resistor value, we will likely have to use several resistors in series and/or parallel combinations to obtain the necessary value. Referring back to Fig. 15.45 in preparation for simulating the circuit using PSpice, however, we notice that no values for R_f or R_1 have been specified.

Although Eq. [20] correctly specifies the relationship between $\mathbf{V}_o$ and $\mathbf{V}_i$, we may also write another equation relating these quantities:

$$0 = \frac{\mathbf{V}_i}{R_1} + \frac{\mathbf{V}_i - \mathbf{V}_o}{R_f}$$

which can be rearranged to obtain

$$\frac{\mathbf{V}_o}{\mathbf{V}_i} = 1 + \frac{R_f}{R_1} \qquad [23]$$

Setting $\omega = 1/RC$ in Eq. [22] results in

$$\frac{\mathbf{V}_i}{\mathbf{V}_o} = \frac{1}{3}$$

Therefore, we need to select values of R_1 and R_f such that $R_f/R_1 = 2$. Unfortunately, if we proceed to perform a transient PSpice analysis on the circuit selecting $R_f = 2$ kΩ and $R_1 = 1$ kΩ, for example, we will likely be disappointed in the outcome. In order to ensure that the circuit is indeed unstable (*a necessary condition for oscillations to begin*), it is necessary to have R_f/R_1 slightly greater than 2. The simulated output of our final design ($R = 7962\ \Omega$, $C = 1\ \mu$F, $R_f = 2.01$ kΩ, and $R_1 = 1$ kΩ) is shown in Fig. 15.46. Note that the magnitude of the oscillations is increasing in the plot; in practice, nonlinear circuit elements are required to stabilize the voltage magnitude of the oscillator circuit.

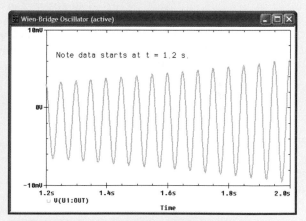

■ **FIGURE 15.46** Simulated output of the Wien-bridge oscillator designed for operation at 20 Hz.

them, and in Sec. 15.7 we have used their poles to establish the form of the natural response.

Now let us see how we might determine a network that can provide a desired transfer function. We consider only a small part of the general problem, working with a transfer function of the form $\mathbf{H(s)} = \mathbf{V}_{\text{out}}(\mathbf{s})/\mathbf{V}_{\text{in}}(\mathbf{s})$, as indicated in Fig. 15.47. For simplicity, we restrict $\mathbf{H(s)}$ to critical frequencies on the negative σ axis (including the origin). Thus, we will consider

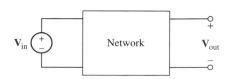

■ **FIGURE 15.47** Given $\mathbf{H(s)} = \mathbf{V}_{\text{out}}/\mathbf{V}_{\text{in}}$, we seek a network having a specified $\mathbf{H(s)}$.

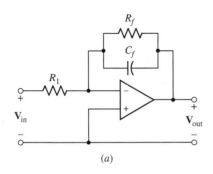

■ **FIGURE 15.48** For an ideal op amp, $H(s) = V_{out}/V_{in} = -Z_f/Z_1$.

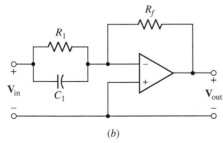

(a)

(b)

■ **FIGURE 15.49** (a) The transfer function $H(s) = V_{out}/V_{in}$ has a pole at $s = -1/R_f C_f$. (b) Here, there is a zero at $s = -1/R_1 C_1$.

transfer functions such as

$$H_1(s) = \frac{10(s+2)}{s+5}$$

or

$$H_2(s) = \frac{-5s}{(s+8)^2}$$

or

$$H_3(s) = 0.1s(s+2)$$

Let us begin by finding the voltage gain of the network of Fig. 15.48, which contains an ideal op amp. The voltage between the two input terminals of the op amp is essentially zero, and the input impedance of the op amp is essentially infinite. We therefore may set the sum of the currents entering the inverting input terminal equal to zero:

$$\frac{V_{in}}{Z_1} + \frac{V_{out}}{Z_f} = 0$$

or

$$\frac{V_{out}}{V_{in}} = -\frac{Z_f}{Z_1}$$

If Z_f and Z_1 are both resistances, the circuit acts as an inverting amplifier, or possibly an ***attenuator*** (if the ratio is less than unity). Our present interest, however, lies with those cases in which one of these impedances is a resistance while the other is an *RC* network.

In Fig. 15.49*a*, we let $Z_1 = R_1$, while Z_f is the parallel combination of R_f and C_f. Therefore,

$$Z_f = \frac{R_f/sC_f}{R_f + (1/sC_f)} = \frac{R_f}{1 + sC_f R_f} = \frac{1/C_f}{s + (1/R_f C_f)}$$

and

$$H(s) = \frac{V_{out}}{V_{in}} = -\frac{Z_f}{Z_1} = -\frac{1/R_1 C_f}{s + (1/R_f C_f)}$$

We have a transfer function with a single (finite) critical frequency, a pole at $s = -1/R_f C_f$.

Moving on to Fig. 15.49*b*, we now let Z_f be resistive while Z_1 is an *RC* parallel combination:

$$Z_1 = \frac{1/C_1}{s + (1/R_1 C_1)}$$

and

$$H(s) = \frac{V_{out}}{V_{in}} = -\frac{Z_f}{Z_1} = -R_f C_1 \left(s + \frac{1}{R_1 C_1} \right)$$

The only finite critical frequency is a zero at $s = -1/R_1 C_1$.

For our ideal op amps, the output or Thévenin impedance is zero and therefore $\mathbf{V}_{out}$ and $\mathbf{V}_{out}/\mathbf{V}_{in}$ arc not functions of any load $\mathbf{Z}_L$ that may be placed across the output terminals. This includes the input to another op amp, as well, and therefore we may connect circuits having poles and zeros at specified locations in *cascade,* where the output of one op amp is connected directly to the input of the next, and thus generates any desired transfer function.

EXAMPLE **15.13**

Synthesize a circuit that will yield the transfer function
H(s) = $\mathbf{V}_{out}/\mathbf{V}_{in} = 10(s + 2)/(s + 5)$.

The pole at $\mathbf{s} = -5$ may be obtained by a network of the form of Fig. 15.49a. Calling this network A, we have $1/R_{fA}C_{fA} = 5$. We arbitrarily select $R_{fA} = 100$ kΩ; therefore, $C_{fA} = 2$ μF. For this portion of the complete circuit,

$$\mathbf{H}_A(\mathbf{s}) = -\frac{1/R_{1A}C_{fA}}{\mathbf{s} + (1/R_{fA}C_{fA})} = -\frac{5 \times 10^5/R_{1A}}{\mathbf{s} + 5}$$

Next, we consider the zero at $\mathbf{s} = -2$. From Fig. 15.49b, $1/R_{1B}C_{1B} = 2$, and, with $R_{1B} = 100$ kΩ, we have $C_{1B} = 5$ μF. Thus

$$\mathbf{H}_B(\mathbf{s}) = -R_{fB}C_{1B}\left(\mathbf{s} + \frac{1}{R_{1B}C_{1B}}\right)$$
$$= -5 \times 10^{-6}R_{fB}(\mathbf{s} + 2)$$

and

$$\mathbf{H}(\mathbf{s}) = \mathbf{H}_A(\mathbf{s})\mathbf{H}_B(\mathbf{s}) = 2.5\frac{R_{fB}}{R_{1A}}\frac{\mathbf{s} + 2}{\mathbf{s} + 5}$$

We complete the design by letting $R_{fB} = 100$ kΩ and $R_{1A} = 25$ kΩ. The result is shown in Fig. 15.50. The capacitors in this circuit are fairly large, but this is a direct consequence of the low frequencies selected for the pole and zero of $\mathbf{H}(\mathbf{s})$. If $\mathbf{H}(\mathbf{s})$ were changed to $10(\mathbf{s} + 2000)/(\mathbf{s} + 5000)$, we could use 2 and 5 nF values.

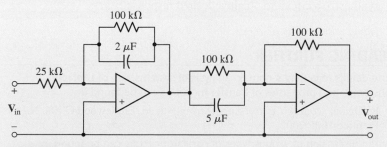

■ **FIGURE 15.50** This network contains two ideal op amps and gives the voltage transfer function H(s) = $V_{out}/V_{in} = 10(s + 2)/(s + 5)$.

PRACTICE

15.17 Specify suitable element values for $\mathbf{Z}_1$ and $\mathbf{Z}_f$ in each of three cascaded stages to realize the transfer function $\mathbf{H}(\mathbf{s}) = -20\mathbf{s}^2/(\mathbf{s} + 1000)$.

Ans: $1\,\mu\text{F} \parallel \infty, 1\,\text{M}\Omega$; $1\,\mu\text{F} \parallel \infty, 1\,\text{M}\Omega$; $100\,\text{k}\Omega \parallel 10\,\text{nF}, 5\,\text{M}\Omega$.

SUMMARY AND REVIEW

These models are summarized in Table 15.1.

❑ Resistors may be represented in the frequency domain by an impedance having the same magnitude.

❑ Inductors may be represented in the frequency domain by an impedance $\mathbf{s}L$. If the initial current is nonzero, then the impedance must be placed in series with a voltage source $-Li(0^-)$ or in parallel with a current source $i(0^-)/\mathbf{s}$.

❑ Capacitors may be represented in the frequency domain by an impedance $1/\mathbf{s}C$. If the initial voltage is nonzero, then the impedance must be placed in series with a voltage source $v(0^-)/\mathbf{s}$ or in parallel with a current source $Cv(0^-)$.

❑ Nodal and mesh analysis in the **s**-domain lead to simultaneous equations in terms of *s*-polynomials. MATLAB is a particularly useful tool for solving such systems of equations.

❑ Superposition, source transformation, and the Thévenin and Norton theorems all apply in the *s*-domain.

❑ A circuit transfer function $\mathbf{H}(\mathbf{s})$ is defined as the ratio of the *s*-domain output to the *s*-domain input. Either quantity may be a voltage or current.

❑ The *zeros* of $\mathbf{H}(\mathbf{s})$ are those values that result in zero magnitude. The *poles* of $\mathbf{H}(\mathbf{s})$ are those values that result in infinite magnitude.

❑ Convolution provides us with both an analytic and a graphical means of determining the output of a circuit from its impulse response $h(t)$.

❑ There are several graphical approaches to representing **s**-domain expressions in terms of poles and zeros. Such plots can be used to synthesize a circuit to obtain a desired response.

READING FURTHER

More details regarding **s**-domain analysis of systems, use of Laplace transforms, and properties of transfer functions can be found in:

K. Ogata, *Modern Control Engineering,* 4th ed. Englewood Cliffs, N.J.: Prentice-Hall, 2002.

A good discussion of various types of oscillator circuits can be found in:

R. Mancini, *Op Amps for Everyone,* 2nd ed. Amsterdam: Newnes, 2003.

and

G. Clayton and S. Winder, *Operational Amplifiers,* 5th ed. Amsterdam: Newnes, 2003.

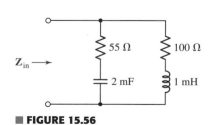

EXERCISES

15.1 Z(s) and Y(s)

1. Draw all possible s-domain equivalents ($t > 0$) of the circuit shown in Fig. 15.51.

2. Draw all possible s-domain equivalents ($t > 0$) of the circuit shown in Fig. 15.52.

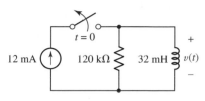

■ **FIGURE 15.51**

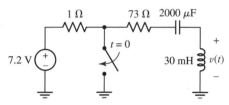

■ **FIGURE 15.52**

3. Refer to Fig. 15.53 and find (a) $\mathbf{Z}_{in}(\mathbf{s})$ as a ratio of two polynomials in $\mathbf{s}$; (b) $\mathbf{Z}_{in}(-80)$; (c) $\mathbf{Z}_{in}(j80)$; (d) the admittance of the parallel RL branch, $\mathbf{Y}_{RL}(\mathbf{s})$, as a ratio of polynomials in $\mathbf{s}$. (e) Repeat for $\mathbf{Y}_{RC}(\mathbf{s})$. (f) Show that $\mathbf{Z}_{in}(\mathbf{s}) = (\mathbf{Y}_{RL} + \mathbf{Y}_{RC})/\mathbf{Y}_{RL}\mathbf{Y}_{RC}$.

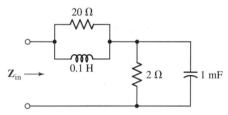

■ **FIGURE 15.53**

4. Find the Thévenin equivalent impedance seen looking into the terminals of the circuit depicted in Fig. 15.54.

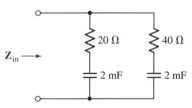

■ **FIGURE 15.54**

5. (a) Find $\mathbf{Z}_{in}(\mathbf{s})$ for the network of Fig. 15.55 as a ratio of two polynomials in $\mathbf{s}$. (b) Find $\mathbf{Z}_{in}(j8)$ in rectangular form. (c) Find $\mathbf{Z}_{in}(-2 + j6)$ in polar form. (d) To what value should the 16 Ω resistor be changed in order that $\mathbf{Z}_{in} = 0$ at $\mathbf{s} = -5 + j0$? (e) To what value should the 16 Ω resistor be changed in order that $\mathbf{Z}_{in} = \infty$ at $\mathbf{s} = -5 + j0$?

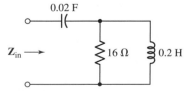

■ **FIGURE 15.55**

6. (a) Find the Thévenin equivalent impedance seen looking into the terminals of the circuit depicted in Fig. 15.56. (b) Plot the magnitude of the impedance as a function of frequency ω for the case of $\sigma = 0$.

■ **FIGURE 15.56**

7. Determine the input impedance $\mathbf{Z}_{in}$ of the circuit shown in Fig. 15.57, a linear circuit model of a common-emitter bipolar junction transistor amplifier valid for frequencies up to several MHz. Express your answer as a ratio of ordered **s**-polynomials.

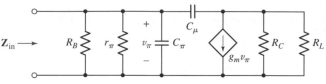

■ **FIGURE 15.57**

8. Find $v(t)$ for the circuit of Fig. 15.58 by initially working in the **s**-domain.

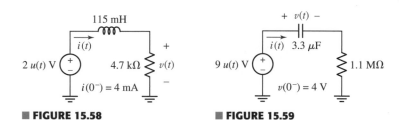

■ **FIGURE 15.58**   ■ **FIGURE 15.59**

■ **FIGURE 15.60**

9. Use **s**-domain analysis techniques to determine the current $i(t)$ through the capacitor of Fig. 15.59.

10. (*a*) Convert the circuit of Fig. 15.60 to an appropriate **s**-domain representation. (*b*) Find an expression for $p(t)$, the power absorbed in the resistor.

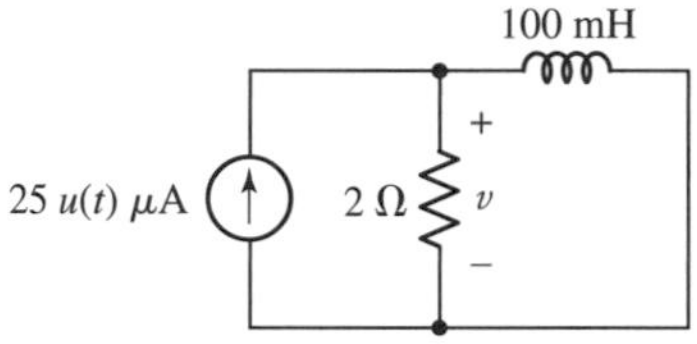

15.2 Nodal and Mesh Analysis in the *s*-Domain

11. Consider the circuit of Fig. 15.61. Using **s**-domain techniques, find the indicated nodal voltages $v_1(t)$ and $v_2(t)$ if $v_1(0^-) = -2$ V.

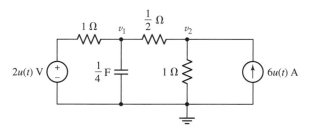

■ **FIGURE 15.61**

12. Consider the circuit of Fig. 15.62. (*a*) Using **s**-domain techniques, find the indicated nodal voltages $v_1(t)$ and $v_2(t)$. (*b*) Sketch $v_1(t)$.

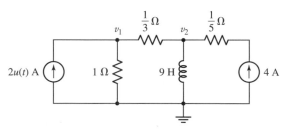

■ **FIGURE 15.62**

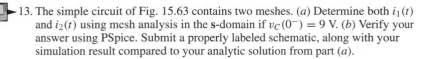

13. The simple circuit of Fig. 15.63 contains two meshes. (*a*) Determine both $i_1(t)$ and $i_2(t)$ using mesh analysis in the **s**-domain if $v_C(0^-) = 9$ V. (*b*) Verify your answer using PSpice. Submit a properly labeled schematic, along with your simulation result compared to your analytic solution from part (*a*).

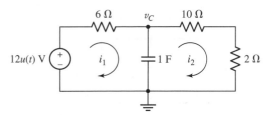

■ **FIGURE 15.63**

14. The simple circuit of Fig. 15.64 contains two meshes. (*a*) Determine both $i_1(t)$ and $i_2(t)$ using mesh analysis in the **s**-domain if $i_1(0^-) - i_2(0^-) = 8$ A. (*b*) Verify your answer using PSpice. Submit a properly labeled schematic, along with your simulation result compared to your analytic solution from part (*a*).

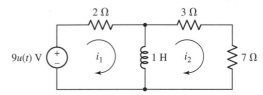

■ **FIGURE 15.64**

15. (*a*) Let $v_s = 10e^{-2t}\cos(10t + 30°)u(t)$ V in the circuit of Fig. 15.65, and work in the frequency domain to find $\mathbf{I}_x$. (*b*) Find $i_x(t)$.

16. Find the nodal voltage $v_1(t)$ in the circuit of Fig. 15.66, assuming zero initial energy.

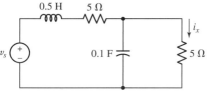

■ **FIGURE 15.65**

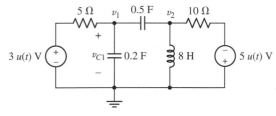

■ **FIGURE 15.66**

17. Determine a time-domain expression for the central mesh current of the circuit in Fig. 15.66, assuming zero initial energy.

18. Find the nodal voltage $v_1(t)$ in the circuit of Fig. 15.66 if the capacitor voltage $v_{C1}(0^-) = 9$ V but no other element is initially storing any energy.

19. Let $i_{s1} = 20e^{-3t}\cos 4t\, u(t)$ A and $i_{s2} = 30e^{-3t}\sin 4t\, u(t)$ A in the circuit of Fig. 15.67. (*a*) Work in the frequency domain to find $\mathbf{V}_x$. (*b*) Find $v_x(t)$.

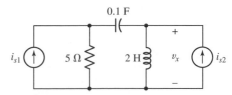

■ **FIGURE 15.67**

20. (*a*) Determine a time-domain expression for $v(t)$ in the circuit shown in Fig. 15.68 if $v(0^-) = 75$ V, and no energy is initially stored in the inductor. (*b*) Use your answer to part (*a*) to determine the steady-state current that flows from the 115 V rms source. (*c*) Verify your answer to part (*b*) using phasor analysis.

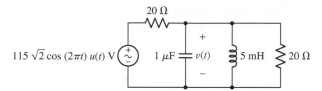

■ **FIGURE 15.68**

21. Determine the mesh currents $i_1(t)$ and $i_2(t)$ in Fig. 15.69 if the current through the 1 mH inductor ($i_2 - i_4$) is 1 A at $t = 0^-$. Verify that your answer approaches the answer obtained using phasor analysis as the circuit response eventually reaches steady state.

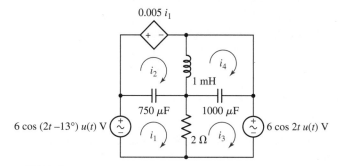

■ **FIGURE 15.69**

22. Assuming no energy initially stored in the circuit of Fig. 15.70, determine the value of v_2 at $t =$: (*a*) 1 ms; (*b*) 100 ms; (*c*) 10 s.

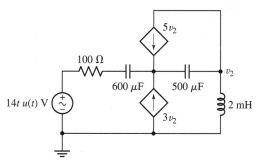

■ **FIGURE 15.70**

23. If the dependent voltage source in the circuit of Fig. 15.71 is damaged by a power surge during a lightning storm so that it no longer functions (i.e., is now

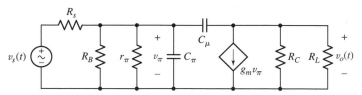

■ **FIGURE 15.71**

an open circuit), find an expression for the power absorbed by the 2 Ω resistor. You should assume that the only energy initially stored in the circuit is in the inductor, such that the current through the 1 mH inductor ($i_2 - i_4$) is 1 mA at $t = 0^-$.

24. (*a*) For the circuit of Fig. 15.71, a linear circuit model of a common-emitter bipolar junction transistor amplifier, determine an expression for the voltage gain $\mathbf{V}_o/\mathbf{V}_s$. You may assume zero energy initially stored in the capacitors; express your answer as a ratio of ordered **s**-polynomials. (*b*) How many poles exist for this transfer function?

15.3 Additional Circuit Analysis Techniques

25. (*a*) Convert the circuit in Fig. 15.72 to an appropriate **s**-domain representation. (*b*) Find the Thévenin equivalent seen by the 1 Ω resistor. (*c*) Analyze the simplified circuit to find an expression for $i(t)$, the instantaneous current through the 1 Ω resistor.

26. Replace the current source in Fig. 15.72 with a voltage source $20u(t)$ V, positive reference at the top. (*a*) Convert the circuit to an appropriate **s**-domain representation. (*b*) Find the Norton equivalent seen by the 1 Ω resistor. (*c*) Analyze the simplified circuit to obtain an expression for $i_C(t)$.

27. For the **s**-domain circuit of Fig. 15.73, determine the Thévenin equivalent seen by the $7s^2$ Ω impedance, and use it to determine the current $\mathbf{I}(\mathbf{s})$.

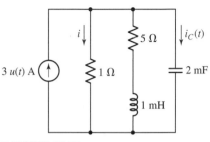

■ **FIGURE 15.72**

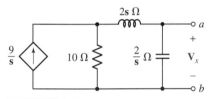

■ **FIGURE 15.73**

28. For the **s**-domain circuit of Fig. 15.74, determine the Thévenin equivalent seen looking into the terminals marked *a* and *b*.

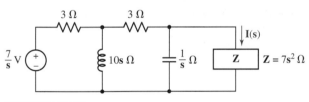

■ **FIGURE 15.74**

29. (*a*) Employ superposition in the **s**-domain to find $\mathbf{V}_1(\mathbf{s})$ and $\mathbf{V}_2(\mathbf{s})$ for the circuit of Fig. 15.75. (*b*) Find $v_1(t)$ and $v_2(t)$.

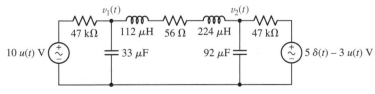

■ **FIGURE 15.75**

30. Determine the power $p(t)$ absorbed by the 56 Ω resistor of Fig. 15.75 by first performing appropriate source transformations in the **s**-domain.

31. (*a*) Find the **s**-domain Norton equivalent seen by the $10\,u(t)$ V source of Fig. 15.75. (*b*) Determine the current flowing out of the $10\,u(t)$ V source at $t = 1.5$ ms.

32. (a) Use superposition in the s-domain to find an expression for $\mathbf{V}_1(s)$ as labeled in Fig. 15.76. (b) Find $v_1(t)$.

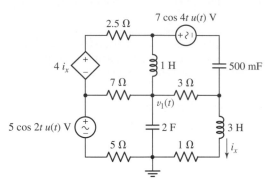

■ **FIGURE 15.76**

33. (a) Employ source transformation in the s-domain to determine $\mathbf{I}(s)$ for the circuit of Fig. 15.77. (b) Find $i(t)$. (c) Find the steady-state value of $i(t)$.

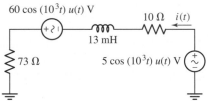

■ **FIGURE 15.77**

15.4 Poles, Zeros, and Transfer Functions

34. Determine the poles and zeros of the following transfer functions:

$$(a)\ \frac{7\mathbf{s}}{\mathbf{s}(3\mathbf{s}^2 - 9\mathbf{s} + 4)}; (b)\ \frac{\mathbf{s}^2 - 1}{(\mathbf{s}^2 + 2\mathbf{s} + 4)(\mathbf{s}^2 + 1)}.$$

35. State all poles and zeros of each of the following s-domain functions:

$$(a)\ \frac{3\mathbf{s}^2}{\mathbf{s}(\mathbf{s}^2 + 4)(\mathbf{s} - 1)}; (b)\ \frac{\mathbf{s}^2 + 2\mathbf{s} - 1}{\mathbf{s}^2(4\mathbf{s}^2 + 2\mathbf{s} + 1)(\mathbf{s}^2 - 1)}.$$

36. The series combination of a 5 Ω resistance and a 0.2 F capacitance is in parallel with the series combination of a 2 Ω resistance and a 5 H inductance. (a) Find the input admittance, $\mathbf{Y}_1(s)$, of this parallel combination as a ratio of two polynomials in s. (b) Identify all the poles and zeros of $\mathbf{Y}_1(s)$. (c) Identify all the poles of the input admittance obtained if a 10 Ω resistance is connected in parallel with $\mathbf{Y}_1(s)$. (d) Identify all the zeros of the input admittance obtained if a 10 Ω resistance is connected in series with $\mathbf{Y}_1(s)$.

37. Determine all the poles and zeros of (a) the input impedance defined in Fig. 15.54; (b) the input impedance defined in Fig. 15.56.

38. An admittance $\mathbf{Y}(s)$ has zeros at $s = 0$ and $s = -10$, and poles at $s = -5$ and $-20\ \mathrm{s}^{-1}$. If $\mathbf{Y}(s) \rightarrow 12$ S as $s \rightarrow \infty$, find (a) $\mathbf{Y}(j10)$; (b) $\mathbf{Y}(-j10)$; (c) $\mathbf{Y}(-15)$; (d) the poles and zeros of $5 + \mathbf{Y}(s)$.

39. (a) Find $\mathbf{Z}_{in}(s)$ for the network shown in Fig. 15.78. (b)Find all the critical frequencies of $\mathbf{Z}_{in}(s)$.

40. A given circuit has a transfer function $\mathbf{H}(s) = (s + 2)/[(s + 5)(s^2 + 6s + 25)]$. Find the s-domain output response if the input is (a) $\delta(t)$; (b) $e^{-4t}u(t)$; (c) $[2\cos 15t]u(t)$; (d) $te^{-t}u(t)$. (e) State the poles and zeros of each output response.

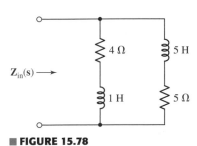

■ **FIGURE 15.78**

15.5 Convolution

41. The impulse response of a certain linear system is $h(t) = 5 \sin \pi t[u(t) - u(t-1)]$. An input signal $x(t) = 2[u(t) - u(t-2)]$ is applied. Use convolution to find and sketch the output $y(t)$.

42. Let $f_1(t) = e^{-5t}u(t)$ and $f_2(t) = (1 - e^{-2t})u(t)$. Find $y(t) = f_1(t) * f_2(t)$ by (a) convolution in the time domain; (b) $\mathcal{L}^{-1}\{\mathbf{F}_1(\mathbf{s})\mathbf{F}_2(\mathbf{s})\}$.

43. When an impulse $\delta(t)$ V is applied to a certain two-port network, the output voltage is $v_o(t) = 4u(t) - 4u(t-2)$ V. Find and sketch $v_o(t)$ if the input voltage is $2u(t-1)$ V.

44. Let $h(t) = 2e^{-3t}u(t)$ and $x(t) = u(t) - \delta(t)$. Find $y(t) = h(t) * x(t)$ by (a) using convolution in the time domain; (b) finding $\mathbf{H}(\mathbf{s})$ and $\mathbf{X}(\mathbf{s})$ and then obtaining $\mathcal{L}^{-1}\{\mathbf{H}(\mathbf{s})\mathbf{X}(\mathbf{s})\}$.

45. The impulse voltage response of a particular circuit is given as $h(t) = 5u(t) - 5u(t-2)$. Find the s-domain and time-domain voltage response if the excitation voltage $v_{\text{in}}(t) =: (a)\ 3\delta(t)$ V; $(b)\ 3u(t)$ V; $(c)\ 3u(t) - 3u(t-2)$ V; $(d)\ 3\cos 3t$ V. (e) Sketch the time-domain voltage response for parts (a)–(d).

46. (a) Determine the impulse response $h(t)$ of the network shown in Fig. 15.79. (b) Use convolution to determine $v_o(t)$ if $v_{\text{in}}(t) = 8u(t)$ V.

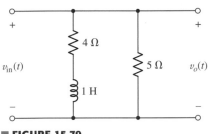

■ **FIGURE 15.79**

47. (a) Determine the impulse response $h(t)$ of the network shown in Fig. 15.80. (b) Use convolution to determine $v_o(t)$ if $v_{\text{in}}(t) = 8e^{-t}u(t)$ V.

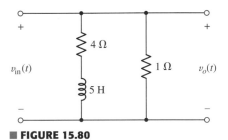

■ **FIGURE 15.80**

15.6 The Complex-Frequency Plane

48. Find $\mathbf{H}(\mathbf{s}) = \mathbf{V}_{\text{out}}/\mathbf{V}_{\text{in}}$ for the network of Fig. 15.81 and locate all its critical frequencies.

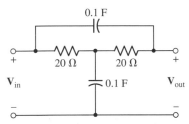

■ **FIGURE 15.81**

49. The pole-zero configuration of $\mathbf{H}(\mathbf{s}) = \mathbf{V}_2(\mathbf{s})/\mathbf{V}_1(\mathbf{s})$ is shown in Fig. 15.82. Let $\mathbf{H}(0) = 1$. Sketch $|\mathbf{H}(\mathbf{s})|$ versus: $(a)\ \sigma$ if $\omega = 0$; $(b)\ \omega$ if $\sigma = 0$. (c) Find $|\mathbf{H}(j\omega)|_{\max}$.

50. A piece of electrical machinery has an input impedance characterized by two zeros at $s = -1$, a pole at $s = -0.5 + j\sqrt{3}/2$, another pole at $s = -0.5 - j\sqrt{3}/2$, and is equal to unity ohms at $s = 0$. (a) Sketch the pole-zero constellation of this impedance. (b) Draw the elastic-sheet model of the

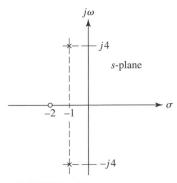

■ **FIGURE 15.82**

impedance magnitude. (*c*) Find a combination of resistors, inductors, and capacitors that have the same impedance. (*Hint: Work backward from the s-domain expression.*)

51. Given the voltage gain $\mathbf{H}(\mathbf{s}) = (10\mathbf{s}^2 + 55\mathbf{s} + 75)/(\mathbf{s}^2 + 16)$: (*a*) indicate the critical frequencies on the **s** plane; (*b*) calculate $\mathbf{H}(0)$ and $\mathbf{H}(\infty)$. (*c*) If a scale model of $|\mathbf{H}(\mathbf{s})|$ has a height of 3 cm at the origin, how high is it at $\mathbf{s} = j3$? (*d*) Roughly sketch $|\mathbf{H}(\sigma)|$ versus σ and $|\mathbf{H}(j\omega)|$ versus ω.

52. In the back corner of a top-secret government laboratory, an odd-shaped metal box is discovered by a researcher whose lunch has been mischievously hidden by colleagues with apparently too much time on their hands. With no sign of food anywhere, the researcher decides to measure the admittance of the box, finding that the admittance can be modeled as $\mathbf{Y}(\mathbf{s}) = (5\mathbf{s}^2 + 5\mathbf{s} + 2)/(5\mathbf{s}^2 + 15\mathbf{s} + 2)$ S. (*a*) Sketch the pole-zero constellation of this admittance. (*b*) Draw the elastic-sheet model of the admittance magnitude. (*c*) Determine the location of the missing lunch if the coefficients of the denominator polynomial correspond to latitude (degrees, minutes, seconds), and the coefficients of the numerator polynomial correspond to longitude (degrees, minutes, seconds). Clearly, the researcher's colleagues have *WAY* too much time on their hands.

53. The pole-zero constellation shown in Fig. 15.83 applies to a current gain $\mathbf{H}(\mathbf{s}) = \mathbf{I}_{out}/\mathbf{I}_{in}$. Let $\mathbf{H}(-2) = 6$. (*a*) Express $\mathbf{H}(\mathbf{s})$ as a ratio of polynomials in **s**. (*b*) Find $\mathbf{H}(0)$ and $\mathbf{H}(\infty)$. (*c*) Determine the magnitude and direction of each arrow from a critical frequency to $\mathbf{s} = j2$.

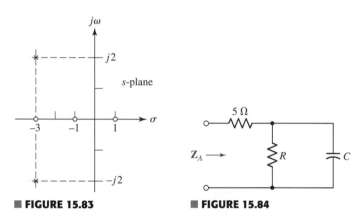

■ **FIGURE 15.83**   ■ **FIGURE 15.84**

54. The three-element network shown in Fig. 15.84 has an input impedance $\mathbf{Z}_A(\mathbf{s})$ that has a zero at $\mathbf{s} = -10 + j0$. If a 20 Ω resistor is placed in series with the network, the zero of the new impedance shifts to $\mathbf{s} = -3.6 + j0$. Find R and C.

55. Let $\mathbf{H}(\mathbf{s}) = 100(\mathbf{s} + 2)/(\mathbf{s}^2 + 2\mathbf{s} + 5)$ and (*a*) show the pole-zero plot for $\mathbf{H}(\mathbf{s})$; (*b*) find $\mathbf{H}(j\omega)$; (*c*) find $|\mathbf{H}(j\omega)|$; (*d*) sketch $|\mathbf{H}(j\omega)|$ versus ω; (*e*) find ω_{max}, the frequency at which $|\mathbf{H}(j\omega)|$ is a maximum.

15.7 Natural Response and the s-Plane

56. Let $\mathbf{Z}_{in}(\mathbf{s}) = (5\mathbf{s} + 20)/(\mathbf{s} + 2)$ Ω for the network shown in Fig. 15.85. Find (*a*) the voltage $v_{ab}(t)$ between the open-circuited terminals if $v_{ab}(0) = 25$ V; (*b*) the current $i_{ab}(t)$ in a short circuit between terminals *a* and *b* if $i_{ab}(0) = 3$ A.

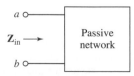

■ **FIGURE 15.85**

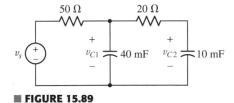

57. Let $\mathbf{Z}_{in}(\mathbf{s}) = 5(s^2 + 4s + 20)/(s + 1)$ Ω for the passive network of Fig. 15.85. Find $i_a(t)$, the instantaneous current entering terminal a, given $v_{ab}(t)$ equal to (a) $160e^{-6t}$ V; (b) $160e^{-6t}u(t)$ V, with $i_a(0) = 0$ and $di_a/dt = 32$ A/s at $t = 0$.

58. (a) Determine $\mathbf{H}(\mathbf{s}) = \mathbf{I}_C/\mathbf{I}_s$ for the circuit shown in Fig. 15.86. (b) Find the poles of $\mathbf{H}(\mathbf{s})$. (c) Find α, ω_0, and ω_d for the *RLC* circuit. (d) Determine the forced response $i_{C,f}(t)$ completely. (e) Give the form of the natural response $i_{C,n}(t)$. (f) Determine values for $i_C(0^+)$ and di_C/dt at $t = 0^+$. (g) Write the complete response, $i_C(t)$.

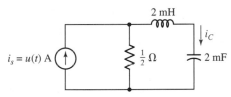

■ **FIGURE 15.86**

59. For the circuit of Fig. 15.87. (a) find the poles of $\mathbf{H}(\mathbf{s}) = \mathbf{I}_{in}/\mathbf{V}_{in}$. (b) Let $i_1(0^+) = 5$ A and $i_2(0^+) = 2$ A, and find $i_{in}(t)$ if $v_{in}(t) = 500u(t)$ V.

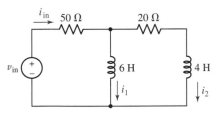

■ **FIGURE 15.87**

60. (a) Find $\mathbf{H}(\mathbf{s}) = \mathbf{V}(\mathbf{s})/\mathbf{I}_s(\mathbf{s})$ for the circuit of Fig. 15.88. Find $v(t)$ if $i_s(t)$ equals (b) $2u(t)$ A; (c) $4e^{-10t}$ A; (d) $4e^{-10t}u(t)$ A.

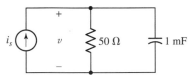

■ **FIGURE 15.88**

61. For the circuit shown in Fig. 15.89. (a) find $\mathbf{H}(\mathbf{s}) = \mathbf{V}_{C2}/\mathbf{V}_s$; (b) let $v_{C1}(0^+) = 0$ and $v_{C2}(0^+) = 0$, and find $v_{C2}(t)$ if $v_s(t) = u(t)$ V.

62. Refer to Fig. 15.90 and find the impedance $\mathbf{Z}_{in}(\mathbf{s})$ seen by the source. Use this expression to help determine $v_{in}(t)$ for $t > 0$.

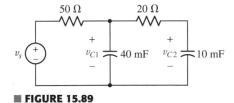

■ **FIGURE 15.89**

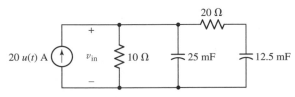

■ **FIGURE 15.90**

15.8 A Technique for Synthesizing the Voltage Ratio H(s) = V$_{out}$/V$_{in}$

63. Find $\mathbf{H(s)} = \mathbf{V}_{out}/\mathbf{V}_{in}$ as a ratio of polynomials in $\mathbf{s}$ for the op-amp circuit of Fig. 15.48, given the impedance values (in Ω): (a) $\mathbf{Z}_1(\mathbf{s}) = 10^3 + (10^8/\mathbf{s})$, $\mathbf{Z}_f(\mathbf{s}) = 5000$; (b) $\mathbf{Z}_1(\mathbf{s}) = 5000$, $\mathbf{Z}_f(\mathbf{s}) = 10^3 + (10^8/\mathbf{s})$; (c) $\mathbf{Z}_1(\mathbf{s}) = 10^3 + (10^8/\mathbf{s})$, $\mathbf{Z}_f(\mathbf{s}) = 10^4 + (10^8/\mathbf{s})$.

64. In the circuit of Fig. 15.49b, let $R_f = 20$ kΩ, and then specify values for R_1 and C_1 so that $\mathbf{H(s)} = \mathbf{V}_{out}/\mathbf{V}_{in}$ equals (a) -50; (b) $-10^{-3}(\mathbf{s} + 10^4)$; (c) $-10^{-4}(\mathbf{s} + 10^3)$; (d) $10^{-3}(\mathbf{s} + 10^5)$, using two stages.

65. In the op-amp circuit of Fig. 15.49a, let $R_f = 20$ kΩ, and then specify values for R_1 and C_f so that $\mathbf{H(s)} = \mathbf{V}_{out}/\mathbf{V}_{in}$ equals (a) -50; (b) $-10^3/(\mathbf{s} + 10^4)$; (c) $-10^4/(\mathbf{s} + 10^3)$; (d) $100/(\mathbf{s} + 10^5)$, using two stages.

66. Use several op amps in cascade to realize the transfer function $\mathbf{H(s)} = \mathbf{V}_{out}/\mathbf{V}_{in} = -10^{-4}\mathbf{s}(\mathbf{s} + 10^2)/(\mathbf{s} + 10^3)$. Use only 10 k$\Omega$ resistors, open circuits, or short circuits, but specify all capacitance values.

67. Design a Wien-bridge oscillator characterized by an oscillation frequency of 1 kHz. Use only the standard resistor values given on the inside cover. Verify your design with an appropriate PSpice simulation.

68. Design a Wien-bridge oscillator with an oscillation frequency of 60 Hz. Verify your design with an appropriate PSpice simulation.

69. Design an oscillator circuit to provide a sinusoidal signal of 440 Hz using only standard resistor values as given on the inside front cover. Verify your design with an appropriate PSpice simulation. What musical note is produced by your circuit?

70. Design a circuit that provides a voltage output composed of a 220 Hz sine wave and a 440 Hz sine wave. Verify your design with an appropriate Pspice simulation. Are the two sine waves in phase with one another?

Frequency Response

Resonant Frequency of
Circuits with Inductors
and Capacitors

Quality Factor

Bandwidth

Frequency and Magnitude
Scaling

Bode Diagram Techniques

Low- and High-Pass Filters

Bandpass Filter Design

Active Filters

INTRODUCTION

Frequency response has appeared in several chapters already, so
that the reader may be wondering why the topic now warrants an
entire chapter. The concept of frequency response is extremely
important in all fields of science and engineering, forming the
foundation for understanding factors that determine the stability
(or instability) of a particular system, be it electrical, mechanical,
chemical, or biological. We will also find that frequency response
concepts are required in many electrical engineering applications
beyond the issue of stability. For instance, in working with communications systems we are often faced with situations that call for
the separation of frequencies (individual radio stations, for example), an operation that can be accomplished once we have a solid
understanding of the frequency response of filtering circuits. In
short, we could easily devote several pages to extolling the virtues
of studying frequency response. However, we prefer to launch into
the subject, beginning with an electrical twist to the concept of
resonance, and culminating in the design of basic filtering circuits
for use in everyday applications such as audio amplifiers.

16.1 • PARALLEL RESONANCE

Why should we be interested in the response to sinusoidal forcing
functions when we so seldom encounter them in practice? The electric power industry is an exception, as the sinusoidal waveform appears throughout, although it is occasionally necessary to consider
other frequencies introduced by the nonlinearity of some devices.
But in most other electrical systems, the forcing functions and

responses are *not* sinusoidal. In any system in which information is to be transmitted, the sinusoid by itself is almost valueless; it contains limited information because its future values are exactly predictable from its past values. Moreover, once one period has been completed, any periodic non-sinusoidal waveform also contains no additional information.

Let us suppose that a certain forcing function is found to contain sinusoidal *components* having frequencies within the range of 10 to 100 Hz. Now let us imagine that this forcing function is applied to a network that has the property that all sinusoidal voltages with frequencies from zero to 200 Hz applied at the input terminals appear doubled in magnitude at the output terminals, with no change in phase angle. The output function is therefore an undistorted facsimile of the input function, but with twice the amplitude. If, however, the network has a frequency response such that the magnitudes of input sinusoids between 10 and 50 Hz are multiplied by a different factor than are those between 50 and 100 Hz, then the output would in general be distorted; it would no longer be a magnified version of the input. This distorted output might be desirable in some cases and undesirable in others. That is, the network frequency response might be chosen *deliberately* to reject some frequency components of the forcing function, or to emphasize others.

Such behavior is characteristic of tuned circuits or resonant circuits, as we will see in this chapter. In discussing resonance we will be able to apply all the methods we have discussed in presenting frequency response.

Resonance

In this section we will begin the study of a very important phenomenon that may occur in circuits that contain both inductors and capacitors. The phenomenon is called **resonance,** and it may be loosely described as the condition existing in any physical system when a fixed-amplitude sinusoidal forcing function produces a response of maximum amplitude. However, we often speak of resonance as occurring even when the forcing function is not sinusoidal. The resonant system may be electrical, mechanical, hydraulic, acoustic, or some other kind, but we will restrict our attention, for the most part, to electrical systems.

Resonance is a familiar phenomenon. Jumping up and down on the bumper of an automobile, for example, can put the vehicle into rather large oscillatory motion if the jumping is done at the proper frequency (about one jump per second), and if the shock absorbers are somewhat decrepit. However, if the jumping frequency is increased or decreased, the vibrational response of the automobile will be considerably less than it was before. A further illustration is furnished in the case of an opera singer who is able to shatter crystal goblets by means of a well-formed note at the proper frequency. In each of these examples, we are thinking of frequency as being adjusted until resonance occurs; it is also possible to adjust the size, shape, and material of the mechanical object being vibrated, but this may not be so easily accomplished physically.

The condition of resonance may or may not be desirable, depending upon the purpose which the physical system is to serve. In the automotive example, a large amplitude of vibration may help to separate locked bumpers, but it would be somewhat disagreeable at 65 mi/h (105 km/h).

Let us now define resonance more carefully. In a two-terminal electrical network containing at least one inductor and one capacitor, we define resonance as the condition which exists when the input impedance of the network is purely resistive. Thus,

a network is in resonance (or resonant) when the voltage and current at the network input terminals are in phase.

We will also find that a maximum-amplitude response is produced in the network when it is in the resonant condition.

We first apply the definition of resonance to a parallel *RLC* network driven by a sinusoidal current source as shown in Fig. 16.1. In many practical situations, this circuit is a very good approximation to the circuit we might build in the laboratory by connecting a physical inductor in parallel with a physical capacitor, where the parallel combination is driven by an energy source having a very high output impedance. The steady-state admittance offered to the ideal current source is

$$\mathbf{Y} = \frac{1}{R} + j\left(\omega C - \frac{1}{\omega L}\right) \qquad [1]$$

Resonance occurs when the voltage and current at the input terminals are in phase. This corresponds to a purely real admittance, so that the necessary condition is given by

$$\omega C - \frac{1}{\omega L} = 0$$

The resonant condition may be achieved by adjusting L, C, or ω; we will devote our attention to the case for which ω is the variable. Hence, the resonant frequency ω_0 is

$$\omega_0 = \frac{1}{\sqrt{LC}} \qquad \text{rad/s} \qquad [2]$$

or

$$f_0 = \frac{1}{2\pi\sqrt{LC}} \qquad \text{Hz} \qquad [3]$$

This resonant frequency ω_0 is identical to the resonant frequency defined in Eq. [10], Chap. 9.

The pole-zero configuration of the admittance function can also be used to considerable advantage here. Given $\mathbf{Y}(s)$,

$$\mathbf{Y}(s) = \frac{1}{R} + \frac{1}{sL} + sC$$

or

$$\mathbf{Y}(s) = C\frac{s^2 + s/RC + 1/LC}{s} \qquad [4]$$

we may display the zeros of $\mathbf{Y}(s)$ by factoring the numerator:

$$\mathbf{Y}(s) = C\frac{(s + \alpha - j\omega_d)(s + \alpha + j\omega_d)}{s}$$

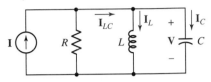

■ **FIGURE 16.1** The parallel combination of a resistor, an inductor, and a capacitor, often referred to as a *parallel resonant circuit.*

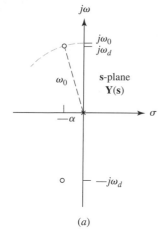

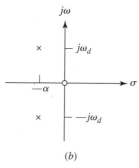

FIGURE 16.2 (*a*) The pole-zero constellation of the input admittance of a parallel resonant circuit is shown on the **s**-plane; $\omega_0^2 = \alpha^2 + \omega_d^2$. (*b*) The pole-zero constellation of the input impedance.

where α and ω_d represent the same quantities that they did when we discussed the natural response of the parallel *RLC* circuit in Sec. 9.4. That is, α is the *exponential damping coefficient*,

$$\alpha = \frac{1}{2RC}$$

and ω_d is the *natural resonant frequency* (not the resonant frequency ω_0),

$$\omega_d = \sqrt{\omega_0^2 - \alpha^2}$$

The pole-zero constellation shown in Fig. 16.2*a* follows directly from the factored form.

In view of the relationship among α, ω_d, and ω_0, it is apparent that the distance from the origin of the **s**-plane to one of the admittance zeros is numerically equal to ω_0. Given the pole-zero configuration, the resonant frequency may therefore be obtained by purely graphical methods. We merely swing an arc, using the origin of the **s**-plane as a center, through one of the zeros. The intersection of this arc and the positive $j\omega$ axis locates the point $\mathbf{s} = j\omega_0$. It is evident that ω_0 is slightly greater than the natural resonant frequency ω_d, but their ratio approaches unity as the ratio of ω_d to α increases.

Resonance and the Voltage Response

Next let us examine the magnitude of the response, the voltage $\mathbf{V(s)}$ indicated in Fig. 16.1, as the frequency ω of the forcing function is varied. If we assume a constant-amplitude sinusoidal current source, the voltage response is proportional to the input impedance. This response can be obtained from the pole-zero plot of the impedance

$$\mathbf{Z(s)} = \frac{\mathbf{s}/C}{(\mathbf{s} + \alpha - j\omega_d)(\mathbf{s} + \alpha + j\omega_d)}$$

shown in Fig. 16.2*b*. The response of course starts at zero, reaches a maximum value in the vicinity of the natural resonant frequency, and then drops again to zero as ω becomes infinite. The frequency response is sketched in Fig. 16.3. The maximum value of the response is indicated as R times the amplitude of the source current, implying that the maximum magnitude of

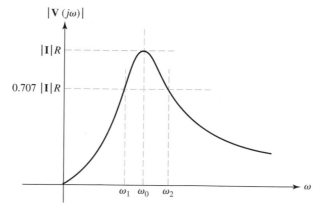

FIGURE 16.3 The magnitude of the voltage response of a parallel resonant circuit is shown as a function of frequency.

the circuit impedance is R; moreover, the response maximum is shown to occur exactly at the resonant frequency ω_0. Two additional frequencies, ω_1 and ω_2, which we will later use as a measure of the width of the response curve, are also identified. Let us first show that the maximum impedance magnitude is R and that this maximum occurs at resonance.

The admittance, as specified by Eq. [1], possesses a constant conductance and a susceptance which has a minimum magnitude (zero) at resonance. The minimum admittance magnitude therefore occurs at resonance, and it is $1/R$. Hence, the maximum impedance magnitude is R, *and it occurs at resonance*.

At the resonant frequency, therefore, the voltage across the parallel resonant circuit of Fig. 16.1 is simply $\mathbf{I}R$, and the *entire* source current $\mathbf{I}$ flows through the resistor. However, current is also present in L and C. For the inductor, $\mathbf{I}_{L,0} = \mathbf{V}_{L,0}/j\omega_0 L = \mathbf{I}R/j\omega_0 L$, and the capacitor current at resonance is $\mathbf{I}_{C,0} = (j\omega_0 C)\mathbf{V}_{C,0} = j\omega_0 C R \mathbf{I}$. Since $1/\omega_0 C = \omega_0 L$ at resonance, we find that

$$\mathbf{I}_{C,0} = -\mathbf{I}_{L,0} = j\omega_0 C R \mathbf{I} \qquad [5]$$

and

$$\mathbf{I}_{C,0} + \mathbf{I}_{L,0} = \mathbf{I}_{LC} = 0$$

Thus, the *net* current flowing into the LC combination is zero. The maximum value of the response magnitude and the frequency at which it occurs are not always found so easily. In less standard resonant circuits, we may find it necessary to express the magnitude of the response in analytical form, usually as the square root of the sum of the real part squared and the imaginary part squared; then we should differentiate this expression with respect to frequency, equate the derivative to zero, solve for the frequency of maximum response, and finally substitute this frequency in the magnitude expression to obtain the maximum-amplitude response. The procedure may be carried out for this simple case merely as a corroborative exercise; but, as we have seen, it is not necessary.

Quality Factor

It should be emphasized that, although the *height* of the response curve of Fig. 16.3 depends only upon the value of R for constant-amplitude excitation, the width of the curve or the steepness of the sides depends upon the other two element values also. We will shortly relate the "width of the response curve" to a more carefully defined quantity, the *bandwidth*, but it is helpful to express this relationship in terms of a very important parameter, the ***quality factor Q***.

We will find that the sharpness of the response curve of any resonant circuit is determined by the maximum amount of energy that can be stored in the circuit, compared with the energy that is lost during one complete period of the response.

We define Q as

$$\boxed{Q = \text{quality factor} = 2\pi \frac{\text{maximum energy stored}}{\text{total energy lost per period}}} \qquad [6]$$

We should be very careful not to confuse the quality factor with charge or reactive power, all of which unfortunately are represented by the letter Q.

The proportionality constant 2π is included in the definition in order to simplify the more useful expressions for Q which we will now obtain. Since energy can be stored only in the inductor and the capacitor, and can be lost only in the resistor, we may express Q in terms of the instantaneous energy associated with each of the reactive elements and the average power P_R dissipated in the resistor:

$$Q = 2\pi \frac{[w_L(t) + w_C(t)]_{\max}}{P_R T}$$

where T is the period of the sinusoidal frequency at which Q is evaluated.

Now let us apply this definition to the parallel RLC circuit of Fig. 16.1 and determine the value of Q at the resonant frequency. This value of Q is denoted by Q_0. We select the current forcing function

$$i(t) = \mathbf{I}_m \cos \omega_0 t$$

and obtain the corresponding voltage response at resonance,

$$v(t) = Ri(t) = R\mathbf{I}_m \cos \omega_0 t$$

The energy stored in the capacitor is then

$$w_C(t) = \frac{1}{2}Cv^2 = \frac{\mathbf{I}_m^2 R^2 C}{2} \cos^2 \omega_0 t$$

and the instantaneous energy stored in the inductor is given by

$$w_L(t) = \frac{1}{2}Li_L^2 = \frac{1}{2}L\left(\frac{1}{L}\int v\,dt\right)^2 = \frac{1}{2L}\left[\frac{R\mathbf{I}_m}{\omega_0}\sin \omega_0 t\right]^2$$

so that

$$w_L(t) = \frac{\mathbf{I}_m^2 R^2 C}{2} \sin^2 \omega_0 t$$

The total *instantaneous* stored energy is therefore constant:

$$w(t) = w_L(t) + w_C(t) = \frac{\mathbf{I}_m^2 R^2 C}{2}$$

and this constant value must also be the maximum value. In order to find the energy lost in the resistor in one period, we take the average power absorbed by the resistor (see Sec. 11.2),

$$P_R = \tfrac{1}{2}\mathbf{I}_m^2 R$$

and multiply by one period, obtaining

$$P_R T = \frac{1}{2f_0}\mathbf{I}_m^2 R$$

We thus find the quality factor at resonance:

$$Q_0 = 2\pi \frac{\mathbf{I}_m^2 R^2 C/2}{\mathbf{I}_m^2 R/2f_0}$$

or

$$Q_0 = 2\pi f_0 RC = \omega_0 RC \qquad [7]$$

This equation (as well as any expression in Eq. [8]) holds only for the simple parallel RLC circuit of Fig. 16.1. Equivalent expressions for Q_0 which are often quite useful may be obtained by simple substitution:

$$Q_0 = R\sqrt{\frac{C}{L}} = \frac{R}{|X_{C,0}|} = \frac{R}{|X_{L,0}|} \qquad [8]$$

So we see that for this specific circuit, decreasing the resistance decreases Q_0; the lower the resistance, the greater the amount of energy lost in the element. Intriguingly, increasing the capacitance *increases* Q_0, but increasing the inductance leads to a *reduction* in Q_0. These statements, of course, apply to operation of the circuit at the resonant frequency.

Other Interpretations of Q

The dimensionless constant Q_0 is a function of all three circuit elements in the parallel resonant circuit. The concept of Q, however, is not limited to electric circuits or even to electrical systems; it is useful in describing *any* resonant phenomenon. For example, let us consider a bouncing golf ball. If we assume a weight W and release the golf ball from a height h_1 above a very hard (lossless) horizontal surface, then the ball rebounds to some lesser height h_2. The energy stored initially is Wh_1, and the energy lost in one period is $W(h_1 - h_2)$. The Q_0 is therefore

$$Q_0 = 2\pi \frac{h_1 W}{(h_1 - h_2)W} = \frac{2\pi h_1}{h_1 - h_2}$$

A perfect golf ball would rebound to its original height and have an infinite Q_0; a more typical value is 35. It should be noted that the Q in this mechanical example has been calculated from the natural response and not from the forced response. The Q of an electric circuit may also be determined from a knowledge of the natural response, as illustrated by Eqs. [10] and [11] in the following discussion.

Another useful interpretation of Q is obtained when we inspect the inductor and capacitor currents at resonance, as given by Eq. [5],

$$\mathbf{I}_{C,0} = -\mathbf{I}_{L,0} = j\omega_0 C R\mathbf{I} = jQ_0\mathbf{I} \qquad [9]$$

Note that each is Q_0 times the source current in amplitude and that each is $180°$ out of phase with the other. Thus, if we apply 2 mA at the resonant frequency to a parallel resonant circuit with a Q_0 of 50, we find 2 mA in the resistor, and 100 mA in both the inductor and the capacitor. A parallel resonant circuit can therefore act as a current amplifier, but not, of course, as a power amplifier, since it is a passive network.

Let us now relate to each other the various parameters which we have associated with a parallel resonant circuit. The three parameters α, ω_d, and ω_0 were introduced much earlier in connection with the natural response. Resonance, by definition, is fundamentally associated with the forced response, since it is defined in terms of a (purely resistive) input impedance, a sinusoidal steady-state concept. The two most important parameters of a resonant circuit are perhaps the resonant frequency ω_0 and the quality factor Q_0. Both the exponential damping coefficient and the natural resonant frequency may be expressed in terms of ω_0 and Q_0:

$$\alpha = \frac{1}{2RC} = \frac{1}{2(Q_0/\omega_0 C)C}$$

or

$$\alpha = \frac{\omega_0}{2Q_0} \qquad [10]$$

and

$$\omega_d = \sqrt{\omega_0^2 - \alpha^2}$$

or

$$\omega_d = \omega_0 \sqrt{1 - \left(\frac{1}{2Q_0}\right)^2} \qquad [11]$$

Damping Factor

For future reference it may be helpful to note one additional relationship involving ω_0 and Q_0. The quadratic factor appearing in the numerator of Eq. [4],

$$s^2 + \frac{1}{RC}s + \frac{1}{LC}$$

may be written in terms of α and ω_0:

$$s^2 + 2\alpha s + \omega_0^2$$

In the field of system theory or automatic control theory, it is traditional to write this factor in a slightly different form that utilizes the dimensionless parameter ζ (zeta), called the **damping factor:**

$$s^2 + 2\zeta\omega_0 s + \omega_0^2$$

Comparison of these expressions allows us to relate ζ to other parameters:

$$\zeta = \frac{\alpha}{\omega_0} = \frac{1}{2Q_0} \qquad [12]$$

EXAMPLE 16.1

Calculate numerical values of ω_0, α, ω_d, and R for a parallel resonant circuit having $L = 2.5$ mH, $Q_0 = 5$, and $C = 0.01$ μF.

From Eq. [2], we see that $\omega_0 = 1/\sqrt{LC} = 200$ krad/s, while $f_0 = \omega_0/2\pi = 31.8$ kHz.

The value of α may be obtained quickly by using Eq. [10],

$$\alpha = \frac{\omega_0}{2Q_0} = \frac{2 \times 10^5}{(2 \times 5)} = 2 \times 10^4 \text{ Np/s}$$

Now we may make use of our old friend from Chap. 9,

$$\omega_d = \sqrt{\omega_0^2 - \alpha^2}$$

to find that

$$\omega_d = \sqrt{(2 \times 10^5)^2 - (2 \times 10^4)^2} = 199.0 \text{ krad/s}$$

Finally, we need a value for the parallel resistance, and Eq. [7] gives us the answer:

$$Q_0 = \omega_0 R C$$

so

$$R = \frac{Q_0}{\omega_0 C} = \frac{5}{(2 \times 10^5 \times 10^{-8})} = 2.50 \text{ k}\Omega$$

PRACTICE

16.1 A parallel resonant circuit is composed of the elements $R = 8 \text{ k}\Omega$, $L = 50 \text{ mH}$, and $C = 80 \text{ nF}$. Find (a) ω_0; (b) Q_0; (c) ω_d; (d) α; (e) ζ.

16.2 Find the values of R, L, and C in a parallel resonant circuit for which $\omega_0 = 1000 \text{ rad/s}$, $\omega_d = 998 \text{ rad/s}$, and $\mathbf{Y}_{\text{in}} = 1 \text{ mS}$ at resonance.

Ans: 16.1: 15.811 krad/s; 10.12; 15.792 krad/s; 781 Np/s; 0.0494. 16.2: 1000 Ω; 126.4 mH; 7.91 μF.

Now let us interpret Q_0 in terms of the pole-zero locations of the admittance $\mathbf{Y}(\mathbf{s})$ of the parallel RLC circuit. We will keep ω_0 constant; this may be done, for example, by changing R while holding L and C constant. As Q_0 is increased, the relationships relating α, Q_0, and ω_0 indicate that the two zeros must move closer to the $j\omega$ axis. These relationships also show that the zeros must simultaneously move away from the σ axis. The exact nature of the movement becomes clearer when we remember that the point at which $\mathbf{s} = j\omega_0$ could be located on the $j\omega$ axis by swinging an arc, centered at the origin, through one of the zeros and over to the positive $j\omega$ axis; since ω_0 is to be held constant, the radius must be constant, and the zeros must therefore move along this arc toward the positive $j\omega$ axis as Q_0 increases.

The two zeros are indicated in Fig. 16.4, and the arrows show the path they take as R increases. When R is infinite, Q_0 is also infinite, and the two zeros are found at $\mathbf{s} = \pm j\omega_0$ on the $j\omega$ axis. As R decreases, the zeros move toward the σ axis along the circular locus, joining to form a double zero on the σ axis at $\mathbf{s} = -\omega_0$ when $R = \frac{1}{2}\sqrt{L/C}$ or $Q_0 = \frac{1}{2}$. This condition may be recalled as that for critical damping, so that $\omega_d = 0$ and $\alpha = \omega_0$. Lower values of R and lower values of Q_0 cause the zeros to separate and move in opposite directions on the negative σ axis, but these low values of Q_0 are not really typical of resonant circuits and we need not track them any further.

Later, we will use the criterion $Q_0 \geq 5$ to describe a high-Q circuit. When $Q_0 = 5$, the zeros are located at $\mathbf{s} = -0.1\omega_0 \pm j0.995\omega_0$, and thus ω_0 and ω_d differ by only one-half of 1 percent.

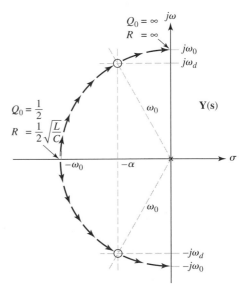

■ **FIGURE 16.4** The two zeros of the admittance $\mathbf{Y}(\mathbf{s})$, located at $\mathbf{s} = -\alpha \pm j\omega_d$, provide a semicircular locus as R increases from $\frac{1}{2}\sqrt{L/C}$ to ∞.

16.2 · BANDWIDTH AND HIGH-Q CIRCUITS

We continue our discussion of parallel resonance by defining half-power frequencies and bandwidth, and then we will make good use of these new concepts in obtaining approximate response data for high-Q circuits. The "width" of a resonance response curve, such as the one shown in Fig. 16.3, may now be defined more carefully and related to Q_0. Let us first define the two half-power frequencies ω_1 and ω_2 as those frequencies at which the magnitude of the input admittance of a parallel resonant circuit is greater than the magnitude at resonance by a factor of $\sqrt{2}$. Since the response curve of Fig. 16.3 displays the voltage produced across the parallel circuit by a sinusoidal current source as a function of frequency, the half-power frequencies also locate those points at which the voltage response is $1/\sqrt{2}$, or 0.707, times its maximum value. A similar relationship holds for the impedance magnitude. We will designate ω_1 as the *lower half-power frequency* and ω_2 as the *upper half-power frequency.*

> These names arise from the fact that a voltage which is $1/\sqrt{2}$ times the resonant voltage is equivalent to a squared voltage which is one-half the squared voltage at resonance. Thus, at the half-power frequencies, the resistor absorbs one-half the power that it does at resonance.

Bandwidth

The (half-power) *bandwidth* of a resonant circuit is defined as the difference of these two half-power frequencies.

$$\mathcal{B} \equiv \omega_2 - \omega_1 \qquad [13]$$

We tend to think of bandwidth as the "width" of the response curve, even though the curve actually extends from $\omega = 0$ to $\omega = \infty$. More exactly, the half-power bandwidth is measured by that portion of the response curve which is equal to or greater than 70.7 percent of the maximum value, as depicted in Fig. 16.5.

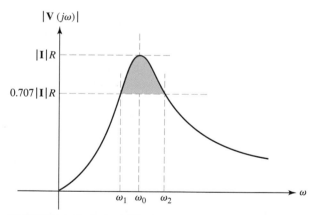

■ **FIGURE 16.5** The bandwidth of the circuit response is highlighted in green; it corresponds to the portion of the response curve greater than or equal to 70.7% of the maximum value.

Now let us express the bandwidth in terms of Q_0 and the resonant frequency. In order to do so, we first express the admittance of the parallel *RLC* circuit,

$$\mathbf{Y} = \frac{1}{R} + j\left(\omega C - \frac{1}{\omega L}\right)$$

in terms of Q_0:

$$\mathbf{Y} = \frac{1}{R} + j\frac{1}{R}\left(\frac{\omega\omega_0 CR}{\omega_0} - \frac{\omega_0 R}{\omega\omega_0 L}\right)$$

or

$$\mathbf{Y} = \frac{1}{R}\left[1 + jQ_0\left(\frac{\omega}{\omega_0} - \frac{\omega_0}{\omega}\right)\right] \qquad [14]$$

We note again that the magnitude of the admittance at resonance is $1/R$, and then realize that an admittance magnitude of $\sqrt{2}/R$ can occur only when a frequency is selected such that the imaginary part of the bracketed quantity has a magnitude of unity. Thus

$$Q_0\left(\frac{\omega_2}{\omega_0} - \frac{\omega_0}{\omega_2}\right) = 1 \qquad \text{and} \qquad Q_0\left(\frac{\omega_1}{\omega_0} - \frac{\omega_0}{\omega_1}\right) = -1$$

> Keep in mind that $\omega_2 > \omega_0$, while $\omega_1 < \omega_0$.

Solving, we have

$$\omega_1 = \omega_0\left[\sqrt{1 + \left(\frac{1}{2Q_0}\right)^2} - \frac{1}{2Q_0}\right] \qquad [15]$$

$$\omega_2 = \omega_0\left[\sqrt{1 + \left(\frac{1}{2Q_0}\right)^2} + \frac{1}{2Q_0}\right] \qquad [16]$$

Although these expressions are somewhat unwieldy, their difference provides a very simple formula for the bandwidth:

$$\mathcal{B} = \omega_2 - \omega_1 = \frac{\omega_0}{Q_0}$$

Equations [15] and [16] may be multiplied by each other to show that ω_0 is exactly equal to the geometric mean of the half-power frequencies:

$$\omega_0^2 = \omega_1\omega_2$$

or

$$\omega_0 = \sqrt{\omega_1\omega_2}$$

Circuits possessing a higher Q_0 have a narrower bandwidth, or a sharper response curve; they have greater **_frequency selectivity,_** or higher quality (factor).

Approximations for High-Q Circuits

Many resonant circuits are deliberately designed to have a large Q_0 in order to take advantage of the narrow bandwidth and high frequency selectivity associated with such circuits. When Q_0 is larger than about 5, it is possible to make some very useful approximations in the expressions for the upper and lower half-power frequencies and in the general expressions for the response in the neighborhood of resonance. Let us arbitrarily refer to a "high-Q circuit" as one for which Q_0 is equal to or greater than 5. The pole-zero configuration of $\mathbf{Y}(\mathbf{s})$ for a parallel RLC circuit having a Q_0 of about 5 is

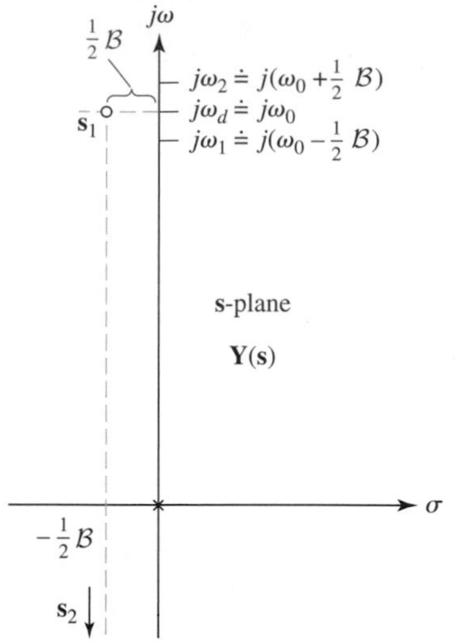

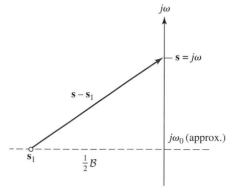

■ FIGURE 16.6 The pole-zero constellation of $Y(s)$ for a parallel *RLC* circuit. The two zeros are exactly $\frac{1}{2}B$ Np/s (or rad/s) to the left of the $j\omega$ axis and approximately $j\omega_0$ rad/s (or Np/s) from the σ axis. The upper and lower half-power frequencies are separated exactly B rad/s, and each is approximately $\frac{1}{2}B$ rad/s away from the resonant frequency and the natural resonant frequency.

shown in Fig. 16.6. Since

$$\alpha = \frac{\omega_0}{2Q_0}$$

then

$$\alpha = \tfrac{1}{2}B$$

and the locations of the two zeros s_1 and s_2 may be approximated:

$$s_{1,2} = -\alpha \pm j\omega_d$$
$$\approx -\tfrac{1}{2}B \pm j\omega_0$$

Moreover, the locations of the two half-power frequencies (on the positive $j\omega$ axis) may also be determined in a concise approximate form:

$$\omega_{1,2} = \omega_0 \left[\sqrt{1 + \left(\frac{1}{2Q_0}\right)^2} \mp \frac{1}{2Q_0}\right] \approx \omega_0\left(1 \mp \frac{1}{2Q_0}\right)$$

or

$$\omega_{1,2} \approx \omega_0 \mp \tfrac{1}{2}B \qquad [17]$$

In a high-Q circuit, therefore, each half-power frequency is located approximately one-half bandwidth from the resonant frequency; this is indicated in Fig. 16.6.

The approximate relationships for ω_1 and ω_2 in Eq. [17] may be added to each other to show that ω_0 is approximately equal to the arithmetic mean of ω_1 and ω_2 in high-Q circuits:

$$\omega_0 \approx \tfrac{1}{2}(\omega_1 + \omega_2)$$

Now let us visualize a test point slightly above $j\omega_0$ on the $j\omega$ axis. In order to determine the admittance offered by the parallel *RLC* network at this frequency, we construct the three vectors from the critical frequencies to the test point. If the test point is close to $j\omega_0$, then the vector from the pole is approximately $j\omega_0$ and that from the lower zero is nearly $j2\omega_0$. The admittance is therefore given approximately by

$$Y(s) \approx C\frac{(j2\omega_0)(s - s_1)}{j\omega_0} \approx 2C(s - s_1) \qquad [18]$$

where C is the capacitance, as shown in Eq. [4]. In order to determine a useful approximation for the vector $(s - s_1)$, let us consider an enlarged view of that portion of the **s**-plane in the neighborhood of the zero s_1 (Fig. 16.7).

In terms of its cartesian components, we see that

$$s - s_1 \approx \tfrac{1}{2}B + j(\omega - \omega_0)$$

where this expression would be exact if ω_0 were replaced by ω_d. We now substitute this equation in the approximation for $Y(s)$, Eq. [18], and factor out $\tfrac{1}{2}B$:

$$Y(s) \approx 2C\left(\tfrac{1}{2}B\right)\left(1 + j\frac{\omega - \omega_0}{\tfrac{1}{2}B}\right)$$

■ FIGURE 16.7 An enlarged portion of the pole-zero constellation for $Y(s)$ of a high-Q_0 parallel *RLC* circuit.

or

$$\mathbf{Y(s)} \approx \frac{1}{R} \left(1 + j\frac{\omega - \omega_0}{\frac{1}{2}\mathcal{B}} \right)$$

The fraction $(\omega - \omega_0)/(\frac{1}{2}\mathcal{B})$ may be interpreted as the "number of half-bandwidths off resonance" and abbreviated by N. Thus,

$$\mathbf{Y(s)} \approx \frac{1}{R}(1 + jN) \qquad [19]$$

where

$$N = \frac{\omega - \omega_0}{\frac{1}{2}\mathcal{B}} \qquad [20]$$

At the upper half-power frequency, $\omega_2 \approx \omega_0 + \frac{1}{2}\mathcal{B}$, $N = +1$, and we are one half-bandwidth above resonance. For the lower half-power frequency, $\omega_1 \approx \omega_0 - \frac{1}{2}\mathcal{B}$, so that $N = -1$, locating us one half-bandwidth below resonance.

Equation [19] is much easier to use than the exact relationships we have had up to now. It shows that the magnitude of the admittance is

$$|\mathbf{Y}(j\omega)| \approx \frac{1}{R}\sqrt{1 + N^2}$$

while the angle of $\mathbf{Y}(j\omega)$ is given by the inverse tangent of N:

$$\text{ang } \mathbf{Y}(j\omega) \approx \tan^{-1} N$$

EXAMPLE 16.2

Determine the approximate value of the admittance of a parallel RLC network for which $R = 40$ kΩ, $L = 1$ H, and $C = \frac{1}{64}$ μF if the operating frequency is $\omega = 8.2$ krad/s.

▶ **Identify the goal of the problem.**
We are asked to determine the approximate value of $\mathbf{Y(s)}$ at $\omega = 8.2$ krad/s for a simple RLC network. This implies that Q_0 must be at least 5, and the operating frequency is not far from the resonant frequency.

▶ **Collect the known information.**
The values for R, L, and C are provided, as well as the frequency at which to evaluate $\mathbf{Y(s)}$. This is sufficient to compute the admittance using either the exact or approximate expressions.

▶ **Devise a plan.**
To use our approximate expression for the admittance, we must first determine Q_0, the quality factor at resonance, as well as the bandwidth.

The resonant frequency ω_0 is given by Eq. [2] as $1/\sqrt{LC} = 8$ krad/s. Thus, $Q_0 = \omega_0 RC = 5$, and the bandwidth is $\omega_0/Q_0 = 1.6$ krad/s. The value of Q_0 for this circuit is sufficient to employ "high-Q" approximations.

(Continued on next page)

▶ **Construct an appropriate set of equations.**
Equation [19] states that

$$\mathbf{Y}(\mathbf{s}) \approx \frac{1}{R}(1 + jN)$$

so

$$|\mathbf{Y}(j\omega)| \approx \frac{1}{R}\sqrt{1 + N^2} \qquad \text{and} \qquad \text{ang } \mathbf{Y}(j\omega) \approx \tan^{-1} N$$

▶ **Determine if additional information is required.**
We still require N, which tells us how many half-bandwidths ω is from the resonant frequency ω_0:

$$N = (8.2 - 8)/0.8 = 0.25$$

▶ **Attempt a solution.**
Now we are ready to employ our approximate relationships for the magnitude and angle of the network admittance,

$$\text{ang } \mathbf{Y} \approx \tan^{-1} 0.25 = 14.04°$$

and

$$|\mathbf{Y}| \approx 25\sqrt{1 + (0.25)^2} = 25.77 \ \mu\text{S}$$

▶ **Verify the solution. Is it reasonable or expected?**
An exact calculation of the admittance using Eq. [1] shows that

$$\mathbf{Y}(j8200) = 25.75\underline{/13.87°} \ \mu\text{S}$$

The approximate method therefore leads to values of admittance magnitude and angle that are reasonably accurate (better than 2 percent) for this frequency.

PRACTICE

16.3 A marginally high-Q parallel resonant circuit has $f_0 = 440$ Hz with $Q_0 = 6$. Use Eqs. [15] and [16] to obtain accurate values for (a) f_1; (b) f_2. Now use Eq. [17] to calculate approximate values for (c) f_1; (d) f_2.

Ans: 404.9 Hz; 478.2 Hz; 403.3 Hz; 476.7 Hz.

Our intention is to use these approximations for high-Q circuits near resonance. We have already agreed that we will let "high-Q" imply $Q_0 \geq 5$, but how near is "near"? It can be shown that the error in magnitude or phase is less than 5 percent if $Q_0 \geq 5$ and $0.9\omega_0 \leq \omega \leq 1.1\omega_0$. Although this narrow band of frequencies may seem to be prohibitively small, it is usually more than sufficient to contain the range of frequencies in which we are most interested. For example, an AM radio usually contains a circuit tuned to a resonant frequency of 455 kHz with a half-power bandwidth of 10 kHz. This circuit must then have a value of 45.5 for Q_0, and the half-power frequencies are about 450 and 460 kHz. Our approximations, however, are valid from 409.5 to 500.5 kHz (with errors less than 5 percent), a range that covers essentially all the peaked portion of the response curve; only in the remote "tails" of the response curve do the approximations lead to unreasonably large errors.[1]

We conclude our coverage of the *parallel* resonant circuit by reviewing some key conclusions we have reached:

- The resonant frequency ω_0 is the frequency at which the imaginary part of the input admittance becomes zero, or the angle of the admittance becomes zero. For this circuit, $\omega_0 = 1/\sqrt{LC}$.

- The circuit's figure of merit Q_0 is defined as 2π times the ratio of the maximum energy stored in the circuit to the energy lost each period in the circuit. For this circuit, $Q_0 = \omega_0 RC$.

- We defined two half-power frequencies, ω_1 and ω_2, as the frequencies at which the admittance magnitude is $\sqrt{2}$ times the minimum admittance magnitude. (These are also the frequencies at which the voltage response is 70.7 percent of the maximum response.)

- The exact expressions for ω_1 and ω_2 are

$$\omega_{1,2} = \omega_0 \left[\sqrt{1 + \left(\frac{1}{2Q_0}\right)^2} \mp \frac{1}{2Q_0} \right]$$

- The approximate (high-Q_0) expressions for ω_1 and ω_2 are

$$\omega_{1,2} \approx \omega_0 \mp \frac{1}{2}\mathcal{B}$$

- The half-power bandwidth $\mathcal{B}$ is given by

$$\mathcal{B} = \omega_2 - \omega_1 = \frac{\omega_0}{Q_0}$$

- The input admittance may also be expressed in approximate form for high-Q circuits:

$$\mathbf{Y} \approx \frac{1}{R}(1 + jN) = \frac{1}{R}\sqrt{1 + N^2}\underline{/\tan^{-1} N}$$

where N is defined as the number of half-bandwidths off resonance, or

$$N = \frac{\omega - \omega_0}{\frac{1}{2}\mathcal{B}}$$

This approximation is valid for $0.9\omega_0 \le \omega \le 1.1\omega_0$.

16.3 SERIES RESONANCE

Although we probably find less use for the series RLC circuit than we do for the parallel RLC circuit, it is still worthy of our attention. We will consider the circuit shown in Fig. 16.8. It should be noted that the various circuit

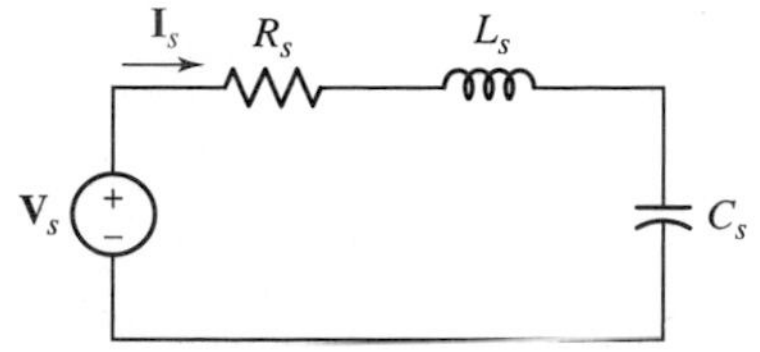

(1) At frequencies remote from resonance, we are often satisfied with very rough results; greater accuracy is not always necessary.

■ **FIGURE 16.8** A series resonant circuit.

elements are given the subscript s (for series) for the time being in order to avoid confusing them with the parallel elements when the circuits are compared.

Our discussion of parallel resonance occupied two sections of considerable length. We could now give the series RLC circuit the same kind of treatment, but it is much cleverer to avoid such needless repetition and use the concept of duality. For simplicity, let us concentrate on the conclusions presented in the last paragraph of the preceding section on parallel resonance. The important results are contained there, and the use of dual language enables us to transcribe this paragraph to present the important results for the series RLC circuit.

"We conclude our coverage of the *series* resonant circuit by reviewing some key conclusions we have reached:

Again, this paragraph is the same as the last paragraph of Sec. 16.2, with the parallel *RLC* language converted to series *RLC* language using duality (hence the quotation marks).

- The resonant frequency ω_0 is the frequency at which the imaginary part of the input impedance becomes zero, or the angle of the impedance becomes zero. For this circuit, $\omega_0 = 1/\sqrt{C_s L_s}$.

- The circuit's figure of merit Q_0 is defined as 2π times the ratio of the maximum energy stored in the circuit to the energy lost each period in the circuit. For this circuit, $Q_0 = \omega_0 L_S / R_S$.

- We defined two half-power frequencies, ω_1 and ω_2, as the frequencies at which the impedance magnitude is $\sqrt{2}$ times the minimum impedance magnitude. (These are also the frequencies at which the current response is 70.7 percent of the maximum response.)

- The exact expressions for ω_1 and ω_2 are

$$\omega_{1,2} = \omega_0 \left[\sqrt{1 + \left(\frac{1}{2Q_0} \right)^2} \mp \frac{1}{2Q_0} \right]$$

- The approximate (high-Q_0) expressions for ω_1 and ω_2 are

$$\omega_{1,2} \approx \omega_0 \mp \frac{1}{2}\mathcal{B}$$

- The half-power bandwidth $\mathcal{B}$ is given by

$$\mathcal{B} = \omega_2 - \omega_1 = \frac{\omega_0}{Q_0}$$

- The input admittance may also be expressed in approximate form for high-Q circuits:

$$\mathbf{Y} \approx \frac{1}{R}(1 + jN) = \frac{1}{R}\sqrt{1 + N^2} \underline{/\tan^{-1} N}$$

where N is defined as the number of half-bandwidths off resonance, or

$$N = \frac{\omega - \omega_0}{\frac{1}{2}\mathcal{B}}$$

This approximation is valid for $0.9\omega_0 \leq \omega \leq 1.1\omega_0$."

From this point on, we will no longer identify series resonant circuits by use of the subscript s, unless clarity requires it.

EXAMPLE **16.3**

The voltage $v_s = 100 \cos \omega t$ mV is applied to a series resonant circuit composed of a 10 Ω resistance, a 200 nF capacitance, and a 2 mH inductance. Use both exact and approximate methods to calculate the current amplitude if $\omega = 48$ krad/s.

The resonant frequency of the circuit is given by

$$\omega_0 = \frac{1}{\sqrt{LC}} = \frac{1}{\sqrt{(2 \times 10^{-3})(200 \times 10^{-9})}} = 50 \text{ krad/s}$$

Since we are operating at $\omega = 48$ krad/s, which is within 10 percent of the resonant frequency, it is reasonable to apply our approximate relationships to estimate the equivalent impedance of the network provided that we find that we are working with a high-Q circuit:

$$\mathbf{Z}_{\text{eq}} \approx R\sqrt{1 + N^2}\underline{/\tan^{-1} N}$$

where N is computed once we determine Q_0. This is a series circuit, so

$$Q_0 = \frac{\omega_0 L}{R} = \frac{(50 \times 10^3)(2 \times 10^{-3})}{10} = 10$$

which qualifies as a high-Q circuit. Thus,

$$\mathcal{B} = \frac{\omega_0}{Q_0} = \frac{50 \times 10^3}{10} = 5 \text{ krad/s}$$

The number of half-bandwidths off-resonance (N) is therefore

$$N = \frac{\omega - \omega_0}{\mathcal{B}/2} = \frac{48 - 50}{2.5} = -0.8$$

Thus,

$$\mathbf{Z}_{\text{eq}} \approx R\sqrt{1 + N^2}\underline{/\tan^{-1} N} = 12.81\underline{/-38.66°} \ \Omega$$

The approximate current magnitude is then

$$\frac{|\mathbf{V}_s|}{|\mathbf{Z}_{\text{eq}}|} = \frac{100}{12.81} = 7.806 \text{ mA}$$

Using the exact expressions, we find that $\mathbf{I} = 7.746\underline{/39.24°}$ mA and thus

$$|\mathbf{I}| = 7.746 \text{ mA}.$$

PRACTICE

16.4 A series resonant circuit has a bandwidth of 100 Hz and contains a 20 mH inductance and a 2 μF capacitance. Determine (a) f_0; (b) Q_0; (c) $\mathbf{Z}_{\text{in}}$ at resonance; (d) f_2.

Ans: 796 Hz; 7.96; 12.57 + $j0$ Ω; 846 Hz (approx.).

The series resonant circuit is characterized by a minimum impedance at resonance, whereas the parallel resonant circuit produces a maximum resonant impedance. The latter circuit provides inductor currents and capacitor currents at resonance which have amplitudes Q_0 times as great as the source current; the series resonant circuit provides inductor voltages and capacitor voltages which are greater than the source voltage by the factor Q_{0s}. The series circuit thus provides voltage amplification at resonance.

A comparison of our results for series and parallel resonance, as well as the exact and approximate expressions we have developed, appears in Table 16.1.

TABLE 16.1 A Short Summary of Resonance

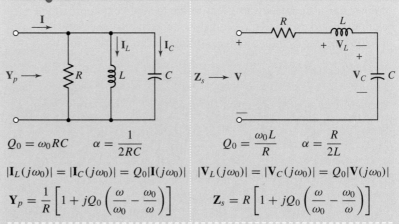

$Q_0 = \omega_0 RC \qquad \alpha = \dfrac{1}{2RC}$ $Q_0 = \dfrac{\omega_0 L}{R} \qquad \alpha = \dfrac{R}{2L}$

$|\mathbf{I}_L(j\omega_0)| = |\mathbf{I}_C(j\omega_0)| = Q_0|\mathbf{I}(j\omega_0)|$ $|\mathbf{V}_L(j\omega_0)| = |\mathbf{V}_C(j\omega_0)| = Q_0|\mathbf{V}(j\omega_0)|$

$\mathbf{Y}_p = \dfrac{1}{R}\left[1 + jQ_0\left(\dfrac{\omega}{\omega_0} - \dfrac{\omega_0}{\omega}\right)\right]$ $\mathbf{Z}_s = R\left[1 + jQ_0\left(\dfrac{\omega}{\omega_0} - \dfrac{\omega_0}{\omega}\right)\right]$

Exact expressions

$$\omega_0 = \frac{1}{\sqrt{LC}} = \sqrt{\omega_1 \omega_2}$$

$$\omega_d = \sqrt{\omega_0^2 - \alpha^2} = \omega_0\sqrt{1 - \left(\frac{1}{2Q_0}\right)^2}$$

$$\omega_{1,2} = \omega_0\left[\sqrt{1 + \left(\frac{1}{2Q_0}\right)^2} \mp \frac{1}{2Q_0}\right]$$

$$N = \frac{\omega - \omega_0}{\frac{1}{2}\mathcal{B}}$$

$$\mathcal{B} = \omega_2 - \omega_1 = \frac{\omega_0}{Q_0} = 2\alpha$$

Approximate expressions

$$(Q_0 \geq 5 \qquad 0.9\omega_0 \leq \omega \leq 1.1\omega_0)$$

$$\omega_d \approx \omega_0$$

$$\omega_{1,2} \approx \omega_0 \mp \tfrac{1}{2}\mathcal{B}$$

$$\omega_0 \approx \tfrac{1}{2}(\omega_1 + \omega_2)$$

$$\mathbf{Y}_p \approx \frac{\sqrt{1 + N^2}}{R}\underline{/\tan^{-1} N}$$

$$\mathbf{Z}_s \approx R\sqrt{1 + N^2}\underline{/\tan^{-1} N}$$

16.4 OTHER RESONANT FORMS

The parallel and series *RLC* circuits of the previous two sections represent *idealized* resonant circuits; they are no more than useful, *approximate* representations of a physical circuit which might be constructed by combining a coil of wire, a carbon resistor, and a tantalum capacitor in parallel or series. The degree of accuracy with which the idealized model fits the *actual* circuit depends on the operating frequency range, the *Q* of the circuit, the materials present in the physical elements, the element sizes, and many other factors. We are not studying the techniques for determining the best model of a given physical circuit, for this requires some knowledge of electromagnetic field theory and the properties of materials; we are, however, concerned with the problem of reducing a more complicated model to one of the two simpler models with which we are more familiar.

The network shown in Fig. 16.9a is a reasonably accurate model for the parallel combination of a physical inductor, capacitor, and resistor. The resistor labeled R_1 is a hypothetical resistor that is included to account for the ohmic, core, and radiation losses of the physical coil. The losses in the dielectric within the physical capacitor, as well as the resistance of the physical resistor in the given *RLC* circuit, are accounted for by the resistor labeled R_2. In this model, *there is no way* to combine elements and produce a simpler model which is equivalent to the original model *for all frequencies*. We will show, however, that a simpler equivalent may be constructed which is valid over a frequency band that is usually large enough to include all frequencies of interest. The equivalent will take the form of the network shown in Fig. 16.9b.

Before we learn how to develop such an equivalent circuit, let us first consider the given circuit, Fig. 16.9a. The resonant radian frequency for this network is *not* $1/\sqrt{LC}$, although if R_1 is sufficiently small it may be very close to this value. The definition of resonance is unchanged, and we may determine the resonant frequency by setting the imaginary part of the input admittance equal to zero:

$$\text{Im}\{\mathbf{Y}(j\omega)\} = \text{Im}\left\{\frac{1}{R_2} + j\omega C + \frac{1}{R_1 + j\omega L}\right\} = 0$$

or

$$\text{Im}\left\{\frac{1}{R_2} + j\omega C + \frac{1}{R_1 + j\omega L}\frac{R_1 - j\omega L}{R_1 - j\omega L}\right\}$$

$$= \text{Im}\left\{\frac{1}{R_2} + j\omega C + \frac{R_1 - j\omega L}{R_1^2 + \omega^2 L^2}\right\} = 0$$

Thus, we have the resonance condition that

$$C = \frac{L}{R_1^2 + \omega^2 L^2}$$

and so

$$\omega_0 = \sqrt{\frac{1}{LC} - \left(\frac{R_1}{L}\right)^2} \qquad [21]$$

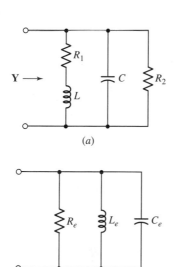

■ FIGURE 16.9 (*a*) A useful model of a physical network which consists of a physical inductor, capacitor, and resistor in parallel. (*b*) A network which can be equivalent to part *a* over a narrow frequency band.

We note that ω_0 is less than $1/\sqrt{LC}$, but sufficiently small values of the ratio R_1/L may result in a negligible difference between ω_0 and $1/\sqrt{LC}$.

The maximum magnitude of the input impedance also deserves consideration. It is not R_2, and it does not occur at ω_0 (or at $\omega = 1/\sqrt{LC}$). The proof of these statements will not be shown, because the expressions soon become algebraically cumbersome; the theory, however, is straightforward. Let us be content with a numerical example.

EXAMPLE 16.4

Using the values $R_1 = 2\ \Omega$, $L = 1$ H, $C = 125$ mF, and $R_2 = 3\ \Omega$ for Fig. 16.9a, determine the resonant frequency and the impedance at resonance.

Substituting the appropriate values in Eq. [21], we find

$$\omega_0 = \sqrt{8 - 2^2} = 2 \text{ rad/s}$$

and this enables us to calculate the input admittance,

$$\mathbf{Y} = \frac{1}{3} + j2\left(\frac{1}{8}\right) + \frac{1}{2 + j(2)(1)} = \frac{1}{3} + \frac{1}{4} = 0.583 \text{ S}$$

and then the input impedance at resonance:

$$\mathbf{Z}(j2) = \frac{1}{0.583} = 1.714\ \Omega$$

At the frequency which would be the resonant frequency if R_1 were zero,

$$\frac{1}{\sqrt{LC}} = 2.83 \text{ rad/s}$$

the input impedance would be

$$\mathbf{Z}(j2.83) = 1.947\underline{/-13.26°}\ \Omega$$

As can be seen in Fig. 16.10, however, the frequency at which the *maximum* impedance magnitude occurs, indicated by ω_m, can be determined to be $\omega_m = 3.26$ rad/s, and the *maximum* impedance magnitude is

$$\mathbf{Z}(j3.26) = 1.980\underline{/-21.4°}\ \Omega$$

The impedance magnitude at resonance and the maximum magnitude differ by about 16 percent. Although it is true that such an error may be neglected occasionally in practice, it is too large to neglect on an exam. The later work in this section will show that the Q of the inductor-resistor combination at 2 rad/s is unity; this low value accounts for the 16 percent discrepancy.

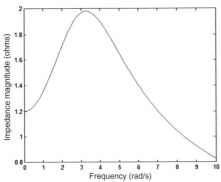

■ **FIGURE 16.10** Plot of |**Z**| vs. ω, generated using the following MATLAB script:

```
EDU» omega = linspace(0,10,100);
EDU» for i = 1:100
Y(i) = 1/3 + j*omega(i)/8 + 1/(2 + j*omega(i));
Z(i) = 1/Y(i);
end
EDU» plot(omega,abs(Z));
EDU» xlabel('frequency (rad/s)');
EDU» ylabel('impedance magnitude (ohms)');
```

PRACTICE

16.5 Referring to the circuit of Fig. 16.9a, let $R_1 = 1$ kΩ and $C = 2.533$ pF. Determine the inductance necessary to select a resonant frequency of 1 MHz. (*Hint:* Recall that $\omega = 2\pi f$.)

Ans: 10 mH.

Equivalent Series and Parallel Combinations

In order to transform the given circuit of Fig. 16.9a into an equivalent of the form shown in Fig. 16.9b, we must discuss the Q of a simple series or parallel combination of a resistor and a reactor (inductor or capacitor). We first consider the series circuit shown in Fig. 16.11a. The Q of this network is again defined as 2π times the ratio of the maximum stored energy to the energy lost each period, but the Q may be evaluated at any frequency we choose. In other words, Q is a function of ω. It is true that we will choose to evaluate it at a frequency which is, or apparently is, the resonant frequency of some network of which the series arm is a part. This frequency, however, is not known until a more complete circuit is available. The reader is encouraged to show that the Q of this series arm is $|X_s|/R_s$, whereas the Q of the parallel network of Fig. 16.11b is $R_p/|X_p|$.

Let us now carry out the details necessary to find values for R_p and X_p so that the parallel network of Fig. 16.11b is equivalent to the series network of Fig. 16.11a at some single specific frequency. We equate $\mathbf{Y}_s$ and $\mathbf{Y}_p$,

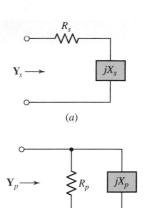

■ **FIGURE 16.11** (a) A series network which consists of a resistance R_s and an inductive or capacitive reactance X_s may be transformed into (b) a parallel network such that $\mathbf{Y}_s = \mathbf{Y}_p$ at one specific frequency. The reverse transformation is equally possible.

$$\mathbf{Y}_s = \frac{1}{R_s + jX_s} = \frac{R_s - jX_s}{R_s^2 + X_s^2}$$

$$= \mathbf{Y}_p = \frac{1}{R_p} - j\frac{1}{X_p}$$

and obtain

$$R_p = \frac{R_s^2 + X_s^2}{R_s}$$

$$X_p = \frac{R_s^2 + X_s^2}{X_s}$$

Dividing these two expressions, we find

$$\frac{R_p}{X_p} = \frac{X_s}{R_s}$$

It follows that the Q's of the series and parallel networks must be equal:

$$Q_p = Q_s = Q$$

The transformation equations may therefore be simplified:

$$R_p = R_s(1 + Q^2) \qquad [22]$$

$$X_p = X_s\left(1 + \frac{1}{Q^2}\right) \qquad [23]$$

R_s and X_s may also be found if R_p and X_p are the given values; the transformation in either direction may be performed.

If $Q \geq 5$, little error is introduced by using the approximate relationships

$$R_p \approx Q^2 R_s \qquad [24]$$

$$X_p \approx X_s \qquad (C_p \approx C_s \quad \text{or} \quad L_p \approx L_s) \qquad [25]$$

EXAMPLE 16.5

Find the parallel equivalent of the series combination of a 100 mH inductor and a 5 Ω resistor at a frequency of 1000 rad/s. Details of the network to which this series combination is connected are unavailable.

At $\omega = 1000$ rad/s, $X_s = 1000(100 \times 10^{-3}) = 100$ Ω. The Q of this series combination is

$$Q = \frac{X_s}{R_s} = \frac{100}{5} = 20$$

Since the Q is sufficiently high (20 is much greater than 5), we use Eqs. [24] and [25] to obtain

$$R_p \approx Q^2 R_s = 2000 \ \Omega \qquad \text{and} \qquad L_p \approx L_s = 100 \text{ mH}$$

Our assertion here is that a 100 mH inductor in series with a 5 Ω resistor provides *essentially the same* input impedance as does a 100 mH inductor in parallel with a 2000 Ω resistor at the frequency 1000 rad/s.

To check the accuracy of the equivalence, let us evaluate the input impedance for each network at 1000 rad/s. We find

$$\mathbf{Z}_s(j1000) = 5 + j100 = 100.1\underline{/87.1°} \ \Omega$$

$$\mathbf{Z}_p(j1000) = \frac{2000(j100)}{2000 + j100} = 99.9\underline{/87.1°} \ \Omega$$

and conclude that the accuracy of our approximation at the transformation frequency is pretty impressive. The accuracy at 900 rad/s is also reasonably good, because

$$\mathbf{Z}_s(j900) = 90.1\underline{/86.8°} \ \Omega$$
$$\mathbf{Z}_p(j900) = 89.9\underline{/87.4°} \ \Omega$$

PRACTICE

16.6 At $\omega = 1000$ rad/s, find a parallel network that is equivalent to the series combination in Fig. 16.12a.

16.7 Find a series equivalent for the parallel network shown in Fig. 16.12b, assuming $\omega = 1000$ rad/s.

Ans: 16.6: 8 H, 640 kΩ; 16.7: 5 H, 250 Ω.

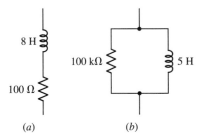

■ **FIGURE 16.12** (a) A series network for which an equivalent parallel network (at $\omega = 1000$ rad/s) is needed. (b) A parallel network for which an equivalent series network (at $\omega = 1000$ rad/s) is needed.

An "ideal" meter is an instrument that measures a particular quantity of interest without disturbing the circuit being tested. Although this is impossible, modern instruments can come very close to being ideal in this respect.

As a further example of the replacement of a more complicated resonant circuit by an equivalent series or parallel *RLC* circuit, let us consider a problem in electronic instrumentation. The simple series *RLC* network in Fig. 16.13a is excited by a sinusoidal voltage source at the network's resonant frequency. The effective (rms) value of the source voltage is 0.5 V, and we wish to measure the effective value of the voltage across the capacitor with an electronic voltmeter (VM) having an internal resistance of 100,000 Ω. That is, an equivalent representation of the voltmeter is an ideal voltmeter in parallel with a 100 kΩ resistor.

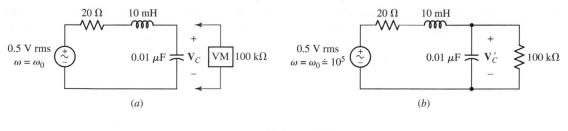

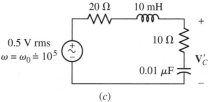

■ **FIGURE 16.13** (*a*) A given series resonant circuit in which the capacitor voltage is to be measured by a nonideal electronic voltmeter. (*b*) The effect of the voltmeter is included in the circuit; it reads V'_c. (*c*) A series resonant circuit is obtained when the parallel *RC* network in part *b* is replaced by the series *RC* network which is equivalent at 10^5 rad/s.

Before the voltmeter is connected, we compute that the resonant frequency is 10^5 rad/s, $Q_0 = 50$, the current is 25 mA, and the rms capacitor voltage is 25 V. (As indicated at the end of Sec. 16.3, this voltage is Q_0 times the applied voltage.) Thus, if the voltmeter were ideal, it would read 25 V when connected across the capacitor.

However, when the actual voltmeter is connected, the circuit shown in Fig. 16.13*b* results. In order to obtain a series *RLC* circuit, it is now necessary to replace the parallel *RC* network with a series *RC* network. Let us assume that the *Q* of this *RC* network is sufficiently high that the equivalent series capacitor will be the same as the given parallel capacitor. We do this in order to approximate the resonant frequency of the final series *RLC* circuit. Thus, if the series *RLC* circuit also contains a 0.01 μF capacitor, the resonant frequency remains 10^5 rad/s. We need to know this estimated resonant frequency in order to calculate the *Q* of the parallel *RC* network; it is

$$Q = \frac{R_p}{|X_p|} = \omega R_p C_p = 10^5 (10^5)(10^{-8}) = 100$$

Since this value is greater than 5, our vicious circle of assumptions is justified, and the equivalent series *RC* network consists of the capacitor $C_s = 0.01\ \mu$F and the resistor

$$R_s \approx \frac{R_p}{Q^2} = 10\ \Omega$$

Hence, the equivalent circuit of Fig. 16.13*c* is obtained. The resonant *Q* of this circuit is now only 33.3, and thus the voltage across the capacitor in the circuit of Fig. 16.13*c* is $16\frac{2}{3}$ V. But we need to find $|\mathbf{V}'_C|$, the voltage across the series *RC* combination; we obtain

$$|\mathbf{V}'_C| = \frac{0.5}{30}|10 - j1000| = 16.67\ \text{V}$$

The capacitor voltage and $|\mathbf{V}'_C|$ are essentially equal, since the voltage across the 10 Ω resistor is quite small.

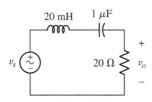

■ **FIGURE 16.14** A first model for a 20 mH inductor, a 1 μF capacitor, and a 20 Ω resistor in series with a voltage generator.

The final conclusion must be that an apparently good voltmeter may still produce a severe effect on the response of a high-Q resonant circuit. A similar effect may occur when a nonideal ammeter is inserted in the circuit.

We wrap up this section with a technical fable.

Once upon a time there was a student named Sean, who had a professor identified simply as Dr. Abel.

In the laboratory one afternoon, Dr. Abel gave Sean three practical circuit devices: a resistor, an inductor, and a capacitor, having nominal element values of 20 Ω, 20 mH, and 1 μF. The student was asked to connect a variable-frequency voltage source to the series combination of these three elements, to measure the resultant voltage across the resistor as a function of frequency, and then to calculate numerical values for the resonant frequency, the Q at resonance, and the half-power bandwidth. The student was also asked to predict the results of the experiment before making the measurements.

Sean, whose normally clear mental processes were sometimes overcome with circuit analysis anxiety, drew an equivalent circuit for this problem that was like the circuit of Fig. 16.14, and then calculated:

$$f_0 = \frac{1}{2\pi \sqrt{LC}} = \frac{1}{2\pi \sqrt{20 \times 10^{-3} \times 10^{-6}}} = 1125 \text{ Hz}$$

$$Q_0 = \frac{\omega_0 L}{R} = 7.07$$

$$\mathcal{B} = \frac{f_0}{Q_0} = 159 \text{ Hz}$$

Next, Sean made the measurements that Dr. Abel requested, compared them with the predicted values, and then felt a strong urge to transfer to the business school. The results were

$$f_0 = 1000 \text{ Hz} \qquad Q_0 = 0.625 \qquad \mathcal{B} = 1600 \text{ Hz}$$

Sean knew that discrepancies of this magnitude could not be characterized as being "within engineering accuracy" or "due to meter errors." Sadly, the results were handed to the professor.

Remembering many past errors in judgment, some of which were even (possibly) self-made, Dr. Abel smiled kindly and called Sean's attention to the Q-meter (or impedance bridge) which is present in most well-equipped laboratories, and suggested that it might be used to find out what these practical circuit elements really looked like at some convenient frequency near resonance—1000 Hz, for example.

Upon doing so, Sean discovered that the resistor had a measured value of 18 Ω and the inductor was 21.4 mH with a Q of 1.2, while the capacitor had a capacitance of 1.41 μF and a dissipation factor (the reciprocal of Q) equal to 0.123.

So, with the hope that springs eternal within the heart of every engineering undergraduate, Sean reasoned that a better model for the practical inductor would be 21.4 mH in series with $\omega L/Q = 112 \ \Omega$, while a more appropriate model for the capacitor would be 1.41 μF in series with $1/\omega C Q = 13.9 \ \Omega$. Using these data, Sean prepared the modified circuit

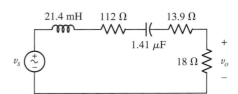

FIGURE 16.15 An improved model in which more accurate values are used and the losses in the inductor and capacitor are acknowledged.

model shown as Fig. 16.15 and calculated a new set of predicted values:

$$f_0 = \frac{1}{2\pi\sqrt{21.4 \times 10^{-3} \times 1.41 \times 10^{-6}}} = 916 \text{ Hz}$$

$$Q_0 = \frac{2\pi \times 916 \times 21.4 \times 10^{-3}}{143.9} = 0.856$$

$$\mathcal{B} = 916/0.856 = 1070 \text{ Hz}$$

Since these results were much closer to the measured values, Sean was much happier. Dr. Abel, however, being a stickler for detail, pondered the differences in the predicted and measured values for both Q_0 and the bandwidth. *"Have you,"* Dr. Abel asked, *"given any consideration to the output impedance of the voltage source?"* *"Not yet,"* said Sean, trotting back to the laboratory bench.

It turned out that the output impedance in question was 50 Ω and so Sean added this value to the circuit diagram, as shown in Fig. 16.16. Using the new equivalent resistance value of 193.9 Ω, improved values for Q_0 and $\mathcal{B}$ were then obtained:

$$Q_0 = 0.635 \qquad \mathcal{B} = 1442 \text{ Hz}$$

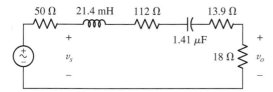

FIGURE 16.16 The final model also contains the output resistance of the voltage source.

Since all the theoretical and experimental values now agreed within 10 percent, Sean was once again an enthusiastic, confident engineering student, motivated to start homework early and read the textbook prior to class.[2] Dr. Abel simply nodded her head agreeably as she moralized:

> *When using real devices,*
> *Watch the models that you choose;*
> *Think well before you calculate,*
> *And mind your Z's and Q's!*

PRACTICE

16.8 The series combination of 10 Ω and 10 nF is in parallel with the series combination of 20 Ω and 10 mH. (*a*) Find the approximate resonant frequency of the parallel network. (*b*) Find the Q of the RC branch. (*c*) Find the Q of the RL branch. (*d*) Find the three-element equivalent of the original network.

Ans: 10^5 rad/s; 100; 50; 10 nF ‖ 10 mH ‖ 33.3 kΩ.

(2) Okay, this last part is a bit much. Sorry about that.

16.5 SCALING

Some of the examples and problems that we have been solving have in-volved circuits containing passive element values ranging around a few ohms, a few henrys, and a few farads. The applied frequencies were a few radians per second. These particular numerical values were used not be-cause they are those commonly met in practice, but because arithmetic ma-nipulations are so much easier than they would be if it were necessary to carry along various powers of 10 throughout the calculations. The scaling procedures that will be discussed in this section enable us to analyze net-works composed of practical-sized elements by scaling the element values to permit more convenient numerical calculations. We will consider both *magnitude scaling* and *frequency scaling.*

Let us select the parallel resonant circuit shown in Fig. 16.17a as our example. The impractical element values lead to the unlikely response curve drawn as Fig. 16.17b; the maximum impedance is 2.5 Ω, the resonant fre-quency is 1 rad/s, Q_0 is 5, and the bandwidth is 0.2 rad/s. These numerical val-ues are much more characteristic of the electrical analog of some mechanical system than they are of any basically electrical device. We have convenient numbers with which to calculate, but an impractical circuit to construct.

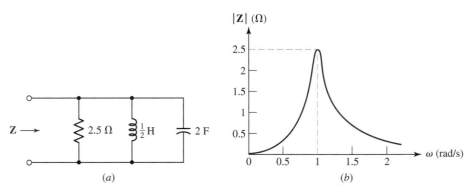

■ **FIGURE 16.17** (*a*) A parallel resonant circuit used as an example to illustrate magnitude and frequency scaling. (*b*) The magnitude of the input impedance is shown as a function of frequency.

Recall that "ordinate" refers to the vertical axis and "abscissa" refers to the horizontal axis.

Let us assume that our goal is to scale this network in such a way as to provide an impedance maximum of 5000 Ω at a resonant frequency of 5×10^6 rad/s, or 796 kHz. In other words, we may use the same response curve shown in Fig. 16.17b if every number on the *ordinate* scale is increased by a factor of 2000 and every number on the *abscissa* scale is increased by a factor of 5×10^6. We will treat this as two problems: (1) scaling in magnitude by a factor of 2000 and (2) scaling in frequency by a factor of 5×10^6.

Magnitude scaling is defined as the process by which the impedance of a two-terminal network is increased by a factor of K_m, the frequency re-maining constant. The factor K_m is real and positive; it may be greater or smaller than unity. We will understand that the shorter statement *"the net-work is scaled in magnitude by a factor of 2"* indicates that the impedance of the new network is to be twice that of the old network at any frequency. Let us now determine how we must scale each type of passive element. To increase the input impedance of a network by a factor of K_m, it is sufficient to increase the impedance of each element in the network by this same factor. Thus, a resistance R must be replaced by a resistance $K_m R$. Each inductance must also exhibit an impedance which is K_m times as great at any frequency. In order to increase an impedance sL by a factor of K_m when **s** remains constant, the inductance L must be replaced by an inductance $K_m L$. In a similar manner, each capacitance C must be replaced by a capacitance C/K_m. In summary, these changes will produce a network which is scaled in magnitude by a factor of K_m:

$$\left. \begin{array}{l} R \to K_m R \\ L \to K_m L \\ C \to \dfrac{C}{K_m} \end{array} \right\} \quad \text{magnitude scaling}$$

When each element in the network of Fig. 16.17a is scaled in magnitude by a factor of 2000, the network shown in Fig. 16.18a results. The response curve shown in Fig. 16.18b indicates that no change in the previously drawn response curve need be made other than a change in the scale of the ordinate.

Let us now take this new network and scale it in frequency. We define frequency scaling as the process by which the frequency at which any impedance occurs is increased by a factor of K_f. Again, we will make use of the shorter expression *"the network is scaled in frequency by a factor of 2"* to indicate that the same impedance is now obtained at a frequency twice as great. Frequency scaling is accomplished by scaling each passive element in frequency. It is apparent that no resistor is affected. The impedance of any inductor is sL, and if this same impedance is to be obtained at a frequency K_f times as great, then the inductance L must be replaced by an inductance of L/K_f. Similarly, a capacitance C is to be replaced by a capacitance C/K_f. Thus, if a network is to be scaled in frequency by a factor of K_f, then the changes necessary in each passive element are

$$\left. \begin{array}{l} R \to R \\ L \to \dfrac{L}{K_f} \\ C \to \dfrac{C}{K_f} \end{array} \right\} \quad \text{frequency scaling}$$

When each element of the magnitude-scaled network of Fig. 16.18a is scaled in frequency by a factor of 5×10^6, the network of Fig. 16.19a is obtained. The corresponding response curve is shown in Fig. 16.19b.

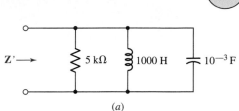

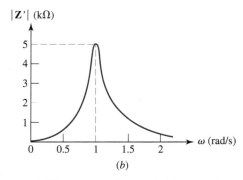

FIGURE 16.18 (a) The network of Fig. 16.17a after being scaled in magnitude by a factor $K_m = 2000$. (b) The corresponding response curve.

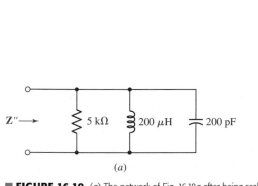

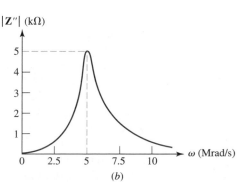

■ **FIGURE 16.19** (a) The network of Fig. 16.18a after being scaled in frequency by a factor $K_f = 5 \times 10^6$. (b) The corresponding response curve.

The circuit elements in this last network have values which are easily achieved in physical circuits; the network can actually be built and tested. It follows that, if the original network of Fig. 16.17a were actually an analog of some mechanical resonant system, we could have scaled this analog in both magnitude and frequency in order to achieve a network which we might construct in the laboratory; tests that are expensive or inconvenient to run on the mechanical system could then be made on the scaled electrical system, and the results should then be "unscaled" and converted into mechanical units to complete the analysis.

An impedance that is given as a function of **s** may also be scaled in magnitude or frequency, and this may be done without any knowledge of the specific elements out of which the two-terminal network is composed. In order to scale $\mathbf{Z}(\mathbf{s})$ in magnitude, the definition of magnitude scaling shows that it is necessary only to multiply $\mathbf{Z}(\mathbf{s})$ by K_m in order to obtain the magnitude-scaled impedance. Thus, the impedance of the parallel resonant circuit shown in Fig. 16.17a is

$$\mathbf{Z}(\mathbf{s}) = \frac{\mathbf{s}}{2\mathbf{s}^2 + 0.4\mathbf{s} + 2}$$

or

$$\mathbf{Z}(\mathbf{s}) = \frac{0.5\mathbf{s}}{(\mathbf{s} + 0.1 + j0.995)(\mathbf{s} + 0.1 - j0.995)}$$

The impedance $\mathbf{Z}'(\mathbf{s})$ of the magnitude-scaled network is

$$\mathbf{Z}'(\mathbf{s}) = K_m \mathbf{Z}(\mathbf{s})$$

If we again select $K_m = 2000$, we have

$$\mathbf{Z}'(\mathbf{s}) = (1000)\frac{\mathbf{s}}{(\mathbf{s} + 0.1 + j0.995)(\mathbf{s} + 0.1 - j0.995)}$$

If $\mathbf{Z}'(\mathbf{s})$ is now to be scaled in frequency by a factor of 5×10^6, then $\mathbf{Z}''(\mathbf{s})$ and $\mathbf{Z}'(\mathbf{s})$ are to provide identical values of impedance if $\mathbf{Z}''(\mathbf{s})$ is evaluated at a frequency K_f times that at which $\mathbf{Z}'(\mathbf{s})$ is evaluated. After some careful cerebral activity, this conclusion may be stated concisely in functional notation:

$$\mathbf{Z}''(\mathbf{s}) = \mathbf{Z}'\left(\frac{\mathbf{s}}{K_f}\right)$$

Note that we obtain $\mathbf{Z}''(\mathbf{s})$ by replacing every **s** in $\mathbf{Z}'(\mathbf{s})$ with $\mathbf{s}/K_f$. The analytic expression for the impedance of the network shown in Fig. 16.19a must therefore be

$$\mathbf{Z}''(\mathbf{s}) = (1000)\frac{\mathbf{s}/(5 \times 10^6)}{[\mathbf{s}/(5 \times 10^6) + 0.1 + j0.995][\mathbf{s}/(5 \times 10^6) + 0.1 - j0.995]}$$

or

$$\mathbf{Z}''(\mathbf{s}) = (1000)\frac{(5 \times 10^6)\mathbf{s}}{[\mathbf{s} + 0.5 \times 10^6 + j4.975 \times 10^6][\mathbf{s} + 0.5 \times 10^6 - j4.975 \times 10^6]}$$

Although scaling is a process normally applied to passive elements, dependent sources may also be scaled in magnitude and frequency. We assume that the output of any source is given as $k_x v_x$ or $k_y i_y$, where k_x has the dimensions of an admittance for a dependent current source and is dimensionless for a dependent voltage source, while k_y has the dimensions of ohms for a dependent voltage source and is dimensionless for a dependent current source. If the network containing the dependent source is scaled in magnitude by K_m, then it is necessary only to treat k_x or k_y as if it were the type of element consistent with its dimensions. That is, if k_x (or k_y) is dimensionless, it is left unchanged; if it is an admittance, it is divided by K_m; and if it is an impedance, it is multiplied by K_m. *Frequency scaling does not affect the dependent sources.*

CAUTION

EXAMPLE **16.6**

Scale the network shown in Fig. 16.20 by $K_m = 20$ and $K_f = 50$, and then find $\mathbf{Z}_{in}(s)$ for the scaled network.

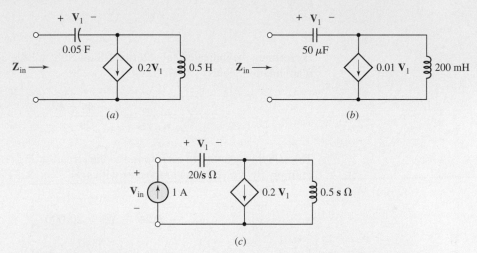

(a)

(b)

(c)

■ **FIGURE 16.20** (a) A network to be magnitude scaled by a factor of 20, and frequency scaled by a factor of 50. (b) The scaled network. (c) A 1 A test source is applied to the input terminals in order to obtain the impedance of the unscaled network in part a.

Magnitude scaling of the capacitor is accomplished by dividing 0.05 F by the scaling factor $K_m = 20$, and frequency scaling is accomplished by dividing by $K_f = 50$. Carrying out both operations simultaneously,

$$C_{scaled} = \frac{0.05}{(20)(50)} = 50 \, \mu F$$

The inductor is also scaled:

$$L_{scaled} = \frac{(20)0.5}{50} = 200 \, mH$$

In scaling the dependent source, only magnitude scaling need be considered, as frequency scaling does not affect dependent sources. Since

(Continued on next page)

this is a *voltage*-controlled *current* source, the multiplying constant 0.2 has units of A/V, or S. Since the factor has units of admittance, we divide by K_m, so that the new term is $0.01\mathbf{V}_1$. The resulting (scaled) network is shown in Fig. 16.20b.

To find the impedance of the new network, we need to apply a 1 A test source at the input terminals. We may work with either circuit; however, let's proceed by first finding the impedance of the *unscaled* network in Fig. 16.20a, and then scaling the result.

Referring to Fig. 16.20c,

$$\mathbf{V}_{in} = \mathbf{V}_1 + 0.5\mathbf{s}(1 - 0.2\mathbf{V}_1)$$

Also,

$$\mathbf{V}_1 = \frac{20}{\mathbf{s}}(1)$$

Performing the indicated substitution followed by a little algebraic manipulation yields

$$\mathbf{Z}_{in} = \frac{\mathbf{V}_{in}}{1} = \frac{\mathbf{s}^2 - 4\mathbf{s} + 40}{2\mathbf{s}}$$

To scale this quantity to correspond to the circuit of Fig. 16.20b we multiply by $K_m = 20$, and replace $\mathbf{s}$ with $\mathbf{s}/\mathbf{K}_f = \mathbf{s}/50$. Thus,

$$\mathbf{Z}_{in_{scaled}} = \frac{0.2\mathbf{s}^2 - 40\mathbf{s} + 20{,}000}{\mathbf{s}} \ \Omega$$

PRACTICE

16.9 A parallel resonant circuit is defined by $C = 0.01$ F, $\mathcal{B} = 2.5$ rad/s, and $\omega_0 = 20$ rad/s. Find the values of R and L if the network is scaled in (a) magnitude by a factor of 800; (b) frequency by a factor of 10^4; (c) magnitude by a factor of 800 and frequency by a factor of 10^4.

Ans: 32 kΩ, 200 H; 40 Ω, 25 μH; 32 kΩ, 20 mH.

16.6 BODE DIAGRAMS

In this section we will discover a quick method of obtaining an *approximate* picture of the amplitude and phase variation of a given transfer function as functions of ω. Accurate curves may, of course, be plotted after calculating values with a programmable calculator or a computer; curves may also be produced directly on the computer. Our object here, however, is to obtain a better picture of the response than we could visualize from a pole-zero plot, but yet not mount an all-out computational offensive.

The Decibel (dB) Scale

The approximate response curve we construct is called an asymptotic plot, or a **Bode plot,** or a **Bode diagram,** after its developer, Hendrik W. Bode, who was an electrical engineer and mathematician with the Bell Telephone Laboratories. Both the magnitude and phase curves are shown using a log-arithmic frequency scale for the abscissa, and the magnitude itself is also shown in logarithmic units called **decibels** (dB). We define the value of $|\mathbf{H}(j\omega)|$ in dB as follows:

$$H_{\mathrm{dB}} = 20 \log |\mathbf{H}(j\omega)|$$

where the common logarithm (base 10) is used. (*A multiplier of 10 instead of 20 is used for power transfer functions, but we will not need it here.*) The inverse operation is

$$|\mathbf{H}(j\omega)| = 10^{(H_{\mathrm{dB}}/20)}$$

Before we actually begin a detailed discussion of the technique for drawing Bode diagrams, it will help to gain some feeling for the size of the decibel unit, to learn a few of its important values, and to recall some of the properties of the logarithm. Since $\log 1 = 0$, $\log 2 = 0.30103$, and $\log 10 = 1$, we note the correspondences:

$$|\mathbf{H}(j\omega)| = 1 \Leftrightarrow H_{\mathrm{dB}} = 0$$
$$|\mathbf{H}(j\omega)| = 2 \Leftrightarrow H_{\mathrm{dB}} \approx 6 \text{ dB}$$
$$|\mathbf{H}(j\omega)| = 10 \Leftrightarrow H_{\mathrm{dB}} = 20 \text{ dB}$$

An increase of $|\mathbf{H}(j\omega)|$ by a factor of 10 corresponds to an increase in H_{dB} by 20 dB. Moreover, $\log 10^n = n$, and thus $10^n \Leftrightarrow 20n$ dB, so that 1000 corresponds to 60 dB, while 0.01 is represented as -40 dB. Using only the values already given, we may also note that $20 \log 5 = 20 \log \frac{10}{2} = 20 \log 10 - 20 \log 2 = 20 - 6 = 14$ dB, and thus $5 \Leftrightarrow 14$ dB. Also, $\log \sqrt{x} = \frac{1}{2} \log x$, and therefore $\sqrt{2} \Leftrightarrow 3$ dB and $1/\sqrt{2} \Leftrightarrow -3$ dB.[3]

We will write our transfer functions in terms of **s**, substituting $\mathbf{s} = j\omega$ when we are ready to find the magnitude or phase angle. If desired, the magnitude may be written in terms of dB at that point.

The decibel is named in honor of Alexander Graham Bell.

PRACTICE

16.10 Calculate H_{dB} at $\omega = 146$ rad/s if $\mathbf{H}(\mathbf{s})$ equals (a) $20/(\mathbf{s} + 100)$; (b) $20(\mathbf{s} + 100)$; (c) $20\mathbf{s}$. Calculate $|\mathbf{H}(j\omega)|$ if H_{dB} equals (d) 29.2 dB; (e) -15.6 dB; (f) -0.318 dB.

Ans: -18.94 dB; 71.0 dB; 69.3 dB; 28.8; 0.1660; 0.964.

Determination of Asymptotes

Our next step is to factor $\mathbf{H}(\mathbf{s})$ to display its poles and zeros. We first con-sider a zero at $\mathbf{s} = -a$, written in a standardized form as

$$\mathbf{H}(\mathbf{s}) = 1 + \frac{\mathbf{s}}{a} \qquad\qquad [26]$$

---

(3) Note that we are being slightly dishonest here by using $20 \log 2 = 6$ dB rather than 6.02 dB. It is customary, however, to represent $\sqrt{2}$ as 3 dB; since the dB scale is inherently logarithmic, the small inaccuracy is seldom significant.

The Bode diagram for this function consists of the two asymptotic curves approached by H_{dB} for very large and very small values of ω. Thus, we begin by finding

$$|\mathbf{H}(j\omega)| = \left|1 + \frac{j\omega}{a}\right| = \sqrt{1 + \frac{\omega^2}{a^2}}$$

and thus

$$H_{dB} = 20\log\left|1 + \frac{j\omega}{a}\right| = 20\log\sqrt{1 + \frac{\omega^2}{a^2}}$$

When $\omega \ll a$,

$$H_{dB} \approx 20\log 1 = 0 \qquad (\omega \ll a)$$

This simple asymptote is shown in Fig. 16.21. It is drawn as a solid line for $\omega < a$, and as a green line for $\omega > a$.

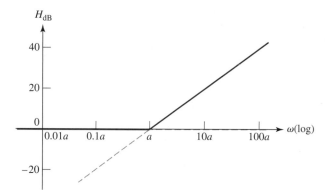

■ **FIGURE 16.21** The Bode amplitude plot for $H(s) = 1 + s/a$ consists of the low- and high-frequency asymptotes, shown as dashed lines. They intersect on the abscissa at the corner frequency. The Bode plot represents the response in terms of two asymptotes, both straight lines and both easily drawn.

When $\omega \gg a$,

$$H_{dB} \approx 20\log\frac{\omega}{a} \qquad (\omega \gg a)$$

At $\omega = a$, $H_{dB} = 0$; at $\omega = 10a$, $H_{dB} = 20$ dB; and at $\omega = 100a$, $H_{dB} = 40$ dB. Thus, the value of H_{dB} increases 20 dB for every 10-fold increase in frequency. The asymptote therefore has a slope of 20 dB/decade. Since H_{dB} increases by 6 dB when ω doubles, an alternate value for the slope is 6 dB/octave. The high-frequency asymptote is also shown in Fig. 16.21, a solid line for $\omega > a$, and a broken line for $\omega < a$. Note that the two asymptotes intersect at $\omega = a$, the frequency of the zero. This frequency is also described as the ***corner, break, 3 dB,*** or ***half-power frequency.***

A ***decade*** refers to a range of frequencies defined by a factor of 10, such as 3 Hz to 30 Hz, or 12.5 MHz to 125 MHz. An ***octave*** refers to a range of frequencies defined by a factor of 2, such as 7 GHz to 14 GHz.

Smoothing Bode Plots

Now let us see how much error is embodied in our asymptotic response curve. At the corner frequency ($\omega = a$),

Note that we continue to abide by the convention of taking $\sqrt{2}$ as corresponding to 3 dB.

$$H_{dB} = 20\log\sqrt{1 + \frac{a^2}{a^2}} = 3 \text{ dB}$$

as compared with an asymptotic value of 0 dB. At $\omega = 0.5a$, we have

$$H_{\text{dB}} = 20 \log \sqrt{1.25} \approx 1 \text{ dB}$$

Thus, the exact response is represented by a smooth curve that lies 3 dB above the asymptotic response at $\omega = a$, and 1 dB above it at $\omega = 0.5a$ (and also at $\omega = 2a$). This information can always be used to smooth out the corner if a more exact result is desired.

Multiple Terms

Most transfer functions will consist of more than a simple zero (or simple pole). This, however, is easily handled by the Bode method, since we are in fact working with logarithms. For example, consider a function

$$\mathbf{H}(\mathbf{s}) = K \left(1 + \frac{\mathbf{s}}{s_1} \right) \left(1 + \frac{\mathbf{s}}{s_2} \right)$$

where $K = $ constant, and $-s_1$ and $-s_2$ represent the two zeros of our function $\mathbf{H}(\mathbf{s})$. H_{dB} for this function may be written as

$$H_{\text{dB}} = 20 \log \left| K \left(1 + \frac{j\omega}{s_1} \right) \left(1 + \frac{j\omega}{s_2} \right) \right|$$

$$= 20 \log \left[K \sqrt{1 + \left(\frac{\omega}{s_1} \right)^2} \sqrt{1 + \left(\frac{\omega}{s_2} \right)^2} \right]$$

or

$$H_{\text{dB}} = 20 \log K + 20 \log \sqrt{1 + \left(\frac{\omega}{s_1} \right)^2} + 20 \log \sqrt{1 + \left(\frac{\omega}{s_2} \right)^2}$$

which is simply the sum of a constant (frequency-independent) term $20 \log K$, and two simple zero terms of the form previously considered. In other words, *we may construct a sketch of H_{dB} by simply graphically adding the plots of the separate terms.* We explore this in the following example.

EXAMPLE 16.7

Obtain the Bode plot of the input impedance of the network shown in Fig. 16.22.

We have the input impedance,

$$\mathbf{Z}_{\text{in}}(\mathbf{s}) = \mathbf{H}(\mathbf{s}) = 20 + 0.2\mathbf{s}$$

Putting this in standard form, we obtain

$$\mathbf{H}(\mathbf{s}) = 20 \left(1 + \frac{\mathbf{s}}{100} \right)$$

(Continued on next page)

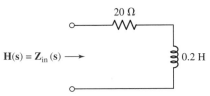

■ FIGURE 16.22 If $\mathbf{H}(\mathbf{s})$ is selected as $\mathbf{Z}_{\text{in}}(\mathbf{s})$ for this network, then the Bode plot for $\mathbf{H}_{\text{dB}}$ is as shown in Fig. 16.23*b*.

The two factors constituting $\mathbf{H}(\mathbf{s})$ are a zero at $\mathbf{s} = -100$, leading to a break frequency of $\omega = 100$ rad/s, and a constant equivalent to $20\log 20 = 26$ dB. Each of these is sketched lightly in Fig. 16.23a. Since we are working with the logarithm of $|\mathbf{H}(j\omega)|$, we next add together the Bode plots corresponding to the individual factors. The resultant magnitude plot appears as Fig. 16.23b. No attempt has been made to smooth out the corner with a $+3$ dB correction at $\omega = 100$ rad/s; this is left to the reader as a quick exercise.

PRACTICE

16.11 Construct a Bode magnitude plot for $\mathbf{H}(\mathbf{s}) = 50 + \mathbf{s}$.

Ans: 34 dB, $\omega < 50$ rad/s; slope $= +20$ dB/decade $\omega > 50$ rad/s.

Phase Response

Returning to the transfer function of Eq. [26], we would now like to determine the *phase response* for the simple zero,

$$\text{ang } \mathbf{H}(j\omega) = \text{ang}\left(1 + \frac{j\omega}{a}\right) = \tan^{-1}\frac{\omega}{a}$$

This expression is also represented by its asymptotes, although three straight-line segments are required. For $\omega \ll a$, ang $\mathbf{H}(j\omega) \approx 0°$, and we use this as our asymptote when $\omega < 0.1a$:

$$\text{ang } \mathbf{H}(j\omega) = 0° \qquad (\omega < 0.1a)$$

At the high end, $\omega \gg a$, we have ang $\mathbf{H}(j\omega) \approx 90°$, and we use this above $\omega = 10a$:

$$\text{ang } \mathbf{H}(j\omega) = 90° \qquad (\omega > 10a)$$

Since the angle is $45°$ at $\omega = a$, we now construct the straight-line asymptote extending from $0°$ at $\omega = 0.1a$, through $45°$ at $\omega = a$, to $90°$ at $\omega = 10a$. This straight line has a slope of $45°$/decade. It is shown as a solid curve in Fig. 16.24, while the exact angle response is shown as a broken line.

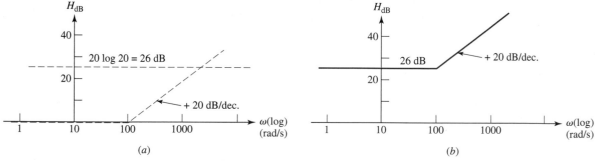

FIGURE 16.23 (*a*) The Bode plots for the factors of $\mathbf{H}(\mathbf{s}) = 20(1 + \mathbf{s}/100)$ are sketched individually. (*b*) The composite Bode plot is shown as the sum of the plots of part *a*.

The maximum differences between the asymptotic and true responses are $\pm 5.71°$ at $\omega = 0.1a$ and $10a$. Errors of $\mp 5.29°$ occur at $\omega = 0.394a$ and $2.54a$; the error is zero at $\omega = 0.159a$, a, and $6.31a$. The phase angle plot is typically left as a straight-line approximation, although smooth curves can also be drawn in a manner similar to that depicted in Fig. 16.24.

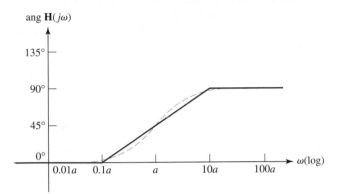

■ **FIGURE 16.24** The asymptotic angle response for $H(s) = 1 + s/a$ is shown as the three straight-line segments in solid color. The endpoints of the ramp are 0° at $0.1a$ and 90° at $10a$. The dashed line represents a more accurate (smoothed) response.

It is worth pausing briefly here to consider what the phase plot is telling us. In the case of a simple zero at $s = a$, we see that for frequencies much less than the corner frequency, the phase of the response function is 0°. For high frequencies, however ($\omega \gg a$), the phase is 90°. In the vicinity of the corner frequency, the phase of the transfer function varies somewhat rapidly. The actual phase angle imparted to the response can therefore be selected through the design of the circuit (which determines a).

PRACTICE

16.12 Draw the Bode phase plot for the transfer function of Example 16.7.

Ans: 0°, $\omega \leq 10$; 90°, $\omega \geq 1000$; 45°, $\omega = 100$; 45°/dec slope, $10 < \omega < 1000$. (ω in rad/s).

Additional Considerations in Creating Bode Plots

We next consider a simple pole,

$$H(s) = \frac{1}{1 + s/a} \qquad [27]$$

Since this is the reciprocal of a zero, the logarithmic operation leads to a Bode plot which is the *negative* of that obtained previously. The amplitude is 0 dB up to $\omega = a$, and then the slope is -20 dB/decade for $\omega > a$. The angle plot is 0° for $\omega < 0.1a$, $-90°$ for $\omega > 10a$, and $-45°$ at $\omega = a$, and it has a slope of $-45°$/decade when $0.1a < \omega < 10a$. The reader is encouraged to generate the Bode plot for this function by working directly with Eq. [27].

Another term that can appear in $\mathbf{H}(\mathbf{s})$ is a factor of $\mathbf{s}$ in the numerator or denominator. If $\mathbf{H}(\mathbf{s}) = \mathbf{s}$, then

$$H_{\mathrm{dB}} = 20 \log |\omega|$$

Thus, we have an infinite straight line passing through 0 dB at $\omega = 1$ and having a slope everywhere of 20 dB/decade. This is shown in Fig. 16.25a. If the $\mathbf{s}$ factor occurs in the denominator, a straight line is obtained having a slope of -20 dB/decade and passing through 0 dB at $\omega = 1$, as shown in Fig. 16.25b.

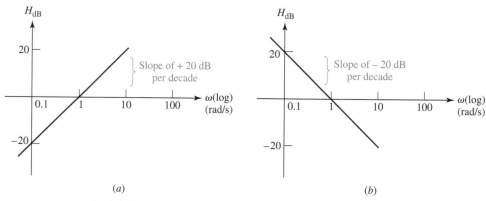

(a) (b)

■ **FIGURE 16.25** The asymptotic diagrams are shown for (a) $\mathbf{H}(\mathbf{s}) = \mathbf{s}$ and (b) $\mathbf{H}(\mathbf{s}) = 1/\mathbf{s}$. Both are infinitely long straight lines passing through 0 dB at $\omega = 1$ and having slopes of ± 20 dB/decade.

Another simple term found in $\mathbf{H}(\mathbf{s})$ is the multiplying constant K. This yields a Bode plot which is a horizontal straight line lying $20 \log |K|$ dB above the abscissa. It will actually be below the abscissa if $|K| < 1$.

EXAMPLE 16.8

Find the Bode plot for the gain of the circuit shown in Fig. 16.26.

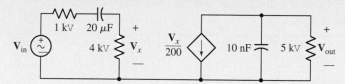

■ **FIGURE 16.26** If $\mathbf{H}(\mathbf{s}) = \mathbf{V}_{\mathrm{out}}/\mathbf{V}_{\mathrm{in}}$, this amplifier is found to have the Bode amplitude plot shown in Fig. 16.27b, and the phase plot shown in Fig. 16.28.

We work from left to right through the circuit and write the expression for the voltage gain,

$$\mathbf{H}(\mathbf{s}) = \frac{\mathbf{V}_{\mathrm{out}}}{\mathbf{V}_{\mathrm{in}}} = \frac{4000}{5000 + 10^6/20\mathbf{s}} \left(-\frac{1}{200}\right) \frac{5000(10^8/\mathbf{s})}{5000 + 10^8/\mathbf{s}}$$

which simplifies (mercifully) to

$$\mathbf{H}(\mathbf{s}) = \frac{-2\mathbf{s}}{(1 + \mathbf{s}/10)(1 + \mathbf{s}/20{,}000)} \qquad [28]$$

We see a constant, $20 \log | -2| = 6$ dB, break points at $\omega = 10$ rad/s and $\omega = 20{,}000$ rad/s, and a linear factor $\mathbf{s}$. Each of these is sketched in Fig. 16.27a, and the four sketches are added to give the Bode magnitude plot in Fig. 16.27b.

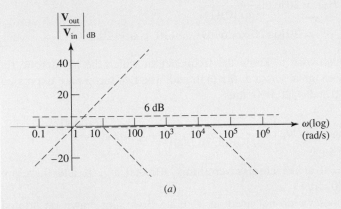

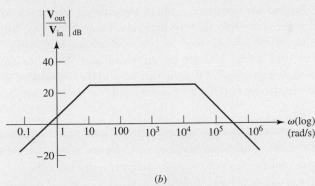

(a)

(b)

■ **FIGURE 16.27** (a) Individual Bode magnitude sketches are made for the factors (-2), $(\mathbf{s})$, $(1 + \mathbf{s}/10)^{-1}$, and $(1 + \mathbf{s}/20{,}000)^{-1}$. (b) The four separate plots of part a are added to give the Bode magnitude plots for the amplifier of Fig. 16.26.

PRACTICE

16.13 Construct a Bode magnitude plot for $\mathbf{H}(\mathbf{s})$ equal to
(a) $50/(\mathbf{s} + 100)$; (b) $(\mathbf{s} + 10)/(\mathbf{s} + 100)$; (c) $(\mathbf{s} + 10)/\mathbf{s}$.

Ans: (a) -6 dB, $\omega < 100$; -20 dB/decade, $\omega > 100$; (b) -20 dB, $\omega < 10$; $+20$ dB/decade, $10 < \omega < 100$; 0 dB, $\omega > 100$; (c) 0 dB, $\omega > 10$; -20 dB/decade, $\omega < 10$.

Before we construct the phase plot for the amplifier of Fig. 16.26, let us take a few moments to investigate several of the details of the magnitude plot.

First, it is wise not to rely too heavily on graphical addition of the individual magnitude plots. Instead, the exact value of the combined magnitude plot may be found easily at selected points by considering the asymptotic

CAUTION

value of each factor of $\mathbf{H}(\mathbf{s})$ at the point in question. For example, in the flat region of Fig. 16.27a between $\omega = 10$ and $\omega = 20{,}000$, we are below the corner at $\omega = 20{,}000$, and so we represent $(1 + \mathbf{s}/20{,}000)$ by 1; but we are above $\omega = 10$, so $(1 + \mathbf{s}/10)$ is represented as $\omega/10$. Hence,

$$H_{\text{dB}} = 20 \log \left| \frac{-2\omega}{(\omega/10)(1)} \right|$$

$$= 20 \log 20 = 26 \text{ dB} \qquad (10 < \omega < 20{,}000)$$

We might also wish to know the frequency at which the asymptotic response crosses the abscissa at the high end. The two factors are expressed here as $\omega/10$ and $\omega/20{,}000$; thus

$$H_{\text{dB}} = 20 \log \left| \frac{-2\omega}{(\omega/10)(\omega/20{,}000)} \right| = 20 \log \left| \frac{400{,}000}{\omega} \right|$$

Since $H_{\text{dB}} = 0$ at the abscissa crossing, $400{,}000/\omega = 1$, and therefore $\omega = 400{,}000$ rad/s.

Many times we do not need an accurate Bode plot drawn on printed semilog paper. Instead we construct a rough logarithmic frequency axis on simple lined paper. After selecting the interval for a decade—say, a distance L extending from $\omega = \omega_1$ to $\omega = 10\omega_1$ (where ω_1 is usually an integral power of 10)—we let x locate the distance that ω lies to the right of ω_1, so that $x/L = \log(\omega/\omega_1)$. Of particular help is the knowledge that $x = 0.3L$ when $\omega = 2\omega_1$, $x = 0.6L$ at $\omega = 4\omega_1$, and $x = 0.7L$ at $\omega = 5\omega_1$.

EXAMPLE 16.9

Draw the phase plot for the transfer function given by Eq. [28], $\mathbf{H}(\mathbf{s}) = -2\mathbf{s}/[(1 + \mathbf{s}/10)(1 + \mathbf{s}/20{,}000)]$.

We begin by inspecting $\mathbf{H}(j\omega)$:

$$\mathbf{H}(j\omega) = \frac{-j2\omega}{(1 + j\omega/10)(1 + j\omega/20{,}000)} \qquad [29]$$

The angle of the numerator is a constant, $-90°$.

The remaining factors are represented as the sum of the angles contributed by breakpoints at $\omega = 10$ and $\omega = 20{,}000$. These three terms appear as broken-line asymptotic curves in Fig. 16.28, and their sum is shown as the solid curve. An equivalent representation is obtained if the curve is shifted upward by $360°$.

Exact values can also be obtained for the asymptotic phase response. For example, at $\omega = 10^4$ rad/s, the angle in Fig. 16.28 is obtained from the numerator and denominator terms in Eq. [29]. The numerator angle is $-90°$. The angle for the pole at $\omega = 10$ is $-90°$, since ω is greater than 10 times the corner frequency. Between 0.1 and 10 times the corner frequency, we recall that the slope is $-45°$ per decade for a simple pole. For the breakpoint at $20{,}000$ rad/s, we therefore calculate the angle, $-45° \log(\omega/0.1a) = -45° \log[10{,}000/(0.1 \times 20{,}000)] = -31.5°$.

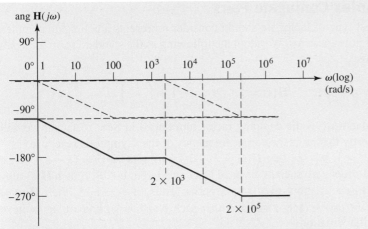

■ **FIGURE 16.28** The solid curve displays the asymptotic phase response of the amplifier shown in Fig. 16.26.

The algebraic sum of these three contributions is $-90° - 90° - 31.5° = -211.5°$, a value that appears to be moderately near the asymptotic phase curve of Fig. 16.28.

PRACTICE
●

16.14 Draw the Bode phase plot for $\mathbf{H(s)}$ equal to (a) $50/(\mathbf{s} + 100)$; (b) $(\mathbf{s} + 10)/(\mathbf{s} + 100)$; (c) $(\mathbf{s} + 10)/\mathbf{s}$.

Ans: (a) $0°$, $\omega < 10$; $-45°/$decade, $10 < \omega < 1000$; $-90°$, $\omega > 1000$; (b) $0°$, $\omega < 1$; $+45°/$decade, $1 < \omega < 10$; $45°$, $10 < \omega < 100$; $-45°/$decade, $100 < \omega < 1000$; $0°$, $\omega > 1000$; (c) $-90°$, $\omega < 1$; $+45°/$decade, $1 < \omega < 100$; $0°$, $\omega > 100$.

Higher-Order Terms

The zeros and poles that we have been considering are all first-order terms, such as $\mathbf{s}^{\pm 1}$, $(1 + 0.2\mathbf{s})^{\pm 1}$, and so forth. We may extend our analysis to higher-order poles and zeros very easily, however. A term $\mathbf{s}^{\pm n}$ yields a magnitude response that passes through $\omega = 1$ with a slope of $\pm 20n$ dB/decade; the phase response is a constant angle of $\pm 90n°$. Also, a multiple zero, $(1 + \mathbf{s}/a)^n$, must represent the sum of n of the magnitude-response curves, or n of the phase-response curves of the simple zero. We therefore obtain an asymptotic magnitude plot that is 0 dB for $\omega < a$ and has a slope of $20n$ dB/decade when $\omega > a$; the error is $-3n$ dB at $\omega = a$, and $-n$ dB at $\omega = 0.5a$ and $2a$. The phase plot is $0°$ for $\omega < 0.1a$, $90n°$ for $\omega > 10a$, $45n°$ at $\omega = a$, and a straight line with a slope of $45n°/$decade for $0.1a < \omega < 10a$, and it has errors as large as $\pm 5.71n°$ at two frequencies.

The asymptotic magnitude and phase curves associated with a factor such as $(1 + \mathbf{s}/20)^{-3}$ may be drawn quickly, but the relatively large errors associated with the higher powers should be kept in mind.

Complex Conjugate Pairs

The last type of factor we should consider represents a conjugate complex pair of poles or zeros. We adopt the following as the standard form for a pair of zeros:

$$\mathbf{H(s)} = 1 + 2\zeta \left(\frac{\mathbf{s}}{\omega_0}\right) + \left(\frac{\mathbf{s}}{\omega_0}\right)^2$$

The quantity ζ is the damping factor introduced in Sec. 16.1, and we will see shortly that ω_0 is the corner frequency of the asymptotic response.

If $\zeta = 1$, we see that $\mathbf{H(s)} = 1 + 2(\mathbf{s}/\omega_0) + (\mathbf{s}/\omega_0)^2 = (1 + \mathbf{s}/\omega_0)^2$, a second-order zero such as we have just considered. If $\zeta > 1$, then $\mathbf{H(s)}$ may be factored to show two simple zeros. Thus, if $\zeta = 1.25$, then $\mathbf{H(s)} = 1 + 2.5(\mathbf{s}/\omega_0) + (\mathbf{s}/\omega_0)^2 = (1 + \mathbf{s}/2\omega_0)(1 + \mathbf{s}/0.5\omega_0)$, and we again have a familiar situation.

A new case arises when $0 \leq \zeta \leq 1$. There is no need to find values for the conjugate complex pair of roots. Instead, we determine the low- and high-frequency asymptotic values for both the magnitude and phase response, and then apply a correction that depends on the value of ζ.

For the magnitude response, we have

$$H_{\mathrm{dB}} = 20 \log |\mathbf{H}(j\omega)| = 20 \log \left| 1 + j2\zeta \left(\frac{\omega}{\omega_0}\right) - \left(\frac{\omega}{\omega_0}\right)^2 \right| \qquad [30]$$

When $\omega \ll \omega_0$, $H_{\mathrm{dB}} = 20 \log |1| = 0$ dB. This is the low-frequency asymptote. Next, if $\omega \gg \omega_0$, only the squared term is important, and $H_{\mathrm{dB}} = 20 \log |-(\omega/\omega_0)^2| = 40 \log(\omega/\omega_0)$. We have a slope of $+40$ dB/decade. This is the high-frequency asymptote, and the two asymptotes intersect at 0 dB, $\omega = \omega_0$. The solid curve in Fig. 16.29 shows this asymptotic representation of the magnitude response. However, a correction must be applied

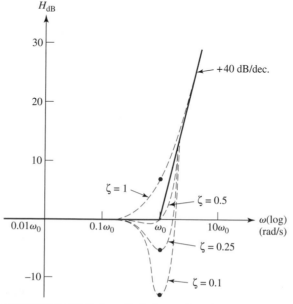

■ **FIGURE 16.29** Bode amplitude plots are shown for $\mathbf{H(s)} = 1 + 2\zeta(\mathbf{s}/\omega_0) + (\mathbf{s}/\omega_0)^2$ for several values of the damping factor ζ.

in the neighborhood of the corner frequency. We let $\omega = \omega_0$ in Eq. [30] and have

$$H_{dB} = 20 \log \left| j2\zeta \left(\frac{\omega}{\omega_0} \right) \right| = 20 \log(2\zeta) \qquad [31]$$

If $\zeta = 1$, a limiting case, the correction is $+6$ dB; for $\zeta = 0.5$, no correction is required; and if $\zeta = 0.1$, the correction is -14 dB. Knowing this one correction value is often sufficient to draw a satisfactory asymptotic magnitude response. Figure 16.29 shows more accurate curves for $\zeta = 1, 0.5, 0.25$, and 0.1, as calculated from Eq. [30]. For example, if $\zeta = 0.25$, then the exact value of H_{dB} at $\omega = 0.5\omega_0$ is

$$H_{dB} = 20 \log |1 + j0.25 - 0.25| = 20 \log \sqrt{0.75^2 + 0.25^2} = -2.0 \text{ dB}$$

The negative peaks do not show a minimum value exactly at $\omega = \omega_0$, as we can see by the curve for $\zeta = 0.5$. The valley is always found at a slightly lower frequency.

If $\zeta = 0$, then $\mathbf{H}(j\omega_0) = 0$ and $H_{dB} = -\infty$. Bode plots are not usually drawn for this situation.

Our last task is to draw the asymptotic phase response for $\mathbf{H}(j\omega) = 1 + j2\zeta(\omega/\omega_0) - (\omega/\omega_0)^2$. Below $\omega = 0.1\omega_0$, we let ang $\mathbf{H}(j\omega) = 0°$; above $\omega = 10\omega_0$, we have ang $\mathbf{H}(j\omega) = \text{ang} [-(\omega/\omega_0)^2] = 180°$. At the corner frequency, ang $\mathbf{H}(j\omega_0) = \text{ang} (j2\zeta) = 90°$. In the interval $0.1\omega_0 < \omega < 10\omega_0$, we begin with the straight line shown as a solid curve in Fig. 16.30. It extends from $(0.1\omega_0, 0°)$, through $(\omega_0, 90°)$, and terminates at $(10\omega_0, 180°)$; it has a slope of $90°/\text{decade}$.

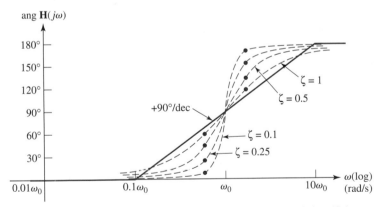

FIGURE 16.30 The straight-line approximation to the phase characteristic for $\mathbf{H}(j\omega) = 1 + j2\zeta(\omega/\omega_0) - (\omega/\omega_0)^2$ is shown as a solid curve, and the true phase response is shown for $\zeta = 1, 0.5, 0.25$, and 0.1 as broken lines.

We must now provide some correction to this basic curve for various values of ζ. From Eq. [30], we have

$$\text{ang } \mathbf{H}(j\omega) = \tan^{-1} \frac{2\zeta(\omega/\omega_0)}{1 - (\omega/\omega_0)^2}$$

One accurate value above and one below $\omega = \omega_0$ may be sufficient to give an approximate shape to the curve. If we take $\omega = 0.5\omega_0$, we find ang $\mathbf{H}(j0.5\omega_0) = \tan^{-1}(4\zeta/3)$, while the angle is $180° - \tan^{-1}(4\zeta/3)$ at

$\omega = 2\omega_0$. Phase curves are shown as broken lines in Fig. 16.30 for $\zeta = 1, 0.5$, 0.25, and 0.1; heavy dots identify accurate values at $\omega = 0.5\omega_0$ and $\omega = 2\omega_0$.

If the quadratic factor appears in the denominator, both the magnitude and phase curves are the *negatives* of those just discussed. We conclude with an example that contains both linear and quadratic factors.

EXAMPLE 16.10

Construct the Bode plot for the transfer function $H(s) = 100,000s/[(s + 1)(10,000 + 20s + s^2)]$.

Let's consider the quadratic factor first and arrange it in a form such that we can see the value of ζ. We begin by dividing the second-order factor by its constant term, 10,000:

$$H(s) = \frac{10s}{(1 + s)(1 + 0.002s + 0.0001s^2)}$$

An inspection of the s^2 term next shows that $\omega_0 = \sqrt{1/0.0001} = 100$. Then the linear term of the quadratic is written to display the factor 2, the factor (s/ω_0), and finally the factor ζ:

$$H(s) = \frac{10s}{(1 + s)[(1 + 2)(0.1)(s/100) + (s/100)^2]}$$

We see that $\zeta = 0.1$.

The asymptotes of the magnitude-response curve are sketched in lightly in Fig. 16.31: 20 dB for the factor of 10, an infinite straight line through $\omega = 1$ with a +20 dB/decade slope for the s factor, a corner at $\omega = 1$ for the simple pole, and a corner at $\omega = 100$ with a slope of -40 dB/decade for the second-order term in the denominator. Adding these four curves and supplying a correction of +14 dB for the quadratic factor leads to the heavy curve of Fig. 16.31.

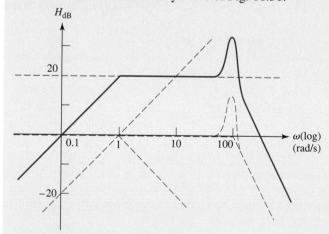

■ **FIGURE 16.31** The Bode magnitude plot of the transfer function $H(s) = \dfrac{100,000s}{(s + 1)(10,000 + 20s + s^2)}$.

The phase response contains three components: $+90°$ for the factor s; $0°$ for $\omega < 0.1$, $-90°$ for $\omega > 10$, and $-45°$/decade for the simple pole; and $0°$ for $\omega < 10$, $-180°$ for $\omega > 1000$, and $-90°$ per decade for the quadratic factor. The addition of these three asymptotes plus some improvement for $\zeta = 0.1$ is shown as the solid curve in Fig. 16.32.

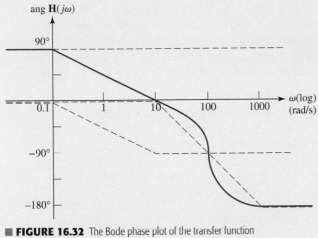

■ **FIGURE 16.32** The Bode phase plot of the transfer function

$$H(s) = \frac{100,000s}{(s+1)(10,000 + 20s + s^2)}.$$

PRACTICE

16.15 If $\mathbf{H}(s) = 1000s^2/(s^2 + 5s + 100)$, sketch the Bode amplitude plot and calculate a value for (a) ω when $H_{dB} = 0$; (b) H_{dB} at $\omega = 1$; (c) H_{dB} as $\omega \to \infty$.

Ans: 0.316 rad/s; 20 dB; 60 dB.

COMPUTER-AIDED ANALYSIS

The technique of generating Bode plots is a valuable one. There are many situations in which an approximate diagram is needed quickly (such as on exams, or when evaluating a particular circuit topology for a specific application), and simply knowing the general shape of the response is adequate. Further, Bode plots can be invaluable when designing filters in terms of enabling us to select factors and coefficient values.

In situations where *exact* response curves are required (such as when verifying a final circuit design), there are several computer-assisted options available to the engineer. The first technique we will consider here is the use of MATLAB to generate a frequency response curve. In order to accomplish this, the circuit must first be analyzed to obtain the correct transfer function. However, it is not necessary to factor or simplify the expression.

Consider the circuit in Fig. 16.26. We previously determined that the transfer function for this circuit can be expressed as

$$\mathbf{H}(s) = \frac{-2s}{(1 + s/10)(1 + s/20,000)}$$

(Continued on next page)

We seek a detailed graph of this function over the frequency range of 100 mrad/s to 1 Mrad/s. Since the final graph will be plotted on a logarithmic scale, there is no need to uniformly space our discrete frequencies. Instead, we use the MATLAB function *logspace*() to generate a frequency vector, where the first two arguments represent the power of 10 for starting and ending frequencies, respectively (-1 and 6 in the present example), and the third argument is the total number of points desired. Thus, our MATLAB script is

```
EDU» w = logspace(−1,6,100);
EDU» denom = (1+j*w/10) .* (1+j*w/20000);
EDU» H = −2*j*w ./ denom;
EDU» Hdb = 20*log10(abs(H));
EDU» semilogx(w,Hdb)
EDU» xlabel('frequency (rad/s)')
EDU» ylabel ('|H(jw)| (dB)')
```

which yields the graph depicted in Fig. 16.33.

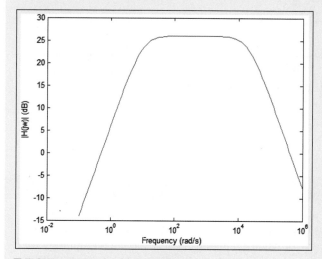

■ **FIGURE 16.33** Plot of H_{dB} generated using MATLAB.

A few comments about the MATLAB code are warranted. First, note that we have substituted $\mathbf{s} = j\omega$ in our expression for $\mathbf{H}(\mathbf{s})$. Also, MATLAB treats the variable w as a vector, or one-dimensional matrix. As such, this variable can cause difficulties in the denominator of an expression as MATLAB will attempt to apply matrix algebra rules to any expression. Thus, the denominator of $\mathbf{H}(j\omega)$ is computed in a separate line, and the operator ".*" is required instead of "*" to multiply the two terms. This new operator is equivalent to the following MATLAB code:

```
EDU» for k = 1:100
denom = (1 + j*w(k)/10) * (1 + j*w(k)/20000);
end
```

In a similar fashion, the new operator "./" is used in the subsequent line of code. The results are desired in dB, so the function *log*10() is invoked; *log*() represents the natural logarithm in MATLAB. Finally, the new plot command *semilogx*() is used to generate a graph with the *x* axis having a logarithmic scale. The reader is encouraged at this point to return to previous examples, and use these techniques to generate exact curves for comparison to the corresponding Bode plots.

PSpice is also commonly used to generate frequency response curves, especially to evaluate a final design. Figure 16.34*a* depicts the circuit of Fig. 16.26, where the voltage across the resistor R3 represents the desired output voltage. The source component

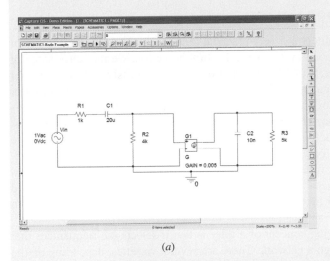

(*a*)

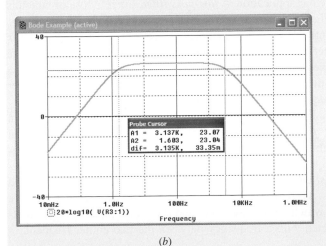

(*b*)

■ **FIGURE 16.34** (*a*) The circuit of Fig. 16.26. (*b*) Frequency response of the circuit plotted in dB scale.

(*Continued on next page*)

VAC has been employed with a fixed voltage of 1 V for convenience. An ac sweep simulation is required to determine the frequency response of our circuit; Fig. 16.34b was generated using 10 points per decade (with Decade selected under **Logarithmic** AC Sweep Type) from 10 mHz to 1 MHz. Note the simulation has been performed in Hz, not rad/s, so the cursor tool is indicating a bandwidth of 3.14 kHz.

Again, the reader is encouraged to simulate example circuits and compare the results to the Bode plots generated previously.

16.7 FILTERS

The design of filters is a very practical (and interesting) subject, worthy of a separate textbook in its own right. In this section, we introduce some of the basic concepts of filtering, and explore both passive and active filter circuits. These circuits may be very simple, consisting of a single capacitor or inductor whose addition to a given network leads to improved performance. They may also be fairly sophisticated, consisting of many resistors, capacitors, inductors, and op amps in order to obtain the precise response curve required for a given application. Filters are used in modern electronics to obtain dc voltages in power supplies, eliminate noise in communication channels, separate radio and television channels from the multiplexed signal provided by antennas, and boost the bass signal in a car stereo, to name just a few applications.

The underlying concept of a filter is that it selects the frequencies that may pass through a network. There are several varieties, depending on the needs of a particular application. A *low-pass filter,* the response of which is illustrated in Fig. 16.35a, passes frequencies below a cutoff frequency, while significantly damping frequencies above that cutoff. A *high-pass filter,* on the other hand, does just the opposite, as shown in Fig. 16.35b. The chief figure of merit of a filter is the sharpness of the cutoff, or the steepness of the curve in the vicinity of the corner frequency. In general, steeper response curves require more complex circuits.

Combining a low-pass and a high-pass filter can lead to what is known as a *bandpass filter,* as illustrated by the response curve shown in Fig. 16.35c. In this type of filter, the region between the two corner frequencies is referred to as the *passband;* the region outside the passband is referred to as the *stopband.* These terms may also be applied to the low- and high-pass filters, as indicated in Fig. 16.35a and b. We can also create a *bandstop filter,* which allows both high and low frequencies to pass but attenuates any signal with a frequency between the two corner frequencies (Fig. 16.35d).

The *notch filter* is a specialized bandstop filter, designed with a narrow response characteristic that blocks a single frequency component of a signal. *Multiband filters* are also possible; these are filter circuits which have multiple passbands and stopbands. The design of such filters is straightforward, but beyond the range of this book.

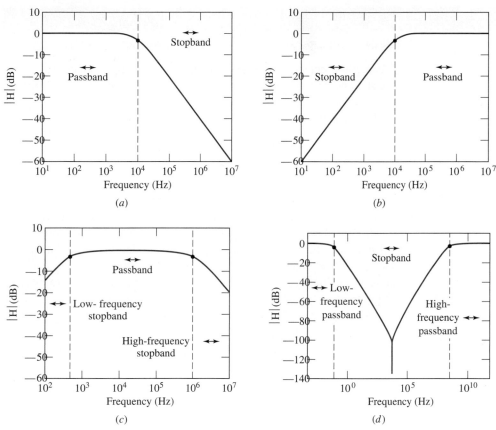

■ **FIGURE 16.35** Frequency response curves for (*a*) a low-pass filter; (*b*) a high-pass filter; (*c*) a bandpass filter; (*d*) a bandstop filter. In each diagram, a solid dot corresponds to −3 dB.

Passive Low-Pass and High-Pass Filters

A filter can be constructed by simply using a single capacitor and a single resistor, as shown in Fig. 16.36*a*. The transfer function for this low-pass filter circuit is

$$\mathbf{H(s)} \equiv \frac{\mathbf{V}_{out}}{\mathbf{V}_{in}} = \frac{1}{1 + RC\mathbf{s}} \qquad [32]$$

$\mathbf{H(s)}$ has a single corner frequency, which occurs at $\omega = 1/RC$, and a zero at $\mathbf{s} = \infty$, leading to its "low-pass" filtering behavior. Low frequencies ($\mathbf{s} \to 0$) result in $|\mathbf{H(s)}|$ near its maximum value (unity, or 0 dB), and high frequencies ($\mathbf{s} \to \infty$) result in $|\mathbf{H(s)}| \to 0$. This behavior can be understood qualitatively by considering the impedance of the capacitor: as the frequency increases, the capacitor begins to act like a short-circuit to ac signals, leading to a reduction in the output voltage. An example response curve for such a filter with $R = 500\ \Omega$ and $C = 2$ nF is shown in Fig. 16.36*b*; the corner frequency of 159 kHz (1 Mrad/s) can be found by moving the cursor to −3 dB. The sharpness of the response curve in the vicinity of the cutoff frequency can be improved by moving to a circuit containing additional reactive (i.e., capacitive and/or inductive) elements.

A high-pass filter can be constructed by simply swapping the locations of the resistor and capacitor in Fig. 16.36*a*, as we see in the next example.

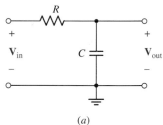

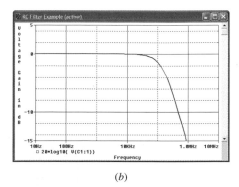

■ **FIGURE 16.36** (*a*) A simple low-pass filter constructed from a resistor-capacitor combination. (*b*) Frequency response of the circuit generated using PSpice.

EXAMPLE 16.11

Design a high-pass filter with a corner frequency of 3 kHz.

We begin by selecting a circuit topology. Since no requirements as to the sharpness of the response are given, we choose the simple circuit of Fig. 16.37.

The transfer function of this circuit is easily found to be

$$\mathbf{H(s)} \equiv \frac{\mathbf{V}_{out}}{\mathbf{V}_x} = \frac{RC\mathbf{s}}{1 + RC\mathbf{s}}$$

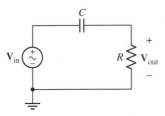

■ **FIGURE 16.37** A simple high-pass filter circuit, for which values for R and C must be selected to obtain a cutoff frequency of 3 kHz.

which has a zero at $\mathbf{s} = 0$ and a pole at $\mathbf{s} = -1/RC$, leading to "high-pass" filter behavior (i.e., $|\mathbf{H}| \to 0$ as $\omega \to \infty$).

The corner frequency of the filter circuit is $\omega_c = 1/RC$, and we seek a value of $\omega_c = 2\pi f_c = 2\pi(3000) = 18.85$ krad/s. Again, we must select a value for either R or C. In practice, our decision would most likely be based on the values of resistors and capacitors at hand, but since no such information has been provided here, we are free to make arbitrary choices.

We therefore choose the standard resistor value 4.7 kΩ for R, leading to a requirement of $C = 11.29$ nF.

The only remaining step is to verify our design with a PSpice simulation; the predicted frequency response curve is shown in Fig. 16.38.

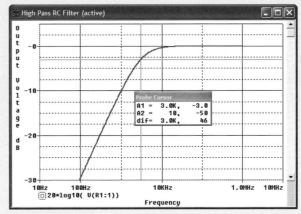

■ **FIGURE 16.38** Simulated frequency response of the final filter design, showing a cutoff (3 dB) frequency of 3 kHZ as expected.

PRACTICE
●
16.16 Design a high-pass filter with a cutoff frequency of 13.56 MHz, a common rf power supply frequency. Verify your design using PSpice.

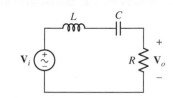

Bandpass Filters

We have already seen several circuits earlier in this chapter which could be classified as "bandpass" filters (e.g., Figs. 16.1 and 16.8). Consider the simple circuit of Fig. 16.39, in which the output is taken across the resistor. The transfer function of this circuit is easily found to be

$$A_V = \frac{sRC}{LCs^2 + RCs + 1} \quad [33]$$

FIGURE 16.39 A simple bandpass filter, constructed using a series *RLC* circuit.

The magnitude of this function is (after a few algebraic maneuvers)

$$|\mathbf{A}_V| = \frac{\omega RC}{\sqrt{(1 - \omega^2 LC)^2 + \omega^2 R^2 C^2}} \quad [34]$$

which, in the limit of $\omega \to 0$, becomes

$$|\mathbf{A}_V| \approx \omega RC \to 0$$

and in the limit of $\omega \to \infty$ becomes

$$|\mathbf{A}_V| \approx \frac{R}{\omega L} \to 0$$

We know from our experience with Bode plots that Eq. [33] represents three critical frequencies: one zero and two poles. In order to obtain a bandpass filter response with a peak value of unity (0 dB), both pole frequencies must be greater than 1 rad/s, the 0 dB crossover frequency of the zero term. These two critical frequencies can be obtained by factoring Eq. [33] or determining the values of ω at which Eq. [34] is equal to $1/\sqrt{2}$. The center frequency of this filter then occurs at $\omega = 1/\sqrt{LC}$. Thus, applying a minor amount of algebraic manipulation after setting Eq. [34] equal to $1/\sqrt{2}$, we find that

$$\left(1 - LC\omega_c^2\right)^2 = \omega_c^2 R^2 C^2 \quad [35]$$

Taking the square root of both sides yields

$$LC\omega_c^2 + RC\omega_c - 1 = 0$$

Applying the quadratic equation, we find that

$$\omega_c = -\frac{R}{L} \pm \frac{\sqrt{R^2 C^2 + 4LC}}{2LC} \quad [36]$$

Negative frequency is a nonphysical solution to our original equation, and so only the positive radicand of Eq. [36] is applicable. However, we may have been a little too hasty in taking the positive square root of both sides of Eq. [35]. Considering the negative square root as well, which is equally valid, we also obtain

$$\omega_c = \frac{R}{L} \pm \frac{\sqrt{R^2 C^2 + 4LC}}{2LC} \quad [37]$$

from which it can be shown that only the positive radicand is physical. Thus, we obtain ω_L from Eq. [36] and ω_H from Eq. [37]; since $\omega_H - \omega_L = \mathcal{B}$, simple algebra shows that $\mathcal{B} = R/L$.

EXAMPLE **16.12**

Design a bandpass filter characterized by a bandwidth of 1 MHz and a high-frequency cutoff of 1.1 MHz.

We choose the circuit topology of Fig. 16.37, and begin by determining the corner frequencies required. The bandwidth is given by $f_H - f_L$, so

$$f_L = 1.1 \times 10^6 - 1 \times 10^6 = 100 \text{ kHz}$$

and

$$\omega_L = 2\pi f_L = 628.3 \text{ krad/s}$$

The high-frequency cutoff (ω_H) is simply 6.912 Mrad/s.

In order to proceed to design a circuit with these characteristics, it is necessary to obtain an expression for each frequency in terms of the variables R, L, and C.

Setting Eq. [37] equal to $2\pi(1.1 \times 10^6)$ allows us to solve for $1/LC$, as we already know that $\mathcal{B} = 2\pi(f_H - f_L) = 6.283 \times 10^6$.

$$\frac{1}{2}\mathcal{B} + \left[\frac{1}{4}\mathcal{B}^2 + \frac{1}{LC}\right]^{1/2} = 2\pi(1.1 \times 10^6)$$

Solving, we find that $1/LC = 4.343 \times 10^{12}$. Arbitrarily selecting $L = 1$ H (a little large, practically speaking), we obtain $R = 6.283$ MΩ and $C = 230.3$ fF. It should be noted that there is no unique solution for this "design" problem—either R, L, or C can be selected as a starting point.

PSpice verification of our design is shown in Fig. 16.40.

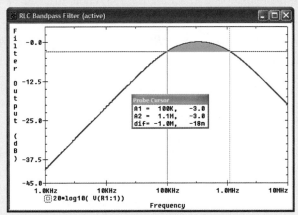

■ **FIGURE 16.40** Simulated response of the bandpass filter design showing a bandwidth of 1 MHz and a high-frequency cutoff of 1.1 MHz as desired. The passband frequencies have been shaded in green.

PRACTICE

16.17 Design a bandpass filter with a low-frequency cutoff of 100 rad/s and a high-frequency cutoff of 10 krad/s.

Ans: One possible answer of many: $R = 990\,\Omega$, $L = 100$ mH, and $C = 10\,\mu$F.

The type of circuit we have been considering is known as a *passive filter,* as it is constructed of only passive components (i.e., no transistors, op amps, or other "active" elements). Although passive filters are relatively common, they are not well suited to all applications. The gain (defined as the output voltage divided by the input voltage) of a passive filter can be difficult to set, and amplification is often desirable in filter circuits.

Active Filters

The use of an active element such as the op amp in filter design can overcome many of the shortcomings of passive filters. As we saw in Chap. 6, op amp circuits can easily be designed to provide gain. Op amp circuits can also exhibit inductor-like behavior through the strategic location of capacitors.

The internal circuitry of an op amp contains very small capacitances (typically on the order of 100 pF), and these limit the maximum frequency at which the op amp will function properly. Thus, any op amp circuit will behave as a low-pass filter, with a cutoff frequency for modern devices of perhaps 20 MHz or more (depending on the circuit gain).

EXAMPLE 16.13

Design an active low-pass filter with a cutoff frequency of 10 kHz and a voltage gain of 40 dB.

For frequencies much less than 10 kHz, we require an amplifier circuit capable of providing a gain of 40 dB, or 100 V/V. This can be accomplished by simply using a noninverting amplifier (such as the one shown in Fig. 16.41a) with

$$\frac{R_f}{R_1} + 1 = 100$$

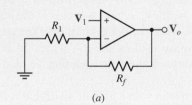

(a)

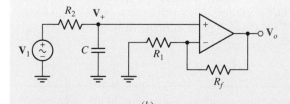

(b)

■ FIGURE 16.41 (a) A simple noninverting op amp circuit. (b) A low-pass filter consisting of a resistor R_2 and a capacitor C has been added to the input.

(Continued on next page)

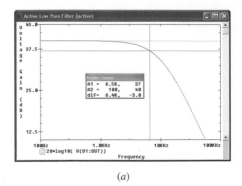

(a)

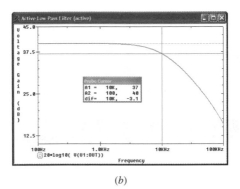

(b)

■ **FIGURE 16.42** (a) Frequency response for filter circuit using a μA741 op amp, showing a corner frequency of 6.4 kHz. (b) Frequency response of the same filter circuit, but using an LF111 op amp instead. The cutoff frequency for this circuit is 10 kHz, the desired value.

To provide a high-frequency corner at 10 kHz, we require a low-pass filter at the input to the op amp (as in Fig. 16.41b). To derive the transfer function, we begin at the noninverting input,

$$\mathbf{V}_+ = \mathbf{V}_i \frac{1/sC}{R_2 + 1/sC} = \mathbf{V}_i \frac{1}{1 + sR_2C}$$

At the inverting input we have

$$\frac{\mathbf{V}_o - \mathbf{V}_+}{R_f} = \frac{\mathbf{V}_+}{R_1}$$

Combining these two equations and solving for $\mathbf{V}_o$, we find that

$$\mathbf{V}_o = \mathbf{V}_i \left(\frac{1}{1 + sR_2C} \right) \left(1 + \frac{R_f}{R_1} \right)$$

The maximum value of the gain $\mathbf{A}_V = \mathbf{V}_o/\mathbf{V}_i$ is $1 + R_f/R_1$, so we set this quantity equal to 100. Since neither resistor appears in the expression for the corner frequency $(R_2C)^{-1}$, either may be selected first. We thus choose $R_1 = 1\,k\Omega$, so that $R_f = 99\,k\Omega$.

Arbitrarily selecting $C = 1\,\mu$F, we find that

$$R_2 = \frac{1}{2\pi(10 \times 10^3)C} = 15.9\,\Omega$$

At this point, our design is complete. Or is it? The simulated frequency response of this circuit is shown in Fig. 16.42a.

It is readily apparent that our design does not in fact meet the 10 kHz cutoff specification. What did we do wrong? A careful check of our algebra does not yield any errors, so an erroneous assumption must have been made somewhere. The simulation was performed using a μA741 op amp, as opposed to the ideal op amp assumed in the derivations. It turns out that this is the source of our discomfort—the same circuit with an LF111 op amp substituted for the μA741 results in a cutoff frequency of 10 kHz as desired; the corresponding simulation result is shown in Fig. 16.42b.

Unfortunately, the μA741 op amp with a gain of 40 dB has a corner frequency in the vicinity of 10 kHz, which cannot be neglected in this instance. The LF111, however, does not reach its first corner frequency until approximately 75 kHz, which is far enough away from 10 kHz that it does not affect our design.

PRACTICE

16.18 Design a low-pass filter circuit with a gain of 30 dB and a cutoff frequency of 1 kHz.

Ans: One possible answer of many: $R_1 = 100\,k\Omega$, $R_f = 3.062\,M\Omega$, $R_2 = 79.58\,\Omega$, and $C = 2\,\mu$F.

PRACTICAL APPLICATION

Bass, Treble, and Midrange Adjustment

The ability to independently adjust the bass, treble, and midrange settings on a sound system is commonly desirable, even in the case of inexpensive equipment. The audio frequency range (at least for the human ear) is commonly accepted to be 20 Hz to 20 kHz, with bass corresponding to lower frequencies (< 500 Hz or so) and treble corresponding to higher frequencies (> 5 kHz or thereabouts).

Designing a simple graphic equalizer is a relatively straightforward endeavor, although a system such as that shown in Fig. 16.43 requires a bit more effort. In the bass, midrange, treble type equalizer common on many portable radios, the main signal (provided by the radio receiver circuit, or perhaps a CD player) consists of a wide spectrum of frequencies having a bandwidth of approximately 20 kHz.

■ FIGURE 16.43 An example of a graphic equalizer.
Courtesy of Alesis

This signal must be sent to three different op amp circuits, each with a different filter at the input. The bass adjustment circuit will require a low-pass filter, the treble adjustment circuit will require a high-pass filter, and the midrange adjustment circuit requires a bandpass filter. The output of each op amp circuit is then fed into a summing amplifier circuit; a block diagram of the complete circuit is shown in Fig. 16.44.

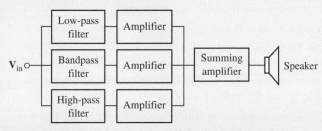

■ FIGURE 16.44 Block diagram of a simple graphic equalizer circuit.

Our basic building block is shown in Fig. 16.45. This circuit consists of a noninverting op amp circuit characterized by a voltage gain of $1 + R_f/R_1$, and a simple low-pass filter composed of a resistor R_2 and a capacitor C. The feedback resistor R_f is a variable resistor (sometimes referred to as a **_potentiometer_**), and allows the gain to be varied through the rotation of a knob; the layperson

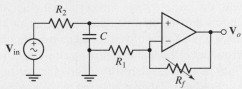

■ FIGURE 16.45 The bass adjustment section of the amplifier circuit.

would call this resistor the volume control. The low-pass filter network restricts the frequencies that will enter the op amp and hence be amplified; the corner frequency is simply $(R_2C)^{-1}$. If the circuit designer needs to allow the user to also select the break frequency for the filter, R_2 may be replaced by a potentiometer, or, alternatively, C could be replaced by a variable capacitor. The remaining stages are constructed in essentially the same way, but with a different filter network at the input.

In order to keep the resistors, capacitors, and op amps separate, we should add an appropriate subscript to each as an indication of the stage to which it belongs (t, m, b). Beginning with the treble stage, we have already encountered problems in using the μA741 in the 10 to 20 kHz range at high gain, so perhaps the LF111 is a better choice here as well. Selecting a treble cutoff frequency of 5 kHz (there is some variation among values selected by different audio circuit designers), we require

$$\frac{1}{R_{2t}C_t} = 2\pi(5 \times 10^3) = 3.142 \times 10^4$$

Arbitrarily selecting $C_t = 1\mu$F results in a required value of 31.83 Ω for R_{2t}. Selecting $C_b = 1\,\mu$F as well (perhaps we can negotiate a quantity discount), we need $R_{2b} = 318.3\ \Omega$ for a bass cutoff frequency of 500 Hz. We leave the design of a suitable bandpass filter for the reader.

The next part of our design is to choose suitable values for R_{1t} and R_{1b}, as well as the corresponding feedback resistors. Without any instructions to the contrary, it is probably simplest to make both stages identical. Therefore, we arbitrarily select both R_{1t} and R_{1b} as 1 kΩ, and R_{ft} and R_{fb} as 10 kΩ potentiometers (meaning that the range will be from 0 to 10 kΩ). This allows the volume of one signal to be up to 11 times louder than the other. In case we need our design to be portable, we select ± 9 V supply voltages, although this can be easily changed if needed.

(Continued on next page)

Now that the design of the filter stage is complete, we are ready to consider the design of the summing stage. For the sake of simplicity, we should power this op amp stage with the same voltage sources as the other stages, which limits the maximum output voltage magnitude to less than 9 V. We use an inverting op amp configuration, with the output of each of the filter op amp stages fed directly into its own 1 kΩ resistor. The other terminal of each 1 kΩ resistor is then connected to the inverting input of the summing amplifier stage. The appropriate potentiometer for the summing amplifier stage must be selected in order to prevent saturation, so that knowledge of both the input voltage range and the output speaker wattage is required.

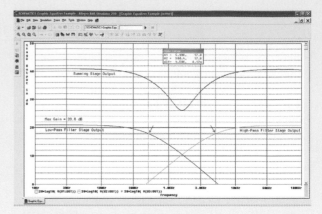

■ **FIGURE 16.46** Simulated frequency response of the equalizer design.

SUMMARY AND REVIEW

- ❑ Resonance is the condition in which a fixed-amplitude sinusoidal forcing function produces a response of maximum amplitude.

- ❑ An electrical network is in resonance when the voltage and current at the network input terminals are in phase.

- ❑ The quality factor is proportional to the maximum energy stored in a network divided by the total energy lost per period.

- ❑ A half-power frequency is defined as the frequency at which the magnitude of a circuit response function is reduced to $1/\sqrt{2}$ times its maximum value.

- ❑ The bandwidth of a resonant circuit is defined as the difference between the upper and lower half-power frequencies.

- ❑ A high-Q circuit is a resonant circuit in which the quality factor is ≥ 5.

- ❑ In a high-Q circuit, each half-power frequency is located approximately one-half bandwidth from the resonant frequency.

- ❑ A series resonant circuit is characterized by a *low* impedance at resonance, whereas a parallel resonant circuit is characterized by a *high* impedance at resonance.

- ❑ A series resonant circuit and a parallel resonant circuit are equivalent if $R_p = R_s(1 + Q^2)$ and $X_p = X_s(1 + Q^{-2})$.

- ❑ Impractical values for components often make design easier. The transfer function of a network may be scaled in magnitude or frequency using appropriate replacement values for components.

- ❑ Bode diagrams allow the rough shape of a transfer function to be plotted quickly from the poles and zeros.

- ❑ The four basic types of filters are low-pass, high-pass, bandpass, and bandstop.

- ❑ Passive filters use only resistors, capacitors, and inductors; active filters are based on op amps or other active elements.

READING FURTHER

A good discussion of a large variety of filters can be found in:

J.T. Taylor and Q. Huang, eds., *CRC Handbook of Electrical Filters*. Boca Raton, Fla: CRC Press, 1997.

A comprehensive compilation of various active filter circuits and design procedures is given in:

D. Lancaster, *Lancaster's Active Filter Cookbook*, 2nd ed. Burlington, Mass.: Newnes, 1996.

EXERCISES

16.1 Parallel Resonance

1. A parallel *RLC* circuit has $R = 1$ kΩ, $C = 47$ μF, and $L = 11$ mH. (*a*) Compute Q_0. (*b*) Determine the resonant frequency (in Hz). (*c*) Sketch the voltage response as a function of frequency if the circuit is excited by a steady-state 1 mA sinusoidal current source.

2. A parallel *RLC* circuit is measured to have a Q_0 of 200. Determine the remaining component value if (*a*) $R = 1$ Ω and $C = 1$ μF; (*b*) $L = 12$ fH and $C = 2.4$ nF; (*c*) $R = 121.7$ kΩ and $L = 100$ pH.

3. A varactor is a semiconductor device whose reactance may be varied by applying a bias voltage. The quality factor can be expressed[4] as

$$Q \approx \frac{\omega C_J R_P}{1 + \omega^2 C_J^2 R_P R_S}$$

where C_J is the junction capacitance (which depends on the voltage applied to the device), R_S is the series resistance of the device, and R_P is an equivalent parallel resistance term. (*a*) If $C_J = 3.77$ pF at 1.5 V, $R_P = 1.5$ MΩ, and $R_S = 2.8$ Ω, plot the quality factor as a function of frequency ω. (*b*) Differentiate the expression for Q to obtain both ω_0 and Q_{max}.

4. Determine Q for (*a*) a ping-pong ball; (*b*) a quarter; (*c*) this textbook. Be sure to provide precise details of the measurement conditions and any observations you make, including averaging or other statistical analysis.

5. A parallel resonant circuit has parameter values of $\alpha = 80$ Np/s and $\omega_d = 1200$ rad/s. If the impedance at $\mathbf{s} = -2\alpha + j\omega_d$ has a magnitude of 400 Ω, calculate Q_0, R, L, and C.

6. Find the resonant frequency of the two-terminal network shown in Fig. 16.47.

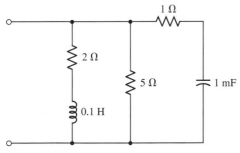

■ FIGURE 16.47

7. Let $R = 1$ MΩ, $L = 1$ H, $C = 1$ μF, and $\mathbf{I} = 10\underline{/0°}$ μA in the circuit of Fig. 16.1. (*a*) Find ω_0 and Q_0. (*b*) Plot $|\mathbf{V}|$ as a function of ω, $995 < \omega < 1005$ rad/s.

(4) S. M. Sze, *Physics of Semiconductor Devices*, 2d ed. New York: Wiley, 1981, p. 116.

8. For the network shown in Fig. 16.48, find (*a*) the resonant frequency ω_0; (*b*) $\mathbf{Z}_{in}(j\omega_0)$.

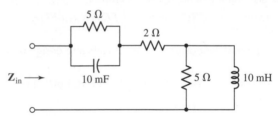

■ **FIGURE 16.48**

9. A parallel resonant circuit has impedance poles at $\mathbf{s} = -50 \pm j1000$ s^{-1}, and a zero at the origin. If $C = 1\,\mu$F: (*a*) find L and R; (*b*) calculate $\mathbf{Z}$ at $\omega = 1000$ rad/s.

D 10. Design a parallel resonant circuit for an AM radio so that a variable inductor can adjust the resonant frequency over the AM broadcast band, 535 to 1605 kHz, with $Q_0 = 45$ at one end of the band and $Q_0 \leq 45$ throughout the band. Let $R = 20$ kΩ, and specify values for C, L_{min}, and L_{max}.

11. (*a*) Find $\mathbf{Y}_{in}$ for the network shown in Fig. 16.49. (*b*) Determine ω_0 and $\mathbf{Z}_{in}(j\omega_0)$ for the network.

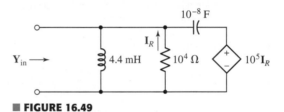

■ **FIGURE 16.49**

12. Determine the resonant frequency for $t > 0$ of the network depicted in Fig. 16.50.

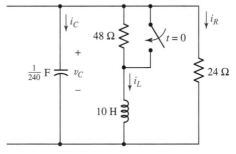

■ **FIGURE 16.50**

13. Determine the resonant frequency for $t > 0$ of the network depicted in Fig. 16.51.

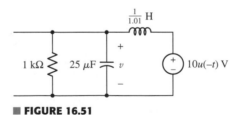

■ **FIGURE 16.51**

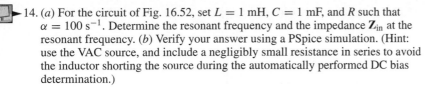

14. (a) For the circuit of Fig. 16.52, set $L = 1$ mH, $C = 1$ mF, and R such that $\alpha = 100$ s^{-1}. Determine the resonant frequency and the impedance $\mathbf{Z}_{in}$ at the resonant frequency. (b) Verify your answer using a PSpice simulation. (Hint: use the VAC source, and include a negligibly small resistance in series to avoid the inductor shorting the source during the automatically performed DC bias determination.)

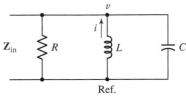

Ref.

■ **FIGURE 16.52**

 15. (a) For the circuit of Fig. 16.52, set $L = 1$ mH, R such that $\alpha = 50$ s^{-1}, and C such that $\omega_d = 5000$ rad/s. Determine the resonant frequency and the impedance $\mathbf{Z}_{in}$ at the resonant frequency. (b) Verify your answer using a PSpice simulation. (Hint: use the VAC source, and include a negligibly small resistance in series to avoid the inductor shorting the source during the automatically performed DC bias determination.)

16.2 Bandwidth and High-Q Circuits

16. A parallel resonant circuit has $\omega_0 = 100$ rad/s, $Q_0 = 80$, and $C = 0.2\,\mu$F. (a) Find R and L. (b) Use approximate methods to plot $|\mathbf{Z}|$ versus ω.

17. Use the exact relationships to find R, L, and C for a parallel resonant circuit that has $\omega_1 = 103$ rad/s, $\omega_2 = 118$ rad/s, and $|\mathbf{Z}(j105)| = 10\,\Omega$.

18. Let $\omega_0 = 30$ krad/s, $Q_0 = 10$, and $R = 600\,\Omega$ for a certain parallel resonant circuit. (a) Find the bandwidth. (b) Calculate N at $\omega = 28$ krad/s. (c) Use approximate methods to determine $\mathbf{Z}_{in}(j28,000)$. (d) Find the true value of $\mathbf{Z}_{in}(j28,000)$. (e) State the percentage error incurred by using the approximate relationships to calculate $|\mathbf{Z}_{in}|$ and ang $\mathbf{Z}_{in}$ at 28 krad/s.

19. A parallel resonant circuit is resonant at 400 Hz with $Q_0 = 8$ and $R = 500\,\Omega$. If a current of 2 mA is applied to the circuit, use approximate methods to find the cyclic frequency of the current if (a) the voltage across the circuit has a magnitude of 0.5 V; (b) the resistor current has a magnitude of 0.5 mA.

20. A parallel resonant circuit has $\omega_0 = 1$ Mrad/s and $Q_0 = 10$. Let $R = 5$ kΩ and find (a) L; (b) the frequency above ω_0 at which $|\mathbf{Z}_{in}| = 2$ kΩ; (c) the frequency at which ang $\mathbf{Z}_{in} = -30°$.

21. Use good approximations on the circuit of Fig. 16.53 to (a) find ω_0; (b) calculate $\mathbf{V}_1$ at the resonant frequency; (c) calculate $\mathbf{V}_1$ at a frequency that is 15 krad/s above resonance.

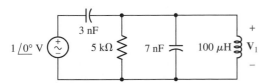

■ **FIGURE 16.53**

22. (a) Apply the definition of resonance to find ω_0 for the network of Fig. 16.54. (b) Find $\mathbf{Z}_{in}(j\omega_0)$.

23. A parallel resonant circuit is characterized by $f_0 = 1000$ Hz, $Q_0 = 40$, and $|\mathbf{Z}_{in}(j\omega_0)| = 2$ kΩ. Use the approximate relationships to find (a) $\mathbf{Z}_{in}$ at 1010 Hz; (b) the frequency range over which the approximations are reasonably accurate.

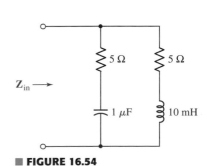

■ **FIGURE 16.54**

24. Find the bandwidth of each of the response curves shown in Fig. 16.55.

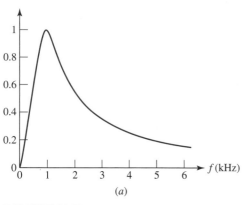

(a)

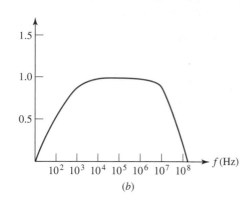

(b)

■ **FIGURE 16.55**

25. A parallel resonant circuit is known to have a bandwidth of 1 MHz, and a lower half-power frequency $f_1 = 5.5$ kHz. (a) What is the upper half-power frequency (in Hz)? (b) What is the circuit's resonant frequency f_0? (c) What is the quality factor of the circuit when operated at its resonant frequency?

26. A parallel resonant circuit is known to have a bandwidth of 1 GHz, and a lower half-power frequency $f_1 = 75.3$ MHz. (a) What is the upper half-power frequency (in Hz)? (b) What is the circuit's resonant frequency f_0? (c) What is the quality factor of the circuit when operated at its resonant frequency?

27. (a) Sketch the voltage response curve for a circuit having a lower half-power frequency of 1000 rad/s, an upper half-power frequency of 4000 rad/s, and a maximum voltage magnitude of 10 V. (b) What is the resonant frequency of the circuit? (c) What is the bandwidth of the circuit? (d) What is the quality factor of the circuit when operated at the resonant frequency?

28. (a) If the 1 μF capacitor of Fig. 16.54 is replaced with a 330 pF capacitor, find the resonant frequency of the new circuit. (b) Verify your answer using PSpice. (Hint: Use the VAC source, and simulate over several decades of frequency.)

 29. Design a parallel *RLC* circuit having a bandwidth of 5.5 kHz and a lower half-power frequency of 500 Hz. Verify your design with an appropriate PSpice simulation.

16.3 Series Resonance

30. A series circuit is constructed from two 5 Ω resistors, four 100 μH inductors, and a 3.3 μF capacitor. (a) Compute the resonant frequency of the circuit. (b) Calculate the quality factor of the circuit when operated at the resonant frequency. (c) Determine the input impedance at the resonant frequency, 0.1 times the resonant frequency, and 10 times the resonant frequency.

31. If a series circuit is known to have a bandwidth of 3 MHz, and a lower half-power frequency $f_1 = 17$ kHz, determine (a) the upper half-power frequency (in Hz); (b) the resonant frequency f_0 of the circuit; (c) the quality factor of the circuit when operated at the resonant frequency.

32. (a) Determine the impedance of a series *RLC* circuit ($R = 1\ \Omega$, $L = 1$ mH, $C = 2$ mF) when operated at the resonant frequency. (b) Verify your solution with an appropriate PSpice simulation. (Hint: A very large resistor in parallel with the capacitor will avoid error messages associated with no dc path to ground.)

33. (a) Determine the impedance of a series *RLC* circuit ($R = 1$ kΩ, $L = 1\ \mu$H, $C = 2\ \mu$F) when operated at the resonant frequency. (b) Verify your solution

with an appropriate PSpice simulation. (Hint: A very large resistor in parallel with the capacitor will avoid error messages associated with no dc path to ground.)

34. (*a*) Use approximate techniques to plot $|\mathbf{V}_{out}|$ versus ω for the circuit shown in Fig. 16.56. (*b*) Find an exact value for $\mathbf{V}_{out}$ at $\omega = 9$ rad/s.

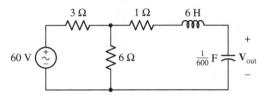

■ **FIGURE 16.56**

35. A series resonant network consists of a 50 Ω resistor, a 4 mH inductor, and a 0.1 μF capacitor. Calculate values for (*a*) ω_0; (*b*) f_0; (*c*) Q_0; (*d*) $\mathcal{B}$; (*e*) ω_1; (*f*) ω_2; (*g*) $\mathbf{Z}_{in}$ at 45 krad/s; (*h*) the ratio of magnitudes of the capacitor impedance to the resistor impedance at 45 krad/s.

36. After deriving $\mathbf{Z}_{in}(\mathbf{s})$ in Fig. 16.57, find: (*a*) ω_0; (*b*) Q_0.

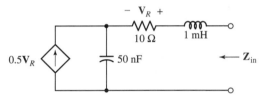

■ **FIGURE 16.57**

37. Inspect the circuit of Fig. 16.58, noting the amplitude of the source voltage. Now decide whether you would be willing to put your bare hands across the capacitor if the circuit were actually built in the lab. Plot $|\mathbf{V}_C|$ versus ω to justify your answer.

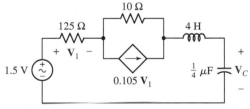

■ **FIGURE 16.58**

38. A certain series resonant circuit has $f_0 = 500$ Hz, $Q_0 = 10$, and $X_L = 500\ \Omega$ at resonance. (*a*) Find R, L, and C. (*b*) If a source $\mathbf{V}_s = 1\underline{/0°}$ V is connected in series with the circuit, find exact values for $|\mathbf{V}_C|$ at the frequencies $f = 450$, 500, and 550 Hz.

39. A three-element network has an input impedance $\mathbf{Z}(\mathbf{s})$ that shows poles at $\mathbf{s} = 0$ and infinity, and a pair of zeros at $\mathbf{s} = -20,000 \pm j80,000\ \text{s}^{-1}$. Specify the three element values if $\mathbf{Z}_{in}(-10,000) = -20 + j0\ \Omega$.

16.4 Other Resonant Forms

40. Make a few reasonable approximations on the network of Fig. 16.59 and obtain values for ω_0, Q_0, $\mathcal{B}$, $\mathbf{Z}_{in}(j\omega_0)$, and $\mathbf{Z}_{in}(j99,000)$.

41. What value of resistance should be connected across the input of the network in Fig. 16.59 to cause it to have a Q_0 of 50?

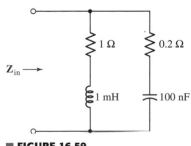

■ **FIGURE 16.59**

42. Refer to the network shown in Fig. 16.60 and use approximate techniques to determine the minimum magnitude of $\mathbf{Z}_{in}$ and the frequency at which it occurs.

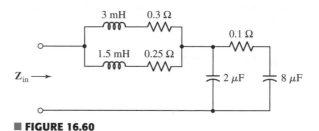

■ FIGURE 16.60

43. For the circuit of Fig. 16.61: (a) prepare an approximate response curve of $|\mathbf{V}|$ versus ω, and (b) calculate the exact value of $\mathbf{V}$ at $\omega = 50$ rad/s.

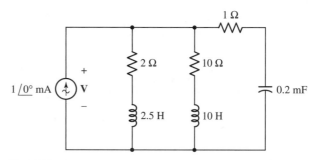

■ FIGURE 16.61

44. (a) Use approximate methods to calculate $|\mathbf{V}_x|$ at $\omega = 2000$ rad/s for the circuit of Fig. 16.62. (b) Obtain the exact value of $|\mathbf{V}_x(j2000)|$.

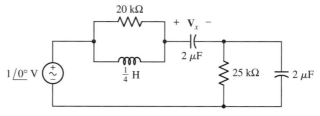

■ FIGURE 16.62

45. A parallel combination of a 5 kΩ resistor and a 1 μF capacitor is constructed. Determine a series-connected equivalent if the operating frequency ω is (a) 10^3 rad/s; (b) 10^4 rad/s; (c) 10^5 rad/s.

46. A series combination of a 5 kΩ resistor and a 1 μF capacitor is constructed. Determine a parallel-connected equivalent if the operating frequency ω is (a) 10^3 rad/s; (b) 10^4 rad/s; (c) 10^5 rad/s.

47. A series combination of a 470 Ω resistor and a 3.3 μH inductor is constructed. Determine a series-connected equivalent if the operating frequency ω is (a) 10^3 rad/s; (b) 10^4 rad/s; (c) 10^5 rad/s.

48. A parallel combination of a 470 Ω resistor and a 3.3 μH inductor is constructed. Determine a parallel-connected equivalent if the operating frequency ω is (a) 10^3 rad/s; (b) 10^4 rad/s; (c) 10^5 rad/s.

49. (a) For the circuit of Fig. 16.63, employ approximate methods to calculate $|\mathbf{V}_x|$ at $f = 1.6$ MHz. (b) Compute the exact value of $|\mathbf{V}_x(j10 \times 10^6)|$. (c) Verify your results with an appropriate PSpice simulation.

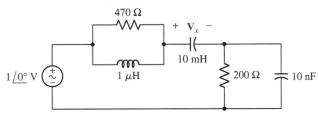

■ FIGURE 16.63

16.5 Scaling

50. The filter shown in Fig. 16.64a has the response curve shown in Fig. 16.64b. (a) Scale the filter so that it operates between a 50 Ω source and a 50 Ω load and has a cutoff frequency of 20 kHz. (b) Draw the new response curve.

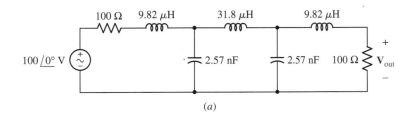

(a)

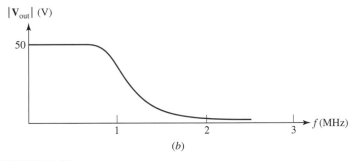

(b)

■ FIGURE 16.64

51. (a) Find $\mathbf{Z}_{in}(\mathbf{s})$ for the network shown in Fig. 16.65. (b) Write an expression for $\mathbf{Z}_{in}(\mathbf{s})$ after it has been scaled by $K_m = 2$, $K_f = 5$. (c) Scale the elements in the network by $K_m = 2$, $K_f = 5$, and draw the new network.

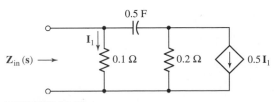

■ FIGURE 16.65

52. (a) Use good approximations to find ω_0 and Q_0 for the circuit of Fig. 16.66. (b) Scale the network to the right of the source so that it is resonant at 1 Mrad/s. (c) Specify ω_0 and $\mathcal{B}$ for the scaled circuit.

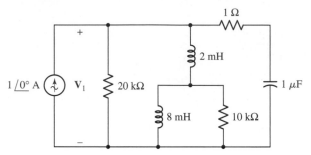

■ **FIGURE 16.66**

53. (a) Draw the new configuration for Fig. 16.67 after the network is scaled by $K_m = 250$ and $K_f = 400$. (b) Determine the Thévenin equivalent of the scaled network at $\omega = 1$ krad/s.

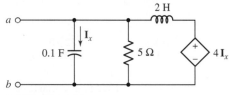

■ **FIGURE 16.67**

54. A network composed entirely of ideal R's, L's, and C's has a pair of input terminals to which a sinusoidal current source $\mathbf{I}_s$ is connected, and a pair of open-circuited output terminals at which a voltage $\mathbf{V}_{\text{out}}$ is defined. If $\mathbf{I}_s = 1\underline{/0°}$ A at $\omega = 50$ rad/s, then $\mathbf{V}_{\text{out}} = 30\underline{/25°}$ V. Specify $\mathbf{V}_{\text{out}}$ for each condition described as follows. If it is impossible to determine the value of $\mathbf{V}_{\text{out}}$, write OTSK.[5] (a) $\mathbf{I}_s = 2\underline{/0°}$ A at $\omega = 50$ rad/s; (b) $\mathbf{I}_s = 2\underline{/40°}$ A at $\omega = 50$ rad/s; (c) $\mathbf{I}_s = 2\underline{/40°}$ A at 200 rad/s; (d) the network is scaled by $K_m = 30$, $\mathbf{I}_s = 2\underline{/40°}$ A, $\omega = 50$ rad/s; (e) $K_m = 30$, $K_f = 4$, $\mathbf{I}_s = 2\underline{/40°}$ A, $\omega = 200$ rad/s.

16.6 Bode Diagrams

55. Find H_{dB} if $\mathbf{H}(\mathbf{s})$ equals (a) 0.2; (b) 50; (c) $12/(\mathbf{s}+2) + 26/(\mathbf{s}+20)$ for $\mathbf{s} = j10$. Find $|\mathbf{H}(\mathbf{s})|$ if H_{dB} equals (d) 37.6 dB; (e) -8 dB; (f) 0.01 dB.

 56. Draw the Bode amplitude plot for (a) $20(\mathbf{s}+1)/(\mathbf{s}+100)$; (b) $2000\mathbf{s}(\mathbf{s}+1)/(\mathbf{s}+100)^2$; (c) $\mathbf{s} + 45 + 200/\mathbf{s}$. (d) Verify your sketches using MATLAB.

57. For Fig. 16.68, prepare Bode amplitude and phase plots for transfer function, $\mathbf{H}(\mathbf{s}) = \mathbf{V}_C/\mathbf{I}_s$.

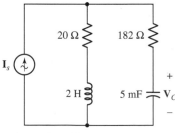

■ **FIGURE 16.68**

(5) Only The Shadow Knows.

58. (a) Using an origin at $\omega = 1$, $H_{dB} = 0$, construct the Bode amplitude plot for $\mathbf{H}(\mathbf{s}) = 5 \times 10^8 \mathbf{s}(\mathbf{s} + 100)/[(\mathbf{s} + 20)(\mathbf{s} + 1000)^3]$. (b) Give the coordinates for all corners and all intercepts on the Bode plot. (c) Give the exact value of $20 \log |\mathbf{H}(j\omega)|$ for each corner frequency in part b.

59. (a) Construct a Bode phase plot for $\mathbf{H}(\mathbf{s}) = 5 \times 10^8 \mathbf{s}(\mathbf{s} + 100)/[(\mathbf{s} + 20)(\mathbf{s} + 1000)^3]$. Place the origin at $\omega = 1$, ang $= 0°$. (b) Give the coordinates for all points on the phase plot at which the slope changes. (c) Give the exact value of ang $\mathbf{H}(j\omega)$ for each frequency listed in part b.

60. (a) Construct a Bode magnitude plot for the transfer function $\mathbf{H}(\mathbf{s}) = 1 + 20/\mathbf{s} + 400/\mathbf{s}^2$. (b) Compare the Bode plot and exact values at $\omega = 5$ and 100 rad/s. (c) Verify your Bode plot using MATLAB.

61. (a) Find $\mathbf{H}(\mathbf{s}) = \mathbf{V}_R/\mathbf{V}_s$ for the circuit shown in Fig. 16.69. (b) Draw Bode amplitude and phase plots for $\mathbf{H}(\mathbf{s})$. (c) Calculate the exact values of H_{dB} and ang $\mathbf{H}(j\omega)$ at $\omega = 20$ rad/s.

62. Construct the Bode amplitude plot for the transfer function $\mathbf{H}(\mathbf{s}) = \mathbf{V}_{out}/\mathbf{V}_{in}$ of the network shown in Fig. 16.70.

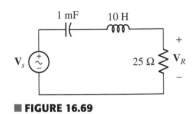

■ FIGURE 16.69

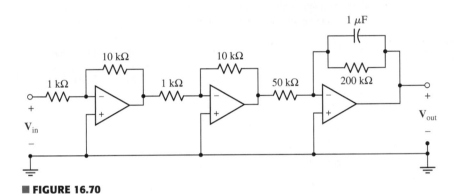

■ FIGURE 16.70

63. For the network of Fig. 16.71: (a) find $\mathbf{H}(\mathbf{s}) = \mathbf{V}_{out}/\mathbf{V}_{in}$; (b) draw the Bode amplitude plot for H_{dB}; (c) draw the Bode phase plot for $\mathbf{H}(j\omega)$.

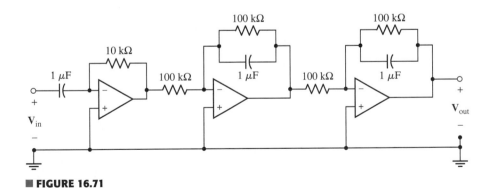

■ FIGURE 16.71

16.7 Filters

64. The audio frequency range of the bottlenose dolphin extends from approximately 250 Hz to 150 kHz. Frequencies between 250 Hz and ~ 50 kHz are thought to be used primarily in social communications, and "clicks" with frequencies above ~ 40 kHz are believed to be primarily used in echolocation.

Design an amplifier circuit that will selectively amplify dolphin social conversations. The microphone can be modeled as a sinusoidal voltage source with peak amplitude less than 15 mV in series with a 1 Ω resistor. The voltage delivered to the 1 kΩ earphone should peak at approximately 1 V.

65. Design a filter circuit that removes the entire range of frequencies audible to the human ear (20 Hz to 20 kHz), while allowing lower- and higher-frequency signals to pass through. Verify your design using PSpice.

66. Design a filter circuit that removes any signals with a frequency greater than or equal to 1 kHz. Verify your design using PSpice.

67. A microphone that is very sensitive to high frequencies is used to detect certain types of imminent jet engine failures, but is unfortunately also picking up low-frequency noise from flap and aileron hydraulic systems, resulting in false alarms. Design a filter circuit to remove the noise signals while selectively amplifying the high-frequency signals by at least a factor of 100. The low-frequency noise signal has its peak energy in the vicinity of 20 Hz, and has fallen to less than 1 percent of its maximum by 1 kHz. The engine failure signals begin in the vicinity of 25 kHz.

68. Complete the design discussed in the Practical Application. (*a*) Begin by designing a suitable midrange stage. (*b*) Simulate the frequency response of your circuit by varying the feedback resistance between its minimum and maximum values.

69. Despite the fact that the human auditory response is commonly accepted to lie within the range of 20 Hz to 20 kHz, the bandwidth of many telephone systems is limited to 3 kHz. Design a filter circuit that will convert 20 kHz bandwidth speech into reduced 3 kHz "telephone bandwidth" speech. The input is a microphone with a maximum voltage of 150 mV and essentially zero series resistance; the output is an 8 Ω speaker. The speech should be amplified by at least a factor of 10. Verify your design using PSpice.

70. Design a circuit that removes 50*n* Hz components from an antenna signal if *n* is an integer in the range of 1 to 4. A good "notch" filter (i.e., a filter that "notches out" a particular frequency) topology is given by the circuit of Fig. 16.39, but with the output now taken across the inductor-capacitor series combination instead of across the resistor. The antenna signal may be modeled as a 1 V peak amplitude time-varying source with zero series resistance.

71. A sensitive piece of monitoring equipment is adversely affected by 60 Hz power line induced noise contaminating the incoming signals. The nature of the signals prevents any type of low-pass, high-pass, or bandpass filters to be used to solve the problem. Design a "notch" filter that selectively removes any 60 Hz signals from the input of the equipment. It may be assumed that the equipment has an essentially infinite Thévenin equivalent resistance. A good "notch" filter topology is given by the circuit of Fig. 16.39, but with the output now taken across the inductor-capacitor series combination instead of across the resistor.

17 Two-Port Networks

KEY CONCEPTS

The Distinction Between
One-Port and Two-Port
Networks

Admittance (y) Parameters

Impedance (z) Parameters

Hybrid (h) Parameters

Transmission (t) Parameters

Transformation Methods
Between y, z, h, and t
Parameters

Circuit Analysis Techniques
Using Network Parameters

INTRODUCTION

A general network having two pairs of terminals, one perhaps labeled the "input terminals" and the other the "output terminals," is a very important building block in electronic systems, communication systems, automatic control systems, transmission and distribution systems, or other systems in which an electrical signal or electric energy enters the input terminals, is acted upon by the network, and leaves via the output terminals. The output terminal pair may very well connect with the input terminal pair of another network. When we studied the concept of Thévenin and Norton equivalent networks in Chap. 5, we were introduced to the idea that it is not always necessary to know the detailed workings of part of a circuit. This chapter extends such concepts to situations where we don't even know the details of the inner workings of our circuit. Armed only with the knowledge that the circuit is linear, and the ability to measure voltages and currents, we will shortly see that it is possible to characterize such a network with a set of parameters that allow us to predict how the network will interact with other networks.

17.1 ONE-PORT NETWORKS

A pair of terminals at which a signal may enter or leave a network is called a *port,* and a network having only one such pair of terminals is called a *one-port network,* or simply a *one-port*. No connections may be made to any other nodes internal to the one-port, and it is therefore evident that i_a must equal i_b in the one-port shown in

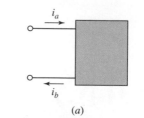

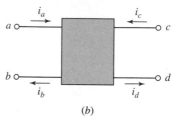

■ **FIGURE 17.1** (*a*) A one-port network. (*b*) A two-port network.

Fig. 17.1*a*. When more than one pair of terminals is present, the network is known as a ***multiport network.*** The two-port network to which this chapter is principally devoted is shown in Fig. 17.1*b*. The currents in the two leads making up each port must be equal, and so it follows that $i_a = i_b$ and $i_c = i_d$ in the two-port shown in Fig. 17.1*b*. Sources and loads must be connected directly across the two terminals of a port if the methods of this chapter are to be used. In other words, each port can be connected only to a one-port network or to a port of another multiport network. For example, no device may be connected between terminals ***a*** and ***c*** of the two-port network in Fig. 17.1*b*. If such a circuit must be analyzed, general loop or nodal equations should be written.

The special methods of analysis which have been developed for two-port networks, or simply two-ports, emphasize the current and voltage relationships at the terminals of the networks and suppress the specific nature of the currents and voltages within the networks. Our introductory study should serve to acquaint us with a number of important parameters and their use in simplifying and systematizing linear two-port network analysis.

Some of the introductory study of one- and two-port networks is accomplished best by using a generalized network notation and the abbreviated nomenclature for determinants introduced in Appendix 2. Thus, if we write a set of loop equations for a passive network,

$$\begin{aligned}
\mathbf{Z}_{11}\mathbf{I}_1 + \mathbf{Z}_{12}\mathbf{I}_2 + \mathbf{Z}_{13}\mathbf{I}_3 + \cdots + \mathbf{Z}_{1N}\mathbf{I}_N &= \mathbf{V}_1 \\
\mathbf{Z}_{21}\mathbf{I}_1 + \mathbf{Z}_{22}\mathbf{I}_2 + \mathbf{Z}_{23}\mathbf{I}_3 + \cdots + \mathbf{Z}_{2N}\mathbf{I}_N &= \mathbf{V}_2 \\
\mathbf{Z}_{31}\mathbf{I}_1 + \mathbf{Z}_{32}\mathbf{I}_2 + \mathbf{Z}_{33}\mathbf{I}_3 + \cdots + \mathbf{Z}_{3N}\mathbf{I}_N &= \mathbf{V}_3 \\
&\cdots \\
\mathbf{Z}_{N1}\mathbf{I}_1 + \mathbf{Z}_{N2}\mathbf{I}_2 + \mathbf{Z}_{N3}\mathbf{I}_3 + \cdots + \mathbf{Z}_{NN}\mathbf{I}_N &= \mathbf{V}_N
\end{aligned} \qquad [1]$$

then the coefficient of each current will be an impedance $\mathbf{Z}_{ij}(\mathbf{s})$, and the circuit determinant, or determinant of the coefficients, is

$$\Delta_{\mathbf{Z}} = \begin{vmatrix}
\mathbf{Z}_{11} & \mathbf{Z}_{12} & \mathbf{Z}_{13} & \cdots & \mathbf{Z}_{1N} \\
\mathbf{Z}_{21} & \mathbf{Z}_{22} & \mathbf{Z}_{23} & \cdots & \mathbf{Z}_{2N} \\
\mathbf{Z}_{31} & \mathbf{Z}_{32} & \mathbf{Z}_{33} & \cdots & \mathbf{Z}_{3N} \\
\cdots & \cdots & \cdots & \cdots & \cdots \\
\mathbf{Z}_{N1} & \mathbf{Z}_{N2} & \mathbf{Z}_{N3} & \cdots & \mathbf{Z}_{NN}
\end{vmatrix} \qquad [2]$$

where N loops have been assumed, the currents appear in subscript order in each equation, and the order of the equations is the same as that of the currents. We also assume that KVL is applied so that the sign of each $\mathbf{Z}_{ii}$ term ($\mathbf{Z}_{11}, \mathbf{Z}_{22}, \ldots, \mathbf{Z}_{NN}$) is positive; the sign of any $\mathbf{Z}_{ij}(i \neq j)$ or mutual term may be either positive or negative, depending on the reference directions assigned to $\mathbf{I}_i$ and $\mathbf{I}_j$.

If there are dependent sources within the network, then it is possible that not all the coefficients in the loop equations must be resistances or impedances. Even so, we will continue to refer to the circuit determinant as $\Delta_{\mathbf{Z}}$.

The use of minor notation (Appendix 2) allows for the input or driving-point impedance at the terminals of a *one-port* network to be expressed very concisely. The result is also applicable to a *two-port* network if one of the two ports is terminated in a passive impedance, including an open or a short circuit.

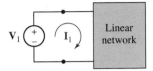

Let us suppose that the one-port network shown in Fig. 17.2 is composed entirely of passive elements and dependent sources; linearity is also assumed. An ideal voltage source $\mathbf{V}_1$ is connected to the port, and the source current is identified as the current in loop 1. Employing the Cramer's rule procedure, then,

Cramer's rule is reviewed in Appendix 2.

$$\mathbf{I}_1 = \frac{\begin{vmatrix} \mathbf{V}_1 & \mathbf{Z}_{12} & \mathbf{Z}_{13} & \cdots & \mathbf{Z}_{1N} \\ 0 & \mathbf{Z}_{22} & \mathbf{Z}_{23} & \cdots & \mathbf{Z}_{2N} \\ 0 & \mathbf{Z}_{32} & \mathbf{Z}_{33} & \cdots & \mathbf{Z}_{3N} \\ \cdots & \cdots & \cdots & & \cdots \\ 0 & \mathbf{Z}_{N2} & \mathbf{Z}_{N3} & \cdots & \mathbf{Z}_{NN} \end{vmatrix}}{\begin{vmatrix} \mathbf{Z}_{11} & \mathbf{Z}_{12} & \mathbf{Z}_{13} & \cdots & \mathbf{Z}_{1N} \\ \mathbf{Z}_{21} & \mathbf{Z}_{22} & \mathbf{Z}_{23} & \cdots & \mathbf{Z}_{2N} \\ \mathbf{Z}_{31} & \mathbf{Z}_{32} & \mathbf{Z}_{33} & \cdots & \mathbf{Z}_{3N} \\ \cdots & \cdots & \cdots & & \cdots \\ \mathbf{Z}_{N1} & \mathbf{Z}_{N2} & \mathbf{Z}_{N3} & \cdots & \mathbf{Z}_{NN} \end{vmatrix}}$$

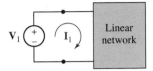

■ **FIGURE 17.2** An ideal voltage source $\mathbf{V}_1$ is connected to the single port of a linear one-port network containing no independent sources; $Z_{in} = \Delta_z/\Delta_{11}$.

or, more concisely,

$$\mathbf{I}_1 = \frac{\mathbf{V}_1 \Delta_{11}}{\Delta_{\mathbf{z}}}$$

Thus,

$$\mathbf{Z}_{in} = \frac{\mathbf{V}_1}{\mathbf{I}_1} = \frac{\Delta_{\mathbf{z}}}{\Delta_{11}} \qquad [3]$$

EXAMPLE **17.1**

Calculate the input impedance for the one-port resistive network shown in Fig. 17.3.

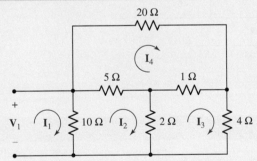

■ **FIGURE 17.3** An example one-port network containing only resistive elements.

We first assign the four mesh currents as shown and write the corresponding mesh equations by inspection:

$$\mathbf{V}_1 = 10\mathbf{I}_1 - 10\mathbf{I}_2$$
$$0 = -10\mathbf{I}_1 + 17\mathbf{I}_2 - 2\mathbf{I}_3 - 5\mathbf{I}_4$$
$$0 = -2\mathbf{I}_2 + 7\mathbf{I}_3 - \mathbf{I}_4$$
$$0 = -5\mathbf{I}_2 - \mathbf{I}_3 + 26\mathbf{I}_4$$

(Continued on next page)

The circuit determinant is then given by

$$\Delta_Z = \begin{vmatrix} 10 & -10 & 0 & 0 \\ -10 & 17 & -2 & -5 \\ 0 & -2 & 7 & -1 \\ 0 & -5 & -1 & 26 \end{vmatrix}$$

and has the value 9680 Ω^4. Eliminating the first row and first column, we have

$$\Delta_{11} = \begin{vmatrix} 17 & -2 & -5 \\ -2 & 7 & -1 \\ -5 & -1 & 26 \end{vmatrix} = 2778 \ \Omega^3$$

Thus, Eq. [3] provides the value of the input impedance,

$$\mathbf{Z}_{\text{in}} = \tfrac{9680}{2778} = 3.485 \ \Omega$$

PRACTICE

17.1 Find the input impedance of the network shown in Fig. 17.4 if it is formed into a one-port network by breaking it at terminals: (a) a and a'; (b) b and b'; (c) c and c'.

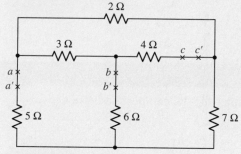

■ **FIGURE 17.4**

Ans: 9.47 Ω; 10.63 Ω; 7.58 Ω.

EXAMPLE 17.2

Find the input impedance of the network shown in Fig. 17.5.

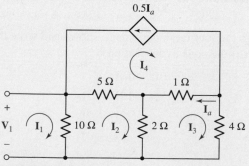

■ **FIGURE 17.5** A one-port network containing a dependent source.

The four mesh equations are written in terms of the four assigned mesh currents:

$$10\mathbf{I}_1 - 10\mathbf{I}_2 \qquad\qquad = \mathbf{V}_1$$
$$-10\mathbf{I}_1 + 17\mathbf{I}_2 - 2\mathbf{I}_3 - 5\mathbf{I}_4 = 0$$
$$- 2\mathbf{I}_2 + 7\mathbf{I}_3 - \mathbf{I}_4 = 0$$

and

$$\mathbf{I}_4 = -0.5\mathbf{I}_a = -0.5(\mathbf{I}_4 - \mathbf{I}_3)$$

or

$$-0.5\mathbf{I}_3 + 1.5\mathbf{I}_4 = 0$$

Thus we can write

$$\Delta_{\mathbf{Z}} = \begin{vmatrix} 10 & -10 & 0 & 0 \\ -10 & 17 & -2 & -5 \\ 0 & -2 & 7 & -1 \\ 0 & 0 & -0.5 & 1.5 \end{vmatrix} = 590\ \Omega^3$$

while

$$\Delta_{11} = \begin{vmatrix} 17 & -2 & -5 \\ -2 & 7 & -1 \\ 0 & -0.5 & 1.5 \end{vmatrix} = 159\ \Omega^2$$

giving

$$\mathbf{Z}_{\text{in}} = \tfrac{590}{159} = 3.711\ \Omega$$

We may also select a similar procedure using nodal equations, yielding the input admittance:

$$\mathbf{Y}_{\text{in}} = \frac{1}{\mathbf{Z}_{\text{in}}} = \frac{\Delta_{\mathbf{Y}}}{\Delta_{11}} \qquad\qquad [4]$$

where Δ_{11} now refers to the minor of $\Delta_{\mathbf{Y}}$.

PRACTICE

17.2 Write a set of nodal equations for the circuit of Fig. 17.6, calculate $\Delta_{\mathbf{Y}}$, and then find the input admittance seen between: (a) node 1 and the reference node; (b) node 2 and the reference.

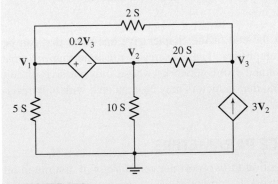

■ FIGURE 17.6

Ans: 10.68 S; 13.16 S.

EXAMPLE 17.3

Use Eq. [4] to again determine the input impedance of the network shown in Fig. 17.3, repeated here as Fig. 17.7.

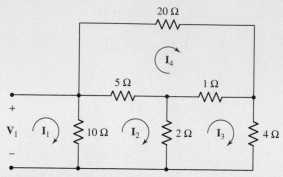

■ **FIGURE 17.7** The circuit from Example 17.1, repeated for convenience.

We first order the node voltages V_1, V_2, and V_3 from left to right, select the reference at the bottom node, and then write the system admittance matrix by inspection:

$$\Delta_Y = \begin{vmatrix} 0.35 & -0.2 & -0.05 \\ -0.2 & 1.7 & -1 \\ -0.05 & -1 & 1.3 \end{vmatrix} = 0.3473 \text{ S}^3$$

$$\Delta_{11} = \begin{vmatrix} 1.7 & -1 \\ -1 & 1.3 \end{vmatrix} = 1.21 \text{ S}^2$$

so that

$$Y_{in} = \frac{0.3473}{1.21} = 0.2870 \text{ S}$$

which corresponds to

$$Z_{in} = \frac{1}{0.287} = 3.484 \ \Omega$$

which agrees with our previous answer to within expected rounding error (we only retained four digits throughout the calculations).

Exercises 8 and 9 at the end of the chapter give one-ports that can be built using operational amplifiers. These exercises illustrate that *negative* resistances may be obtained from networks whose only passive circuit elements are resistors, and that inductors may be simulated with only resistors and capacitors.

17.2 • ADMITTANCE PARAMETERS

Let us now turn our attention to two-port networks. We will assume in all that follows that the network is composed of linear elements and contains no independent sources; dependent sources *are* permissible, however. Further conditions will also be placed on the network in some special cases.

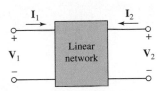

We will consider the two-port as it is shown in Fig. 17.8; the voltage and current at the input terminals are $\mathbf{V}_1$ and $\mathbf{I}_1$, and $\mathbf{V}_2$ and $\mathbf{I}_2$ are specified at the output port. The directions of $\mathbf{I}_1$ and $\mathbf{I}_2$ are both customarily selected as *into* the network at the upper conductors (and out at the lower conductors). Since the network is linear and contains no independent sources within it, $\mathbf{I}_1$ may be considered to be the superposition of two components, one caused by $\mathbf{V}_1$ and the other by $\mathbf{V}_2$. When the same argument is applied to $\mathbf{I}_2$, we may begin with the set of equations

$$\mathbf{I}_1 = \mathbf{y}_{11}\mathbf{V}_1 + \mathbf{y}_{12}\mathbf{V}_2 \qquad [5]$$

$$\mathbf{I}_2 = \mathbf{y}_{21}\mathbf{V}_1 + \mathbf{y}_{22}\mathbf{V}_2 \qquad [6]$$

where the $\mathbf{y}$'s are no more than proportionality constants, or unknown coefficients, for the present. However, it should be clear that their dimensions must be A/V, or S. They are therefore called the $\mathbf{y}$ (or admittance) parameters, and are defined by Eqs. [5] and [6].

The $\mathbf{y}$ parameters, as well as other sets of parameters we will define later in the chapter, are represented concisely as matrices. Here, we define the (2×1) column matrix $\mathbf{I}$,

$$\mathbf{I} = \begin{bmatrix} \mathbf{I}_1 \\ \mathbf{I}_2 \end{bmatrix} \qquad [7]$$

the (2×2) square matrix of the $\mathbf{y}$ parameters,

$$\mathbf{y} = \begin{bmatrix} \mathbf{y}_{11} & \mathbf{y}_{12} \\ \mathbf{y}_{21} & \mathbf{y}_{22} \end{bmatrix} \qquad [8]$$

and the (2×1) column matrix $\mathbf{V}$,

$$\mathbf{V} = \begin{bmatrix} \mathbf{V}_1 \\ \mathbf{V}_2 \end{bmatrix} \qquad [9]$$

Thus, we may write the matrix equation $\mathbf{I} = \mathbf{y}\mathbf{V}$, or

$$\begin{bmatrix} \mathbf{I}_1 \\ \mathbf{I}_2 \end{bmatrix} = \begin{bmatrix} \mathbf{y}_{11} & \mathbf{y}_{12} \\ \mathbf{y}_{21} & \mathbf{y}_{22} \end{bmatrix} \begin{bmatrix} \mathbf{V}_1 \\ \mathbf{V}_2 \end{bmatrix}$$

and matrix multiplication of the right-hand side gives us the equality

$$\begin{bmatrix} \mathbf{I}_1 \\ \mathbf{I}_2 \end{bmatrix} = \begin{bmatrix} \mathbf{y}_{11}\mathbf{V}_1 + \mathbf{y}_{12}\mathbf{V}_2 \\ \mathbf{y}_{21}\mathbf{V}_1 + \mathbf{y}_{22}\mathbf{V}_2 \end{bmatrix}$$

These (2×1) matrices must be equal, element by element, and thus we are led to the defining equations, [5] and [6].

The most useful and informative way to attach a physical meaning to the $\mathbf{y}$ parameters is through a direct inspection of Eqs. [5] and [6]. Consider Eq. [5], for example; if we let $\mathbf{V}_2$ be zero, then we see that $\mathbf{y}_{11}$ must be given by the ratio of $\mathbf{I}_1$ to $\mathbf{V}_1$. We therefore describe $\mathbf{y}_{11}$ as the admittance measured at the input terminals with the output terminals *short-circuited* ($\mathbf{V}_2 = 0$). Since there can be no question which terminals are short-circuited, $\mathbf{y}_{11}$ is best described as the *short-circuit input admittance*. Alternatively, we might describe $\mathbf{y}_{11}$ as the reciprocal of the input impedance measured with the output terminals short-circuited, but a description as an admittance is obviously more direct. It is not the *name* of the parameter that is important. Rather, it

■ FIGURE 17.8 A general two-port with terminal voltages and currents specified. The two-port is composed of linear elements, possibly including dependent sources, but not containing any independent sources.

The notation adopted in this text to represent a matrix is standard, but also can be easily confused with our previous notation for phasors or general complex quantities. The nature of any such symbol should be clear from the context in which it is used.

is the conditions which must be applied to Eq. [5] or [6], and hence to the network, that are most meaningful; when the conditions are determined, the parameter can be found directly from an analysis of the circuit (or by experiment on the physical circuit). Each of the $\mathbf{y}$ parameters may be described as a current-voltage ratio with either $\mathbf{V}_1 = 0$ (the input terminals short-circuited) or $\mathbf{V}_2 = 0$ (the output terminals short-circuited):

$$\mathbf{y}_{11} = \frac{\mathbf{I}_1}{\mathbf{V}_1}\bigg|_{\mathbf{V}_2=0} \tag{10}$$

$$\mathbf{y}_{12} = \frac{\mathbf{I}_1}{\mathbf{V}_2}\bigg|_{\mathbf{V}_1=0} \tag{11}$$

$$\mathbf{y}_{21} = \frac{\mathbf{I}_2}{\mathbf{V}_1}\bigg|_{\mathbf{V}_2=0} \tag{12}$$

$$\mathbf{y}_{22} = \frac{\mathbf{I}_2}{\mathbf{V}_2}\bigg|_{\mathbf{V}_1=0} \tag{13}$$

Because each parameter is an admittance which is obtained by short-circuiting either the output or the input port, the $\mathbf{y}$ parameters are known as the ***short-circuit admittance parameters.*** The specific name of $\mathbf{y}_{11}$ is the ***short-circuit input admittance,*** $\mathbf{y}_{22}$ is the ***short-circuit output admittance,*** and $\mathbf{y}_{12}$ and $\mathbf{y}_{21}$ are the ***short-circuit transfer admittances.***

EXAMPLE 17.4

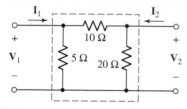

■ **FIGURE 17.9** A resistive two-port.

Find the four short-circuit admittance parameters for the resistive two-port shown in Fig. 17.9.

The values of the parameters may be easily established by applying Eqs. [10] to [13], which we obtained directly from the defining equations, [5] and [6]. To determine $\mathbf{y}_{11}$, we short-circuit the output and find the ratio of $\mathbf{I}_1$ to $\mathbf{V}_1$. This may be done by letting $\mathbf{V}_1 = 1$ V, for then $\mathbf{y}_{11} = \mathbf{I}_1$. By inspection of Fig. 17.9, it is apparent that 1 V applied at the input with the output short-circuited will cause an input current of $(\frac{1}{5} + \frac{1}{10})$, or 0.3 A. Hence,

$$\mathbf{y}_{11} = 0.3 \text{ S}$$

In order to find $\mathbf{y}_{12}$, we short-circuit the input terminals and apply 1 V at the output terminals. The input current flows through the short circuit and is $\mathbf{I}_1 = -\frac{1}{10}$ A. Thus

$$\mathbf{y}_{12} = -0.1 \text{ S}$$

By similar methods,

$$\mathbf{y}_{21} = -0.1 \text{ S} \qquad \mathbf{y}_{22} = 0.15 \text{ S}$$

The describing equations for this two-port in terms of the admittance parameters are, therefore,

$$\mathbf{I}_1 = 0.3\mathbf{V}_1 - 0.1\mathbf{V}_2 \tag{14}$$

$$\mathbf{I}_2 = -0.1\mathbf{V}_1 + 0.15\mathbf{V}_2 \tag{15}$$

and

$$y = \begin{bmatrix} 0.3 & -0.1 \\ -0.1 & 0.15 \end{bmatrix} \quad \text{(all S)}$$

It is not necessary to find these parameters one at a time by using Eqs. [10] to [13], however. We may find them all at once.

EXAMPLE 17.5

Assign node voltages V_1 and V_2 in the two-port of Fig. 17.9 and write the expressions for I_1 and I_2 in terms of them.

We have

$$I_1 = \frac{V_1}{5} + \frac{V_1 - V_2}{10} = 0.3V_1 - 0.1V_2$$

and

$$I_2 = \frac{V_2 - V_1}{10} + \frac{V_2}{20} = -0.1V_1 + 0.15V_2$$

These equations are identical with Eqs. [14] and [15], and the four y parameters may be read from them *directly*.

PRACTICE

17.3 By applying the appropriate 1 V sources and short circuits to the circuit shown in Fig. 17.10, find (a) y_{11}; (b) y_{21}; (c) y_{22}; (d) y_{12}.

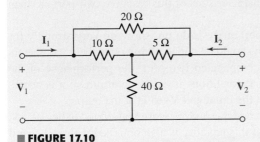

■ FIGURE 17.10

Ans: 0.1192 S; −0.1115 S; 0.1269 S; −0.1115 S.

In general, it is easier to use Eqs. [10], [11], [12], or [13] when only one parameter is desired. If we need all of them, however, it is usually easier to assign V_1 and V_2 to the input and output nodes, to assign other node-to-reference voltages at any interior nodes, and then to carry through with the general solution.

In order to see what use might be made of such a system of equations, let us now terminate each port with some specific one-port network.

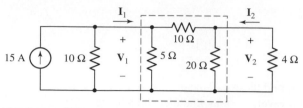

■ **FIGURE 17.11** The resistive two-port network of Fig. 17.9, terminated with specific one-port networks.

Consider the simple two-port network of Example 17.4, shown in Fig. 17.11 with a practical current source connected to the input port and a resistive load connected to the output port. A relationship must now exist between V_1 and I_1 that is independent of the two-port network. This relationship may be determined solely from this external circuit. If we apply KCL (or write a single nodal equation) at the input,

$$I_1 = 15 - 0.1V_1$$

For the output, Ohm's law yields

$$I_2 = -0.25V_2$$

Substituting these expressions for I_1 and I_2 in Eqs. [14] and [15], we have

$$15 = \quad 0.4V_1 - 0.1V_2$$
$$0 = -0.1V_1 + 0.4V_2$$

from which are obtained

$$V_1 = 40 \text{ V} \qquad V_2 = 10 \text{ V}$$

The input and output currents are also easily found:

$$I_1 = 11 \text{ A} \qquad I_2 = -2.5 \text{ A}$$

and the complete terminal characteristics of this resistive two-port are then known.

The advantages of two-port analysis do not show up very strongly for such a simple example, but it should be apparent that once the **y** parameters are determined for a more complicated two-port, the performance of the two-port for different terminal conditions is easily determined; it is necessary only to relate V_1 to I_1 at the input and V_2 to I_2 at the output.

In the example just concluded, y_{12} and y_{21} were both found to be -0.1 S. It is not difficult to show that this equality is also obtained if three general impedances Z_A, Z_B, and Z_C are contained in this Π network. It is somewhat more difficult to determine the specific conditions which are necessary in order that $y_{12} = y_{21}$, but the use of determinant notation is of some help. Let us see if the relationships of Eqs. [10] to [13] can be expressed in terms of the impedance determinant and its minors.

Since our concern is with the two-port and not with the specific networks with which it is terminated, we will let V_1 and V_2 be represented by two ideal voltage sources. Equation [10] is applied by letting $V_2 = 0$ (thus short-circuiting the output) and finding the input admittance. The network now, however, is simply a one-port, and the input impedance of a one-port was found in Sec. 17.1. We select loop **1** to include the input terminals, and let I_1 be that loop's current; we identify $(-I_2)$ as the loop current in loop **2** and assign the remaining loop currents in any convenient manner. Thus,

$$\mathbf{Z}_{\text{in}}|_{\mathbf{V}_2=0} = \frac{\Delta_{\mathbf{Z}}}{\Delta_{11}}$$

and, therefore,

$$\mathbf{y}_{11} = \frac{\Delta_{11}}{\Delta_{\mathbf{Z}}}$$

Similarly,

$$\mathbf{y}_{22} = \frac{\Delta_{22}}{\Delta_{\mathbf{Z}}}$$

In order to find $\mathbf{y}_{12}$, we let $\mathbf{V}_1 = 0$ and find $\mathbf{I}_1$ as a function of $\mathbf{V}_2$. We find that $\mathbf{I}_1$ is given by the ratio

$$\mathbf{I}_1 = \frac{\begin{vmatrix} 0 & \mathbf{Z}_{12} & \cdots & \mathbf{Z}_{1N} \\ -\mathbf{V}_2 & \mathbf{Z}_{22} & \cdots & \mathbf{Z}_{2N} \\ 0 & \mathbf{Z}_{32} & \cdots & \mathbf{Z}_{3N} \\ \cdots & \cdots & \cdots & \cdots \\ 0 & \mathbf{Z}_{N2} & \cdots & \mathbf{Z}_{NN} \end{vmatrix}}{\begin{vmatrix} \mathbf{Z}_{11} & \mathbf{Z}_{12} & \cdots & \mathbf{Z}_{1N} \\ \mathbf{Z}_{21} & \mathbf{Z}_{22} & \cdots & \mathbf{Z}_{2N} \\ \mathbf{Z}_{31} & \mathbf{Z}_{32} & \cdots & \mathbf{Z}_{3N} \\ \cdots & \cdots & \cdots & \cdots \\ \mathbf{Z}_{N1} & \mathbf{Z}_{N2} & \cdots & \mathbf{Z}_{NN} \end{vmatrix}}$$

Thus,

$$\mathbf{I}_1 = -\frac{(-\mathbf{V}_2)\Delta_{21}}{\Delta_{\mathbf{Z}}}$$

and

$$\mathbf{y}_{12} = \frac{\Delta_{21}}{\Delta_{\mathbf{Z}}}$$

In a similar manner, we may show that

$$\mathbf{y}_{21} = \frac{\Delta_{12}}{\Delta_{\mathbf{Z}}}$$

The equality of $\mathbf{y}_{12}$ and $\mathbf{y}_{21}$ is thus contingent on the equality of the two minors of $\Delta_{\mathbf{Z}}$, Δ_{12}, and Δ_{21}. These two minors are

$$\Delta_{21} = \begin{vmatrix} \mathbf{Z}_{12} & \mathbf{Z}_{13} & \mathbf{Z}_{14} & \cdots & \mathbf{Z}_{1N} \\ \mathbf{Z}_{32} & \mathbf{Z}_{33} & \mathbf{Z}_{34} & \cdots & \mathbf{Z}_{3N} \\ \mathbf{Z}_{42} & \mathbf{Z}_{43} & \mathbf{Z}_{44} & \cdots & \mathbf{Z}_{4N} \\ \cdots & \cdots & \cdots & \cdots & \cdots \\ \mathbf{Z}_{N2} & \mathbf{Z}_{N3} & \mathbf{Z}_{N4} & \cdots & \mathbf{Z}_{NN} \end{vmatrix}$$

and

$$\Delta_{12} = \begin{vmatrix} \mathbf{Z}_{21} & \mathbf{Z}_{23} & \mathbf{Z}_{24} & \cdots & \mathbf{Z}_{2N} \\ \mathbf{Z}_{31} & \mathbf{Z}_{33} & \mathbf{Z}_{34} & \cdots & \mathbf{Z}_{3N} \\ \mathbf{Z}_{41} & \mathbf{Z}_{43} & \mathbf{Z}_{44} & \cdots & \mathbf{Z}_{4N} \\ \cdots & \cdots & \cdots & \cdots & \cdots \\ \mathbf{Z}_{N1} & \mathbf{Z}_{N3} & \mathbf{Z}_{N4} & \cdots & \mathbf{Z}_{NN} \end{vmatrix}$$

Their equality is shown by first interchanging the rows and columns of one minor (for example, Δ_{21}), an operation which any college algebra book proves is valid, and then letting every mutual impedance $\mathbf{Z}_{ij}$ be replaced by $\mathbf{Z}_{ji}$. Thus, we set

$$\mathbf{Z}_{12} = \mathbf{Z}_{21} \qquad \mathbf{Z}_{23} = \mathbf{Z}_{32} \qquad \text{etc.}$$

This equality of $\mathbf{Z}_{ij}$ and $\mathbf{Z}_{ji}$ is evident for the three familiar passive elements, the resistor, capacitor, and inductor, and it is also true for mutual inductance. However, it is *not* true for *every* type of device which we may wish to include inside a two-port network. Specifically, it is not true in general for a dependent source, and it is not true for the gyrator, a useful model for Hall-effect devices and for waveguide sections containing ferrites. Over a narrow range of radian frequencies, the gyrator provides an additional phase shift of 180° for a signal passing from the output to the input over that for a signal in the forward direction, and thus $\mathbf{y}_{12} = -\mathbf{y}_{21}$. A common type of passive element leading to the inequality of $\mathbf{Z}_{ij}$ and $\mathbf{Z}_{ji}$, however, is a nonlinear element.

Any device for which $\mathbf{Z}_{ij} = \mathbf{Z}_{ji}$ is called a *bilateral element,* and a circuit which contains only bilateral elements is called a *bilateral circuit.* We have therefore shown that an important property of a bilateral two-port is

$$\mathbf{y}_{12} = \mathbf{y}_{21}$$

and this property is glorified by stating it as the *reciprocity theorem:*

> A simple way of stating the theorem is to say that the interchange of an ideal voltage source and an ideal ammeter in any passive, linear, bilateral circuit will not change the ammeter reading.

In any passive linear bilateral network, if the single voltage source $\mathbf{V}_x$ in branch x produces the current response $\mathbf{I}_y$ in branch y, then the removal of the voltage source from branch x and its insertion in branch y will produce the current response $\mathbf{I}_y$ in branch x.

If we had been working with the admittance determinant of the circuit and had proved that the minors Δ_{21} and Δ_{12} of the admittance determinant $\Delta_{\mathbf{Y}}$ were equal, then we should have obtained the reciprocity theorem in its dual form:

> In other words, the interchange of an ideal current source and an ideal voltmeter in any passive linear bilateral circuit will not change the voltmeter reading.

In any passive linear bilateral network, if the single current source $\mathbf{I}_x$ between nodes x and x' produces the voltage response $\mathbf{V}_y$ between nodes y and y', then the removal of the current source from nodes x and x' and its insertion between nodes y and y' will produce the voltage response $\mathbf{V}_y$ between nodes x and x'.

Two-ports containing dependent sources receive emphasis in Sec. 17.3.

PRACTICE

17.4 In the circuit of Fig. 17.10, let $\mathbf{I}_1$ and $\mathbf{I}_2$ represent ideal current sources. Assign the node voltage $\mathbf{V}_1$ at the input, $\mathbf{V}_2$ at the output, and $\mathbf{V}_x$ from the central node to the reference node. Write three nodal equations, eliminate $\mathbf{V}_x$ to obtain two equations, and then rearrange these equations into the form of Eqs. [5] and [6] so that all four $\mathbf{y}$ parameters may be read directly from the equations.

17.5 Find **y** for the two-port shown in Fig. 17.12.

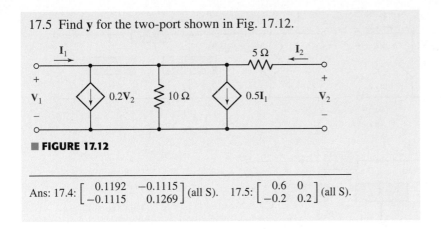

■ **FIGURE 17.12**

Ans: 17.4: $\begin{bmatrix} 0.1192 & -0.1115 \\ -0.1115 & 0.1269 \end{bmatrix}$ (all S). 17.5: $\begin{bmatrix} 0.6 & 0 \\ -0.2 & 0.2 \end{bmatrix}$ (all S).

17.3 • SOME EQUIVALENT NETWORKS

When analyzing electronic circuits, it is usually necessary to replace the active device (and perhaps some of its associated passive circuitry) with an equivalent two-port containing only three or four impedances. The validity of the equivalent may be restricted to small signal amplitudes and a single frequency, or perhaps a limited range of frequencies. The equivalent is also a linear approximation of a nonlinear circuit. However, if we are faced with a network containing a number of resistors, capacitors, and inductors, plus a transistor labeled 2N3823, then we cannot analyze the circuit by any of the techniques we have studied previously; the transistor must first be replaced by a linear model, just as we replaced the op amp by a linear model in Chap. 6. The **y** parameters provide one such model in the form of a two-port network that is often used at high frequencies. Another common linear model for a transistor appears in Sec. 17.5.

The two equations that determine the short-circuit admittance parameters,

$$\mathbf{I}_1 = \mathbf{y}_{11}\mathbf{V}_1 + \mathbf{y}_{12}\mathbf{V}_2 \qquad [16]$$

$$\mathbf{I}_2 = \mathbf{y}_{21}\mathbf{V}_1 + \mathbf{y}_{22}\mathbf{V}_2 \qquad [17]$$

have the form of a pair of nodal equations written for a circuit containing two nonreference nodes. The determination of an equivalent circuit that leads to Eqs. [16] and [17] is made more difficult by the inequality, in general, of $\mathbf{y}_{12}$ and $\mathbf{y}_{21}$; it helps to resort to a little trickery in order to obtain a pair of equations that possess equal mutual coefficients. Let us both add and subtract $\mathbf{y}_{12}\mathbf{V}_1$ (the term we would like to see present on the right side of Eq. [17]):

$$\mathbf{I}_2 = \mathbf{y}_{12}\mathbf{V}_1 + \mathbf{y}_{22}\mathbf{V}_2 + (\mathbf{y}_{21} - \mathbf{y}_{12})\mathbf{V}_1 \qquad [18]$$

or

$$\mathbf{I}_2 - (\mathbf{y}_{21} - \mathbf{y}_{12})\mathbf{V}_1 = \mathbf{y}_{12}\mathbf{V}_1 + \mathbf{y}_{22}\mathbf{V}_2 \qquad [19]$$

The right-hand sides of Eqs. [16] and [19] now show the proper symmetry for a bilateral circuit; the left-hand side of Eq. [19] may be interpreted as the algebraic sum of two current sources, one an independent source $\mathbf{I}_2$ entering node **2**, and the other a dependent source $(\mathbf{y}_{21} - \mathbf{y}_{12})\mathbf{V}_1$ leaving node **2**.

Let us now "read" the equivalent network from Eqs. [16] and [19]. We first provide a reference node, and then a node labeled $\mathbf{V}_1$ and one labeled $\mathbf{V}_2$.

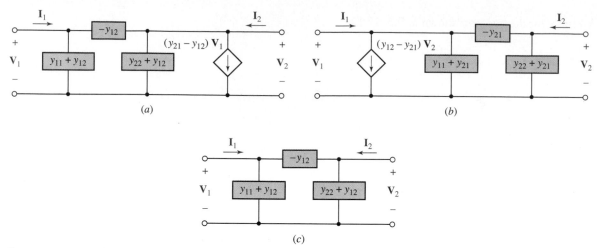

■ **FIGURE 17.13** (*a, b*) Two-ports which are equivalent to any general linear two-port. The dependent source in part *a* depends on V₁, and that in part *b* depends on V₂. (*c*) An equivalent for a bilateral network.

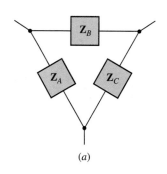

(*a*)

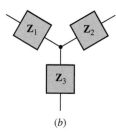

(*b*)

■ **FIGURE 17.14** The three-terminal Δ network (*a*) and the three-terminal Y network (*b*) are equivalent if the six impedances satisfy the conditions of the Y-Δ (or Π-T) transformation, Eqs. [20] to [25].

From Eq. [16], we establish the current $\mathbf{I}_1$ flowing into node **1**, we supply a mutual admittance $(-\mathbf{y}_{12})$ between nodes **1** and **2**, and we supply an admittance of $(\mathbf{y}_{11} + \mathbf{y}_{12})$ between node **1** and the reference node. With $\mathbf{V}_2 = 0$, the ratio of $\mathbf{I}_1$ to $\mathbf{V}_1$ is then $\mathbf{y}_{11}$, as it should be. Now consider Eq. [19]; we cause the current $\mathbf{I}_2$ to flow into the second node, we cause the current $(\mathbf{y}_{21} - \mathbf{y}_{12})\mathbf{V}_1$ to leave the node, we note that the proper admittance $(-\mathbf{y}_{12})$ exists between the nodes, and we complete the circuit by installing the admittance $(\mathbf{y}_{22} + \mathbf{y}_{12})$ from node **2** to the reference node. The completed circuit is shown in Fig. 17.13*a*.

Another form of equivalent network is obtained by subtracting and adding $\mathbf{y}_{21}\mathbf{V}_2$ in Eq. [16]; this equivalent circuit is shown in Fig. 17.13*b*.

If the two-port is bilateral, then $\mathbf{y}_{12} = \mathbf{y}_{21}$, and either of the equivalents reduces to a simple passive Π network. The dependent source disappears. This equivalent of the bilateral two-port is shown in Fig. 17.13*c*.

There are several uses to which these equivalent circuits may be put. In the first place, we have succeeded in showing that an equivalent of any complicated linear two-port *exists*. It does not matter how many nodes or loops are contained within the network; the equivalent is no more complex than the circuits of Fig. 17.13. One of these may be much simpler to use than the given circuit if we are interested only in the terminal characteristics of the given network.

The three-terminal network shown in Fig. 17.14*a* is often referred to as a Δ of impedances, while that in Fig. 17.14*b* is called a Y. One network may be replaced by the other if certain specific relationships between the impedances are satisfied, and these interrelationships may be established by use of the **y** parameters. We find that

$$\mathbf{y}_{11} = \frac{1}{\mathbf{Z}_A} + \frac{1}{\mathbf{Z}_B} = \frac{1}{\mathbf{Z}_1 + \mathbf{Z}_2\mathbf{Z}_3/(\mathbf{Z}_2 + \mathbf{Z}_3)}$$

$$\mathbf{y}_{12} = \mathbf{y}_{21} = -\frac{1}{\mathbf{Z}_B} = \frac{-\mathbf{Z}_3}{\mathbf{Z}_1\mathbf{Z}_2 + \mathbf{Z}_2\mathbf{Z}_3 + \mathbf{Z}_3\mathbf{Z}_1}$$

$$\mathbf{y}_{22} = \frac{1}{\mathbf{Z}_C} + \frac{1}{\mathbf{Z}_B} = \frac{1}{\mathbf{Z}_2 + \mathbf{Z}_1\mathbf{Z}_3/(\mathbf{Z}_1 + \mathbf{Z}_3)}$$

These equations may be solved for $\mathbf{Z}_A$, $\mathbf{Z}_B$, and $\mathbf{Z}_C$ in terms of $\mathbf{Z}_1$, $\mathbf{Z}_2$, and $\mathbf{Z}_3$:

$$\mathbf{Z}_A = \frac{\mathbf{Z}_1\mathbf{Z}_2 + \mathbf{Z}_2\mathbf{Z}_3 + \mathbf{Z}_3\mathbf{Z}_1}{\mathbf{Z}_2} \qquad [20]$$

$$\mathbf{Z}_B = \frac{\mathbf{Z}_1\mathbf{Z}_2 + \mathbf{Z}_2\mathbf{Z}_3 + \mathbf{Z}_3\mathbf{Z}_1}{\mathbf{Z}_3} \qquad [21]$$

$$\mathbf{Z}_C = \frac{\mathbf{Z}_1\mathbf{Z}_2 + \mathbf{Z}_2\mathbf{Z}_3 + \mathbf{Z}_3\mathbf{Z}_1}{\mathbf{Z}_1} \qquad [22]$$

or, for the inverse relationships:

$$\mathbf{Z}_1 = \frac{\mathbf{Z}_A\mathbf{Z}_B}{\mathbf{Z}_A + \mathbf{Z}_B + \mathbf{Z}_C} \qquad [23]$$

$$\mathbf{Z}_2 = \frac{\mathbf{Z}_B\mathbf{Z}_C}{\mathbf{Z}_A + \mathbf{Z}_B + \mathbf{Z}_C} \qquad [24]$$

$$\mathbf{Z}_3 = \frac{\mathbf{Z}_C\mathbf{Z}_A}{\mathbf{Z}_A + \mathbf{Z}_B + \mathbf{Z}_C} \qquad [25]$$

> The reader may recall these useful relationships from Chap. 5, where their derivation was described.

These equations enable us to transform easily between the equivalent Y and Δ networks, a process known as the Y-Δ transformation (or Π-T transformation if the networks are drawn in the forms of those letters). In going from Y to Δ, Eqs. [20] to [22], first find the value of the common numerator as the sum of the products of the impedances in the Y taken two at a time. Each impedance in the Δ is then found by dividing the numerator by the impedance of that element in the Y which has no common node with the desired Δ element. Conversely, given the Δ, first take the sum of the three impedances around the Δ; then divide the product of the two Δ impedances having a common node with the desired Y element by that sum.

These transformations are often useful in simplifying passive networks, particularly resistive ones, thus avoiding the need for any mesh or nodal analysis.

EXAMPLE 17.6

Find the input resistance of the circuit shown in Fig. 17.15a.

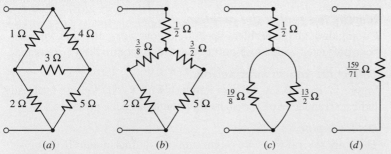

■ **FIGURE 17.15** (*a*) A resistive network whose input resistance is desired. This example is repeated from Chap. 5. (*b*) The upper Δ is replaced by an equivalent Y. (*c*, *d*) Series and parallel combinations give the equivalent input resistance $\frac{159}{71}$ Ω.

(Continued on next page)

We first make a Δ-Y transformation on the upper Δ appearing in Fig. 17.15*a*. The sum of the three resistances forming this Δ is $1 + 4 + 3 = 8 \, \Omega$. The product of the two resistors connected to the top node is $1 \times 4 = 4 \, \Omega^2$. Thus, the upper resistor of the Y is $\frac{4}{8}$, or $\frac{1}{2} \, \Omega$. Repeating this procedure for the other two resistors, we obtain the network shown in Fig. 17.15*b*.

We next make the series and parallel combinations indicated, obtaining in succession Fig. 17.15*c* and *d*. Thus, the input resistance of the circuit in Fig. 17.15*a* is found to be $\frac{159}{71}$, or 2.24 Ω.

Now let us tackle a slightly more complicated example, shown as Fig. 17.16. We note that the circuit contains a dependent source, and thus the Y-Δ transformation is not applicable.

EXAMPLE 17.7

The circuit shown in Fig. 17.16 is an approximate linear equivalent of a transistor amplifier in which the emitter terminal is the bottom node, the base terminal is the upper input node, and the collector terminal is the upper output node. A 2000 Ω resistor is connected between collector and base for some special application and makes the analysis of the circuit more difficult. Determine the y parameters for this circuit.

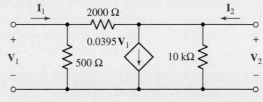

■ **FIGURE 17.16** The linear equivalent circuit of a transistor in common-emitter configuration with resistive feedback between collector and base.

▶ **Identify the goal of the problem.**
Cutting through the problem-specific jargon, we realize that we have been presented with a two-port network and require the **y** parameters.

▶ **Collect the known information.**
Figure 17.16 shows a two-port network with $\mathbf{V}_1$, $\mathbf{I}_1$, $\mathbf{V}_2$, and $\mathbf{I}_2$ already indicated, and a value for each component has been provided.

▶ **Devise a plan.**
There are several ways we might think about this circuit. If we recognize it as being in the form of the equivalent circuit shown in Fig. 17.13*a*, then we may immediately determine the values of the **y** parameters. If recognition is not immediate, then the **y** parameters

may be determined for the two-port by applying the relationships of Eqs. [10] to [13]. We also might avoid any use of two-port analysis methods and write equations directly for the circuit as it stands. The first option seems best in this case.

▶ **Construct an appropriate set of equations.**

By inspection, we find that $-\mathbf{y}_{21}$ corresponds to the admittance of our 2 kΩ resistor, that $\mathbf{y}_{11} + \mathbf{y}_{12}$ corresponds to the admittance of the 500 Ω resistor, the gain of the dependent current source corresponds to $\mathbf{y}_{21} - \mathbf{y}_{12}$, and finally that $\mathbf{y}_{22} + \mathbf{y}_{12}$ corresponds to the admittance of the 10 kΩ resistor. Hence we may write

$$\mathbf{y}_{12} = -\tfrac{1}{2000}$$

$$\mathbf{y}_{11} = \tfrac{1}{500} - \mathbf{y}_{12}$$

$$\mathbf{y}_{21} = 0.0395 + \mathbf{y}_{12}$$

$$\mathbf{y}_{22} = \tfrac{1}{10,000} - \mathbf{y}_{12}$$

▶ **Determine if additional information is required.**

With the equations written as they are, we see that once $\mathbf{y}_{12}$ is computed, the remaining $\mathbf{y}$ parameters may also be obtained.

▶ **Attempt a solution.**

Plugging the numbers into a calculator, we find that

$$\mathbf{y}_{12} = -\tfrac{1}{2000} = -0.5 \text{ mS}$$

$$\mathbf{y}_{11} = \tfrac{1}{500} - \left(-\tfrac{1}{2000}\right) = 2.5 \text{ mS}$$

$$\mathbf{y}_{22} = \tfrac{1}{10,000} - \left(-\tfrac{1}{2000}\right) = 0.6 \text{ mS}$$

and

$$\mathbf{y}_{21} = 0.0395 + \left(-\tfrac{1}{2000}\right) = 39 \text{ mS}$$

The following equations must then apply:

$$\mathbf{I}_1 = 2.5\mathbf{V}_1 - 0.5\mathbf{V}_2 \qquad\qquad [26]$$

$$\mathbf{I}_2 = 39\mathbf{V}_1 + 0.6\mathbf{V}_2 \qquad\qquad [27]$$

where we are now using units of mA, V, and mS or kΩ.

▶ **Verify the solution. Is it reasonable or expected?**

Writing two nodal equations directly from the circuit, we find

$$\mathbf{I}_1 = \frac{\mathbf{V}_1 - \mathbf{V}_2}{2} + \frac{\mathbf{V}_1}{0.5} \qquad \text{or} \qquad \mathbf{I}_1 = 2.5\mathbf{V}_1 - 0.5\mathbf{V}_2$$

and

$$-39.5\mathbf{V}_1 + \mathbf{I}_2 = \frac{\mathbf{V}_2 - \mathbf{V}_1}{2} + \frac{\mathbf{V}_2}{10} \qquad \text{or} \qquad \mathbf{I}_2 = 39\mathbf{V}_1 + 0.6\mathbf{V}_2$$

which agree with Eqs. [26] and [27] obtained directly from the $\mathbf{y}$ parameters.

Now let us make use of Eqs. [26] and [27] by analyzing the performance of the two-port in Fig. 17.16 under several different operating conditions. We first provide a current source of $1\underline{/0°}$ mA at the input and connect a 0.5 kΩ (2 mS) load to the output. The terminating networks are therefore both one-ports and give us the following specific information relating $\mathbf{I}_1$ to $\mathbf{V}_1$ and $\mathbf{I}_2$ to $\mathbf{V}_2$:

$$\mathbf{I}_1 = 1 \text{ (for any } \mathbf{V}_1) \qquad \mathbf{I}_2 = -2\mathbf{V}_2$$

We now have four equations in the four variables, $\mathbf{V}_1$, $\mathbf{V}_2$, $\mathbf{I}_1$, and $\mathbf{I}_2$. Substituting the two one-port relationships in Eqs. [26] and [27], we obtain two equations relating $\mathbf{V}_1$ and $\mathbf{V}_2$:

$$1 = 2.5\mathbf{V}_1 - 0.5\mathbf{V}_2 \qquad 0 = 39\mathbf{V}_1 + 2.6\mathbf{V}_2$$

Solving, we find that

$$\mathbf{V}_1 = 0.1 \text{ V} \quad \mathbf{V}_2 = -1.5 \text{ V}$$
$$\mathbf{I}_1 = 1 \text{ mA} \quad \mathbf{I}_2 = 3 \text{ mA}$$

These four values apply to the two-port operating with a prescribed input ($\mathbf{I}_1 = 1$ mA) and a specified load ($R_L = 0.5$ kΩ).

The performance of an amplifier is often described by giving a few specific values. Let us calculate four of these values for this two-port with its terminations. We will define and evaluate the voltage gain, the current gain, the power gain, and the input impedance.

The *voltage gain* $\mathbf{G}_V$ is

$$\mathbf{G}_V = \frac{\mathbf{V}_2}{\mathbf{V}_1}$$

From the numerical results, it is easy to see that $\mathbf{G}_V = -15$.

The *current gain* $\mathbf{G}_I$ is defined as

$$\mathbf{G}_I = \frac{\mathbf{I}_2}{\mathbf{I}_1}$$

and we have

$$\mathbf{G}_I = 3$$

Let us define and calculate the *power gain* G_P for an assumed sinusoidal excitation. We have

$$G_P = \frac{P_{\text{out}}}{P_{\text{in}}} = \frac{\text{Re}\left[-\frac{1}{2}\mathbf{V}_2\mathbf{I}_2^*\right]}{\text{Re}\left[\frac{1}{2}\mathbf{V}_1\mathbf{I}_1^*\right]} = 45$$

The device might be termed either a voltage, a current, or a power amplifier, since all the gains are greater than unity. If the 2 kΩ resistor were removed, the power gain would rise to 354.

The input and output impedances of the amplifier are often desired in order that maximum power transfer may be achieved to or from an adjacent two-port. We define the *input impedance* $\mathbf{Z}_{\text{in}}$ as the ratio of input voltage to current:

$$\mathbf{Z}_{\text{in}} = \frac{\mathbf{V}_1}{\mathbf{I}_1} = 0.1 \text{ kΩ}$$

This is the impedance offered to the current source when the 500 Ω load is connected to the output. (With the output short-circuited, the input impedance is necessarily $1/\mathbf{y}_{11}$, or 400 Ω.)

It should be noted that the input impedance *cannot* be determined by replacing every source with its internal impedance and then combining resistances or conductances. In the given circuit, such a procedure would yield a value of 416 Ω. The error, of course, comes from treating the *dependent* source as an *independent* source. If we think of the input impedance as being numerically equal to the input voltage produced by an input current of 1 A, the application of the 1 A source produces some input voltage $\mathbf{V}_1$, and the strength of the dependent source $(0.0395\mathbf{V}_1)$ cannot be zero. We should recall that when we obtain the Thévenin equivalent impedance of a circuit containing a dependent source along with one or more independent sources, we must replace the independent sources with short circuits or open circuits, but a dependent source must not be killed. Of course, if the voltage or current on which the dependent source depends is zero, then the dependent source will itself be inactive; occasionally a circuit may be simplified by recognizing such an occurrence.

Besides $\mathbf{G}_V$, $\mathbf{G}_I$, G_P, and $\mathbf{Z}_{\text{in}}$, there is one other performance parameter that is quite useful. This is the *output impedance* $\mathbf{Z}_{\text{out}}$, and it is determined for a different circuit configuration.

The output impedance is just another term for the Thévenin impedance appearing in the Thévenin equivalent circuit of that portion of the network faced by the load. In our circuit, which we have assumed is driven by a $1\underline{/0^\circ}$ mA current source, we therefore replace this independent source with an open circuit, leave the dependent source alone, and seek the *input* impedance seen looking to the left from the output terminals (with the load removed). Thus, we define

$$\mathbf{Z}_{\text{out}} = \mathbf{V}_2|_{\mathbf{I}_2=1\text{ A with all other independent sources killed }and\ R_L\text{ removed}}$$

We therefore remove the load resistor, apply $1\underline{/0^\circ}$ mA (since we are working in V, mA, and kΩ) at the output terminals, and determine $\mathbf{V}_2$. We place these requirements on Eqs. [26] and [27], and obtain

$$0 = 2.5\mathbf{V}_1 - 0.5\mathbf{V}_2 \qquad 1 = 39\mathbf{V}_1 + 0.6\mathbf{V}_2$$

Solving,

$$\mathbf{V}_2 = 0.1190\text{ V}$$

and thus

$$\mathbf{Z}_{\text{out}} = 0.1190\text{ k}\Omega$$

An alternative procedure might be to find the open-circuit output voltage and the short-circuit output current. That is, the Thévenin impedance is the output impedance:

$$\mathbf{Z}_{\text{out}} = \mathbf{Z}_{\text{th}} = -\frac{\mathbf{V}_{2\text{oc}}}{\mathbf{I}_{2\text{sc}}}$$

Carrying out this procedure, we first rekindle the independent source so that $\mathbf{I}_1 = 1$ mA, and then open-circuit the load so that $\mathbf{I}_2 = 0$. We have

$$1 = 2.5\mathbf{V}_1 - 0.5\mathbf{V}_2 \qquad 0 = 39\mathbf{V}_1 + 0.6\mathbf{V}_2$$

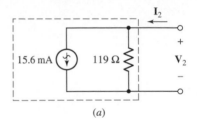

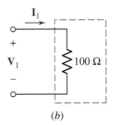

■ **FIGURE 17.17** (*a*) The Norton equivalent of the network in Fig. 17.16 to the left of the output terminal, $I_1 = 1\underline{/0^\circ}$ mA. (*b*) The Thévenin equivalent of that portion of the network to the right of the input terminals, if $I_2 = -2V_2$ mA.

and thus

$$V_{2oc} = -1.857 \text{ V}$$

Next, we apply short-circuit conditions by setting $V_2 = 0$ and again let $I_1 = 1$ mA. We find that

$$I_1 = 1 = 2.5V_1 - 0 \qquad I_2 = 39V_1 + 0$$

and thus

$$I_{2sc} = 15.6 \text{ mA}$$

The assumed directions of V_2 and I_2 therefore result in a Thévenin or output impedance

$$Z_{out} = -\frac{V_{2oc}}{I_{2sc}} = -\frac{-1.857}{15.6} = 0.1190 \text{ k}\Omega$$

as before.

We now have enough information to enable us to draw the Thévenin or Norton equivalent of the two-port of Fig. 17.16 when it is driven by a $1\underline{/0^\circ}$ mA current source and terminated in a 500 Ω load. Thus, the Norton equivalent presented to the load must contain a current source equal to the short-circuit current I_{2sc} in parallel with the output impedance; this equivalent is shown in Fig. 17.17*a*. Also, the Thévenin equivalent offered to the $1\underline{/0^\circ}$ mA input source must consist solely of the input impedance, as drawn in Fig. 17.17*b*.

Before leaving the **y** parameters, we should recognize their usefulness in describing the parallel connection of two-ports, as indicated in Fig. 17.18. When we first defined a port in Sec. 17.1, we noted that the currents entering and leaving the two terminals of a port had to be equal, and there could be no external connections made that bridged between ports. Apparently the parallel connection shown in Fig. 17.18 violates this condition. However, if each two-port has a reference node that is common to its input and output port, and if the two-ports are connected in parallel so that they have a common reference node, then all ports remain ports after the connection. Thus, for the *A* network,

$$I_A = y_A V_A$$

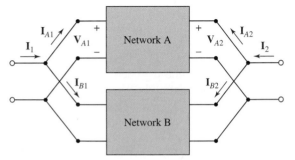

■ **FIGURE 17.18** The parallel connection of two two-port networks. If both inputs and outputs have the same reference node, then the admittance matrix $y = y_A + y_B$.

where

$$\mathbf{I}_A = \begin{bmatrix} \mathbf{I}_{A1} \\ \mathbf{I}_{A2} \end{bmatrix} \quad \text{and} \quad \mathbf{V}_A = \begin{bmatrix} \mathbf{V}_{A1} \\ \mathbf{V}_{A2} \end{bmatrix}$$

and for the B network

$$\mathbf{I}_B = \mathbf{y}_B \mathbf{V}_B$$

But

$$\mathbf{V}_A = \mathbf{V}_B = \mathbf{V} \quad \text{and} \quad \mathbf{I} = \mathbf{I}_A + \mathbf{I}_B$$

Thus,

$$\mathbf{I} = (\mathbf{y}_A + \mathbf{y}_B)\mathbf{V}$$

and we see that each **y** parameter of the parallel network is given as the sum of the corresponding parameters of the individual networks,

$$\mathbf{y} = \mathbf{y}_A + \mathbf{y}_B \tag{28}$$

This may be extended to any number of two-ports connected in parallel.

PRACTICE

17.6 Find **y** and $\mathbf{Z}_{\text{out}}$ for the terminated two-port shown in Fig. 17.19.

17.7 Use Δ-Y and Y-Δ transformations to determine R_{in} for the network shown in (a) Fig. 17.20a; (b) Fig. 17.20b.

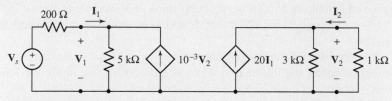

■ **FIGURE 17.19**

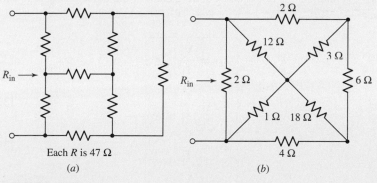

Each R is 47 Ω

(a)

(b)

■ **FIGURE 17.20**

Ans: 17.6: $\begin{bmatrix} 2 \times 10^{-4} & -10^{-3} \\ -4 \times 10^{-3} & 20.3 \times 10^{-3} \end{bmatrix}$ (S); 51.1 Ω. 17.7: 53.71 Ω, 1.311 Ω.

17.4   IMPEDANCE PARAMETERS

The concept of two-port parameters has been introduced in terms of the short-circuit admittance parameters. There are other sets of parameters, however, and each set is associated with a particular class of networks for which its use provides the simplest analysis. We will consider three other types of parameters, the open-circuit impedance parameters, which are the subject of this section; and the hybrid and the transmission parameters, which are discussed in following sections.

We begin again with a general linear two-port that does not contain any independent sources; the currents and voltages are assigned as before (Fig. 17.8). Now let us consider the voltage $\mathbf{V}_1$ as the response produced by two current sources $\mathbf{I}_1$ and $\mathbf{I}_2$. We thus write for $\mathbf{V}_1$

$$\mathbf{V}_1 = \mathbf{z}_{11}\mathbf{I}_1 + \mathbf{z}_{12}\mathbf{I}_2 \qquad [29]$$

and for $\mathbf{V}_2$

$$\mathbf{V}_2 = \mathbf{z}_{21}\mathbf{I}_1 + \mathbf{z}_{22}\mathbf{I}_2 \qquad [30]$$

or

$$\mathbf{V} = \begin{bmatrix} \mathbf{V}_1 \\ \mathbf{V}_2 \end{bmatrix} = \mathbf{z}\mathbf{I} = \begin{bmatrix} \mathbf{z}_{11} & \mathbf{z}_{12} \\ \mathbf{z}_{21} & \mathbf{z}_{22} \end{bmatrix} \begin{bmatrix} \mathbf{I}_1 \\ \mathbf{I}_2 \end{bmatrix} \qquad [31]$$

Of course, in using these equations it is not necessary that $\mathbf{I}_1$ and $\mathbf{I}_2$ be current sources; nor is it necessary that $\mathbf{V}_1$ and $\mathbf{V}_2$ be voltage sources. In general, we may have any networks terminating the two-port at either end. As the equations are written, we probably think of $\mathbf{V}_1$ and $\mathbf{V}_2$ as given quantities, or independent variables, and $\mathbf{I}_1$ and $\mathbf{I}_2$ as unknowns, or dependent variables.

The six ways in which two equations may be written to relate these four quantities define the different systems of parameters. We study the four most important of these six systems of parameters.

The most informative description of the $\mathbf{z}$ parameters, defined in Eqs. [29] and [30], is obtained by setting each of the currents equal to zero. Thus

$$\mathbf{z}_{11} = \left. \frac{\mathbf{V}_1}{\mathbf{I}_1} \right|_{\mathbf{I}_2=0} \qquad [32]$$

$$\mathbf{z}_{12} = \left. \frac{\mathbf{V}_1}{\mathbf{I}_2} \right|_{\mathbf{I}_1=0} \qquad [33]$$

$$\mathbf{z}_{21} = \left. \frac{\mathbf{V}_2}{\mathbf{I}_1} \right|_{\mathbf{I}_2=0} \qquad [34]$$

$$\mathbf{z}_{22} = \left. \frac{\mathbf{V}_2}{\mathbf{I}_2} \right|_{\mathbf{I}_1=0} \qquad [35]$$

Since zero current results from an open-circuit termination, the $\mathbf{z}$ parameters are known as the *open-circuit impedance parameters*. They are easily related to the short-circuit admittance parameters by solving Eqs. [29] and

[30] for $\mathbf{I}_1$ and $\mathbf{I}_2$:

$$\mathbf{I}_1 = \frac{\begin{vmatrix} \mathbf{V}_1 & \mathbf{z}_{12} \\ \mathbf{V}_2 & \mathbf{z}_{22} \end{vmatrix}}{\begin{vmatrix} \mathbf{z}_{11} & \mathbf{z}_{12} \\ \mathbf{z}_{21} & \mathbf{z}_{22} \end{vmatrix}}$$

or

$$\mathbf{I}_1 = \left(\frac{\mathbf{z}_{22}}{\mathbf{z}_{11}\mathbf{z}_{22} - \mathbf{z}_{12}\mathbf{z}_{21}} \right) \mathbf{V}_1 - \left(\frac{\mathbf{z}_{12}}{\mathbf{z}_{11}\mathbf{z}_{22} - \mathbf{z}_{12}\mathbf{z}_{21}} \right) \mathbf{V}_2$$

Using determinant notation, and being careful that the subscript is a lower-case $\mathbf{z}$, we assume that $\Delta_{\mathbf{z}} \neq 0$ and obtain

$$\mathbf{y}_{11} = \frac{\Delta_{11}}{\Delta_{\mathbf{z}}} = \frac{\mathbf{z}_{22}}{\Delta_{\mathbf{z}}} \qquad \mathbf{y}_{12} = -\frac{\Delta_{21}}{\Delta_{\mathbf{z}}} = -\frac{\mathbf{z}_{12}}{\Delta_{\mathbf{z}}}$$

and from solving for $\mathbf{I}_2$,

$$\mathbf{y}_{21} = -\frac{\Delta_{12}}{\Delta_{\mathbf{z}}} = -\frac{\mathbf{z}_{21}}{\Delta_{\mathbf{z}}} \qquad \mathbf{y}_{22} = \frac{\Delta_{22}}{\Delta_{\mathbf{z}}} = \frac{\mathbf{z}_{11}}{\Delta_{\mathbf{z}}}$$

In a similar manner, the $\mathbf{z}$ parameters may be expressed in terms of the admittance parameters. Transformations of this nature are possible between any of the various parameter systems, and quite a collection of occasionally useful formulas may be obtained. Transformations between the $\mathbf{y}$ and $\mathbf{z}$ parameters (as well as the $\mathbf{h}$ and $\mathbf{t}$ parameters which we will consider in the following sections) are given in Table 17.1 as a helpful reference.

TABLE 17.1 Transformations Between **y**, **z**, **h**, and **t** Parameters

	y		**z**		**h**		**t**	
y	$\mathbf{y}_{11}$	$\mathbf{y}_{12}$	$\dfrac{\mathbf{z}_{22}}{\Delta_{\mathbf{z}}}$	$\dfrac{-\mathbf{z}_{12}}{\Delta_{\mathbf{z}}}$	$\dfrac{1}{\mathbf{h}_{11}}$	$\dfrac{-\mathbf{h}_{12}}{\mathbf{h}_{11}}$	$\dfrac{\mathbf{t}_{22}}{\mathbf{t}_{12}}$	$\dfrac{-\Delta_{\mathbf{t}}}{\mathbf{t}_{12}}$
	$\mathbf{y}_{21}$	$\mathbf{y}_{22}$	$\dfrac{-\mathbf{z}_{21}}{\Delta_{\mathbf{z}}}$	$\dfrac{\mathbf{z}_{11}}{\Delta_{\mathbf{z}}}$	$\dfrac{\mathbf{h}_{21}}{\mathbf{h}_{11}}$	$\dfrac{\Delta_{\mathbf{h}}}{\mathbf{h}_{11}}$	$\dfrac{-1}{\mathbf{t}_{12}}$	$\dfrac{\mathbf{t}_{11}}{\mathbf{t}_{12}}$
z	$\dfrac{\mathbf{y}_{22}}{\Delta_{\mathbf{y}}}$	$\dfrac{-\mathbf{y}_{12}}{\Delta_{\mathbf{y}}}$	$\mathbf{z}_{11}$	$\mathbf{z}_{12}$	$\dfrac{\Delta_{\mathbf{h}}}{\mathbf{h}_{22}}$	$\dfrac{\mathbf{h}_{12}}{\mathbf{h}_{22}}$	$\dfrac{\mathbf{t}_{11}}{\mathbf{t}_{21}}$	$\dfrac{\Delta_{\mathbf{t}}}{\mathbf{t}_{21}}$
	$\dfrac{-\mathbf{y}_{21}}{\Delta_{\mathbf{y}}}$	$\dfrac{\mathbf{y}_{11}}{\Delta_{\mathbf{y}}}$	$\mathbf{z}_{21}$	$\mathbf{z}_{22}$	$\dfrac{-\mathbf{h}_{21}}{\mathbf{h}_{22}}$	$\dfrac{1}{\mathbf{h}_{22}}$	$\dfrac{1}{\mathbf{t}_{21}}$	$\dfrac{\mathbf{t}_{22}}{\mathbf{t}_{21}}$
h	$\dfrac{1}{\mathbf{y}_{11}}$	$\dfrac{-\mathbf{y}_{12}}{\mathbf{y}_{11}}$	$\dfrac{\Delta_{\mathbf{z}}}{\mathbf{z}_{22}}$	$\dfrac{\mathbf{z}_{12}}{\mathbf{z}_{22}}$	$\mathbf{h}_{11}$	$\mathbf{h}_{12}$	$\dfrac{\mathbf{t}_{12}}{\mathbf{t}_{22}}$	$\dfrac{\Delta_{\mathbf{t}}}{\mathbf{t}_{22}}$
	$\dfrac{\mathbf{y}_{21}}{\mathbf{y}_{11}}$	$\dfrac{\Delta_{\mathbf{y}}}{\mathbf{y}_{11}}$	$\dfrac{-\mathbf{z}_{21}}{\mathbf{z}_{22}}$	$\dfrac{1}{\mathbf{z}_{22}}$	$\mathbf{h}_{21}$	$\mathbf{h}_{22}$	$\dfrac{-1}{\mathbf{t}_{22}}$	$\dfrac{\mathbf{t}_{21}}{\mathbf{t}_{22}}$
t	$\dfrac{-\mathbf{y}_{22}}{\mathbf{y}_{21}}$	$\dfrac{-1}{\mathbf{y}_{21}}$	$\dfrac{\mathbf{z}_{11}}{\mathbf{z}_{21}}$	$\dfrac{\Delta_{\mathbf{z}}}{\mathbf{z}_{21}}$	$\dfrac{-\Delta_{\mathbf{h}}}{\mathbf{h}_{21}}$	$\dfrac{-\mathbf{h}_{11}}{\mathbf{h}_{21}}$	$\mathbf{t}_{11}$	$\mathbf{t}_{12}$
	$\dfrac{-\Delta_{\mathbf{y}}}{\mathbf{y}_{21}}$	$\dfrac{-\mathbf{y}_{11}}{\mathbf{y}_{21}}$	$\dfrac{1}{\mathbf{z}_{21}}$	$\dfrac{\mathbf{z}_{22}}{\mathbf{z}_{21}}$	$\dfrac{-\mathbf{h}_{22}}{\mathbf{h}_{21}}$	$\dfrac{-1}{\mathbf{h}_{21}}$	$\mathbf{t}_{21}$	$\mathbf{t}_{22}$

For all parameter sets: $\Delta_{\mathbf{p}} = \mathbf{p}_{11}\mathbf{p}_{22} - \mathbf{p}_{12}\mathbf{p}_{21}$.

If the two-port is a bilateral network, reciprocity is present; it is easy to show that this results in the equality of z_{12} and z_{21}.

Equivalent circuits may again be obtained from an inspection of Eqs. [29] and [30]; their construction is facilitated by adding and subtracting either $z_{12}I_1$ in Eq. [30] or $z_{21}I_2$ in Eq. [29]. Each of these equivalent circuits contains a dependent voltage source.

Let us leave the derivation of such an equivalent to some leisure moment, and consider next an example of a rather general nature. Can we construct a general Thévenin equivalent of the two-port, as viewed from the output terminals? It is necessary first to assume a specific input circuit configuration, and we will select an independent voltage source $\mathbf{V}_s$ (positive sign at top) in series with a generator impedance $\mathbf{Z}_g$. Thus

$$\mathbf{V}_s = \mathbf{V}_1 + \mathbf{I}_1 \mathbf{Z}_g$$

Combining this result with Eqs. [29] and [30], we may eliminate $\mathbf{V}_1$ and $\mathbf{I}_1$ and obtain

$$\mathbf{V}_2 = \frac{\mathbf{z}_{21}}{\mathbf{z}_{11} + \mathbf{Z}_g}\mathbf{V}_s + \left(\mathbf{z}_{22} - \frac{\mathbf{z}_{12}\mathbf{z}_{21}}{\mathbf{z}_{11} + \mathbf{Z}_g}\right)\mathbf{I}_2$$

The Thévenin equivalent circuit may be drawn directly from this equation; it is shown in Fig. 17.21. The output impedance, expressed in terms of the $\mathbf{z}$ parameters, is

$$\mathbf{Z}_{\text{out}} = \mathbf{z}_{22} - \frac{\mathbf{z}_{12}\mathbf{z}_{21}}{\mathbf{z}_{11} + \mathbf{Z}_g}$$

If the generator impedance is zero, the simpler expression

$$\mathbf{Z}_{\text{out}} = \frac{\mathbf{z}_{11}\mathbf{z}_{22} - \mathbf{z}_{12}\mathbf{z}_{21}}{\mathbf{z}_{11}} = \frac{\Delta_{\mathbf{z}}}{\Delta_{22}} = \frac{1}{\mathbf{y}_{22}} \qquad (\mathbf{Z}_g = 0)$$

is obtained. For this special case, the output *admittance* is identical to $\mathbf{y}_{22}$, as indicated by the basic relationship of Eq. [13].

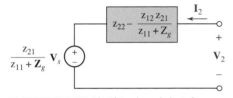

■ **FIGURE 17.21** The Thévenin equivalent of a general two-port, as viewed from the output terminals, expressed in terms of the open-circuit impedance parameters.

EXAMPLE 17.8

Given the set of impedance parameters

$$\mathbf{z} = \begin{bmatrix} 10^3 & 10 \\ -10^6 & 10^4 \end{bmatrix} \qquad \text{(all } \Omega\text{)}$$

which is representative of a transistor operating in the common-emitter configuration, determine the voltage, current, and power gains, as well as the input and output impedances. The two-port may be viewed as driven by an ideal sinusoidal voltage source $\mathbf{V}_s$ in series with a 500 Ω resistor, and terminated in a 10 kΩ load resistor.

The two describing equations for the two-port are

$$\mathbf{V}_1 = 10^3 \mathbf{I}_1 + 10 \mathbf{I}_2 \tag{36}$$

$$\mathbf{V}_2 = -10^6 \mathbf{I}_1 + 10^4 \mathbf{I}_2 \tag{37}$$

and the characterizing equations of the input and output networks are

$$\mathbf{V}_s = 500\mathbf{I}_1 + \mathbf{V}_1 \qquad [38]$$

$$\mathbf{V}_2 = -10^4\mathbf{I}_2 \qquad [39]$$

From these last four equations, we may easily obtain expressions for $\mathbf{V}_1$, $\mathbf{I}_1$, $\mathbf{V}_2$, and $\mathbf{I}_2$ in terms of $\mathbf{V}_s$:

$$\mathbf{V}_1 = 0.75\mathbf{V}_s \qquad \mathbf{I}_1 = \frac{\mathbf{V}_s}{2000}$$

$$\mathbf{V}_2 = -250\mathbf{V}_s \qquad \mathbf{I}_2 = \frac{\mathbf{V}_s}{40}$$

From this information, it is simple to determine the voltage gain,

$$\mathbf{G}_V = \frac{\mathbf{V}_2}{\mathbf{V}_1} = -333$$

the current gain,

$$\mathbf{G}_I = \frac{\mathbf{I}_2}{\mathbf{I}_1} = 50$$

the power gain,

$$G_P = \frac{\mathrm{Re}\left[-\frac{1}{2}\mathbf{V}_2\mathbf{I}_2^*\right]}{\mathrm{Re}\left[\frac{1}{2}\mathbf{V}_1\mathbf{I}_1^*\right]} = 16{,}670$$

and the input impedance,

$$\mathbf{Z}_{\mathrm{in}} = \frac{\mathbf{V}_1}{\mathbf{I}_1} = 1500 \; \Omega$$

The output impedance may be obtained by referring to Fig. 17.21:

$$\mathbf{Z}_{\mathrm{out}} = \mathbf{z}_{22} - \frac{\mathbf{z}_{12}\mathbf{z}_{21}}{\mathbf{z}_{11} + \mathbf{Z}_g} = 16.67 \; \mathrm{k}\Omega$$

In accordance with the predictions of the maximum power transfer theorem, the power gain reaches a maximum value when $\mathbf{Z}_L = \mathbf{Z}_{\mathrm{out}}^* = 16.67$ kΩ; that maximum value is 17,045.

The **y** parameters are useful when two-ports are interconnected in parallel, and, in a dual manner, the **z** parameters simplify the problem of a series connection of networks, shown in Fig. 17.22. Note that the series connection is *not* the same as the cascade connection that we will discuss later in connection with the transmission parameters. If each two-port has a common reference node for its input and output, and if the references are connected together as indicated in Fig. 17.22, then $\mathbf{I}_1$ flows through the input ports of the two networks in series. A similar statement holds for $\mathbf{I}_2$. Thus, ports remain ports after the interconnection. It follows that $\mathbf{I} = \mathbf{I}_A = \mathbf{I}_B$ and

$$\mathbf{V} = \mathbf{V}_A + \mathbf{V}_B = \mathbf{z}_A\mathbf{I}_A + \mathbf{z}_B\mathbf{I}_B$$
$$= (\mathbf{z}_A + \mathbf{z}_B)\mathbf{I} = \mathbf{z}\mathbf{I}$$

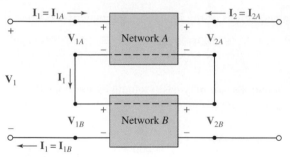

FIGURE 17.22 The series connection of two two-port networks is made by connecting the four common reference nodes together; then the matrix $\mathbf{z} = \mathbf{z}_A + \mathbf{z}_B$.

where

$$\mathbf{z} = \mathbf{z}_A + \mathbf{z}_B$$

so that $\mathbf{z}_{11} = \mathbf{z}_{11A} + \mathbf{z}_{11B}$, and so forth.

PRACTICE

17.8 Find **z** for the two-port shown in (*a*) Fig. 17.23*a*; (*b*) Fig. 17.23*b*.
17.9 Find **z** for the two-port shown in Fig. 17.23*c*.

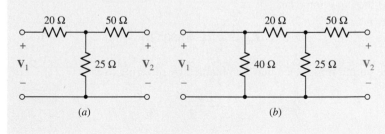

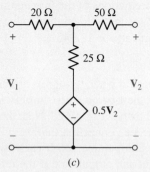

FIGURE 17.23

Ans: 17.8: $\begin{bmatrix} 45 & 25 \\ 25 & 75 \end{bmatrix}$ (Ω), $\begin{bmatrix} 21.2 & 11.76 \\ 11.76 & 67.6 \end{bmatrix}$ (Ω).   17.9: $\begin{bmatrix} 70 & 100 \\ 50 & 150 \end{bmatrix}$ (Ω).

PRACTICAL APPLICATION

Characterizing Transistors

Parameter values for bipolar junction transistors are commonly quoted in terms of **h** parameters. Invented in the late 1940s by researchers at Bell Laboratories (Fig. 17.24), the transistor is a nonlinear semiconductor device that forms the basis for almost all amplifiers and digital logic circuits.

■ **FIGURE 17.24** Photograph of the first demonstrated bipolar junction transistor ("bjt").
Lucent Technologies Inc./Bell Labs

The three terminals of a transistor are labeled the *base* (*b*), *collector* (*c*), and *emitter* (*e*) as shown in Fig. 17.25, and are named after their roles in the transport of charge within the device. The **h** parameters of a bipolar junction transistor are typically measured with the emitter terminal grounded, also known as the *common-emitter* configuration; the base is then designated as the input and the collector as the output. As mentioned previously, however, the transistor is a nonlinear device, and so definition of **h** parameters which

are valid for all voltages and currents is not possible. Therefore, it is common practice to quote **h** parameters at a specific value of collector current I_C and collector-emitter voltage V_{CE}. Another consequence of the nonlinearity of the device is that ac **h** parameters and dc **h** parameters are often quite different in value.

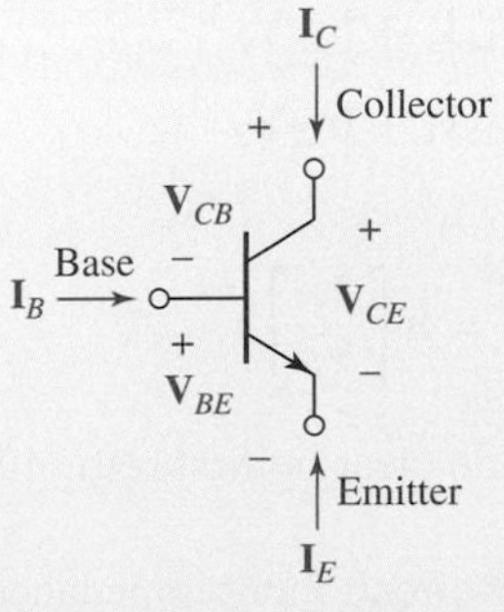

There are many types of instruments which may be employed to obtain the **h** parameters for a particular transistor. One example is a semiconductor parameter analyzer, shown in Fig. 17.26. This instrument sweeps the desired current (plotted on the vertical axis) against a specified voltage (plotted on the horizontal axis). A "family" of curves is produced by varying a third parameter, often the base current, in discrete steps.

As an example, the manufacturer of the 2N3904 NPN silicon transistor quotes **h** parameters as indicated in Table 17.2; note that the specific parameters are given alternative designations (h_{ie}, h_{re}, etc.) by transistor engineers. The measurements were made at $I_C = 1.0$ mA, $V_{CE} = 10$ Vdc, and $f = 1.0$ kHz.

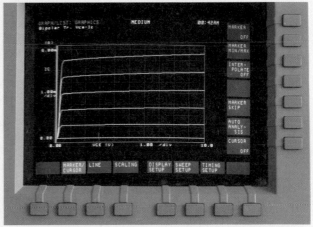

■ **FIGURE 17.26** Display snapshot of an HP 4155A Semiconductor Parameter Analyzer used to measure the **h** parameters of a 2N3904 bipolar junction transistor (bjt).

Just for fun, one of the authors and a friend decided to measure these parameters for themselves. Grabbing an inexpensive device off the shelf and using the instrument in Fig. 17.26, they found

$$h_{oe} = 3.3 \ \mu\text{mhos} \qquad h_{fe} = 109$$
$$h_{ie} = 3.02 \ \text{k}\Omega \qquad h_{re} = 4 \times 10^{-3}$$

(Continued on next page)

■ **FIGURE 17.25** Schematic of a bjt showing currents and voltages defined using the IEEE convention.

TABLE 17.2 Summary of 2N3904 ac Parameters

Parameter	Name	Specification	Units
h_{ie} (h_{11})	Input impedance	1.0–10	kΩ
h_{re} (h_{12})	Voltage feedback ratio	$0.5\text{–}8.0 \times 10^{-4}$	–
h_{fe} (h_{21})	Small-signal current gain	100–400	–
h_{oe} (h_{22})	Output admittance	1.0–40	μmhos

the first three of which were all well within the manufacturer's published tolerances, although much closer to the minimum values than to the maximum values. The value for h_{re}, however, was an order of magnitude larger than the maximum value specified by the manufacturer's datasheet! This was rather disconcerting, as we thought we were doing pretty well up to that point.

Upon further reflection, we realized that the experimental setup allowed the device to heat up during the measurement, as we were sweeping below and above $I_C = 1$ mA. Transistors, unfortunately, can change their properties rather dramatically as a function of temperature; the manufacturer values were specifically for 25°C. Once the sweep was changed to minimize device heating, we obtained a value of 2.0×10^{-4} for h_{re}. Linear circuits are by far much easier to work with, but nonlinear circuits can be much more interesting!

17.5 • HYBRID PARAMETERS

The difficulty in measuring quantities such as the open-circuit impedance parameters arises when a parameter such as $\mathbf{z}_{21}$ must be measured. A known sinusoidal current is easily supplied at the input terminals, but because of the exceedingly high output impedance of the transistor circuit, it is difficult to open-circuit the output terminals and yet supply the necessary dc biasing voltages and measure the sinusoidal output voltage. A short-circuit current measurement at the output terminals is much simpler to implement.

The hybrid parameters are defined by writing the pair of equations relating $\mathbf{V}_1$, $\mathbf{I}_1$, $\mathbf{V}_2$, and $\mathbf{I}_2$ as if $\mathbf{V}_1$ and $\mathbf{I}_2$ were the independent variables:

$$\mathbf{V}_1 = \mathbf{h}_{11}\mathbf{I}_1 + \mathbf{h}_{12}\mathbf{V}_2 \qquad [40]$$

$$\mathbf{I}_2 = \mathbf{h}_{21}\mathbf{I}_1 + \mathbf{h}_{22}\mathbf{V}_2 \qquad [41]$$

or

$$\begin{bmatrix} \mathbf{V}_1 \\ \mathbf{I}_2 \end{bmatrix} = \mathbf{h} \begin{bmatrix} \mathbf{I}_1 \\ \mathbf{V}_2 \end{bmatrix} \qquad [42]$$

The nature of the parameters is made clear by first setting $\mathbf{V}_2 = 0$. Thus,

$$\mathbf{h}_{11} = \left.\frac{\mathbf{V}_1}{\mathbf{I}_1}\right|_{\mathbf{V}_2=0} = \text{short-circuit input impedance}$$

$$\mathbf{h}_{21} = \left.\frac{\mathbf{I}_2}{\mathbf{I}_1}\right|_{\mathbf{V}_2=0} = \text{short-circuit forward current gain}$$

Letting $\mathbf{I}_1 = 0$, we obtain

$$\mathbf{h}_{12} = \left.\frac{\mathbf{V}_1}{\mathbf{V}_2}\right|_{\mathbf{I}_1=0} = \text{open-circuit reverse voltage gain}$$

$$\mathbf{h}_{22} = \left.\frac{\mathbf{I}_2}{\mathbf{V}_2}\right|_{\mathbf{I}_1=0} = \text{open-circuit output admittance}$$

Since the parameters represent an impedance, an admittance, a voltage gain, and a current gain, they are called the "hybrid" parameters.

The subscript designations for these parameters are often simplified when they are applied to transistors. Thus, $\mathbf{h}_{11}$, $\mathbf{h}_{12}$, $\mathbf{h}_{21}$, and $\mathbf{h}_{22}$ become $\mathbf{h}_i$, $\mathbf{h}_r$, $\mathbf{h}_f$, and $\mathbf{h}_o$, respectively, where the subscripts denote input, reverse, forward, and output.

EXAMPLE **17.9**

Find h for the bilateral resistive circuit drawn in Fig. 17.27.

With the output short-circuited ($\mathbf{V}_2 = 0$), the application of a 1 A source at the input ($\mathbf{I}_1 = 1$ A) produces an input voltage of 3.4 V ($\mathbf{V}_1 = 3.4$ V); hence, $\mathbf{h}_{11} = 3.4\ \Omega$. Under these same conditions, the output current is easily obtained by current division: $\mathbf{I}_2 = -0.4$ A; thus, $\mathbf{h}_{21} = -0.4$.

The remaining two parameters are obtained with the input open-circuited ($\mathbf{I}_1 = 0$). We apply 1 V to the output terminals ($\mathbf{V}_2 = 1$ V). The response at the input terminals is 0.4 V ($\mathbf{V}_1 = 0.4$ V), and thus $\mathbf{h}_{12} = 0.4$. The current delivered by this source at the output terminals is 0.1 A ($\mathbf{I}_2 = 0.1$ A), and therefore $\mathbf{h}_{22} = 0.1$ S.

We therefore have $\mathbf{h} = \begin{bmatrix} 3.4\ \Omega & 0.4 \\ -0.4 & 0.1\ \text{S} \end{bmatrix}$. It is a consequence of the reciprocity theorem that $\mathbf{h}_{12} = -\mathbf{h}_{21}$ for a bilateral network.

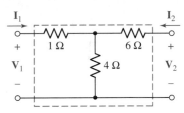

■ **FIGURE 17.27** A bilateral network for which the **h** parameters are found: $\mathbf{h}_{12} = -\mathbf{h}_{21}$.

PRACTICE

17.10 Find **h** for the two-port shown in (*a*) Fig. 17.28*a*; (*b*) Fig. 17.28*b*.

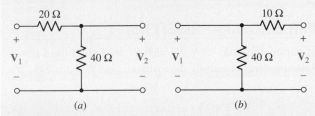

(a)   (b)

■ **FIGURE 17.28**

17.11 If $\mathbf{h} = \begin{bmatrix} 5\ \Omega & 2 \\ -0.5 & 0.1\ \text{S} \end{bmatrix}$, find (*a*) **y**; (*b*) **z**.

Ans: 17.10: $\begin{bmatrix} 20\ \Omega & 1 \\ -1 & 25\ \text{ms} \end{bmatrix}$, $\begin{bmatrix} 8\ \Omega & 0.8 \\ -0.8 & 20\ \text{ms} \end{bmatrix}$. 17.11: $\begin{bmatrix} 0.2 & -0.4 \\ -0.1 & 0.3 \end{bmatrix}$ (S),

$\begin{bmatrix} 15 & 20 \\ 5 & 10 \end{bmatrix}$ (Ω).

The circuit shown in Fig. 17.29 is a direct translation of the two defining equations, [40] and [41]. The first represents KVL about the input loop, while the second is obtained from KCL at the upper output node. This circuit is also a popular transistor equivalent circuit. Let us assume some reasonable values for the common-emitter configuration: $\mathbf{h}_{11} = 1200\ \Omega$, $\mathbf{h}_{12} = 2 \times 10^{-4}$, $\mathbf{h}_{21} = 50$, $\mathbf{h}_{22} = 50 \times 10^{-6}$ S, a voltage generator of $1\underline{/0°}$ mV in series with 800 Ω, and a 5 kΩ load. For the input,

$$10^{-3} = (1200 + 800)\mathbf{I}_1 + 2 \times 10^{-4}\mathbf{V}_2$$

and at the output,

$$\mathbf{I}_2 = -2 \times 10^{-4}\mathbf{V}_2 = 50\mathbf{I}_1 + 50 \times 10^{-6}\mathbf{V}_2$$

Solving,

$$\mathbf{I}_1 = 0.510\ \mu A \qquad \mathbf{V}_1 = 0.592\ mV$$
$$\mathbf{I}_2 = 20.4\ \mu A \qquad \mathbf{V}_2 = -102\ mV$$

Through the transistor we have a current gain of 40, a voltage gain of -172, and a power gain of 6880. The input impedance to the transistor is 1160 Ω, and a few more calculations show that the output impedance is 22.2 kΩ.

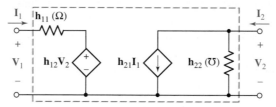

FIGURE 17.29 The four **h** parameters are referred to a two-port. The pertinent equations are $\mathbf{V}_1 = \mathbf{h}_{11}\mathbf{I}_1 + \mathbf{h}_{12}\mathbf{V}_2$ and $\mathbf{I}_2 = \mathbf{h}_{21}\mathbf{I}_1 + \mathbf{h}_{22}\mathbf{V}_2$.

Hybrid parameters may be added directly when two-ports are connected in series at the input and in parallel at the output. This is called a series-parallel interconnection, and it is not used very often.

17.6 TRANSMISSION PARAMETERS

The last two-port parameters that we will consider are called the **t** *parameters,* the *ABCD parameters,* or simply the *transmission parameters.* They are defined by

$$\mathbf{V}_1 = \mathbf{t}_{11}\mathbf{V}_2 - \mathbf{t}_{12}\mathbf{I}_2 \tag{43}$$

and

$$\mathbf{I}_1 = \mathbf{t}_{21}\mathbf{V}_2 - \mathbf{t}_{22}\mathbf{I}_2 \tag{44}$$

or

$$\begin{bmatrix} \mathbf{V}_1 \\ \mathbf{I}_1 \end{bmatrix} = \mathbf{t} \begin{bmatrix} \mathbf{V}_2 \\ -\mathbf{I}_2 \end{bmatrix} \tag{45}$$

where $\mathbf{V}_1$, $\mathbf{V}_2$, $\mathbf{I}_1$, and $\mathbf{I}_2$ are defined as usual (Fig. 17.8). The minus signs that appear in Eqs. [43] and [44] should be associated with the output current, as $(-\mathbf{I}_2)$. Thus, both $\mathbf{I}_1$ and $-\mathbf{I}_2$ are directed to the right, the direction of energy or signal transmission.

Other widely used nomenclature for this set of parameters is

$$\begin{bmatrix} \mathbf{t}_{11} & \mathbf{t}_{12} \\ \mathbf{t}_{21} & \mathbf{t}_{22} \end{bmatrix} = \begin{bmatrix} \mathbf{A} & \mathbf{B} \\ \mathbf{C} & \mathbf{D} \end{bmatrix} \qquad [46]$$

Note that there are no minus signs in the $\mathbf{t}$ or $\mathbf{ABCD}$ matrices.

Looking again at Eqs. [43] to [45], we see that the quantities on the left, often thought of as the given or independent variables, are the input voltage and current, $\mathbf{V}_1$ and $\mathbf{I}_1$; the dependent variables, $\mathbf{V}_2$ and $\mathbf{I}_2$, are the output quantities. Thus, the transmission parameters provide a direct relationship between input and output. Their major use arises in transmission-line analysis and in cascaded networks.

Let us find the $\mathbf{t}$ parameters for the bilateral resistive two-port of Fig. 17.30a. To illustrate one possible procedure for finding a single parameter, consider

$$\mathbf{t}_{12} = \left. \frac{\mathbf{V}_1}{-\mathbf{I}_2} \right|_{\mathbf{V}_2 = 0}$$

We therefore short-circuit the output $(\mathbf{V}_2 = 0)$ and set $\mathbf{V}_1 = 1$ V, as shown in Fig. 17.30b. Note that we cannot set the denominator equal to unity by placing a 1 A current source at the output; we already have a short circuit there. The equivalent resistance offered to the 1 V source is $R_{eq} = 2 + (4\|10)$ Ω, and we then use current division to get

$$-\mathbf{I}_2 = \frac{1}{2 + (4\|10)} \times \frac{10}{10 + 4} = \frac{5}{34} \text{ A}$$

Hence,

$$\mathbf{t}_{12} = \frac{1}{-\mathbf{I}_2} = \frac{34}{5} = 6.8 \ \Omega$$

If it is necessary to find all four parameters, we write any convenient pair of equations using all four terminal quantities, $\mathbf{V}_1$, $\mathbf{V}_2$, $\mathbf{I}_1$, and $\mathbf{I}_2$. From Fig. 17.30a, we have two mesh equations:

$$\mathbf{V}_1 = 12\mathbf{I}_1 + 10\mathbf{I}_2 \qquad [47]$$

$$\mathbf{V}_2 = 10\mathbf{I}_1 + 14\mathbf{I}_2 \qquad [48]$$

Solving Eq. [48] for $\mathbf{I}_1$, we get

$$\mathbf{I}_1 = 0.1\mathbf{V}_2 - 1.4\mathbf{I}_2$$

so that $\mathbf{t}_{21} = 0.1$ S and $\mathbf{t}_{22} = 1.4$. Substituting the expression for $\mathbf{I}_1$ in Eq. [47], we find

$$\mathbf{V}_1 = 12(0.1\mathbf{V}_2 - 1.4\mathbf{I}_2) + 10\mathbf{I}_2 = 1.2\mathbf{V}_2 - 6.8\mathbf{I}_2$$

and $\mathbf{t}_{11} = 1.2$ and $\mathbf{t}_{12} = 6.8$ Ω, once again.

For reciprocal networks, the determinant of the $\mathbf{t}$ matrix is equal to unity:

$$\Delta_\mathbf{t} = \mathbf{t}_{11}\mathbf{t}_{22} - \mathbf{t}_{12}\mathbf{t}_{21} = 1$$

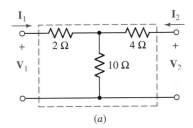

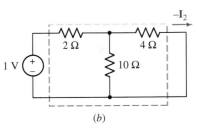

■ **FIGURE 17.30** (a) A two-port resistive network for which the $\mathbf{t}$ parameters are to be found. (b) To find $\mathbf{t}_{12}$, set $\mathbf{V}_1 = 1$ V with $\mathbf{V}_2 = 0$; then $\mathbf{t}_{12} = 1/(-\mathbf{I}_2) = 6.8\ \Omega$.

In the resistive example of Fig. 17.30, $\Delta_t = 1.2 \times 1.4 - 6.8 \times 0.1 = 1$. Good!

We conclude our two-port discussion by connecting two two-ports in cascade, as illustrated for two networks in Fig. 17.31. Terminal voltages and currents are indicated for each two-port, and the corresponding **t** parameter relationships are, for network A,

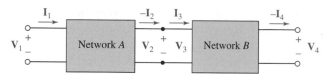

FIGURE 17.31 When two-port networks A and B are cascaded, the **t** parameter matrix for the combined network is given by the matrix product $\mathbf{t} = \mathbf{t}_A \mathbf{t}_B$.

$$\begin{bmatrix} \mathbf{V}_1 \\ \mathbf{I}_1 \end{bmatrix} = \mathbf{t}_A \begin{bmatrix} \mathbf{V}_2 \\ -\mathbf{I}_2 \end{bmatrix} = \mathbf{t}_A \begin{bmatrix} \mathbf{V}_3 \\ \mathbf{I}_3 \end{bmatrix}$$

and for network B,

$$\begin{bmatrix} \mathbf{V}_3 \\ \mathbf{I}_3 \end{bmatrix} = \mathbf{t}_B \begin{bmatrix} \mathbf{V}_4 \\ -\mathbf{I}_4 \end{bmatrix}$$

Combining these results, we have

$$\begin{bmatrix} \mathbf{V}_1 \\ \mathbf{I}_1 \end{bmatrix} = \mathbf{t}_A \mathbf{t}_B \begin{bmatrix} \mathbf{V}_4 \\ -\mathbf{I}_4 \end{bmatrix}$$

Therefore, the **t** parameters for the cascaded networks are found by the matrix product,

$$\mathbf{t} = \mathbf{t}_A \mathbf{t}_B$$

This product is *not* obtained by multiplying corresponding elements in the two matrices. If necessary, review the correct procedure for matrix multiplication in Appendix 2.

EXAMPLE 17.10

Find the t parameters for the cascaded networks shown in Fig. 17.32.

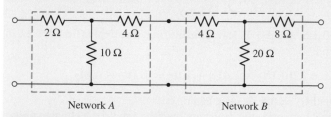

FIGURE 17.32 A cascaded connection.

Network A is the two-port of Fig. 17.32, and, therefore

$$\mathbf{t}_A = \begin{bmatrix} 1.2 & 6.8\ \Omega \\ 0.1\ \text{S} & 1.4 \end{bmatrix}$$

while network B has resistance values twice as large, so that

$$\mathbf{t}_B = \begin{bmatrix} 1.2 & 13.6\ \Omega \\ 0.05\ \text{S} & 1.4 \end{bmatrix}$$

For the combined network,

$$\mathbf{t} = \mathbf{t}_A \mathbf{t}_B = \begin{bmatrix} 1.2 & 6.8 \\ 0.1 & 1.4 \end{bmatrix} \begin{bmatrix} 1.2 & 13.6 \\ 0.05 & 1.4 \end{bmatrix}$$

$$= \begin{bmatrix} 1.2 \times 1.2 + 6.8 \times 0.05 & 1.2 \times 13.6 + 6.8 \times 1.4 \\ 0.1 \times 1.2 + 1.4 \times 0.05 & 0.1 \times 13.6 + 1.4 \times 1.4 \end{bmatrix}$$

and

$$\mathbf{t} = \begin{bmatrix} 1.78 & 25.84\ \Omega \\ 0.19\ \text{S} & 3.32 \end{bmatrix}$$

PRACTICE

17.12 Given $\mathbf{t} = \begin{bmatrix} 3.2 & 8\ \Omega \\ 0.2\ \text{S} & 4 \end{bmatrix}$, find ($a$) $\mathbf{z}$; (b) $\mathbf{t}$ for two identical networks in cascade; (c) $\mathbf{z}$ for two identical networks in cascade.

Ans: $\begin{bmatrix} 16 & 56 \\ 5 & 20 \end{bmatrix}$ (Ω); $\begin{bmatrix} 11.84 & 57.6\ \Omega \\ 1.44\ \text{S} & 17.6 \end{bmatrix}$; $\begin{bmatrix} 8.22 & 87.1 \\ 0.694 & 12.22 \end{bmatrix}$ (Ω).

COMPUTER-AIDED ANALYSIS

The characterization of two-port networks using $\mathbf{t}$ parameters creates the opportunity for vastly simplified analysis of cascaded two-port network circuits. As seen in this section, where, for example,

$$\mathbf{t}_A = \begin{bmatrix} 1.2 & 6.8\ \Omega \\ 0.1\ \text{S} & 1.4 \end{bmatrix}$$

and

$$\mathbf{t}_B = \begin{bmatrix} 1.2 & 13.6\ \Omega \\ 0.05\ \text{S} & 1.4 \end{bmatrix}$$

we found that the $\mathbf{t}$ parameters characterizing the cascaded network can be found by simply multiplying $\mathbf{t}_A$ and $\mathbf{t}_B$:

$$\mathbf{t} = \mathbf{t}_A \cdot \mathbf{t}_B$$

Such matrix operations are easily carried out using scientific calculators or software packages such as MATLAB. The MATLAB script,

(Continued on next page)

for example, would be

> EDU» tA = [1.2 6.8; 0.1 1.4];
> EDU» tB = [1.2 13.6; 0.05 1.4];
> EDU» t = tA*tB

t =

1.7800	25.8700
0.1900	3.3200

as we found in Example 17.10.

In terms of entering matrices in MATLAB, each has a case-sensitive variable name (tA, tB, and t in this example). Matrix elements are entered a row at a time, beginning with the top row; rows are separated by a semicolon. Again, the reader should always be careful to remember that the order to operations is critical when performing matrix algebra. For example, tB*tA results in a totally different matrix than the one we sought:

$$\mathbf{t}_B \cdot \mathbf{t}_A = \begin{bmatrix} 2.8 & 27.2 \\ 0.2 & 2.3 \end{bmatrix}$$

For simple matrices such as seen in this example, a scientific calculator is just as handy (if not more so). However, larger cascaded networks are more easily handled on a computer, where it is more convenient to see all arrays on the screen simultaneously.

SUMMARY AND REVIEW

❑ In order to employ the analysis methods described in this chapter, it is critical to remember that each port can only be connected to either a one-port network or a port of another multiport network.

❑ The defining equations for analyzing a two-port network in terms of its admittance (**y**) parameters are:

$$\mathbf{I}_1 = \mathbf{y}_{11}\mathbf{V}_1 + \mathbf{y}_{12}\mathbf{V}_2 \qquad \text{and} \qquad \mathbf{I}_2 = \mathbf{y}_{21}\mathbf{V}_1 + \mathbf{y}_{22}\mathbf{V}_2$$

where

$$\mathbf{y}_{11} = \left.\frac{\mathbf{I}_1}{\mathbf{V}_1}\right|_{\mathbf{V}_2=0} \qquad\qquad \mathbf{y}_{12} = \left.\frac{\mathbf{I}_1}{\mathbf{V}_2}\right|_{\mathbf{V}_1=0}$$

$$\mathbf{y}_{21} = \left.\frac{\mathbf{I}_2}{\mathbf{V}_1}\right|_{\mathbf{V}_2=0} \qquad \text{and} \qquad \mathbf{y}_{22} = \left.\frac{\mathbf{I}_2}{\mathbf{V}_2}\right|_{\mathbf{V}_1=0}$$

❑ The defining equations for analyzing a two-port network in terms of its impedance (**z**) parameters are:

$$\mathbf{V}_1 = \mathbf{z}_{11}\mathbf{I}_1 + \mathbf{z}_{12}\mathbf{I}_2 \qquad \text{and} \qquad \mathbf{V}_2 = \mathbf{z}_{21}\mathbf{I}_1 + \mathbf{z}_{22}\mathbf{I}_2$$

❑ The defining equations for analyzing a two-port network in terms of its hybrid (**h**) parameters are:

$$\mathbf{V}_1 = \mathbf{h}_{11}\mathbf{I}_1 + \mathbf{h}_{12}\mathbf{V}_2 \qquad \text{and} \qquad \mathbf{I}_2 = \mathbf{h}_{21}\mathbf{I}_1 + \mathbf{h}_{22}\mathbf{V}_2$$

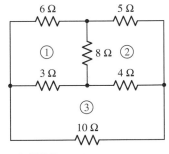

❑ The defining equations for analyzing a two-port network in terms of its transmission (**t**) parameters (also called the **ABCD** parameters) are:

$$\mathbf{V}_1 = \mathbf{t}_{11}\mathbf{V}_2 - \mathbf{t}_{12}\mathbf{I}_2 \qquad \text{and} \qquad \mathbf{I}_1 = \mathbf{t}_{21}\mathbf{V}_2 - \mathbf{t}_{22}\mathbf{I}_2$$

❑ It is straightforward to convert between **h**, **z**, **t**, and **y** parameters, depending on circuit analysis needs; the transformations are summarized in Table 17.1.

READING FURTHER

Further details of matrix methods for circuit analysis can be found in:

R. A. DeCarlo and P. M. Lin, *Linear Circuit Analysis,* 2nd ed. New York: Oxford University Press, 2001.

Analysis of transistor circuits using network parameters is described in:

W. H. Hayt, Jr. and G. W. Neudeck, *Electronic Circuit Analysis and Design,* 2nd ed. New York: Wiley, 1995.

EXERCISES

17.1 One-Port Networks

1. Consider the following set of equations:

$$4\mathbf{I}_1 - 8\mathbf{I}_2 + 9\mathbf{I}_3 = 12$$
$$5\mathbf{I}_1 \qquad\quad - 7\mathbf{I}_3 = 4$$
$$7\mathbf{I}_1 + 3\mathbf{I}_2 + \ \mathbf{I}_3 = 0$$

(*a*) Write this set of equations in matrix form. (*b*) Determine $\Delta_{\mathbf{Z}}$. (*c*) Determine Δ_{11}. (*d*) Calculate $\mathbf{I}_1$. (*e*) Calculate $\mathbf{I}_3$.

2. Find $\Delta_{\mathbf{Z}}$ for the network shown in Fig. 17.33, and then use it as a help in finding the power generated by a 100 V dc source inserted in the outside branch of mesh: (*a*) 1; (*b*) 2; (*c*) 3.

3. Find $\Delta_{\mathbf{Y}}$ for the network shown in Fig. 17.34, and then use it as a help in finding the power generated by a 10 A dc source inserted between the reference node and node: (*a*) 1; (*b*) 2; (*c*) 3.

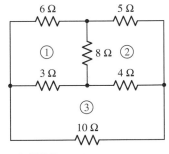

■ **FIGURE 17.33**

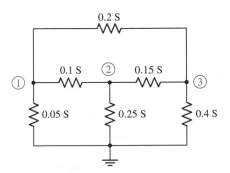

■ **FIGURE 17.34**

4. The resistance matrix of a certain one-port network is given as Fig. 17.35. Find R_{in} for a source inserted only in mesh 1.

$$[\mathbf{R}] = \begin{bmatrix} 3 & -1 & -2 & 0 \\ -1 & 4 & 1 & 3 \\ -2 & 2 & 5 & 2 \\ 0 & 3 & -2 & 6 \end{bmatrix} (\Omega)$$

■ **FIGURE 17.35**

5. Find the Thévenin equivalent impedance $\mathbf{Z}_{th}(\mathbf{s})$ for the one-port of Fig. 17.36.

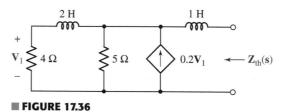

■ **FIGURE 17.36**

6. Find $\mathbf{Z}_{in}$ for the one-port shown in Fig. 17.37 by (a) finding $\Delta_{\mathbf{Z}}$; (b) finding $\Delta_{\mathbf{Y}}$ and $\mathbf{Y}_{in}$ first, and then $\mathbf{Z}_{in}$.

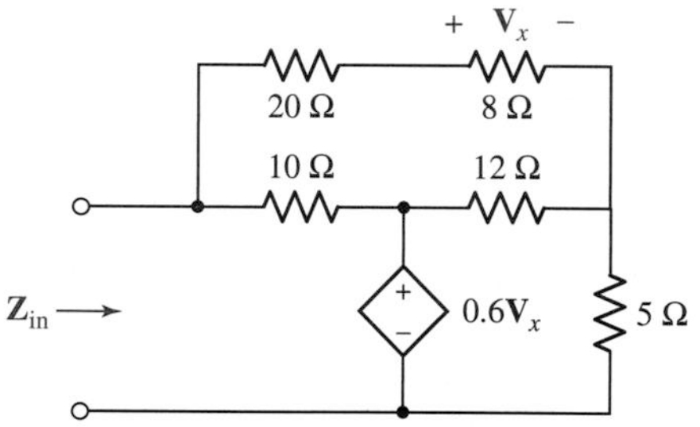

■ **FIGURE 17.37**

7. Find the output impedance for the network of Fig. 17.38, as a function of **s**.

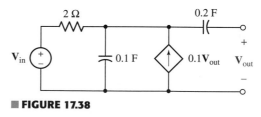

■ **FIGURE 17.38**

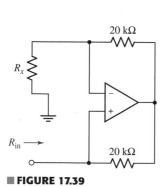

■ **FIGURE 17.39**

8. If the op amp shown in Fig. 17.39 is assumed to be ideal ($R_i = \infty$, $R_o = 0$, and $A = \infty$), find R_{in}.

9. (a) If both the op amps shown in the circuit of Fig. 17.40 are assumed to be ideal ($R_i = \infty$, $R_o = 0$, and $A = \infty$), find $\mathbf{Z}_{in}$. (b) $R_1 = 4$ kΩ, $R_2 = 10$ kΩ,

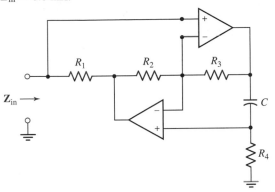

$R_3 = 10 \text{ k}\Omega$, $R_4 = 1 \text{ k}\Omega$, and $C = 200 \text{ pF}$, show that $\mathbf{Z}_{in} = j\omega L_{in}$, where $L_{in} = 0.8 \text{ mH}$.

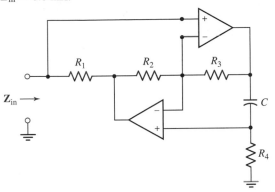

■ **FIGURE 17.40**

17.2 Admittance Parameters

10. For the linear network depicted in Fig. 17.8, find

(a) $\mathbf{I}_2$ if $\mathbf{y} = \begin{bmatrix} 0.01 & 0.3 \\ 0.3 & -0.02 \end{bmatrix}$ (S) and $\mathbf{V} = \begin{bmatrix} 9 \\ -3.5 \end{bmatrix}$ (V);

(b) $\mathbf{V}_1$ if $\mathbf{y} = \begin{bmatrix} -0.1 & 0.15 \\ 0.15 & 0.8 \end{bmatrix}$ (S) and $\mathbf{I} = \begin{bmatrix} 0.001 \\ 0.02 \end{bmatrix}$ (A).

11. Find $\mathbf{y}_{11}$ and $\mathbf{y}_{12}$ for the two-port shown in Fig. 17.41.

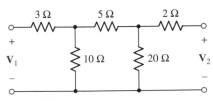

■ **FIGURE 17.41**

12. If the two-port shown in Fig. 17.42 has the parameter values $\mathbf{y}_{11} = 10$, $\mathbf{y}_{12} = -5$, $\mathbf{y}_{21} = 50$, and $\mathbf{y}_{22} = 20$, all in mS, find $\mathbf{V}_1$ and $\mathbf{V}_2$ when $\mathbf{V}_s = 100 \text{ V}$, $R_s = 25 \text{ }\Omega$, and $R_L = 100 \text{ }\Omega$.

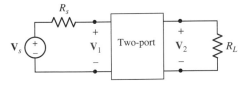

■ **FIGURE 17.42**

13. Find the four **y** parameters for the network of Fig. 17.43.

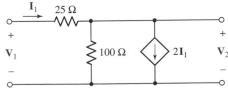

■ **FIGURE 17.43**

14. Find **y** for the two-port shown in Fig. 17.44.

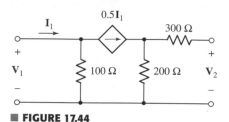

■ **FIGURE 17.44**

15. Let $\mathbf{y} = \begin{bmatrix} 0.1 & -0.0025 \\ -8 & 0.05 \end{bmatrix}$ (S) for the two-port of Fig. 17.45. (*a*) Find values

for the ratios $\mathbf{V}_2/\mathbf{V}_1$, $\mathbf{I}_2/\mathbf{I}_1$, and $\mathbf{V}_1/\mathbf{I}_1$. (*b*) Remove the 5 Ω resistor, set the 1 V source equal to zero, and find $\mathbf{V}_2/\mathbf{I}_2$.

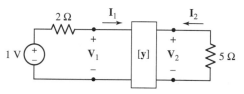

■ **FIGURE 17.45**

16. The admittance parameters of a certain two-port are $\mathbf{y} = \begin{bmatrix} 10 & -5 \\ -20 & 2 \end{bmatrix}$ (mS).

Find the new **y** if a 100 Ω resistor is connected: (*a*) in series with one of the input leads; (*b*) in series with one of the output leads.

17. Complete the table given as part of Fig. 17.46, and also give values for the **y** parameters.

18. For the general linear network depicted in Fig. 17.8, find

(*a*) $\mathbf{I}_2$ if $\mathbf{y} = \begin{bmatrix} 10^{-3} & j0.01 \\ j0.01 & -j0.005 \end{bmatrix}$ (S) and $\mathbf{V} = \begin{bmatrix} 12\underline{/43°} \\ 2\underline{/0°} \end{bmatrix}$ (V);

(*b*) $\mathbf{V}_2$ if $\mathbf{y} = \begin{bmatrix} -j5 & 10 \\ 4 & j10 \end{bmatrix}$ (S) and $\mathbf{I} = \begin{bmatrix} 120\underline{/30°} \\ 88\underline{/45°} \end{bmatrix}$ (A).

19. The metal-oxide-semiconductor field effect transistor (MOSFET), a three-terminal nonlinear element used in many electronics applications, is often specified in terms of its **y** parameters. The ac parameters are strongly dependent on the measurement conditions, and commonly named y_{is}, y_{rs}, y_{fs}, and y_{os}, as in:

$$I_g = y_{is} V_{gs} + y_{rs} V_{ds} \qquad [49]$$

$$I_d = y_{fs} V_{gs} + y_{os} V_{ds} \qquad [50]$$

where I_g is the transistor gate current, I_d is the transistor drain current, and the third terminal (the source) is common to the input and output during the measurement. Thus, V_{gs} is the voltage between the gate and the source, and V_{ds} is the voltage between the drain and the source. The typical high-frequency model used to approximate the behavior of a MOSFET is shown in Fig. 17.47.

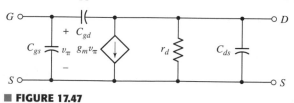

■ **FIGURE 17.47**

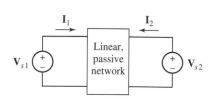

	$\mathbf{V}_{s1}$ (V)	$\mathbf{V}_{s2}$ (V)	$\mathbf{I}_1$ (A)	$\mathbf{I}_2$ (A)
Exp't #1	100	50	5	−32.5
Exp't #2	50	100	−20	−5
Exp't #3	20	0		
Exp't #4			5	0
Exp't #5			5	15

■ **FIGURE 17.46**

(*a*) For the configuration stated above, which transistor terminal is used as the input, and which terminal is used as the output? (*b*) Derive expressions for the parameters y_{is}, y_{rs}, y_{fs}, and y_{os} defined in Eqs. [49] and [50], in terms of the model parameters C_{gs}, C_{gd}, g_m, r_d, and C_{ds} of Fig. 17.47. (*c*) Compute y_{is}, y_{rs}, y_{fs}, and y_{os} if $g_m = 4.7$ mS, $C_{gs} = 3.4$ pF, $C_{gd} = 1.4$ pF, $C_{ds} = 0.4$ pF, and $r_d = 10$ kΩ.

17.3 Some Equivalent Networks

20. Convert the Δ network of Fig. 17.48 to a Y-connected network.

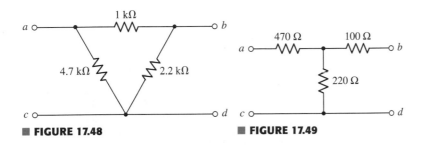

■ **FIGURE 17.48**   ■ **FIGURE 17.49**

21. Convert the Y network of Fig. 17.49 to a Δ-connected network.
22. Find R_{in} for the one-port shown in Fig. 17.50 by using Y-Δ and Δ-Y transformations as appropriate.

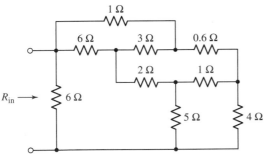

■ **FIGURE 17.50**

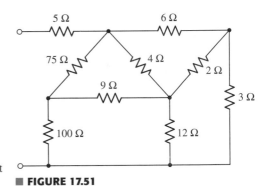

■ **FIGURE 17.51**

23. Use Y-Δ and Δ-Y transformations to find the input resistance of the one-port shown in Fig. 17.51.
24. Find $\mathbf{Z}_{in}$ for the network of Fig. 17.52.

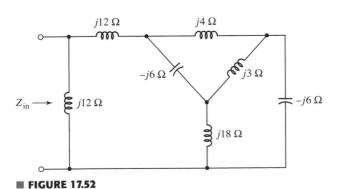

■ **FIGURE 17.52**

25. Let $\mathbf{y} = \begin{bmatrix} 0.4 & -0.002 \\ -5 & 0.04 \end{bmatrix}$ (S) for the two-port of Fig. 17.53, and find (a) $\mathbf{G}_V$; (b) $\mathbf{G}_I$; (c) G_P; (d) $\mathbf{Z}_{\text{in}}$; (e) $\mathbf{Z}_{\text{out}}$.

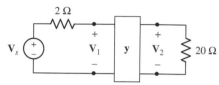

■ **FIGURE 17.53**

26. Let $\mathbf{y} = \begin{bmatrix} 0.1 & -0.05 \\ -0.5 & 0.2 \end{bmatrix}$ (S) for the two-port of Fig. 17.54. Find (a) $\mathbf{G}_V$; (b) $\mathbf{G}_I$; (c) G_P; (d) $\mathbf{Z}_{\text{in}}$; (e) $\mathbf{Z}_{\text{out}}$. (f) If the reverse voltage gain $\mathbf{G}_{V,\text{rev}}$ is defined as $\mathbf{V}_1/\mathbf{V}_2$ with $\mathbf{V}_s = 0$ and R_L removed, calculate $\mathbf{G}_{V,\text{rev}}$. (g) If the insertion power gain G_{ins} is defined as the ratio of $P_{5\Omega}$ with the two-port in place to $P_{5\Omega}$ with the two-port replaced by jumpers connecting each input terminal to the corresponding output terminal, calculate G_{ins}.

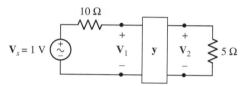

■ **FIGURE 17.54**

27. (a) Draw an equivalent circuit in the form of Fig. 17.13b for which $\mathbf{y} = \begin{bmatrix} 1.5 & -1 \\ 4 & 3 \end{bmatrix}$ (mS). (b) If two of these two-ports are connected in parallel, draw the new equivalent circuit and show that $\mathbf{y}_{\text{new}} = 2\mathbf{y}$.

28. (a) Find $\mathbf{y}_a$ for the two-port of Fig. 17.55a. (b) Find $\mathbf{y}_b$ for Fig. 17.55b. (c) Draw the network that is obtained when these two-ports are connected in parallel, and show that $\mathbf{y}$ for this network is equal to $\mathbf{y}_a + \mathbf{y}_b$.

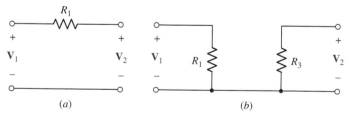

(a)   (b)

■ **FIGURE 17.55**

17.4 Impedance Parameters

29. For the linear network depicted in Fig. 17.8,

(a) find $\mathbf{V}_1$ if $\mathbf{z} = \begin{bmatrix} 4.7 & 2.2 \\ 2.2 & 3.3 \end{bmatrix}$ (kΩ) and $\mathbf{I} = \begin{bmatrix} 1.5 \\ -2.5 \end{bmatrix}$ (mA); (b) $\mathbf{I}_2$ if $\mathbf{z} = \begin{bmatrix} -10 & 15 \\ 15 & 6 \end{bmatrix}$ (kΩ) and $\mathbf{V} = \begin{bmatrix} 1 \\ -2 \end{bmatrix}$ (V).

30. Consider the general linear network of Fig. 17.8. Calculate

(a) $\mathbf{V}_2$ if $\mathbf{z} = \begin{bmatrix} 5 & j \\ j & -j2 \end{bmatrix}$ (Ω) and $\mathbf{I} = \begin{bmatrix} 2\underline{/20°} \\ 2\underline{/0°} \end{bmatrix}$ (A);

(b) $\mathbf{I}_1$ if $\mathbf{z} = \begin{bmatrix} -j & 2 \\ 4 & j4 \end{bmatrix}$ (Ω) and $\mathbf{V} = \begin{bmatrix} 137\underline{/30°} \\ 105\underline{/45°} \end{bmatrix}$ (V).

31. Find $\mathbf{z}$ for the two-port shown in Fig. 17.56.

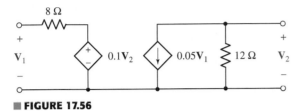

■ **FIGURE 17.56**

32. (a) Find $\mathbf{z}$ for the two-port of Fig. 17.57. (b) If $\mathbf{I}_1 = \mathbf{I}_2 = 1$ A, find the voltage gain $\mathbf{G}_V$.

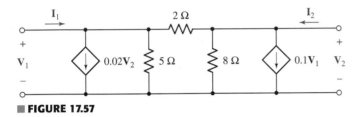

■ **FIGURE 17.57**

33. A certain two-port is described by $\mathbf{z} = \begin{bmatrix} 4 & 1.5 \\ 10 & 3 \end{bmatrix}$ (Ω). The input consists of a source $\mathbf{V}_s$ in series with 5 Ω, while the output is $R_L = 2$ Ω. Find (a) $\mathbf{G}_I$; (b) $\mathbf{G}_V$; (c) G_P; (d) $\mathbf{Z}_{\text{in}}$; (e) $\mathbf{Z}_{\text{out}}$.

34. Let $[\mathbf{z}] = \begin{bmatrix} 1000 & 100 \\ -2000 & 400 \end{bmatrix}$ (Ω) for the two-port of Fig. 17.58. Find the average power delivered to the (a) 200 Ω resistor; (b) 500 Ω resistor; (c) two-port.

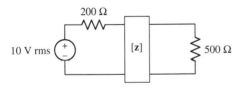

■ **FIGURE 17.58**

35. Find the four $\mathbf{z}$ parameters at $\omega = 10^8$ rad/s for the transistor high-frequency equivalent circuit shown in Fig. 17.59.

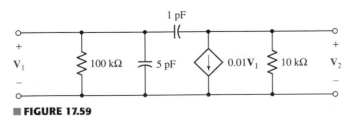

■ **FIGURE 17.59**

36. A two-port for which $\mathbf{z} = \begin{bmatrix} 20 & 2 \\ 40 & 10 \end{bmatrix}$ (Ω) is driven by a source $\mathbf{V}_s = 100\underline{/0°}$ V in series with 5 Ω, and terminated in a 25 Ω resistor. Find the Thévenin-equivalent circuit presented to the 25 Ω resistor.

17.5 Hybrid Parameters

37. The **h** parameters for a certain two-port are $\mathbf{h} = \begin{bmatrix} 9\,\Omega & -2 \\ 20 & 0.2\text{ S} \end{bmatrix}$. Find the new **h** that results if a 1 Ω resistor is connected in series with (a) the input; (b) the output.

38. Find $\mathbf{Z}_{\text{in}}$ and $\mathbf{Z}_{\text{out}}$ for a two-port driven by a source having $R_s = 100$ Ω and terminated with $R_L = 500$ Ω, if $\mathbf{h} = \begin{bmatrix} 100\,\Omega & 0.01 \\ 20 & 1\text{ mS} \end{bmatrix}$.

39. Refer to the two-port shown in Fig. 17.60 and find (a) $\mathbf{h}_{12}$; (b) $\mathbf{z}_{12}$; (c) $\mathbf{y}_{12}$.

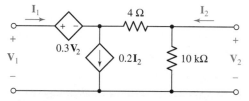

■ **FIGURE 17.60**

40. Let $\mathbf{h}_{11} = 1$ kΩ, $\mathbf{h}_{12} = -1$, $\mathbf{h}_{21} = 4$, and $\mathbf{h}_{22} = 500$ μS for the two-port shown in Fig. 17.61. Find the average power delivered to (a) $R_s = 200$ Ω; (b) $R_L = 1$ kΩ; (c) the entire two-port.

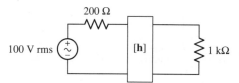

■ **FIGURE 17.61**

41. (a) Find **h** for the two-port of Fig. 17.62. (b) Find $\mathbf{Z}_{\text{out}}$ if the input contains $\mathbf{V}_s$ in series with $R_s = 200$.

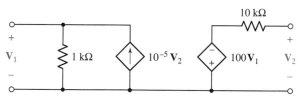

■ **FIGURE 17.62**

42. Find **y**, **z**, and **h** for both of the two-ports shown in Fig. 17.63. If any parameter is infinite, skip that parameter set.

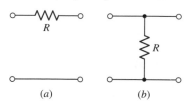

(a)　　　　(b)

■ **FIGURE 17.63**

43. Figure 17.64 depicts a commonly used high-frequency bipolar junction transistor (bjt) model, which is valid for small ac signal magnitudes. If the emitter terminal (labeled E) is common to the input and output, and the base terminal (labeled B) is used as the input, derive an expression in terms of r_x, r_π, C_π, C_μ, g_m, and r_d for (a) h_{oe}; (b) h_{fe}; (c) h_{ie}; and (d) h_{re}.

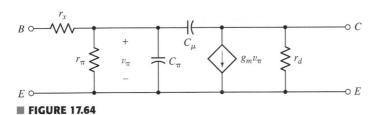

■ **FIGURE 17.64**

17.6 Transmission Parameters

44. Given $\mathbf{y} = \begin{bmatrix} 1 & -2 \\ 3 & 4 \end{bmatrix}$, $\mathbf{b} = \begin{bmatrix} 4 & 6 \\ -1 & 5 \end{bmatrix}$, $\mathbf{c} = \begin{bmatrix} 3 & 2 & 4 & -1 \\ -2 & 3 & 5 & 0 \end{bmatrix}$, and

$\mathbf{d} = \begin{bmatrix} 1 & 2 & -1 \\ 3 & 0 & 5 \\ -2 & -3 & 1 \\ 4 & -4 & 2 \end{bmatrix}$, calculate: (a) $\mathbf{y} \cdot \mathbf{b}$; (b) $\mathbf{b} \cdot \mathbf{y}$; (c) $\mathbf{b} \cdot \mathbf{c}$; (d) $\mathbf{c} \cdot \mathbf{d}$;

(e) $\mathbf{y} \cdot \mathbf{b} \cdot \mathbf{c} \cdot \mathbf{d}$.

45. (a) Find $\mathbf{t}$ for the two-port shown in Fig. 17.65. (b) Calculate $\mathbf{Z}_{out}$ for this two-port if $R_s = 15\ \Omega$ for the source.

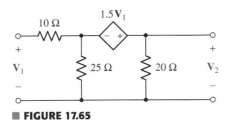

■ **FIGURE 17.65**

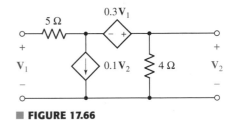

■ **FIGURE 17.66**

46. Find $\mathbf{t}$ for the two-port shown in Fig. 17.66.

47. (a) Find $\mathbf{t}_A$, $\mathbf{t}_B$, and $\mathbf{t}_C$ for the cascaded two-ports of Fig. 17.67. (b) Find $\mathbf{t}$ for the six-resistor two-port.

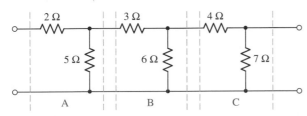

■ **FIGURE 17.67**

2 Ω

$\mathbf{t}_A$

■ **FIGURE 17.68**

48. (*a*) Find $\mathbf{t}_A$ for the single 2 Ω resistor of Fig. 17.68. (*b*) Show that $\mathbf{t}$ for a single 10 Ω resistor can be obtained by $(\mathbf{t}_A)^5$.

49. (*a*) Find $\mathbf{t}_a$, $\mathbf{t}_b$, and $\mathbf{t}_c$ for the networks shown in Fig. 17.69*a*, *b*, and *c*. (*b*) By using the rules for interconnecting two-ports in cascade, find $\mathbf{t}$ for the network of Fig. 17.69*d*.

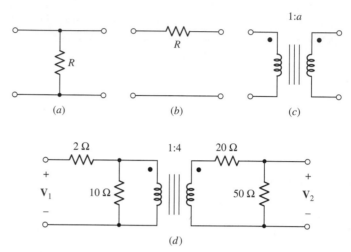

■ **FIGURE 17.69**

50. (*a*) Find $\mathbf{t}$ for the two-port shown in Fig. 17.70. (*b*) Use the techniques of cascading two-ports to find $\mathbf{t}_{new}$ if a 20 Ω resistor is connected across the output.

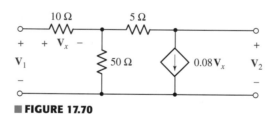

■ **FIGURE 17.70**

Fourier Circuit Analysis

INTRODUCTION

In this chapter we continue our introduction to circuit analysis by studying periodic functions in both the time and frequency domains. Specifically, we will consider forcing functions which are *periodic* and have functional natures which satisfy certain mathematical restrictions that are characteristic of any function which we can generate in the laboratory. Such functions may be represented as the sum of an infinite number of sine and cosine functions which are harmonically related. Therefore, since the forced response to each sinusoidal component can be determined easily by sinusoidal steady-state analysis, the response of the linear network to the general periodic forcing function may be obtained by superposing the partial responses.

The topic of Fourier series is of vital importance in a number of fields, particularly communications. The use of Fourier-based techniques to assist in circuit analysis, however, had been slowly falling out of fashion for a number of years. Now as we face an increasingly larger fraction of global power usage coming from equipment employing pulse-modulated power supplies (e.g., computers), the subject of harmonics in power systems and power electronics is rapidly becoming a serious problem in even large-scale generation plants. It is only with Fourier-based analysis that the underlying problems and possible solutions can be understood.

18.1 TRIGONOMETRIC FORM OF THE FOURIER SERIES

We know that the complete response of a linear circuit to an arbitrary forcing function is composed of the sum of a *forced response* and a *natural response*. The natural response has been considered both in the time domain (Chaps. 7, 8, and 9) and in the frequency domain (Chaps. 14 and 15). The forced response has also been considered from several perspectives, including the phasor-based techniques of Chap. 10. As we have discovered, in some cases we need *both* components of the total response of a particular circuit, while in others we need only the natural or the forced response. In this section, we refocus our attention on forcing functions that are *sinusoidal* in nature, and discover how to write a general periodic function as a *sum* of such functions—leading us into a discussion of a new set of circuit analysis procedures.

Harmonics

Some feeling for the validity of representing a general *periodic* function by an infinite sum of sine and cosine functions may be gained by considering a simple example. Let us first assume a cosine function of radian frequency ω_0,

$$v_1(t) = 2\cos \omega_0 t$$

where

$$\omega_0 = 2\pi f_0$$

and the period T is

$$T = \frac{1}{f_0} = \frac{2\pi}{\omega_0}$$

Although T does not usually carry a zero subscript, it is the period of the fundamental frequency. The ***harmonics*** of this sinusoid have frequencies $n\omega_0$, where ω_0 is the fundamental frequency and $n = 1, 2, 3, \dots$. The frequency of the first harmonic is the ***fundamental frequency.***

Next let us select a third-harmonic voltage

$$v_{3a}(t) = \cos 3\omega_0 t$$

The fundamental $v_1(t)$, the third harmonic $v_{3a}(t)$, and the sum of these two waves are shown as functions of time in Fig. 18.1*a*. It should be noted that the sum is periodic, with period $T = 2\pi/\omega_0$.

The form of the resultant periodic function changes as the phase and amplitude of the third-harmonic component change. Thus, Fig. 18.1*b* shows the effect of combining $v_1(t)$ and a third harmonic of slightly larger amplitude,

$$v_{3b}(t) = 1.5\cos 3\omega_0 t$$

By shifting the phase of the third harmonic by 90 degrees to give

$$v_{3c}(t) = \sin 3\omega_0 t$$

the sum, shown in Fig. 18.1*c*, takes on a still different character. In all cases, the period of the resultant waveform is the same as the period of the fundamental waveform. The nature of the waveform depends on the amplitude and phase of every possible harmonic component, and we will find that we are able to generate waveforms which have extremely nonsinusoidal characteristics by an appropriate combination of sinusoidal functions.

After we have become familiar with the use of the sum of an infinite number of sine and cosine functions to represent a periodic waveform, we will consider the frequency-domain representation of a general nonperiodic waveform in a manner similar to the Laplace transform.

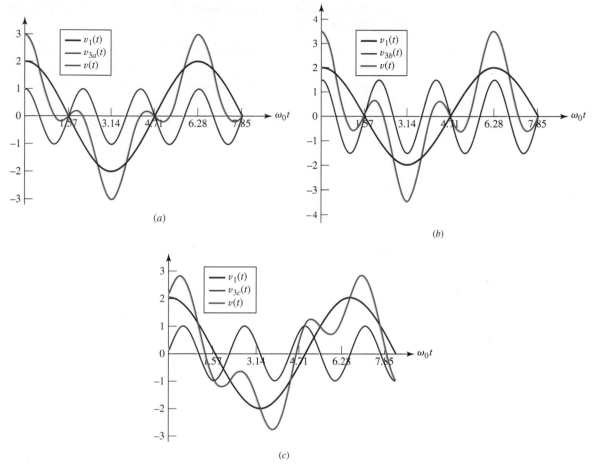

FIGURE 18.1 Several of the infinite number of different waveforms which may be obtained by combining a fundamental and a third harmonic. The fundamental is $v_1 = 2 \cos \omega_0 t$, and the third harmonic is: (a) $v_{3a} = \cos 3\omega_0 t$; (b) $v_{3b} = 1.5 \cos 3\omega_0 t$; (c) $v_{3c} = \sin 3\omega_0 t$.

PRACTICE

18.1 Let a third-harmonic voltage be added to the fundamental to yield $v = 2 \cos \omega_0 t + V_{m3} \sin 3\omega_0 t$, the waveform shown in Fig. 18.1c for $V_{m3} = 1$. (a) Find the value of V_{m3} so that $v(t)$ will have zero slope at $\omega_0 t = 2\pi/3$. (b) Evaluate $v(t)$ at $\omega_0 t = 2\pi/3$.

Ans: 0.577; −1.000.

The Fourier Series

We first consider a *periodic* function $f(t)$, defined in Sec. 11.2 by the functional relationship

$$f(t) = f(t + T)$$

where T is the period. We further assume that the function $f(t)$ satisfies the following properties:

1. $f(t)$ is single-valued everywhere; that is, $f(t)$ satisfies the mathematical definition of a function.

2. The integral $\int_{t_0}^{t_0+T} |f(t)|\, dt$ exists (i.e., is not infinite) for any choice of t_0.

3. $f(t)$ has a finite number of discontinuities in any one period.

4. $f(t)$ has a finite number of maxima and minima in any one period.

We will take f(t) to represent either a voltage or a current waveform, and any such waveform which we can actually produce must satisfy these four conditions; perhaps it should be noted, however, that certain mathematical functions do exist for which these four conditions are not satisfied.

Given such a periodic function $f(t)$, the Fourier theorem states that $f(t)$ may be represented by the infinite series

$$
\begin{aligned}
f(t) &= a_0 + a_1 \cos \omega_0 t + a_2 \cos 2\omega_0 t + \cdots \\
&\quad + b_1 \sin \omega_0 t + b_2 \sin 2\omega_0 t + \cdots \\
&= a_0 + \sum_{n=1}^{\infty} (a_n \cos n\omega_0 t + b_n \sin n\omega_0 t)
\end{aligned}
\tag{1}
$$

where the fundamental frequency ω_0 is related to the period T by

$$
\omega_0 = \frac{2\pi}{T}
$$

and where a_0, a_n, and b_n are constants that depend upon n and $f(t)$. Equation [1] is the trigonometric form of the **_Fourier series_** for $f(t)$, and the process of determining the values of the constants a_0, a_n, and b_n is called _Fourier analysis_. Our object is not the proof of this theorem, but only a simple development of the procedures of Fourier analysis and a feeling that the theorem is plausible.

Some Useful Trigonometric Integrals

Before we discuss the evaluation of the constants appearing in the Fourier series, let us collect a set of useful trigonometric integrals. We let both n and k represent any element of the set of integers 1, 2, 3, ... In the following integrals, 0 and T are used as the integration limits, but it is understood that any interval of one period is equally correct. Since the average value of a sinusoid over one period is zero,

$$
\int_0^T \sin n\omega_0 t\, dt = 0
\tag{2}
$$

and

$$
\int_0^T \cos n\omega_0 t\, dt = 0
\tag{3}
$$

It is also a simple matter to show that the following three definite integrals are zero:

$$
\int_0^T \sin k\omega_0 t \cos n\omega_0 t\, dt = 0
\tag{4}
$$

$$\int_0^T \sin k\omega_0 t \, \sin n\omega_0 t \, dt = 0 \qquad (k \neq n) \qquad [5]$$

$$\int_0^T \cos k\omega_0 t \, \cos n\omega_0 t \, dt = 0 \qquad (k \neq n) \qquad [6]$$

Those cases which are excepted in Eqs. [5] and [6] are also easily evaluated; we obtain

$$\int_0^T \sin^2 n\omega_0 t \, dt = \frac{T}{2} \qquad [7]$$

$$\int_0^T \cos^2 n\omega_0 t \, dt = \frac{T}{2} \qquad [8]$$

Evaluation of the Fourier Coefficients

The evaluation of the unknown constants in the Fourier series may now be accomplished readily. We first attack a_0. If we integrate each side of Eq. [1] over a full period, we obtain

$$\int_0^T f(t) \, dt = \int_0^T a_0 \, dt + \int_0^T \sum_{n=1}^{\infty} (a_n \cos n\omega_0 t + b_n \sin n\omega_0 t) \, dt$$

But every term in the summation is of the form of Eq. [2] or [3], and thus

$$\int_0^T f(t) \, dt = a_0 T$$

or

$$a_0 = \frac{1}{T} \int_0^T f(t) \, dt \qquad [9]$$

This constant a_0 is simply the average value of $f(t)$ over a period, and we therefore describe it as the dc component of $f(t)$.

To evaluate one of the cosine coefficients—say, a_k, the coefficient of $\cos k\omega_0 t$—we first multiply each side of Eq. [1] by $\cos k\omega_0 t$ and then integrate both sides of the equation over a full period:

$$\int_0^T f(t) \cos k\omega_0 t \, dt = \int_0^T a_0 \cos k\omega_0 t \, dt$$

$$+ \int_0^T \sum_{n=1}^{\infty} a_n \cos k\omega_0 t \, \cos n\omega_0 t \, dt$$

$$+ \int_0^T \sum_{n=1}^{\infty} b_n \cos k\omega_0 t \, \sin n\omega_0 t \, dt$$

From Eqs. [3], [4], and [6] we note that every term on the right-hand side of this equation is zero except for the single a_n term where $k = n$. We evaluate that term using Eq. [8], and in so doing we find a_k, or a_n:

$$a_n = \frac{2}{T} \int_0^T f(t) \cos n\omega_0 t \, dt \qquad [10]$$

This result is *twice* the average value of the product $f(t) \cos n\omega_0 t$ over a period.

In a similar way, we obtain b_k by multiplying by $\sin k\omega_0 t$, integrating over a period, noting that all but one of the terms on the right-hand side are zero, and performing that single integration by Eq. [7]. The result is

$$b_n = \frac{2}{T} \int_0^T f(t) \sin n\omega_0 t \, dt \qquad [11]$$

which is *twice* the average value of $f(t) \sin n\omega_0 t$ over a period.

Equations [9] to [11] now enable us to determine values for a_0 and all the a_n and b_n in the Fourier series, Eq. [1], as summarized below:

$$f(t) = a_0 + \sum_{n=1}^{\infty} (a_n \cos n\omega_0 t + b_n \sin n\omega_0 t) \qquad [1]$$

$$\omega_0 = \frac{2\pi}{T} = 2\pi f_0$$

$$a_0 = \frac{1}{T} \int_0^T f(t) \, dt \qquad [9]$$

$$a_n = \frac{2}{T} \int_0^T f(t) \cos n\omega_0 t \, dt \qquad [10]$$

$$b_n = \frac{2}{T} \int_0^T f(t) \sin n\omega_0 t \, dt \qquad [11]$$

EXAMPLE 18.1

The "half-sinusoidal" waveform shown in Fig. 18.2 represents the voltage response obtained at the output of a half-wave rectifier circuit, a nonlinear circuit whose purpose is to convert a sinusoidal input voltage to a (pulsating) dc output voltage. Find the Fourier series representation of this waveform.

▶ **Identify the goal of the problem.**
We are presented with a periodic function that partially resembles a sinusoidal waveform, and are asked to find the Fourier series representation. If not for the removal of all negative voltages the problem would be trivial, as only *one* sinusoid would be required.

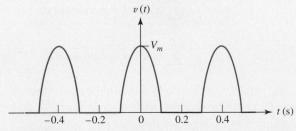

■ **FIGURE 18.2** The output of a half-wave rectifier to which a sinusoidal input is applied.

▶ *Collect the known information.*

In order to represent this voltage as a Fourier series, we must first determine the period and then express the graphical voltage as an analytical function of time. From the graph, the period is seen to be

$$T = 0.4 \text{ s}$$

and thus

$$f_0 = 2.5 \text{ Hz}$$

and

$$\omega_0 = 5\pi \text{ rad/s}$$

▶ *Devise a plan.*

The most straightforward approach is to apply Eqs. 9–11 to calculate the set of coefficients a_0, a_n, and b_n. To do this, we need a functional expression for $v(t)$, the most straightforward being defined over the interval $t = 0$ to $t = 0.4$ as:

$$v(t) = \begin{cases} V_m \cos 5\pi t & 0 \le t \le 0.1 \\ 0 & 0.1 \le t \le 0.3 \\ V_m \cos 5\pi t & 0.3 \le t \le 0.4 \end{cases}$$

However, choosing the period to extend from $t = -0.1$ to $t = 0.3$ will result in fewer equations and, hence, fewer integrals:

$$v(t) = \begin{cases} V_m \cos 5\pi t & -0.1 \le t \le 0.1 \\ 0 & 0.1 \le t \le 0.3 \end{cases} \qquad [12]$$

This form is preferable, although either description will yield the correct results.

▶ *Construct an appropriate set of equations.*

The zero-frequency component is easily obtained:

$$a_0 = \frac{1}{0.4} \int_{-0.1}^{0.3} v(t)\, dt = \frac{1}{0.4} \left[\int_{-0.1}^{0.1} V_m \cos 5\pi t\, dt + \int_{0.1}^{0.3} (0)\, dt \right]$$

The amplitude of a *general* cosine term is

$$a_n = \frac{2}{0.4} \int_{-0.1}^{0.1} V_m \cos 5\pi t \cos 5\pi n t\, dt$$

and the amplitude of a general sine term is

$$\frac{2}{0.4} \int_{-0.1}^{0.1} V_m \cos 5\pi t \sin 5\pi n t\, dt$$

which, in fact, is always zero, and hence will not be considered further.

▶ *Determine if additional information is required.*

The form of the function we obtain upon integrating is different when n is unity than it is for any other choice of n. If $n = 1$, we have

$$a_1 = 5V_m \int_{-0.1}^{0.1} \cos^2 5\pi t\, dt = \frac{V_m}{2} \qquad [13]$$

Notice that integration over an entire period must be broken up into subintervals of the period, in each of which the functional form of $v(t)$ is known.

(Continued on next page)

whereas if n is not equal to unity, we find

$$a_n = 5V_m \int_{-0.1}^{0.1} \cos 5\pi t \cos 5\pi nt \, dt$$

▶ **Attempt a solution.**
Solving, we find that

$$a_0 = \frac{V_m}{\pi} \qquad [14]$$

$$a_n = 5V_m \int_{-0.1}^{0.1} \frac{1}{2}[\cos 5\pi(1+n)t + \cos 5\pi(1-n)t] \, dt$$

or

$$a_n = \frac{2V_m}{\pi} \frac{\cos(\pi n/2)}{1 - n^2} \qquad (n \neq 1) \qquad [15]$$

> It should be pointed out, incidentally, that the expression for a_n when $n \neq 1$ will yield the correct result for $n = 1$ in the limit as $n \to 1$.

(A similar integration shows that $b_n = 0$ for any value of n, and the Fourier series thus contains no sine terms.) The Fourier series is therefore obtained from Eqs. [1], [13], [14], and [15]:

$$v(t) = \frac{V_m}{\pi} + \frac{V_m}{2}\cos 5\pi t + \frac{2V_m}{3\pi}\cos 10\pi t - \frac{2V_m}{15\pi}\cos 20\pi t$$

$$+ \frac{2V_m}{35\pi}\cos 30\pi t - \cdots \qquad [16]$$

▶ **Verify the solution. Is it reasonable or expected?**
Our solution can be checked by plugging values into Eq. [16] and truncating after a specific number of terms. Another approach, however, is to plot the function as shown in Fig. 18.3 for $n = 1$, 2, and 6.

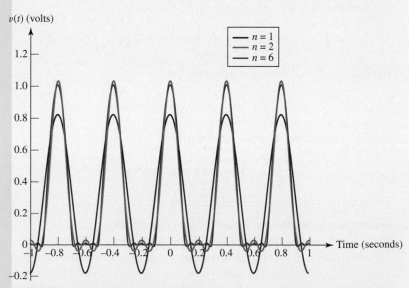

■ **FIGURE 18.3** Equation [16] truncated after $n = 1$ term; $n = 2$ term and $n = 6$ term, showing convergence to the half-sinusoid $v(t)$. A magnitude of $V_m = 1$ has been chosen for convenience.

As can be seen, as more terms are included, the more the plot resembles that of Fig. 18.2.

PRACTICE

18.2 A periodic waveform $f(t)$ is described as follows: $f(t) = -4$, $0 < t < 0.3$; $f(t) = 6, 0.3 < t < 0.4$; $f(t) = 0, 0.4 < t < 0.5$; $T = 0.5$. Evaluate: (a) a_0; (b) a_3; (c) b_1.

18.3 Write the Fourier series for the three voltage waveforms shown in Fig. 18.4.

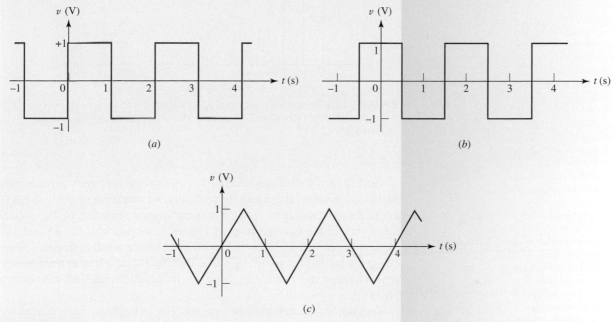

(a)

(b)

(c)

■ FIGURE 18.4

Ans: 18.2: $-1.200; 1.383; -4.44.$ 18.3: $(4/\pi)(\sin \pi t + \frac{1}{3} \sin 3\pi t + \frac{1}{5} \sin 5\pi t + \cdots)$ V; $(4/\pi)(\cos \pi t - \frac{1}{3} \cos 3\pi t + \frac{1}{5} \cos 5\pi t - \cdots)$V;$(8/\pi^2)(\sin \pi t - \frac{1}{9} \sin 3\pi t + \frac{1}{25} \sin 5\pi t - \cdots).$

Line and Phase Spectra

We depicted the function $v(t)$ of Example 18.1 graphically in Fig. 18.2, and analytically in Eq. [12]—both representations being in the time domain. The Fourier series representation of $v(t)$ given in Eq. [16] is also a time-domain expression, but may be transformed into a *frequency-domain* representation as well. For example, Fig. 18.5 shows the amplitude of each frequency component of $v(t)$, a type of plot known as a ***line spectrum.*** Here, the magnitude of each frequency component (i.e., $|a_0|$, $|a_1|$, etc.) is indicated by the length of the vertical line at the corresponding frequency (f_0, f_1, etc.); for the sake of convenience, we have taken $V_m = 1$. Given a different value of V_m, we simply scale the y axis values by the new value.

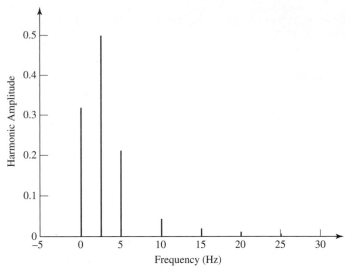

■ FIGURE 18.5 The discrete line spectrum of $v(t)$ as represented in Eq. [16], showing the first seven frequency components. A magnitude of $V_m = 1$ has been chosen for convenience.

Such a plot, sometimes referred to as a *discrete spectrum,* gives a great deal of information at a glance. In particular, we can see how many terms of the series are required to obtain a reasonable approximation of the original waveform. In the line spectrum of Fig. 18.5, we note that the 8th and 10th harmonics (20 and 25 Hz, respectively) add only a small correction. Truncating the series after the 6th harmonic therefore should lead to a reasonable approximation; the reader can judge this for herself/himself by considering Fig. 18.3.

One note of caution must be injected. The example we have considered contains no sine terms, and the amplitude of the nth harmonic is therefore $|a_n|$. If b_n is not zero, then the amplitude of the component at a frequency $n\omega_0$ must be $\sqrt{a_n^2 + b_n^2}$. This is the general quantity which we must show in a line spectrum. When we discuss the complex form of the Fourier series, we will see that this amplitude is obtained more directly.

In addition to the amplitude spectrum, we also may construct a discrete **phase spectrum.** At any frequency $n\omega_0$, we combine the cosine and sine terms to determine the phase angle ϕ_n:

$$a_n \cos n\omega_0 t + b_n \sin n\omega_0 t = \sqrt{a_n^2 + b_n^2} \cos\left(n\omega_0 t + \tan^{-1}\frac{-b_n}{a_n}\right)$$

$$= \sqrt{a_n^2 + b_n^2} \cos(n\omega_0 t + \phi_n)$$

or

$$\phi_n = \tan^{-1}\frac{-b_n}{a_n}$$

In Eq. [16], $\phi_n = 0°$ or $180°$ for every n.

The Fourier series obtained for this example includes no sine terms and no odd harmonics (except the fundamental) among the cosine terms. It is possible to anticipate the absence of certain terms in a Fourier series, before any integrations are performed, by an inspection of the symmetry of the given time function. We will investigate the use of symmetry in the following section.

18.2 • THE USE OF SYMMETRY

Even and Odd Symmetry

The two types of symmetry which are most readily recognized are *even-function symmetry* and *odd-function symmetry*, or simply *even symmetry* and *odd symmetry*. We say that $f(t)$ possesses the property of even symmetry if

$$f(t) = f(-t) \qquad [17]$$

Such functions as t^2, $\cos 3t$, $\ln(\cos t)$, $\sin^2 7t$, and a constant C all possess even symmetry; the replacement of t by $(-t)$ does not change the value of any of these functions. This type of symmetry may also be recognized graphically, for if $f(t) = f(-t)$ then mirror symmetry exists about the $f(t)$ axis. The function shown in Fig. 18.6a possesses even symmetry; if the figure were to be folded along the $f(t)$ axis, then the portions of the graph for positive and negative time would fit exactly, one on top of the other.

We define odd symmetry by stating that if odd symmetry is a property of $f(t)$, then

$$f(t) = -f(-t) \qquad [18]$$

In other words, if t is replaced by $(-t)$, then the negative of the given function is obtained; for example, t, $\sin t$, $t \cos 70t$, $t\sqrt{1+t^2}$, and the function sketched in Fig. 18.6b are all odd functions and possess odd symmetry. The graphical characteristics of odd symmetry are apparent if the portion of $f(t)$ for $t > 0$ is rotated about the positive t axis and the resultant figure is then rotated about the $f(t)$ axis; the two curves will fit exactly, one on top of the other. That is, we now have symmetry about the origin, rather than about the $f(t)$ axis as we did for even functions.

Having definitions for even and odd symmetry, we should note that the product of two functions with even symmetry, or of two functions with odd symmetry, yields a function with even symmetry. Furthermore, the product of an even and an odd function gives a function with odd symmetry.

Symmetry and Fourier Series Terms

Now let us investigate the effect that even symmetry produces in a Fourier series. If we think of the expression which equates an even function $f(t)$ and the sum of an infinite number of sine and cosine functions, then it is apparent that the sum must also be an even function. A sine wave, however, is an odd function, and *no sum of sine waves can produce any even function other than zero* (which is both even and odd). It is thus plausible that the Fourier series of any even function is composed of only a constant and cosine functions. Let us now show carefully that $b_n = 0$. We have

$$b_n = \frac{2}{T} \int_{-T/2}^{T/2} f(t) \sin n\omega_0 t \, dt$$

$$= \frac{2}{T} \left[\int_{-T/2}^{0} f(t) \sin n\omega_0 t \, dt + \int_{0}^{T/2} f(t) \sin n\omega_0 t \, dt \right]$$

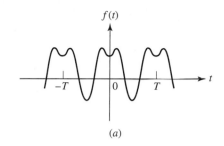

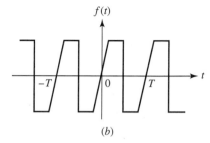

■ **FIGURE 18.6** (a) A waveform showing even symmetry. (b) A waveform showing odd symmetry.

Now let us replace the variable t in the first integral by $-\tau$, or $\tau = -t$, and make use of the fact that $f(t) = f(-t) = f(\tau)$:

$$b_n = \frac{2}{T}\left[\int_{T/2}^{0} f(-\tau)\sin(-n\omega_0\tau)(-d\tau) + \int_{0}^{T/2} f(t)\sin n\omega_0 t\,dt\right]$$

$$= \frac{2}{T}\left[-\int_{0}^{T/2} f(\tau)\sin n\omega_0\tau\,d\tau + \int_{0}^{T/2} f(t)\sin n\omega_0 t\,dt\right]$$

But the symbol we use to identify the variable of integration cannot affect the value of the integral. Thus,

$$\int_{0}^{T/2} f(\tau)\sin n\omega_0\tau\,d\tau = \int_{0}^{T/2} f(t)\sin n\omega_0 t\,dt$$

and

$$b_n = 0 \qquad \text{(even sym.)} \qquad [19]$$

No sine terms are present. Therefore, if $f(t)$ shows even symmetry, then $b_n = 0$; conversely, if $b_n = 0$, then $f(t)$ must have even symmetry.

A similar examination of the expression for a_n leads to an integral over the *half period* extending from $t = 0$ to $t = \frac{1}{2}T$:

$$a_n = \frac{4}{T}\int_{0}^{T/2} f(t)\cos n\omega_0 t\,dt \qquad \text{(even sym.)} \qquad [20]$$

The fact that a_n may be obtained for an even function by taking "twice the integral over half the range" should seem logical.

A function having odd symmetry can contain no constant term or cosine terms in its Fourier expansion. Let us prove the second part of this statement. We have

$$a_n = \frac{2}{T}\int_{-T/2}^{T/2} f(t)\cos n\omega_0 t\,dt$$

$$= \frac{2}{T}\left[\int_{-T/2}^{0} f(t)\cos n\omega_0 t\,dt + \int_{0}^{T/2} f(t)\cos n\omega_0 t\,dt\right]$$

and we now let $t = -\tau$ in the first integral:

$$a_n = \frac{2}{T}\left[\int_{T/2}^{0} f(-\tau)\cos(-n\omega_0\tau)(-d\tau) + \int_{0}^{T/2} f(t)\cos n\omega_0 t\,dt\right]$$

$$= \frac{2}{T}\left[\int_{0}^{T/2} f(-\tau)\cos n\omega_0\tau\,d\tau + \int_{0}^{T/2} f(t)\cos n\omega_0 t\,dt\right]$$

But $f(-\tau) = -f(\tau)$, and therefore

$$a_n = 0 \qquad \text{(odd sym.)} \qquad [21]$$

A similar, but simpler, proof shows that

$$a_0 = 0 \qquad \text{(odd sym.)}$$

With odd symmetry, therefore, $a_n = 0$ and $a_0 = 0$; conversely, if $a_n = 0$ and $a_0 = 0$, odd symmetry is present.

The values of b_n may again be obtained by integrating over half the range:

$$b_n = \frac{4}{T} \int_0^{T/2} f(t) \sin n\omega_0 t \, dt \qquad \text{(odd sym.)} \qquad [22]$$

Examples of even and odd symmetry were afforded by Practice Problem 18.3, preceding this section. In both parts a and b, a square wave of the same amplitude and period is the given function. The time origin, however, is selected to provide odd symmetry in part a and even symmetry in part b, and the resultant series contain, respectively, only sine terms and only cosine terms. It is also noteworthy that the point at which $t = 0$ could be selected to provide neither even nor odd symmetry; the determination of the coefficients of the terms in the Fourier series would then take at least twice as long.

Half-Wave Symmetry

The Fourier series for both of these square waves have one other interesting characteristic: neither contains any even *harmonics*.[1] That is, the only frequency components present in the series have frequencies which are odd multiples of the fundamental frequency; a_n and b_n are zero for even values of n. This result is caused by another type of symmetry, called half-wave symmetry. We will say that $f(t)$ possesses *half-wave symmetry* if

$$f(t) = -f\left(t - \tfrac{1}{2}T\right)$$

or the equivalent expression,

$$f(t) = -f\left(t + \tfrac{1}{2}T\right)$$

Except for a change of sign, each half cycle is like the adjacent half cycles. Half-wave symmetry, unlike even and odd symmetry, is not a function of the choice of the point $t = 0$. Thus, we can state that the square wave (Fig. 18.4a or b) shows half-wave symmetry. Neither waveform shown in Fig. 18.6 has half-wave symmetry, but the two somewhat similar functions plotted in Fig. 18.7 do possess half-wave symmetry.

It may be shown that the Fourier series of any function which has half-wave symmetry contains only odd harmonics. Let us consider the coefficients a_n. We have again

$$a_n = \frac{2}{T} \int_{-T/2}^{T/2} f(t) \cos n\omega_0 t \, dt$$

$$= \frac{2}{T} \left[\int_{-T/2}^{0} f(t) \cos n\omega_0 t \, dt + \int_{0}^{T/2} f(t) \cos n\omega_0 t \, dt \right]$$

which we may represent as

$$a_n = \frac{2}{T}(I_1 + I_2)$$

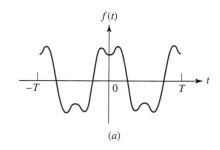

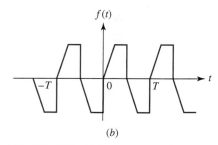

■ **FIGURE 18.7** (*a*) A waveform somewhat similar to the one shown in Fig. 18.6*a* but possessing half-wave symmetry. (*b*) A waveform somewhat similar to the one shown in Fig. 18.6*b* but possessing half-wave symmetry.

(1) Constant vigilance is required to avoid confusion between an even function and an even harmonic, or between an odd function and an odd harmonic. For example, b_{10} is the coefficient of an even harmonic, and it is zero if $f(t)$ is an even function.

Now we substitute the new variable $\tau = t + \frac{1}{2}T$ in the integral I_1:

$$I_1 = \int_0^{T/2} f\left(\tau - \frac{1}{2}T\right) \cos n\omega_0 \left(\tau - \frac{1}{2}T\right) d\tau$$

$$= \int_0^{T/2} -f(\tau)\left(\cos n\omega_0\tau \cos \frac{n\omega_0 T}{2} + \sin n\omega_0\tau \sin \frac{n\omega_0 T}{2}\right) d\tau$$

But $\omega_0 T$ is 2π, and thus

$$\sin \frac{n\omega_0 T}{2} = \sin n\pi = 0$$

Hence

$$I_1 = -\cos n\pi \int_0^{T/2} f(\tau) \cos n\omega_0\tau \, d\tau$$

After noting the form of I_2, we therefore may write

$$a_n = \frac{2}{T}(1 - \cos n\pi) \int_0^{T/2} f(t) \cos n\omega_0 t \, dt$$

The factor $(1 - \cos n\pi)$ indicates that a_n is zero if n is even. Thus,

$$a_n = \begin{cases} \dfrac{4}{T} \displaystyle\int_0^{T/2} f(t) \cos n\omega_0 t \, dt & n \text{ odd} \\ 0 & n \text{ even} \end{cases} \qquad (\tfrac{1}{2}\text{-wave sym.}) \quad [23]$$

A similar investigation shows that b_n is also zero for all even n, and therefore

$$b_n = \begin{cases} \dfrac{4}{T} \displaystyle\int_0^{T/2} f(t) \sin n\omega_0 t \, dt & n \text{ odd} \\ 0 & n \text{ even} \end{cases} \qquad (\tfrac{1}{2}\text{-wave sym.}) \quad [24]$$

It should be noted that half-wave symmetry may be present in a waveform which also shows odd symmetry or even symmetry. The waveform sketched in Fig. 18.7a, for example, possesses both even symmetry and half-wave symmetry. When a waveform possesses half-wave symmetry and either even or odd symmetry, then it is possible to reconstruct the waveform if the function is known over any quarter-period interval. The value of a_n or b_n may also be found by integrating over any quarter period. Thus,

$$\left.\begin{aligned} a_n &= \frac{8}{T} \int_0^{T/4} f(t) \cos n\omega_0 t \, dt && n \text{ odd} \\ a_n &= 0 && n \text{ even} \\ b_n &= 0 && \text{all } n \end{aligned}\right\} \quad \begin{aligned} &(\tfrac{1}{2}\text{-wave and even sym.}) \\ &\qquad\qquad\qquad\qquad [25] \end{aligned}$$

$$\left.\begin{aligned} a_n &= 0 && \text{all } n \\ b_n &= \frac{8}{T} \int_0^{T/4} f(t) \sin \omega_0 t \, dt && n \text{ odd} \\ b_n &= 0 && n \text{ even} \end{aligned}\right\} \quad \begin{aligned} &(\tfrac{1}{2}\text{-wave and odd sym.}) \\ &\qquad\qquad\qquad\qquad [26] \end{aligned}$$

Table 18.1 provides a short summary of the simplifications arising from the various types of symmetry discussed.

It is *always* worthwhile spending a few moments investigating the symmetry of a function for which a Fourier series is to be determined.

TABLE 18.1 Summary of Symmetry-Based Simplifications in Fourier Series

Symmetry Type	Characteristic	Simplification
Even	$f(t) = -f(t)$	$b_n = 0$
Odd	$f(t) = -f(-t)$	$a_n = 0$
Half-Wave	$f(t) = -f\left(t - \dfrac{T}{2}\right)$ or $f(t) = -f\left(t + \dfrac{T}{2}\right)$	$a_n = \begin{cases} \dfrac{4}{T}\displaystyle\int_0^{T/2} f(t)\cos n\omega_0 t\, dt & n \text{ odd} \\ 0 & n \text{ even} \end{cases}$ $b_n = \begin{cases} \dfrac{4}{T}\displaystyle\int_0^{T/2} f(t)\sin n\omega_0 t\, dt & n \text{ odd} \\ 0 & n \text{ even} \end{cases}$
Half-Wave and Even	$f(t) = -f\left(t - \dfrac{T}{2}\right)$ and $f(t) = -f(t)$ or $f(t) = -f\left(t + \dfrac{T}{2}\right)$ and $f(t) = -f(t)$	$a_n = \begin{cases} \dfrac{8}{T}\displaystyle\int_0^{T/4} f(t)\cos n\omega_0 t\, dt & n \text{ odd} \\ 0 & n \text{ even} \end{cases}$ $b_n = 0 \qquad \text{all } n$
Half-Wave and Odd	$f(t) = -f\left(t - \dfrac{T}{2}\right)$ and $f(t) = -f(-t)$ or $f(t) = -f\left(t + \dfrac{T}{2}\right)$ and $f(t) = -f(-t)$	$a_n = 0 \qquad \text{all } n$ $b_n = \begin{cases} \dfrac{8}{T}\displaystyle\int_0^{T/4} f(t)\sin n\omega_0 t\, dt & n \text{ odd} \\ 0 & n \text{ even} \end{cases}$

PRACTICE

18.4 Sketch each of the functions described, state whether or not even symmetry, odd symmetry, and half-wave symmetry are present, and give the period: (a) $v = 0, -2 < t < 0$ and $2 < t < 4; v = 5,$ $0 < t < 2; v = -5, 4 < t < 6$; repeats; (b) $v = 10, 1 < t < 3; v = 0,$ $3 < t < 7; v = -10, 7 < t < 9$; repeats; (c) $v = 8t, -1 < t < 1;$ $v = 0, 1 < t < 3$; repeats.

18.5 Determine the Fourier series for the waveforms of Practice Problem 18.4a and b.

Ans: 18.4: No, no, yes, 8; no, no, no, 8; no, yes, no, 4.

18.5: $\displaystyle\sum_{n=1(\text{odd})}^{\infty} \frac{10}{n\pi}\left(\sin\frac{n\pi}{2}\cos\frac{n\pi t}{4} + \sin\frac{n\pi t}{4}\right)$;

$\displaystyle\sum_{n=1}^{\infty} \frac{10}{n\pi}\left[\left(\sin\frac{3n\pi}{4} - 3\sin\frac{n\pi}{4}\right)\cos\frac{n\pi t}{4} + \left(\cos\frac{n\pi}{4} - \cos\frac{3n\pi}{4}\right)\sin\frac{n\pi t}{4}\right].$

18.3 • COMPLETE RESPONSE TO PERIODIC FORCING FUNCTIONS

Through the use of the Fourier series, we may now express an arbitrary periodic forcing function as the sum of an infinite number of sinusoidal forcing functions. The forced response to each of these functions may be determined by conventional steady-state analysis, and the form of the natural response may be determined from the poles of an appropriate network transfer function. The initial conditions existing throughout the network, including the initial value of the forced response, enable the amplitude of the natural response to be selected; the complete response is then obtained as the sum of the forced and natural responses.

EXAMPLE 18.2

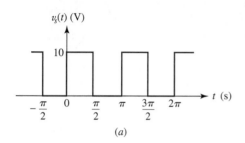

(a)

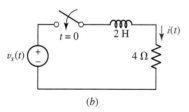

(b)

■ **FIGURE 18.8** (a) A square-wave voltage forcing function. (b) The forcing function of part a is applied to this series *RL* circuit at $t = 0$; the complete response $i(t)$ is desired.

Recall that $V_m \sin \omega t$ is equal to $V_m \cos(\omega t - 90°)$, corresponding to $V_m \underline{/-90°} = -jV_m$.

Find the periodic response obtained when the square wave of Fig. 18.8a, including its dc component, is applied to the series _RL_ circuit shown in Fig. 18.8b. The forcing function is applied at $t = 0$, and the current is the desired response. Its initial value is zero.

The forcing function has a fundamental frequency $\omega_0 = 2$ rad/s, and its Fourier series may be written down by comparison with the Fourier series developed for the waveform of Fig. 18.4a in the solution of Practice Problem 18.3,

$$v_s(t) = 5 + \frac{20}{\pi} \sum_{n=1(\text{odd})}^{\infty} \frac{\sin 2nt}{n}$$

We will find the forced response for the *n*th harmonic by working in the frequency domain. Thus,

$$v_{sn}(t) = \frac{20}{n\pi} \sin 2nt$$

and

$$\mathbf{V}_{sn} = \frac{20}{n\pi} \underline{/-90°} = -j\frac{20}{n\pi}$$

The impedance offered by the *RL* circuit at this frequency is

$$\mathbf{Z}_n = 4 + j(2n)2 = 4 + j4n$$

and thus the component of the forced response at this frequency is

$$\mathbf{I}_{fn} = \frac{\mathbf{V}_{sn}}{\mathbf{Z}_n} = \frac{-j5}{n\pi(1 + jn)}$$

Transforming to the time domain, we have

$$i_{fn} = \frac{5}{n\pi} \frac{1}{\sqrt{1 + n^2}} \cos(2nt - 90° - \tan^{-1} n)$$

$$= \frac{5}{\pi(1 + n^2)} \left(\frac{\sin 2nt}{n} - \cos 2nt \right)$$

Since the response to the dc component is simply $5\text{ V}/4\text{ }\Omega = 1.25\text{ A}$, the forced response may be expressed as the summation

$$i_f(t) = 1.25 + \frac{5}{\pi} \sum_{n=1(\text{odd})}^{\infty} \left[\frac{\sin 2nt}{n(1+n^2)} - \frac{\cos 2nt}{1+n^2} \right]$$

The familiar natural response of this simple circuit is the single exponential term [characterizing the single pole of the transfer function, $\mathbf{I}_f/\mathbf{V}_s = 1/(4+2\mathbf{s})$]

$$i_n(t) = Ae^{-2t}$$

The *complete* response is therefore the sum

$$i(t) = i_f(t) + i_n(t)$$

Letting $t = 0$, we find A using $i(0) = 0$:

$$A = -1.25 + \frac{5}{\pi} \sum_{n=1(\text{odd})}^{\infty} \frac{1}{1+n^2}$$

Although correct, it is more convenient to use the numerical value of the summation. The sum of the first five terms of $\Sigma\, 1/(1+n^2)$ is 0.671, the sum of the first ten terms is 0.695, the sum of the first twenty terms is 0.708, and the exact sum is 0.720 to three significant figures. Thus

$$A = -1.25 + \frac{5}{\pi}(0.720) = -0.104$$

and

$$i(t) = -0.104e^{-2t} + 1.25$$

$$+ \frac{5}{\pi} \sum_{n=1(\text{odd})}^{\infty} \left[\frac{\sin 2nt}{n(1+n^2)} - \frac{\cos 2nt}{1+n^2} \right] \quad \text{amperes}$$

In obtaining this solution, we have had to use many of the most general concepts introduced in this and the preceding 17 chapters. Some we did not have to use because of the simple nature of this particular circuit, but their places in the general analysis were indicated. In this sense, we may look upon the solution of this problem as a significant achievement in our introductory study of circuit analysis. In spite of this glorious feeling of accomplishment, however, it must be pointed out that the complete response, as obtained in Example 18.2 in analytical form, is not of much value as it stands; it furnishes no clear picture of the nature of the response. What we really need is a sketch of $i(t)$ as a function of time. This may be obtained by a laborious calculation at a sufficient number of instants of time; a desktop computer or a programmable calculator can be of great assistance here. The sketch may be approximated by the graphical addition of the natural response, the dc term, and the first few harmonics; this is an unrewarding task.

When all is said and done, the most informative solution of this problem is probably obtained by making a repeated transient analysis. That is, the form of the response can certainly be calculated in the interval from $t = 0$ to $t = \pi/2$ s; it is an exponential rising toward 2.5 A. After determining the value at the end of this first interval, we have an initial condition for the next $(\pi/2)$-second interval. The process is repeated until the response assumes a generally periodic nature. The method is eminently suitable to this example, for there is negligible change in the current waveform in the successive periods $\pi/2 < t < 3\pi/2$ and $3\pi/2 < t < 5\pi/2$. The complete current response is sketched in Fig. 18.9.

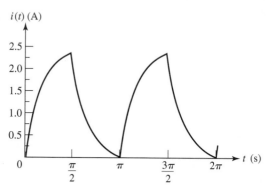

■ **FIGURE 18.9** The initial portion of the complete response of the circuit of Fig. 18.8*b* to the forcing function of Fig. 18.8*a*.

PRACTICE

18.6 Use the methods of Chap. 8 to determine the value of the current sketched in Fig. 18.9 at t equal to (*a*) $\pi/2$; (*b*) π; (*c*) $3\pi/2$.

Ans: 2.392 A; 0.1034 A; 2.396 A.

18.4 COMPLEX FORM OF THE FOURIER SERIES

In obtaining a frequency spectrum, we have seen that the amplitude of each frequency component depends on both a_n and b_n; that is, the sine term and the cosine term both contribute to the amplitude. The exact expression for this amplitude is $\sqrt{a_n^2 + b_n^2}$. It is also possible to obtain the amplitude directly by using a form of Fourier series in which each term is a cosine function with a phase angle; the amplitude and phase angle are functions of $f(t)$ and n.

An even more convenient and concise form of the Fourier series is obtained if the sines and cosines are expressed as exponential functions with complex multiplying constants.

Let us first take the trigonometric form of the Fourier series:

$$f(t) = a_0 + \sum_{n=1}^{\infty}(a_n \cos n\omega_0 t + b_n \sin n\omega_0 t)$$

and then substitute the exponential forms for the sine and cosine. After rearranging,

$$f(t) = a_0 + \sum_{n=1}^{\infty}\left(e^{jn\omega_0 t}\frac{a_n - jb_n}{2} + e^{-jn\omega_0 t}\frac{a_n + jb_n}{2}\right)$$

The reader may recall the identities

$$\sin \alpha = \frac{e^{j\alpha} - e^{-j\alpha}}{j2}$$

and

$$\cos \alpha = \frac{e^{j\alpha} + e^{-j\alpha}}{2}$$

We now define a complex constant $\mathbf{c}_n$:

$$\mathbf{c}_n = \tfrac{1}{2}(a_n - jb_n) \qquad (n = 1, 2, 3, \ldots) \qquad [27]$$

The values of a_n, b_n, and $\mathbf{c}_n$ all depend on n and $f(t)$. Suppose we now replace n with $(-n)$; how do the values of the constants change? The coefficients a_n and b_n are defined by Eqs. [10] and [11], and it is evident that

$$a_{-n} = a_n$$

but

$$b_{-n} = -b_n$$

From Eq. [27], then,

$$\mathbf{c}_{-n} = \tfrac{1}{2}(a_n + jb_n) \qquad (n = 1, 2, 3, \ldots) \qquad [28]$$

Thus,

$$\mathbf{c}_n = \mathbf{c}_{-n}^*$$

We also let

$$\mathbf{c}_0 = a_0$$

We may therefore express $f(t)$ as

$$f(t) = \mathbf{c}_0 + \sum_{n=1}^{\infty} \mathbf{c}_n e^{jn\omega_0 t} + \sum_{n=1}^{\infty} \mathbf{c}_{-n} e^{-jn\omega_0 t}$$

or

$$f(t) = \sum_{n=0}^{\infty} \mathbf{c}_n e^{jn\omega_0 t} + \sum_{n=1}^{\infty} \mathbf{c}_{-n} e^{-jn\omega_0 t}$$

Finally, instead of summing the second series over the positive integers from 1 to ∞, let us sum over the negative integers from -1 to $-\infty$:

$$f(t) = \sum_{n=0}^{\infty} \mathbf{c}_n e^{jn\omega_0 t} + \sum_{n=-1}^{-\infty} \mathbf{c}_n e^{jn\omega_0 t}$$

or

$$\boxed{f(t) = \sum_{n=-\infty}^{\infty} \mathbf{c}_n e^{jn\omega_0 t}} \qquad [29]$$

By agreement, a summation from $-\infty$ to ∞ is understood to include a term for $n = 0$.

Equation [29] is the *complex form* of the Fourier series for $f(t)$; its conciseness is one of the most important reasons for its use. In order to obtain the expression by which a particular complex coefficient $\mathbf{c}_n$ may be evaluated, we substitute Eqs. [10] and [11] in Eq. [27]:

$$\mathbf{c}_n = \frac{1}{T} \int_{-T/2}^{T/2} f(t) \cos n\omega_0 t \, dt - j \frac{1}{T} \int_{-T/2}^{T/2} f(t) \sin n\omega_0 t \, dt$$

and then we use the exponential equivalents of the sine and cosine and simplify:

$$\mathbf{c}_n = \frac{1}{T} \int_{-T/2}^{T/2} f(t) e^{-jn\omega_0 t} \, dt \qquad\qquad [30]$$

Thus, a single concise equation serves to replace the two equations required for the trigonometric form of the Fourier series. Instead of evaluating two integrals to find the Fourier coefficients, only one integration is required; moreover, it is almost always a simpler integration. It should be noted that the integral of Eq. [30] contains the multiplying factor $1/T$, whereas the integrals for a_n and b_n both contain the factor $2/T$.

Collecting the two basic relationships for the exponential form of the Fourier series, we have

$$f(t) = \sum_{n=-\infty}^{\infty} \mathbf{c}_n e^{jn\omega_0 t} \qquad\qquad [29]$$

$$\mathbf{c}_n = \frac{1}{T} \int_{-T/2}^{T/2} f(t) e^{-jn\omega_0 t} \, dt \qquad\qquad [30]$$

where $\omega_0 = 2\pi/T$ as usual.

The amplitude of the component of the exponential Fourier series at $\omega = n\omega_0$, where $n = 0, \pm 1, \pm 2, \ldots$, is $|\mathbf{c}_n|$. We may plot a discrete frequency spectrum giving $|\mathbf{c}_n|$ versus $n\omega_0$ or nf_0, using an abscissa that shows both positive and negative values; and when we do this, the graph is symmetrical about the origin, since Eqs. [27] and [28] show that $|\mathbf{c}_n| = |\mathbf{c}_{-n}|$.

We note also from Eqs. [29] and [30] that the amplitude of the sinusoidal component at $\omega = n\omega_0$, where $n = 1, 2, 3, \ldots$, is $\sqrt{a_n^2 + b_n^2} = 2|\mathbf{c}_n| = 2|\mathbf{c}_{-n}| = |\mathbf{c}_n| + |\mathbf{c}_{-n}|$. For the dc component, $a_0 = \mathbf{c}_0$.

The exponential Fourier coefficients, given by Eq. [30], are also affected by the presence of certain symmetries in $f(t)$. Thus, appropriate expressions for $\mathbf{c}_n$ are

$$\mathbf{c}_n = \frac{2}{T} \int_0^{T/2} f(t) \cos n\omega_0 t \, dt \qquad \text{(even sym.)} \qquad\qquad [31]$$

$$\mathbf{c}_n = \frac{-j2}{T} \int_0^{T/2} f(t) \sin n\omega_0 t \, dt \qquad \text{(odd sym.)} \qquad\qquad [32]$$

$$\mathbf{c}_n = \begin{cases} \dfrac{2}{T} \displaystyle\int_0^{T/2} f(t) e^{-jn\omega_0 t} \, dt & \left(n \text{ odd}, \tfrac{1}{2}\text{-wave sym.}\right) \qquad [33a] \\[4mm] 0 & \left(n \text{ even}, \tfrac{1}{2}\text{-wave sym.}\right) \qquad [33b] \end{cases}$$

$$\mathbf{c}_n = \begin{cases} \dfrac{4}{T} \displaystyle\int_0^{T/4} f(t) \cos n\omega_0 t \, dt & \left(n \text{ odd}, \tfrac{1}{2}\text{-wave and even sym.}\right) \quad [34a] \\[4mm] 0 & \left(n \text{ even}, \tfrac{1}{2}\text{-wave and even sym.}\right) \quad [34b] \end{cases}$$

$$\mathbf{c}_n = \begin{cases} \dfrac{-j4}{T} \displaystyle\int_0^{T/4} f(t) \sin n\omega_0 t \, dt & \left(n \text{ odd}, \tfrac{1}{2}\text{-wave and odd sym.}\right) \quad [35a] \\[4mm] 0 & \left(n \text{ even}, \tfrac{1}{2}\text{-wave and odd sym.}\right) \quad [35b] \end{cases}$$

EXAMPLE 18.3

Determine c_n for the square wave of Fig. 18.10.

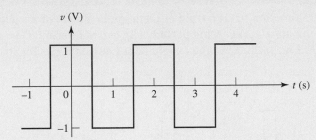

■ **FIGURE 18.10** A square wave function possessing both even and half-wave symmetry.

This square wave possesses both even and half-wave symmetry. If we ignore the symmetry and use our general equation [30], with $T = 2$ and $\omega_0 = 2\pi/2 = \pi$, we have

$$c_n = \frac{1}{T} \int_{-T/2}^{T/2} f(t)e^{-jn\omega_0 t}\, dt$$

$$= \frac{1}{2}\left[\int_{-1}^{-0.5} -e^{-jn\pi t}\, dt + \int_{-0.5}^{0.5} e^{-jn\pi t}\, dt - \int_{0.5}^{1} e^{-jn\pi t}\, dt \right]$$

$$= \frac{1}{2}\left[\frac{-1}{-jn\pi}(e^{-jn\pi t})_{-1}^{-0.5} + \frac{1}{-jn\pi}(e^{-jn\pi t})_{-0.5}^{0.5} + \frac{-1}{-jn\pi}(e^{-jn\pi t})_{0.5}^{1} \right]$$

$$= \frac{1}{j2n\pi}(e^{jn\pi/2} - e^{jn\pi} - e^{-jn\pi/2} + e^{jn\pi/2} + e^{-jn\pi} - e^{-jn\pi/2})$$

$$= \frac{1}{j2n\pi}(2e^{jn\pi/2} - 2e^{-jn\pi/2}) = \frac{2}{n\pi}\sin\frac{n\pi}{2}$$

We thus find that $c_0 = 0$, $c_1 = 2/\pi$, $c_2 = 0$, $c_3 = -2/3\pi$, $c_4 = 0$, $c_5 = 2/5\pi$, and so forth. These values agree with the trigonometric Fourier series given as the answer we obtained in Practice Problem 18.3 for the same waveform shown in Fig. 18.4b if we remember that $a_n = 2c_n$ when $b_n = 0$.

Utilizing the symmetry of the waveform (even and half-wave), there is less work when we apply Eqs. [34a] and [34b], leading to

$$c_n = \frac{4}{T} \int_{0}^{T/4} f(t)\cos n\omega_0 t\, dt$$

$$= \frac{4}{2} \int_{0}^{0.5} \cos n\pi t\, dt = \frac{2}{n\pi}(\sin n\pi t)\Big|_{0}^{0.5}$$

$$= \begin{cases} \dfrac{2}{n\pi}\sin\dfrac{n\pi}{2} & (n\text{ odd}) \\ 0 & (n\text{ even}) \end{cases}$$

These results are the same as those we just obtained when we did not take the symmetry of the waveform into account.

Now let us consider a more difficult, more interesting example.

EXAMPLE **18.4**

A certain function $f(t)$ is a train of rectangular pulses of amplitude V_0 and duration τ, recurring periodically every T seconds, as shown in Fig. 18.11. Find the exponential Fourier series for $f(t)$.

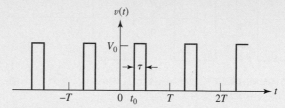

■ **FIGURE 18.11** A periodic sequence of rectangular pulses.

The fundamental frequency is $f_0 = 1/T$. No symmetry is present, and the value of a general complex coefficient is found from Eq. [30]:

$$
\mathbf{c}_n = \frac{1}{T} \int_{-T/2}^{T/2} f(t) e^{-jn\omega_0 t}\, dt = \frac{V_0}{T} \int_{t_0}^{t_0+\tau} e^{-jn\omega_0 t}\, dt
$$

$$
= \frac{V_0}{-jn\omega_0 T} (e^{-jn\omega_0 (t_0+\tau)} - e^{-jn\omega_0 t_0})
$$

$$
= \frac{2V_0}{n\omega_0 T} e^{-jn\omega_0 (t_0+\tau/2)} \sin\left(\frac{1}{2} n\omega_0 \tau\right)
$$

$$
= \frac{V_0 \tau}{T} \frac{\sin\left(\frac{1}{2} n\omega_0 \tau\right)}{\frac{1}{2} n\omega_0 \tau} e^{-jn\omega_0 (t_0+\tau/2)}
$$

The magnitude of $\mathbf{c}_n$ is therefore

$$
|\mathbf{c}_n| = \frac{V_0 \tau}{T} \left| \frac{\sin\left(\frac{1}{2} n\omega_0 \tau\right)}{\frac{1}{2} n\omega_0 \tau} \right| \tag{36}
$$

and the angle of $\mathbf{c}_n$ is

$$
\text{ang } \mathbf{c}_n = -n\omega_0 \left(t_0 + \frac{\tau}{2}\right) \qquad \text{(possibly plus } 180°\text{)} \tag{37}
$$

Equations [36] and [37] represent our solution to this exponential Fourier series problem.

The Sampling Function

The trigonometric factor in Eq. [36] occurs frequently in modern communication theory, and it is called the *sampling function.* The "sampling" refers to the time function of Fig. 18.11 from which the sampling function is derived. The product of this sequence of pulses and any other function $f(t)$

represents *samples* of $f(t)$ every T seconds if τ is small and $V_0 = 1$. We define

$$\text{Sa}(x) = \frac{\sin x}{x}$$

Because of the way in which it helps to determine the amplitude of the various frequency components in $f(t)$, it is worth our while to discover the important characteristics of this function. First, we note that $\text{Sa}(x)$ is zero whenever x is an integral multiple of π; that is,

$$\text{Sa}(n\pi) = 0 \qquad n = 1, 2, 3, \ldots$$

When x is zero, the function is indeterminate, but it is easy to show that its value is unity:

$$\text{Sa}(0) = 1$$

The magnitude of $\text{Sa}(x)$ therefore decreases from unity at $x = 0$ to zero at $x = \pi$. As x increases from π to 2π, $|\text{Sa}(x)|$ increases from zero to a maximum less than unity, and then decreases to zero once again. As x continues to increase, the successive maxima continually become smaller because the numerator of $\text{Sa}(x)$ cannot exceed unity and the denominator is continually increasing. Also, $\text{Sa}(x)$ shows even symmetry.

Now let us construct the line spectrum. We first consider $|\mathbf{c}_n|$, writing Eq. [36] in terms of the fundamental cyclic frequency f_0:

$$|\mathbf{c}_n| = \frac{V_0 \tau}{T} \left| \frac{\sin(n\pi f_0 \tau)}{n\pi f_0 \tau} \right| \qquad [38]$$

The amplitude of any $\mathbf{c}_n$ is obtained from Eq. [38] by using the known values τ and $T = 1/f_0$ and selecting the desired value of n, $n = 0, \pm 1, \pm 2, \ldots$. Instead of evaluating Eq. [38] at these discrete frequencies, let us sketch the *envelope* of $|\mathbf{c}_n|$ by considering the frequency nf_0 to be a continuous variable. That is, f, which is nf_0, can actually take on only the discrete values of the harmonic frequencies $0, \pm f_0, \pm 2f_0, \pm 3f_0$, and so forth, but we may think of n for the moment as a continuous variable. When f is zero, $|\mathbf{c}_n|$ is evidently $V_0 \tau / T$, and when f has increased to $1/\tau$, $|\mathbf{c}_n|$ is zero. The resultant envelope is first sketched as in Fig. 18.12a. The line spectrum is then obtained by simply erecting a vertical line at each harmonic frequency, as shown in the sketch. The amplitudes shown are those of the $\mathbf{c}_n$. The particular case sketched applies to the case where $\tau / T = 1/(1.5\pi) = 0.212$. In this example, it happens that there is no harmonic exactly at that frequency at which the envelope amplitude is zero; another choice of τ or T could produce such an occurrence, however.

In Fig. 18.12b, the amplitude of the sinusoidal component is plotted as a function of frequency. Note again that $a_0 = \mathbf{c}_0$ and $\sqrt{a_n^2 + b_n^2} = |\mathbf{c}_n| + |\mathbf{c}_{-n}|$.

There are several observations and conclusions which we may make about the line spectrum of a periodic sequence of rectangular pulses, as given in Fig. 18.12b. With respect to the envelope of the discrete spectrum, it is evident that the "width" of the envelope depends upon τ, and not upon T. As a matter of fact, the shape of the envelope is not a function of T. It follows that the bandwidth of a filter which is designed to pass the periodic pulses is a function of the pulse width τ, but not of the pulse period T; an

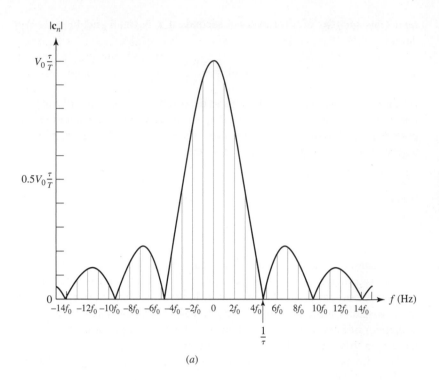

(a)

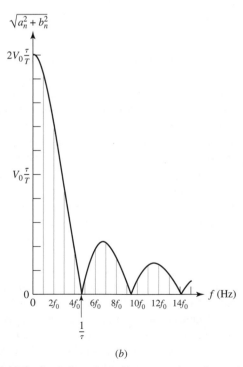

(b)

■ **FIGURE 18.12** (a) The discrete line spectrum of $|c_n|$ versus $f = nf_0, n = 0, \pm 1, \pm 2, \ldots$ corresponding to the pulse train shown in Fig. 18.11. (b) $\sqrt{a^2 + b^2}$ versus $f = nf_0, n = 0, 1, 2, \ldots$ for the same pulse train.

inspection of Fig. 18.12b indicates that the required bandwidth is about $1/\tau$ Hz. If the pulse period T is increased (or the pulse repetition frequency f_0 is decreased), the bandwidth $1/\tau$ does not change, but the number of spectral lines between zero frequency and $1/\tau$ Hz increases, albeit discontinuously; the amplitude of each line is inversely proportional to T. Finally, a shift in the time origin does not change the line spectrum; that is, $|\mathbf{c}_n|$ is not a function of t_0. The relative phases of the frequency components do change with the choice of t_0.

PRACTICE

18.7 Determine the general coefficient $\mathbf{c}_n$ in the complex Fourier series for the waveform shown in Fig.: (a) 18.4a; (b) 18.4c.

Ans: $-j2/(n\pi)$ for n odd, 0 for n even; $-j[4/(n^2\pi^2)]\sin n\pi/2$ for all n.

18.5 DEFINITION OF THE FOURIER TRANSFORM

Now that we are familiar with the basic concepts of the Fourier series representation of periodic functions, let us proceed to define the Fourier transform by first recalling the spectrum of the periodic train of rectangular pulses we obtained in Sec. 18.4. That was a *discrete* line spectrum, which is the type that we must always obtain for periodic functions of time. The spectrum was discrete in the sense that it was not a smooth or continuous function of frequency; instead, it had nonzero values only at specific frequencies.

There are many important forcing functions, however, that are not periodic functions of time, such as a single rectangular pulse, a step function, a ramp function, or the somewhat strange type of function called the *impulse function* defined in Chap. 14. Frequency spectra may be obtained for such nonperiodic functions, but they will be *continuous* spectra in which some energy, in general, may be found in any nonzero frequency interval, no matter how small.

We will develop this concept by beginning with a periodic function and then letting the period become infinite. Our experience with periodic rectangular pulses should indicate that the envelope will decrease in amplitude without otherwise changing shape, and that more and more frequency components will be found in any given frequency interval. In the limit, we should expect an envelope of vanishingly small amplitude, filled with an infinite number of frequency components separated by vanishingly small frequency intervals. The number of frequency components between 0 and 100 Hz, for example, becomes infinite, but the amplitude of each one approaches zero. At first thought, a spectrum of zero amplitude is a puzzling concept. We know that the line spectrum of a periodic forcing function shows the amplitude of each frequency component. But what does the zero-amplitude continuous spectrum of a nonperiodic forcing function signify? That question will be answered in the following section; now we proceed to carry out the limiting procedure just suggested.

We begin with the exponential form of the Fourier series:

$$f(t) = \sum_{n=-\infty}^{\infty} \mathbf{c}_n e^{jn\omega_0 t} \qquad [39]$$

where

$$\mathbf{c}_n = \frac{1}{T} \int_{-T/2}^{T/2} f(t) e^{-jn\omega_0 t}\, dt \qquad [40]$$

and

$$\omega_0 = \frac{2\pi}{T} \qquad [41]$$

We now let

$$T \to \infty$$

and thus, from Eq. [41], ω_0 must become vanishingly small. We represent this limit by a differential:

$$\omega_0 \to d\omega$$

Thus

$$\frac{1}{T} = \frac{\omega_0}{2\pi} \to \frac{d\omega}{2\pi} \qquad [42]$$

Finally, the frequency of any "harmonic" $n\omega_0$ must now correspond to the general frequency variable which describes the continuous spectrum. In other words, n must tend to infinity as ω_0 approaches zero, so that the product is finite:

$$n\omega_0 \to \omega \qquad [43]$$

When these four limiting operations are applied to Eq. [40], we find that $\mathbf{c}_n$ must approach zero, as we had previously presumed. If we multiply each side of Eq. [40] by the period T and then undertake the limiting process, a nontrivial result is obtained:

$$\mathbf{c}_n T \to \int_{-\infty}^{\infty} f(t) e^{-j\omega t}\, dt$$

The right-hand side of this expression is a function of ω (and *not* of t), and we represent it by $\mathbf{F}(j\omega)$:

$$\mathbf{F}(j\omega) = \int_{-\infty}^{\infty} f(t) e^{-j\omega t}\, dt \qquad [44]$$

Now let us apply the limiting process to Eq. [39]. We begin by multiplying and dividing the summation by T,

$$f(t) = \sum_{n=-\infty}^{\infty} \mathbf{c}_n T e^{jn\omega_0 t} \frac{1}{T}$$

next replacing $\mathbf{c}_n T$ with the new quantity $\mathbf{F}(j\omega)$, and then making use of expressions [42] and [43]. In the limit, the summation becomes an integral, and

$$f(t) = \frac{1}{2\pi} \int_{-\infty}^{\infty} \mathbf{F}(j\omega) e^{j\omega t}\, d\omega \qquad [45]$$

Equations [44] and [45] are collectively called the *Fourier transform pair*. The function $\mathbf{F}(j\omega)$ is the *Fourier transform* of $f(t)$, and $f(t)$ is the *inverse Fourier transform* of $\mathbf{F}(j\omega)$.

This transform-pair relationship is most important! We should memorize it, draw arrows pointing to it, and mentally keep it on the conscious level henceforth and forevermore. We emphasize the importance of these relations by repeating them in boxed form:

$$\mathbf{F}(j\omega) = \int_{-\infty}^{\infty} e^{-j\omega t} f(t)\, dt \qquad [46a]$$

$$f(t) = \frac{1}{2\pi} \int_{-\infty}^{\infty} e^{j\omega t} \mathbf{F}(j\omega)\, d\omega \qquad [46b]$$

The exponential terms in these two equations carry opposite signs for the exponents. To keep them straight, it may help to note that the positive sign is associated with the expression for $f(t)$, as it is with the complex Fourier series, Eq. [39].

It is appropriate to raise one question at this time. For the Fourier transform relationships of Eq. [46], can we obtain the Fourier transform of *any* arbitrarily chosen $f(t)$? It turns out that the answer is affirmative for almost any voltage or current that we can actually produce. A sufficient condition for the existence of $\mathbf{F}(j\omega)$ is that

$$\int_{-\infty}^{\infty} |f(t)| dt < \infty$$

This condition is not *necessary,* however, because some functions that do not meet it still have a Fourier transform; the step function is one such example. Furthermore, we will see later that $f(t)$ does not even need to be nonperiodic in order to have a Fourier transform; the Fourier series representation for a periodic time function is just a special case of the more general Fourier transform representation.

As we indicated earlier, the Fourier transform-pair relationship is unique. For a given $f(t)$ there is one specific $\mathbf{F}(j\omega)$; and for a given $\mathbf{F}(j\omega)$ there is one specific $f(t)$.

The reader may have already noticed a few similarities between the Fourier transform and the Laplace transform. Key differences between the two include the fact that initial energy storage is not easily incorporated in circuit analysis using Fourier transforms while it is very easily incorporated in the case of Laplace transforms. Also, there are several time functions (e.g., the *increasing* exponential), for which a Fourier transform does not exist. However, if it is spectral information as opposed to transient response in which we are primarily concerned, the Fourier transform is the ticket.

EXAMPLE 18.5

Use the Fourier transform to obtain the continuous spectrum of the single rectangular pulse Fig. 18.13*a*.

The pulse is a truncated version of the sequence considered previously in Fig. 18.11, and is described by

$$f(t) = \begin{cases} V_0 & t_0 < t < t_0 + \tau \\ 0 & t < t_0 \text{ and } t > t_0 + \tau \end{cases}$$

(Continued on next page)

The Fourier transform of $f(t)$ is found from Eq. [46a]:

$$\mathbf{F}(j\omega) = \int_{t_0}^{t_0+\tau} V_0 e^{-j\omega t} \, dt$$

and this may be easily integrated and simplified:

$$\mathbf{F}(j\omega) = V_0 \tau \frac{\sin \frac{1}{2}\omega\tau}{\frac{1}{2}\omega\tau} e^{-j\omega(t_0+\tau/2)}$$

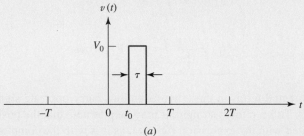

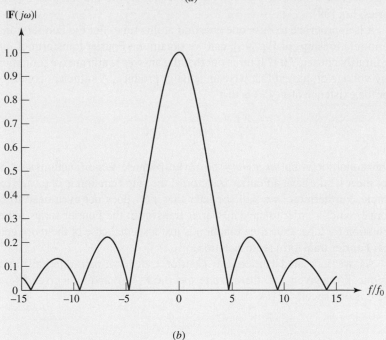

■ **FIGURE 18.13** (*a*) A single rectangular pulse identical to those of the sequence in Fig. 18.11. (*b*) A plot of $|\mathbf{F}(j\omega)|$ corresponding to the pulse, with $V_0 = 1$, $\tau = 1$, and $t_0 = 0$. The frequency axis has been normalized to the value of $f_0 = 1/1.5\,\pi$ corresponding to Fig. 18.12*a* to allow comparison; note that f_0 has no meaning or relevance in the context of $\mathbf{F}(j\omega)$.

The magnitude of $\mathbf{F}(j\omega)$ yields the continuous frequency spectrum, and it is obviously of the form of the sampling function. The value of $\mathbf{F}(0)$ is $V_0\tau$. The shape of the spectrum is identical with the envelope in Fig. 18.12*b*. A plot of $|\mathbf{F}(j\omega)|$ as a function of ω does *not* indicate the magnitude of the voltage present at any given frequency. What is it, then? Examination of Eq. [45] shows that, if $f(t)$ is a voltage wave-form, then $\mathbf{F}(j\omega)$ is dimensionally "volts per unit frequency," a concept that was introduced in Sec. 15.1.

PRACTICE

18.8 If $f(t) = -10$ V, $-0.2 < t < -0.1$s, $f(t) = 10$ V, $0.1 < t < 0.2$s, and $f(t) = 0$ for all other t, evaluate $\mathbf{F}(j\omega)$ for ω equal to (a) 0; (b) 10π rad/s; (c) -10π rad/s; (d) 15π rad/s; (e) -20π rad/s.
18.9 If $\mathbf{F}(j\omega) = -10$ V/(rad/s) for $-4 < \omega < -2$ rad/s, $+10$ V/(rad/s) for $2 < \omega < 4$ rad/s, and 0 for all other ω, find the numerical value of $f(t)$ at t equal to (a) 10^{-4} s; (b) 10^{-2} s; (c) $\pi/4$ s; (d) $\pi/2$ s; (e) π s.

Ans: 18.8: 0; $j1.273$ V/(rad/s); $-j1.273$ V/(rad/s); $-j0.424$ V/(rad/s); 0.
18.9: $j1.9099 \times 10^{-3}$ V; $j0.1910$ V; $j4.05$ V; $-j4.05$ V; 0.

18.6 SOME PROPERTIES OF THE FOURIER TRANSFORM

Our object in this section is to establish several of the mathematical proper-ties of the Fourier transform and, even more important, to understand its physical significance. We begin by using Euler's identity to replace $e^{-j\omega t}$ in Eq. [46a]:

$$\mathbf{F}(j\omega) = \int_{-\infty}^{\infty} f(t) \cos \omega t \, dt - j \int_{-\infty}^{\infty} f(t) \sin \omega t \, dt \qquad [47]$$

Since $f(t)$, $\cos \omega t$, and $\sin \omega t$ are all real functions of time, both the integrals in Eq. [47] are real functions of ω. Thus, by letting

$$\mathbf{F}(j\omega) = A(\omega) + jB(\omega) = |\mathbf{F}(j\omega)|e^{j\phi(\omega)} \qquad [48]$$

we have

$$A(\omega) = \int_{-\infty}^{\infty} f(t) \cos \omega t \, dt \qquad [49]$$

$$B(\omega) = - \int_{-\infty}^{\infty} f(t) \sin \omega t \, dt \qquad [50]$$

$$|\mathbf{F}(j\omega)| = \sqrt{A^2(\omega) + B^2(\omega)} \qquad [51]$$

and

$$\phi(\omega) = \tan^{-1} \frac{B(\omega)}{A(\omega)} \qquad [52]$$

Replacing ω by $-\omega$ shows that $A(\omega)$ and $|\mathbf{F}(j\omega)|$ are both even functions of ω, while $B(\omega)$ and $\phi(\omega)$ are both odd functions of ω.

Now, if $f(t)$ is an even function of t, then the integrand of Eq. [50] is an odd function of t, and the symmetrical limits force $B(\omega)$ to be zero; thus, if $f(t)$ is even, its Fourier transform $\mathbf{F}(j\omega)$ is a real, even function of ω, and the phase function $\phi(\omega)$ is zero or π for all ω. However, if $f(t)$ is an odd function of t, then $A(\omega) = 0$ and $\mathbf{F}(j\omega)$ is both odd and a pure imaginary function of ω; $\phi(\omega)$ is $\pm\pi/2$. In general, however, $\mathbf{F}(j\omega)$ is a complex function of ω.

Finally, we note that the replacement of ω by $-\omega$ in Eq. [47] forms the *conjugate* of $\mathbf{F}(j\omega)$. Thus,

$$\mathbf{F}(-j\omega) = A(\omega) - jB(\omega) = \mathbf{F}^*(j\omega)$$

and we have

$$\mathbf{F}(j\omega)\mathbf{F}(-j\omega) = \mathbf{F}(j\omega)\mathbf{F}^*(j\omega) = A^2(\omega) + B^2(\omega) = |\mathbf{F}(j\omega)|^2$$

Physical Significance of the Fourier Transform

With these basic mathematical properties of the Fourier transform in mind, we are now ready to consider its physical significance. Let us suppose that $f(t)$ is either the voltage across or the current through a 1 Ω resistor, so that $f^2(t)$ is the instantaneous power delivered to the 1 Ω resistor by $f(t)$. Integrating this power over all time, we obtain the total energy delivered by $f(t)$ to the 1 Ω resistor,

$$W_{1\Omega} = \int_{-\infty}^{\infty} f^2(t)\, dt \qquad [53]$$

Now let us resort to a little trickery. Thinking of the integrand in Eq. [53] as $f(t)$ times itself, we replace one of those functions with Eq. [46b]:

$$W_{1\Omega} = \int_{-\infty}^{\infty} f(t) \left[\frac{1}{2\pi} \int_{-\infty}^{\infty} e^{j\omega t} \mathbf{F}(j\omega)\, d\omega \right] dt$$

Since $f(t)$ is not a function of the variable of integration ω, we may move it inside the bracketed integral and then interchange the order of integration:

$$W_{1\Omega} = \frac{1}{2\pi} \int_{-\infty}^{\infty} \left[\int_{-\infty}^{\infty} \mathbf{F}(j\omega)e^{j\omega t} f(t)\, dt \right] d\omega$$

Next we shift $\mathbf{F}(j\omega)$ outside the inner integral, causing that integral to become $\mathbf{F}(-j\omega)$:

$$W_{1\Omega} = \frac{1}{2\pi} \int_{-\infty}^{\infty} \mathbf{F}(j\omega)\mathbf{F}(-j\omega)\, d\omega = \frac{1}{2\pi} \int_{-\infty}^{\infty} |\mathbf{F}(j\omega)|^2\, d\omega$$

Collecting these results,

$$\int_{-\infty}^{\infty} f^2(t)\, dt = \frac{1}{2\pi} \int_{-\infty}^{\infty} |\mathbf{F}(j\omega)|^2\, d\omega \qquad [54]$$

Marc Antoine Parseval-Deschenes was a rather obscure French mathematician, geographer, and occasional poet who published these results in 1805, seventeen years before Fourier published his theorem.

Equation [54] is a very useful expression known as Parseval's theorem. This theorem, along with Eq. [53], tells us that the energy associated with $f(t)$ can be obtained either from an integration over all time in the time domain or by $1/(2\pi)$ times an integration over all (radian) frequency in the frequency domain.

Parseval's theorem also leads us to a greater understanding and interpretation of the meaning of the Fourier transform. Consider a voltage $v(t)$ with Fourier transform $\mathbf{F}_v(j\omega)$ and 1 Ω energy $W_{1\Omega}$:

$$W_{1\Omega} = \frac{1}{2\pi} \int_{-\infty}^{\infty} |\mathbf{F}_v(j\omega)|^2\, d\omega = \frac{1}{\pi} \int_{0}^{\infty} |\mathbf{F}_v(j\omega)|^2\, d\omega$$

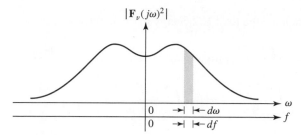

■ **FIGURE 18.14** The area of the slice $|F_v(j\omega)|^2$ is the 1 Ω energy associated with $v(t)$ lying in the bandwidth df.

where the rightmost equality follows from the fact that $|\mathbf{F}_v(j\omega)|^2$ is an even function of ω. Then, since $\omega = 2\pi f$, we can write

$$W_{1\Omega} = \int_{-\infty}^{\infty} |\mathbf{F}_v(j\omega)|^2 \, df = 2 \int_{0}^{\infty} |\mathbf{F}_v(j\omega)|^2 \, df \qquad [55]$$

Figure 18.14 illustrates a typical plot of $|\mathbf{F}_v(j\omega)|^2$ as a function of both ω and f. If we divide the frequency scale up into vanishingly small increments df, Eq. [55] shows us that the area of a differential slice under the $|\mathbf{F}_v(j\omega)|^2$ curve, having a width df, is $|\mathbf{F}_v(j\omega)|^2 \, df$. This area is shown shaded. The sum of all such areas, as f ranges from minus to plus infinity, is the total 1 Ω energy contained in $v(t)$. Thus, $|\mathbf{F}_v(j\omega)|^2$ is the (1 Ω) **energy density** or energy per unit bandwidth (J/Hz) of $v(t)$, and this energy density is always a real, even, nonnegative function of ω. By integrating $|\mathbf{F}_v(j\omega)|^2$ over an appropriate frequency interval, we are able to calculate that portion of the total energy lying within the chosen interval. Note that the energy density is not a function of the phase of $\mathbf{F}_v(j\omega)$, and thus there are an infinite number of time functions and Fourier transforms that possess identical energy-density functions.

EXAMPLE **18.6**

The one-sided [i.e., $v(t) = 0$ for $t < 0$] exponential pulse

$$v(t) = 4e^{-3t} u(t) \text{ V}$$

is applied to the input of an ideal bandpass filter. If the filter passband is defined by $1 < |f| < 2$ Hz, calculate the total output energy.

We call the filter output voltage $v_o(t)$. The energy in $v_o(t)$ will therefore be equal to the energy of that part of $v(t)$ having frequency components in the intervals $1 < f < 2$ and $-2 < f < -1$. We determine the Fourier transform of $v(t)$,

$$\mathbf{F}_v(j\omega) = 4 \int_{-\infty}^{\infty} e^{-j\omega t} e^{-3t} u(t) \, dt$$

$$= 4 \int_{0}^{\infty} e^{-(3+j\omega)t} \, dt = \frac{4}{3 + j\omega}$$

(Continued on next page)

and then we may calculate the total 1 Ω energy in the input signal by either

$$W_{1\Omega} = \frac{1}{2\pi} \int_{-\infty}^{\infty} |\mathbf{F}_v(j\omega)|^2 \, d\omega$$

$$= \frac{8}{\pi} \int_{-\infty}^{\infty} \frac{d\omega}{9 + \omega^2} = \frac{16}{\pi} \int_0^{\infty} \frac{d\omega}{9 + \omega^2} = \frac{8}{3} \text{ J}$$

or

$$W_{1\Omega} = \int_{-\infty}^{\infty} v^2(t) \, dt = 16 \int_0^{\infty} e^{-6t} \, dt = \frac{8}{3} \text{ J}$$

The total energy in $v_o(t)$, however, is smaller:

$$W_{o1} = \frac{1}{2\pi} \int_{-4\pi}^{-2\pi} \frac{16 \, d\omega}{9 + \omega^2} + \frac{1}{2\pi} \int_{2\pi}^{4\pi} \frac{16 \, d\omega}{9 + \omega^2}$$

$$= \frac{16}{\pi} \int_{2\pi}^{4\pi} \frac{d\omega}{9 + \omega^2} = \frac{16}{3\pi} \left(\tan^{-1} \frac{4\pi}{3} - \tan^{-1} \frac{2\pi}{3} \right) = 358 \text{ mJ}$$

In general, we see that an ideal bandpass filter enables us to remove energy from prescribed frequency ranges while still retaining the energy contained in other frequency ranges. The Fourier transform helps us to describe the filtering action quantitatively without actually evaluating $v_o(t)$, although we will see later that the Fourier transform can also be used to obtain the expression for $v_o(t)$ if we wish to do so.

PRACTICE

18.10 If $i(t) = 10e^{20t}[u(t + 0.1) - u(t - 0.1)]$ A, find (a) $\mathbf{F}_i(j0)$; (b) $\mathbf{F}_i(j10)$; (c) $A_i(10)$; (d) $B_i(10)$; (e) $\phi_i(10)$.

18.11 Find the 1 Ω energy associated with the current $i(t) = 20e^{-10t}u(t)$ A in the interval: (a) $-0.1 < t < 0.1$ s; (b) $-10 < \omega < 10$ rad/s; (c) $10 < \omega < \infty$ rad/s.

Ans: 18.10: 3.63 A/(rad/s); 3.33$\underline{/-31.7°}$ A/(rad/s); 2.83 A/(rad/s); −1.749 A/(rad/s); −31.7°. 18.11: 17.29 J; 10 J; 5 J.

18.7 FOURIER TRANSFORM PAIRS FOR SOME SIMPLE TIME FUNCTIONS

The Unit-Impulse Function

We now seek the Fourier transform of the unit impulse $\delta(t - t_0)$, a function we introduced in Sec. 14.4. That is, we are interested in the spectral properties or frequency-domain description of this singularity function. If we use the notation $\mathcal{F}\{\}$ to symbolize "Fourier transform of $\{\}$," then

$$\mathcal{F}\{\delta(t - t_0)\} = \int_{-\infty}^{\infty} e^{-j\omega t} \delta(t - t_0) \, dt$$

From our earlier discussion of this type of integral, we have

$$\mathcal{F}\{\delta(t - t_0)\} = e^{-j\omega t_0} = \cos \omega t_0 - j \sin \omega t_0$$

This complex function of ω leads to the 1 Ω energy-density function,

$$|\mathcal{F}\{\delta(t - t_0)\}|^2 = \cos^2 \omega t_0 + \sin^2 \omega t_0 = 1$$

This remarkable result says that the (1 Ω) energy per unit bandwidth is unity *at all frequencies,* and that the total energy in the unit impulse is infinitely large. No wonder, then, that we must conclude that the unit impulse is "impractical" in the sense that it cannot be generated in the laboratory. Moreover, even if one were available to us, it must appear distorted after being subjected to the finite bandwidth of any practical laboratory instrument.

Since there is a unique one-to-one correspondence between a time function and its Fourier transform, we can say that the inverse Fourier transform of $e^{-j\omega t_0}$ is $\delta(t - t_0)$. Utilizing the symbol $\mathcal{F}^{-1}\{\ \}$ for the inverse transform, we have

$$\mathcal{F}^{-1}\{e^{-j\omega t_0}\} = \delta(t - t_0)$$

Thus, we now know that

$$\frac{1}{2\pi} \int_{-\infty}^{\infty} e^{j\omega t} e^{-j\omega t_0} d\omega = \delta(t - t_0)$$

even though we would fail in an attempt at the direct evaluation of this improper integral. Symbolically, we may write

$$\delta(t - t_0) \Leftrightarrow e^{-j\omega t_0} \qquad [56]$$

where $\Leftrightarrow$ indicates that the two functions constitute a Fourier transform pair.

Continuing with our consideration of the unit-impulse function, let us consider a Fourier transform in that form,

$$\mathbf{F}(j\omega) = \delta(\omega - \omega_0)$$

which is a unit impulse *in the frequency domain* located at $\omega = \omega_0$. Then $f(t)$ must be

$$f(t) = \mathcal{F}^{-1}\{\mathbf{F}(j\omega)\} = \frac{1}{2\pi} \int_{-\infty}^{\infty} e^{j\omega t} \delta(\omega - \omega_0) \, d\omega = \frac{1}{2\pi} e^{j\omega_0 t}$$

where we have used the sifting property of the unit impulse. Thus we may now write

$$\frac{1}{2\pi} e^{j\omega_0 t} \Leftrightarrow \delta(\omega - \omega_0)$$

or

$$e^{j\omega_0 t} \Leftrightarrow 2\pi \, \delta(\omega - \omega_0) \qquad [57]$$

Also, by a simple sign change we obtain

$$e^{-j\omega_0 t} \Leftrightarrow 2\pi \, \delta(\omega + \omega_0) \qquad [58]$$

Clearly, the time function is complex in both expressions [57] and [58], and does not exist in the real world of the laboratory. Time functions such as $\cos \omega_0 t$, for example, can be produced with laboratory equipment, but a function like $e^{-j\omega_0 t}$ cannot.

However, we know that

$$\cos \omega_0 t = \tfrac{1}{2} e^{j\omega_0 t} + \tfrac{1}{2} e^{-j\omega_0 t}$$

and it is easily seen from the definition of the Fourier transform that

$$\mathcal{F}\{f_1(t)\} + \mathcal{F}\{f_2(t)\} = \mathcal{F}\{f_1(t) + f_2(t)\} \qquad [59]$$

Therefore,

$$\mathcal{F}\{\cos \omega_0 t\} = \mathcal{F}\{\tfrac{1}{2}e^{j\omega_0 t}\} + \mathcal{F}\{\tfrac{1}{2}e^{-j\omega_0 t}\}$$
$$= \pi\,\delta(\omega - \omega_0) + \pi\,\delta(\omega + \omega_0)$$

which indicates that the frequency-domain description of $\cos \omega_0 t$ shows a *pair* of impulses, located at $\omega = \pm\omega_0$. This should not be a great surprise, for in our first discussion of complex frequency in Chap. 14, we noted that a sinusoidal function of time was always represented by a pair of imaginary frequencies located at $\mathbf{s} = \pm j\omega_0$. We have, therefore,

$$\cos \omega_0 t \Leftrightarrow \pi[\delta(\omega + \omega_0) + \delta(\omega - \omega_0)] \qquad [60]$$

The Constant Forcing Function

The first forcing function that we considered many chapters ago was a dc voltage or current. To find the Fourier transform of a constant function of time, $f(t) = K$, our first inclination might be to substitute this constant in the defining equation for the Fourier transform and evaluate the resulting integral. If we did, we would find ourselves with an indeterminate expression on our hands. Fortunately, however, we have already solved this problem, for from expression [58],

$$e^{-j\omega_0 t} \Leftrightarrow 2\pi\,\delta(\omega + \omega_0)$$

We see that if we simply let $\omega_0 = 0$, then the resulting transform pair is

$$1 \Leftrightarrow 2\pi\,\delta(\omega) \qquad [61]$$

from which it follows that

$$K \Leftrightarrow 2\pi K\,\delta(\omega) \qquad [62]$$

and our problem is solved. The frequency spectrum of a constant function of time consists only of a component at $\omega = 0$, which we knew all along.

The Signum Function

As another example, let us obtain the Fourier transform of a singularity function known as the ***signum function,*** sgn(t), defined by

$$\text{sgn}(t) = \begin{cases} -1 & t < 0 \\ 1 & t > 0 \end{cases} \qquad [63]$$

or

$$\text{sgn}(t) = u(t) - u(-t)$$

Again, if we should try to substitute this time function in the defining equation for the Fourier transform, we would face an indeterminate expression upon substitution of the limits of integration. This same problem will arise every time we attempt to obtain the Fourier transform of a time function that does not approach zero as $|t|$ approaches infinity. Fortunately, we can avoid this situation by using the *Laplace transform,* as it contains a built-in convergence factor that cures many of the inconvenient ills associated with the evaluation of certain Fourier transforms.

Along those lines, the signum function under consideration can be written as

$$\text{sgn}(t) = \lim_{a \to 0}[e^{-at}u(t) - e^{at}u(-t)]$$

Notice that the expression within the brackets *does* approach zero as $|t|$ gets very large. Using the definition of the Fourier transform, we obtain

$$\mathcal{F}\{\text{sgn}(t)\} = \lim_{a \to 0}\left[\int_0^\infty e^{-j\omega t}e^{-at}\,dt - \int_{-\infty}^0 e^{-j\omega t}e^{at}\,dt\right]$$

$$= \lim_{a \to 0}\frac{-j2\omega}{\omega^2 + a^2} = \frac{2}{j\omega}$$

The real component is zero, since $\text{sgn}(t)$ is an odd function of t. Thus,

$$\text{sgn}(t) \Leftrightarrow \frac{2}{j\omega} \qquad [64]$$

The Unit Step Function

As a final example in this section, let us look at the familiar unit-step function, $u(t)$. Making use of our work on the signum function in the preceding paragraphs, we represent the unit step by

$$u(t) = \tfrac{1}{2} + \tfrac{1}{2}\text{sgn}(t)$$

and obtain the Fourier transform pair

$$u(t) \Leftrightarrow \left[\pi\,\delta(\omega) + \frac{1}{j\omega}\right] \qquad [65]$$

Table 18.2 presents the conclusions drawn from the examples discussed in this section, along with a few others that have not been detailed here.

EXAMPLE 18.7

Use Table 18.2 to find the Fourier transform of the time function $3e^{-t}\cos 4t\,u(t)$.

From the next to the last entry in the table, we have

$$e^{-\alpha t}\cos\omega_d t\,u(t) \Leftrightarrow \frac{\alpha + j\omega}{(\alpha + j\omega)^2 + \omega_d^2}$$

We therefore identify α as 1 and ω_d as 4, and have

$$\mathbf{F}(j\omega) = (3)\frac{1 + j\omega}{(1 + j\omega)^2 + 16}$$

PRACTICE

18.12 Evaluate the Fourier transform at $\omega = 12$ for the time function:
(a) $4u(t) - 10\delta(t)$; (b) $5e^{-8t}u(t)$; (c) $4\cos 8tu(t)$; (d) $-4\,\text{sgn}(t)$.
18.13 Find $f(t)$ at $t = 2$ if $\mathbf{F}(j\omega)$ is: (a) $5e^{-j3\omega} - j(4/\omega)$;
(b) $8[\delta(\omega - 3) + \delta(\omega + 3)]$; (c) $(8/\omega)\sin 5\omega$.

Ans: 18.12: $10.01\underline{/-178.1°}$; $0.347\underline{/-56.3°}$; $-j0.6$; $j0.667$.   18.13: 2.00; 2.45; 4.00.

TABLE 18.2 A Summary of Some Fourier Transform Pairs

$f(t)$	$f(t)$	$\mathcal{F}\{f(t)\} = F(j\omega)$	$|F(j\omega)|$
	$\delta(t - t_0)$	$e^{-j\omega t_0}$	
	$e^{j\omega_0 t}$	$2\pi\delta(\omega - \omega_0)$	
	$\cos\omega_0 t$	$\pi[\delta(\omega + \omega_0) + \delta(\omega - \omega_0)]$	
	1	$2\pi\delta(\omega)$	
	$\mathrm{sgn}(t)$	$\dfrac{2}{j\omega}$	
	$u(t)$	$\pi\delta(\omega) + \dfrac{1}{j\omega}$	
	$e^{-\alpha t}u(t)$	$\dfrac{1}{\alpha + j\omega}$	
	$[e^{-\alpha t}\cos\omega_d t]u(t)$	$\dfrac{\alpha + j\omega}{(\alpha + j\omega)^2 + \omega_d^2}$	
	$u(t + \tfrac{1}{2}T) - u(t - \tfrac{1}{2}T)$	$T\,\dfrac{\sin\frac{\omega T}{2}}{\frac{\omega T}{2}}$	

18.8 • THE FOURIER TRANSFORM OF A GENERAL PERIODIC TIME FUNCTION

In Sec. 18.5 we remarked that we would be able to show that periodic time functions, as well as nonperiodic functions, possess Fourier transforms. Let us now establish this fact on a rigorous basis. Consider a periodic time function $f(t)$ with period T and Fourier series expansion, as outlined by Eqs. [39], [40], and [41], repeated here for convenience:

$$f(t) = \sum_{n=-\infty}^{\infty} \mathbf{c}_n e^{jn\omega_0 t} \qquad [39]$$

$$\mathbf{c}_n = \frac{1}{T} \int_{-T/2}^{T/2} f(t) e^{-jn\omega_0 t}\, dt \qquad [40]$$

and

$$\omega_0 = \frac{2\pi}{T} \qquad [41]$$

Bearing in mind that the Fourier transform of a sum is just the sum of the transforms of the terms in the sum, and that $\mathbf{c}_n$ is not a function of time, we can write

$$\mathcal{F}\{f(t)\} = \mathcal{F}\left\{ \sum_{n=-\infty}^{\infty} \mathbf{c}_n e^{jn\omega_0 t} \right\} = \sum_{n=-\infty}^{\infty} \mathbf{c}_n \mathcal{F}\{e^{jn\omega_0 t}\}$$

After obtaining the transform of $e^{jn\omega_0 t}$ from expression [57], we have

$$f(t) \Leftrightarrow 2\pi \sum_{n=-\infty}^{\infty} \mathbf{c}_n \delta(\omega - n\omega_0) \qquad [66]$$

This shows that $f(t)$ has a discrete spectrum consisting of impulses located at points on the ω axis given by $\omega = n\omega_0$, $n = \dots, -2, -1, 0, 1, \dots$. The strength of each impulse is 2π times the value of the corresponding Fourier coefficient appearing in the complex form of the Fourier series expansion for $f(t)$.

As a check on our work, let us see whether the inverse Fourier transform of the right side of expression [66] is once again $f(t)$. This inverse transform can be written as

$$\mathcal{F}^{-1}\{\mathbf{F}(j\omega)\} = \frac{1}{2\pi} \int_{-\infty}^{\infty} e^{j\omega t} \left[2\pi \sum_{n=-\infty}^{\infty} \mathbf{c}_n \delta(\omega - n\omega_0) \right] d\omega \stackrel{?}{=} f(t)$$

Since the exponential term does not contain the index of summation n, we can interchange the order of the integration and summation operations:

$$\mathcal{F}^{-1}\{\mathbf{F}(j\omega)\} = \sum_{n=-\infty}^{\infty} \int_{-\infty}^{\infty} \mathbf{c}_n e^{j\omega t} \delta(\omega - n\omega_0)\, d\omega \stackrel{?}{=} f(t)$$

Because it is not a function of the variable of integration, $\mathbf{c}_n$ can be treated as a constant. Then, using the sifting property of the impulse, we obtain

$$\mathcal{F}^{-1}\{\mathbf{F}(j\omega)\} = \sum_{n=-\infty}^{\infty} \mathbf{c}_n e^{jn\omega_0 t} \stackrel{?}{=} f(t)$$

which is exactly the same as Eq. [39], the complex Fourier series expansion for $f(t)$. The question marks in the preceding equations can now be removed, and the existence of the Fourier transform for a periodic time function is established. This should come as no great surprise, however. In the last section we evaluated the Fourier transform of a cosine function, which is certainly periodic, although we made no direct reference to its periodicity. However, we did use a backhanded approach in getting the transform. But now we have a mathematical tool by which the transform can be obtained more directly. To demonstrate this procedure, consider $f(t) = \cos \omega_0 t$ once more. First we evaluate the Fourier coefficients c_n:

$$\mathbf{c}_n = \frac{1}{T} \int_{-T/2}^{T/2} \cos \omega_0 t \, e^{-jn\omega_0 t} \, dt = \begin{cases} \frac{1}{2} & n = \pm 1 \\ 0 & \text{otherwise} \end{cases}$$

Then

$$\mathcal{F}\{f(t)\} = 2\pi \sum_{n=-\infty}^{\infty} \mathbf{c}_n \delta(\omega - n\omega_0)$$

This expression has values that are nonzero only when $n = \pm 1$, and it follows, therefore, that the entire summation reduces to

$$\mathcal{F}\{\cos \omega_0 t\} = \pi[\delta(\omega - \omega_0) + \delta(\omega + \omega_0)]$$

which is precisely the expression that we obtained before. What a relief!

PRACTICE

18.14 Find (a) $\mathcal{F}\{5 \sin^2 3t\}$; (b) $\mathcal{F}\{A \sin \omega_0 t\}$; (c) $\mathcal{F}\{6 \cos(8t + 0.1\pi)\}$.

Ans: $2.5\pi[2\delta(\omega) - \delta(\omega + 6) - \delta(\omega - 6)]$; $j\pi A[\delta(\omega + \omega_0) - \delta(\omega - \omega_0)]$; $[18.85\underline{/18°}]\delta(\omega - 8) + [18.85\underline{/-18°}]\delta(\omega + 8)$.

18.9 THE SYSTEM FUNCTION AND RESPONSE IN THE FREQUENCY DOMAIN

In Sec. 15.5, the problem of determining the output of a physical system in terms of the input and the impulse response was solved by using the convolution integral and initially working in the time domain. The input, the output, and the impulse response are all time functions. Subsequently, we found that it was often more convenient to perform such operations in the frequency domain, as the Laplace transform of the convolution of two functions is simply the product of each function in the frequency domain. Along the same lines, we find the same is true when working with Fourier transforms.

To do this we examine the Fourier transform of the system output. Assuming arbitrarily that the input and output are voltages, we apply the basic definition of the Fourier transform and express the output by the convolution integral:

$$\mathcal{F}\{v_0(t)\} = \mathbf{F}_0(j\omega) = \int_{-\infty}^{\infty} e^{-j\omega t} \left[\int_{-\infty}^{\infty} v_i(t-z)h(z)\,dz \right] dt$$

where we again assume no initial energy storage. At first glance this expression may seem rather formidable, but it can be reduced to a result that is surprisingly simple. We may move the exponential term inside the inner integral because it does not contain the variable of integration z. Next we reverse the order of integration, obtaining

$$\mathbf{F}_0(j\omega) = \int_{-\infty}^{\infty} \left[\int_{-\infty}^{\infty} e^{-j\omega t} v_i(t-z)h(z)\,dt \right] dz$$

Since it is not a function of t, we can extract $h(z)$ from the inner integral and simplify the integration with respect to t by a change of variable, $t - z = x$:

$$\mathbf{F}_0(j\omega) = \int_{-\infty}^{\infty} h(z) \left[\int_{-\infty}^{\infty} e^{-j\omega(x+z)} v_i(x)\,dx \right] dz$$

$$= \int_{-\infty}^{\infty} e^{-j\omega z} h(z) \left[\int_{-\infty}^{\infty} e^{-j\omega x} v_i(x)\,dx \right] dz$$

But now the sun is starting to break through, for the inner integral is merely the Fourier transform of $v_i(t)$. Furthermore, it contains no z terms and can be treated as a constant in any integration involving z. Thus, we can move this transform, $\mathbf{F}_i(j\omega)$, completely outside all the integral signs:

$$\mathbf{F}_0(j\omega) = \mathbf{F}_i(j\omega) \int_{-\infty}^{\infty} e^{-j\omega z} h(z)\,dz$$

Finally, the remaining integral exhibits our old friend once more, another Fourier transform! This one is the Fourier transform of the impulse response, which we will designate by the notation $\mathbf{H}(j\omega)$. Therefore, all our work has boiled down to the simple result:

$$\mathbf{F}_0(j\omega) = \mathbf{F}_i(j\omega)\mathbf{H}(j\omega) = \mathbf{F}_i(j\omega)\mathcal{F}\{h(t)\}$$

This is another important result: it defines the *system function* $\mathbf{H}(j\omega)$ as the ratio of the Fourier transform of the response function to the Fourier transform of the forcing function. Moreover, the system function and the impulse response constitute a Fourier transform pair:

$$h(t) \Leftrightarrow \mathbf{H}(j\omega) \qquad [67]$$

The development in the preceding paragraph also serves to prove the general statement that the Fourier transform of the convolution of two time functions is the product of their Fourier transforms,

$$\boxed{\mathcal{F}\{f(t) * g(t)\} = \mathbf{F}_f(j\omega)\mathbf{F}_g(j\omega)} \qquad [68]$$

To recapitulate, if we know the Fourier transforms of the forcing function and the impulse response, then the Fourier transform of the response function can be obtained as their product. The result is a description of the response function in the frequency domain; the time-domain description of the response function is obtained by simply taking the inverse Fourier transform. Thus we see that the process of convolution in the time domain is equivalent to the relatively simple operation of multiplication in the frequency domain.

The foregoing comments might make us wonder once again why we would ever choose to work in the time domain at all, but we must always remember that we seldom get something for nothing. A poet once said, *"Our sincerest laughter/with some pain is fraught."*[2] The pain herein is the occasional difficulty in obtaining the inverse Fourier transform of a response function, for reasons of mathematical complexity. On the other hand, a simple desktop computer can convolve two time functions with magnificent celerity. For that matter, it can also obtain an FFT (fast Fourier transform) quite rapidly. Consequently there is no clear-cut advantage between working in the time domain and in the frequency domain. A decision must be made each time a new problem arises; it should be based on the information available and on the computational facilities at hand.

Consider a forcing function of the form

$$v_i(t) = u(t) - u(t-1)$$

and a unit-impulse response defined by

$$h(t) = 2e^{-t}u(t)$$

We first obtain the corresponding Fourier transforms. The forcing function is the difference between two unit-step functions. These two functions are identical, except that one is initiated 1 s after the other. We will evaluate the response due to $u(t)$; the response due to $u(t-1)$ is the same, but delayed in time by 1 s. The difference between these two partial responses will be the total response due to $v_i(t)$.

The Fourier transform of $u(t)$ was obtained in Sec. 18.7:

$$\mathcal{F}\{u(t)\} = \pi\delta(\omega) + \frac{1}{j\omega}$$

The system function is obtained by taking the Fourier transform of $h(t)$, listed in Table 18.2,

$$\mathcal{F}\{h(t)\} = \mathbf{H}(j\omega) = \mathcal{F}\{2e^{-t}u(t)\} = \frac{2}{1+j\omega}$$

The inverse transform of the product of these two functions yields that component of $v_o(t)$ caused by $u(t)$,

$$v_{o1}(t) = \mathcal{F}^{-1}\left\{\frac{2\pi\delta(\omega)}{1+j\omega} + \frac{2}{j\omega(1+j\omega)}\right\}$$

Using the sifting property of the unit impulse, the inverse transform of the first term is just a constant equal to unity. Thus,

$$v_{o1}(t) = 1 + \mathcal{F}^{-1}\left\{\frac{2}{j\omega(1+j\omega)}\right\}$$

The second term contains a product of terms in the denominator, each of the form $(\alpha + j\omega)$, and its inverse transform is found most easily by making use of the partial-fraction expansion that we developed in Sec. 14.5. Let us

(2) P.B. Shelley, "To a Skylark," 1821.

select a technique for obtaining a partial-fraction expansion that has one big advantage—it always works, although faster methods are usually available for most situations. We assign an unknown quantity in the numerator of each fraction, here two in number,

$$\frac{2}{j\omega(1+j\omega)} = \frac{A}{j\omega} + \frac{B}{1+j\omega}$$

and then substitute a corresponding number of simple values for $j\omega$. Here we let $j\omega = 1$:

$$1 = A + \frac{B}{2}$$

and then let $j\omega = -2$:

$$1 = -\frac{A}{2} - B$$

This leads to $A = 2$ and $B = -2$. Thus,

$$\mathcal{F}^{-1}\left\{\frac{2}{j\omega(1+j\omega)}\right\} = \mathcal{F}^{-1}\left\{\frac{2}{j\omega} - \frac{2}{1+j\omega}\right\} = \text{sgn}(t) - 2e^{-t}u(t)$$

so that

$$v_{o1}(t) = 1 + \text{sgn}(t) - 2e^{-t}u(t)$$
$$= 2u(t) - 2e^{-t}u(t)$$
$$= 2(1 - e^{-t})u(t)$$

It follows that $v_{o2}(t)$, the component of $v_o(t)$ produced by $u(t-1)$, is

$$v_{o2}(t) = 2(1 - e^{-(t-1)})u(t-1)$$

Therefore,

$$v_o(t) = v_{o1}(t) - v_{o2}(t)$$
$$= 2(1 - e^{-t})u(t) - 2(1 - e^{-t+1})u(t-1)$$

The discontinuities at $t = 0$ and $t = 1$ dictate a separation into three time intervals:

$$v_o(t) = \begin{cases} 0 & t < 0 \\ 2(1 - e^{-t}) & 0 < t < 1 \\ 2(e-1)e^{-t} & t > 1 \end{cases}$$

PRACTICE

18.15 The impulse response of a certain linear network is $h(t) = 6e^{-20t}u(t)$. The input signal is $3e^{-6t}u(t)$ V. Find (a) $\mathbf{H}(j\omega)$; (b) $\mathbf{V}_i(j\omega)$; (c) $\mathbf{V}_o(j\omega)$; (d) $v_o(0.1)$; (e) $v_o(0.3)$; (f) $v_{o,\text{max}}$.

Ans: $6/(20 + j\omega)$; $3/(6 + j\omega)$; $18/[(20 + j\omega)(6 + j\omega)]$; 0.532 V; 0.209 V; 0.5372.

COMPUTER-AIDED ANALYSIS

The material presented in this chapter forms the foundation for many advanced fields of study, including signal processing, communications, and controls. We are only able to introduce some of the more fundamental concepts within the context of an introductory circuits text, but even at this point some of the power of Fourier-based analysis can be brought to bear. As a first example, consider the op amp circuit of Fig. 18.15, constructed in PSpice using a μA741 operational amplifier.

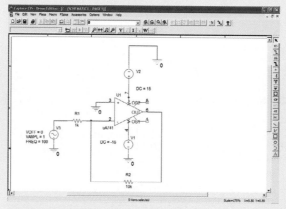

■ **FIGURE 18.15** An inverting amplifier circuit with a voltage gain of −10, driven by a sinusoidal input operating at 100 Hz.

The circuit has a voltage gain of −10, and so we would expect a sinusoidal output of 10 V amplitude. This is indeed what we obtain from a transient analysis of the circuit, as shown in Fig. 18.16.

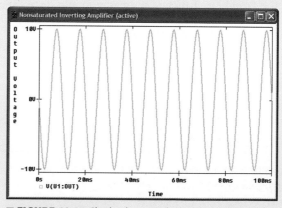

■ **FIGURE 18.16** Simulated output voltage of the amplifier circuit shown in Fig. 18.15.

PSpice allows us to determine the frequency spectrum of the output voltage through what is known as a fast Fourier transform (FFT), a discrete-time approximation to the exact Fourier transform of the signal. From within Probe, we select **Fourier** under the **Trace** menu; the result is the plot shown in Fig. 18.17. As expected, the line spectrum for the output voltage of this amplifier circuit consists of a single feature at a frequency of 100 Hz.

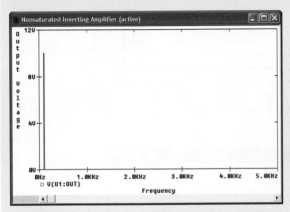

■ **FIGURE 18.17** Discrete approximation to the Fourier transform of Fig. 18.16.

As the input voltage magnitude is increased, the output of the amplifier approaches the saturation condition determined by the positive and negative dc supply voltages (±15 V in this example). This behavior is evident in the simulation result of Fig. 18.18, which corresponds to an input voltage magnitude of 1.8 V. A key feature of interest is that the output voltage waveform is no longer a pure sinusoid. As a result, we expect nonzero values at harmonic frequencies to appear in the

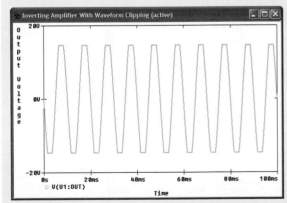

■ **FIGURE 18.18** Transient analysis simulation results for the amplifier circuit when the input voltage magnitude is increased to 1.8 V. Saturation effects manifest themselves in the plot as clipped waveform extrema.

(Continued on next page)

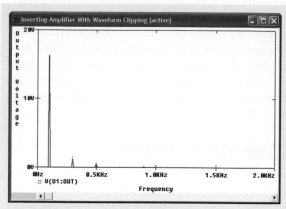

■ **FIGURE 18.19** Frequency spectrum of the waveform depicted in Fig. 18.18, showing the presence of several harmonic components in addition to the fundamental frequency. The finite width of the features is an artifact of the numerical discretization (a set of discrete time values was used).

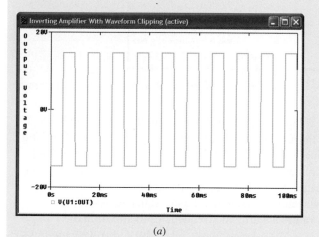

(a)

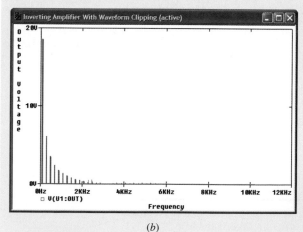

(b)

■ **FIGURE 18.20** (a) Severe effects of amplifier saturation are observed in the simulated response to a 15 V sinusoidal input. (b) An FFT of the waveform shows a significant increase in the fraction of energy present in harmonics as opposed to the fundamental frequency of 100 Hz.

frequency spectrum of the function, as is the case in Fig. 18.19. The effect of reaching saturation in the amplifier circuit is a distortion of the signal; if connected to a speaker, we do not hear a "clean" 100 Hz waveform. Instead, we now hear a superposition of waveforms which include not only the 100 Hz fundamental frequency, but significant harmonic components at 300 and 500 Hz as well. Further distortion of the waveform would increase the amount of energy in harmonic frequencies, so that contributions from higher-frequency harmonics would become more significant. This is evident in the simulation results of Fig. 18.20a and b, which show the output voltage in the time and frequency domains, respectively.

18.10 • THE PHYSICAL SIGNIFICANCE OF THE SYSTEM FUNCTION

In this section we will try to connect several aspects of the Fourier transform with work we completed in earlier chapters.

Given a general linear two-port network N without any initial energy storage, we assume sinusoidal forcing and response functions, arbitrarily taken to be voltages, as shown in Fig. 18.21. We let the input voltage be simply $A \cos(\omega_x t + \theta)$, and the output can be described in general terms as $v_o(t) = B \cos(\omega_x t + \phi)$, where the amplitude B and phase angle ϕ are functions of ω_x. In phasor form, we can write the forcing and response functions as $\mathbf{V}_i = A e^{j\theta}$ and $\mathbf{V}_o = B e^{j\phi}$. The ratio of the phasor response to the phasor forcing function is a complex number that is a function of ω_x:

$$\frac{\mathbf{V}_o}{\mathbf{V}_i} = \mathbf{G}(\omega_x) = \frac{B}{A} e^{j(\phi - \theta)}$$

where B/A is the amplitude of $\mathbf{G}$ and $\phi - \theta$ is its phase angle. This transfer function $\mathbf{G}(\omega_x)$ could be obtained in the laboratory by varying ω_x over a large range of values and measuring the amplitude B/A and phase $\phi - \theta$ for each value of ω_x. If we then plotted each of these parameters as a function of frequency, the resultant pair of curves would completely describe the transfer function.

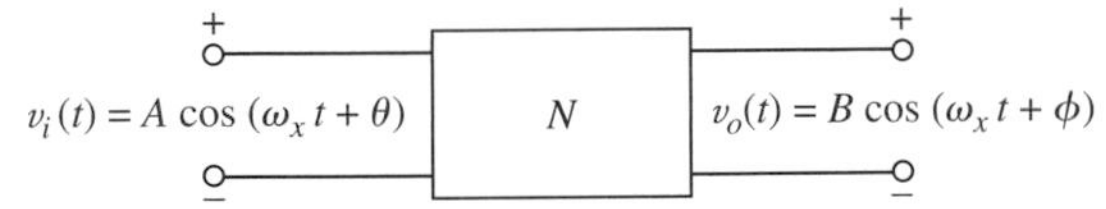

■ FIGURE 18.21 Sinusoidal analysis can be used to determine the transfer function $H(j\omega_x) = (B/A)e^{j(\phi-\theta)}$, where B and ϕ are functions of ω_x.

Now let us hold these comments in the backs of our minds for a moment as we consider a slightly different aspect of the same analysis problem.

For the circuit with sinusoidal input and output shown in Fig. 18.21, what is the system function $\mathbf{H}(j\omega)$? To answer this question, we begin with the definition of $\mathbf{H}(j\omega)$ as the ratio of the Fourier transforms of the output and the input. Both of these time functions involve the functional form

$\cos(\omega_x t + \beta)$, whose Fourier transform we have not evaluated as yet, although we can handle $\cos \omega_x t$. The transform we need is

$$\mathcal{F}\{\cos(\omega_x t + \beta)\} = \int_{-\infty}^{\infty} e^{-j\omega t} \cos(\omega_x t + \beta)\, dt$$

If we make the substitution $\omega_x t + \beta = \omega_x \tau$, then

$$\mathcal{F}\{\cos(\omega_x t + \beta)\} = \int_{-\infty}^{\infty} e^{-j\omega\tau + j\omega\beta/\omega_x} \cos \omega_x \tau \, d\tau$$

$$= e^{j\omega\beta/\omega_x} \mathcal{F}\{\cos \omega_x t\}$$

$$= \pi e^{j\omega\beta/\omega_x} [\delta(\omega - \omega_x) + \delta(\omega + \omega_x)]$$

This is a new Fourier transform pair,

$$\cos(\omega_x t + \beta) \Leftrightarrow \pi e^{j\omega\beta/\omega_x} [\delta(\omega - \omega_x) + \delta(\omega + \omega_x)] \qquad [69]$$

which we can now use to evaluate the desired system function,

$$\mathbf{H}(j\omega) = \frac{\mathcal{F}\{B \cos(\omega_x t + \phi)\}}{\mathcal{F}\{A \cos(\omega_x t + \theta)\}}$$

$$= \frac{\pi B e^{j\omega\phi/\omega_x} [\delta(\omega - \omega_x) + \delta(\omega + \omega_x)]}{\pi A e^{j\omega\theta/\omega_x} [\delta(\omega - \omega_x) + \delta(\omega + \omega_x)]}$$

$$= \frac{B}{A} e^{j\omega(\phi-\theta)/\omega_x}$$

Now we recall the expression for $\mathbf{G}(\omega_x)$,

$$\mathbf{G}(\omega_x) = \frac{B}{A} e^{j(\phi-\theta)}$$

where B and ϕ were evaluated at $\omega = \omega_x$, and we see that evaluating $\mathbf{H}(j\omega)$ at $\omega = \omega_x$ gives

$$\mathbf{H}(\omega_x) = \mathbf{G}(\omega_x) = \frac{B}{A} e^{j(\phi-\theta)}$$

Since there is nothing special about the x subscript, we conclude that the system function and the transfer function are identical:

$$\mathbf{H}(j\omega) = \mathbf{G}(\omega) \qquad [70]$$

The fact that one argument is ω while the other is indicated by $j\omega$ is immaterial and arbitrary; the j merely makes possible a more direct comparison between the Fourier and Laplace transforms.

Equation [70] represents a direct connection between Fourier transform techniques and sinusoidal steady-state analysis. Our previous work on steady-state sinusoidal analysis using phasors was but a special case of the more general techniques of Fourier transform analysis. It was "special" in the sense that the inputs and outputs were sinusoids, whereas the use of Fourier transforms and system functions enables us to handle nonsinusoidal forcing functions and responses.

Thus, to find the system function $\mathbf{H}(j\omega)$ for a network, all we need to do is to determine the corresponding sinusoidal transfer function as a function of ω (or $j\omega$).

EXAMPLE 18.8

Find the voltage across the inductor of the circuit shown in Fig. 18.22a when the input voltage is a simple exponentially decaying pulse, as indicated.

We need the system function; but it is not necessary to apply an impulse, find the impulse response, and then determine its inverse transform. Instead we use Eq. [70] to obtain the system function $\mathbf{H}(j\omega)$ by assuming that the input and output voltages are both sinusoids described by their corresponding phasors, as shown in Fig. 18.22b. Using voltage division, we have

$$\mathbf{H}(j\omega) = \frac{\mathbf{V}_o}{\mathbf{V}_i} = \frac{j2\omega}{4 + j2\omega}$$

The transform of the forcing function is

$$\mathcal{F}\{v_i(t)\} = \frac{5}{3 + j\omega}$$

and thus the transform of $v_o(t)$ is given as

$$\mathcal{F}\{v_o(t)\} = \mathbf{H}(j\omega)\mathcal{F}\{v_i(t)\}$$

$$= \frac{j2\omega}{4 + j2\omega}\frac{5}{3 + j\omega}$$

$$= \frac{15}{3 + j\omega} - \frac{10}{2 + j\omega}$$

where the partial fractions appearing in the last step help to determine the inverse Fourier transform

$$v_o(t) = \mathcal{F}^{-1}\left\{\frac{15}{3 + j\omega} - \frac{10}{2 + j\omega}\right\}$$

$$= 15e^{-3t}u(t) - 10e^{-2t}u(t)$$

$$= 5(3e^{-3t} - 2e^{-2t})u(t)$$

Our problem is completed without fuss, convolution, or differential equations.

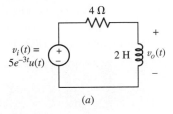

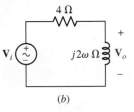

■ **FIGURE 18.22** (a) The response $v_o(t)$ caused by $v_i(t)$ is desired. (b) The system function $\mathsf{H}(j\omega)$ may be determined by sinusoidal steady-state analysis: $\mathsf{H}(j\omega) = \mathsf{V}_o/\mathsf{V}_i$.

PRACTICE

18.16 Use Fourier-transform techniques on the circuit of Fig. 18.23 to find $i_1(t)$ at $t = 1.5$ ms if i_s equals (a) $\delta(t)$ A; (b) $u(t)$ A; (c) $\cos 500t$ A.

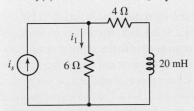

■ **FIGURE 18.23**

Ans: -141.7 A; 0.683 A; 0.308 A.

Image Processing

Although a great deal of progress has been made toward developing a complete understanding of the function of muscle, there remain many open questions. A great deal of research in this field has been carried out using vertebrate skeletal muscle, in particular the *sartorius* or leg muscle of the frog (Fig. 18.24).

■ FIGURE 18.24 A face only a biologist could love. (© Geostock/Getty Images.)

Of the many analytical techniques scientists use, one of the most common is electron microscopy. Fig. 18.25

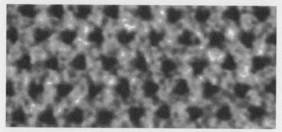

■ FIGURE 18.25 Electron micrograph of a region of frog sartorius muscle tissue. False color has been employed for clarity.

shows an electron micrograph of frog sartorius muscle tissue, sectioned in such a fashion as to highlight the regular arrangement of *myosin,* a filamentary type of contractile protein. Of interest to structural biologists is the periodicity and disorder of these proteins over a large area of muscle tissue. In order to develop a model for these characteristics, a numerical approach is preferable, where the analysis of such images can be automated. As can be seen in the figure, however, the image produced by the electron microscope can be contaminated by a high level of background noise, making automated identification of the myosin filaments prone to error.

Introduced with the intent of ultimately assisting us in the analysis of time-varying linear circuits, the Fourier-based techniques of this chapter are in fact very powerful general methods which find application in many other situations. Among these, the field of *image processing* makes frequent use of Fourier techniques, especially through the fast Fourier transform (FFT) and related numerical methods. The image of Fig. 18.25 can be described by a spatial function $f(x, y)$ where $f(x, y) = 0$ corresponds to white, $f(x, y) = 1$ corresponds to red, and (x, y) denotes a pixel location in the image. Defining a filter function $h(x, y)$ that has the appearance of Fig. 18.26a, the convolution operation

$$g(x, y) = f(x, y) * h(x, y)$$

results in the image of Fig. 18.26b in which the myosin filaments (viewed on end) are more clearly identifiable.

In practice, this image processing is performed in the frequency domain, where the FFT of both f and h are calculated, and the resulting matrices multiplied together.

Epilogue

Returning again to Eq. [70], the identity between the system function $\mathbf{H}(j\omega)$ and the sinusoidal steady-state transfer function $\mathbf{G}(\omega)$, we may now consider the system function as the ratio of the output phasor to the input phasor. Suppose that we hold the input-phasor amplitude at unity and the phase angle at zero. Then the output phasor is $\mathbf{H}(j\omega)$. Under these conditions, if we record the output amplitude and phase as functions of ω, for all ω, we have recorded the system function $\mathbf{H}(j\omega)$ as a function of ω, for all ω. We thus have examined the system response under the condition that an infinite number of sinusoids, all with unity amplitude and zero phase, were successively applied at the input. Now suppose that our input is a single unit impulse, and look at the

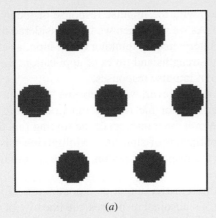

(a)

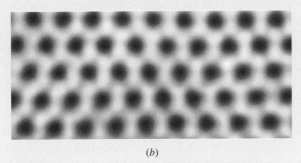

(b)

■ FIGURE 18.26 (a) Spatial filter having hexagonal symmetry. (b) Image after convolution and inverse discrete Fourier transform are performed, showing a reduction in background noise.

An inverse FFT operation then produces the filtered image of Fig. 18.26b. Why does this convolution equate to a filtering operation? The myosin filament arrangement possesses hexagonal symmetry, as does the filter function $h(x, y)$—in a sense, both the myosin filament arrangement and the filter function possess the same

spatial frequencies. The convolution of f with h results in a reinforcement of the hexagonal pattern within the original image, and the removal of noise pixels (which do not possess hexagonal symmetry). This can be understood qualitatively if we model a horizontal row of Fig. 18.25 as a sinusoidal function $f(x) = \cos \omega_0 t$, which has the Fourier transform shown in Fig. 18.27a—a matched pair of impulse functions separated by $2\omega_0$. If we convolve this function with a filter function $h(x) = \cos \omega_1 t$, the Fourier transform of which is depicted in Fig. 18.27b, we get zero if $\omega_1 \neq \omega_0$; the frequencies (periodicities) of the two functions do not match. If, instead, we choose a filter function with the same frequency as $f(x)$, the convolution has a nonzero value at $\omega = \pm\omega_0$.

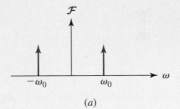

(a)

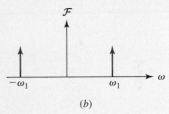

(b)

■ FIGURE 18.27 (a) Fourier transform of $f(x) = \cos \omega_0 t$. (b) Fourier transform of $h(x) = \cos \omega_1 t$.

impulse response $h(t)$. Is the information we examine really any different from that we just obtained? The Fourier transform of the unit impulse is a constant equal to unity, indicating that all frequency components are present, all with the same magnitude, and all with zero phase. Our system response is the sum of the responses to all these components. The result might be viewed at the output on a cathode-ray oscilloscope. It is evident that the system function and the impulse-response function contain equivalent information regarding the response of the system.

We therefore have two different methods of describing the response of a system to a general forcing function; one is a time-domain description, and the other a frequency-domain description. Working in the time domain, we convolve the forcing function with the impulse response of the system to obtain the response function. As we saw when we first considered convolution, this procedure may be interpreted by thinking of the input as a continuum of impulses of different strengths and times of application; the output which results is a continuum of impulse responses.

In the frequency domain, however, we determine the response by multiplying the Fourier transform of the forcing function by the system function. In this case we interpret the transform of the forcing function as a frequency spectrum, or a continuum of sinusoids. Multiplying this by the system function, we obtain the response function, also as a continuum of sinusoids.

Whether we choose to think of the output as a continuum of impulse responses or as a continuum of sinusoidal responses, the linearity of the network and the superposition principle enable us to determine the total output as a time function by summing over all frequencies (the inverse Fourier transform), or as a frequency function by summing over all time (the Fourier transform).

Unfortunately, both of these techniques have some difficulties or limitations associated with their use. In using convolution, the integral itself can often be rather difficult to evaluate when complicated forcing functions or impulse response functions are present. Furthermore, from the experimental point of view, we cannot really measure the impulse response of a system because we cannot actually generate an impulse. Even if we approximate the impulse by a narrow high-amplitude pulse, we would probably drive our system into saturation and out of its linear operating range.

With regard to the frequency domain, we encounter one absolute limitation in that we may easily hypothesize forcing functions that we would like to apply theoretically that do not possess Fourier transforms. Moreover, if we wish to find the time-domain description of the response function, we must evaluate an inverse Fourier transform, and some of these inversions can be extremely difficult.

Finally, neither of these techniques offers a very convenient method of handling initial conditions. For this, the Laplace transform is clearly superior.

The greatest benefits derived from the use of the Fourier transform arise through the abundance of useful information it provides about the spectral properties of a signal, particularly the energy or power per unit bandwidth. Some of this information is also easily obtained through the Laplace transform; we must leave a detailed discussion of the relative merits of each to more advanced signals and systems courses.

So, why has this all been withheld until now? The best answer is probably that these powerful techniques can overcomplicate the solution of simple problems and tend to obscure the physical interpretation of the performance of the simpler networks. For example, if we are interested only in the forced response, then there is little point in using the Laplace transform and obtaining both the forced and natural response after laboring through a difficult inverse transform operation.

Well, we could go on, but all good things must come to an end. Best of luck to you in your future studies.

SUMMARY AND REVIEW

❏ The harmonic frequencies of a sinusoid having the fundamental frequency ω_0 are $n\omega_0$, where n is an integer.

❏ The Fourier theorem states that provided a function $f(t)$ satisfies certain key properties, it may be represented by the infinite series $a_0 + \sum_{n=1}^{\infty}(a_n \cos n\omega_0 t + b_n \sin n\omega_0 t)$, where $a_0 = (1/T) \int_0^T f(t)\, dt$, $a_n = (2/T) \int_0^T f(t) \cos n\omega_0 t\, dt$, and $b_n = (2/T) \int_0^T f(t) \sin n\omega_0 t\, dt$.

❏ A function $f(t)$ possesses *even* symmetry if $f(t) = f(-t)$.

❏ A function $f(t)$ possesses *odd* symmetry if $f(t) = -f(-t)$.

❏ A function $f(t)$ possesses *half-wave* symmetry if $f(t) = -f(t - \frac{1}{2}T)$.

❏ The Fourier series of an even function is composed of only a constant and cosine functions.

❏ The Fourier series of an odd function is composed of only sine functions.

❏ The Fourier series of any function possessing half-wave symmetry contains only odd harmonics.

❏ The Fourier series of a function may also be expressed in complex or exponential form, where $f(t) = \sum_{n=-\infty}^{\infty} \mathbf{c}_n e^{jn\omega_0 t}$ and $\mathbf{c}_n = (1/T) \int_{-T/2}^{T/2} f(t) e^{-jn\omega_0 t}\, dt$.

❏ The Fourier transform allows us to represent time-varying functions in the frequency domain, in a manner similar to that of the Laplace transform. The defining equations are $\mathbf{F}(j\omega) = \int_{-\infty}^{\infty} e^{-j\omega t} f(t)\, dt$ and $f(t) = (1/2\pi) \int_{-\infty}^{\infty} e^{j\omega t} \mathbf{F}(j\omega)\, d\omega$.

READING FURTHER

A very readable treatment of Fourier analysis can be found in

A. Pinkus and S. Zafrany, *Fourier Series and Integral Transforms*. Cambridge: Cambridge University Press, 1997.

Finally, for those interested in learning more about muscle research, including electron microscopy of tissue, an excellent treatment can be found in

J. Squire, *The Structural Basis of Muscular Contraction*. New York: Plenum Press, 1981.

EXERCISES

18.1 Trigonometric Form of the Fourier Series

1. Find the first five harmonic frequencies ($n = 1 - 5$) of the following waveforms: (a) $v_1(t) = 77 \cos(5t)$ V; (b) $i(t) = 32 \sin(5t)$ nA; (c) $q(t) = 4\cos(90t - 85°)$ C.

2. State the period and the fundamental frequency of each of the following: (a) $q(t) = 8.5 \cos(2\pi t)$ nC; (b) $v(t) = 9 \sin(5.95t)$ MV; (c) $i(t) = 1.113 \cos(t - 45°)$ pA.

3. Let $v(t) = 3 - 3\cos(100\pi t - 40°) + 4\sin(200\pi t - 10°) + 2.5 \cos 300\pi t$ V. Find (a) V_{av}; (b) V_{eff}; (c) T; (d) $v(18\text{ ms})$.

4. (a) Make a sketch of the voltage waveform $v(t) = 2 \cos 2\pi t + 1.8 \sin 4\pi t$ in the interval $0 < t < T$. (b) Find the maximum value of $v(t)$ in this interval. (c) Find the magnitude of the most negative value of $v(t)$ in this interval.

5. Calculate a_0 for the following functions: (a) $5 \cos 100t$; (b) $5 \sin 100t$; (c) $5 + \cos 100t$; (d) $5 + \sin 100t$.

6. Calculate a_0 for the following functions: (a) $100 \cos(5t - 18°)$; (b) $100 \sin(5t - 18°)$; (c) $100 + 100 \cos(5t - 18°)$; (d) $100 + 100 \sin(5t - 18°)$.

7. Calculate a_0, a_1, a_2, b_1, and b_2 for $f(t) = $ (a) 3; (b) $3 \cos 3t$; (c) $3 \sin 3t$; (d) $3 \cos(3t - 10°)$.

8. Calculate a_0, a_1, a_2, b_1, and b_2 for $f(t) = 5u(t - 1) - 5u(t - 2) + 5u(t - 3) - 5u(t - 4) + \ldots$.

9. Calculate a_0, a_1, a_2, a_3, b_1, b_2, and b_3 for $g(t) = 2u(t) - 2u(t - 2) + 2u(t - 3) - 2u(t - 5) + \ldots$.

10. Determine numerical values for a_0, a_1, a_2, a_3, b_1, b_2, and b_3 for $h(t) = -3 + 8 \sin 2t + f(t)$, where $f(t) = u(t - 1) - u(t - 2) + u(t - 3) - u(t - 4) + \ldots$.

11. The waveform shown in Fig. 18.28 is periodic with $T = 10$ s. Find (a) the average value; (b) the effective value; (c) the value of a_3.

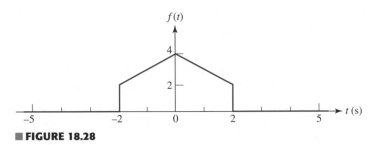

■ **FIGURE 18.28**

12. For the periodic waveform illustrated in Fig. 18.29, find (a) T; (b) f_0; (c) ω_0; (d) a_0; (e) b_2.

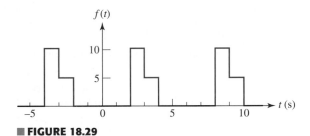

■ **FIGURE 18.29**

13. Find a_3, b_3, and $\sqrt{a_3^2 + b_3^2}$ for the waveform shown in Fig. 18.29.

14. Obtain the trigonometric form of the Fourier series, give the value of T, and determine the average value for each of these periodic functions of time: (a) $3.8 \cos^2 80\pi t$; (b) $3.8 \cos^3 80\pi t$; (c) $3.8 \cos 79\pi t - 3.8 \sin 80\pi t$.

15. A periodic function of time with $T = 2$ s has the following values: $f(t) = 0$, $-1 < t < 0$; $f(t) = 1$, $0 < t < t_1$; and $f(t) = 0$, $t_1 < t < 1$. (a) What value of t_1 will maximize b_4? (b) Find $b_{4,\max}$.

16. Let an electrical signal be described by $g(t) = -5 + 8 \cos 10t - 5 \cos 15t + 3 \cos 20t - 8 \sin 10t - 4 \sin 15t + 2 \sin 20t$. Find (a) the period of $g(t)$; (b) the bandwidth (in hertz) of the signal; (c) the average value of $g(t)$; (d) the effective value of $g(t)$; (e) the discrete amplitude and phase spectra of the signal.

17. The waveform of Example 18.1 (shown in Fig. 18.2) is the output of a half-wave rectifier. If the half sinusoids occupy all the intervals $-0.5 < t < -0.3$, $-0.3 < t < -0.1$, $-0.1 < t < 0.1$, and so forth, then the output is that of a full-wave rectifier. Find the trigonometric Fourier series for this case.

18.2 The Use of Symmetry

18. (a) Specify the types of symmetry present in the waveform of Fig. 18.30.
    (b) Which of the a_n, b_n, or a_0 are zero? (c) Calculate a_1, b_1, a_2, b_2, a_3, and b_3.

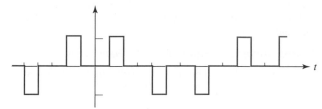

■ **FIGURE 18.30**

19. The periodic function $y(t)$ is known to have odd symmetry and the amplitude spectrum shown in Fig. 18.31. If all the a_n and b_n are nonnegative: (a) determine the Fourier series for $y(t)$; (b) find the effective value of $y(t)$; (c) calculate the value of $y(0.2 \text{ ms})$.

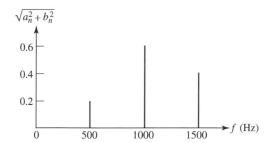

■ **FIGURE 18.31**

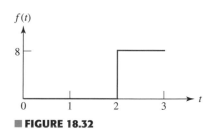

■ **FIGURE 18.32**

20. Use the waveform given for $f(t)$ over the interval $0 < t < 3$ in Fig. 18.32 to sketch a new function $g(t)$ that is equal to $f(t)$ for $0 < t < 3$ but also has (a) $T = 6$ and even symmetry; (b) $T = 6$ and odd symmetry; (c) $T = 12$, and even and half-wave symmetry; (d) $T = 12$, and odd and half-wave symmetry. (e) Evaluate a_5 and b_5 for each case.

21. The waveform shown in Fig. 18.33 repeats every 4 ms. (a) Find the dc component a_0. (b) Specify the values of a_1 and b_1. (c) Specify a function $f_x(t)$ that equals $f(t)$ in the 4 ms interval shown, but has a period of 8 ms and shows even symmetry. (d) Find a_1 and b_1 for $f_x(t)$.

22. Make use of symmetry as much as possible to obtain numerical values for a_0, a_n, and b_n, $1 \le n \le 10$, for the waveform shown in Fig. 18.34.

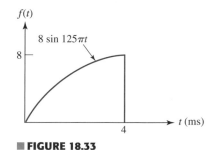

■ **FIGURE 18.33**

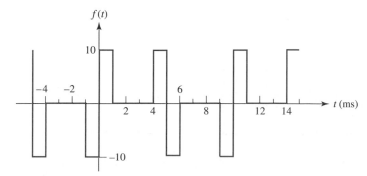

■ **FIGURE 18.34**

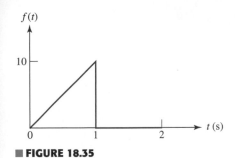

■ FIGURE 18.35

23. A function $f(t)$ has both odd and half-wave symmetry. The period is 8 ms. It is also known that $f(t) = 10^3 t$, $0 < t < 1$ ms, and $f(t) = 0$, $1 < t < 2$ ms. Find values for b_n, $1 \le n \le 5$.

24. A portion of $f(t)$ is shown in Fig. 18.35. Show $f(t)$ over the interval $0 < t < 8$ s if $f(t)$ has (a) odd symmetry and $T = 4$ s; (b) even symmetry and $T = 4$ s; (c) odd and half-wave symmetry and $T = 8$ s; (d) even and half-wave symmetry and $T = 8$ s.

18.3 Complete Response to Periodic Forcing Functions

25. Replace the square wave of Fig. 18.8a with that shown in Fig. 18.36 and repeat the analysis of Example 18.2 to obtain a new expression for (a) $i_f(t)$; (b) $i(t)$.

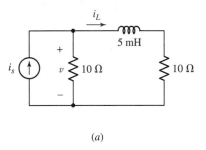

■ FIGURE 18.36

26. The waveform for $v_s(t)$ shown in Fig. 18.36 is applied to the circuit of Fig. 18.8b. Use the standard methods of transient analysis to calculate $i(t)$ at t equal to (a) 0.2π s; (b) 0.4π s; (c) 0.6π s.

27. An ideal voltage source v_s, an open switch, a 2 Ω resistor, and a 2 F capacitor are in series. The source voltage is shown in Fig. 18.36. The switch closes at $t = 0$ and the capacitor voltage is the desired response. (a) Work in the frequency domain of the nth harmonic to find the forced response as a trigonometric Fourier series. (b) Specify the functional form of the natural response. (c) Determine the complete response.

28. The circuit of Fig. 18.37a is subjected to the waveform depicted in Fig. 18.37b. Determine the steady-state voltage $v(t)$.

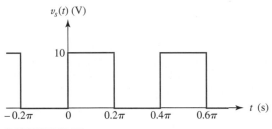

(a)

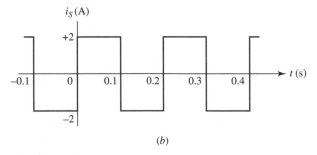

(b)

■ FIGURE 18.37

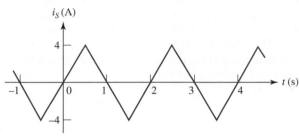

29. The circuit of Fig. 18.37a is subjected to the periodic waveform shown in Fig. 18.38. Determine the steady-state current $i_L(t)$.

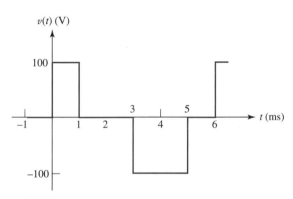

■ **FIGURE 18.38**

18.4 Complex Form of the Fourier Series

30. Let $T = 6$ ms for the periodic waveform shown in Fig. 18.39. Find $\mathbf{c}_3$, $\mathbf{c}_{-3}$, $|\mathbf{c}_3|$, a_3, b_3, and $\sqrt{a_3^2 + b_3^2}$.

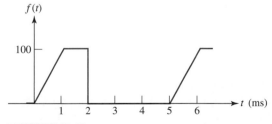

■ **FIGURE 18.39**

31. (a) Find the complex Fourier series for the periodic waveform shown in Fig. 18.40. (b) Give numerical values for $\mathbf{c}_n$, $n = 0, \pm 1$, and ± 2.

■ **FIGURE 18.40**

32. The pulse shown in Fig. 18.11 has an amplitude of 8 V, a duration of 0.2 μs, and a repetition rate of 6000 pulses per second. (a) Find the frequency at which the envelope of the frequency spectrum has an amplitude of zero. (b) Determine the frequency separation of the spectral lines. (c) Find $|c_n|$ for that spectral component closest to 20 kHz. (d) ... closest to 2 MHz. (e) Specify the

nominal bandwidth that an amplifier should have to transmit this pulse train with reasonable fidelity. (*f*) State the number of spectral components in the frequency range $2 < \omega < 2.2$ Mrad/s. (*g*) Calculate the amplitude of c_{227} and state its frequency.

33. A voltage waveform has a period $T = 5$ ms and complex coefficient values: $c_0 = 1$, $c_1 = 0.2 - j0.2$, $c_2 = 0.5 + j0.25$, $c_3 = -1 - j2$, and $c_n = 0$ for $|n| \geq 4$. (*a*) Find $v(t)$. (*b*) Calculate $v(1 \text{ ms})$.

34. A pulse sequence has a period of 5 μs, an amplitude of unity for $-0.6 < t < -0.4$ μs and for $0.4 < t < 0.6$ μs, and zero amplitude elsewhere in the period interval. This series of pulses might represent the decimal number 3 being transmitted in binary form by a digital computer. (*a*) Find c_n. (*b*) Evaluate c_4. (*c*) Evaluate c_0. (*d*) Find $|c_n|_{max}$. (*e*) Find N so that $|c_n| \leq 0.1|c_n|_{max}$ for all $n > N$. (*f*) What bandwidth is required to transmit this portion of the spectrum?

35. Let a periodic voltage $v_s(t) = 40$ V for $0 < t < \frac{1}{96}$ s, and 0 for $\frac{1}{96} < t < \frac{1}{16}$ s. If $T = \frac{1}{16}$ s, find (*a*) c_3; (*b*) the power delivered to the load in the circuit of Fig. 18.41.

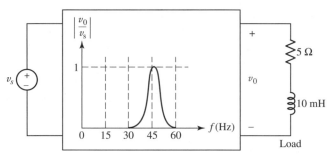

■ **FIGURE 18.41**

18.5 Definition of the Fourier Transform

36. Given the time function $f(t) = 5[u(t+3) + u(t+2) - u(t-2) - u(t-3)]$: (*a*) sketch $f(t)$; (*b*) use the definition of the Fourier transform to find $\mathbf{F}(j\omega)$.

37. Use the defining equations for the Fourier transform to find $\mathbf{F}(j\omega)$ if $f(t)$ equals (*a*) $e^{-at}u(t)$, $a > 0$; (*b*) $e^{-a(t-t_0)}u(t-t_0)$, $a > 0$; (*c*) $te^{-at}u(t)$, $a > 0$.

38. Find the Fourier transform of the single triangular pulse in Fig. 18.42.

39. Find the Fourier transform of the single sinusoidal pulse in Fig. 18.43.

40. Let $f(t) = (8\cos t)[u(t+0.5\pi) - u(t-0.5\pi)]$. Calculate $\mathbf{F}(j\omega)$ for ω equal to (*a*) 0; (*b*) 0.8; (*c*) 3.1.

41. Use the defining equations for the inverse Fourier transform to find $f(t)$, and then evaluate it at $t = 0.8$ for $\mathbf{F}(j\omega)$ equal to (*a*) $4[u(\omega+2) - u(\omega-2)]$; (*b*) $4e^{-2|\omega|}$; (*c*) $(4\cos\pi\omega)[u(\omega+0.5) - u(\omega-0.5)]$.

18.6 Some Properties of the Fourier Transform

42. Given the voltage $v(t) = 20e^{1.5t}u(-t-2)$ V, find (*a*) $\mathbf{F}_v(j0)$; (*b*) $A_v(2)$; (*c*) $B_v(2)$; (*d*) $|\mathbf{F}_v(j2)|$; (*e*) $\phi_v(2)$.

43. Let $i(t)$ be the time-varying current through a 4 Ω resistor. If the magnitude of the Fourier transform of $i(t)$ is known to be $|\mathbf{I}(j\omega)| = (3\cos 10\omega) [u(\omega + 0.05\pi) - u(\omega - 0.05\pi)]$ A/(rad/s), find (*a*) the total energy present in the signal; (*b*) the frequency ω_x such that half the total energy lies in the range $|\omega| < \omega_x$.

44. Let $f(t) = 10te^{-4t}u(t)$, and find (*a*) the 1 Ω energy represented by that signal; (*b*) $|\mathbf{F}(j\omega)|$; (*c*) the energy density at $\omega = 0$ and $\omega = 4$ rad/s.

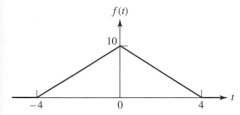

■ **FIGURE 18.42**

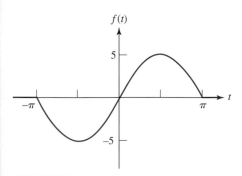

■ **FIGURE 18.43**

45. If $v(t) = 8e^{-2|t|}$ V, find (a) the 1 Ω energy associated with this signal; (b) $|\mathbf{F}_v(j\omega)|$; (c) the frequency range $|\omega| < \omega_1$ in which 90 percent of the 1 Ω energy lies.

18.7 Fourier Transform Pairs for Some Simple Time Functions

46. Use the definition of the Fourier transform to prove the following results, where $\mathcal{F}\{f(t)\} = \mathbf{F}(j\omega)$: (a) $\mathcal{F}\{f(t - t_0)\} = e^{-j\omega t_0}\mathcal{F}\{f(t)\}$; (b) $\mathcal{F}\{df(t)/dt\} = j\omega\mathcal{F}\{f(t)\}$; (c) $\mathcal{F}\{f(kt)\} = (1/|k|)\mathbf{F}(j\omega/k)$; (d) $\mathcal{F}\{f(-t)\} = \mathbf{F}(-j\omega)$; (e) $\mathcal{F}\{tf(t)\} = j\,d[\mathbf{F}(j\omega)]/d\omega$.

47. Find $\mathcal{F}\{f(t)\}$ if $f(t)$ is given by (a) $4[\text{sgn}(t)]\delta(t - 1)$; (b) $4[\text{sgn}(t - 1)]\delta(t)$; (c) $4[\sin(10t - 30°)]$.

48. Find $\mathbf{F}(j\omega)$ if $f(t)$ equals (a) $A\cos(\omega_0 t + \phi)$; (b) $3\,\text{sgn}(t - 2) - 2\delta(t) - u(t - 1)$; (c) $(\sinh kt)u(t)$.

49. Find $f(t)$ at $t = 5$ if $\mathbf{F}(j\omega)$ equals (a) $3u(\omega + 3) - 3u(\omega - 1)$; (b) $3u(-3 - \omega) + 3u(\omega - 1)$; (c) $2\delta(\omega) + 3u(-3 - \omega) + 3u(\omega - 1)$.

50. Find $f(t)$ if $\mathbf{F}(j\omega)$ equals (a) $3/(1 + j\omega) + 3/j\omega + 3 + 3\delta(\omega - 1)$; (b) $(5\sin 4\omega)/\omega$; (c) $6(3 + j\omega)/[(3 + j\omega)^2 + 4]$.

18.8 The Fourier Transform of a General Periodic Time Function

51. Find the Fourier transform of the periodic time function shown in Fig. 18.44.

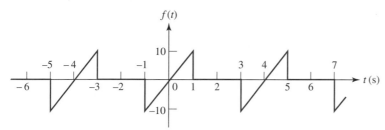

■ **FIGURE 18.44**

52. The periodic function $f(t)$ is defined over the period $0 < t < 4$ ms by $f_1(t) = 10u(t) - 6u(t - 0.001) - 4u(t - 0.003)$. Find $\mathbf{F}(j\omega)$.

53. If $\mathbf{F}(j\omega) = 20\sum_{n=1}^{\infty}[1/(|n|! + 1)]\delta(\omega - 20n)$, find the value of $f(0.05)$.

54. Given an input $x(t) = 5[u(t) - u(t - 1)]$, use convolution to find the output $y(t)$ if $h(t)$ equals (a) $2u(t)$; (b) $2u(t - 1)$; (c) $2u(t - 2)$.

55. Let $x(t) = 5[u(t) - u(t - 2)]$ and $h(t) = 2[u(t - 1) - u(t - 2)]$. Find $y(t)$ at $t = -0.4, 0.4, 1.4, 2.4, 3.4$, and 4.4 using convolution.

18.9 The System Function and Response in the Frequency Domain

56. The impulse response of a certain linear system is $h(t) = 3(e^{-t} - e^{-2t})$. Given the input $x(t) = u(t)$, find the output for $t > 0$.

57. The unit-impulse response and the input to a certain linear system are shown in Fig. 18.45. (a) Obtain an integral expression for the output that is valid in the interval $4 < t < 6$ and does not contain any singularity functions. (b) Evaluate the output at $t = 5$.

58. Given an input signal $x(t) = 5e^{-(t-2)}u(t - 2)$ and the impulse response $h(t) = (4t - 16)[u(t - 4) - u(t - 7)]$, find the value of the output signal at (a) $t = 5$; (b) $t = 8$; (c) $t = 10$.

59. When an input $\delta(t)$ is applied to a linear system, the ouput is $\sin t$ for $0 < t < \pi$, and zero elsewhere. Now, if the input $e^{-t}u(t)$ is applied, specify the numerical value of the output at t equal to (a) 1; (b) 2.5; (c) 4.

60. Let $x(t) = 0.8(t - 1)[u(t - 1) - u(t - 3)]$ and $h(t) = 0.2(t - 2)[u(t - 2) - u(t - 3)]$. Evaluate $y(t)$ for (a) $t = 3.8$; (b) $t = 4.8$.

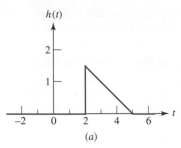
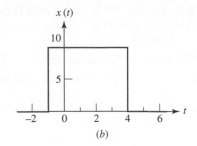

(a)   (b)

■ **FIGURE 18.45**

61. A signal $x(t) = 10e^{-2t}u(t)$ is applied to a linear system for which the impulse response is $h(t) = 10e^{-2t}u(t)$. Find the output $y(t)$.

62. An impulse is applied to a linear system, generating the output $h(t) = 5e^{-4t}\,u(t)$ V. What percentage of the 1 Ω energy in this response: (a) occurs during the time interval $0.1 < t < 0.8$ s? (b) lies in the frequency band $-2 < \omega < 2$ rad/s?

63. If $\mathbf{F}(j\omega) = 2/[(1 + j\omega)(2 + j\omega)]$, find (a) the total 1 Ω energy present in the signal, and (b) the maximum value of $f(t)$.

64. Find $\mathcal{F}^{-1}[\mathbf{F}(j\omega)]$ if $\mathbf{F}(j\omega)$ equals (a) $1/[(j\omega)(2 + j\omega)(3 + j\omega)]$; (b) $(1 + j\omega)/[(j\omega)(2 + j\omega)(3 + j\omega)]$; (c) $(1 + j\omega)^2/[(j\omega)(2 + j\omega)(3 + j\omega)]$; (d) $(1 + j\omega)^3/[(j\omega)(2 + j\omega)(3 + j\omega)]$.

65. Let's develop a network having an impulse response $h(t) = 2e^{-t}u(t)$.
(a) Determine $\mathbf{H}(j\omega) = \mathbf{V}_o(j\omega)/\mathbf{V}_i(j\omega)$. (b) By inspecting either $h(t)$ or $\mathbf{H}(j\omega)$, note that the network has a single energy-storage element. Arbitrarily selecting an RC circuit with $R = 1$ Ω, $C = 1$ F to provide the necessary time constant, determine the form of the circuit to give $\frac{1}{2}h(t)$ or $\frac{1}{2}\mathbf{H}(j\omega)$. (c) Place an ideal voltage amplifier in cascade with the network to provide the proper multiplicative constant. What is the gain of the amplifier?

18.10 The Physical Significance of the System Function

66. Find $v_o(t)$ for the circuit of Fig. 18.46.

67. Find $v_C(t)$ for the circuit illustrated in Fig. 18.47.

68. Let $f(t) = 5e^{-2t}u(t)$ and $g(t) = 4e^{-3t}u(t)$. (a) Find $f(t) * g(t)$ by working in the time domain. (b) Find $f(t) * g(t)$ by using multiplication in the frequency domain.

69. The voltage source in Fig. 18.22 is replaced by $v_i(t) = 12$ sgn(t). Determine $v_o(t)$, the voltage across the inductor, using Fourier transform techniques.

70. A particular system has impulse response $h(t) = 2e^{-t}\cos 4t$. Find the output if the input is (a) 2; (b) $2\delta(t - 1)$; (c) $2u(t + 0.25) - 2u(t - 0.25)$.

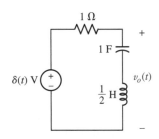

■ **FIGURE 18.46**

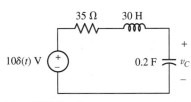

■ **FIGURE 18.47**

AN INTRODUCTION TO NETWORK TOPOLOGY

After working many circuits problems, it slowly becomes evident that many of the circuits we see have quite a bit in common, at least in terms of the arrangement of components. From this realization, it is possible to create a more abstract view of circuits which we call *network topology*. This appendix presents an introduction to several basic concepts of network topology; implementation is left to the reader's discretion.

A1.1 TREES AND GENERAL NODAL ANALYSIS

We now plan to generalize the method of nodal analysis that we have come to know and love. Since nodal analysis is applicable to any network, we cannot promise that we will be able to solve a wider class of circuit problems. We can, however, look forward to being able to select a general nodal analysis method for any particular problem that may result in fewer equations and less work.

We must first extend our list of definitions relating to network topology. We begin by defining *topology* itself as a branch of geometry which is concerned with those properties of a geometrical figure which are unchanged when the figure is twisted, bent, folded, stretched, squeezed, or tied in knots, with the provision that no parts of the figure are to be cut apart or to be joined together. A sphere and a tetrahedron are topologically identical, as are a square and a circle. In terms of electric circuits, then, we are not now concerned with the particular types of elements appearing in the circuit, but only with the way in which branches and nodes are arranged. As a matter of fact, we usually suppress the nature of the elements and simplify the drawing of the circuit by showing the elements as lines. The resultant drawing is called a linear graph, or simply a graph. A circuit and its graph are shown in Fig. A1.1. Note that all nodes are identified by heavy dots in the graph.

Since the topological properties of the circuit or its graph are unchanged when it is distorted, the three graphs shown in Fig. A1.2 are all topologically identical with the circuit and graph of Fig. A1.1.

Topological terms that we already know and have been using correctly are

Node: A point at which two or more elements have a common connection.
Path: A set of elements that may be traversed in order without passing through the same node twice.
Branch: A single path, containing one simple element, which connects one node to any other node.

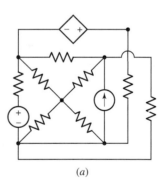

(a)

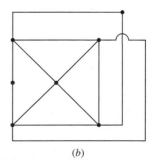

(b)

■ **FIGURE A1.1** (*a*) A given circuit. (*b*) The linear graph of this circuit.

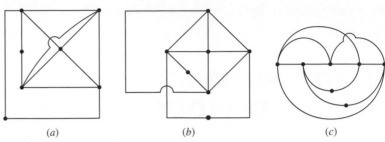

■ FIGURE A1.2 (*a*, *b*, *c*) Alternative linear graphs of the circuit of Fig. A1.1.

Loop: A closed path.

Mesh: A loop which does not contain any other loops within it.

Planar circuit: A circuit which may be drawn on a plane surface in such a way that no branch passes over or under any other branch.

Nonplanar circuit: Any circuit which is not planar.

The graphs of Fig. A1.2 each contain 12 branches and 7 nodes.

Three new properties of a linear graph must now be defined—a *tree,* a *cotree,* and a *link.* We define a *tree* as any set of branches which does not contain any loops and yet connects every node to every other node, not necessarily directly. There are usually a number of different trees which may be drawn for a network, and the number increases rapidly as the complexity of the network increases. The simple graph shown in Fig. A1.3*a* has eight possible trees, four of which are shown by heavy lines in Fig. A1.3*b*, *c*, *d*, and *e*.

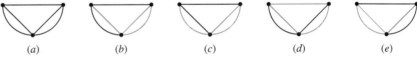

■ FIGURE A1.3 (*a*) The linear graph of a three-node network. (*b*, *c*, *d*, *e*) Four of the eight different trees which may be drawn for this graph are shown by the black lines.

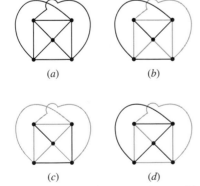

(*a*)   (*b*)

(*c*)   (*d*)

■ FIGURE A1.4 (*a*) A linear graph. (*b*) A possible tree for this graph. (*c*, *d*) These sets of branches do not satisfy the definition of a tree.

In Fig. A1.4*a* a more complex graph is shown. Figure A1.4*b* shows one possible tree, and Fig. A1.4*c* and *d* show sets of branches which are not trees because neither set satisfies the definition.

After a tree has been specified, those branches that are not part of the tree form the *cotree,* or complement of the tree. The lightly drawn branches in Fig. A1.3*b* to *e* show the cotrees that correspond to the heavier trees.

Once we understand the construction of a tree and its cotree, the concept of the *link* is very simple, for a link is any branch belonging to the cotree. It is evident that any particular branch may or may not be a link, depending on the particular tree which is selected.

The number of links in a graph may easily be related to the number of branches and nodes. If the graph has N nodes, then exactly $(N - 1)$ branches are required to construct a tree because the first branch chosen connects two nodes and each additional branch includes one more node.

Thus, given B branches, the number of links L must be

$$L = B - (N - 1)$$

or

$$L = B - N + 1 \qquad [1]$$

There are L branches in the cotree and $(N - 1)$ branches in the tree.

In any of the graphs shown in Fig. A1.3, we note that $3 = 5 - 3 + 1$, and in the graph of Fig. A1.4b, $6 = 10 - 5 + 1$. A network may be in several disconnected parts, and Eq. [1] may be made more general by replacing $+1$ with $+S$, where S is the number of separate parts. However, it is also possible to connect two separate parts by a single conductor, thus causing two nodes to form one node; no current can flow through this single conductor. This process may be used to join any number of separate parts, and thus we will not suffer any loss of generality if we restrict our attention to circuits for which $S = 1$.

We are now ready to discuss a method by which we may write a set of nodal equations that are independent and sufficient. The method will enable us to obtain many different sets of equations for the same network, and all the sets will be valid. However, the method does not provide us with every possible set of equations. Let us first describe the procedure, illustrate it by three examples, and then point out the reason that the equations are independent and sufficient.

Given a network, we should:

1. Draw a graph and then identify a tree.
2. Place all voltage sources in the tree.
3. Place all current sources in the cotree.
4. Place all control-voltage branches for voltage-controlled dependent sources in the tree, if possible.
5. Place all control-current branches for current-controlled dependent sources in the cotree, if possible.

These last four steps effectively associate voltages with the tree and currents with the cotree.

We now assign a voltage variable (with its plus-minus pair) across each of the $(N - 1)$ branches in the tree. A branch containing a voltage source (dependent or independent) should be assigned that source voltage, and a branch containing a controlling voltage should be assigned that controlling voltage. The number of new variables that we have introduced is therefore equal to the number of branches in the tree $(N - 1)$, reduced by the number of voltage sources in the tree, and reduced also by the number of control voltages we were able to locate in the tree. In Example A1.3, we will find that the number of new variables required may be zero.

Having a set of variables, we now need to write a set of equations that are sufficient to determine these variables. The equations are obtained through the application of KCL. Voltage sources are handled in the same way that they were in our earlier attack on nodal analysis; each voltage source and the two nodes at its terminals constitute a supernode or a part of a supernode. Kirchhoff's current law is then applied at all but one of the remaining nodes and supernodes. We set the sum of the currents leaving the node in all of the branches connected to it equal to zero. Each current is expressed in terms of the voltage variables we just assigned. One node may be ignored, just as was the case earlier for the reference node. Finally, in case there are current-controlled dependent sources, we must write an equation for each control current that relates it to the voltage variables; this also is no different from the procedure used before with nodal analysis.

Let us try out this process on the circuit shown in Fig. A1.5*a*. It contains four nodes and five branches, and its graph is shown in Fig. A1.5*b*.

EXAMPLE **A1.1**

Find the value of v_x in the circuit of Fig. A1.5*a*.

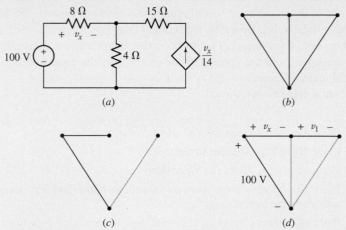

■ **FIGURE A1.5** (*a*) A circuit used as an example for general nodal analysis. (*b*) The graph of the given circuit. (*c*) The voltage source and the control voltage are placed in the tree, while the current source goes in the cotree. (*d*) The tree is completed and a voltage is assigned across each tree branch.

In accordance with steps 2 and 3 of the tree-drawing procedure, we place the voltage source in the tree and the current source in the cotree. Following step 4, we see that the v_x branch may also be placed in the tree, since it does not form any loop which would violate the definition of a tree. We have now arrived at the two tree branches and the single link shown in Fig. A1.5*c*, and we see that we do not yet have a tree, since the right node is not connected to the others by a path through tree branches. The only possible way to complete the tree is shown in Fig. A1.5*d*. The 100-V source voltage, the control voltage v_x, and a new voltage variable v_1 are next assigned to the three tree branches as shown.

We therefore have two unknowns, v_x and v_1, and we need to obtain two equations in terms of them. There are four nodes, but the presence of the voltage source causes two of them to form a single supernode. Kirchhoff's current law may be applied at any two of the three remaining nodes or supernodes. Let's attack the right node first. The current leaving to the left is $-v_1/15$, while that leaving downward is $-v_x/14$. Thus, our first equation is

$$-\frac{v_1}{15} + \frac{-v_x}{14} = 0$$

The central node at the top looks easier than the supernode, and so we set the sum of the current to the left $(-v_x/8)$, the current to the right $(v_1/15)$, and the downward current through the 4-Ω resistor equal to zero. This latter current is given by the voltage across the resistor divided by 4 Ω, but there is no voltage labeled on that link. However, when a tree is constructed according to the definition, there is a path through it from any node to any other node. Then, since every branch in the tree is assigned a voltage, we may express the voltage across any link in terms of the tree-branch voltages. This downward current is therefore $(-v_x + 100)/4$, and we have the second equation,

$$-\frac{v_x}{8} + \frac{v_1}{15} + \frac{-v_x + 100}{4} = 0$$

The simultaneous solution of these two nodal equations gives

$$v_1 = -60 \text{ V} \qquad v_x = 56 \text{ V}$$

EXAMPLE **A1.2**

Find the values of v_x and v_y in the circuit of Fig. A1.6a.

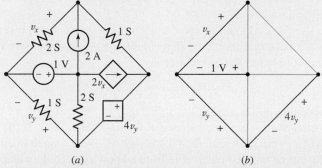

■ **FIGURE A1.6** (a) A circuit with 5 nodes. (b) A tree is chosen such that both voltage sources and both control voltages are tree branches.

We draw a tree so that both voltage sources and both control voltages appear as tree-branch voltages and, hence, as assigned variables. As it happens, these four branches constitute a tree, Fig. A1.6b, and tree-branch voltages v_x, 1, v_y, and $4v_y$ are chosen, as shown.

(Continued on next page)

Both voltage sources define supernodes, and we apply KCL twice, once to the top node,

$$2v_x + 1(v_x - v_y - 4v_y) = 2$$

and once to the supernode consisting of the right node, the bottom node, and the dependent voltage source,

$$1v_y + 2(v_y - 1) + 1(4v_y + v_y - v_x) = 2v_x$$

Instead of the four equations we would expect using previously studied techniques, we have only two, and we find easily that $v_x = \frac{26}{9}$ V and $v_y = \frac{4}{3}$ V.

EXAMPLE A1.3

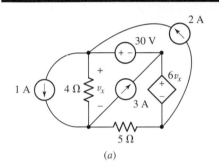

(a)

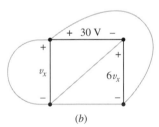

(b)

■ **FIGURE A1.7** (a) A circuit for which only one general nodal equation need be written. (b) The tree and the tree-branch voltages used.

Find the value of v_x in the circuit of Fig. A1.7a.

The two voltage sources and the control voltage establish the three-branch tree shown in Fig. A1.7b. Since the two upper nodes and the lower right node all join to form one supernode, we need write only one KCL equation. Selecting the lower left node, we have

$$-1 - \frac{v_x}{4} + 3 + \frac{-v_x + 30 + 6v_x}{5} = 0$$

and it follows that $v_x = -\frac{32}{3}$ V. In spite of the apparent complexity of this circuit, the use of general nodal analysis has led to an easy solution. Employing mesh currents or node-to-reference voltages would require more equations and more effort.

We will discuss the problem of finding the best analysis scheme in the following section.

If we should need to know some other voltage, current, or power in the previous example, one additional step would give the answer. For example, the power provided by the 3-A source is

$$3\left(-30 - \tfrac{32}{3}\right) = -122 \text{ W}$$

Let us conclude by discussing the sufficiency of the assumed set of tree-branch voltages and the independence of the nodal equations. If these tree-branch voltages are sufficient, then the voltage of every branch in either the tree or the cotree must be obtainable from a knowledge of the values of all the tree-branch voltages. This is certainly true for those branches in the tree. For the links we know that each link extends between two nodes, and, by definition, the tree must also connect those two nodes. Hence, every link voltage may also be established in terms of the tree-branch voltages.

Once the voltage across every branch in the circuit is known, then all the currents may be found by using either the given value of the current if the branch consists of a current source, by using Ohm's law if it is a resistive branch, or by using KCL and these current values if the branch happens to be a voltage source. Thus, all the voltages and currents are determined and sufficiency is demonstrated.

To demonstrate independency, let us satisfy ourselves by assuming the situation where the only sources in the network are independent current sources. As we have noticed earlier, independent voltage sources in the circuit result in fewer equations, while dependent sources usually necessitate a greater number of equations. With independent current sources only, there will then be precisely $(N - 1)$ nodal equations written in terms of $(N - 1)$ tree-branch voltages. To show that these $(N - 1)$ equations are independent, visualize the application of KCL to the $(N - 1)$ different nodes. Each time we write the KCL equation, there is a new tree branch involved—the one which connects that node to the remainder of the tree. Since that circuit element has not appeared in any previous equation, we must obtain an independent equation. This is true for each of the $(N - 1)$ nodes in turn, and hence we have $(N - 1)$ independent equations.

PRACTICE

A1.1 (*a*) How many trees may be constructed for the circuit of Fig. A1.8 that follow all five of the tree-drawing suggestions listed earlier? (*b*) Draw a suitable tree, write two equations in two unknowns, and find i_3. (*c*) What power is supplied by the dependent source?

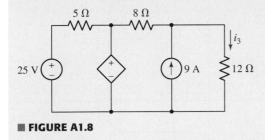

■ **FIGURE A1.8**

Ans: 1; 7.2 A; 547 W.

A1.2 LINKS AND LOOP ANALYSIS

Now we will consider the use of a tree to obtain a suitable set of loop equations. In some respects this is the *dual* of the method of writing nodal equations. Again it should be pointed out that, although we are able to guarantee that any set of equations we write will be both sufficient and independent, we should not expect that the method will lead directly to every possible set of equations.

We again begin by constructing a tree, and we use the same set of rules as we did for general nodal analysis. The objective for either nodal or loop analysis is to place voltages in the tree and currents in the cotree; this is a mandatory rule for sources and a desirable rule for controlling quantities.

Now, however, instead of assigning a voltage to each branch in the tree, we assign a current (including reference arrow, of course) to each element in the cotree or to each link. If there were 10 links, we would assign exactly 10 link currents. Any link that contains a current source is assigned that source current as the link current. Note that each link current may also be thought of as a loop current, for the link must extend between two specific nodes, and there must also be a path between those same two nodes through the tree. Thus, with each link there is associated a single specific loop that includes that one link and a unique path through the tree. It is evident that the assigned current may be thought of either as a loop current or as a link current. The link connotation is most helpful at the time the currents are being defined, for one must be established for each link; the loop interpretation is more convenient at equation-writing time, because we will apply KVL around each loop.

Let us try out this process of defining link currents by considering the circuit shown in Fig. A1.9*a*. The tree selected is one of several that might be constructed for which the voltage source is in a tree branch and the current source is in a link. Let us first consider the link containing the current source. The loop associated with this link is the left-hand mesh, and so we show our link current flowing about the perimeter of this mesh (Fig. A1.9*b*). An obvious choice for the symbol for this link current is "7 A." Remember that no other current can flow through this particular link, and thus its value must be exactly the strength of the current source.

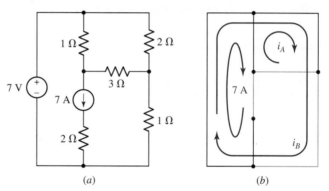

FIGURE A1.9 (*a*) A simple circuit. (*b*) A tree is chosen such that the current source is in a link and the voltage source is in a tree branch.

We next turn our attention to the 3-Ω resistor link. The loop associated with it is the upper right-hand mesh, and this loop (or mesh) current is defined as i_A and also shown in Fig. A1.9*b*. The last link is the lower 1-Ω resistor, and the only path between its terminals through the tree is around the perimeter of the circuit. That link current is called i_B, and the arrow indicating its path and reference direction appears in Fig. A1.9*b*. It is not a mesh current.

Note that each link has only one current present in it, but a tree branch may have any number from 1 to the total number of link currents assigned. The use of long, almost closed, arrows to indicate the loops helps to indicate which loop currents flow through which tree branch and what their reference directions are.

A KVL equation must now be written around each of these loops. The variables used are the assigned link currents. Since the voltage across a current source cannot be expressed in terms of the source current, and since we have already used the value of the source current as the link current, we discard any loop containing a current source.

For the example of Fig. A1.9, find the values of i_A and i_B.

We first traverse the i_A loop, proceeding clockwise from its lower left corner. The current going our way in the 1-Ω resistor is $(i_A - 7)$, in the 2-Ω element it is $(i_A + i_B)$, and in the link it is simply i_B. Thus

$$1(i_A - 7) + 2(i_A + i_B) + 3i_A = 0$$

For the i_B link, clockwise travel from the lower left corner leads to

$$-7 + 2(i_A + i_B) + 1i_B = 0$$

Traversal of the loop defined by the 7-A link is not required. Solving, we have $i_A = 0.5$ A, $i_B = 2$ A, once again. The solution has been achieved with one less equation than before!

Evaluate i_1 in the circuit shown in Fig. A1.10a.

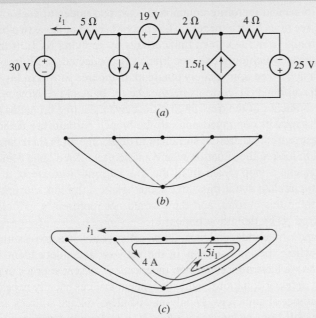

■ **FIGURE A1.10** (a) A circuit for which i_1 may be found with one equation using general loop analysis. (b) The only tree that satisfies the rules outlined in Sec. A1.1. (c) The three link currents are shown with their loops.

This circuit contains six nodes, and its tree therefore must have five branches. Since there are eight elements in the network, there are three links in the cotree. If we place the three voltage sources in the tree and the two current sources and the current control in the cotree, we are led to the tree shown in Fig. A1.10b. The source current of 4 A defines a

(Continued on next page)

loop as shown in Fig. A1.10c. The dependent source establishes the loop current $1.5i_1$ around the right mesh, and the control current i_1 gives us the remaining loop current about the perimeter of the circuit. Note that all three currents flow through the 4-Ω resistor.

We have only one unknown quantity, i_1, and after discarding the loops defined by the two current sources, we apply KVL around the outside of the circuit:

$$-30 + 5(-i_1) + 19 + 2(-i_1 - 4) + 4(-i_1 - 4 + 1.5i_1) - 25 = 0$$

Besides the three voltage sources, there are three resistors in this loop. The 5-Ω resistor has one loop current in it, since it is also a link; the 2-Ω resistor contains two loop currents; and the 4-Ω resistor has three. A carefully drawn set of loop currents is a necessity if errors in skipping currents, utilizing extra ones, or erring in choosing the correct direction are to be avoided. The foregoing equation is guaranteed, however, and it leads to $i_1 = -12$ A.

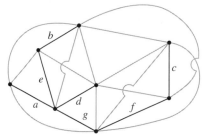

■ **FIGURE A1.11** A tree that is used as an example to illustrate the sufficiency of the link currents.

How may we demonstrate sufficiency? Let us visualize a tree. It contains no loops and therefore contains at least two nodes to each of which only one tree branch is connected. The current in each of these two branches is easily found from the known link currents by applying KCL. If there are other nodes at which only one tree branch is connected, these tree-branch currents may also be immediately obtained. In the tree shown in Fig. A1.11, we thus have found the currents in branches **a**, **b**, **c**, and **d**. Now we move along the branches of the tree, finding the currents in tree branches **e** and **f**; the process may be continued until all the branch currents are determined. The link currents are therefore sufficient to determine all branch currents. It is helpful to look at the situation where an incorrect "tree" has been drawn which contains a loop. Even if all the link currents were zero, a current might still circulate about this "tree loop." Hence, the link currents could not determine this current, and they would not represent a sufficient set. Such a "tree" is by definition impossible.

To demonstrate independence, let us satisfy ourselves by assuming the situation where the only sources in the network are independent voltage sources. As we have noticed earlier, independent current sources in the circuit result in fewer equations, while dependent sources usually necessitate a greater number of equations. If only independent voltage sources are present, there will then be precisely $(B - N + 1)$ loop equations written in terms of the $(B - N + 1)$ link currents. To show that these $(B - N + 1)$ loop equations are independent, it is necessary only to point out that each represents the application of KVL around a loop which contains one link not appearing in any other equation. We might visualize a different resistance $R_1, R_2, \ldots, R_{B-N+1}$ in each of these links, and it is then apparent that one equation can never be obtained from the others, since each contains one coefficient not appearing in any other equation.

Hence, the link currents are sufficient to enable a complete solution to be obtained, and the set of loop equations which we use to find the link currents is a set of independent equations.

Having looked at both general nodal analysis and loop analysis, we should now consider the advantages and disadvantages of each method so that an intelligent choice of a plan of attack can be made on a given analysis problem.

The nodal method in general requires $(N - 1)$ equations, but this number is reduced by 1 for each independent or dependent voltage source in a tree branch, and increased by 1 for each dependent source that is voltage-controlled by a link voltage, or current-controlled.

The loop method basically involves $(B - N + 1)$ equations. However, each independent or dependent current source in a link reduces this number by 1, while each dependent source that is current-controlled by a tree-branch current, or is voltage-controlled, increases the number by 1.

As a grand finale for this discussion, let us inspect the T-equivalent-circuit model for a transistor shown in Fig. A1.12, to which is connected a sinusoidal source, $4 \sin 1000t$ mV, and a 10-kΩ load.

EXAMPLE A1.6

Find the input (emitter) current i_e and the load voltage v_L in the circuit of Fig. A1.12, assuming typical values for the emitter resistance $r_e = 50$ Ω; the base resistance $r_b = 500$ Ω; the collector resistance $r_c = 20$ kΩ; and the common-base forward-current-transfer ratio $\alpha = 0.99$.

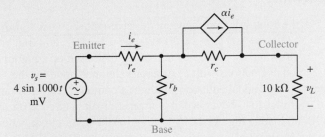

■ **FIGURE A1.12** A sinusoidal voltage source and a 10-kΩ load are connected to the T-equivalent circuit of a transistor. The common connection between the input and output is at the base terminal of the transistor, and the arrangement is called the *common-base* configuration.

Although the details are requested in the practice problems that follow, we should see readily that the analysis of this circuit might be accomplished by drawing trees requiring three general nodal equations $(N - 1 - 1 + 1)$ or two loop equations $(B - N + 1 - 1)$. We might also note that three equations are required in terms of node-to-reference voltages, as are three mesh equations.

No matter which method we choose, these results are obtained for this specific circuit:

$$i_e = 18.42 \sin 1000t \ \mu A$$
$$v_L = 122.6 \sin 1000t \ mV$$

(Continued on next page)

and we therefore find that this transistor circuit provides a voltage gain (v_L/v_s) of 30.6, a current gain $(v_L/10{,}000i_e)$ of 0.666, and a power gain equal to the product $30.6(0.666) = 20.4$. Higher gains could be secured by operating this transistor in a common-emitter configuration.

PRACTICE

A1.2 Draw a suitable tree and use general loop analysis to find i_{10} in the circuit of (a) Fig. A1.13a by writing just one equation with i_{10} as the variable; (b) Fig. A1.13b by writing just two equations with i_{10} and i_3 as the variables.

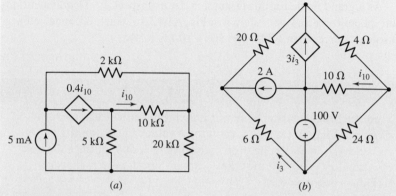

(a)   (b)

■ **FIGURE A1.13**

A1.3 For the transistor amplifier equivalent circuit shown in Fig. A1.12, let $r_e = 50\ \Omega$, $r_b = 500\ \Omega$, $r_c = 20\ \text{k}\Omega$, and $\alpha = 0.99$, and find both i_e and v_L by drawing a suitable tree and using (a) two loop equations; (b) three nodal equations with a common reference node for the voltage; (c) three nodal equations without a common reference node.

A1.4 Determine the Thévenin and Norton equivalent circuits presented to the 10-kΩ load in Fig. A1.12 by finding (a) the open-circuit value of v_L; (b) the (downward) short-circuit current; (c) the Thévenin equivalent resistance. All circuit values are given in Practice Problem A1.3.

Ans: A1.2: -4.00 mA; 4.69 A.   A1.3: $18.42 \sin 1000t\ \mu\text{A}$; $122.6 \sin 1000t$ mV. A1.4: $147.6 \sin 1000t$ mV; $72.2 \sin 1000t\ \mu\text{A}$; 2.05 kΩ.

SOLUTION OF SIMULTANEOUS EQUATIONS

Consider the simple system of equations

$$7v_1 - 3v_2 - 4v_3 = -11 \qquad [1]$$

$$-3v_1 + 6v_2 - 2v_3 = 3 \qquad [2]$$

$$-4v_1 - 2v_2 + 11v_3 = 25 \qquad [3]$$

This set of equations *could* be solved by a systematic elimination of the variables. Such a procedure is lengthy, however, and may never yield answers if done unsystematically for a greater number of simultaneous equations. Fortunately, there are many more options available to us, some of which we will explore in this appendix.

The Scientific Calculator

Perhaps the most straightforward approach when faced with a system of equations such as Eqs. [1] to [3], in which we have numerical coefficients and are only interested in the specific values of our unknowns (as opposed to algebraic relationships), is to employ any of the various scientific calculators presently on the market. For example, on a Texas Instruments *TI*-86, we access this feature by pressing 2nd SIMULT . The resulting display is

SIMULT
Number =

to which we will respond with the keystroke sequence 3 ENTER . The calculator then displays

a1, 1x1...a1, 3x3=b1

a1, 1=

a1, 2=

a1, 3=

b1=

indicating that we may begin entering the numeric data for Eq. [1]. Note that prior to beginning such an operation, we should take the time required to write the system of equations in an ordered fashion in order to avoid confusing our coefficients. We respond with the keystroke sequence 7 ENTER (−) 3 ENTER (−) 4 ENTER (−) 11 ENTER . We then are presented with a new display for our second equation.

After entering the information for all three equations, we press the key to instruct the *TI*-86 to solve for the unknowns $x1$, $x2$, and $x3$ (as it names our v_1, v_2, and v_3). The calculator display then shows

$$x1 = 1.000$$
$$x2 = 2.000$$
$$x3 = 3.000$$

It should be noted that each calculator capable of solving simultaneous equations has its own procedure for entering the required information—therefore, it is a good idea not to throw away anything marked "Owner's Manual" or "Instructions," no matter how tempting such an action might be.

Matrices

Another powerful approach to the solution of a system of equations is based on the concept of matrices. Consider Eqs. [1], [2], and [3]. The array of the constant coefficients of the equations,

$$\mathbf{G} = \begin{bmatrix} 7 & -3 & -4 \\ -3 & 6 & -2 \\ -4 & -2 & 11 \end{bmatrix}$$

is called a ***matrix;*** the symbol $\mathbf{G}$ has been selected since each element of the matrix is a conductance value. A matrix itself has no "value"; it is merely an ordered array of elements. We use a letter in boldface type to represent a matrix, and we enclose the array itself by square brackets.

A matrix having m rows and n columns is called an $(m \times n)$ (pronounced "m by n") matrix. Thus,

$$\mathbf{A} = \begin{bmatrix} 2 & 0 & 5 \\ -1 & 6 & 3 \end{bmatrix}$$

is a (2×3) matrix, and the $\mathbf{G}$ matrix of our example is a (3×3) matrix. An $(n \times n)$ matrix is a ***square matrix*** of order n.

An $(m \times 1)$ matrix is called a *column matrix,* or a ***vector.*** Thus,

$$\mathbf{V} = \begin{bmatrix} \mathbf{V}_1 \\ \mathbf{V}_2 \end{bmatrix}$$

is a (2×1) column matrix of phasor voltages, and

$$\mathbf{I} = \begin{bmatrix} \mathbf{I}_1 \\ \mathbf{I}_2 \end{bmatrix}$$

is a (2×1) phasor-current vector. A $(1 \times n)$ matrix is known as a *row vector.*

Two $(m \times n)$ matrices are equal if their corresponding elements are equal. Thus, if a_{jk} is that element of $\mathbf{A}$ located in row j and column k and b_{jk} is the element at row j and column k in matrix $\mathbf{B}$, then $\mathbf{A} = \mathbf{B}$ *if and only if* $a_{jk} = b_{jk}$ for all $1 \leq j \leq m$ and $1 \leq k \leq n$. Thus, if

$$\begin{bmatrix} \mathbf{V}_1 \\ \mathbf{V}_2 \end{bmatrix} = \begin{bmatrix} \mathbf{z}_{11}\mathbf{I}_1 + \mathbf{z}_{12}\mathbf{I}_2 \\ \mathbf{z}_{21}\mathbf{I}_1 + \mathbf{z}_{22}\mathbf{I}_2 \end{bmatrix}$$

then $V_1 = z_{11}I_1 + z_{12}I_2$ and $V_2 = z_{21}I_1 + z_{22}I_2$.

Two $(m \times n)$ matrices may be added by adding corresponding elements. Thus,

$$\begin{bmatrix} 2 & 0 & 5 \\ -1 & 6 & 3 \end{bmatrix} + \begin{bmatrix} 1 & 2 & 3 \\ -3 & -2 & -1 \end{bmatrix} = \begin{bmatrix} 3 & 2 & 8 \\ -4 & 4 & 2 \end{bmatrix}$$

Next let us consider the matrix product **AB**, where **A** is an $(m \times n)$ matrix and **B** is a $(p \times q)$ matrix. If $n = p$, the matrices are said to be *conformal*, and their product exists. That is, matrix multiplication is defined only for the case where the number of columns of the first matrix in the product is equal to the number of rows in the second matrix.

The formal definition of matrix multiplication states that the product of the $(m \times n)$ matrix **A** and the $(n \times q)$ matrix **B** is an $(m \times q)$ matrix having elements c_{jk}, $1 \le j \le m$ and $1 \le k \le q$, where

$$c_{jk} = a_{j1}b_{1k} + a_{j2}b_{2k} + \cdots + a_{jn}b_{nk}$$

That is, to find the element in the second row and third column of the product, we multiply each of the elements in the second row of **A** by the corresponding element in the third column of **B** and then add the n results. For example, given the (2×3) matrix **A** and the (3×2) matrix **B**,

$$\begin{bmatrix} a_{11} & a_{12} & a_{13} \\ a_{21} & a_{22} & a_{23} \end{bmatrix} \begin{bmatrix} b_{11} & b_{12} \\ b_{21} & b_{22} \\ b_{31} & b_{32} \end{bmatrix} =$$

$$\begin{bmatrix} (a_{11}b_{11} + a_{12}b_{21} + a_{13}b_{31}) & (a_{11}b_{12} + a_{12}b_{22} + a_{13}b_{32}) \\ (a_{21}b_{11} + a_{22}b_{21} + a_{23}b_{31}) & (a_{21}b_{12} + a_{22}b_{22} + a_{23}b_{32}) \end{bmatrix}$$

The result is a (2×2) matrix.

As a numerical example of matrix multiplication, we take

$$\begin{bmatrix} 3 & 2 & 1 \\ -2 & -2 & 4 \end{bmatrix} \begin{bmatrix} 2 & 3 \\ -2 & -1 \\ 4 & -3 \end{bmatrix} = \begin{bmatrix} 6 & 4 \\ 16 & -16 \end{bmatrix}$$

where $6 = (3)(2) + (2)(-2) + (1)(4)$, $4 = (3)(3) + (2)(-1) + (1)(-3)$, and so forth.

Matrix multiplication is not commutative. For example, given the (3×2) matrix **C** and the (2×1) matrix **D**, it is evident that the product **CD** may be calculated, but the product **DC** is not even defined.

As a final example, let

$$\mathbf{t}_A = \begin{bmatrix} 2 & 3 \\ -1 & 4 \end{bmatrix}$$

and

$$\mathbf{t}_B = \begin{bmatrix} 3 & 1 \\ 5 & 0 \end{bmatrix}$$

so that both $\mathbf{t}_A\mathbf{t}_B$ and $\mathbf{t}_B\mathbf{t}_A$ are defined. However,

$$\mathbf{t}_A\mathbf{t}_B = \begin{bmatrix} 21 & 2 \\ 17 & -1 \end{bmatrix}$$

while

$$\mathbf{t}_B\mathbf{t}_A = \begin{bmatrix} 5 & 13 \\ 10 & 15 \end{bmatrix}$$

PRACTICE

A2.1 Given $\mathbf{A} = \begin{bmatrix} 1 & -3 \\ 3 & 5 \end{bmatrix}$, $\mathbf{B} = \begin{bmatrix} 4 & -1 \\ -2 & 3 \end{bmatrix}$, $\mathbf{C} = \begin{bmatrix} 50 \\ 30 \end{bmatrix}$, and $\mathbf{V} = \begin{bmatrix} V_1 \\ V_2 \end{bmatrix}$, find (a) $\mathbf{A} + \mathbf{B}$; (b) $\mathbf{AB}$; (c) $\mathbf{BA}$; (d) $\mathbf{AV} + \mathbf{BC}$; (e) $\mathbf{A}^2 = \mathbf{AA}$.

Ans: $\begin{bmatrix} 5 & -4 \\ 1 & 8 \end{bmatrix}$; $\begin{bmatrix} 10 & -10 \\ 2 & 12 \end{bmatrix}$; $\begin{bmatrix} 1 & -17 \\ 7 & 21 \end{bmatrix}$; $\begin{bmatrix} V_1 - 3V_2 + 170 \\ 3V_1 + 5V_2 - 10 \end{bmatrix}$; $\begin{bmatrix} -8 & -18 \\ 18 & 16 \end{bmatrix}$.

Matrix Inversion

If we write our system of equations using matrix notation,

$$\begin{bmatrix} 7 & -3 & -4 \\ -3 & 6 & -2 \\ -4 & -2 & 11 \end{bmatrix} \begin{bmatrix} v_1 \\ v_2 \\ v_3 \end{bmatrix} = \begin{bmatrix} -11 \\ 3 \\ 25 \end{bmatrix} \qquad [4]$$

we may solve for the voltage vector by multiplying both sides of Eq. [4] by the inverse of our matrix $\mathbf{G}$:

$$\mathbf{G}^{-1}\begin{bmatrix} 7 & -3 & -4 \\ -3 & 6 & -2 \\ -4 & -2 & 11 \end{bmatrix} \begin{bmatrix} v_1 \\ v_2 \\ v_3 \end{bmatrix} = \mathbf{G}^{-1}\begin{bmatrix} -11 \\ 3 \\ 25 \end{bmatrix} \qquad [5]$$

This procedure makes use of the identity $\mathbf{G}^{-1}\mathbf{G} = \mathbf{I}$, where $\mathbf{I}$ is the identity matrix, a square matrix of the same size as $\mathbf{G}$, with zeros everywhere except the diagonal. Each element on the diagonal of an identity matrix is unity. Thus, Eq. [5] becomes

$$\begin{bmatrix} 1 & 0 & 0 \\ 0 & 1 & 0 \\ 0 & 0 & 1 \end{bmatrix} \begin{bmatrix} v_1 \\ v_2 \\ v_3 \end{bmatrix} = \mathbf{G}^{-1}\begin{bmatrix} -11 \\ 3 \\ 25 \end{bmatrix}$$

which may be simplified to

$$\begin{bmatrix} v_1 \\ v_2 \\ v_3 \end{bmatrix} = \mathbf{G}^{-1}\begin{bmatrix} -11 \\ 3 \\ 25 \end{bmatrix}$$

since the identity matrix times any vector is simply equal to that vector (the proof is left to the reader as a 30-second exercise). The solution of our system of equations has therefore been transformed into the problem of obtaining the inverse matrix of $\mathbf{G}$. Many scientific calculators provide the means of performing matrix algebra.

Once again using the *TI-86* as an example, we enter the keystroke sequence 2nd MATRX , and see the display shown in Fig. A2.1.

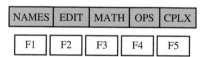

NAMES	EDIT	MATH	OPS	CPLX
F1	F2	F3	F4	F5

■ **FIGURE A2.1** Matrix manipulation display of the *TI*-86, with the corresponding function key shown beneath each menu item.

To create a new matrix called **G**, we press $\boxed{\text{F2}}$, resulting in the display

MATRX
Name=

We next press $\boxed{\text{G}}$ $\boxed{\text{ENTER}}$ and the display reads

MATRX:G   1   x1
[0   ]

We next type $\boxed{3}$ $\boxed{\text{ENTER}}$ twice to define **G** as a 3×3 matrix, and the display changes to

MATRX:G   3 x3
[0   0   0]
[0   0   0]
[0   0   0]

1,   1=0

We respond by typing

$\boxed{7}$ $\boxed{\text{ENTER}}$ $\boxed{(-)}$ $\boxed{3}$ $\boxed{\text{ENTER}}$ $\boxed{(-)}$ $\boxed{4}$ $\boxed{\text{ENTER}}$

and continue until we have entered each coefficient, then pressing $\boxed{\text{EXIT}}$. We next create a current vector **I** by once again invoking the matrix menu, and creating a matrix **I** with the dimensions 3×1. We proceed by entering the values -11, 3, and 25. We can check the values we have entered by typing $\boxed{\text{ALPHA}}$ $\boxed{\text{G}}$ $\boxed{\text{ENTER}}$, or $\boxed{\text{ALPHA}}$ $\boxed{\text{I}}$ $\boxed{\text{ENTER}}$.

The calculator is now prepared to manipulate our arrays to solve the system of equations. We need only type

$\boxed{\text{ALPHA}}$ $\boxed{\text{G}}$ $\boxed{x^{-1}}$ $\boxed{\text{I}}$ $\boxed{\text{ENTER}}$

Again, the reader is cautioned to refer to a specific calculator owner's manual for details.

Determinants

Although a matrix *itself* has no "value," the ***determinant*** of a square matrix *does* have a value. To be precise, we should say that the determinant of a matrix is a value, but common usage enables us to speak of both the array itself and its value as the determinant. We shall symbolize a determinant by Δ, and employ a suitable subscript to denote the matrix to which the determinant refers. Thus,

$$\Delta_G = \begin{vmatrix} 7 & -3 & -4 \\ -3 & 6 & -2 \\ -4 & -2 & 11 \end{vmatrix}$$

Note that simple vertical lines are used to enclose the determinant.

The value of any determinant is obtained by expanding it in terms of its minors. To do this, we select any row j or any column k, multiply each element in that row or column by its minor and by $(-1)^{j+k}$, and then add the products. The minor of the element appearing in both row j and column k is the determinant obtained when row j and column k are removed; it is indicated by Δ_{jk}.

As an example, let us expand the determinant Δ_G along column 3. We first multiply the (-4) at the top of this column by $(-1)^{1+3} = 1$ and then by its minor:

$$(-4)(-1)^{1+3}\begin{vmatrix} -3 & 6 \\ -4 & -2 \end{vmatrix}$$

and then repeat for the other two elements in column 3, adding the results:

$$-4\begin{vmatrix} -3 & 6 \\ -4 & -2 \end{vmatrix} + 2\begin{vmatrix} 7 & -3 \\ -4 & -2 \end{vmatrix} + 11\begin{vmatrix} 7 & -3 \\ -3 & 6 \end{vmatrix}$$

The minors contain only two rows and two columns. They are of order 2, and their values are easily determined by expanding in terms of minors again, here a trivial operation. Thus, for the first determinant, we expand along the first column by multiplying (-3) by $(-1)^{1+1}$ and its minor, which is merely the element (-2), and then multiplying (-4) by $(-1)^{2+1}$ and by 6. Thus,

$$\begin{vmatrix} -3 & 6 \\ -4 & -2 \end{vmatrix} = (-3)(-2) - 4(-6) = 30$$

It is usually easier to remember the result for a second-order determinant as "upper left times lower right minus upper right times lower left." Finally,

$$\begin{aligned}
\Delta_G &= -4[(-3)(-2) - 6(-4)] \\
&\quad + 2[(7)(-2) - (-3)(-4)] \\
&\quad + 11[(7)(6) - (-3)(-3)] \\
&= -4(30) + 2(-26) + 11(33) \\
&= 191
\end{aligned}$$

For practice, let us expand this same determinant along the first row:

$$\begin{aligned}
\Delta_G &= 7\begin{vmatrix} 6 & -2 \\ -2 & 11 \end{vmatrix} - (-3)\begin{vmatrix} -3 & -2 \\ -4 & 11 \end{vmatrix} + (-4)\begin{vmatrix} -3 & 6 \\ -4 & -2 \end{vmatrix} \\
&= 7(62) + 3(-41) - 4(30) \\
&= 191
\end{aligned}$$

The expansion by minors is valid for a determinant of any order.

Repeating these rules for evaluating a determinant in more general terms, we would say, given a matrix $\mathbf{a}$,

$$\mathbf{a} = \begin{bmatrix} a_{11} & a_{12} & \cdots & a_{1N} \\ a_{21} & a_{22} & \cdots & a_{2N} \\ \cdots & \cdots & \cdots & \cdots \\ a_{N1} & a_{N2} & \cdots & a_{NN} \end{bmatrix}$$

that Δ_a may be obtained by expansion in terms of minors along any row j:

$$\begin{aligned}
\Delta_a &= a_{j1}(-1)^{j+1}\Delta_{j1} + a_{j2}(-1)^{j+2}\Delta_{j2} + \cdots + a_{jN}(-1)^{j+N}\Delta_{jN} \\
&= \sum_{n=1}^{N} a_{jn}(-1)^{j+n}\Delta_{jn}
\end{aligned}$$

or along any column k:

$$\Delta_a = a_{1k}(-1)^{1+k}\Delta_{1k} + a_{2k}(-1)^{2+k}\Delta_{2k} + \cdots + a_{Nk}(-1)^{N+k}\Delta_{Nk}$$

$$= \sum_{n=1}^{N} a_{nk}(-1)^{n+k}\Delta_{nk}$$

The cofactor C_{jk} of the element appearing in both row j and column k is simply $(-1)^{j+k}$ times the minor Δ_{jk}. Thus, $C_{11} = \Delta_{11}$, but $C_{12} = -\Delta_{12}$. We may now write

$$\Delta_a = \sum_{n=1}^{N} a_{jn}C_{jn} = \sum_{n=1}^{N} a_{nk}C_{nk}$$

As an example, let us consider this fourth-order determinant:

$$\Delta = \begin{vmatrix} 2 & -1 & -2 & 0 \\ -1 & 4 & 2 & -3 \\ -2 & -1 & 5 & -1 \\ 0 & -3 & 3 & 2 \end{vmatrix}$$

We find

$$\Delta_{11} = \begin{vmatrix} 4 & 2 & -3 \\ -1 & 5 & -1 \\ -3 & 3 & 2 \end{vmatrix} = 4(10+3) + 1(4+9) - 3(-2+15) = 26$$

$$\Delta_{12} = \begin{vmatrix} -1 & 2 & -3 \\ -2 & 5 & -1 \\ 0 & 3 & 2 \end{vmatrix} = -1(10+3) + 2(4+9) + 0 = 13$$

and $C_{11} = 26$, whereas $C_{12} = -13$. Finding the value of Δ for practice, we have

$$\Delta = 2C_{11} + (-1)C_{12} + (-2)C_{13} + 0$$
$$= 2(26) + (-1)(-13) + (-2)(3) + 0 = 59$$

Cramer's Rule

We next consider Cramer's rule, which enables us to find the values of the unknown variables. It is also useful in solving systems of equations where numerical coefficients have not yet been specified, thus confounding our calculators. Let us again consider Eqs. [1], [2], and [3]; we define the determinant Δ_1 as that determinant which is obtained when the first column of Δ_G is replaced by the three constants on the right-hand sides of the three equations. Thus,

$$\Delta_1 = \begin{vmatrix} -11 & -3 & -4 \\ 3 & 6 & -2 \\ 25 & -2 & 11 \end{vmatrix}$$

We expand along the first column:

$$\Delta_1 = -11 \begin{vmatrix} 6 & -2 \\ -2 & 11 \end{vmatrix} - 3 \begin{vmatrix} -3 & -4 \\ -2 & 11 \end{vmatrix} + 25 \begin{vmatrix} -3 & -4 \\ 6 & -2 \end{vmatrix}$$
$$= -682 + 123 + 750 = 191$$

Cramer's rule then states that

$$v_1 = \frac{\Delta_1}{\Delta_G} = \frac{191}{191} = 1 \text{ V}$$

and

$$v_2 = \frac{\Delta_2}{\Delta_G} = \begin{vmatrix} 7 & -11 & -4 \\ -3 & 3 & -2 \\ -4 & 25 & 11 \end{vmatrix} = \frac{581 - 63 - 136}{191} = 2 \text{ V}$$

and finally,

$$v_3 = \frac{\Delta_3}{\Delta_G} = \begin{vmatrix} 7 & -3 & -11 \\ -3 & 6 & 3 \\ -4 & -2 & 25 \end{vmatrix} = \frac{1092 - 291 - 228}{191} = 3 \text{ V}$$

Cramer's rule is applicable to a system of N simultaneous linear equations in N unknowns; for the ith variable v_i:

$$v_i = \frac{\Delta_i}{\Delta_G}$$

PRACTICE

A2.2 Evaluate:

(a) $\begin{vmatrix} 2 & -3 \\ -2 & 5 \end{vmatrix}$; (b) $\begin{vmatrix} 1 & -1 & 0 \\ 4 & 2 & -3 \\ 3 & -2 & 5 \end{vmatrix}$; (c) $\begin{vmatrix} 2 & -3 & 1 & 5 \\ -3 & 1 & -1 & 0 \\ 0 & 4 & 2 & -3 \\ 6 & 3 & -2 & 5 \end{vmatrix}$;

(d) Find i_2 if $5i_1 - 2i_2 - i_3 = 100$, $-2i_1 + 6i_2 - 3i_3 - i_4 = 0$, $-i_1 - 3i_2 + 4i_3 - i_4 = 0$, and $-i_2 - i_3 = 0$.

Ans: 4; 33; −411; 1.266.

A PROOF OF THÉVENIN'S THEOREM

We shall prove Thévenin's theorem in the same form in which it is stated in Sec. 5.4 of Chap. 5, repeated here for reference:

> Given any linear circuit, rearrange it in the form of two networks A and B connected by two wires. Define a voltage v_{oc} as the open-circuit voltage which appears across the terminals of A when B is disconnected. Then all currents and voltages in B will remain unchanged if all *independent* voltage and current sources in A are "killed" or "zeroed out," and an independent voltage source v_{oc} is connected, with proper polarity, in series with the dead (inactive) A network.

We will effect our proof by showing that the original A network and the Thévenin equivalent of the A network both cause the same current to flow into the terminals of the B network. If the currents are the same, then the voltages must be the same; in other words, if we apply a certain current, which we might think of as a current source, to the B network, then the current source and the B network constitute a circuit that has a specific input voltage as a response. Thus, the current determines the voltage. Alternatively we could, if we wished, show that the terminal voltage at B is unchanged, because the voltage also determines the current uniquely. If the input voltage and current to the B network are unchanged, then it follows that the currents and voltages *throughout* the B network are also unchanged.

Let us first prove the theorem for the case where the B network is inactive (no independent sources). After this step has been accomplished, we may then use the superposition principle to extend the theorem to include B networks that contain independent sources. Each network may contain dependent sources, provided that their control variables are in the same network.

The current i, flowing in the upper conductor from the A network to the B network in Fig. A3.1a, is therefore caused entirely by the independent

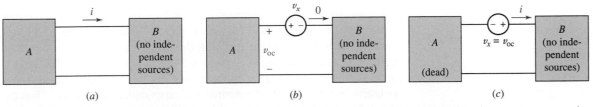

■ **FIGURE A3.1** (a) A general linear network A and a network B that contains no independent sources. Controls for dependent sources must appear in the same part of the network. (b) The Thévenin source is inserted in the circuit and adjusted until $i = 0$. No voltage appears across network B and thus $v_x = v_{oc}$. The Thévenin source thus produces a current $-i$ while network A provides i. (c) The Thévenin source is reversed and network A is killed. The current is therefore i.

sources present in the *A* network. Suppose now that we add an additional voltage source v_x, which we shall call the Thévenin source, in the conductor in which i is measured, as shown in Fig. A3.1*b*, and then adjust the magnitude and time variation of v_x until the current is reduced to zero. By our definition of v_{oc}, then, the voltage across the terminals of *A* must be v_{oc}, since $i = 0$. Network *B* contains no independent sources, and no current is entering its terminals; therefore, there is no voltage across the terminals of the *B* network, and by Kirchhoff's voltage law the voltage of the Thévenin source is v_{oc} volts, $v_x = v_{oc}$. Moreover, since the Thévenin source and the *A* network jointly deliver no current to *B*, and since the *A* network by itself delivers a current i, superposition requires that the Thévenin source acting by itself must deliver a current of $-i$ to *B*. The source acting alone in a reversed direction, as shown in Fig. A3.1*c*, therefore produces a current i in the upper lead. This situation, however, is the same as the conclusion reached by Thévenin's theorem: the Thévenin source v_{oc} acting in series with the inactive *A* network is equivalent to the given network.

Now let us consider the case where the *B* network may be an active network. We now think of the current i, flowing from the *A* network to the *B* network in the upper conductor, as being composed of two parts, i_A and i_B, where i_A is the current produced by *A* acting alone and the current i_B is due to *B* acting alone. Our ability to divide the current into these two components is a direct consequence of the applicability of the superposition principle to these two *linear* networks; the complete response and the two partial responses are indicated by the diagrams of Fig. A3.2.

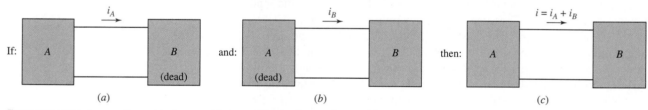

■ **FIGURE A3.2** Superposition enables the current *i* to be considered as the sum of two partial responses.

The partial response i_A has already been considered; if network *B* is inactive, we know that network *A* may be replaced by the Thévenin source and the inactive *A* network. In other words, of the three sources which we must keep in mind—those in *A*, those in *B*, and the Thévenin source—the partial response i_A occurs when *A* and *B* are dead and the Thévenin source is active. Preparing for the use of superposition, we now let *A* remain inactive, but turn on *B* and turn off the Thévenin source; by definition, the partial response i_B is obtained. Superimposing the results, the response when *A* is dead and both the Thévenin source and *B* are active is $i_A + i_B$. This sum is the original current i, and the situation wherein the Thévenin source and *B* are active but *A* is dead is the desired Thévenin equivalent circuit. Thus the active network *A* may be replaced by its Thévenin source, the open-circuit voltage, in series with the inactive *A* network, regardless of the status of the *B* network; it may be either active or inactive.

A PSpice® TUTORIAL

SPICE is an acronym for Simulation Program with Integrated Circuit Emphasis. A very powerful program, it is an industry standard and used throughout the world for a variety of circuit analysis applications. SPICE was originally developed in the early 1970s by Donald O. Peterson and coworkers at the University of California at Berkeley. Interestingly, Peterson advocated free and unhindered distribution of knowledge created in university labs, choosing to make an impact as opposed to profiting financially. In 1984, MicroSim Corporation introduced a PC version of SPICE called PSpice®, which built intuitive graphical interfaces around the core SPICE software routines. There are now several variations of SPICE available commercially, as well as competing software products.

The goal of this appendix is to simply introduce the basics of computer-aided circuit analysis; more details are presented in the main text as well as in the references listed under Reading Further. Advanced topics covered in the references include how to determine the sensitivity of an output variable to changes in a specific component value; how to obtain plots of the output versus a source value; determining ac output as a function of source frequency; methods for performing noise and distortion analyses; nonlinear component models; and how to model temperature effects on specific types of circuits.

The acquisition of MicroSim by OrCAD, and the subsequent acquisition of OrCAD by Cadence, has led to quite a few changes in this popular circuit simulation package. At the time of this writing, OrCAD 10.3 is the current professional release, retailing for approximately US$1000; a scaled-back version called OrCAD 10.0 Lite is available for free download (www.cadence.com). This new version replaces the popular PSpice Student Release 9.1, and although slightly different, particularly in terms of the schematic editing, should seem generally familiar to users of previous PSpice releases.

The documentation which accompanies the Demo version OrCAD 10.0 Lite lists several restrictions that do not apply to the professional (commercially available) version. The most significant is that only circuits having 60 or fewer parts may be saved and simulated; larger circuits can be drawn and viewed, however. We have chosen to work with the OrCAD Capture schematics editor, as the current version is very similar fundamentally to the PSpice A/D Schematic Capture editor. Although at present Cadence also provides PSpice A/D for download, it is no longer supported.

Getting Started

A computer-aided circuit analysis consists of three separate steps: (1) drawing the schematic; (2) simulating the circuit; and (3) extracting the desired information from the simulation output. The OrCAD Capture schematic

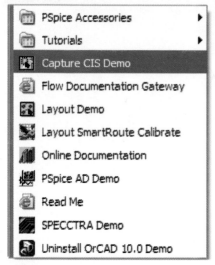

■ **FIGURE A4.1** Orcad Demo programs menu.

editor is launched through the Windows programs list found under the ▮ **start** menu; a menu similar to the one shown in Fig. A4.1 should appear. Upon selecting Capture CIS Demo, the schematics editor opens, as shown in Fig. A4.2.

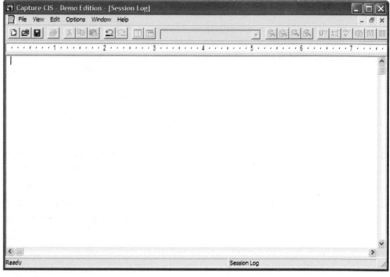

■ **FIGURE A4.2** Capture CIS Demo window.

Under the **File** menu, select **New,** then **Project;** the window of Fig. A4.3*a* appears. After providing a simulation filename and a directory path, the window of Fig. A4.3*b* appears (simply select the "blank project"

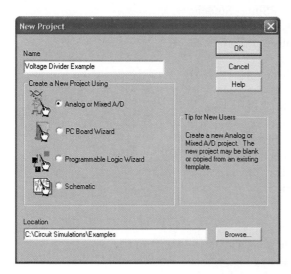

(*a*)

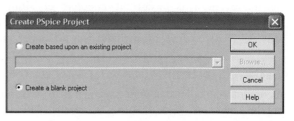

(*b*)

■ **FIGURE A4.3** (*a*) New Project window. (*b*) Create PSpice Project window.

option). We are now presented with the main schematics editor window, as in Fig. A4.4.

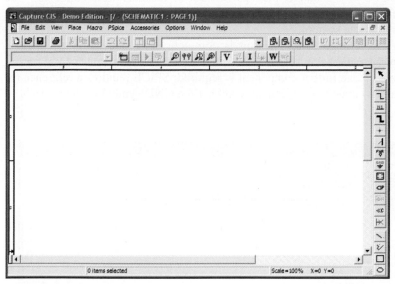

■ **FIGURE A4.4** Main Capture CIS Demo schematics page.

At this point, we're ready to draw a circuit, so let's try a simple voltage divider for purposes of illustration. We will first place the necessary components on the grid, and then wire them together.

Pulling down the **Place** menu, we choose **Part,** resulting in the window shown in Fig. A4.5. Typing a lower case "r" as shown, we click OK and are now able to move a resistor symbol around the schematic window using the mouse. A single left click places a resistor (named R1) at the mouse location; a second left click places a second resistor on our schematic (named R2). A single right click and selecting **End Mode** cancels further resistor placements. The second resistor does not have the appropriate orientation, but is easily manipulated by highlighting it with a single left click, then

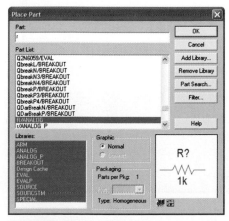

■ **FIGURE A4.5** Place Part menu.

typing *Ctrl* + R. If we do not know the name of the desired part, we can browse through the library of parts provided. If 1 kΩ resistors are not desired—for example, perhaps two 500 Ω resistors were called for—we change the default values simply by double-clicking the "1k" next to the appropriate symbol.

No voltage divider circuit is complete of course without a voltage source. Double-clicking the 0Vdc default, we choose a value of 9 V for our source. One further component is required: SPICE requires a reference (or ground) node to be specified. Clicking the GND symbol to the far right of

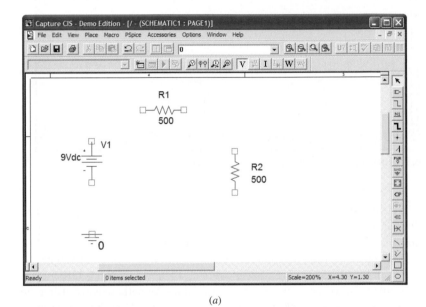

(*a*)

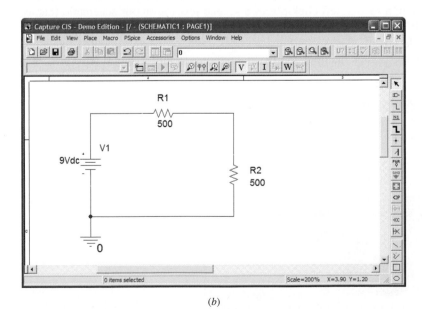

(*b*)

■ **FIGURE A4.6** (*a*) Parts placed on the grid. (*b*) Fully wired circuit, ready to simulate.

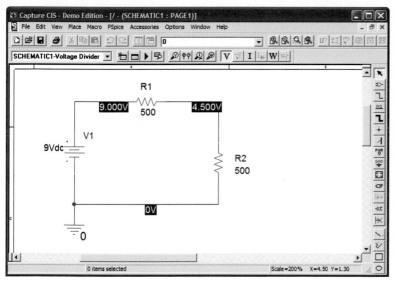

the schematic window, we choose 0/Source from the options. Our progress so far is shown in Fig. A4.6a; all that remains is to wire the components together. This is accomplished by pulling down the **Place** menu and selecting **Wire.** The left and right mouse keys control each wire (some experimenting is called for here—afterwards, select any unwanted wire segments and hit the Delete key). Our final circuit is shown in Fig. A4.6b. It is worth noting that the editor will allow the user to wire through a resistor (thus shorting it out), which can be difficult to see. Generally a warning symbol appears before wiring to an inappropriate location.

Prior to simulating our circuit, we save it by clicking the save icon or selecting **Save** from the **File** menu. From the **PSpice** menu, we select **New Simulation Profile,** and type Voltage Divider in the dialog box that appears. The Simulation Settings dialog box that appears allows us to set parameters for a variety of types of simulations, but for the present example we need only select OK. Once again pulling down the **PSpice** menu, we select **Run.** The simulation results are shown in Fig. A4.7.

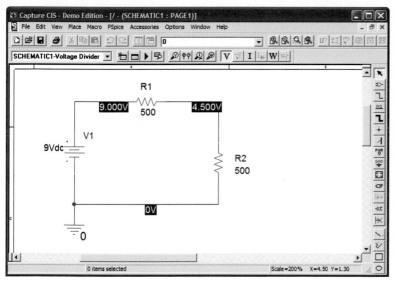

■ FIGURE A4.7 Simulation results.

Fortunately, our simulation yields the expected result—an even split of our source voltage across the two equal-valued resistors. We can also view the simulation results by selecting **View Output File** under the **PSpice** menu. Scrolling down to the end of this file, we see the following lines:

NODE	VOLTAGE	NODE	VOLTAGE
(N00157)	9.0000	(N00166)	4.5000

where node 157 is the positive reference of our voltage source, and node 166 is the junction between the two resistors. This information is available at the top of the file.

READING FURTHER

Two very good books devoted to SPICE and PSpice simulation are:

P. W. Tuinenga, *SPICE: A Guide to Circuit Simulation and Analysis Using PSpice.* Englewood Cliffs, N.J.: Prentice-Hall, 1995.

R. W. Goody, *OrCAD PSpice for Windows Volume 1: DC and AC Circuits,* 3rd ed. Englewood Cliffs, N.J.: Prentice-Hall, 2001.

An interesting history of circuit simulators, as well as Donald Peterson's contributions to the field, can be found in

T. Perry, "Donald O. Peterson [electronic engineering biography]," *IEEE Spectrum* **35** (1998) 22–27.

COMPLEX NUMBERS

This appendix includes sections covering the definition of a complex number, the basic arithmetic operations for complex numbers, Euler's identity, and the exponential and polar forms of the complex number. We first introduce the concept of a complex number.

A5.1 THE COMPLEX NUMBER

Our early training in mathematics dealt exclusively with real numbers, such as 4, $-\frac{2}{7}$, and π. Soon, however, we began to encounter algebraic equations, such as $x^2 = -3$, which could not be satisfied by any real number. Such an equation can be solved only through the introduction of the *imaginary unit,* or the *imaginary operator,* which we shall designate by the symbol j. By definition, $j^2 = -1$, and thus $j = \sqrt{-1}$, $j^3 = -j$, $j^4 = 1$, and so forth. The product of a real number and the imaginary operator is called an *imaginary number,* and the sum of a real number and an imaginary number is called a *complex number*. Thus, a number having the form $a + jb$, where a and b are real numbers, is a complex number.

We shall designate a complex number by means of a special single symbol; thus, $\mathbf{A} = a + jb$. The complex nature of the number is indicated by the use of boldface type; in handwritten material, a bar over the letter is customary. The complex number $\mathbf{A}$ just shown is described as having a *real component* or *real part a* and an *imaginary component* or *imaginary part b*. This is also expressed as

$$\text{Re}\{\mathbf{A}\} = a \qquad \text{Im}\{\mathbf{A}\} = b$$

The imaginary component of $\mathbf{A}$ is *not jb*. By definition, the imaginary component is a real number.

It should be noted that all real numbers may be regarded as complex numbers having imaginary parts equal to zero. The real numbers are therefore included in the system of complex numbers, and we may now consider them as a special case. When we define the fundamental arithmetic operations for complex numbers, we should therefore expect them to reduce to the corresponding definitions for real numbers if the imaginary part of every complex number is set equal to zero.

Since any complex number is completely characterized by a pair of real numbers, such as a and b in the previous example, we can obtain some visual assistance by representing a complex number graphically on a rectangular, or Cartesian, coordinate system. By providing ourselves with a real axis and an imaginary axis, as shown in Fig. A5.1, we form a *complex plane,* or *Argand diagram,* on which any complex number can be represented as a single point. The complex numbers $\mathbf{M} = 3 + j1$ and $\mathbf{N} = 2 - j2$

> Mathematicians designate the imaginary operator by the symbol i, but it is customary to use j in electrical engineering in order to avoid confusion with the symbol for current.

> The choice of the words *imaginary* and *complex* is unfortunate. They are used here and in the mathematical literature as technical terms to designate a class of numbers. To interpret imaginary as "not pertaining to the physical world" or complex as "complicated" is neither justified nor intended.

are indicated. It is important to understand that this complex plane is only a visual aid; it is not at all essential to the mathematical statements which follow.

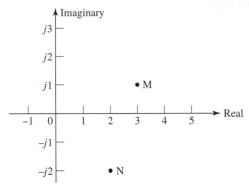

■ **FIGURE A5.1** The complex numbers $M = 3 + j1$ and $N = 2 - j2$ are shown on the complex plane.

We shall define two complex numbers as being equal if, and only if, their real parts are equal and their imaginary parts are equal. Graphically, then, to each point in the complex plane there corresponds only one complex number, and conversely, to each complex number there corresponds only one point in the complex plane. Thus, suppose we are given the two complex numbers:

$$\mathbf{A} = a + jb \qquad \text{and} \qquad \mathbf{B} = c + jd$$

Then, if

$$\mathbf{A} = \mathbf{B}$$

it is necessary that

$$a = c \qquad \text{and} \qquad b = d$$

A complex number expressed as the sum of a real number and an imaginary number, such as $\mathbf{A} = a + jb$, is said to be in *rectangular* or *cartesian* form. Other forms for a complex number will appear shortly.

Let us now define the fundamental operations of addition, subtraction, multiplication, and division for complex numbers. The sum of two complex numbers is defined as the complex number whose real part is the sum of the real parts of the two complex numbers and whose imaginary part is the sum of the imaginary parts of the two complex numbers. Thus,

$$(a + jb) + (c + jd) = (a + c) + j(b + d)$$

For example,

$$(3 + j4) + (4 - j2) = 7 + j2$$

The difference of two complex numbers is taken in a similar manner; for example,

$$(3 + j4) - (4 - j2) = -1 + j6$$

Addition and subtraction of complex numbers may also be accomplished graphically on the complex plane. Each complex number is represented as a vector, or directed line segment, and the sum is obtained by completing the parallelogram, illustrated by Fig. A5.2*a*, or by connecting the vectors in a head-to-tail manner, as shown in Fig. A5.2*b*. A graphical sketch is often useful as a check for a more exact numerical solution.

The product of two complex numbers is defined by

$$(a + jb)(c + jd) = (ac - bd) + j(bc + ad)$$

This result may be easily obtained by a direct multiplication of the two binomial terms, using the rules of the algebra of real numbers, and then simplifying the result by letting $j^2 = -1$. For example,

$$(3 + j4)(4 - j2) = 12 - j6 + j16 - 8j^2$$
$$= 12 + j10 + 8$$
$$= 20 + j10$$

It is easier to multiply the complex numbers by this method, particularly if we immediately replace j^2 by -1, than it is to substitute in the general formula that defines the multiplication.

Before defining the operation of division for complex numbers, we should define the conjugate of a complex number. The *conjugate* of the complex number $\mathbf{A} = a + jb$ is $a - jb$ and is represented as $\mathbf{A}^*$. The conjugate of any complex number is therefore easily obtained by merely changing the sign of the imaginary part of the complex number. Thus, if

$$\mathbf{A} = 5 + j3$$

then

$$\mathbf{A}^* = 5 - j3$$

It is evident that the conjugate of any complicated complex expression may be found by replacing every complex term in the expression by its conjugate, which may be obtained by replacing every j in the expression by $-j$.

The definitions of addition, subtraction, and multiplication show that the following statements are true: the sum of a complex number and its conjugate is a real number; the difference of a complex number and its conjugate is an imaginary number; and the product of a complex number and its conjugate is a real number. It is also evident that if $\mathbf{A}^*$ is the conjugate of $\mathbf{A}$, then $\mathbf{A}$ is the conjugate of $\mathbf{A}^*$; in other words, $\mathbf{A} = (\mathbf{A}^*)^*$. A complex number and its conjugate are said to form a *conjugate complex pair* of numbers.

We now define the quotient of two complex numbers:

$$\frac{\mathbf{A}}{\mathbf{B}} = \frac{(\mathbf{A})(\mathbf{B}^*)}{(\mathbf{B})(\mathbf{B}^*)}$$

and thus

$$\frac{a + jb}{c + jd} = \frac{(ac + bd) + j(bc - ad)}{c^2 + d^2}$$

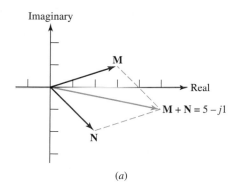

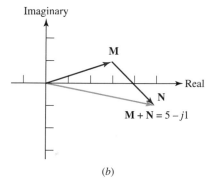

■ FIGURE A5.2 (*a*) The sum of the complex numbers $\mathbf{M} = 3 + j1$ and $\mathbf{N} = 2 - j2$ is obtained by constructing a parallelogram. (*b*) The sum of the same two complex numbers is found by a head-to-tail combination.

Inevitably in a physical problem a complex number is somehow accompanied by its conjugate.

We multiply numerator and denominator by the conjugate of the denominator in order to obtain a denominator which is real; this process is called *rationalizing the denominator.* As a numerical example,

$$\frac{3 + j4}{4 - j2} = \frac{(3 + j4)(4 + j2)}{(4 - j2)(4 + j2)}$$

$$= \frac{4 + j22}{16 + 4} = 0.2 + j1.1$$

The addition or subtraction of two complex numbers which are each expressed in rectangular form is a relatively simple operation; multiplication or division of two complex numbers in rectangular form, however, is a rather unwieldy process. These latter two operations will be found to be much simpler when the complex numbers are given in either exponential or polar form. These forms will be introduced in Secs. A5.3 and A5.4.

PRACTICE

A5.1 Let $\mathbf{A} = -4 + j5$, $\mathbf{B} = 3 - j2$, and $\mathbf{C} = -6 - j5$, and find
(a) $\mathbf{C} - \mathbf{B}$; (b) $2\mathbf{A} - 3\mathbf{B} + 5\mathbf{C}$; (c) $j^5\mathbf{C}^2(\mathbf{A} + \mathbf{B})$; (d) $\mathbf{B}\,\mathrm{Re}[\mathbf{A}] + \mathbf{A}\,\mathrm{Re}[\mathbf{B}]$.
A5.2 Using the same values for $\mathbf{A}$, $\mathbf{B}$, and $\mathbf{C}$ as in the previous
problem, find (a) $[(\mathbf{A} - \mathbf{A}^*)(\mathbf{B} + \mathbf{B}^*)^*]^*$; (b) $(1/\mathbf{C}) - (1/\mathbf{B})^*$;
(c) $(\mathbf{B} + \mathbf{C})/(2\mathbf{BC})$.

Ans: A5.1: $-9 - j3$; $-47 - j9$; $27 - j191$; $-24 + j23$.   A5.2: $-j60$;
$-0.329 + j0.236$; $0.0662 + j0.1179$.

A5.2 EULER'S IDENTITY

In Chap. 9 we encounter functions of time which contain complex numbers, and we are concerned with the differentiation and integration of these functions with respect to the real variable t. We differentiate and integrate such functions with respect to t by exactly the same procedures we use for real functions of time. That is, the complex constants are treated just as though they were real constants when performing the operations of differentiation or integration. If $\mathbf{f}(t)$ is a complex function of time, such as

$$\mathbf{f}(t) = a \cos ct + jb \sin ct$$

then

$$\frac{d\mathbf{f}(t)}{dt} = -ac \sin ct + jbc \cos ct$$

and

$$\int \mathbf{f}(t)\,dt = \frac{a}{c} \sin ct - j\frac{b}{c} \cos ct + \mathbf{C}$$

where the constant of integration $\mathbf{C}$ is a complex number in general.

It is sometimes necessary to differentiate or integrate a function of a complex variable with respect to that complex variable. In general, the successful accomplishment of either of these operations requires that the function which is to be differentiated or integrated satisfy certain conditions. All our functions do meet these conditions, and integration or differentiation with respect to a complex variable is achieved by using methods identical to those used for real variables.

At this time we must make use of a very important fundamental relationship known as Euler's identity (pronounced "oilers"). We shall prove this identity, for it is extremely useful in representing a complex number in a form other than rectangular form.

The proof is based on the power series expansions of $\cos\theta$, $\sin\theta$, and e^z, given toward the back of your favorite college calculus text:

$$\cos\theta = 1 - \frac{\theta^2}{2!} + \frac{\theta^4}{4!} - \frac{\theta^6}{6!} + \cdots$$

$$\sin\theta = \theta - \frac{\theta^3}{3!} + \frac{\theta^5}{5!} - \frac{\theta^7}{7!} + \cdots$$

or

$$\cos\theta + j\sin\theta = 1 + j\theta - \frac{\theta^2}{2!} - j\frac{\theta^3}{3!} + \frac{\theta^4}{4!} + j\frac{\theta^5}{5!} - \cdots$$

and

$$e^z = 1 + z + \frac{z^2}{2!} + \frac{z^3}{3!} + \frac{z^4}{4!} + \frac{z^5}{5!} + \cdots$$

so that

$$e^{j\theta} = 1 + j\theta - \frac{\theta^2}{2!} - j\frac{\theta^3}{3!} + \frac{\theta^4}{4!} + \cdots$$

We conclude that

$$e^{j\theta} = \cos\theta + j\sin\theta \qquad [1]$$

or, if we let $z = -j\theta$, we find that

$$e^{-j\theta} = \cos\theta - j\sin\theta \qquad [2]$$

By adding and subtracting Eqs. [1] and [2], we obtain the two expressions which we used without proof in our study of the underdamped natural response of the parallel and series *RLC* circuits,

$$\cos\theta = \tfrac{1}{2}(e^{j\theta} + e^{-j\theta}) \qquad [3]$$

$$\sin\theta = -j\tfrac{1}{2}(e^{j\theta} - e^{-j\theta}) \qquad [4]$$

PRACTICE

A5.3 Use Eqs. [1] through [4] to evaluate: (*a*) e^{-j1}; (*b*) e^{1-j1}; (*c*) $\cos(-j1)$; (*d*) $\sin(-j1)$.

A5.4 Evaluate at $t = 0.5$: (*a*) $(d/dt)(3\cos 2t - j2\sin 2t)$; (*b*) $\int_0^t (3\cos 2t - j2\sin 2t)\, dt$; Evaluate at $\mathbf{s} = 1 + j2$: (*c*) $\int_{\mathbf{s}}^{\infty} \mathbf{s}^{-3}\, d\mathbf{s}$; (*d*) $(d/d\mathbf{s})[3/(\mathbf{s}+2)]$.

Ans: A5.3: $0.540 - j0.841$; $1.469 - j2.29$; 1.543; $-j1.175$.   A5.4: $-5.05 - j2.16$; $1.262 - j0.460$; $-0.06 - j0.08$; $-0.0888 + j0.213$.

A5.3 THE EXPONENTIAL FORM

Let us now take Euler's identity

$$e^{j\theta} = \cos\theta + j\sin\theta$$

and multiply each side by the real positive number C:

$$Ce^{j\theta} = C\cos\theta + jC\sin\theta \qquad [5]$$

The right-hand side of Eq. [5] consists of the sum of a real number and an imaginary number and thus represents a complex number in rectangular form; let us call this complex number $\mathbf{A}$, where $\mathbf{A} = a + jb$. By equating the real parts

$$a = C\cos\theta \qquad [6]$$

and the imaginary parts

$$b = C\sin\theta \qquad [7]$$

then squaring and adding Eqs. [6] and [7],

$$a^2 + b^2 = C^2$$

or

$$C = +\sqrt{a^2 + b^2} \qquad [8]$$

and dividing Eq. [7] by Eq. [6]:

$$\frac{b}{a} = \tan\theta$$

or

$$\theta = \tan^{-1}\frac{b}{a} \qquad [9]$$

we obtain the relationships of Eqs. [8] and [9], which enable us to determine C and θ from a knowledge of a and b. For example, if $\mathbf{A} = 4 + j2$, then we identify a as 4 and b as 2 and find C and θ:

$$C = \sqrt{4^2 + 2^2} = 4.47$$
$$\theta = \tan^{-1}\tfrac{2}{4} = 26.6°$$

We could use this new information to write $\mathbf{A}$ in the form

$$\mathbf{A} = 4.47\cos 26.6° + j4.47\sin 26.6°$$

but it is the form of the left-hand side of Eq. [5] which will prove to be the more useful:

$$\mathbf{A} = Ce^{j\theta} = 4.47e^{j26.6°}$$

A complex number expressed in this manner is said to be in *exponential form*. The real positive multiplying factor C is known as the *amplitude* or *magnitude,* and the real quantity θ appearing in the exponent is called the *argument* or *angle.* A mathematician would always express θ in radians and would write

$$\mathbf{A} = 4.47e^{j0.464}$$

but engineers customarily work in terms of degrees. The use of the degree symbol (°) in the exponent should make confusion impossible.

To recapitulate, if we have a complex number which is given in rectangular form,

$$\mathbf{A} = a + jb$$

and wish to express it in exponential form,

$$\mathbf{A} = Ce^{j\theta}$$

we may find C and θ by Eqs. [8] and [9]. If we are given the complex number in exponential form, then we may find a and b by Eqs. [6] and [7].

When $\mathbf{A}$ is expressed in terms of numerical values, the transformation between exponential (or polar) and rectangular forms is available as a built-in operation on most hand-held scientific calculators.

One question will be found to arise in the determination of the angle θ by using the arctangent relationship of Eq. [9]. This function is multivalued, and an appropriate angle must be selected from various possibilities. One method by which the choice may be made is to select an angle for which the sine and cosine have the proper signs to produce the required values of a and b from Eqs. [6] and [7]. For example, let us convert

$$\mathbf{V} = 4 - j3$$

to exponential form. The amplitude is

$$C = \sqrt{4^2 + (-3)^2} = 5$$

and the angle is

$$\theta = \tan^{-1} \frac{-3}{4} \qquad [10]$$

A value of θ has to be selected which leads to a positive value for $\cos\theta$, since $4 = 5\cos\theta$, and a negative value for $\sin\theta$, since $-3 = 5\sin\theta$. We therefore obtain $\theta = -36.9°$, $323.1°$, $-396.9°$, and so forth. Any of these angles is correct, and we usually select that one which is the simplest, here, $-36.9°$. We should note that the alternative solution of Eq. [10], $\theta = 143.1°$, is not correct, because $\cos\theta$ is negative and $\sin\theta$ is positive.

A simpler method of selecting the correct angle is available if we represent the complex number graphically in the complex plane. Let us first select a complex number, given in rectangular form, $\mathbf{A} = a + jb$, which lies in the first quadrant of the complex plane, as illustrated in Fig. A5.3. If we draw a line from the origin to the point which represents the complex number, we shall have constructed a right triangle whose hypotenuse is evidently the amplitude of the exponential representation of the complex number. In other words, $C = \sqrt{a^2 + b^2}$. Moreover, the counterclockwise angle which the line makes with the positive real axis is seen to be the angle θ of the exponential representation, because $a = C\cos\theta$ and $b = C\sin\theta$. Now if we are given the rectangular form of a complex number which lies in another quadrant, such as $\mathbf{V} = 4 - j3$, which is depicted in Fig. A5.4, the correct angle is graphically evident, either $-36.9°$ or $323.1°$ for this example. The sketch may often be visualized and need not be drawn.

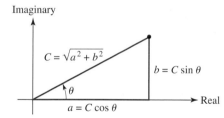

■ **FIGURE A5.3** A complex number may be represented by a point in the complex plane through choosing the correct real and imaginary parts from the rectangular form, or by selecting the magnitude and angle from the exponential form.

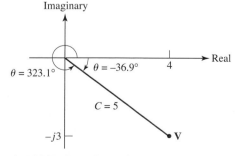

■ **FIGURE A5.4** The complex number $\mathbf{V} = 4 - j3 = 5e^{-j36.9°}$ is represented in the complex plane.

If the rectangular form of the complex number has a negative real part, it is often easier to work with the negative of the complex number, thus avoiding angles greater than 90° in magnitude. For example, given

$$\mathbf{I} = -5 + j2$$

we write

$$\mathbf{I} = -(5 - j2)$$

and then transform $(5 - j2)$ to exponential form:

$$\mathbf{I} = -Ce^{j\theta}$$

where

$$C = \sqrt{29} = 5.39 \qquad \text{and} \qquad \theta = \tan^{-1}\tfrac{-2}{5} = -21.8°$$

We therefore have

$$\mathbf{I} = -5.39e^{-j21.8°}$$

The negative sign may be removed from the complex number by increasing or decreasing the angle by 180°, as shown by reference to a sketch in the complex plane. Thus, the result may be expressed in exponential form as

$$\mathbf{I} = 5.39e^{j158.2°} \qquad \text{or} \qquad \mathbf{I} = 5.39e^{-j201.8°}$$

Note that use of an electronic calculator in the inverse tangent mode always yields angles having magnitudes less than 90°. Thus, both $\tan^{-1}[(-3)/4]$ and $\tan^{-1}[3/(-4)]$ come out as $-36.9°$. Calculators that provide rectangular-to-polar conversion, however, give the correct angle in all cases.

One last remark about the exponential representation of a complex number should be made. Two complex numbers, both written in exponential form, are equal if, and only if, their amplitudes are equal and their angles are equivalent. Equivalent angles are those which differ by multiples of 360°. For example, if $\mathbf{A} = Ce^{j\theta}$ and $\mathbf{B} = De^{j\phi}$, then if $\mathbf{A} = \mathbf{B}$, it is necessary that $C = D$ and $\theta = \phi \pm (360°)n$, where $n = 0, 1, 2, 3, \ldots$.

PRACTICE

A5.5 Express each of the following complex numbers in exponential form, using an angle lying in the range $-180° < \theta \le 180°$;
(a) $-18.5 - j26.1$; (b) $17.9 - j12.2$; (c) $-21.6 + j31.2$.
A5.6 Express each of these complex numbers in rectangular form:
(a) $61.2e^{-j111.1°}$; (b) $-36.2e^{j108°}$; (c) $5e^{-j2.5}$.

Ans: A5.5: $32.0e^{-j125.3°}$; $21.7e^{-j34.3°}$; $37.9e^{j124.7°}$.   A5.6: $-22.0 - j57.1$; $11.19 - j34.4$; $-4.01 - j2.99$.

A5.4 THE POLAR FORM

The third (and last) form in which we may represent a complex number is essentially the same as the exponential form, except for a slight difference in symbolism. We use an angle sign ($\underline{/}$) to replace the combination e^j. Thus, the exponential representation of a complex number **A**,

$$\mathbf{A} = Ce^{j\theta}$$

may be written somewhat more concisely as

$$\mathbf{A} = C\underline{/\theta}$$

The complex number is now said to be expressed in *polar* form, a name which suggests the representation of a point in a (complex) plane through the use of polar coordinates.

It is apparent that transformation from rectangular to polar form or from polar form to rectangular form is basically the same as transformation between rectangular and exponential form. The same relationships exist between C, θ, a, and b.

The complex number

$$\mathbf{A} = -2 + j5$$

is thus written in exponential form as

$$\mathbf{A} = 5.39e^{j111.8°}$$

and in polar form as

$$\mathbf{A} = 5.39\underline{/111.8°}$$

In order to appreciate the utility of the exponential and polar forms, let us consider the multiplication and division of two complex numbers represented in exponential or polar form. If we are given

$$\mathbf{A} = 5\underline{/53.1°} \qquad \text{and} \qquad \mathbf{B} = 15\underline{/-36.9°}$$

then the expression of these two complex numbers in exponential form

$$\mathbf{A} = 5e^{j53.1°} \qquad \text{and} \qquad \mathbf{B} = 15e^{-j36.9°}$$

enables us to write the product as a complex number in exponential form whose amplitude is the product of the amplitudes and whose angle is the algebraic sum of the angles, in accordance with the normal rules for multiplying two exponential quantities:

$$(\mathbf{A})(\mathbf{B}) = (5)(15)e^{j(53.1°-36.9°)}$$

or

$$\mathbf{AB} = 75e^{j16.2°} = 75\underline{/16.2°}$$

From the definition of the polar form, it is evident that

$$\frac{\mathbf{A}}{\mathbf{B}} = 0.333\underline{/90°}$$

Addition and subtraction of complex numbers are accomplished most easily by operating on complex numbers in rectangular form, and the addition or subtraction of two complex numbers given in exponential or polar form should begin with the conversion of the two complex numbers to rectangular form. The reverse situation applies to multiplication and division; two numbers given in rectangular form should be transformed to polar form, unless the numbers happen to be small integers. For example, if we wish to multiply $(1 - j3)$ by $(2 + j1)$, it is easier to multiply them directly as they stand and obtain $(5 - j5)$. If the numbers can be multiplied mentally, then time is wasted in transforming them to polar form.

We should now endeavor to become familiar with the three different forms in which complex numbers may be expressed and with the rapid conversion from one form to another. The relationships among the three forms seem almost endless, and the following lengthy equation summarizes the various interrelationships

$$\mathbf{A} = a + jb = \text{Re}[\mathbf{A}] + j\text{Im}[\mathbf{A}] = Ce^{j\theta} = \sqrt{a^2 + b^2}\,e^{j\,\tan^{-1}(b/a)}$$
$$= \sqrt{a^2 + b^2}\,\underline{/\tan^{-1}(b/a)}$$

Most of the conversions from one form to another can be done quickly with the help of a calculator, and many calculators are equipped to solve linear equations with complex numbers.

We shall find that complex numbers are a convenient mathematical artifice which facilitates the analysis of real physical situations.

PRACTICE ●

A5.7 Express the result of each of these complex-number manipulations in polar form, using six significant figures just for the pure joy of calculating: (*a*) $[2 - (1\underline{/-41°})]/(0.3\underline{/41°})$; (*b*) $50/(2.87\underline{/83.6°} + 5.16\underline{/63.2°})$; (*c*) $4\underline{/18°} - 6\underline{/-75°} + 5\underline{/28°}$.

A5.8 Find $\mathbf{Z}$ in rectangular form if (*a*) $\mathbf{Z} + j2 = 3/\mathbf{Z}$; (*b*) $\mathbf{Z} = 2\ln(2 - j3)$; (*c*) $\sin\mathbf{Z} = 3$.

Ans: A5.7: $4.69179\underline{/-13.2183°}$; $6.318\,33\underline{/-70.4626°}$; $11.5066\underline{/54.5969°}$.
A5.8: $\pm1.414 - j1$; $2.56 - j1.966$; $1.571 \pm j1.763$.

A BRIEF MATLAB® TUTORIAL

The intention of this tutorial is to provide a very brief introduction to some basic concepts required to use a powerful software package known as MATLAB. The use of MATLAB is a completely optional part of the material in this textbook, but as it is becoming an increasingly more common tool in all areas of electrical engineering, we felt that it was worthwhile to provide students with the opportunity to begin exploring some of the features of this software, particularly in plotting 2D and 3D functions, performing matrix operations, solving simultaneous equations, and manipulating algebraic expressions. Many institutions now provide the full version of MATLAB for their students, but at the time of this writing, a student version is available at significantly reduced cost from The MathWorks, Inc. (http://www.mathworks.com/academia/student_version/).

Getting Started

MATLAB is launched by clicking on the program icon; the typical opening window is shown in Fig. A6.1. Programs may be run from files or by directly entering commands in the window. MATLAB also has extensive online help resources, useful for both beginners and advanced users alike. Typical MATLAB programs very much resemble programs written in C, although familiarity with this language is by no means required.

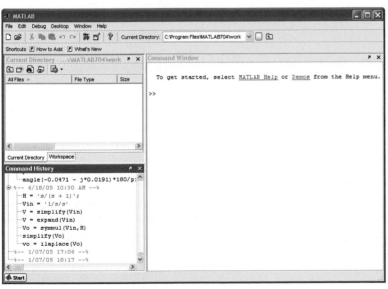

■ **FIGURE A6.1** MATLAB command window upon startup.

Variables and Mathematical Operations

MATLAB makes a great deal more sense once the user realizes that all variables are matrices, even if simply 1×1 matrices. Variable names can be up to 19 characters in length, which is extremely useful in constructing programs with adequate readability. The first character must be a letter, but all remaining characters can be any letter or number; the underscore (_) character may also be used. Variable names in MATLAB are case-sensitive. MATLAB includes several predefined variables. Relevant predefined variables for the material presented in this text include:

eps	The machine accuracy
realmin	The smallest (positive) floating point number handled by the computer
realmax	The largest floating point number handled by the computer
inf	Infinity (defined as $1/0$)
NaN	Literally, "Not a Number." This includes situations such as 0/0
pi	π (3.14159....)
i, j	Both are initially defined as $\sqrt{-1}$. They may be assigned other values by the user

A complete list of currently defined variables can be obtained with the command *who*. Variables are assigned by using an equal sign (=). If the statement is terminated with a semicolon (;), then another prompt appears. If the statement is simply terminated by a carriage return (i.e., by pressing the Enter key), then the variable is repeated. For example,

EDU» input_voltage = 5;

EDU» input_current = 1e−3

input_current =

1.0000e−003

EDU»

Complex variables are easy to define in MATLAB: for example,

EDU» s = 9 + j*5;

creates a complex variable **s** with value $9 + j5$.

A matrix other than a 1×1 matrix is defined using brackets. For example, we would express the matrix $\mathbf{t} = \begin{bmatrix} 2 & -1 \\ 3 & 0 \end{bmatrix}$ in MATLAB as

EDU» t = [2   − 1; 3   0];

Note that the matrix elements are entered a row at a time; row elements are separated by a space, and rows are separated by a semicolon (;). The same arithmetic operations are available for matrices, so, for example, we may find $t + t$ as

EDU» t + t

ans =

    4   −2
    6   0

Arithmetic operators include:

∧	power	\	left division
*	multiplication	+	addition
/	right (ordinary) division	−	subtraction

The order of operations is important. The order of precedence is power, then multiplication and division, then addition and subtraction.

$$\text{EDU» } x = 1 + 5 \wedge 2 * 3$$

$$x =$$

$$76$$

The concept of left division may seem strange at first, but is very useful in matrix algebra. For example,

$$\text{EDU» } 1/5$$

$$\text{ans} =$$
$$0.2000$$
$$\text{EDU» } 1\backslash5$$
$$\text{ans} =$$
$$5$$
$$\text{EDU» } 5\backslash1$$
$$\text{ans} =$$
$$0.2000$$

And, in the case of the matrix equation $\mathbf{Ax} = \mathbf{B}$, where

$$\mathbf{A} = \begin{bmatrix} 2 & 4 \\ 1 & 6 \end{bmatrix} \quad \text{and} \quad \mathbf{B} = \begin{bmatrix} -1 \\ 2 \end{bmatrix}, \text{ we find } \mathbf{x} \text{ with}$$

$$\text{EDU» } A = [2\ 4;\ 1\ 6];$$
$$\text{EDU» } B = [-1;\ 2];$$
$$\text{EDU» } x = A\backslash B$$

$$x =$$

$$-1.7500$$
$$0.6250$$

Alternatively, we can also write

$$\text{EDU» } x = A\wedge-1*B$$

$$x =$$

$$-1.7500$$
$$0.6250$$

or

$$\text{EDU}\text{»}\ \text{inv(A)*B}$$

$$\text{ans} =$$

$$-1.7500$$
$$0.6250$$

When in doubt, parentheses can help a great deal.

Some Useful Functions

Space requirements prevent us from listing every function contained in MATLAB. Some of the more basic ones include:

abs(x)	$\lvert x\rvert$	log 10(x)	$\log_{10} x$		
exp(x)	e^x	sin(x)	$\sin x$	asin(x)	$\sin^{-1} x$
sqrt(x)	$\sqrt{x}$	cos(x)	$\cos x$	acos(x)	$\cos^{-1} x$
log(x)	$\ln x$	tan(x)	$\tan x$	atan(x)	$\tan^{-1} x$

Functions useful for manipulating complex variables include:

real(s)	$\text{Re}\{s\}$
imag(s)	$\text{Im}\{s\}$
abs(s)	$\sqrt{a^2 + b^2}$, where $\mathbf{s} \equiv a + jb$
angle(s)	$\tan^{-1}(b/a)$, where $\mathbf{s} \equiv a + jb$
conj(s)	complex conjugate of $\mathbf{s}$

Another extremely useful command, often forgotten, is simply *help*.

Occasionally we require a vector, such as when we plan to create a plot. The command *linspace*(min, max, number of points) is invaluable in such instances:

$$\text{EDU}\text{»}\ \text{frequency} = \text{linspace(0,10,5)}$$

$$\text{frequency} =$$

$$0 \quad 2.5000 \quad 5.0000 \quad 7.5000 \quad 10.0000$$

A useful cousin is the command *logspace*().

Generating Plots

Plotting with MATLAB is extremely easy. For example, Fig. A6.2 shows the result of executing the following MATLAB program:

$$\text{EDU}\text{»}\ \text{x} = \text{linspace(0,2*pi,100)};$$
$$\text{EDU}\text{»}\ \text{y} = \text{sin(x)};$$
$$\text{EDU}\text{»}\ \text{plot(x,y)};$$
$$\text{EDU}\text{»}\ \text{xlabel('Angle (radians)')};$$
$$\text{EDU}\text{»}\ \text{ylabel('f(x)')};$$

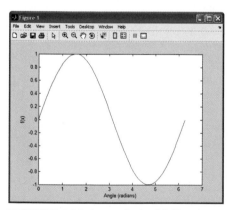

■ **FIGURE A6.2** An example plot of sin(x), $0 < x < 2\pi$, generated using MATLAB. The variable x is a vector comprised of 100 equally spaced elements.

Writing Programs

Although the MATLAB examples in this text are presented as lines typed into the Command Window, it is possible (and often prudent, if repetition is an issue) to write a program so that calculations are more convenient. This is accomplished in MATLAB by writing what is termed an m-file. This is simply a text file saved with a ".m" extension (for example, first_program.m). In a nod to Kernighan and Ritchie, we pull down **New M-File** under the **File** menu, which opens up the m-file editor. (Note that you can use another editor, for example WordPad, if you prefer.)

We type in

$$r = input('Hello, World')$$

as shown in Fig. A6.3.

FIGURE A6.3 Example m-file created in m-file editor.

We next save it as first_program.m in an appropriate directory, then close the editor. Under the **File** menu, we select **Open,** and find first_program.m. This reopens the editor (so we could have skipped closing it earlier). We run our program by hitting f5 or selecting **Run** under the **Debug** menu. In the Command Window, we see our greeting; MATLAB is waiting for a keyboard response, so just hit the Enter key.

Let's expand a previous example to allow the magnitude to be user-selected as in Fig. A6.4. We are now allowed to enter an arbitrary amplitude for our plot.

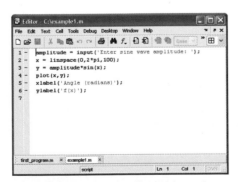

FIGURE A6.4 Example m-file named example1.m for generating sine wave plot.

We leave it to the reader to choose when to write a program/m-file and when to simply use the Command Window directly.

READING FURTHER

There are a large number of excellent MATLAB references available, with new titles appearing regularly. Two worth looking at are:

D. C. Hanselman and B. L. Littlefield, *Mastering MATLAB 7*. Upper Saddle River, N.J.: Prentice-Hall, 2005.

W. J. Palm III, *Introduction to MATLAB 7 for Engineers,* 2nd ed. New York: McGraw-Hill, 2005.

ADDITIONAL LAPLACE TRANSFORM THEOREMS

In this appendix, we briefly present several Laplace transform theorems typically used in more advanced situations in addition to those described in Chap. 14.

Transforms of Periodic Time Functions

The time-shift theorem is very useful in evaluating the transform of periodic time functions. Suppose that $f(t)$ is periodic with a period T for positive values of t. The behavior of $f(t)$ for $t < 0$ has no effect on the (one-sided) Laplace transform, as we know. Thus, $f(t)$ can be written as

$$f(t) = f(t - nT) \qquad n = 0, 1, 2, \ldots$$

If we now define a new time function which is nonzero only in the first period of $f(t)$,

$$f_1(t) = [u(t) - u(t - T)]f(t)$$

then the original $f(t)$ can be represented as the sum of an infinite number of such functions, delayed by integral multiples of T. That is,

$$f(t) = [u(t) - u(t - T)]f(t) + [u(t - T) - u(t - 2T)]f(t)$$
$$+ [u(t - 2T) - u(t - 3T)]f(t) + \cdots$$
$$= f_1(t) + f_1(t - T) + f_1(t - 2T) + \cdots$$

or

$$f(t) = \sum_{n=0}^{\infty} f_1(t - nT)$$

The Laplace transform of this sum is just the sum of the transforms,

$$\mathbf{F}(\mathbf{s}) = \sum_{n=0}^{\infty} \mathcal{L}\{f_1(t - nT)\}$$

so that the time-shift theorem leads to

$$\mathbf{F}(\mathbf{s}) = \sum_{n=0}^{\infty} e^{-nT\mathbf{s}}\mathbf{F}_1(\mathbf{s})$$

where

$$\mathbf{F}_1(\mathbf{s}) = \mathcal{L}\{f_1(t)\} = \int_{0^-}^{T} e^{-\mathbf{s}t} f(t)\, dt$$

Since $\mathbf{F}_1(\mathbf{s})$ is not a function of n, it can be removed from the summation, and $\mathbf{F}(\mathbf{s})$ becomes

$$\mathbf{F}(\mathbf{s}) = \mathbf{F}_1(\mathbf{s})[1 + e^{-T\mathbf{s}} + e^{-2T\mathbf{s}} + \cdots]$$

When we apply the binomial theorem to the bracketed expression, it simplifies to $1/(1 - e^{-T\mathbf{s}})$. Thus, we conclude that the periodic function $f(t)$, with period T, has a Laplace transform expressed by

$$\mathbf{F}(\mathbf{s}) = \frac{\mathbf{F}_1(\mathbf{s})}{1 - e^{-T\mathbf{s}}} \qquad [1]$$

where

$$\mathbf{F}_1(\mathbf{s}) = \mathcal{L}\{[u(t) - u(t - T)]f(t)\} \qquad [2]$$

is the transform of the first period of the time function.

To illustrate the use of this transform theorem for periodic functions, let us apply it to the familiar rectangular pulse train, Fig. A7.1. We may describe this periodic function analytically:

$$v(t) = \sum_{n=0}^{\infty} V_0[u(t - nT) - u(t - nT - \tau)] \qquad t > 0$$

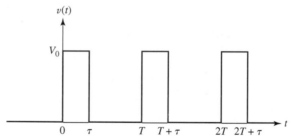

■ **FIGURE A7.1** A periodic train of rectangular pulses for which $F(s) = (V_0/s)(1 - e^{-s\tau})/(1 - e^{-sT})$.

The function $\mathbf{V}_1(\mathbf{s})$ is simple to calculate:

$$\mathbf{V}_1(\mathbf{s}) = V_0 \int_{0^-}^{\tau} e^{-st}\, dt = \frac{V_0}{\mathbf{s}}(1 - e^{-s\tau})$$

Now, to obtain the desired transform, we just divide by $(1 - e^{-s T})$:

$$\mathbf{V}(\mathbf{s}) = \frac{V_0}{\mathbf{s}} \frac{(1 - e^{-s\tau})}{(1 - e^{-s T})} \qquad [3]$$

We should note how several different theorems show up in the transform in Eq. [3]. The $(1 - e^{-s T})$ factor in the denominator accounts for the periodicity of the function, the $e^{-s\tau}$ term in the numerator arises from the time delay of the negative square wave that turns off the pulse, and the $V_0/\mathbf{s}$ factor is, of course, the transform of the step functions involved in $v(t)$.

Determine the transform of the periodic function of Fig. A7.2.

We begin by writing an equation which describes $f(t)$, a function composed of alternating positive and negative impulse functions.

$$f(t) = 2\delta(t-1) - 2\delta(t-3) + 2\delta(t-5) - 2\delta(t-7) + \cdots$$

Defining a new function f_1 and recognizing a period $T = 4$ s,

$$f_1(t) = 2[\delta(t-1) - \delta(t-3)]$$

we can make use of the time periodicity operation as listed in Table 14.2 to find $\mathbf{F(s)}$

$$\mathbf{F(s)} = \frac{1}{1 - e^{-T\mathbf{s}}}\mathbf{F_1(s)} \tag{4}$$

where

$$\mathbf{F_1(s)} = \int_{0^-}^{T} f(t)e^{-\mathbf{s}t}\, dt = \int_{0^-}^{4} f_1(t)e^{-\mathbf{s}t}\, dt$$

There are several ways to evaluate this integral. The easiest is to recognize that its value will remain the same if the upper limit is increased to ∞, allowing us to make use of the time shift theorem. Thus,

$$\mathbf{F_1(s)} = 2[e^{-\mathbf{s}} - e^{-3\mathbf{s}}] \tag{5}$$

Our example is completed by multiplying Eq. [5] by the factor indicated in Eq. [4], so that

$$\mathbf{F(s)} = \frac{2}{1 - e^{-4\mathbf{s}}}(e^{-\mathbf{s}} - e^{-3\mathbf{s}}) = \frac{2e^{-\mathbf{s}}}{1 + e^{-2\mathbf{s}}}$$

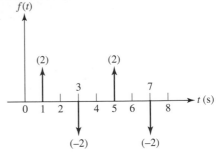

■ **FIGURE A7.2** A periodic function based on unit impulse functions.

PRACTICE

A7.1 Determine the Laplace transform of the periodic function shown in Fig. A7.3.

Ans: $\left(\dfrac{8}{s^2 + \pi^2/4}\right)\dfrac{s + (\pi/2)e^{-s} + (\pi/2)e^{-3s} - se^{-4s}}{1 - e^{-4s}}$.

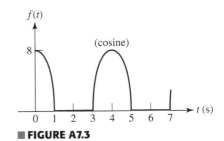

■ **FIGURE A7.3**

Frequency Shifting

The next new theorem establishes a relationship between $\mathbf{F(s)} = \mathcal{L}\{f(t)\}$ and $\mathbf{F(s} + a)$. We consider the Laplace transform of $e^{-at}f(t)$,

$$\mathcal{L}\{e^{-at}f(t)\} = \int_{0^-}^{\infty} e^{-\mathbf{s}t}e^{-at}f(t)\, dt = \int_{0^-}^{\infty} e^{-(\mathbf{s}+a)t}f(t)\, dt$$

Looking carefully at this result, we note that the integral on the right is identical to that defining $\mathbf{F(s)}$ with one exception: $(\mathbf{s} + a)$ appears in place of $\mathbf{s}$. Thus,

$$e^{-at}f(t) \Leftrightarrow \mathbf{F(s} + a) \tag{6}$$

We conclude that replacing **s** by $(\mathbf{s} + a)$ in the frequency domain corresponds to multiplication by e^{-at} in the time domain. This is known as the *frequency-shift* theorem. It can be put to immediate use in evaluating the transform of the exponentially damped cosine function that we used extensively in previous work. Beginning with the known transform of the cosine function,

$$\mathcal{L}\{\cos \omega_0 t\} = \mathbf{F(s)} = \frac{\mathbf{s}}{\mathbf{s}^2 + \omega_0^2}$$

then the transform of $e^{-at} \cos \omega_0 t$ must be $\mathbf{F(s} + a)$:

$$\mathcal{L}\{e^{-at} \cos \omega_0 t\} = \mathbf{F(s} + a) = \frac{\mathbf{s} + a}{(\mathbf{s} + a)^2 + \omega_0^2} \qquad [7]$$

PRACTICE

A7.2 Find $\mathcal{L}\{e^{-2t} \sin(5t + 0.2\pi)u(t)\}$.

Ans: $(0.588\mathbf{s} + 4.05)/(\mathbf{s}^2 + 4\mathbf{s} + 29)$.

Differentiation in the Frequency Domain

Next let us examine the consequences of differentiating $\mathbf{F(s)}$ with respect to **s**. The result is

$$\frac{d}{d\mathbf{s}}\mathbf{F(s)} = \frac{d}{d\mathbf{s}} \int_{0^-}^{\infty} e^{-\mathbf{s}t} f(t)\, dt$$

$$= \int_{0^-}^{\infty} -te^{-\mathbf{s}t} f(t)\, dt = \int_{0^-}^{\infty} e^{-\mathbf{s}t}[-tf(t)]\, dt$$

which is simply the Laplace transform of $[-tf(t)]$. We therefore conclude that differentiation with respect to **s** in the frequency domain results in multiplication by $-t$ in the time domain, or

$$-tf(t) \Leftrightarrow \frac{d}{d\mathbf{s}}\mathbf{F(s)} \qquad [8]$$

Suppose now that $f(t)$ is the unit-ramp function $tu(t)$, whose transform we know is $1/\mathbf{s}^2$. We can use our newly acquired frequency-differentiation theorem to determine the inverse transform of $1/\mathbf{s}^3$ as follows:

$$\frac{d}{d\mathbf{s}}\left(\frac{1}{\mathbf{s}^2}\right) = -\frac{2}{\mathbf{s}^3} \Leftrightarrow -t\mathcal{L}^{-1}\left\{\frac{1}{\mathbf{s}^2}\right\} = -t^2 u(t)$$

and

$$\frac{t^2 u(t)}{2} \Leftrightarrow \frac{1}{\mathbf{s}^3} \qquad [9]$$

Continuing with the same procedure, we find

$$\frac{t^3}{3!}u(t) \Leftrightarrow \frac{1}{\mathbf{s}^4} \qquad [10]$$

and in general

$$\frac{t^{(n-1)}}{(n-1)!}u(t) \Leftrightarrow \frac{1}{\mathbf{s}^n} \qquad [11]$$

PRACTICE

A7.3 Find $\mathcal{L}\{t\sin(5t + 0.2\pi)u(t)\}$.

Ans: $(0.588s^2 + 8.09s - 14.69)/(s^2 + 25)^2$.

Integration in the Frequency Domain

The effect on $f(t)$ of integrating $\mathbf{F}(\mathbf{s})$ with respect to $\mathbf{s}$ may be shown by beginning with the definition once more,

$$\mathbf{F}(\mathbf{s}) = \int_{0^-}^{\infty} e^{-st} f(t)\, dt$$

performing the frequency integration from $\mathbf{s}$ to ∞,

$$\int_{\mathbf{s}}^{\infty} \mathbf{F}(\mathbf{s})\, d\mathbf{s} = \int_{\mathbf{s}}^{\infty} \left[\int_{0^-}^{\infty} e^{-st} f(t)\, dt\right] d\mathbf{s}$$

interchanging the order of integration,

$$\int_{\mathbf{s}}^{\infty} \mathbf{F}(\mathbf{s})\, d\mathbf{s} = \int_{0^-}^{\infty} \left[\int_{\mathbf{s}}^{\infty} e^{-st}\, d\mathbf{s}\right] f(t)\, dt$$

and performing the inner integration,

$$\int_{\mathbf{s}}^{\infty} \mathbf{F}(\mathbf{s})\, d\mathbf{s} = \int_{0^-}^{\infty} \left[-\frac{1}{t} e^{-st}\right]_{\mathbf{s}}^{\infty} f(t)\, dt = \int_{0^-}^{\infty} \frac{f(t)}{t} e^{-st}\, dt$$

Thus,

$$\frac{f(t)}{t} \Leftrightarrow \int_{\mathbf{s}}^{\infty} \mathbf{F}(\mathbf{s})\, d\mathbf{s} \qquad [12]$$

For example, we have already established the transform pair

$$\sin\omega_0 t u(t) \Leftrightarrow \frac{\omega_0}{s^2 + \omega_0^2}$$

Therefore,

$$\mathcal{L}\left\{\frac{\sin\omega_0 t u(t)}{t}\right\} = \int_{\mathbf{s}}^{\infty} \frac{\omega_0\, d\mathbf{s}}{s^2 + \omega_0^2} = \left. \tan^{-1}\frac{\mathbf{s}}{\omega_0}\right|_{\mathbf{s}}^{\infty}$$

and we have

$$\frac{\sin\omega_0 t u(t)}{t} \Leftrightarrow \frac{\pi}{2} - \tan^{-1}\frac{\mathbf{s}}{\omega_0} \qquad [13]$$

PRACTICE

A7.4 Find $\mathcal{L}\{\sin^2 5t u(t)/t\}$.

Ans: $\frac{1}{4}\ln[(s^2 + 100)/s^2]$.

The Time-Scaling Theorem

We next develop the time-scaling theorem of Laplace transform theory by evaluating the transform of $f(at)$, assuming that $\mathcal{L}\{f(t)\}$ is known. The procedure is very simple:

$$\mathcal{L}\{f(at)\} = \int_{0^-}^{\infty} e^{-st} f(at)\, dt = \frac{1}{a} \int_{0^-}^{\infty} e^{-(s/a)\lambda} f(\lambda)\, d\lambda$$

where the change of variable $at = \lambda$ has been employed. The last integral is recognizable as $1/a$ times the Laplace transform of $f(t)$, except that **s** is replaced by **s**/a in the transform. It follows that

$$f(at) \Leftrightarrow \frac{1}{a}\mathbf{F}\left(\frac{\mathbf{s}}{a}\right) \tag{14}$$

As an elementary example of the use of this time-scaling theorem, consider the determination of the transform of a 1-kHz cosine wave. Assuming we know the transform of a 1-rad/s cosine wave,

$$\cos t\, u(t) \Leftrightarrow \frac{\mathbf{s}}{\mathbf{s}^2 + 1}$$

the result is

$$\mathcal{L}\{\cos 2000\pi t\, u(t)\} = \frac{1}{2000\pi} \frac{\mathbf{s}/2000\pi}{(\mathbf{s}/2000\pi)^2 + 1} = \frac{\mathbf{s}}{\mathbf{s}^2 + (2000\pi)^2}$$

PRACTICE
●

A7.5 Find $\mathcal{L}\{\sin^2 5t\, u(t)\}$.

Ans: $50/[\mathbf{s}(\mathbf{s}^2 + 100)]$.

INDEX

A

A_1 and A_2 values
    critical damping and, 333
    overdamped parallel *RLC* circuit,
        324–325
abc phase sequence, 464–465
ABCD parameters, two-port
        networks, 720–724
Absorbed power, 16, 19, 45
    by element, 44
    in resistors, 23–27
ac circuit analysis, 3, 4. *See also* ac circuit
        power analysis; Circuit
        analysis
ac circuit power analysis, 419–456.
        See also Complex power
    apparent power/power factor,
        437–439
    average power. *See* Average power
    instantaneous power, 420–422, 445
    maximum average power, 429
    RMS values of current/voltage,
        432–437, 445
        average power computations, 434
        multiple-frequency circuits,
          434–435
        periodic waveform values,
          432–433
        sinusoidal waveform values,
          433–434
    sinusoidal excitation, instantaneous
        power, 421
    sinusoidal steady state theorem,
        428–429
Active element, 215
Active filters, 677–678
Active network, 21
Addition, Laplace transform
        operation, 561
Additive fluxes, 495
Additive property, of the Laplace
        transform, 546

AD549K op amp, 191, 193
Admittance, 236–237, 572
    parameters. *See* Two-port
        networks
    in sinusoidal steady-state, 392–393
AD622 op amp, 204
Algebraic alternatives, complex forcing
        functions, 378–379
American Wire Gauge (AWG), 26
Ampère, A.M., 12
Amperes, 10, 11, 12
Amplifiers, equivalent networks
        and, 708–710
Amplitude
    exponential form of complex number,
        826–828
    of response, proportional forcing
        function, 374
    of sinusoids, 369
Analysis
    of circuits. *See* Circuit analysis
    computer-aided. *See* Computer-aided
        analysis
    defined, 5–6
    Fourier circuit. *See* Fourier circuit
        analysis
    mesh. *See* Nodal and mesh analysis
    nodal. *See* Nodal and mesh analysis
    power. *See* ac circuit power analysis
    PSpice Type command, 105
    sinusoidal steady-state. *See* Sinusoidal
        steady-state analysis
    transient, 3, 4, 264–266
Analytical Engine, 6
Angles, exponential complex
        numbers, 826–828
Angular frequency, of
        sinusoids, 369
Anode, 187
Apparent power, 440, 441, 445
    power factor and, 437–439
Argand diagram, 821–822

Argument
    exponential form of complex number,
        826–828
    of sinusoids, 369
Arrows, for current, 9, 13
Asymptotes, Bode diagrams
        and, 657–658
Attenuator, 176, 614
Automotive suspensions, modeling, 356
Auxiliary equation, 321
Average power, 441, 445
    ac circuits, 422–432, 445
    ideal resistor absorption of, 426
    maximum, 429
    maximum transfer of, 428–430
    nonperiodic functions, 430–432
    periodic waveforms, 423–424
    reactive element absorption of, 427
    RMS value and, 434
    in the sinusoidal steady state, 424–425
    superposition and, 431

B

Babbage, Charles, 6
Balanced load, 458
Balanced three-phase system, 458
B_1 and B_2 values, 337–338
Bandpass filters, 672, 675–676
Bandstop filters, 672
Bandwidth, and high-Q circuits, 636–641
Base, of transistors, 717
Basic components and electric
        circuits, 9–34
    charge, 11–12
    current. *See* Current
    Ohm's law. *See* Ohm's law
    power. *See* Power
    units and scales, 9–11
    voltage. *See* Voltage
Bass, treble, and midrange filters,
        679–680
Beaty, H. Wayne, 28

Bias Point command (PSpice), 105
Bilateral circuit, 702
Bilateral element, 702
Bode, Hendrik W., 657
Bode diagrams/plots, 656–672
    additional considerations, 661–664
    asymptotes, determining, 657–658
    complex conjugate pairs, 666–668
    computer-aided analysis for, 669–672
    decibel (dB) scale, 657
    higher-order terms and, 665
    multiple terms in, 659–660
    phase response and, 660–661
    smoothing of, 658–659
Bossanyi, E., 485
Boyce, W. E., 302
Branch current, 94
Branches, defined, 793
Break frequency, 658
Buffer design, 178
Burton, T., 485
Butterworth filters, 678

C
Candela, 10
Capacitors, 215–224
    defined, 216
    duality. *See* Duality
    electrochemical, 223
    energy storage, 220–222
    ideal, 215–218, 224
    integral voltage-current relationships,
        218–220
    linearity, consequences of, 235–238
    modeling
        of ideal capacitors, 215–218
        with PSpice, 243–245
        in the *s*-domain, 575–576
    op amp circuits with, 238–239
    in parallel, 234
    phasor relationships for, 385–386
    practical application, 223
    s-domain circuits and, 575–577
    in series, 234
    ultracapacitor, 223
Cartesian form, complex numbers, 822
Cascaded op amps, 182–185, 615
Cathode, 187
Cavendish, Henry, 22
cba phase sequence, 464–465

Characteristic equation, 259, 321
Charge, 11–12
    conservation of, 11, 155
    distance and, 5
Chassis ground, 61–62
Chebyshev filters, 678
Circuit analysis. *See also* Circuit analysis
        techniques
    engineering and, 4–5
    linear. *See* Linear circuits
    nonlinear. *See* Nonlinear circuit
        analysis
    in the *s*-domain. *See* *s*-domain circuit
        analysis
    software, 7. *See also* Computer-aided
        analysis
Circuit analysis techniques, 121–172
    delta-wye (Δ-Y) conversion, 152–154
    linearity and superposition, 121–131
    maximum power transfer, 150–152
    Norton equivalent circuits.
        See Thévenin/Norton
        equivalent circuits
    selection process for, 155–156
    source transformations. *See* Source
        transformations
    superposition. *See* Superposition
    Thévenin equivalent circuits.
        See Thévenin/Norton
        equivalent circuits
Circuits
    analysis of. *See* Circuit analysis
    components of. *See* Basic components
        and electric circuits
    elements of, 17–18, 21
    networks and, 21–22
    response résumé, source-free series
        RLC, 344–345
    transfer functions for, 497
Clayton, G., 616
Closed-loop operation, op amps, 201
Closed-loop voltage gain, 191
Closed paths, 38, 92
Coefficient of mutual inductance, 492
Coils, in wattmeters, 476–477
Collectors, 717
Column matrix, 806
Common-emitter configuration, 717
Common mode rejection, op amps,
    193–194

Comparators, 201–202
Complementary function, source-free
    RL circuits, 256
Complementary solution. *See* Natural
    responses
Complete response, 735–736
    driven *RL* circuits, 285–289
    to periodic forcing functions, 750–752
    of *RLC* circuits. *See RLC* circuits
Complex conjugate pairs, Bode diagrams
    and, 666–668
Complex forcing function. *See* Sinusoidal
    steady-state analysis
Complex form, of Fourier series,
    752–759
Complex frequency, 322
    dc case, 535
    defined, 533–537
    exponential case, 535
    exponentially damped sinusoids, 536
    general form, 534–535
    neper frequency, 534, 537
    radian frequency, 537
    s-domain circuit analysis
        and, 598–607
        at complex frequencies, 608
        frequency dependence,
            magnitude/phase angle,
            604–607
        graphing and, 600–602
        magnitude frequency dependence,
            604–607
        natural response and, 607–611
            general perspective, 609–610
            special case, 610
        operating at complex
            frequencies, 608
        phase angle frequency
            dependence, 604–607
        pole-zero constellations, 602–604
        response as a function of σ, 599
        response as a function
            of ω, 599–600
    s in relation to reality, 536–537
    sinusoidal case, 535
Complex numbers, 821–830
    arithmetic operations for, 822–824
    described, 821–822
    Euler's identity, 824–825
    exponential form of, 826–828

imaginary unit (operator), 821
    polar form of, 828–830
    rectangular (cartesian) form of, 822
Complex plane, 821–822
    s-domain circuit analysis and.
        See Complex frequency
Complex power, 440–445
    apparent power, 440, 441, 445
        and power factor, 437–439
    average power, 441
    complex power, 441
    formula, 441
    measuring, 441
    power factor, 437–439
        correction, 442
        lagging, 441
        leading, 441
    power triangle, 441
    quadrature component, 441
    quadrature power, 441
    reactive power, 440, 441, 445
    terminology, 445
    volt-ampere (VA), 441
    volt-ampere-reactive (VAR)
        units, 440, 441
    watt (W), 441
Complex representation, phasor as
        abbreviation for, 381
Components. *See* Basic components and
        electric circuits
Computer-aided analysis, 6–7, 128–130.
        See also MATLAB; PSpice
    Bode diagrams and, 669–672
    fast Fourier Transform, 776–779
    Laplace transforms and, 551–553
    magnetically coupled circuits,
        508–509
    nodal and mesh analysis, 103–108,
        578–580
    op amps, 198–201
    s-domain nodal and mesh analysis,
        578–580
    sinusoidal steady-state analysis,
        402–403
    source-free parallel *RLC* circuits,
        342–343
    source-free *RL* circuits, 264–266
    system function, 776–779
    for two-port networks, 723–724
Conductance, 27–28, 392

Conformal matrices, 807
Conservation of charge, 11, 155
Conservation of energy, 14, 44, 155
Constant charge, 12
Controlled sources, of voltage/current,
        18, 19–21
Convolution
    Laplace transform operation, 561,
        595–596
    s-domain circuit analysis
        and, 589–598
        convolution integral, 591
        four-step process for analysis, 589
        graphical methods of, 592–593
        impulse response, 589–590
        Laplace transform and, 595–596
        realizable systems and, 591–592
        transfer function comments, 597
Cooper, George R., 544
Corner frequency, 397, 658
Cosines, sines converted to, 371
Cotree, 794–795
Coulomb, 11
Coupling coefficient, 502
Cramer's rule, 84, 811–812
Create command (PSpice), 105
Critical frequencies, *s*-domain circuit
        analysis, 588
Critically damped response,
        RLC circuits
    form of, 332–333
    graphical representation, 334–335
    source-free circuits
        parallel, 323, 345
        series, 344–345
Current, 9, 11, 12–13
    actual direction vs. convention, 13
    branch current, 94
    capacitor voltage-current
        relationships, 218–220
    coil, 476
    current-controlled current source,
        18, 19–21
    current-controlled voltage source,
        18, 19–21
    gain, amplifiers, 708
    graphical symbols for, 13
    laws. *See* Voltage and current laws
    mesh, 92, 93–95, 503
    response, resonance and, 631

sources
    controlled, 18, 19–21
    practical, 133, 137–138
    reliable, op amps, 188–190
    series/parallel connections, 49–51
    and voltage. *See* Voltage
    superposition applicable to, 431
    types of, 13
    and voltage division, 57–60
Cutoff frequency, transistor amplifier,
        396–397
Cutoff voltage, 223

D

Damped sinusoidal forcing function,
        537–540
Damped sinusoidal response, 336
Damping factor, parallel resonance and,
        634–635
Damping out, of transients, 330
Davies, B., 564
3 dB frequency, 658
dc (direct current)
    analysis, 3
    case, complex frequency, 535
    current source, 19
    parameter sweep, 128–130
    short circuits to, 225
    sources, 19, 173, 289
Dead network, 142, 145
Decade (of frequencies), 658
DeCarlo, R. A., 108, 156, 407, 725
Decibel (dB) scale, Bode diagrams, 657
Delivered power, 19
Delta (Δ) connection, 470–476
    connected sources, 473–476
    Y-connected loads vs., 473
Delta (Δ) of impedances, equivalent
        networks, 704–705
Delta-wye (Δ-Y) conversion, 152–154
Dependent sources
    linear, 122
    Thévenin/Norton equivalent
        circuits, 145–147
    of voltage/current, 18, 19–21
Derivative-of-the-current voltage, 18
Design, defined, 5–6
Determinants, 809–811
Difference amplifier, 179–182, 193–194
    summary, 180

Difference Engine, 6
Differential equations
    algebraic alternative, sinusoidal
        steady-state, 378–379
    homogeneous linear, 255–256
    for source-free parallel *RLC* circuits,
        320–322
Differential input voltage, 193
Digital cellular devices, 223
Digital integrated circuits, frequency
    limits in, 300
Digital multimeter (DMM), 148–149
DiPrima, R. C., 302
Direct approach, source-free
    RL circuits, 256–257
Direction of travel, current, 12
Direct procedure, driven *RL* circuits,
    281–282
Discrete spectrum, 744
Dissipation of power, 45
Distance, charge and, 5
Distinct poles, method of residues and,
    548–549
Distributed-parameter networks, 35
Dot convention
    circuit transfer function, 497
    mutual inductance, 493–497
    physical basis of, 495–497
    power gain, 497
Double-subscript notation, polyphase
    circuits, 459–460
Drexler, H. B., 246
Driven *RC* circuits, 289–294
Driven *RL* circuits, 280–283
    complete response determination,
        285–289
    direct procedure, 281–282
    intuitive understanding of, 283
    natural and forced response,
        282, 283–289
    response from dc sources,
        summarized, 289
Duality, 232, 240–242

E

Earth ground, 61–62
Edison, Thomas, 457
Effective (RMS) value. *See* RMS value
Electrical networks, behavior of, 607
Electric circuits. *See* Circuits

Electrochemical capacitor, 223
Emitters, 717
Energy, 14
    accounting, source-free *RL*
        circuits, 261
    conservation of, 14, 44, 155
    density, 765
    instantaneous, stored, 632
    magnetically coupled circuits.
        See Magnetically coupled
        circuits
    storage capacitors, 220–222
    storage inductors, 230–232
    work units, 10
Engineering, circuit analysis and, 4–5
Engineering systems, behavior of, 607
Engineering units, 11
ENIAC, 6
Equivalent circuits, ideal transformers,
    518–520
Equivalent combinations, frequency
    response and, 647–651
Equivalent networks, two-port. *See* Two-
    port networks
Equivalent practical sources, 133–136
Equivalent resistance, 52, 142
Equivalent voltage sources, 131
Even functions, 747n
Even harmonics, 747, 747n
Even symmetry, Fourier series
    analysis, 745, 749
Exponential case, complex frequency, 535
Exponential damping coefficient,
    322, 630
Exponential form, complex numbers,
    826–828
Exponential function $e^{-\alpha t}$, 545
Exponentially damped sinusoids, 536
Exponential response, *RL* circuits,
    262–266

F

Fairchild Corp., 173
Fall time, of waveforms, 294
Faraday, Michael, 216n, 224, 225
farad (F), 216
Fast Fourier transform (FFT),
    774, 776–779
    image processing example, 782
Feedback control, 5

Feynman, R., 63
Fiber Optic intercom, 181
Filters (frequency), 672–680
    active, 677–678
    bandpass, 672, 675–676
    bandstop, 672
    bass/treble/midrange adjustment,
        679–680
    Butterworth, 678
    Chebyshev, 678
    high-pass, 672, 673–674
    low-pass, 672, 673–674
    maximally flat, 678
    multiband, 672
    notch, 672
    passive
        defined, 677
        low-pass and high-pass,
            673–674
    practical application, 679–680
Final-value, Laplace transforms,
    561–563
Finite resistance, underdamped
    source-free parallel *RLC*,
    338–340
Finite wire impedance, 461
Fink, Donald G., 28
Flowchart, for problem-solving, 8
Force, voltage and, 5
Forced responses, 369, 735–736
    driven *RL* circuits, 282, 283–289
    to sinusoids. *See* Sinusoidal
        steady-state analysis
    source-free *RL* circuits, 256
Forcing functions, 122
    sinusoidal waveform as, 369
    source-free *RL* circuits, 256
Forms of responses
    critically damped *RLC* circuits,
        332–333
    underdamped source-free parallel
        RLC circuits, 336–337
Fourier circuit analysis, 4, 735–792.
        See also Fourier series;
        Fourier transform
    complete response to periodic forcing
        functions, 750–752
    epilogue, 782–784
    image processing, 782–783
    practical application, 782–783

Fourier series
  coefficients, 739–740
  complex form, 752–759
  sampling function, 756–759
  symmetry, use of, 745–749
    even and odd symmetry, 745, 749
    Fourier terms and, 745–747
    half-wave symmetry, 747–748, 749
    for simplification purposes, 749t
  trigonometric form of, 735–745
    coefficients, evaluating, 739–740
    derived, 737–738
    equation for, 738
    harmonics, 736–737
    integrals, useful, 738–739
    line spectra, 743–744
    phase spectra, 744–745
Fourier transform. *See also* Fourier
    transform pairs
  defined, 759–763
  fast Fourier transform (FFT), 774,
    776–779
    image processing example, 782
  of general periodic time function,
    771–772
  physical significance of, 764–765
  properties of, 763–766
  system function, frequency domain.
    See System function
Fourier transform pairs, 760, 761
  for constant forcing function, 768
  for signum function, 768–769
  summary of, 770
  for unit-impulse function, 766–768
  for unit step function, 769
Free response, source-free
    RL circuits, 256
Frequency
  angular, of sinusoids, 369
  complex. *See* Complex frequency
  corner, 397
  cutoff, transistor amplifier, 396–397
  dependence, *s* plane, 604–607
  differentiation, Laplace transforms,
    561, 840–841
  domain. *See* Frequency domain
  fundamental frequency, 736
  integration, Laplace transforms,
    561, 841
  limits, digital integrated circuits, 300

  multiple, RMS value with,
    434–435
  natural resonant, 336–337
  op amps and, 197–198
  radian, of sinusoids, 369
  response. *See* Frequency response
  scaling, 652–656
  selectivity, parallel resonance and, 637
  shift, Laplace transforms, 561,
    839–840
  of sinusoids, 370–371
  source-free parallel *RLC* circuits,
    322–323
  unit definitions for, 322
Frequency domain
  phasor representation, 382
  system function and, 772–779
  time domain converted to, 539
  V-I expressions, phasor relationships
    and, 385
Frequency response, 3, 4, 627–690
  Bode diagrams. *See* Bode
    diagrams/plots
  equivalent series/parallel
    combinations, 647–651
  filters. *See* Filters (frequency)
  parallel resonance. *See* Parallel
    resonance
  resonant forms, other, 645–651
  scaling, 652–656
  series resonance, 641–644
Friction coefficient, 5
Fundamental frequency, 736

G

Gain, of op amps, 612
General Conference on Weights
    and Measures, 10
General form, complex frequency,
    534–535
General practical voltage source, 132
General *RC* circuits, 273–276
General *RL* circuits, 269–270
General solution, source-free
    RL circuits, 258–261
George A. Philbrick Researches,
    Inc., 205
Global positioning systems
    (GPS), 612
Goody, R. W., 360, 820

Graphics/Graphing
  on complex-frequency *(s)* plane,
    600–602
  of convolution, *s*-domain analysis,
    592–593
  of critically damped response, *RLC*
    circuits, 334–335
  of current, symbols for, 13
  overdamped response, *RLC* circuits,
    329–330
  underdamped response, *RLC*
    circuits, 338
Ground (neutral) connection, 61–62, 458
Groups, of independent sources, 123

H

Half-power frequency, 658
Half-wave symmetry, Fourier,
    747–748, 749
Hanselman, D. C., 836
Harmonics, Fourier, 736–737
Hayt, W. H., Jr., 204, 407, 725
Heathcote, M., 520
Henry, Joseph, 224
henry (H), 224
Higher-order terms, Bode diagrams, 665
High-pass filters, 672
  passive, 673–674
High-*Q* circuits, bandwidth and, 636–641
Homogeneity property, Laplace
    transforms, 546
Homogeneous linear differential
    equations, 255–256
$H(s) = V_{out}/V_{in}$, synthesizing, 612–616
Huang, Q., 681
Hybrid parameters, two-port networks,
    718–720

I

Ideal capacitor model, 215–218
Ideal inductor model, 224–227
Ideal operational amplifiers.
    See Operational amplifiers
Ideal resistor, average power
    absorption, 426
Ideal sources, of voltage, 18
Ideal transformers, 510–520
  equivalent circuits, 518–520
  for impedance matching, 512–513
  step-down transformers, 514

Ideal transformers—*Cont.*
  step-up transformers, 514
  turns ratio of, 510–512
  for voltage level adjustment, 513–514
  voltage relationship in the time
        domain, 515–518
Ideal voltage sources, 131–133
Image processing, Fourier analysis
        and, 782–783
Imaginary sources → imaginary
        responses, 377–378
Imaginary unit (operator)/component, 821
  of complex forcing function, 376
  of complex power, 440
  imaginary sources → imaginary
        responses, 377–378
Immittance, 392–393
Impedance, 236–237, 571
  input, 587
  matching, 512–513
  sinusoidal steady-state, 387–391
    defined, 387
    parallel impedance combinations,
        387–388
    reactance and, 388
    resistance and, 388
    series impedance combinations,
        387
Impulse response, convolution and,
        589–590
Inactive network, 145
Independent current sources, 18, 19
Independent voltage sources, 18–19
Inductors/Inductance, 224–232, 491
  characteristics, ideal, 232
  defined, 224
  duality. *See* Duality
  energy storage, 230–232
  in the frequency domain, 572, 577
  ideal inductor model, 224–227
  inductive reactance, 374
  infinite voltage spikes, 227
  integral voltage-current relationships,
        228–230
  linearity, consequences of, 235–238
  modeled, 243–245, 572–575
  in parallel, 233–234
  phasor relationships for, 384–385
  in series, 232–233
  in the time domain, 577

Infinite voltage spikes, inductors and, 227
Initial value, Laplace transforms, 561–563
In-phase sinusoids, 370–371
Input bias, 193
Input impedance, 587
  amplifiers, 708–710
  one-port networks, 693–696
Input offset voltage, op amps, 196
Instantaneous charge, 12
Instantaneous power, 420–422, 445
Instantaneous stored energy, parallel
        resonance and, 632
Instrumentation amplifier, 202–204
Integral-of-the-current voltage, 18
Integral voltage-current relationships
  capacitors, 218–220
  inductors, 228–230
Internal generated voltage, 474
Internal resistance, 132
International System of Units (SI), 10
Intuitive understanding,
        driven *RL* circuits, 283
Inverse transforms. *See* Laplace
        transform(s)
Inverting amplifier, 175, 180
Inverting input, 174
IN750 Zener diode, 188

J

Jaeger, R., 22
Jenkins, N., 485
Joules, 10
Jung, W. G., 204, 246

K

Kaiser, C.J., 246
kelvin, 10
Lord Kelvin, 10
Kennedy, B. K., 521
Kilograms, 10
Kilowatthour (kWh), 437
Kirchhoff, Gustav Robert, 36
Kirchhoff's laws
  current law (KCL), 35, 36–38
    nodal analysis and, 80, 155
    phasors and, 386
  voltage law (KVL), 35, 38–42
    circuit analysis and, 155
    in mesh analysis, 98
    order of elements and, 52

L

Lagging power factor, 441
Lagging sinusoids, 370–371
Lancaster, D., 681
Laplace analysis, 4
Laplace transform(s), 533–570
  computer-aided analysis,
        551–553
  convolution and, 595–596
  damped sinusoidal forcing
        function, 537–540
  defined, 540–543
  for exponential function $e^{-\alpha t}$, 545
  frequency-differentiation theorem,
        840–841
  frequency-integration theorem, 841
  frequency-shift theorem, 839–840
  initial-value/final-value theorems,
        561–563
  inverse transform techniques, 546–551
    distinct poles/method of residues,
        548–549
    linearity theorem, 546–547
    for rational functions, 547–548
    repeated poles, 550
  one-sided, 542–543
  operations, table of, 561
  pairs, 559
  of periodic time functions, 837–839
  for ramp function $tu(t)$, 545
  sifting property, 544–545
  of simple time functions, 543–545
  sinusoid theorem, 558
  system stability theorem, 560
  theorems for, 553–561
  time differentiation theorem, 553–554
  time-integration theorem, 555–556
  time-scaling theorem, 842
  time-shift theorem, 558, 837–839
  two-sided inverse Laplace
        transform, 542
  two-sided Laplace transform, 541
  for unit-impulse function $\delta(t - t_0)$,
        544–545
  for unit-step function $u(t)$, 544
LC circuit, lossless, 357–359
Leading sinusoids, 370–371
Leighton, R. B., 63
LF411 op amp, 191, 198
Lin, P. M., 108, 156, 407, 725

Linden, D., 156
Linear circuits, 2–4
  ac analysis, 3, 4
  complex forcing functions, 377–378
  conservation laws, 155
  dc analysis, 3
  frequency response analysis, 3, 4
  linear voltage-current relationships, 121–122
  transient analysis, 3, 4
Linear dependent source, 122
Linear elements, 121–122
Linear homogeneous differential equations, 255–256
Linearity, 121–122
  consequences, capacitors/inductors, 235–238
  inverse transform theorem, 546–547
Linear resistor, 23
Linear transformers, 503–509
  primary mesh current, 503
  reflected impedance, 503–504
  secondary mesh current, 503
  T and Π equivalent networks, 505–508
Linear voltage-current relationship, 121–122
Line spectra, Fourier series analysis, 743–744
Line terminals, 464
Line-to-line voltages, three-phase Y-Y connection, 465–466
Links, 794–795
  loop analysis and, 799–804
Littlefield, B. L., 836
LMC6035 op amp, 174
LM8272 dual op amp, 174
LM324 op amp, 191
LM741 op amp, 198
Loop
  analysis, links and, 799–804
  defined, 794
  mesh analysis and, 92
Lossless LC circuit, 357–359
Lower half-power frequency, 636
Low-pass filters, 672
  passive, 673–674
Lumped-parameter networks, 35

M
M, upper limit for, 501
McGillem, Clare D., 544
McLyman, W. T., 521
McPartland, B. J., 63
McPartland, J. F., 63
Magnetically coupled circuits, 491–532. *See also* Transformers
  computer-aided analysis, 508–509
  coupling coefficient, 502
  energy considerations, 499–502
  equality of M_{12} and M_{21}, 500–501
  ideal transformers. *See* Ideal transformers
  linear transformers, 503–509
  magnetic flux, 491, 492, 495
  mutual inductance. *See* Mutual inductance
  upper limit for M, establishing, 501
Magnetic flux, 491, 492, 495
Magnitude
  exponential form of complex number, 826–828
  frequency dependence and (s) plane, 604–607
  scaling, 652–656
Mancini, R., 204, 246, 616
MATLAB, 85, 551–553
  tutorial, 831–836
Matrices
  determinants of, 809–811
  inversion of, 808–809
  matrix form of equations, 85
  simultaneous equations, solving, 806–812
Maximally flat filters, 678
Maximum average power, 429
Maximum power transfer, 150–152, 428–430
Maxwell, James Clerk, 216
Mesh. *See* Nodal and mesh analysis
Meters, 10
Method of residues, 548–549
Metric system of units, 10
Microfarads (μF), 217
MicroSim Corporation, 103
Midrange filters, 679–680
Models/Modeling, 3
  of automotive suspension systems, 356

of ideal capacitors, 215–218
of inductors
  ideal inductors, 224–227
  with PSpice, 243–245
  in the s-domain, 572–575
of op amps, detailed, 190–192
Moles, 10
M_{12}/M_{21} equality, magnetically coupled circuits, 500–501
μA741 op amp, 191–192, 193, 196
Multiband filters, 672
Multiple-frequency circuits, RMS value with, 434–435
Multiple terms, in Bode diagrams, 659–660
Multiport network, 692. *See also* Two-port networks
Mutual inductance, 491–499
  additive fluxes, 495
  coefficient of, 492
  dot convention, 493–497
    additive fluxes, 495
    circuit transfer function, 497
    physical basis of, 495–497
    power gain, 497
  magnetic flux, 491, 492, 495
  self inductance added to, 494

N
Napier, John, 534
NASA Marshall Space Flight Center, 5
National Bureau of Standards, 9–10
National Semiconductor Corp., 174, 198
Natural resonant frequency, 336–337, 630
Natural responses, 276, 369, 372, 735–736
  and the complex-frequency (s) plane, 607–611
  driven RL circuits, 282, 283–289
  source-free RL circuits, 256
Negative (absorbed) power, 16, 19
Negative charge, 11
Negative feedback
  op amps, 194–195
  path, 612
Negative phase sequence, 464–465
Negative resistances, 696
Neper frequency, 537
  defined, 322

Nepers (Np), 534
Networks, 21–22
   active, 21
   passive, 21
   topology. *See* Network topology
   two-port. *See* Two-port networks
Network topology, 793–804
   links and loop analysis, 799–804
   trees and general nodal analysis,
     793–799
Neudeck, G. W., 204, 407, 725
Neutral (ground) connection, 458, 464
New Simulation Profile command
     (PSpice), 104–105
Nodal and mesh analysis, 3, 79–120
   compared, 101–103
   computer-aided, 103–108, 578–580
   location of sources and, 101
   mesh analysis, 92–98, 155
     Kirchhoff's voltage law applied
       to, 98
     mesh current, 92, 93–95, 503
     mesh defined, 794
     procedure, summarized, 98
     supermesh, 98, 100–101
   nodal analysis, 3, 80–89, 155
     basic procedure, summary, 88–89
     Kirchhoff's current law and, 80
     nodes defined, 36, 793
     procedure, summarized, 98
     reference node, 80
     sinusoidal steady-state analysis,
       393–395
     supermesh, 98, 100–101
     supernodes, 89–91
     trees and, 793–799
     voltage source effects, 89–91
   node-base PSpice schematics,
     106–107
   s-domain circuit analysis and,
     578–584
     computer-aided, 578–580
     of sinusoidal steady-states,
       393–395
Noninverting amplifier circuit, 180
   output waveform, 176–177
Noninverting input, 174
Nonlinear circuit analysis, 2
Nonperiodic functions, average power
     for, 430–432

Nonplanar circuit, defined, 794
Norton, E. L., 139
Norton equivalents. *See* Thévenin/Norton
     equivalent circuits
Notch filters, 672
Number systems, units and scales, 9
Numerical value, of current, 12
1N750 Zener diode, 187–188

O

Octave (of frequencies), 658
Odd functions, 747n
Odd harmonics, 747n
Odd symmetry, Fourier series analysis,
     745, 749
Ogata, K., 564, 616
Ohm, Georg Simon, 22
Ohms (Ω), 22
Ohm's law, 22–28
   conductance, 27–28
   defined, 22
   power absorption in resistors,
     23–27
   practical application, 25–26
   resistance units defined, 22
One-port networks, 691–696
   input impedance calculations for,
     693–696
One-sided Laplace transform,
     542–543
Op amps. *See* Operational amplifiers
OPA690 op amp, 191, 197
Open circuit, 27–28
   to dc, 217
   impedance parameters, 712–713
Open-loop
   configuration, op amps, 201
   voltage gain, 190–191
Operating at complex frequencies, 608
Operational amplifiers, 173–214
   AD549K op amp, 191, 193
   AD622 op amp, 204
   capacitors with, 238–239
   cascaded stages, 182–185
   common mode rejection, 193–194
   comparators, 201–202
   computer-aided analysis, 198–201
   frequency and, 197–198
   ideal, 174–182
     derivation of, 192–193

     difference amplifier, 179–182,
      193–194
     inverting amplifier, 175, 180
     noninverting amplifier circuit,
      176–177, 180
     rules, 174, 175
     summary, 180
     summing amplifier, 178–179, 180
     voltage follower circuit, 177, 180
   input offset voltage, 196
   instrumentation amplifier, 202–204
   LF411 op amp, 191, 198
   LM324 op amp, 191
   LM741 op amp, 198
   LM8272 dual op amp, 174
   LMC6035 op amp, 174
   modeling, 190–192
   μA741 op amp, 191–192, 193, 196
   negative feedback, 194–195
   OPA690 op amp, 191, 197
   outputs depend on inputs, 174
   packaging, 198
   parameter values, typical, 191
   Philbrick K2-W op amp, 174
   positive feedback, 195
   practical considerations, 190–201
   reliable current sources, 188–190
   reliable voltage sources, 186–188
   saturation, 195–196
   slew rate, 197–198
   tank pressure monitoring system,
     184–185
Operations, Laplace transform, table
     of, 561
Order of elements, KVL and, 52
Ørsted, Hans Christian, 224
Oscillator, 612
   circuit design, 612–613
   function, 338
Out-of-phase sinusoids, 370–371
Output impedance, amplifiers, 709
Output resistance, 132
Overdamped response
   source-free parallel *RLC* circuits, 323,
     324–331, 345
     A_1 and A_2 values, finding, 324–325
     graphical representation of,
      329–330
   source-free series *RLC* circuits,
     344–345

P

Packages, op amp, 198
Pairs, Laplace transform, 559
Palm, W. J., III, 836
Π and T equivalent networks, 505–508
Parallel element combinations, 45
   capacitors, 234
   impedance combinations, 387–388
   inductors, 233–234
   series/parallel combination
      equivalents, 647–651
Parallel resonance, 627–641, 644
   bandwidth and high-Q circuits,
      636–641
   key conclusions on, 641
   current response and, 631
   damping
      exponential coefficient, 630
      factor, 634–635
   defined, 628–631
   frequency selectivity, 637
   instantaneous stored energy, 632
   natural resonant frequency, 630
   quality factor (Q), 631–641
      bandwidth and, 636–641
      damping factor and, 634–635
      other interpretations of Q, 633–634
   summary of, 644
   voltage response and, 630–631
Parameter values, op amps, 191
Parseval-Deschenes, Marc Antione, 764
Particular integral, 285
Particular solution, 285
   source-free RL circuits, 256
Passband, 672
Passive element, 215
Passive filters
   defined, 677
   low-pass and high-pass, 673–674
Passive network, 21
Passive sign convention, 16
Path
   defined, 793
   mesh analysis, 92
   voltage, 14
Periodic functions/waveforms, 431.
      See also Sinusoidal
      steady-state analysis;
      Sinusoidal waveforms
   ac average power of, 295, 423–424

complete response to, 750–752
   fall time of, 294
   as forcing functions, 369
   Laplace transforms of, 837–839
   as output, noninverting amplifiers,
      176–177
   period T of, 295, 369–370
   pulse width of, 295
   rise time of, 294
   RMS values for, 432–434
   time delay of, 295
Perry, T., 820
Peterson, Donald O., 815
Phase angle θ, 370, 604–607
Phase comparison, sinusoidal
      waves, 371
Phase response, Bode diagrams and,
      660–661
Phase spectra, Fourier series analysis,
      744–745
Phase voltages, 464
Phasor(s), 4, 382, 571. *See also* Phasor
      relationships for R, L, and C
   diagrams, sinusoidal steady-states,
      404–407
Phasor relationships for R, L, and C
   as abbreviated complex
      representation, 381
   capacitors, 385–386
   frequency-domain representation, 382
   frequency-domain V-I
      expressions, 385
   impedance defined from. *See*
      Sinusoidal steady-state
      analysis
   inductors, 384–385
   Kirchhoff's laws using, 386
   phasor representation, 382
   resistors, 383–384
   time-domain representation, 382
   time-domain V-I expressions, 385
Philbrick, George A., 205
Philbrick K2-W op amp, 174
Philbrick Researches, Inc., 173
Physically realizable systems, 591–592
Physical significance, of Fourier
      transforms, 764–765
Physical sources, unit-step function and,
      278–279
Pinkus, A., 564, 785

Planar circuit, 92, 101
   defined, 794
Polar form, of complex numbers, 828–830
Poles, 547
   method of residues and, 548–549
   pole-zero constellations, 602–604
   repeated, inverse transforms, 550
   zeros, and transfer functions, 588
Polya, G., 8
Polyphase circuits, 457–490
   delta (Δ) connection, 470–476
      of sources, 473–476
      Y-connected loads vs., 473
   double-subscript notation, 459–460
   polyphase systems, 458–460
   single-phase three-wire systems,
      460–464
   three-phase Y-Y connection. *See*
      Three-phase Y-Y connection
Port, 691
Positive charge, 11
Positive feedback, 195, 612
Positive phase sequence, 464–465
Positive power, 16, 19
Potential coil, 476
Potential difference, 14
Power, 9, 15–17. *See also* ac circuit
      power analysis
   absorbed. *See* Absorbed power
   average. *See* Average power
   dissipation, 45
   expression for, 15
   factor. *See* Power factor
   gain, 497, 708
   generating systems, 474–475
   maximum transfer of, 150–152
   measuring. *See* Power measurement
   negative. *See* Absorbed power
   positive, 16, 19
   reactive, 440, 441, 445
   superposition applicable to, 431
   terminology recap, 445
   triangle, 441
   units, 10
Power factor, 445
   apparent power and, 437–439
   complex power, 437–439
   correction, 442
   lagging, 441
   leading, 441

Power measurement, 441
  three-phase systems, 476–484
  two-wattmeter method, 481–483
  wattmeters, use of, 476–478
  wattmeter theory and formulas,
    478–481
Practical current sources, 133,
    137–138
Practical voltage sources, 131–133,
    137–138
Prefixes, SI, 10–11
Primary mesh current, 503
Prime mover, 474
Probe software, 342–343
Problem-solving strategies, 7–8
PSpice, 103, 104–107, 128–130
  Bias Point command, 105
  capacitors modeled with, 243–245
  Create command, 105
  inductors modeled with, 243–245
  New Simulation Profile command,
    104–105
  node-base schematics, 106–107
  Run command, 105
  for sinusoidal steady-state analysis,
    402–403
  for transient analysis, 264–266
  tutorial, 815–820
  Type command, 105
Pulse width (PW), of waveforms, 295
Purely reactive elements, average power
    absorption, 427

Q

Quadrature power, 441
Quality factor (Q). *See* Parallel
    resonance

R

Radian frequency, 369, 537
Ragazzini, J. R., 205
Ramp function $tu(t)$, Laplace transform
    for, 545
Randall, R. M., 205
Rational functions, inverse transforms for,
    547–548
Rawlins, C. B., 25, 26
RC circuits
  driven, 289–294
  general, 273–276

sequentially switched, 294–299
  I: time to fully charge/fully
    discharge, 296–297, 298
  II: time to fully charge but not fully
    discharge, 297, 298
  III: no time to fully charge but time
    to fully discharge, 297, 298
  IV: no time to fully charge or fully
    discharge, 298–299
source-free, 266–269
time constant (τ), 267–268
unit-step function, 276–280
Reactance
  impedance and, 388
  inductive, 374
  synchronous, 474
Reactive elements, average power
    absorption, 427
Reactive power, 440, 441, 445
Realizable systems, *s*-domain analysis,
    591–592
Real portion, of complex forcing
    function, 376
Real sources → real responses, complex
    forcing functions, 377–378
Receiving mode, 223
Reciprocity theorem, 702
Rectangular form, complex
    numbers, 822
Rectangular pulse function, 279–280
Rectifiers, 459
Reference node, 80
Reflected impedance, 503–504
Reliable current sources, op amps,
    188–190
Reliable voltage sources, op amps,
    186–188
Repeated poles, inverse transform
    techniques, 550
Resistance/Resistors/Resistivity, 9, 25.
    See also Ohm's law
  equivalent, 52
  in the frequency domain, 571–572
  ideal, average power
    absorption, 426
  impedance and, 388, 389
  internal, 132
  linear, 23
  output, 132
  phasor relationships for, 383–384

in *s*-domain circuit analysis,
    571–572, 577
  in series and parallel, 51–57
  in the time domain, 577
Resonance, 322
  current response and, 631
  parallel. *See* Parallel resonance
  series, 641–644
  summary table for, 644
  voltage response and, 630–631
Resonant frequency, 322
Response, 121
  in the frequency domain, 772–779
  as a function of σ, *s*-domain, 599
  as a function of ω, *s*-domain,
    599–600
  functions, 122
  source-free series *RLC* circuits,
    344–345
Rise time (TR), of waveforms, 294
RLC circuits, 319–368
  automotive suspensions
    modeled, 356
  complete response of, 349–357
    complicated part, 350–355
    uncomplicated part, 349–350
  lossless *LC* circuit, 357–359
  phasor relationships for. *See* Phasor
    relationships for *R*, *L*, and *C*
  solution process summary, 355–357
  source-free critical damping, 332–336
    A_1 and A_2 values, 333
    form of critically damped response,
      332–333
    graphical representation of,
      334–335
  source-free parallel circuits, 319–323
    computer-aided analysis, 342–343
    critically damped response,
      323, 345
    differential equation for, 320–322
    equations summary, 345
    frequency terms defined, 322–323
    overdamped response, 323,
      324–331, 345
      A_1 and A_2 values, 324–325
      graphical representation,
        329–330
    underdamped response, 323,
      336–343, 345

B_1 and B_2 values, 337–338
finite resistance, role of, 338–340
form of, 336–337
graphical representation, 338
source-free series circuits, 343–349
circuit response résumé, 344–345
critically damped response, 344–345
equations summary, 345
overdamped response, 344–345
underdamped response, 344–345
RL circuits
driven. *See* Driven *RL* circuits
exponential response properties, 262–266
exponential response time constant (τ), 262–264
general, 269–270
natural response. *See* Natural responses
sequentially switched, 294–299
I: time to fully charge/fully discharge, 296–297, 298
II: time to fully charge but not fully discharge, 297, 298
III: no time to fully charge but time to fully discharge, 297, 298
IV: no time to fully charge or fully discharge, 298–299
slicing thinly: 0^+ vs. 0^-, 270–273
source-free, 255–261
alternative approach, 258
complementary function, 256
computer-aided analysis, 264–266
direct approach, 256–257
energy, accounting for, 261
forced response, 256
forcing function, 256
free response, 256
general solution approach, 258–261
natural response, 256
the particular solution, 256
the steady-state response, 256
transient response, 256
unit-step function, 276–280
RMS value
for average power, 434

for current and voltage, 432–437, 445
with multiple-frequency circuits, 434–435
for periodic waveforms, 432–433
for sinusoidal waveforms, 433–434
Robotic manipulator, 5
Root-mean-square (RMS) value. *See* RMS value
Rotor, 474
Row vector, 806
Run command (PSpice), 105
Russell, F. A., 205

S

s, defined, 536–537
Sampling function, Fourier series, 756–759
Sands, M. L., 63
Satellite system telephones, 223
Saturation, op amp, 195–196
Scalar multiplication, 561
Scales, units and, 9–11
Scaling
and frequency response, 652–656
Laplace transform operation, 561
Scientific calculators, 805–806
s-domain circuit analysis, 571–626
additional techniques, 585–588
complex frequency and. *See* Complex frequency
convolution and. *See* Convolution
$\mathbf{H}(s) = \mathbf{V_{out}}/\mathbf{V_{in}}$ voltage ratio, synthesized, 612–616
nodal and mesh analysis in, 578–584
computer-aided analysis, 578–580
poles, zeros, and transfer functions, 588
Thévenin equivalent technique, 587–588
Z(s) and Y(s), 571–577
capacitors
in frequency domain, 577
modeled in the **s** domain, 575–576
in time domain, 577
inductors
in frequency domain, 572, 577
modeled in the **s** domain, 572–575
in time domain, 577

resistors
in frequency domain, 571–572, 577
in time domain, 577
summary of element representations, 577
Secondary mesh current, 503
Seconds, 10
Self-inductance, 491
added to mutual inductance, 494
Sequentially switched *RL* or *RC* circuits. *See RC* circuits; *RL* circuits
Series connections, 42
capacitors, 234
impedance combinations, 387
inductors in, 232–233
and parallel combinations. *See also* Source transformations
connected sources, 49–51, 137–138
other resonant forms, 647–651
Series resonance, 641–644
Settling time, 330
Sharpe, D., 485
Short circuit(s), 27–28
admittance and, 712–713
for equivalent networks, 703–704
input admittance, 697–698
output admittance, 698
transfer admittance, 698
two-port networks, 698
to dc, 225
SI base units, 10
siemen (S), 572
Sifting property, 544–545
Signal ground, 61–62
Signs
passive convention, 16
for voltages, 9, 14
Simon, Paul-René, 28
Simple time functions, Laplace transforms of, 543–545
Simulation Program with Integrated Circuit Emphasis, 103
Simultaneous equations, solving, 805–812
Cramer's rule, 811–812
determinants and, 809–811
matrices, 806–812
scientific calculators and, 805–806

Sines, converted to cosines, 371
Single-loop circuit, 42–45
Single-node-pair circuit, 45–49
Single-phase three-wire systems,
        460–464
Singularity functions, 277
Sinusoids
    complex frequency case, 535
    as forcing functions, 627–628
    Laplace transforms of, 558
Sinusoidal steady-state analysis,
        369–418
    ac circuit average power, 424–425
    admittance, 392–393
    amplitude, 369
    angular frequency, 369
    argument, 369
    characteristics of sinusoids,
        369–371
    complex forcing function, 376–380
        algebraic alternative to differential
            equations, 378–379
        applying, 377–378
        imaginary part, 376
        imaginary sources → imaginary
            responses, 377–378
        real part, 376
        real sources → real responses,
            377–378
        superposition theorem, 377–378
    computer-aided analysis, 402–403
    conductance, 392
    cutoff frequency, transistor amplifier,
        396–397
    forced responses to sinusoids, 369,
        372–376
        alternative form of, 373–374
        amplitude, response vs. forcing
            function, 374
        steady-state, 372–373
    frequency, 370–371
    immittance, 392–393
    impedance. See Impedance
    lagging and leading, 370–371
    natural response, 369
    nodal and mesh analysis, 393–395
    out-of-phase, 370–371
    period, 369–370
    in phase, 370–371
    phase comparison requirements, 371

phasor diagrams, 404–407
phasor relationships and. See Phasor
        relationships for R, L, and C
radian frequency, 369
sines converted to cosines, 371
sinusoidal waveform forcing
        function, 369
superposition, source transformations,
        and, 396–403
susceptance, 392
Sinusoidal waveforms
    as forcing functions, 369
    oscillator circuit design and, 612–613
    phase comparison, 371
    RMS values of current/voltage,
        433–434
SI prefixes, 10–11
Slew rate, op amps, 197–198
Slicing thinly: 0^+ vs. 0^-, RL circuits,
        270–273
Smoothing, of Bode diagrams, 658–659
Solve( ) routine, 86
Source-free RC circuits, 266–269
Source-free RLC circuits. See RLC
        circuits
Source-free RL circuits. See RL circuits
Source transformations, 3, 131–138, 155
    equivalent practical sources, 133–136
    key concept requirements, 137–138
    practical current sources, 133, 137–138
    practical voltage sources, 131–133,
        137–138
    and sinusoidal steady-state analysis,
        396–403
    summary, 138
SPICE, 103. See also PSpice
Square matrix, 806
Squire, J., 785
Stability, of a system, 560
Standby mode, 223
Stator, 474
Steady-state analysis/response, 285.
        See also Sinusoidal
        steady-state analysis
    source-free RL circuits, 256
Step-down transformers, 514
Step-up transformers, 514
Stopband, 672
Structure (programming), 86
Summing amplifier, 178–179, 180

Superconducting transformers, 516–517
Supermesh, 98, 100–101
Supernodes, 89–91
Superposition, 3, 121–131, 155,
        377–378
    basic procedure, 128
    applicable to current, 431
    applicable to power, 431
    limitations of, 131
    sinusoidal steady-state analysis,
        396–403
    superposition theorem, 123
Supplied power, 16
    equaling absorbed power, 44
Susceptance, 392
Suspension systems, automotive,
        modeling of, 356
Symmetrical components, 470
Symmetry, use of, Fourier series analysis,
        745–749
Synchronous generator, 474
Synchronous reactance, 474
System function, 589
    computer-aided analysis, 776–779
    fast Fourier transform (FFT), 774,
        776–779
        image processing example, 782
    physical significance of, 779–781
    response, in frequency domain,
        772–779
Systems, stability of, 560
Szwarc, Joseph, 28

T

T and Π equivalent networks, 505–508
Tank pressure monitoring system,
        184–185
Taylor, Barry N., 28
Taylor, J. T., 681
Tesla, Nikola, 457
Thévenin, M. L., 139
Thévenin/Norton equivalent circuits, 3,
        139–149, 155–156
    when dependent sources are present,
        145–147
    Norton's theorem, 3, 143–145,
        155–156
    linearity for
        capacitors/inductors, 238
    resistance, 142, 155–156

s-domain circuit analysis, 587–588
Thévenin's theorem, 3, 139, 141–143, 155–156
  linearity for capacitors/inductors, 238
  proof of, 813–814
  and sinusoidal steady-state analysis, 396–403
  two-port networks, 709–710
Three-phase system, balanced, 458
Three-phase Y-Y connection, 464–470
  abc phase sequence, 464–465
  cba phase sequence, 464–465
  Delta (Δ) connection vs., 473
  line-to-line voltages, 465–466
  negative phase sequence, 464–465
  positive phase sequence, 464–465
  power measurement in. *See* Power measurement
  total instantaneous power, 467–468
  with unbalanced load, 470
Tightly coupled coils, 502
Time constant (τ)
  exponential response of *RL* circuits, 262–264
  RC circuits, 267–268
Time delay (TD) of waveforms, 295
Time differentiation, Laplace transforms and, 553–554, 561
Time domain
  capacitors in, 577
  converted to frequency domain, 539
  ideal transformer voltage relationships in, 515–518
  inductors in, 577
  representation, phasors, 382
  resistors in, 577
  V-I expressions, phasor relationships and, 385
Time functions, simple, Laplace transforms of, 543–545
Time integration, Laplace transforms and, 555–556, 561
Time periodicity, Laplace transforms and, 561, 837–839
Time-scaling theorem, Laplace transforms and, 842
Time shift, Laplace transforms and, 558, 561, 837–839

Topology, 793. *See also* Network topology
Total instantaneous power, three-phase, 458, 467–468
T parameters, two-port networks, 720–724
Transconductance, 21
Transfer functions, 497, 588, 597
Transfer of charge, 12
Transformations
  source. *See* Source transformations
  between y, z, h, and t parameters, 713
Transformers. *See also* Magnetically coupled circuits
  ideal. *See* Ideal transformers
  linear. *See* Linear transformers
  superconducting, 516
Transient analysis, 3, 4
  PSpice capability for, 264–266
Transient response, 283
  source-free *RL* circuits, 256
Transistors, 22, 396–397, 717–718
Transmission parameters, two-port networks, 720–724
Transmitting mode, 223
Treble filters, 679–680
Trees, 793–799
Trigonometric form, of Fourier series. *See* Fourier series
Trigonometric integrals, Fourier series analysis, 738–739
Tuinenga, P., 108, 820
Turns ratio, ideal transformers, 510–512
12AX7A vacuum tube, 174
Two-port networks, 691–734
  ABCD parameters, 720–724
  admittance parameters, 696–703
    bilateral circuit, 702
    bilateral element, 702
    reciprocity theorem, 702
    short-circuit admittance parameters, 698
    short-circuit input admittance, 697–698
    short-circuit output admittance, 698
    short-circuit transfer admittance, 698
    y parameters, 697–699, 710–711

  computer-aided analysis for, 723–724
  equivalent networks, 703–711
    amplifiers, 708–710
    Δ of impedances method, 704–705
    Norton equivalent method, 709–710
    short-circuit admittance method, 703–704
    Thévenin equivalent method, 709–710
    Y-Δ not applicable, 706–707
    yv subtraction/addition method, 704
  hybrid parameters, 718–720
  impedance parameters, 712–716
  one-port networks. *See* one-port networks
  t parameters, 720–724
  transistors, characterizing, 717–718
  transmission parameters, 720–724
Two-sided inverse Laplace transform, 542
Two-sided Laplace transform, 541

U

Ultracapacitor, 223
Unbalance Y-connected loads, 470
Underdamped response
  source-free parallel *RLC* circuits. *See RLC* circuits
  source-free series *RLC* circuits, 344–345
Unit-impulse function, 277
  Laplace transform for, 544–545
Units and scales, 9–11
Unit-step function *u(t)*, 276–280
  Fourier transform pairs for, 769
  Laplace transforms for, 544
  and physical sources, 278–279
  RC circuits, 276–280
  rectangular, 279–280
  RL circuits, 276–280
Unity gain amplifier, 177, 180
Upper half-power frequency, 636

V

Vectors, 85, 806
Volta, Alessandro Giuseppe Antonio Anastasio, 14n

Voltage, 9, 14–15
  actual polarity vs. convention, 14
  current sources and, 17–22, 49–51
    active elements, 21
    circuit element, 21
    dependent sources of
        voltage/current, 18, 19–21
    derivative-of-the-current
        voltage, 18
    independent current sources, 19
    independent voltage sources, 18–19
    integral-of-the-current voltage, 18
    networks and circuits, 21–22
    passive elements, 21
  cutoff, 223
  force and, 5
  input offset, op amps, 196
  integral voltage-current relationships,
      for capacitors, 218–220
  internally generated, 474
  laws. *See* Voltage and current laws
  sources. *See* Voltage sources
  voltage and current division, 57–60
Voltage amplifier, 176
Voltage and current division, 57–60
Voltage and current laws, 35–78
  branches, 35–36
  equivalent resistance, 52
  Kirchhoff's current law (KCL), 35,
      36–38
  Kirchhoff's voltage law (KVL), 35,
      38–42
    order of elements and, 52
  loops, 35–36
  nodes, 35–36
  paths, 35–36

resistors in series and parallel, 51–57
series and parallel connected sources,
    49–51
single-loop circuit, 42–45
single-node-pair circuit, 45–49
voltage and current division, 57–60
Voltage coil, 476
Voltage-controlled current source, 19
Voltage-controlled voltage source, 19
Voltage follower circuit, 177, 180
Voltage gain, amplifiers, 708
Voltage level adjustment, ideal
    transformers for, 513–514
Voltage ratio $H(s) = V_{out}/V_{in}$,
    synthesizing, 612–616
Voltage regulation, 475
Voltage relationship, ideal transformers,
    time domain, 515–518
Voltage response, resonance and, 630–631
Voltage sources
  ideal, 131–133
  practical, 131–133
  reliable, op amps, 186–188
  series and parallel connected sources,
      49–51
  source effects, nodal and mesh
      analysis, 89–91
Volt-ampere-reactive (VAR) units, 441
  complex power, 440
Volt-amperes (VA), 438, 441

W

Wattmeters, for three-phase systems
  theory and formulas, 478–481
  two wattmeter method, 481–483
  use, 476–478

Watts (W), 10, 441
Weber, E., 302, 360
Weedy, B. M., 446, 485
Westinghouse, George, 457
Wheatstone bridge, 74
Wheeler, H. A., 534
Wien-bridge oscillator, 612
Winder, S., 616
Wire gauges, 25–26
Work (energy) units, 10

Y

Y parameters, two-port networks,
    697–699, 710–711
Y(s) and Z(s). *See* **s**-domain circuit
    analysis
Yv method, for equivalent
    networks, 704

Z

Zafrany, S., 564, 785
Zandman, Felix, 28
Zener diode, 186–188
Zener voltage, 187
Zeros, 547
  s-domain circuit analysis
      pole-zero constellations,
          602–604
      zeros, poles, and transfer
          functions, 588
Zero$^+$ vs. Zero$^-$, slicing thinly:
    RL circuits, 270–273
Zeta (ζ) damping factor, 634
Z parameters, 712–716
Z(s), Y(s) and. *See* **s**-domain circuit
    analysis

A Short Table of Integrals

$$\int \sin^2 ax \, dx = \frac{x}{2} - \frac{\sin 2ax}{4a}$$

$$\int \cos^2 ax \, dx = \frac{x}{2} + \frac{\sin 2ax}{4a}$$

$$\int x \sin ax \, dx = \frac{1}{a^2}(\sin ax - ax \cos ax)$$

$$\int x^2 \sin ax \, dx = \frac{1}{a^3}(2ax \sin ax + 2 \cos ax - a^2x^2 \cos ax)$$

$$\int x \cos ax \, dx = \frac{1}{a^2}(\cos ax + ax \sin ax)$$

$$\int x^2 \cos ax \, dx = \frac{1}{a^3}(2ax \cos ax - 2 \sin ax + a^2x^2 \sin ax)$$

$$\int \sin ax \sin bx \, dx = \frac{\sin(a-b)x}{2(a-b)} - \frac{\sin(a+b)x}{2(a+b)}; a^2 \neq b^2$$

$$\int \sin ax \cos bx \, dx = -\frac{\cos(a-b)x}{2(a-b)} - \frac{\cos(a+b)x}{2(a+b)}; a^2 \neq b^2$$

$$\int \cos ax \cos bx \, dx = \frac{\sin(a-b)x}{2(a-b)} + \frac{\sin(a+b)x}{2(a+b)}; a^2 \neq b^2$$

$$\int x e^{ax} \, dx = \frac{e^{ax}}{a^2}(ax - 1)$$

$$\int x^2 e^{ax} \, dx = \frac{e^{ax}}{a^3}(a^2x^2 - 2ax + 2)$$

$$\int e^{ax} \sin bx \, dx = \frac{e^{ax}}{a^2 + b^2}(a \sin bx - b \cos bx)$$

$$\int e^{ax} \cos bx \, dx = \frac{e^{ax}}{a^2 + b^2}(a \cos bx + b \sin bx)$$

$$\int \frac{dx}{a^2 + x^2} = \frac{1}{a} \tan^{-1} \frac{x}{a}$$